AF576055

# Advances in Biocompatible and Biodegradable Polymers II

# Advances in Biocompatible and Biodegradable Polymers II


Editors

**José Miguel Ferri**
**Vicent Fombuena Borràs**
**Miguel Fernando Aldás Carrasco**

Basel • Beijing • Wuhan • Barcelona • Belgrade • Novi Sad • Cluj • Manchester

*Editors*
José Miguel Ferri
Department of Nuclear and Chemical Engineering
Universitat Politècnica de València
Alcoy
Spain

Vicent Fombuena Borràs
Department of Nuclear and Chemical Engineering
Universitat Politècnica de València
Alcoy
Spain

Miguel Fernando Aldás Carrasco
Center of Polymers Applied Research (CIAP)
Escuela Politécnica Nacional
Quito
Ecuador

*Editorial Office*
MDPI
St. Alban-Anlage 66
4052 Basel, Switzerland

This is a reprint of articles from the Special Issue published online in the open access journal *Polymers* (ISSN 2073-4360) (available at: www.mdpi.com/journal/polymers/special_issues/biocompat_biodegrad_polym_II).

For citation purposes, cite each article independently as indicated on the article page online and as indicated below:

Lastname, A.A.; Lastname, B.B. Article Title. *Journal Name* **Year**, *Volume Number*, Page Range.

**ISBN 978-3-0365-9063-9 (Hbk)**
**ISBN 978-3-0365-9062-2 (PDF)**
**doi.org/10.3390/books978-3-0365-9062-2**

# Contents

# About the Editors

**José Miguel Ferri**

Dr. José Miguel Ferri (h-index 16, Scopus) works at the Department of Nuclear and Chemical Engineering of the Escuela Politécnica Superior de Alcoy (EPSA) of the Universitat Politécnica de Valéncia (UPV) in Alcoy, Spain. He holds degrees in technical industrial mechanical engineering and materials engineering. He joined the Technological Institute of Materials (ITM) in 2014 as a research technician and obtained an international PhD in engineering and industrial production in 2017. His field of research is related to the development of new metallic materials (Mg foams for the medical sector), polymeric materials (improvement of thermal, mechanical, and processing properties of biodegradable polymers), and composite materials (metallic matrix for the electronics and aeronautics sectors, and polymeric matrix for the aeronautics and construction sectors).

**Vicent Fombuena Borràs**

Dr. Vicent Fombuena (h-index 21, Scopus) works at the Department of Nuclear and Chemical Engineering of the Escuela Politécnica Superior de Alcoy (EPSA) of the Universitat Politécnica de Valéncia (UPV) in Alcoy, Spain. He holds degrees in technical industrial chemical engineering and materials engineering. He joined the Technological Institute of Materials (ITM) in 2008 as a research technician and obtained an international PhD in engineering and industrial production in 2012. His research work has focused on the implementation of a circular economy model in the polymer industry. He has focused his efforts on obtaining multiple active compounds from seeds and agroforestry residues to be applied to thermoplastic and thermosetting polymers. He has extensive experience in the techniques of chemical, thermal, and mechanical characterization of materials.

**Miguel Fernando Aldás Carrasco**

Miguel Aldás (h-index 12, Scopus) has been a lecturer and researcher at the Escuela Politécnica Nacional (Quito, Ecuador) since 2008. He collaborates with the Centro de Investigaciones Aplicadas a Polímeros (Center of Polymers Applied Research) (CIAP) and the Faculty of Chemical Engineering and Agroindustry. He holds a degree in chemical engineering from the Escuela Politécnica Nacional (Quito, Ecuador), a Master's degree in Innovative Materials from the Université Claude Bernard Lyon 1 (Lyon, France), and a PhD degree in Engineering from the Universitat Politécnica de Valéncia (Valencia, Spain). He worked under the direction of Prof. Juan López Martínez and Dr. Marina P. Arrieta, being part of the team of researchers at the Materials Technology Institute (ITM-UPV) in the Alcoy campus within the Development of Materials for Sustainable Structures (DEMES) group. His field of research focuses on the use of natural additives for biopolymers and biodegradable materials, thermoplastic starch-based materials, the study of the degradation of synthetic and biodegradable plastic materials, and the technology of processing techniques.

# Preface

Among the strategies for reducing the negative effects on the environment affected by the uncontrolled consumption and low potential for the recovery of conventional plastics, the synthesis of new biodegradable and recyclable plastics represents one of the most promising methods for minimizing the negative effects of conventional non-biodegradable plastics. The spectrum of existing biodegradable materials is still very narrow. Therefore, to achieve greater applicability, research is being carried out on biodegradable polymer mixtures, the synthesis of new polymers, and the incorporation of new stabilizers for thermal degradation, alongside the use of other additives such as antibacterials or new and more sustainable plasticizers. Some studies analyze direct applications, such as shape memory foams, new cartilage implants, drug release, etc. The reader can find several studies on the degradation of biodegradable polymers under composting conditions. However, novel bacteria that degrade polymers considered non-biodegradable in other, unusual conditions (such as conditions of high salinity) are also presented.

**José Miguel Ferri, Vicent Fombuena Borràs, and Miguel Fernando Aldás Carrasco**
*Editors*

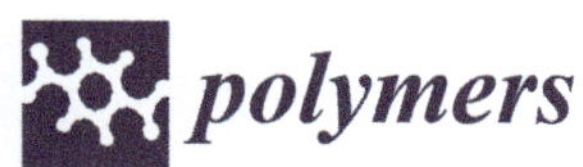 

*Article*

# Morphology and Selected Properties of Modified Potato Thermoplastic Starch

**Regina Jeziorska [1,*], Agnieszka Szadkowska [1], Maciej Studzinski [1], Michal Chmielarek [2] and Ewa Spasowka [1]**

[1] Łukasiewicz Network-Industrial Chemistry Institute, Rydygiera 8, 01-793 Warsaw, Poland
[2] Department of High-Energetic Materials, Faculty of Chemistry, Warsaw University of Technology, Noakowskiego 3, 00-664 Warsaw, Poland
* Correspondence: regina.jeziorska@ichp.lukasiewicz.gov.pl; Tel.: +48-664-010-897

**Abstract:** Potato thermoplastic starch (TPS) containing 1 wt.% of pure halloysite (HNT), glycerol-modified halloysite (G-HNT) or polyester plasticizer-modified halloysite (PP-HNT) was prepared by melt-extrusion. Halloysites were characterized by FTIR, SEM, TGA, and DSC. Interactions between TPS and halloysites were studied by FTIR, SEM, and DMTA. The Vicat softening temperature, tensile, and flexural properties were also determined. FTIR proved the interactions between halloysite and the organic compound as well as between starch, plasticizers and halloysites. Pure HNT had the best thermal stability, but PP-HNT showed better thermal stability than G-HNT. The addition of HNT and G-HNT improved the TPS's thermal stability, as evidenced by significantly higher $T_{5\%}$. Modified TPS showed higher a Vicat softening point, suggesting better hot water resistance. Halloysite improved TPS stiffness due to higher storage modulus. However, TPS/PP-HNT had the lowest stiffness, and TPS/HNT the highest. Halloysite increased $T_\alpha$ and lowered $T_\beta$ due to its simultaneous reinforcing and plasticizing effect. TPS/HNT showed an additional β-relaxation peak, suggesting the formation of a new crystalline phase. The mechanical properties of TPS were also improved in the presence of both pure and modified halloysites.

**Keywords:** starch; halloysite; hot water resistance

**Citation:** Jeziorska, R.; Szadkowska, A.; Studzinski, M.; Chmielarek, M.; Spasowka, E. Morphology and Selected Properties of Modified Potato Thermoplastic Starch. *Polymers* **2023**, *15*, 1762. https://doi.org/10.3390/polym15071762

Academic Editors: José Miguel Ferri, Vicent Fombuena Borràs and Miguel Fernando Aldás Carrasco

Received: 7 March 2023
Revised: 23 March 2023
Accepted: 24 March 2023
Published: 1 April 2023

## 1. Introduction

As the materials of the millennium, plastics have become synonymous with technical and economic progress [1,2]. Unfortunately, the growing amount of polymer waste collected in landfills has a negative impact on the plants, animals, and people's living conditions [3,4]. To reduce the negative effects on the environment, research is being conducted on the production of polymers derived from renewable and biodegradable raw materials [5–7]. Currently, the share of biodegradable plastics in the global plastics market is small, but shows an upward trend. In the modern world, caring about the natural environment is becoming not only an obligation, but also a necessity [8–10]. According to the EU Directive (SUP) on the ban on the production (from 2021) of certain single-use plastic products (cutlery, plates, stirrers, expanded polystyrene packaging), manufacturers of these products are obliged to replace them with biodegradable materials [4,6,11].

One of the most widespread natural polymers is starch [12–14]. Its biodegradability and bio renewability make this polymer "double green", and it meets the principles of "green chemistry". Great hopes for solving the problem of waste, especially packaging waste, are associated with the use of thermoplastic starch (TPS), which can be processed using traditional techniques such as extrusion, injection, pressing, etc. The advantages of starch are its low price and universal availability. The disadvantages, limiting its widespread use in industry, are the values of its glass transition and melting point (230–240 °C), which are higher than the thermal decomposition temperature (220 °C) [15].

Starch is produced from many different renewable sources. It is a cheap and inherently biodegradable material. It can be completely converted by microorganisms into carbon dioxide, water, minerals, and biomass, without negatively impacting the environment [16]. Starch grains come in all shapes and sizes, ranging from 2 μm to over 100 μm, depending on their botanical origin and type [17,18].

Starch contains mainly of 15–35 wt.% linear amylose and 65–85 wt.% non-linear amylopectin. The association of amylose and amylopectin, in the native state, gives a semicrystalline structure [2,13]. It is well known that native starch is brittle, and its mechanical properties are very poor. Lack of water causes thermal degradation to occur below the glass transition temperature [19]. Thanks to plasticizers, thermal processes, and shear stress, it is possible to convert native starch into a thermoplastic polymer [9,14,20]. However, due to poor mechanical properties and high hygroscopicity, plasticized starch cannot yet compete with polymers from non-renewable raw materials. To overcome these problems, nanocellulose [21], montmorillonite [22,23], kaolin [24], bentonite [25], nano silica [26], chitosan [27,28] and halloysite [29,30] have been used.

Halloysite, due to its structure consisting of hollow nanotubes and multiple rolled aluminosilicate layers, is physically and chemically similar to kaolinite [30–32]. The nanotubes' outer diameter is 50–200 nm, their inner diameter is 5–30 nm, and their length is 0.5–25 μm. The negatively charged $SiO_2$ forms the outer surface and the positively charged $Al_2O_3$ forms the inner surface of the HNT. Halloysite, thanks to high specific surface area, good thermal stability, low cost, and unique surface properties, is one of the most important clay minerals [33,34].

Schmitt et al. [35] found that the addition of 2–8 wt.% unmodified and modified (quaternary ammonium salt with benzalkonium chloride) halloysite nanotubes improved thermal stability and tensile properties of starch. The modified halloysite had a much higher Young's modulus than the unmodified one.

Therefore, it can be expected that even a small addition of HNT will increase thermoplastic starch's resistance to hot water and improve its mechanical properties. As far as we know, no comprehensive research on the use of pure halloysite and halloysite containing glycerol or polyester plasticizer for the modification of thermoplastic potato starch has been described yet. It was expected that HNT would act as a plasticizer and a reinforcing agent simultaneously.

Therefore, this article presents the effect of unmodified and modified with glycerol or polyester plasticizer halloysite on the thermal, mechanical, and morphological properties of thermoplastic potato starch. Before modification, pure HNT was subjected to ultrasonication to introduce more organic compound into its porous structure and improve molecular interactions between TPS components. The structure and thermal properties were analyzed by Fourier infrared spectroscopy (FTIR), scanning electron microscopy (SEM), dynamic thermomechanical analysis (DMTA), thermogravimetric thermal analysis (TGA), differential scanning calorimetry (DSC) and Vicat softening temperature. Moreover, its tensile and flexural properties were determined.

## 2. Materials and Methods

### *2.1. Materials*

Native potato starch with amylose content of 21%, pH 5.5–7.5 was obtained from B.E.S.T. Company, Parczew, Poland. Before processing, the starch was conditioned at 23 °C and 50% relative humidity for 48 h. Under these conditions, native starch contained 11% moisture. Glycerol (99% purity) (Sigma-Aldrich, Munich, Germany) and sorbitol (Sigma-Aldrich) were used as plasticizers. Pure (HNT) and modified with glycerol (G-HNT) or polyester plasticizer (PP-HNT) halloysite nanotubes were used as a plasticizer and reinforcer, simultaneously (Table 1). Pure halloysite was delivered from the open pit mine "Dunino", Intermark Company, Gliwice, Poland. The polyester plasticizer was obtained by the method described in [36]. Briefly, the plasticizer was obtained in a three-stage process. In the first stage, 2-butyl-2-ethyl-1,3-propanediol (BEPD), 2-methyl-1,3-propanediol (MPD), ethylene glycol (EG), and propylene glycol (PG) were placed into the reactor in appropriate proportions. After starting the stirrer and heating, a Fascat 4100 catalyst and succinic acid were added. The synthesis was carried out in an inert gas atmosphere at a temperature below 177 °C. After collecting about 81% of the theoretical condensate amount (water with a small glycols content), the process was continued under reduced pressure at a temperature of 175−180 °C until the acid number (AV) and hydroxyl number (HV) of 111.3 mg KOH/g and 1.7 mg KOH/g, respectively, were obtained. In the second step, Fascat 4100 catalyst and 2-ethylhexyl alcohol (EHA) were added. The process was carried out in an inert gas atmosphere at a temperature of 160−186 °C to obtain AV of 5.2 mg KOH/g. In the third stage, distillation of unreacted EHA was carried out at a temperature of 180−189 °C under reduced pressure. After obtaining an AV of 3.6 mg KOH/g and volatility of 0.38 wt.%, the process was continued for 3 h. A plasticizer with the following parameters was obtained: AV 2.2 mg KOH/g; HV 3.6 mg KOH/g; viscosity (25 °C) 2550 mPa·s; volatility (2 h/140 °C) 0.26 wt.%; $M_w$ 2340 g/mol.

**Table 1.** Pure and modified halloysite characteristic.

| Halloysite | BET Surface Area ($m^2$/g) | Average Pore Size (nm) | Average Particle Size (nm) |
|---|---|---|---|
| Pure HNT | 61 | 11 | 93 |
| HNT after ultra-sonication | 35 | 17 | 173 |
| PP-HNT | 15 | 24 | 414 |
| G-HNT | 8 | 37 | 774 |

### 2.2. Halloysite Modification and Characterization

Modified halloysite was obtained in a two-stage process published elsewhere [37]. In the first stage, the suspension of native halloysite in demineralized water was subjected to an ultrasonic field with a power of 250–350 W and frequency of 20–40 kHz for 2–3 h. After ultra-sonication, a solution of modifying agent in ethyl alcohol was added to the suspension of halloysite in water and stirred for 3 h using a mechanical stirrer in an ultrasonic field. After evaporation of the alcohol, the precipitate was ground in a ball mill.

A BET-N2 sorption method (Tri Star II 3010, Micromeritics, Norcross, GE, USA) was used to determine the specific surface area of halloysite nanoparticles. The morphology of pure and modified halloysite nanotubes was studied by scanning electron microscopy (SEM) at 20 kV, using a Jeol JSM-6490LV microscope (Japan).

### 2.3. Thermoplastic Starch Processing

Plasticized starch was obtained in a two-stage process published elsewhere [37]. At first, 69 wt.% native starch, 10 wt.% glycerol and 20 wt.% sorbitol, and 1 wt.% of halloysite were blended in a laboratory mixer (LMX5-S-VS, Labtech Engineering Co. Ltd., Thailand) for 2 min at a speed of 1000 rpm. Then, the stirrer speed was reduced to 400 rpm and glycerol was introduced, after which the stirrer speed was increased to 1300 rpm. After reaching a temperature of 65 °C, the process was terminated. The homogenized starch was dried at 110 °C for 2 h (to a water content of about 10%), and then plasticized in a KraussMaffei Berstorff (Munich, Germany) twin-screw extruder with a screw diameter of 25 mm and a length of 51D. The process was carried out at a temperature of 40–160 °C and a screw speed of 100 rpm. Test samples were injection molded at 155–175 °C using an Arburg 420 M single screw injection machine (Allrounder 1000-250, Lossburg, Germany). The mold temperature was 20 °C.

### 2.4. Methods

Fourier infrared spectroscopy (FTIR) (Thermo Fisher Scientific, model Nicolet 6700, Waltham, MA, USA) was used to analyze the chemical structure. The spectra were recorded using at least 64 scans with 2 $cm^{-1}$ resolution, in the spectral range of 4000–500 $cm^{-1}$, using a KBr pellets technique. At least three samples of each halloysite and starch were evaluated.

The starch morphology was characterized by SEM in high vacuum using a Joel JSM 6100 microscope (Tokyo, Japan) operating at 20 kV. The impact fractured surface of the injection molded samples was coated with a thin gold layer to avoid charging and increase image contrast. At least three samples of each starch were evaluated.

The viscoelastic properties were studied by torsion dynamic thermomechanical analysis (DMTA) using a dynamic analyzer (Rheometrics RDS 2, Rheometric Scientific Inc., Piscataway, NJ, USA) at a frequency of 1 Hz. The strain level was 0.1%. Data were collected from −150 to 100 °C at a heating rate of 3 °C/min. The specimens (38 mm × 10 mm × 2 mm) were cut from injection molded samples. At least three samples of each starch were evaluated.

Thermal stability was determined by thermogravimetric analysis (TGA) using a TGA/SDTA 851e thermogravimetric analyzer (Metler Toledo, Greifensee, Switzerland) in a nitrogen atmosphere at heating rate of 10 °C/min, from 25 to 700 °C. The temperatures of 5% weight loss ($T_{5\%}$), 10% weight loss ($T_{10\%}$), and total weight loss at 700 °C were determined. In addition, the maximum decomposition rate temperature ($T_{max}$) was determined from the differential thermogravimetric curve, which is the first derivative of the TG curve. The minima visible on DTG curve correspond to $T_{max}$. At least three samples of each halloysite and starch were evaluated.

Differential scanning calorimetry (DSC) analysis was carried out using a DSC 822e apparatus (Metler Toledo, Switzerland) at a temperature from −50 °C to 200 °C in an argon atmosphere. The heating rate was 10 °C/min, and the argon flow rate was 50 mL/min. At least three samples of each halloysite and starch were tested.

The Vicat softening point was determined using HV3 apparatus by CEAST, Italy, according to ISO 306, method A (load 10 N, heating rate 50 °C/h).

The tensile and flexural properties were measured using an Instron 5500R universal testing machine (Massachusetts, UK) according to ISO 527 and ISO 178, respectively. The crosshead speeds for tensile and flexural tests were 5 and 2 mm/min, respectively. The gage length for tensile tests was 50 mm. The average of five measurements was taken as the result.

## 3. Results and Discussion

### 3.1. Halloysite Characterization

BET confirmed that before and after ultra-sonication, pure HNT had a porous structure with an average pore size of 11 nm and 17 nm, respectively (Table 1). At the same time, the average particle size increased from 93 nm to 173 nm. PP-HNT and G-HNT also showed a porous structure, and their average pore size gradually increased from 11 nm in HNT to 24 nm and 37 nm, respectively. A significant reduction in the specific surface area of the halloysite after modification and an increase in the size of pores and particles compared to pure halloysite indicated the build-up of the applied organic compound on the HNT surface (Table 1), confirmed by SEM, as shown in Figure 1. These observations are consistent with other studies [38,39].

#### 3.1.1. Infrared Spectroscopy (FTIR)

FTIR spectra of pure and modified halloysite are presented in Figures 2 and 3. In the hydroxyl region (4000–3000 $cm^{-1}$), the HNT spectrum shows two distinct peaks at 3690 and 3618 $cm^{-1}$, corresponding to the stretching vibration of the $Al_2$-OH groups' inner-surface and the deformation vibration of interlayer water, respectively [31]. The weak peak at 1630 $cm^{-1}$ is related to the OH stretching and deformation vibration of the adsorbed water, respectively. The bands at wavenumbers 1113 and 1023 $cm^{-1}$ are assigned to the Si-O in-plane stretching vibration and perpendicular Si-O-Si stretching vibration, respectively. The band at wavenumber 907 $cm^{-1}$ corresponds to the O-H deformation of the internal hydroxyl groups, and that at 787 $cm^{-1}$ is due to a symmetrical Si-O stretching vibration. The bands at 744 and 687 $cm^{-1}$ are attributed to the perpendicular Si-O stretching vibration. All these observations suggest the presence of more than one type of water in the halloysite structure, and are consistent with previous reports on halloysite [31,40–44].

**Figure 1.** SEM images of pure HNT (**a**), PP-HNT (**b**) and G-HNT (**c**).

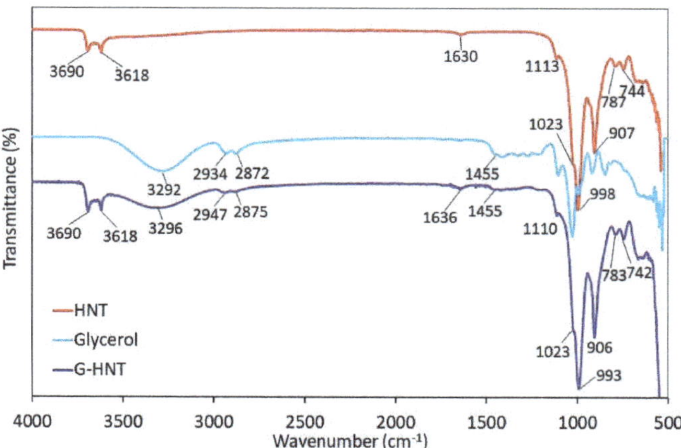


**Figure 2.** FTIR spectra of HNT, glycerol and G-HNT in the 4000–500 $cm^{-1}$ region.

In the case of PP-HNT and G-HNT, it was clear that the characteristic peaks of HNT were maintained. In the G-HNT spectrum, the presence of new peaks at the wavenumbers 3296 $cm^{-1}$ (OH vibration), 2947 $cm^{-1}$ (asymmetrical CH stretching vibration), 2875 $cm^{-1}$ (symmetrical CH stretching vibration), and 1455 $cm^{-1}$ (CH bending vibration), belonging to glycerol, prove the interaction of glycerol with HNT. In addition, these bands were slightly shifted compared to glycerol, and their intensity decreased significantly. The band of adsorbed water at 1630 $cm^{-1}$ shifted to 1636 $cm^{-1}$. The bands at 1113, 998, and 787$cm^{-1}$ shifted slightly to 1110, 993 and 783 $cm^{-1}$, respectively. Compared to the pure HNT, the spectrum of PP-HNT (Figure 3) shows new peaks at 2954 $cm^{-1}$ (C−H stretching vibration), 2926 $cm^{-1}$ (asymmetrical C−H stretching vibration), 2862 $cm^{-1}$ ($CH_2$ stretching vibration), and 1636 $cm^{-1}$ (C=O stretching vibration), belonging to succinic acid, proving the interaction of polyester plasticizer with HNT. The intensity of the bands decreased significantly compared to polyester plasticizer. The band at 787 $cm^{-1}$ shifted slightly to 791 $cm^{-1}$. All these observations suggest the interactions between halloysite and the organic compounds (glycerol, polyester plasticizer).

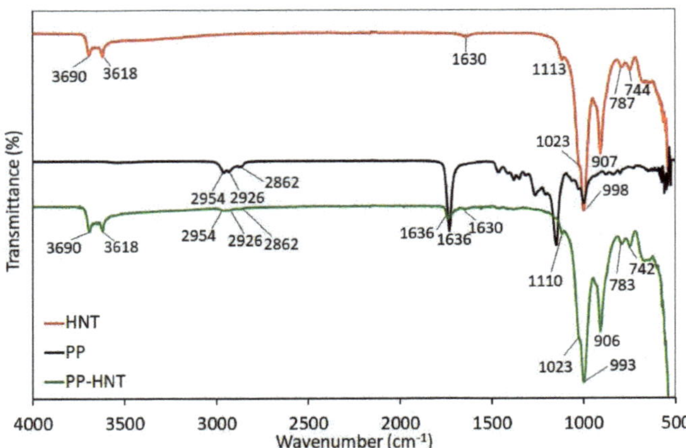


**Figure 3.** FTIR spectra of HNT, polyester plasticizer and PP-HNT in the 4000–500 $cm^{-1}$ region.

3.1.2. Thermal Properties

The thermal decomposition of glycerol, polyester plasticizer, and pure and modified halloysite is summarized in Table 2 and shown in Figure 4. The TG curve of pure HNT shows two phases of weight loss (Figure 4a,b). The mass loss from 25 °C to 150 °C is due to the removal of physically adsorbed water, while the mass loss in the temperature range of 150–403 °C is related to the removal of interlayer residual water. The mass loss above 403 °C is due to Al-OH and Si-OH groups dehydroxylation [45,46]. In contrast, PP-HNT and G-HNT show three weight loss phases with a gradual loss up to 262 °C and 164 °C, respectively, followed by two steeper weight losses (Figure 4a,b). The first mass loss up to about 100 °C is attributed to the adsorbed water loss, the second loss is due to the glycerol or polyester plasticizer loss (Figure 4c,d), and the third is due to the halloysite dehydroxylation. There was a significant decrease in the 5% weight loss temperature ($T_{5\%}$) compared to pure HNT results from the presence of an organic compound (glycerol, polyester plasticizer) in the halloysite structure, which degrades at a much lower temperature (Figure 4c,d, Table 2). The HNT modification caused a decrease in a maximum rate of decomposition temperature ($T_{max3}$) compared to the $T_{max}$ of pure HNT (Figure 4b).

**Table 2.** TGA data of glycerol, polyester plasticizer, pure and modified halloysite.

| Halloysite | $T_{5\%}$ (°C) | $T_{max}$ (°C) | Residue (%) |
|---|---|---|---|
| Glycerol | 182 | 258 | 0 |
| Polyester plasticizer | 261 | 275, 395, 407 | 3 |
| Pure HNT | 403 | 58, 491 | 86 |
| PP-HNT | 262 | 56, 268, 483 | 79 |
| G-HNT | 164 | 58, 203, 482 | 72 |

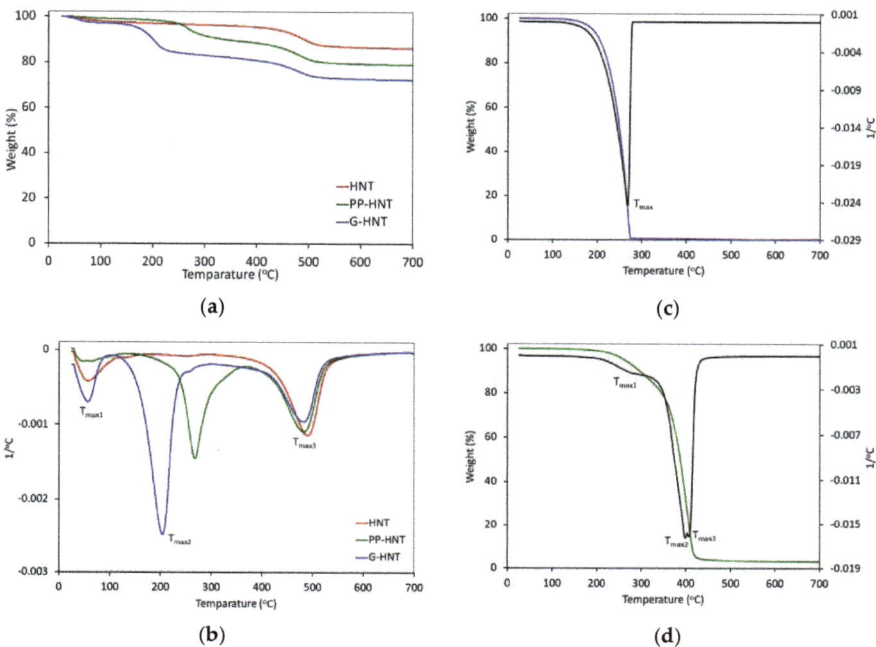


**Figure 4.** TG and DTG curves of halloysites (**a**,**b**), glycerol (**c**), and polyester plasticizer (**d**).

The DSC curves of pure and modified HNT show phase change as a function of temperature (Figure 5). Pure HNT and G-HNT clearly show typical endothermic peaks at 106 °C and 114 °C, respectively, where absorbed water melts. In contrast, PP-HNT shows two endothermic peaks at 56 °C and 88 °C, which can also be attributed to the melting of absorbed water. These results are consistent with the FTIR observations and other studies [47,48].

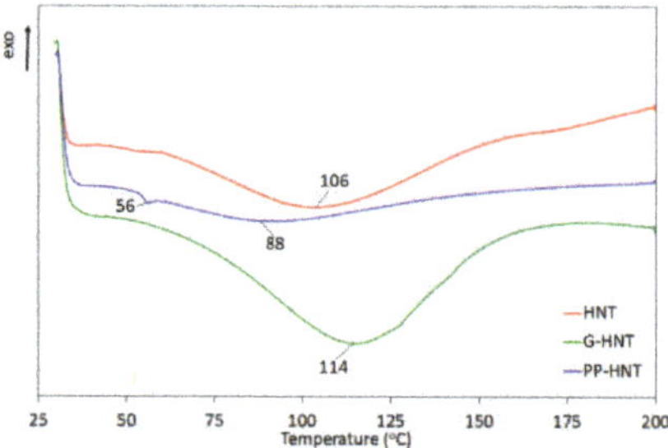


**Figure 5.** DSC curves of pure and modified HNT.

### 3.2. *Thermoplastic Starch*

#### 3.2.1. Chemical Structure and Molecular Interactions

The chemical structure and molecular interactions of native starch, plasticizers and halloysite were studied by FTIR spectroscopy. Unmodified and modified TPS show similar FTIR absorption bands (Figure 6). The broad band between 3660 cm$^{-1}$ and 2990 cm$^{-1}$ is associated with complex vibrational stresses of free, inter-, and intramolecular bonds of hydroxyl groups [49,50]. The peak at 2936 cm$^{-1}$ is related to C–H stretching (–$CH_2$) of the anhydrous glucose ring [51], and the peak at 1646 cm$^{-1}$ to water resulting from the starch hygroscopicity [49]. It is known that the water present in starch is strongly and directly bound to the starch molecules through the ion-dipole interaction, which is stronger than the normal water-water bond [51]. The peaks between 1149 cm$^{-1}$ and 1012 cm$^{-1}$ indicate the interactions between starch and plasticizers molecules [51]. The peak at 998 cm$^{-1}$ is attributed to intramolecular hydrogen bonding of hydroxyl groups or the water plasticizing effect [50]. Moreover, the intensity of the peaks with maximum at 1012 cm$^{-1}$ and 998 cm$^{-1}$ is lower for TPS/HNT and TPS/G-HNT compared to TPS, but higher for TPS/PP-HNT, demonstrating the molecular interactions between starch, plasticizers (glycerol/sorbitol) and halloysite. Similar observations on the interactions between starch and plasticizers have been reported in other studies [52,53].

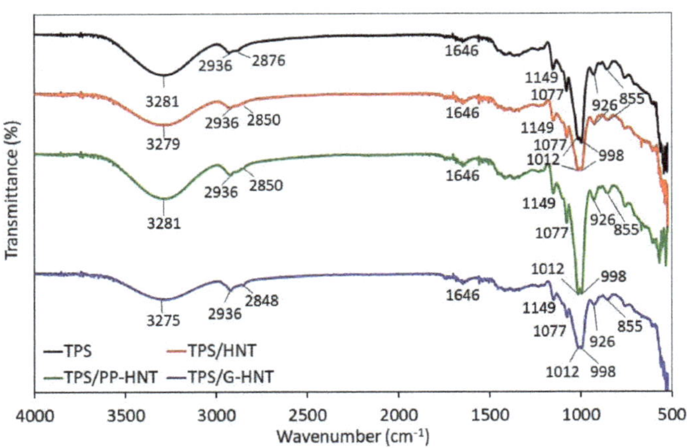


**Figure 6.** FTIR spectra of unmodified and halloysite-modified TPS in the 4000–500 cm$^{-1}$ region.

### 3.2.2. Morphology

The morphology of native and plasticized starch was observed by SEM, as illustrated in Figure 7. TPS shows a uniform morphology with no visible remains of starch granules, which suggests that the plasticizers (glycerol, sorbitol), together with heat and shear stress in the extrusion process, effectively destroyed the starch granules. SEM images indicate that booth pure and modified HNTs are evenly distributed in TPS matrix (Figure 7b–e). TPS, TPS/PP-HNT and TPS/G-HNT exhibit rough fracture surfaces (Figure 7b,d,e). However, the roughest surface is observed for TPS/PP-HNT (Figure 7d), while TPS/HNT shows the smoothest surface (Figure 7c). It is worth noticing to note that the addition of HNT and G-HNT improves TPS homogeneity, confirming the greater plasticizing effect. Based on the SEM results, it can be concluded that TPS/PP-HNT is the most flexible, while TPS/HNT the most brittle, which is also confirmed by the mechanical properties.

### 3.2.3. Viscoelastic Behavior

Dynamic thermomechanical analysis (DMA) was performed to investigate the interfacial interactions between starch and halloysite. Two distinct decreases in the storage modulus (G′) can be observed in Figure 8a. The first one at −50 °C corresponds to the β relaxation related to the crystalline structure of thermoplastic starch. The second decrease occurs in the temperature range from 30 to 70 °C and is associated with the α relaxation transition ($T_\alpha$), which probably corresponds to the melting process of thermoplastic starch. Compared to TPS, the storage modulus of TPS modified with halloysite shows a significant increase in the following order: TPS/G-HNT < TPS/PP-HNT < TPS/HNT, which indicates higher stiffness. This is due to the limitations of the starch chains' segmental movement [9,54] and the presence of nanotubes with a high modulus and high aspect ratio [55]. This phenomenon is particularly evident above the glass transition of the starch-rich phase ($T_\alpha$).

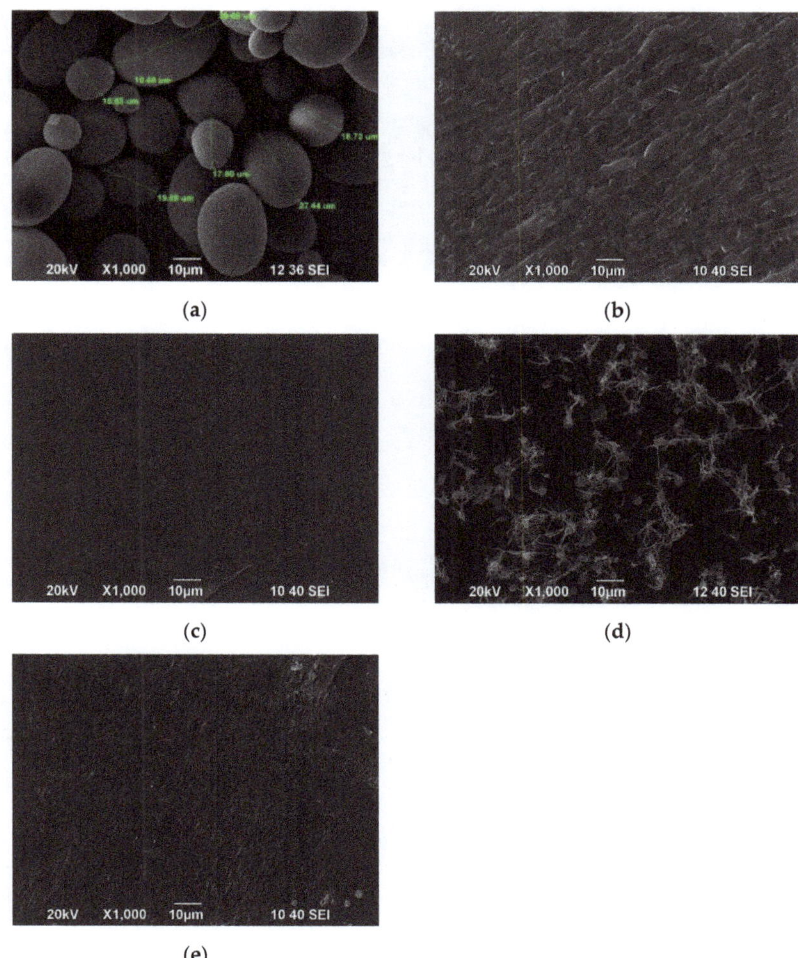


**Figure 7.** SEM images of native starch (**a**), TPS (**b**), TPS/HNT (**c**), TPS/PP-HNT (**d**) and TPS/G-HNT (**e**).

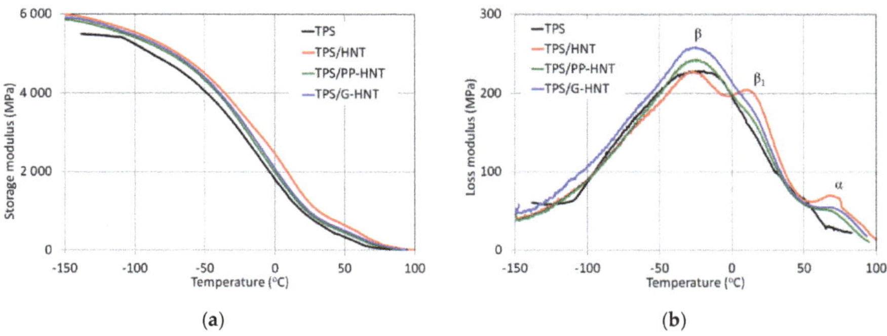


**Figure 8.** Storage modulus (**a**) and loss modulus (**b**) of unmodified and halloysite-modified TPS.

In the case of TPS, the loss modulus (G″) shows two peaks at −21 °C and 50 °C (Figure 8b). The G″ peaks are known as polymer relaxations related to glass transition temperature ($T_g$) or secondary transformations. The upper transition is attributed to the glass transition of the starch-rich phase ($T_\alpha$), while the lower one is attributed to the glass transition of the glycerol-rich phase ($T_\beta$) [9,26,55]. An additional, a β relaxation peak is observed only for pure HNT, which suggests the formation of a new crystalline phase. All modified TPS show lower $T_\beta$ and higher $T_\alpha$ (Table 3). A decrease in $T_\beta$ suggests the presence of more flexible regions or greater mobility of the starch chains fragments in the crystalline phase, indicating a greater plasticizing effect. In contrast, an increase in $T_\alpha$ suggests the presence of a stiffer regions or less mobility of the starch chain fragments in the amorphous phase, confirming a bigger reinforcing effect. It can be concluded that HNTs behave simultaneously as a plasticizing and reinforcing agent. A similar phenomenon has been reported in other studies [9,26].

**Table 3.** Storage modulus and relaxation temperature of unmodified and halloysite-modified TPS.

| Sample | Storage Modulus (MPa) | | Relaxation Temperature (°C) | | |
|---|---|---|---|---|---|
| | −21 °C | 23 °C | $T_\alpha$ | $T_\beta$ | $T_{\beta1}$ |
| TPS | 2820 | 900 | 50 | −22 | – |
| TPS/HNT | 3400 | 1380 | 70 | −27 | 12 |
| TPS/PP-HNT | 3100 | 1040 | 69 | −25 | – |
| TPS/G-HNT | 3090 | 1120 | 72 | −25 | – |

3.2.4. Thermal Properties

Thermogravimetric analysis (TGA) was used to determine the temperature above which the degradation and destruction of thermoplastic starch begins. The addition of 1 wt.% halloysite affects the thermal stability of TPS, depending on its modification (Figure 9a, Table 4). The highest thermal stability was obtained for TPS/HNT, as evidenced by an increase in $T_{5\%}$ and $T_{10\%}$ by 19 °C and 17 °C, respectively, compared to unmodified TPS. The incorporation of G-HNT into TPS also increases the decomposition onset temperature (9 °C) and 10% mass loss temperature (3 °C). Thus, due to high thermal stability and good interaction with native starch, glycerol and sorbitol, HNT and G-HNT improve the thermal stability of starch [30,35]. However, as expected, PP-HNT slightly decreases the thermal resistance of TPS, as evidenced by lower thermal decomposition temperatures. These results are consistent with morphology analysis (Figure 7) and mechanical properties (Figure 10).

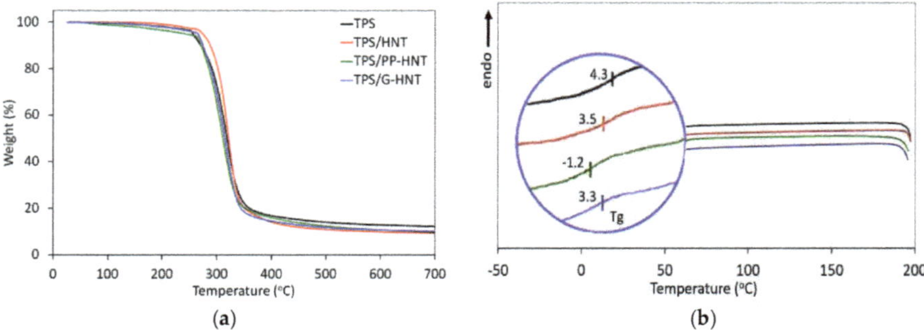


**Figure 9.** TGA (**a**) and DSC (**b**) thermograms of unmodified and modified TPS.

**Table 4.** Thermal properties of unmodified and modified TPS.

| Sample | $T_{5\%}$ (°C) | $T_{10\%}$ (°C) | Residue (%) | $T_{max}$ (°C) | $T_g$ (°C) | Vicat (°C) |
|---|---|---|---|---|---|---|
| TPS | 253 | 268 | 12 | 313 | 4.3 | 65 |
| TPS/HNT | 272 | 285 | 10 | 315 | 3.5 | 76 |
| TPS/PP-HNT | 243 | 268 | 10 | 307 | −1.2 | 72 |
| TPS/G-HNT | 262 | 271 | 10 | 313 | 3.3 | 67 |

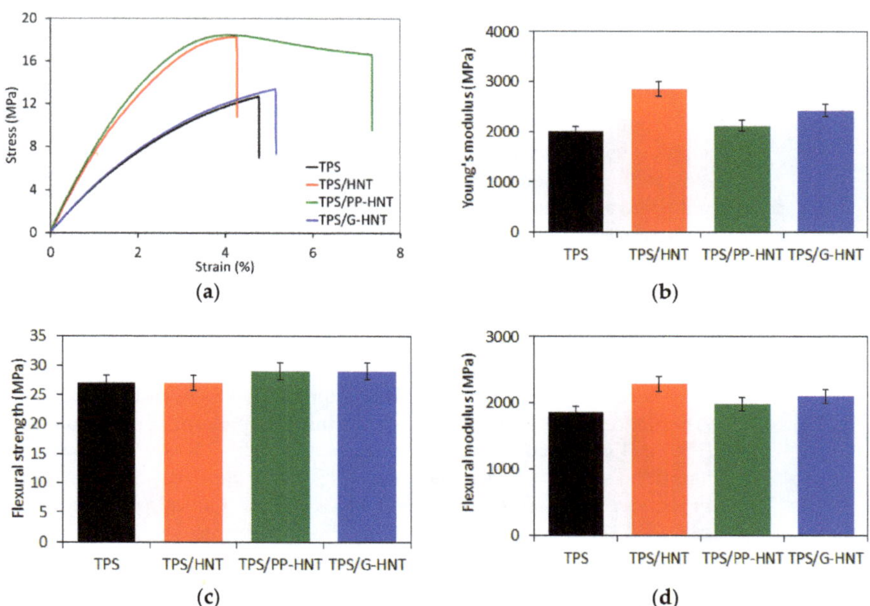


**Figure 10.** Stress–strain curves (**a**), Young's modulus (**b**), flexural strength (**c**) and flexural modulus (**d**) of unmodified and halloysite-modified TPS.

The DSC curves of unmodified and halloysite-modified TPS show no phase change as a function of temperature (Figure 9b), suggesting that TPS is fully amorphous. In general, halloysite slightly reduced the glass transition temperature ($T_g$) of TPS. $T_g$ was observed from −1.2 °C to 4.3 °C, depending on halloysite modification. TPS/PP-HNT showed the lowest $T_g$, suggesting better flexibility, which was confirmed by mechanical properties (Figure 10) and SEM observations (Figure 7).

An important parameter indicating resistance to hot water is the Vicat softening point. The addition of halloysite causes a significant increase in the Vicat temperature by 3–11 °C, suggesting better hot water resistance. The highest Vicat point was obtained for starch containing pure HNT, and the lowest for starch with the addition of G-HNT.

#### 3.2.5. Mechanical Properties

The tensile and flexural properties of unmodified and halloysite-modified TPS are presented in Figure 10. It is clear from Figure 10a that HNT reduces the elongation at break while increasing the tensile strength. In contrast, modified halloysite improves elongation. The highest elongation and tensile strength were obtained using PP-HNT, which confirms the plasticizing and reinforcing effect of halloysite. Moreover, TPS had significantly lower flexural strength and Young's modulus than the modified with halloysite. In the case of TPS/HNT, the tensile strength and Young's modulus increased up to 39% and 45%, respectively. After adding the modified halloysite, the tensile and flexural modulus of TPS decreased while the flexural strength increased, suggesting that the starch became more flexible compared to TPS/HNT. This explains the reduction in starch–starch intermolecular interactions, which results in an increase in the free volume and mobility of the starch chains. These findings are consistent with previous morphology analysis and DMTA results. The improvement of tensile and flexural properties results from the high aspect ratio and high mechanical properties of halloysites in combination with good interactions between starch, plasticizers and halloysite. Indeed, the good interfacial adhesion between the halloysites and the starch together with the uniform dispersion of the halloysites in the starch matrix enables effective stress transfer from the matrix to the reinforcement, resulting in an increase in tensile and flexural strength. Similar results were reported in other studies [35,52].

## 4. Conclusions

Potato thermoplastic starch containing pure halloysite, glycerol-modified halloysite or polyester plasticizer-modified halloysite was obtained in a twin-screw extruder. Ultrasonication was used to introduce more organic compound into the porous structure of halloysite and improve molecular interactions between native starch, plasticizers and halloysite. BET, FTIR and SEM confirmed successive incorporation of glycerol and polyester plasticizer into halloysite. However, modified halloysite showed lower thermal stability then pure HNT due to the presence of the organic compound (glycerol, polyester plasticizer) in the halloysite structure, which degrades at a much lower temperature. FTIR confirmed the interactions between starch, plasticizers (glycerol, sorbitol) and halloysite. HNT and G-HNT, due to high thermal stability and good interaction with TPS components, increased its heat resistance. Moreover, HNTs improved stiffness of thermoplastic potato starch, as evidenced by the higher storage modulus, depending on the halloysite modification. TPS/HNT exhibited the highest stiffness, while TPS/PP-HNT the lowest stiffness. DMTA confirmed simultaneous reinforcing and plasticizing effect of halloysite, as evidenced by an increase in $T_\alpha$ and a decrease in $T_\beta$. The mechanical properties were significantly improved after addition of pure and modified halloysites into starch matrix due to uniform distribution of halloysite in the starch matrix. Pure HNT showed a significantly higher Young's modulus and flexural modulus than modified halloysites, but a lower elongation at break, tensile and flexural strength compared to PP-HNT. In addition, hot water resistance was improved due to a significantly higher Vicat softening point.

Potato starch is a cheap, renewable, easily available, and particularly good raw material for both physical and chemical modification, while halloysite is a natural and biocompatible raw material. Therefore, biodegradable starch-based materials modified with halloysite can replace conventional non-degradable polymers in single-use plastic products. Such studies are currently being conducted by our research group and will be the subject of a future publications.

## 5. Patents

The results of the work are protected by the Polish patent application P 441 782 (2022).

**Author Contributions:** Conceptualization, R.J.; methodology, R.J. and A.S.; software, A.S.; validation, R.J. and A.S.; formal analysis, R.J.; investigation, R.J., A.S., M.S., M.C. and E.S.; resources, R.J.; data curation, R.J.; writing—original draft preparation, R.J.; writing—review and editing, R.J.; visualization, A.S.; super-vision, R.J.; project administration, R.J.; funding acquisition, R.J. All authors have read and agreed to the published version of the manuscript.

**Funding:** This research received no external funding.

**Institutional Review Board Statement:** Not applicable.

**Informed Consent Statement:** Not applicable.

**Data Availability Statement:** The data that supported the findings of this study are available from the corresponding author upon request.

**Conflicts of Interest:** The authors declare no conflict of interest.

## References


1. Yottakot, S.; Leelavatcharamas, V. Isolation and optimization of polylactic acid (PLA)-packaging-degrading Actinomycete for PLA-packaging degradation. *Pertanika J. Trop. Agric. Sci.* **2019**, *42*, 1111–1129.
2. Srivastava, A.; Srivasatva, A.K.; Singh, A.; Singh, P.; Verma, S.; Vats, M.; Sagadevan, S. Biopolymers as renewable polymeric materials for sustainable development-an overview. *Polimery* **2022**, *67*, 185–197. [CrossRef]
3. Halimatul, M.J.; Sapuan, S.M.; Jawaid, M.; Ishak, M.R.; Ilyas, R.A. Effect of sago starch and plasticizer content on the properties of thermoplastic films: Mechanical testing and cyclic soaking-drying. *Polimery* **2019**, *64*, 422–431. [CrossRef]
4. Borkowski, K. Pollution of seas and oceans with plastic waste as an accelerator of new legal regulations in this area (in Polish). *Polimery* **2019**, *64*, 759–763. [CrossRef]
5. Bangar, S.P.; Whiteside, W.S.; Ashogbon, A.O.; Kumar, M. Recent advances in thermoplastic starches for food packaging: A review. *Food Packag. Shelf Life* **2021**, *30*, 1007–1043. [CrossRef]
6. de Kock, L.; Sadan, Z.; Arp, R.; Upadhyaya, P. A Circular economy response to plastic pollution: Current policy landscape and consumer perception. *S. Afr. J. Sci.* **2020**, *116*, 1–2. [CrossRef]
7. Razak, N.S.; Mohamed, R. Antimicrobial sustainable biopolymers for biomedical plastics applications–an overview. *Polimery* **2021**, *66*, 574–583. [CrossRef]
8. Haghighi, H.; Licciardello, F.; Fava, P.; Siesler, H.W.; Pulvirenti, A. Recent advances on chitosan-based films for sustainable food packaging applications. *Food Packag. Shelf Life* **2020**, *26*, 1005–1051. [CrossRef]
9. Swierz-Motysia, B.; Jeziórska, R.; Szadkowska, A.; Piotrowska, M. Preparation and properties of biodegradable polylactide and thermoplastic starch blends (in Polish). *Polimery* **2011**, *56*, 271–280. [CrossRef]
10. Sin, L.T.; Tueen, B.S. *Polylactic Acid: A Practical Guide for the Processing, Manufacturing, and Applications of PLA*, 2nd ed.; William Andrew: Oxford, UK, 2019.
11. Directive (EU) 2019/904 of the European Parliament and of the Council of 5 June 2019 on the Reduction of the Impact of Certain Plastic Products on the Environment. Available online: https://eur-lex.europa.eu/eli/dir/2019/904/oj (accessed on 7 March 2023).
12. Ai, Y.; Jane, J. Understanding starch properties and functionality. In *Starch Food*, 2nd ed.; Sjöö, M., Nilsson, L., Eds.; Woodhead Publishing: Oxford, UK, 2018; pp. 151–178.
13. Agarwal, S. Major factors affecting the characteristics of starch based biopolymer films. *Eur. Polym. J.* **2021**, *160*, 110788. [CrossRef]
14. Mlynarczyk, K.; Longwic, F.; Podkoscielna, B.; Klepka, T. Influence on natural fillers on the thermal and mechanical properties of epoxy resin composites. *Polimery* **2022**, *67*, 102–109. [CrossRef]
15. Zhang, Y.; Rempel, C.; McLaren, D. Thermoplastic starch. In *Innovation in Food Packaging*, 2nd ed.; Han, J., Ed.; Academic Press: Cambridge, MA, USA, 2014; pp. 391–412.
16. Ross, A.S. Starch in foods. In *Food Carbohydrate Chemistry*; Wrolstad, R.E., Ed.; John Wiley & Sons Ltd.: Oxford, UK, 2012; pp. 107–133.
17. Perez, S.; Baldwin, P.M.; Gallant, D.J. Structural features of starch granules. In *Starch Chemistry and Technology*; BeMiller, J., Whistler, R., Eds.; Academic Press: New York, NY, USA, 2009; pp. 149–192.
18. Zhou, Y.; Zhao, D.; Winkworth-Smith, C.G.; Wang, Y.; Liang, J.; Foster, T.J.; Cheng, Y. Effect of a small amount of sodium carbonate on konjac glucomannan-induced changes in wheat starch gel. *Carbohydr. Polym.* **2014**, *114*, 357–364. [CrossRef]
19. Ma, X.F.; Yu, J.G.; Wang, N. Fly ash reinforced thermoplastic starch composites. *Carbohydr. Polym.* **2007**, *67*, 32–39. [CrossRef]
20. Stepto, R.F.T. The processing of starch as a thermoplastic. *Macromol. Symp.* **2003**, *201*, 203–212. [CrossRef]
21. Tabassi, N.; Moghbeli, M.R.; Ghasemi, I. Thermoplastic starch/cellulose nanocrystal green composites prepared in an internal mixer. *Iran. Polym. J.* **2016**, *25*, 45–57. [CrossRef]
22. Huang, L.; Han, X.; Chen, H.; An, S.; Zhao, H.; Xu, H.; Huang, C.; Wang, S.; Liu, Y. Preparation and barrier performance of layer-modified soil-stripping/cassava starch composite films. *Polymers* **2020**, *12*, 1611. [CrossRef] [PubMed]
23. Mansour, G.; Zoumaki, M.; Marinopoulou, A.; Tzetzis, D.; Prevezanos, M.; Raphaelides, S.N. Characterization and properties of non-granular thermoplastic starch-Clay biocomposite films. *Carbohydr. Polym.* **2020**, *245*, 116629. [CrossRef]
24. Ashaduzzaman, M.; Saha, D.; Rashid, M.M. Mechanical and thermal properties of self-assembled kaolin-doped starch-based environment-friendly nanocomposite films. *J. Compos. Sci.* **2020**, *4*, 38. [CrossRef]
25. Lai, D.S.; Osman, A.F.; Adnan, S.A.; Ibrahim, I.; Alrashdi, A.A.; Ahmad Salimi, M.N.; Ul-Hamid, A. On the use of OPEFB-derived microcrystalline cellulose and nano-bentonite for development of thermoplastic starch hybrid bio-composites with improved performance. *Polymers* **2021**, *13*, 897. [CrossRef]
26. Jeziorska, R.; Szadkowska, A.; Spasowka, E.; Lukomska, A.; Chmielarek, M. Characteristics of biodegradable polylactide/thermoplastic starch/nanosilica composites: Effects of plasticizer and nanosilica functionality. *Adv. Mater. Sci. Eng.* **2018**, 15. [CrossRef]
27. Adewale, P.; Yancheshmeh, M.S.; Lam, E. Starch modification for non-food, industrial applications: Market intelligence and critical review. *Carbohydr. Polym.* **2022**, *291*, 119590. [CrossRef]
28. da Costa, J.C.M.; Miki, K.S.L.; da Silva Ramos, A.; Teixeira-Costa, B.E. Development of biodegradable films based on purple yam starch/chitosan for food application. *Heliyon* **2020**, *6*, e03718. [CrossRef] [PubMed]

29. He, Y.; Kong, W.; Wang, W.; Liu, T.; Liu, Y.; Gong, Q.; Gao, J. Modified natural halloysite/potato starch composite films. *Carbohydr. Polym.* **2012**, *87*, 2706–2711. [CrossRef]
30. Xie, Y.; Chang, P.R.; Wang, S.; Yu, J.; Ma, X. Preparation, and properties of halloysite nanotubes/plasticized *Dioscorea opposite Thunb.* starch composites. *Carbohydr. Polym.* **2011**, *83*, 186–191. [CrossRef]
31. Zhang, A.B.; Pan, L.; Zhang, H.Y.; Liu, S.T.; Ye, Y.; Xia, M.S.; Chen, X.G. Effects of acid treatment on the physico-chemical and pore characteristics of halloysite. *Colloids Surf. A Physicochem. Eng. Asp.* **2012**, *396*, 182–188. [CrossRef]
32. Yendluri, R.; Lvov, Y.; de Villiers, M.M.; Vinokurov, V.; Naumenko, E.; Tarasova, E.; Fakhrullin, R. Paclitaxel encapsulated in halloysite clay nanotubes for intestinal and intracellular delivery. *J. Pharm. Sci.* **2017**, *106*, 3131–3139. [CrossRef] [PubMed]
33. Liu, M.; Jia, Z.; Jia, D.; Zhou, C. Recent advance in research on halloysite nanotubes polymer nanocomposite. *Prog. Polym. Sci.* **2014**, *39*, 1498–1525. [CrossRef]
34. Balikile, R.D.; Roy, A.S.; Nagaraju, S.C.; Ramgopal, G. Conductivity properties of hollow $ZnFe_2O_4$ nanofibers doped polyaniline nanocomposites. *J. Mater. Sci. Mater. Electron.* **2017**, *28*, 7368–7375. [CrossRef]
35. Schmitt, H.; Prashantha, K.; Soulestin, J.; Lacrampe, M.F.; Krawczak, P. Preparation, and properties of novel melt-blended halloysite nanotubes/wheat starch nanocomposites. *Carbohydr. Polym.* **2012**, *89*, 920–927. [CrossRef]
36. Wardzinska-Jarmulska, E.; Szczepaniak, B.; Szczepankowska, B.; Modzelewska, A.; Stanecka, J.; Badowska, A.; Potajczuk-Czaja, K.; Grzybek, R. Method of Obtaining Unplasticized Polyester Plasticizer. Polish Patent 236 221, 13 August 2020.
37. Jeziorska, R.; Legocka, I.; Szadkowska, A.; Spasowka, E.; Zubrowska, M.; Studzinski, M.; Wierzbicka, E.; Dzierżawski, J.; Kolasa, J.; Rucinski, J. Method of Producing Modified Thermoplastic Starch and Biodegradable Composites Containing Modified Thermoplastic Starch. Polish Patent Application 441 782, 19 July 2022.
38. Li, Y.; Yuan, X.; Guan, X.; Bai, J.; Wang, H. One-pot synthesis of siliceous ferrihydrite–coated halloysite nanorods in alkaline medium: Structure, properties, and cadmium adsorption performance. *J. Colloid Interface Sci.* **2023**, *636*, 435–444. [CrossRef]
39. Sid, D.; Baitiche, M.; Bourzami, R.; Merir, R.; Djerboua, F.; Gil, A.; Boutahala, M. Experimental and theoretical studies of the interaction of ketoprofen in halloysite nanotubes. *Colloids Surfaces A Physicochem. Eng. Asp.* **2021**, *627*, 127136. [CrossRef]
40. Joussein, E.; Petit, S.; Churchman, J.; Theng, B.; Righi, D.; Delvaux, B. Halloysite clay minerals—A review. *Clay Miner.* **2005**, *40*, 383–426. [CrossRef]
41. Mellouk, S.; Cherifi, S.; Sassi, M.; Marouf-Khelifa, K.; Bengueddach, A.; Schott, J.; Khelifa, A. Intercalation of halloysite from Djebel Debagh (Algeria) and adsorption of copper ions. *Appl. Clay Sci.* **2009**, *44*, 230–236. [CrossRef]
42. Frost, R.L.; Kristof, J.; Schmidt, J.M.; Kloprogge, J.T. Raman spectroscopy of potassium acetate-intercalated kaolinites at liquid nitrogen temperature. *Spectrochim. Acta Part A Mol. Biomol. Spectrosc.* **2001**, *57*, 603–609. [CrossRef]
43. Cheng, H.; Frost, R.L.; Yang, J.; Liu, Q.; He, J. Infrared, and infrared emission spectroscopic study of typical Chinese kaolinite and halloysite. *Spectrochim. Acta Part A Mol. Biomol. Spectrosc.* **2010**, *77*, 1014–1020. [CrossRef]
44. Horvath, E.; Kristov, J.; Frost, R.L. Vibrational spectroscopy of intercalated kaolinites. Part I. *Appl. Spectrosc. Rev.* **2010**, *45*, 130–147. [CrossRef]
45. Hendessi, S.; Sevinis, E.B.; Unal, S.; Cebeci, F.C.; Menceloglu, Y.Z.; Unal, H. Antibacterialsustained-release coatings from halloysite nanotubes/water borne polyurethanes. *Prog. Org. Coat.* **2016**, *101*, 253–261. [CrossRef]
46. Yuan, P.; Tan, D.; Annabi-Bergaya, F.; Yan, W.; Fan, M.; Liu, D.; He, H. Changes in structure, morphology, porosity, and surface activity of mesoporous halloysite nanotubes under heating. *Clays Clay Miner.* **2012**, *60*, 561–573. [CrossRef]
47. Kittaka, S.; Ishimaru, S.; Kuranishi, M.; Matsuda, T.; Yamaguchi, T. Enthalpy and interfacial free energy changes of water capillary condensed in mesoporous silica, MCM-41, and SBA-15. *Phys. Chem. Chem. Phys.* **2006**, *8*, 3223–3231. [CrossRef]
48. Jahnert, S.; Vaca Chavez, F.; Schaumann, G.E.; Schreiber, A.; Schonhoff, M.; Findenegg, G.H. Melting and freezing of water in cylindrical silica nanopores. *Phys. Chem. Chem. Phys.* **2008**, *10*, 6039–6051. [CrossRef]
49. Zhang, Y.; Han, J.H. Plasticization of pea starch films with monosaccharides and polyols. *J. Food Sci.* **2006**, *71*, 253–261. [CrossRef]
50. Muscat, D.; Adhikari, B.; Adhikari, R.; Chaudhary, D.S. Comparative study of film forming behaviour of low and high amylose starches using glycerol and xylitol as plasticizers. *J. Food Eng.* **2012**, *109*, 189–201. [CrossRef]
51. Zullo, J.R.; Iannace, S. The effects of different starch sources and plasticizers on film blowing of thermoplastic starch: Correlation among process, elongational properties and macromolecular structure. *Carbohydr. Polym.* **2009**, *77*, 376–383. [CrossRef]
52. Dang, K.M.; Yoksan, R. Thermoplastic starch blown films with improved mechanical and barrier properties. *Int. J. Biol. Macromol.* **2021**, *188*, 290–299. [CrossRef] [PubMed]
53. Tiefenbacher, K.F. Chapter six-wafer sheet manufacturing: Technology and products. In *Wafer and Waffle Processing and Manufacturing*, 1st ed.; Tiefenbacher, K.F., Ed.; Academic Press: London, UK, 2017; pp. 405–486.
54. Zhou, W.Y.; Guo, B.; Liu, M.; Liao, R.; Bakr, A.; Jia, D. Poly(vinyl alcohol)/halloysite nanotubes bionanocomposite films: Properties and in vitro osteoblasts and fibroblasts response. *J. Biomed. Mater. Res. Part A* **2009**, *93A*, 1574–1587. [CrossRef]
55. Guimaraes, L.; Enyashin, A.N.; Seifert, G.; Duarte, H.A. Structural, electronic, and mechanical properties of single-walled halloysite nanotube models. *J. Phys. Chem. C* **2010**, *114*, 11358–11363. [CrossRef]

 MDPI

Article

# Starch-Derived Superabsorbent Polymer in Remediation of Solid Waste Sludge Based on Water–Polymer Interaction

**Juan Matmin [1,2,*], Salizatul Ilyana Ibrahim [3], Mohd Hayrie Mohd Hatta [4], Raidah Ricky Marzuki [1], Khairulazhar Jumbri [5] and Nik Ahmad Nizam Nik Malek [2,6]**

1 Department of Chemistry, Faculty of Science, Universiti Teknologi Malaysia UTM, Johor Bahru 81310, Johor, Malaysia
2 Centre for Sustainable Nanomaterials, Ibnu Sina Institute for Scientific and Industrial Research, Universiti Teknologi Malaysia UTM, Johor Bahru 81310, Johor, Malaysia
3 Centre of Foundation Studies, Universiti Teknologi MARA Cawangan Selangor, Kampus Dengkil, Dengkil 43800, Selangor, Malaysia
4 Centre for Research and Development, Asia Metropolitan University, Johor Bahru 81750, Johor, Malaysia
5 Department of Fundamental and Applied Sciences, Universiti Teknologi PETRONAS, Seri Iskandar 32610, Perak, Malaysia
6 Department of Biosciences, Faculty of Science, Universiti Teknologi Malaysia UTM, Johor Bahru 81310, Johor, Malaysia
* Correspondence: juanmatmin@utm.my; Tel.: +60-7-5535581

**Abstract:** The purpose of this study is to assess water–polymer interaction in synthesized starch-derived superabsorbent polymer (S-SAP) for the treatment of solid waste sludge. While S-SAP for solid waste sludge treatment is still rare, it offers a lower cost for the safe disposal of sludge into the environment and recycling of treated solid as crop fertilizer. For that to be possible, the water–polymer interaction on S-SAP must first be fully comprehended. In this study, the S-SAP was prepared through graft polymerization of poly (methacrylic acid-co-sodium methacrylate) on the starch backbone. By analyzing the amylose unit, it was possible to avoid the complexity of polymer networks when considering S-SAP using molecular dynamics (MD) simulations and density functional theory (DFT). Through the simulations, formation of hydrogen bonding between starch and water on the H06 of amylose was assessed for its flexibility and less steric hindrance. Meanwhile, water penetration into S-SAP was recorded by the specific radial distribution function (RDF) of atom–molecule interaction in the amylose. The experimental evaluation of S-SAP correlated with high water capacity by measuring up to 500% of distilled water within 80 min and more than 195% of the water from solid waste sludge for 7 days. In addition, the S-SAP swelling showed a notable performance of a 77 g/g swelling ratio within 160 min, while a water retention test showed that S-SAP was capable of retaining more than 50% of the absorbed water within 5 h of heating at 60 °C. The water retention of S-SAP adheres to pseudo-second-order kinetics for chemisorption reactions. Therefore, the prepared S-SAP might have potential applications as a natural superabsorbent, especially for the development of sludge water removal technology.

**Keywords:** superabsorbent polymer; water–polymer interaction; solid waste sludge treatment; starch biopolymer; water absorbency



**Citation:** Matmin, J.; Ibrahim, S.I.; Mohd Hatta, M.H.; Ricky Marzuki, R.; Jumbri, K.; Nik Malek, N.A.N. Starch-Derived Superabsorbent Polymer in Remediation of Solid Waste Sludge Based on Water–Polymer Interaction. *Polymers* **2023**, *15*, 1471. https://doi.org/10.3390/polym15061471

Academic Editors: José Miguel Ferri, Ali Reza Zanjanijam, Vicent Fombuena Borràs and Miguel Fernando Aldás Carrasco

Received: 20 January 2023
Revised: 6 March 2023
Accepted: 8 March 2023
Published: 16 March 2023



## 1. Introduction

Over the past decades, industrial solid waste sludge has primarily been disposed of in open landfills, causing hazardous contamination of water resources with toxic pollutants [1,2]. When solid waste sludge is exposed to rain, the environmental problem becomes severe. This is because toxic pollutants start to leach out of the sludge, spreading into soil and underground water [3–5]. To reduce the increasing number of open landfills for waste treatment, the use of incinerator systems is a more preferable method of getting rid of industrial wastes [6]. In the process, ash, flue gas, and heat are produced, causing air pollution as a secondary environmental issue [6–9]. Another method for treating solid waste sludge is membrane technologies, which require a long operating time and very high maintenance costs [10–13]. With prolonged use, the performance of the membrane system will decrease because of certain drawbacks such as compaction, fouling and scaling [14]. In light of sustainability and cost-effectiveness issues, the development of a more efficient solid waste sludge treatment system is urgently needed.

For more cost-effective waste management in industry, the removal of water and toxic metals from sludge to a dried non-hazardous solid is highly desirable to reduce environmental impact. In addition, treated sludge can be safely disposed of into the environment and reused as fertilizers [15]. In view of this, superabsorbent polymers (SAPs) have attracted attention for the treatment of industrial waste due to their high water absorption capacity [16,17]. For example, a petroleum-based SAP composed of acrylic acid and acrylamide is used to remove heavy metals from wastewater [18,19]. To form natural-based SAPs, grafting copolymerization strategies onto carboxymethylcellulose (CMC) [20–22] and the deacetylation process of chitin from crab, crayfish, and shrimp shells [23–25] have extensively been studied. Recently, Menceloğlu et al. (2022) developed SAP–halloysite nanotube (HNT) via free radical polymerization to enhance water retention capacities and rheological characteristics [26]. Nevertheless, the HNT filler in the SAP limits their application to solid waste sludge treatment and reduces their biodegradability. In comparison to petroleum-based SAPs, biopolymer SAPs, derived from natural resources such as cellulose, chitosan and carbohydrates, are more environmentally friendly and biodegradable [27,28].

In 2021, Chang and co-workers studied SAPs carbohydrates originating from various starches to demonstrate decent water capacity in agriculture usage [29]. Similarly, Sanders et al. (2021) used a combination of MD simulations and the grand canonical Monte Carlo (GCMC) method to study the effect of water on starch [30]. In this technique, the emulsifiers' processes are limited to single-molecule host–guest binding in dilute solutions. Meanwhile, Zhiguang et al. (2022) demonstrated an unwound double-amylose helix, and enhanced the bending degree of starch molecules using MD simulation [31]. However, the MD simulation could only be performed at high temperatures that affect the conformation of the starch molecules. Evidently, the interaction between water and polymer on biopolymer SAP, particularly starch, can be difficult to correlate, either experimentally or through simulation.

Nonetheless, it is necessary to design an experiment to improve the water retention capacities of SAPs. The main problems with finding suitable biopolymers to prepare SAPs lie in the complexity of polymer networks. To tackle these problems, this work is the first investigation to be conducted on the starch-derived SAP (S-SAP) using MD simulations and density functional theory (DFT), based on their simplest polymer backbone, the amylose unit, to determine the water–polymer interactions. To support the simulation findings, the water retentions of S-SAP in distilled water and solid waste sludge were evaluated and confirmed using Fourier transform infrared (FTIR) spectroscopy. In order to fully explain the effect of water variations on hydrogen bond formation, starch's involvement in producing a high water retention capacity is crucial.

To the best of our knowledge, the study of this type of S-SAP for the treatment of solid waste sludge has not been reported elsewhere. It may be possible to use MD simulation to predict the specific interactions between water and other biopolymer-SAPs based on the original works presented here. By adopting these approaches, a brand new biopolymer backbone can be functionalized or grafted into excellent SAP materials without any risks of unwanted molecular interference and complicated experimental design.

## 2. Materials and Methods

### 2.1. Materials

The starch $(C_6H_{10}O_5)_n$, (p.a. 99%; CAS number: 9005-25-8 from Sigma Aldrich, St. Louis, MO, USA), methacrylic acid $C_4H_6O_2$, (p.a. 99%; CAS number: 79-41-4 from Sigma

Aldrich, St. Louis, MO, USA), ammonium persulfate $(NH_4)_2S_2O_8$, (p.a. 99%; CAS number: 7727-54-0 from Sigma Aldrich, St. Louis, MO, USA), N,N′-methylenebisacrylamide $C_7H_{10}N_2O_2$, (p.a. 99%; CAS number: 110-26-9 from Merck KGaA, Darmstadt, Germany), sodium dodecyl sulfate $(CH_3(CH_2)_{11}SO_4Na$, (p.a. 99%; CAS number: 151-21-3 from Sigma Aldrich, St. Louis, MO, USA), and p-octyl poly(ethylene glycol)phenyl ether (p.a. 99%; CAS number: 48145-04-6 from Sigma Aldrich, St. Louis, MO, USA) were purchased and used for the preparation of superabsorbent. The starch was used without further purification, while methacrylic acid was distilled under reduced pressure prior to usage. Meanwhile, the ammonium persulfate (APS) was recrystallized from water before usage. All other chemicals are of analytical grade and the solutions were prepared with distilled water.

In this study, the solid waste sludge is classified as the primary solid waste (SW 204) obtained as an offering from an undisclosed disposal facility of a polymer production plant located at Pasir Gudang, Johor, Malaysia. Prior to the collection of solid waste sludge, the sample was thickened in the waste treatment ponds by adding cationic polyacrylamide as the organic flocculants to remove water. Later, the separated solid waste sludge was allowed to flow over the filtration conveyor on a gravity belt, while the drained wastewater was discharged through the permeable slit of the belt. Consequently, the solid waste sludge entrapped high amounts of water primarily due to the presence of organic and inorganic mixture. The basic composition and physical characteristics of the solid waste sludge for the test are shown in Table 1. The sludge obtained from solid waste is used directly without any pretreatment.

**Table 1.** Sample description of solid waste sludge.

| Parameters | Description |
| --- | --- |
| Code | SW204 |
| Origin | Pasir Gudang, Johor |
| Condition | Sludges containing one or several metals |
| Source | Polymer production plant |
| Crystallinity | Highly amorphous |
| Moisture contents | 60–70 wt.% |

### *2.2. Preparation of Starch-Derived Superabsorbent Polymer (S-SAP)*

1.44 g of starch was added to 30 mL of distilled water in a flask equipped with a stirrer, condenser, and thermometer. The slurry was stirred continuously and gelatinized at 90 °C for 30 min, before being cooled to 40 °C. Afterward, 1.5 mg of sodium dodecyl sulfate (SDS) was dissolved in 2.0 mL of water, while 9.4 mg of organophosphate was added into the gelatinized starch solution. Upon vigorously stirring for 15 min, the volume of the slurry increased greatly, and foaming starch was obtained. In a different flask, 7.2 g of methacrylic acid was partially neutralized with 12.78 mL of 5 M NaOH solution, followed by the additions of 19.2 mg of N,N′-methylenebisacrylamide (MBA) and 124.8 mg of APS to the mixture. Subsequently, the solubilised mixture was poured into the foaming starch slurry. To complete the polymerisation reaction, the water bath was heated at 55 °C for 3 h. The formed gels were dewatered by immersing them into 100 mL of ethanol for 48 h. Finally, the excess ethanol was removed from the surface using a filter paper, and the samples were spread on a Petri dish for an overnight drying process at room temperature.

### *2.3. Characterization*

Morphological properties were characterized using a field emission scanning electron microscope (FESEM), Hitachi SU8020 and 20 keV. Prior to FESEM imaging, the samples were coated with platinum (Pt) using a vacuum sputter coater. The elemental compositions in the prepared samples were analyzed by using energy dispersive X-ray (EDX) (Hitachi) analysis. The structural properties of the samples were characterized by X-ray diffraction spectroscopy (XRD) using a Rigaku with Cu anode (PAN analytical Co. X'Pert PRO) at 40 kV and 30 mA. The Fourier transform infrared (FTIR) spectroscopy was performed to investigate the functional groups and chemical bonding using an attenuated total reflectance Perkin Elmer spectrophotometer, and the wavelengths ranged from 4000 $cm^{-1}$ to 650 $cm^{-1}$. Both the solid and liquid samples had preset scan numbers of 8 and 16, respectively. A Mettler-Toledo model DSC STAR system was used to perform differential scanning calorimetry (DSC) thermogram. A certain amount of the sample was heated up to 400 °C at a heating rate of 10 °C/min. A Philips PW 1400 WD-XRF spectrometer, equipped with a W anticathode X-ray tube (50 kV, 25 mA), scintillation gas proportional counters in tandem and a LiF [200] crystal, was used for the X-ray fluorescence (XRF) analysis.

### 2.4. Density Functional Theory

For visualization of the compounds involved, GaussView 5 software was employed in this study. Meanwhile, calculation of the studied compound was performed by applying the density functional theory method on the B3LYP functional and 6-31G (d,p) basis set using Gaussian 16 software obtained from the Centre for Information Communication and Technology (CICT), Universiti Teknologi Malaysia.

### 2.5. Molecular Dynamics

To simulate the water–polymer condition, the model structure of the simplest form of starch based on linear polymer amylose was obtained from the Research Collaboratory for Structural Bioinformatics Protein Data Bank RCSB-PDB, through http://www.rcsb.org/ (accessed on 16 June 2020), with a PDB ID 5JIW. A cubic simulation box was used and the volume of the box was calculated based on a cut-off distance of 12 Å. In this molecular study, fifty molecules of amylose were randomly placed in a 64 $nm^3$ simulation box with periodic boundary conditions (PBC) applied in all directions. The box was then filled with the requisite number of SPC models of water molecules. A GROMOS 56A6CARBO Force Field was deployed to represent the intra–inter-molecular potential of polymer and water. Figure 1 depicts the cubic simulation box filled with amylose and water molecules.

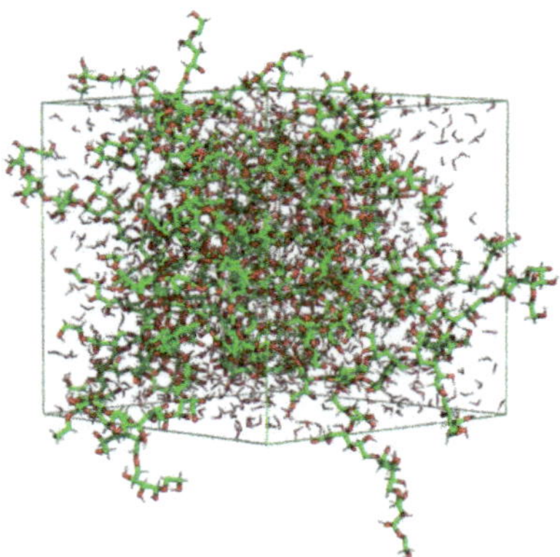

**Figure 1.** The cubic simulation box filled with amylose and water molecules.

In computing the MD simulation, the following parameters were used. Primarily, the integration step of 2.0 fs was used. The non-bonded interactions were calculated up to 12 Å, and the long-range electrostatic interactions were adopted using particle mesh Ewald (PME) with a grid spacing of 1.2 Å and fourth-order interpolation. Neighbour searching was carried out up to 12 Å and updated every five steps. The bond lengths were constrained using LINCS. Temperature and pressure controls were implemented using the Berendsen thermostat and Berendsen barostat, respectively. The reference pressure was set to 1.0 bar, and a relaxation time of 2.0 ps was applied. The isothermal compressibility for pressure control was set to $4.5 \times 10^{-5}$ $bar^{-1}$. In addition, the heat was separated in two heat baths with a temperature coupling constant of 0.1 ps, while two-step energy minimization was performed with each of the respective energy systems minimized at 10,000 steps of the steepest descent, followed by 10,000 steps of conjugate gradients. The system was then introduced in the canonical ensemble (NVT) for 2000 ps. Two different temperatures, which were 298 and 323 K, were deployed for 10 ns during the production simulation.

The average root mean square deviation (RMSD) was calculated by fitting the simulated amylose against the initial X-ray crystal structure, while the radial distribution function (RDF) was determined between the residues' center-of-mass (RES-COM) of water molecules around the amylose. The presence of the formation of hydrogen bonding between the two molecules can be confirmed, provided that the distance between the hydrogen atom and the acceptor is less than 0.35 nm and the angle formed by acceptor–donor-hydrogen is less than 30°. The hydrogen bonding interactions were calculated as an average. All the pictures shown were created using PyMol.

### 2.6. Absorption Properties

The absorption studies are conducted for the absorbency test on dry S-SAP in distilled water (as control) and the absorption of wastewater from sludge. In the respective absorption test, three replicates of ca. 1 g of dry S-SAP were prepared, placed in a nylon tea bag, and immersed in a 250 mL beaker with a sufficient amount of distilled water at room temperature to reach the saturation point of S-SAP. The difference in absorbency ($Q_{water}$) of the SAPs was calculated using Equation (1).

$$Q_{water} = \frac{m_f - m_i}{m_i} \times 100\% \quad (1)$$

where $m_i$ is the initial weight of the S-SAP and $m_f$ is the final weight of the S-SAP.

For the wastewater absorption from sludge, three replicates of approximately 2 g of the dry S-SAP composites were prepared and placed in a nylon tea bag. In this part, a large amount of dry S-SAP was used in order to provide more surface area for the absorption of wastewater from solid sludge. Subsequently, the prepared tea bags were immersed in a beaker with a sufficient amount of SW 204 sludge at room temperature to reach swelling equilibrium. To calculate the absorbed wastewater from sludge, Equation (1) was used.

### 2.7. Swelling Properties

Some 1 g of S-SAP was transferred into a nylon tea bag before immersing it in a beaker with a sufficient amount of distilled water at room temperature to reach the swelling equilibrium. The amount of water entrapped by S-SAP was weighed upon removing the excess water. The weight of the swollen S-SAP was recorded based on the swelling ratio, which corresponded with the size of the swollen S-SAP. The swelling ratio ($q$) was determined by using Equation (2):

$$q = \frac{M_t - M_o}{M_o} \quad (2)$$

where $M_t$ is the weight of the swollen SAP and $M_o$ is the weight of the dry S-SAP. The water retention of the swollen S-SAP was determined by heating them at different temperatures (60 °C, 70 °C, 80 °C, 90 °C and 100 °C) for 5 h.

## 3. Results and Discussion

### 3.1. Synthesis of Starch-Derived Superabsorbent Polymer (S-SAP)

In this study, the S-SAP was formed by graft polymerisation of poly(methacrylic acid -co-sodium methacrylate) as a monomer on the starch, in the presence of methylenebis-acrylamide (MBA) as a crosslinker and ammonium persulfate (APS) as a radical initiator. The neutralization of methacrylic acid took place when the monomer was grafted onto the starch network. At high temperatures, the initiator (APS) was notably present in the solution to decompose and produce sulphate anion radicals (SAR). This SAR would initiate the polymerization process by activating the monomer, and the active sites in the monomer would then act as free radical donors to the adjacent monomers prior to the propagation of the homo polymer. During the polymerization process, a polymer chain was created, with the end vinyl group in the cross-linker producing inter-penetrating, three-dimensional S-SAP containing an abundant number of free carboxyl groups. These reactions are shown schematically in Figure 2.

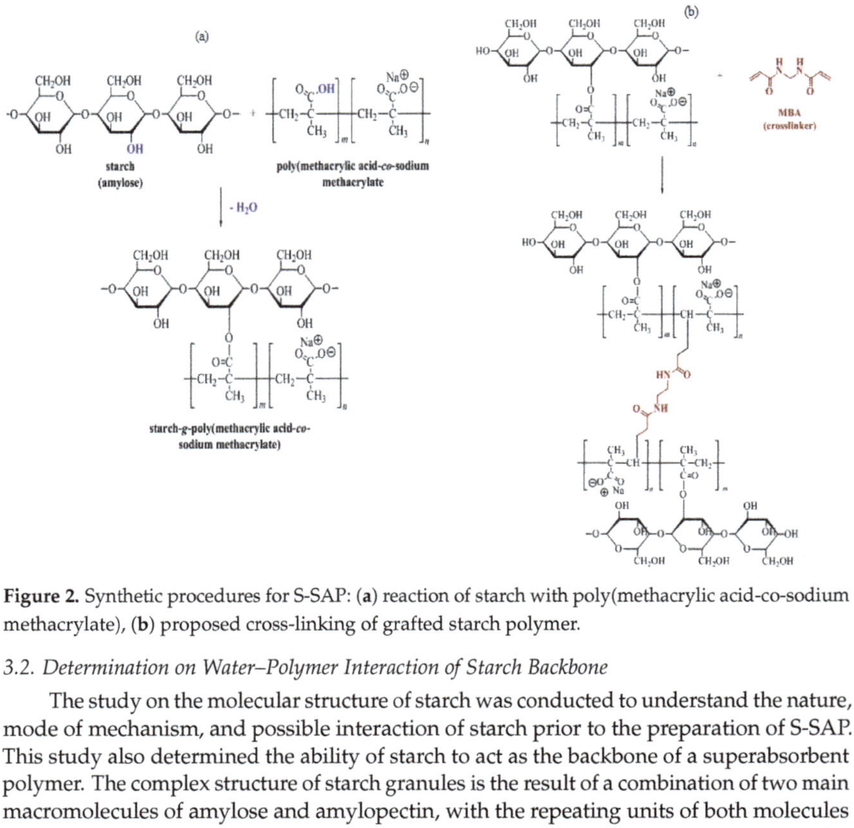

**Figure 2.** Synthetic procedures for S-SAP: (**a**) reaction of starch with poly(methacrylic acid-co-sodium methacrylate), (**b**) proposed cross-linking of grafted starch polymer.

*3.2. Determination on Water–Polymer Interaction of Starch Backbone*

The study on the molecular structure of starch was conducted to understand the nature, mode of mechanism, and possible interaction of starch prior to the preparation of S-SAP. This study also determined the ability of starch to act as the backbone of a superabsorbent polymer. The complex structure of starch granules is the result of a combination of two main macromolecules of amylose and amylopectin, with the repeating units of both molecules known as D-glucan and glucopyranose. Notably, starch exists in the form of granules of different sizes and shapes. Typically, this type of starch demonstrates round and oval granules of various sizes, as shown in Figure 3a. In particular, starch possesses a semi-crystalline morphology with different degrees of crystallinity depending on their classes. Based on the XRD pattern shown in Figure 3b, the starch shows several intense peaks at 2θ of 5.5°, 15° and 17°. A few less intense peaks were also observed at 2θ of 23° and 24°. The observed diffraction patterns suggest that the respective starch has semi-crystalline properties. In addition, the broad diffraction peaks observed at the range of ca. 2θ of 10°–17° indicate the presence of small crystallite sizes.

To determine the functional groups and effects of water–polymer interactions on the starch granules, FTIR spectroscopy measurements were carried out, as shown in Figure 4. The appearance of a wide peak at the region of 1200–800 $cm^{-1}$ corresponded to the vibration band of glucan ring, which overlapped with the stretching mode of C-OH and the bending vibration of the C-O-C glycosidic band [32]. On the other hand, the peaks that appeared at the wavelengths of 1149 $cm^{-1}$, 1077 $cm^{-1}$ and 997 $cm^{-1}$, [32,33] were attributed to the formation of C-O and C-C stretching, as well as the formation of C-OH [33,34]. Meanwhile, the peak observed at ca. 1638 $cm^{-1}$ corresponded to the bending mode of H-O-H, which resulted from the consecutive addition of water. The presence of the respective peak is in agreement with the findings reported in the literature [34,35].

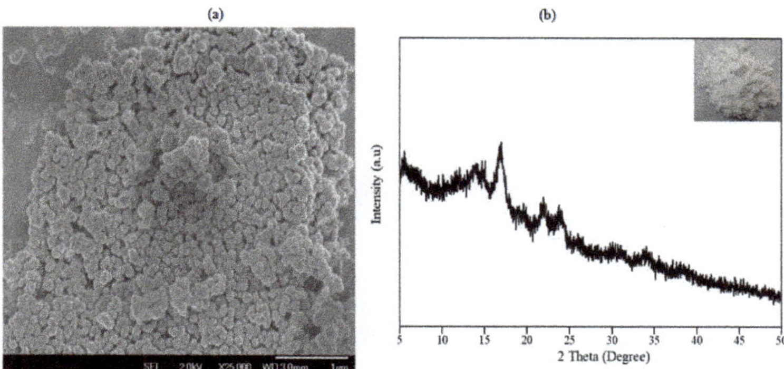


**Figure 3.** (**a**) The FESEM micrograph of potato starch; and (**b**) The XRD diffractogram of potato starch. Inset represents a photograph image of starch powder.

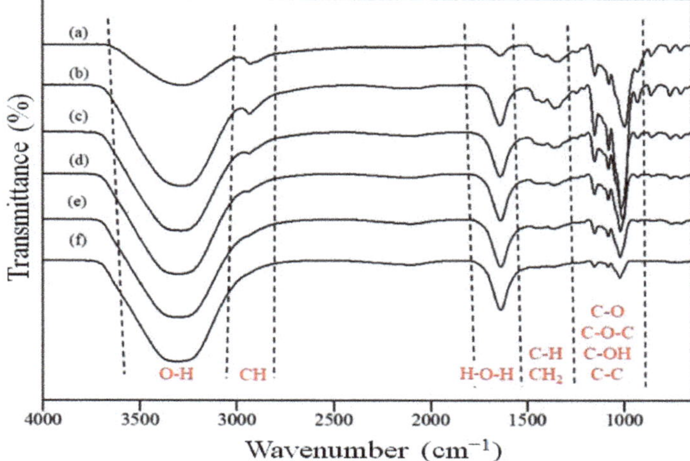


**Figure 4.** The FTIR spectra of starch in with the presence of different concentrations of water; (a) 0 mol, (b) 1 mol, (c) 2 mol, (d) 3 mol, (e) 4 mol and (f) 5 mol.

In this process, the formation of hydrogel is initiated as starch starts to gelatinize into a stable three-dimensional network. This is a crucial step to disrupt the crystalline structure of starch and to enable the strong interaction of the water–polymer chain. Moreover, the ability of starch to form hydrogen bonds with water is a crucial indication that the starch is capable of absorbing water efficiently. Basically, when starch is treated with water in excess, the crystalline structure of the starch's double helices is affected due to the breakage of internal hydrogen bonding. Therefore, the water molecules will form a bonding with the exposed hydroxyl groups of amylose and amylopectin. During this process, the granular structure of starch begins to swell and its solubility in water increases. Consequently, the stretching vibration appears at ca. 1200–800 $cm^{-1}$, which refers to the crystalline domains that contain the double helices of amylopectin side chains [36] becoming less significant as more water is added consecutively. However, it was observed that the FTIR spectra of respective crystalline domains became more intense and significant, as demonstrated by the stretching vibration at ca. 1200–800 $cm^{-1}$. The significant increase might be due to the retrogradation process in starch, in which the broken double helical crystalline structure in amorphous starch molecules reformed after losing their water binding ability [37].

To further confirm the presence of the hydrogen bonding interaction between starch and water, the density functional theory (DFT) was also performed to model the dynamic structure of starch and water. From Figure S1a, the bond length of O11 and H11 after optimization using a basic DFT/6-31G (d,p) level provides the value of 1.85 Å, indicating the hydrogen bonding interaction [38]. As can be observed in Figure S1, the hydrogen bond is highly affected by the dipole forces. Based on the DFT calculation, the obtained value showed an increment from 2.991912 D (amylose without water) to 3.193061 D (amylose with water). This change indicates some kind of dependency on the nature of the surrounding atom, in which the presence of an electronegative oxygen atom of O11 in the system yields a stronger inductive effect on the hydrogen atom of H111. On the other hand, the decreasing value of Mulliken charge for the oxygen atom of O11 in Figure S1b, and in the structure of amylose with and without water in Figure S1c, are −0.522e and −0.561e, respectively, proving the existence of a hydrogen interaction with H111 [39].

Furthermore, the interaction of starch with water was also analyzed using the molecular dynamics of both starch and water. The root mean square displacement (RMSD) measures the average atomic displacement between the initial and final conformations of the simulation trajectory. In this measurement, the initial conformation value was taken from the RCSB Protein Data bank (http://www.rcsb.org/ (accessed on 16 June 2020)), while the coordinate of the final position was obtained from the NPT production simulation. The structural stability of amylose solvated in water molecules was obtained from the last 2 ns, and the result is shown in Figure 5a. The analysis clearly shows that the average RMSD is about 0.18 Å, and this value is in good agreement with the report of a previous study [40]. This shows that the conformation structure of amylose had swelled and extended in the absence of an extrinsic mixture. It has been reported that in the presence of amylose–linoleic acid complexes, the ligand was able to preserve the helical conformation of the polymer [40,41].

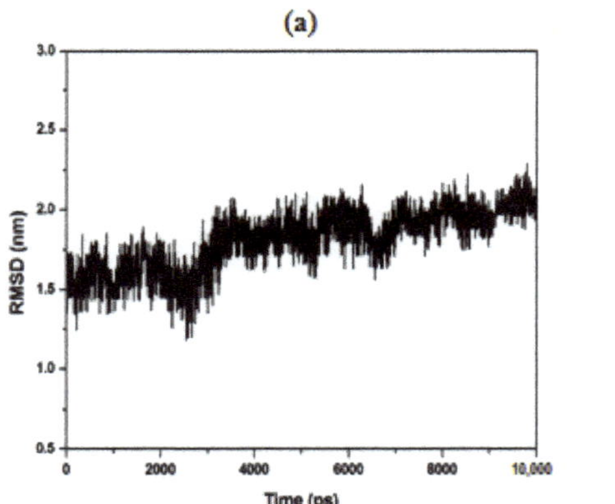


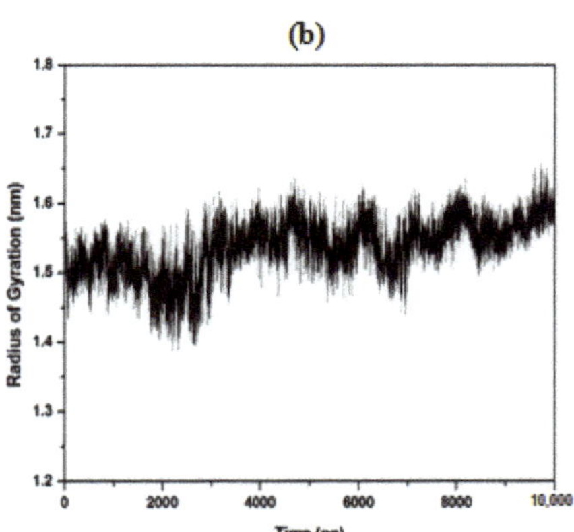


**Figure 5.** Molecular dynamics of (**a**) RMSD of amylose solvated in SPC water molecules at 298 K for 10 ns. The trajectory for analysis was taken from the last 2 ns of the NPT production simulation. (**b**) Average radius of gyration (RoG) of 50 molecules of amylose, representing the amylose region solvated in SPC water molecules at 298 K and 1.0 bar.

Another analysis technique that can be used to study the formation of a hydrogen bond between starch and water is the radius of gyration (RoG). In this study, the RoG refers to the distribution of the atomic structure against its initial position. Figure 5b represents the RoG of amylose in water with an average value of 0.15 Å, which is slightly higher in comparison to the value reported by Gao et al. (2021) [42,43], who observed the RoG value of 0.13 Å for a single amylose strand. The obtained result indicated that the amylose fluctuated greatly and reduced its compactness in water at 298 K. When carbohydrates, which have hydrophilic properties, strongly interact with water via intermolecular bonding, the strong attraction between the carbohydrates and water will reduce the compactness of the amylose complex, since water molecules are highly mobile.

The radial distribution function (RDF) describes the probability density of finding a molecule from a reference particle. In this study, the water density around the amylose molecule was investigated. Figure 6a shows that the RDF of the amylose water complex distance is smaller, with a distance value of about 2.6 Å. This validates our prediction on the existence of a hydrogen bonding interaction between the polar site of amylose with water molecules. The result indicates that water can penetrate the polymer networks, leading to a high probability of interaction with amylose molecules. This penetration causes the amylose structure to stretch out, thus decreasing its helical stability. The specific RDF atom–molecule interaction was also carried out to determine the parts in the amylose site wherein the most intermolecular hydrogen bonding interaction occurs. In general, amylose chains have a strong affinity towards themselves and towards materials that contain hydroxyl groups. The amylose chains are intra-connected via intra-molecular OH-O-type hydrogen bonds to form fleet sheets with CH-O hydrogen bonds [43]. The formation of intermolecular interaction is strongly dependent on the absorption sites: OH moieties pointing out of the amylose or above the equatorial hydroxyls. As shown in Figure 6b, the polar hydroxyl site of H06 was found to interact more with water molecules due to its flexibility and less steric hindrance as compared to other sites.

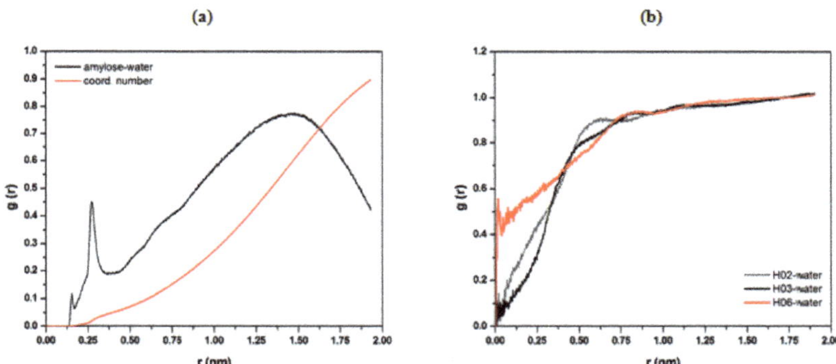


**Figure 6.** Center-of-mass radial distribution function (COM-RDF) of water molecules around the amylose structure (**a**) and specific COM-RDF of a hydrogen atom in amylose toward water molecules' interaction (**b**). The g(r) represents the probability of finding an atom or molecule around a distance r of another chosen atom or molecule as a reference point.

The average number of intermolecular hydrogen bonds between amylose and water molecules was determined from a production simulation trajectory. Based on the trajectory, a hydrogen bond is considered to exist in one conformation, provided that the distance between the hydrogen atom and the acceptor is less than 0.135 nm, and the angle formed by the acceptor–donor hydrogen is less than 30° [43–45]. In this study, the average number of hydrogen bonds calculated from the molecular dynamic (MD) simulation is 13, and this non-bonded interaction was expected to exist due to the polar OH group of amylose that provides sites for the hydrogen bond donor and acceptor. Most of the time, it can be assumed that the amount of hydrogen bonding was maintained throughout the simulation.

The number of density maps for water molecules residing in a polymeric cavity of amylose was obtained by scanning the axial space to compute the density distributions. The density distributions were projected onto the x-y plane and the simulation time was averaged for 2 ns. Based on the density maps shown in Figure 7, it was observed that the molecules were able to penetrate the amylose complex by splitting into smaller conglomerates, as indicated by the orange region. In addition, these density maps further confirm that the water molecules dissociated with the solvent as they interacted with amylose chains, thus reducing the hydrogen bonds among the amylose molecules.

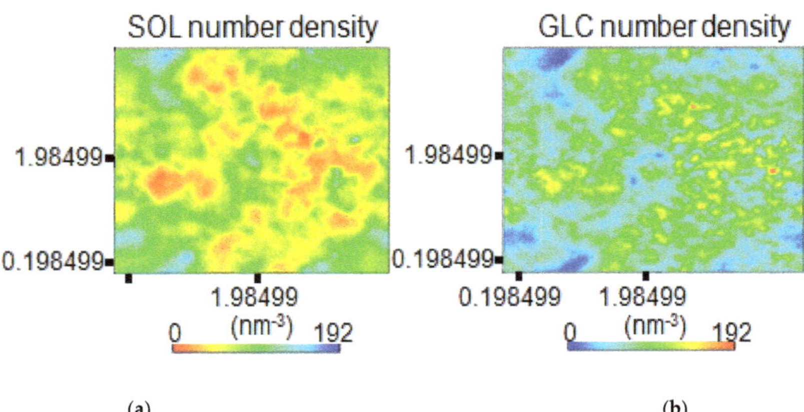


(**a**) (**b**)

**Figure 7.** Density maps (number of molecules/nm$^3$) showing (**a**) water molecule distributions and (**b**) amylose at P = 1.0 bar and T = 198.15 K (in the interlayer at d = 001 spacing) for the amylose–$H_2O$ system. The results were obtained by averaging over 2.0 ns of simulation time.

### *3.3. Physiochemical Properties of S-SAP*

The physicochemical properties S-SAP were studied using the FESEM, EDX, XRD and DSC analyses, as presented in Figure 8. Based on the micrograph shown in Figure 8a, it was observed that the morphology of S-SAP displayed continuous agglomeration of loose particles of homogeneous S-SAP. Based on the loose structure, it can be noted that the foaming process of the starch slurry assists in improving the SAP's absorption characteristics. This finding is also in a good agreement with those reported by Saha et al. (2020) [46], in which the grafting of a compound onto the smooth and tight surface of the poly(acrylic acid) derivatives led to a heterogeneous and loose structure with several cavities present in the starch.

The elemental composition in the S-SAP was characterized using the EDX spectroscopy in Figure 8b. The successful formation of poly(methacrylic acid-co-sodium methacrylate) in the grafted S-SAP can be proven by the structural composition, as part of the methacrylic acid was neutralized after the addition of NaOH to form sodium methacrylate. The EDX analysis detected the presence of elements C, O and Na, with 52.3, 30.2 and 17.5 wt.%, respectively. These corresponded to the chemical composition of S-SAP, namely starch g-poly(methacrylic acid–co–sodium methacrylate). The diffraction pattern illustrated in Figure 8c shows that S-SAP was highly amorphous based on the appearance of wide and broad diffraction peaks. In Figure 8c, the appearance of wide and broad peaks at 2θ of 31.5° suggests that the prepared S-SAP was in an amorphous phase. In addition, the wide and low intense peaks at 2θ of 17° indicate the disappearance of the crystalline phase of starch [46]. The lack of intense sharp peaks suggests that the crystalline phase of starch was completely replaced during the graft polymerisation process, which involved gelatinization and the addition of poly(methacrylate-co-sodium methacrylate) blocks.

The thermal stability and glass transition temperature (Tg) of S-SAP was studied using differential scanning calorimetry (DSC). The thermogram for SAP shown in Figure 8d was obtained using DSC that ramped up to 400 °C and a heating rate of 10 °C/min. As depicted in Figure 8d, the curves show three characteristic stages which denote the glass transition, melting and fusion of crystallite. The Tg occurred at 131.59 °C, followed by two endothermic curves which corresponded to endothermic transition at a peak temperature of 183.29 °C, and the fusion of crystalline at 274.23 °C with enthalpy of 889.19 J/g and 3.31 J/g, respectively. Figure 8d illustrates that the first endothermic curve in the DSC thermogram can be classified as an endothermic transition, which can be associated with the decreasing heat capacity of the sample. Moreover, weight loss was due to the loss of water that was tightly bound to the starch. Furthermore, the second endothermic peak, observed at 274.23 °C with enthalpy of 3.31 J/g, was detected as the fusion of crystallite, in which disintegrated S-SAP changed into small flakes and could be fused and combined together.

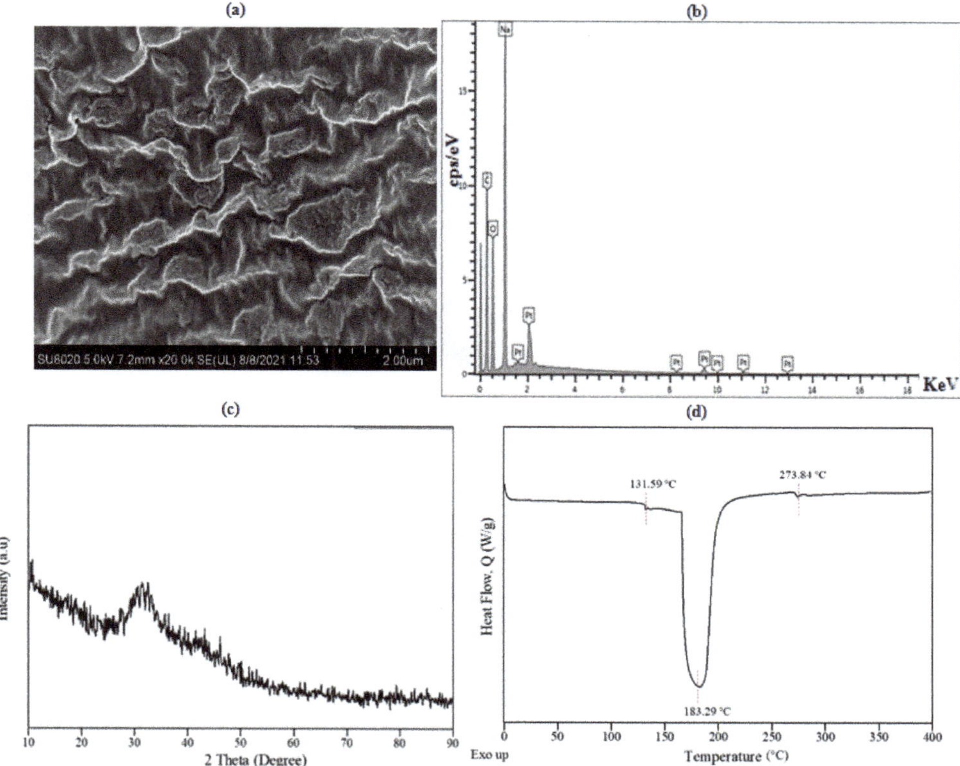


**Figure 8.** Analysis of S-SAP showing (**a**) FESEM micrographs at 20.0 k magnification, (**b**) EDX spectra for elemental composition of S-SAP, (**c**) XRD diffractogram, and (**d**) DSC thermogram.

The FTIR analysis was performed to identify the bonds and functional groups in the S-SAP ranging from 4000 to 650 cm$^{-1}$. For comparison purposes, the FTIR analysis was also determined for methacrylic acid, starch, and poly(methacrylic acid-co-sodium methacrylate), along with SAP, to confirm the grafting of starch into poly(methacrylic acid-co-sodium methacrylate). Based on Figure 9, the peak observed at 1690 cm$^{-1}$ in the methacrylic acid spectrum and at 1642 cm$^{-1}$ in the SAP spectrum corresponded to the stretching vibration of the COOH group, thereby confirming the presence of poly(methacrylic acid-co-sodium methacrylate) in S-SAP. Meanwhile, the peaks at 1632, 1532, and 1541 cm$^{-1}$ for the sample of methacrylic acid, poly(methacrylic acid-co-sodium methacrylate) and S-SAP, respectively, were attributed to C=C stretching vibrations [47]. Although the characteristic peaks of semi-crystalline pure starch around 1200–800 cm$^{-1}$ had weakened, they were still present in SAP, confirming that the starch had grafted onto the poly(methacrylic acid-co-sodium methacrylate) networks.

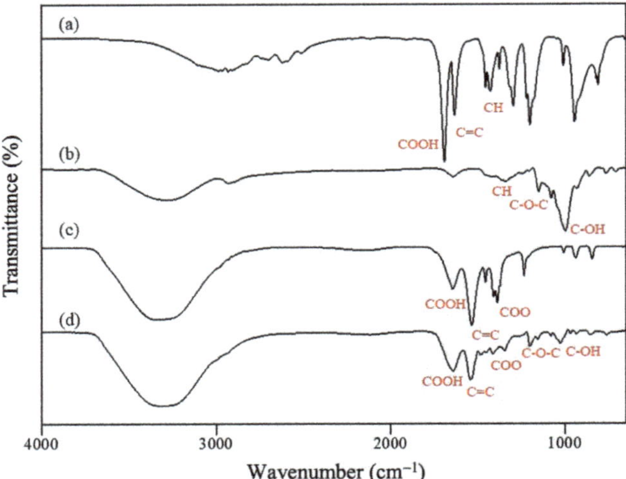


**Figure 9.** FTIR spectra of (a) methacrylic acid (b) starch (c) poly(methacrylic acid- co-sodium methacrylate) and (d) S-SAP.

*3.4. The Performance of S-SAP*

The absorption properties of the S-SAP were investigated to determine the maximum absorbed water capacity and the performance of the polymeric materials. Figure 10 shows the absorption properties of S-SAP in both distilled water (as control) and solid waste sludge based on the differences in absorbance over time. As shown in Figure 10a, the water absorbency of S-SAP composite reflects three stages. The first stage (0–80 min) showed a rapid increase in the amount of water with time, while the second stage (80–160 min) revealed a decreasing absorption of distilled water by the superabsorbent. The final stage (165–180 min) depicted the absorbency almost in an equilibrium state, as no increase in water absorbency over time was recorded. The water absorbency of the S-SAP reached 500% after 80 min, but the absorption decreased subsequently and gradually in the second stage, i.e., from 500% to 100% for another 80 min (80–160 min) before reaching the equilibrium state at 0%.

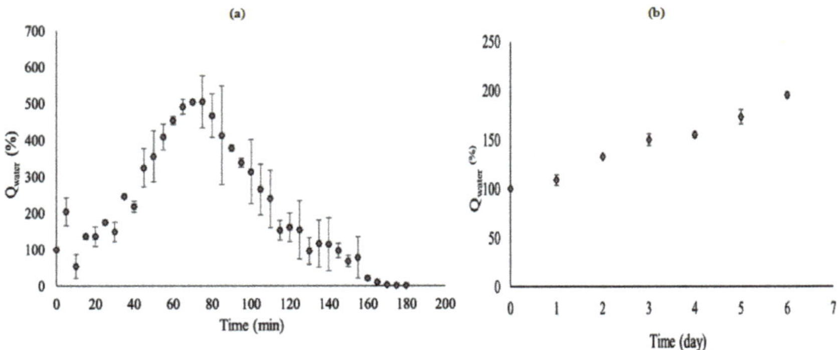


**Figure 10.** Difference in absorbency, $Q_{water}$, of water versus time for (**a**) distilled water and (**b**) solid waste sludge.

The rapid absorption of water in the first stage was due to the presence of numerous free OH groups in the polymeric network of the superabsorbent that are capable of forming hydrogen bonding with water. In the second stage, the OH group in the polymeric network was almost fully bonded with the free water molecules and thus was unable to absorb more water. This resulted in a decrease in the absorption efficiency of SAP. In the third stage, as all the OH groups in S-SAP were fully bonded to the water molecules, no hydrogen bond with water could be formed. Therefore, 0% of absorbency was plotted because the saturation point equilibrium was already achieved [48].

Figure 10b shows the absorption properties of S-SAP in solid waste sludge. The first stage of the absorption trend demonstrates that the amount of wastewater absorbed was increasing steadily up to 195% over 6 days. Compared to the absorption in distilled water, the absorption trend of wastewater from sludge was slower, since S-SAP absorbed the entrapped wastewater from damp sludge for the absorption to happen. In addition, the increased ionic strength of the wastewater due to its basic pH also caused a rapid decrease in ion osmotic pressure, which contributed to the slow uptake of wastewater from damp sludge by S-SAP. Based on the absorption trend of wastewater from the solid waste sludge, it is expected that the absorption of wastewater from sludge will continue until the saturation point of the S-SAP reaches 500%.

A preliminary study using the XRF spectroscopy was conducted to determine the efficiency of S-SAP in absorbing metal contaminants from solid waste sludge, as shown in Figure S2. The results show the dominant proportion of Na metal, with a concentration of 951000 ppm; this can be attributed to the sodium methacrylate at the end of the structure in SAP. In addition, ten metal elements were also detected alongside Na: Zn, Al, Si, S, Cl, K, Ca, Fe, Cu, and Zr. The preliminary results show that all the absorbed metals are positively charged, revealing that the absorption process occurred due to the electrostatic attraction between the positive metal ions and the negatively charged functional groups abundantly found in the S-SAP. The specific concentrations and details on the removal mechanism of the respective elements will be described elsewhere.

Figure 11a illustrates that the swelling of S-SAP occurred via two-stage processes of rapid swelling and steady-state/equilibrium state. During the rapid swelling process, the swelling ratio of S-SAP increased steadily at 160 min up to 77 g/g. At 160 min and above, the swelling ratio showed no changes, indicating that the maximum swelling capacity was achieved. The high swelling ratio of S-SAP observed early in the reaction can be explained through the available vacant area in the polymeric network that enables rapid penetration of the solvent into the superabsorbent. Therefore, the swelling ratio increased until the maximum capacity was achieved. Additionally, when a dry hydrogel is in contact with thermodynamically compatible solvent molecules, it will attack the hydrogel surface and penetrate into the polymeric network of the S-SAP [49]. More specifically, the expansion of the rubbery phase's network resulted from the separation of the unsolvated glassy phase from the rubbery hydrogel region with a moving boundary. Hence, the vacant area was created during the expansion of the networks of the rubbery phase, thus enabling the molecules of solvent to enter the superabsorbent network.

Upon reaching the second phase (>120 min), the void in the polymeric network of S-SAP was almost filled with solvent molecules and completely filled when the saturation point of the S-SAP was reached (160 min). The stable three-dimensional structure of the prepared S-SAP resulted in a low swelling ratio, which was caused by high crosslinking density of the polymers. In particular, a higher concentration of crosslinkers produces a large number of growing polymeric chains that generate an additional network. Consequently, this results in a decreased swelling ratio when the crosslink density exceeds a certain degree, due to the decrease in space between the crosslink branching. However, the highly crosslinked S-SAP has a higher dry-state gel strength and can maintain its shape even after modest pressure. Ironically, the S-SAP does not have a good re-swelling capacity, which may be attributed to the irreversible collapse of the S-SAP during the re-swelling process at 110 °C for 90 min, in which the collapsed polymeric network suggests a low mechanical property.

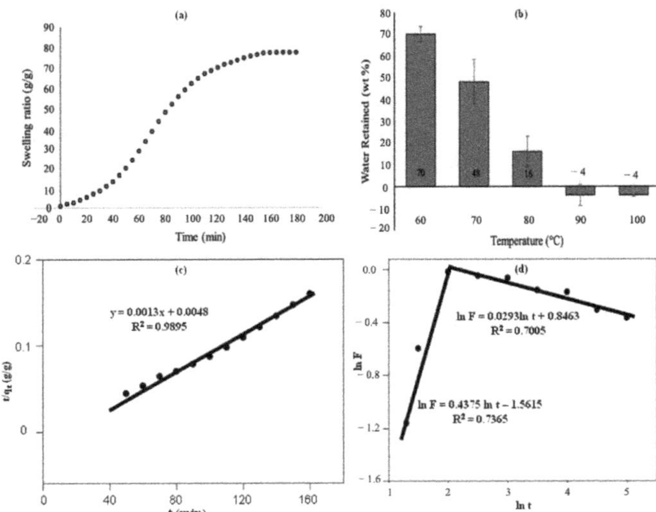


**Figure 11.** S-SAP performance for (**a**) swelling ratio (q) over time and (**b**) water retention at different temperatures, (**c**,**d**) swelling kinetics.

The ability of the superabsorbent to retain water content at a certain temperature was investigated by a water retention test. The test was performed by heating the swollen-S-SAP at different temperatures. The water retention capacity of S-SAP is presented in Figure 11b. The results show that the swollen superabsorbent has a decreasing tendency to retain water with increasing temperature. During the absorption process, the combination of the polymeric network of S-SAP opening up and the water molecules from the surrounding sticking up to it by forming hydrogen bonding causes the solid surface of the superabsorbent to become thicker and more viscous, thereby making the appearance of the superabsorbent sticky and swollen.

However, the hydrogen bonding formed between water molecules and the polymeric network is disrupted when the swollen polymer is exposed to high temperature. About 70% of water could be held at 60 °C after 5 h of the water retention test. Likewise, about 48% and 16% were retained at 70 °C and 80 °C after 5 h of heating. This shows that the S-SAP can hold water upon heating at 5 h, and implies that the superabsorbent possesses a good water retention ability and stable polymeric network even after the oven heating process. Moreover, the hydrogen bonding formed by the surrounding distilled water molecules with superabsorbent did not break down during the process. When the absorption process was conducted at humid temperature, the water would not easily release, even after being absorbed by the superabsorbent.

According to previous works, the swelling behaviour of S-SAP was evaluated by the pseudo-second-order swelling kinetics model using Equation (3), and the Ritger–Peppas model using Equation (4):

$$\frac{t}{q_t} = \frac{1}{k_2 q_{e^2}} + \frac{t}{q_t} \tag{3}$$

$$\ln F = \ln q_t - \ln q_e = \ln k + n \ln t \tag{4}$$

where $q_e$ (g/g) is the water absorbency at equilibrium and $q_t$ (g/g) is the water absorbency at time $t$. While $k_2$ (g·mg$^{-1}$ min$^{-1}$) is the rate constant, $F$ is the fractional uptake at time $t$, $n$ is the swelling exponent, and $k$ is the structural parameter.

As shown in Figure 11c, the relationship between $\frac{t}{q_t}$ and $t$ can be considered linear, with a satisfactory correlation coefficient ($R^2$ = 0.9895). The swelling kinetics of the S-SAP fit well with the pseudo-second-order relationship. Figure 11d shows an acceptable correlation coefficient of the Ritger–Peppas model which is almost comparable to similar works [49,50]. Based on this result, the chemical interactions via chemisorptions on S-SAP dominate the water absorption process.

In the case of high temperature heating at above 90 °C, the water retention value was negative, suggesting that the water released by heating is more than the absorbed water. This could be due to the partial vaporization of water content from the dry superabsorbent, along with the absorbed water upon 5 h of heating. Water molecules start to vaporize as they reach the boiling point of water. It should be noted that high temperature provides high kinetic energy to water molecules to break down the hydrogen bonding. In addition, the synthesized S-SAP could retain more than 50% of distilled water after 5 h of heating at 60 °C. For reusability, it is possible to determine the most suitable temperature for carrying out the desorption process and removing the water content from the S-SAP based on this information.

## 4. Conclusions

In summary, a new approach of using S-SAP for cost-effective solid waste sludge remediation is presented in this work. Firstly, the S-SAP was successfully prepared via graft polymerisation of poly(methacrylic acid-co-sodium methacrylate) as a monomer on the starch, using MBA as the crosslinker and APS as the radical initiator. The structural analysis showed that starch granules are semi-crystalline, as confirmed based on the XRD patterns. On consecutive addition of water to starch, the FTIR measured an intensified bending mode of H-O-H, suggesting water penetration into the starch polymer networks. Furthermore, the MD simulations assessed the formation of hydrogen bonding between starch and water on the polar hydroxyl site of H06 of amylose due to its flexibility and less steric hindrance as compared to other sites. Meanwhile, the DFT and MD simulation analyses showed that the starch backbone is highly capable of absorbing water efficiently into its polymeric network. Based on the MD simulation, the specific RDF atom–molecule interaction in the amylose backbone was able to record a high degree of water penetration into the S-SAP. By using DFT, the presence of hydrogen bonds can be confirmed by the changes in the value of Mulliken charge of atom O11. To support the simulation findings, the water retention of S-SAP was evaluated according to pseudo-second-order kinetics for chemisorption reactions.

The FESEM showed that the morphology of S-SAP possesses a continuous agglomeration of loose particles of homogeneous polymeric networks. Meanwhile, as confirmed by FTIR, the successful grafting of starch onto the PMA network might have decreased the crystallinity of semi-crystalline starch. The low crystallinity of S-SAP is in agreement with the XRD analysis. Absorption studies using water have shown that the S-SAP was capable of absorbing distilled water up to 500% (i.e., the saturation point) within 80 min. With respect to the absorption of water from solid sludge, the high amount of water absorption (up to 195%) and the significant water retention capacity were measured. In future, the S-SAP might have many potential applications as a low cost natural superabsorbent with a high water retention capacity, based on the presented water–polymer interaction.

**Supplementary Materials:** The following supporting information can be downloaded at: https://www.mdpi.com/article/10.3390/polym15061471/s1. Figure S1 DFT analysis on (a) H-bond of O11 with H111. The Mulliken Charge of amylose without (b) water and (c) with water. Figure S2 XRF spectra for absorption of metals contaminants from sludge by NSS using (a) RX9 source, (b) Cu source, and (c) Mo source.

**Author Contributions:** Conceptualization, J.M.; Methodology, J.M.; Software, K.J.; Investigation, J.M.; Resources, J.M., S.I.I. and N.A.N.N.M.; Data curation, M.H.M.H.; Writing—original draft, J.M., R.R.M. and K.J.; Writing—review & editing, J.M., M.H.M.H. and N.A.N.N.M.; Visualization, R.R.M. and K.J.; Supervision, J.M.; Project administration, J.M.; Funding acquisition, Salizatul Ilyana Ibrahim. All authors have read and agreed to the published version of the manuscript.

**Funding:** This research was funded by UTMER, grant number Q.J130000.2654.18J18 and the APC was funded by Universiti Teknologi MARA through GERAN PENYELIDIKAN KHAS 2020 (GPK), Research Grant Number 600-RMC/GPK 5/3 (204/2020).

**Institutional Review Board Statement:** Not applicable.

**Data Availability Statement:** Not available.

**Conflicts of Interest:** The authors declare no conflict of interest.

## References


1. Del Río-Gamero, B.; Perez-Baez, S.O.; Gómez-Gotor, A. Calculation of greenhouse gas emissions in Canary Islands wastewater treatment plants. *Desalin. Water Treat* **2020**, *197*, 101–111. [CrossRef]
2. Farzaneh, G.; Khorasani, N.; Ghodousi, J.; Panahi, M. Assessment of surface and groundwater resources quality close to municipal solid waste landfill using multiple indicators and multivariate statistical methods. *Int. J. Environ. Res.* **2021**, *15*, 383–394. [CrossRef]
3. Lin, Y.T.; Zhuang, G.L.; Wey, M.Y.; Tseng, H.H. The viable fabrication of gas separation membrane used by reclaimed rubber from waste tires. *Polymers* **2020**, *12*, 2540. [CrossRef] [PubMed]
4. Gopinath, A.; Divyapriya, G.; Srivastava, V.; Laiju, A.R.; Nidheesh, P.V.; Kumar, M.S. Conversion of sewage sludge into biochar: A potential resource in water and wastewater treatment. *Environ. Res.* **2021**, *194*, 110656. [CrossRef] [PubMed]
5. Dogaris, I.; Ammar, E.; Philippidis, G.P. Prospects of integrating algae technologies into landfill leachate treatment. *World J. Microbiol. Biotechnol.* **2020**, *36*, 39. [CrossRef]
6. Kajiwara, N.; Noma, Y.; Tamiya, M.; Teranishi, T.; Kato, Y.; Ito, Y.; Sakai, S. Destruction of decabromodiphenyl ether during incineration of plastic television housing waste at commercial-scale industrial waste incineration plants. *J. Environ. Chem. Eng.* **2021**, *9*, 105172. [CrossRef]
7. Xue, Y.; Liu, X. Detoxification, solidification and recycling of municipal solid waste incineration fly ash: A review. *J. Chem. Eng.* **2021**, *420*, 130349. [CrossRef]
8. Liang, Y.; Donghai, X.; Peng, F.; Botian, H.; Yang, G.; Shuzhong, W. Municipal sewage sludge incineration and its air pollution control. *J. Clean. Prod.* **2021**, *295*, 126456. [CrossRef]
9. Vattanapuripakorn, W.; Khannam, K.; Sonsupap, S.; Tongsantia, U.; Sarasamkan, J.; Bubphachot, B. Treatment of flue gas from an infectious waste incinerator using the ozone system. *Environ. Nat. Resour. J.* **2021**, *19*, 348–357. [CrossRef]
10. Aldalbahi, A.; El-Naggar, M.; Khattab, T.; Abdelrahman, M.; Rahaman, M.; Alrehaili, A.; El-Newehy, M. Development of Green and Sustainable Cellulose Acetate/Graphene Oxide Nanocomposite Films as Efficient Adsorbents for Wastewater Treatment. *Polymers* **2020**, *12*, 2501. [CrossRef]
11. Cai, Y.H.; Burkhardt, C.J.; Schäfer, A.I. Renewable energy powered membrane technology: Impact of osmotic backwash on organic fouling during solar irradiance fluctuation. *J. Membr. Sci.* **2022**, *647*, 120286. [CrossRef]
12. Kamali, M.; Suhas, D.P.; Costa, M.E.; Capela, I.; Aminabhavi, T.M. Sustainability considerations in membrane-based technologies for industrial effluents treatment. *J. Chem. Eng.* **2019**, *368*, 474–494. [CrossRef]
13. Khan, S.B.; Irfan, S.; Lam, S.S.; Sun, X.; Chen, S. 3D printed nanofiltration membrane technology for waste water distillation. *J. Water Process. Eng.* **2022**, *49*, 102958. [CrossRef]
14. Coha, M.; Farinelli, G.; Tiraferri, A.; Minella, M.; Vione, D. Advanced oxidation processes in the removal of organic substances from produced water: Potential, configurations, and research needs. *J. Chem. Eng.* **2021**, *414*, 128668. [CrossRef]
15. Elmi, A.; AlOlayan, M. Sewage sludge land application: Balancing act between agronomic benefits and environmental concerns. *J. Clean. Prod.* **2020**, *250*, 119512. [CrossRef]
16. Oladosu, Y.; Rafii, M.Y.; Arolu, F.; Chukwu, S.C.; Salisu, M.A.; Fagbohun, I.K.; Haliru, B.S. Superabsorbent polymer hydrogels for sustainable agriculture: A review. *Horticulturae* **2022**, *8*, 605. [CrossRef]
17. Behera, S.; Mahanwar, P.A. Superabsorbent polymers in agriculture and other applications: A review. *Polym. Plast. Technol. Eng.* **2020**, *59*, 341–356. [CrossRef]
18. Chen, J.; Wu, J.; Raffa, P.; Picchioni, F.; Koning, C.E. Superabsorbent polymers: From long-established, microplastics generating systems, to sustainable, biodegradable and future proof alternatives. *Prog. Polym. Sci.* **2022**, *125*, 101475. [CrossRef]
19. El Idrissi, A.; El Gharrak, A.; Achagri, G.; Essamlali, Y.; Amadine, O.; Akil, A.; Zahouily, M. Synthesis of urea-containing sodium alginate-g-poly (acrylic acid-co-acrylamide) superabsorbent-fertilizer hydrogel reinforced with carboxylated cellulose nanocrystals for efficient water and nitrogen utilization. *J. Environ. Chem. Eng.* **2022**, *10*, 108282. [CrossRef]
20. Kowalski, G.; Ptaszek, P.; Kuterasiński, L. Swelling of hydrogels based on carboxymethylated starch and poly(acrylic acid): Nonlinear rheological approach. *Polymers* **2020**, *12*, 2564. [CrossRef]
21. Mushtaq, F.; Raza, Z.A.; Batool, S.R.; Zahid, M.; Onder, O.C.; Rafique, A.; Nazeer, M.A. Preparation, properties, and applications of gelatin-based hydrogels (GHs) in the environmental, technological, and biomedical sectors. *Int. J. Biol. Macromol.* **2022**, *218*, 601–633. [CrossRef] [PubMed]
22. Meshram, I.; Kanade, V.; Nandanwar, N.; Ingle, P. Super-absorbent polymer: A review on the characteristics and application. *Int. J. Adv. Res. Chem. Sci.* **2020**, *7*, 8–21.

23. Wanyan, Q.; Qiu, Y.; Xie, W.; Wu, D. Tuning degradation and mechanical properties of Poly (l-lactic acid) with biomass-Derived Poly (l-malic acid). *J. Polym. Environ.* **2020**, *28*, 884–891. [CrossRef]
24. Sand, A.; Vyas, A. Superabsorbent polymer based on guar gum-graft-acrylamide: Synthesis and characterization. *J. Polym. Res.* **2020**, *27*, 43. [CrossRef]
25. Menceloğlu, Y.; Menceloğlu, Y.Z.; Seven, S.A. Triblock superabsorbent polymer nanocomposites with enhanced water retention capacities and rheological characteristics. *ACS Omega* **2022**, *7*, 20486–20494. [CrossRef]
26. Metin, C.; Alparslan, Y.; Baygar, T.; Baygar, T. Physicochemical, microstructural and thermal characterization of chitosan from blue crab shell waste and its bioactivity characteristics. *J. Polym. Environ.* **2019**, *27*, 2552–2561. [CrossRef]
27. Díez-Pascual, A.M. Synthesis and applications of biopolymer composites. *Int. J. Mol. Sci.* **2019**, *20*, 2321. [CrossRef]
28. Bachra, Y.; Grouli, A.; Damiri, F.; Talbi, M.; Berrada, M. A novel superabsorbent polymer from crosslinked carboxymethyl tragacanth gum with glutaraldehyde: Synthesis, characterization, and swelling properties. *Int. J. Biomater.* **2021**, *2021*, 5008833. [CrossRef] [PubMed]
29. Chang, L.; Xu, L.; Liu, Y.; Qiu, D. Superabsorbent polymers used for agricultural water retention. *Polym. Test.* **2021**, *94*, 107021. [CrossRef]
30. Sanders, J.M.; Misra, M.; Mustard, T.J.; Giesen, D.J.; Zhang, T.; Shelley, J.; Halls, M.D. Characterizing moisture uptake and plasticization effects of water on amorphous amylose starch models using molecular dynamics methods. *Carbohydr. Polym.* **2021**, *15*, 117161. [CrossRef]
31. Zhiguang, C.; Junrong, H.; Huayin, P.; Keipper, W. The effects of temperature on starch molecular conformation and hydrogen bonding. *Starch* **2022**, *74*, 2100288. [CrossRef]
32. Fang, S.; Wang, G.; Xing, R.; Chen, X.; Liu, S.; Qin, Y.; Li, P. Synthesis of superabsorbent polymers based on chitosan derivative graft acrylic acid-co-acrylamide and its property testing. *Int. J. Biol. Macromol.* **2019**, *132*, 575–584. [CrossRef] [PubMed]
33. Gieroba, B.; Sroka-Bartnicka, A.; Kazimierczak, P.; Kalisz, G.; Pieta, I.S.; Nowakowski, R.; Przekora, A. Effect of gelation temperature on the molecular structure and physicochemical properties of the curdlan matrix: Spectroscopic and microscopic analyses. *Int. J. Mol. Sci.* **2020**, *21*, 6154. [CrossRef] [PubMed]
34. Guo, J.; Kong, L.; Du, B.; Xu, B. Morphological and physicochemical characterization of starches isolated from chestnuts cultivated in different regions of China. *Int. J. Biol. Macromol.* **2019**, *130*, 357–368. [CrossRef] [PubMed]
35. Ji, Y.; Lin, X.; Yu, J. Preparation and characterization of oxidized starch-chitosan complexes for adsorption of procyanidins. *LWT* **2020**, *117*, 108610. [CrossRef]
36. Cheng, Y.; Zhang, X.; Zhang, J.; He, Z.; Wang, Y.; Wang, J.; Zhang, J. Hygroscopic hydrophobic coatings from cellulose: Manipulation of the aggregation morphology of water. *J. Chem. Eng.* **2022**, *441*, 136016. [CrossRef]
37. Liu, Y.; Chen, L.; Xu, H.; Liang, Y.; Zheng, B. Understanding the digestibility of rice starch-gallic acid complexes formed by high pressure homogenization. *Int. J. Biol. Macromol.* **2019**, *134*, 856–863. [CrossRef]
38. Wang, L.; Zhang, L.; Wang, H.; Ai, L.; Xiong, W. Insight into protein-starch ratio on the gelatinization and retrogradation characteristics of reconstituted rice flour. *Int. J. Biol. Macromol.* **2020**, *146*, 524–529. [CrossRef] [PubMed]
39. Chen, X.; Cui, F.; Zi, H.; Zhou, Y.; Liu, H.; Xiao, J. Development and characterization of a hydroxypropyl starch/zein bilayer edible film. *Int. J. Biol. Macromol.* **2019**, *141*, 1175–1182. [CrossRef]
40. Ding, N.; Zhao, B.; Han, X.; Li, C.; Gu, Z.; Li, Z. Starch-binding domain modulates the specificity of maltopentaose production at moderate temperatures. *J. Agric. Food Chem.* **2022**, *70*, 9057–9065. [CrossRef]
41. Marvi, P.K.; Amjad-Iranagh, S.; Halladj, R. Molecular Dynamics Assessment of Doxorubicin Adsorption on Surface-Modified Boron Nitride Nanotubes (BNNTs). *J. Phys. Chem. B* **2021**, *125*, 13168–13180. [CrossRef] [PubMed]
42. Wang, C.; Chao, C.; Yu, J.; Copeland, L.; Huang, Y.; Wang, S. Mechanisms underlying the formation of Amylose–Lauric Acid–β-Lactoglobulin complexes: Experimental and molecular dynamics studies. *J. Agric. Food Chem.* **2022**, *70*, 10635–10643. [CrossRef] [PubMed]
43. Gao, Q.; Bie, P.; Tong, X.; Zhang, B.; Fu, X.; Huang, Q. Complexation between high-amylose starch and binary aroma compounds of decanal and thymol: Cooperativity or competition? *J. Agric. Food Chem.* **2021**, *69*, 11665–11675. [CrossRef]
44. Liu, S.; Sun, Y.; Guo, D.; Lu, R.; Mao, Y.; Ou, H. Porous boronate imprinted microsphere prepared based on new RAFT functioned cellulose nanocrystalline with multiple H-bonding at the emulsion droplet interface for highly specific separation of Naringin. *J. Chem. Eng.* **2023**, *452*, 139294. [CrossRef]
45. Siemers, M.; Lazaratos, M.; Karathanou, K.; Guerra, F.; Brown, L.S.; Bondar, A.N. Bridge: A graph-based algorithm to analyze dynamic H-bond networks in membrane proteins. *J. Chem. Theory Comput.* **2019**, *15*, 6781–6798. [CrossRef] [PubMed]
46. Saha, A.; Sekharan, S.; Manna, U.; Sahoo, L. Transformation of non-water sorbing fly ash to a water sorbing material for drought management. *Sci. Rep.* **2020**, *10*, 18664. [CrossRef] [PubMed]
47. Todica, M.; Nagy, E.M.; Niculaescu, C.; Stan, O.; Cioica, N.; Pop, C.V. XRD investigation of some thermal degraded starch- based materials. *J. Spectrosc.* **2016**, *2016*, 9605312. [CrossRef]
48. Gao, L.; Luo, H.; Wang, Q.; Hu, G.; Xiong, Y. Synergistic effect of hydrogen bonds and chemical bonds to construct a starch-based water-absorbing/retaining hydrogel composite reinforced with cellulose and poly (ethylene glycol). *ACS Omega* **2021**, *6*, 35039–35049. [CrossRef]

49. Czarnecka, E.; Nowaczyk, J. Synthesis and characterization superabsorbent polymers made of starch, acrylic acid, acrylamide, poly (Vinyl alcohol), 2-hydroxyethyl methacrylate, 2-acrylamido-2-methylpropane sulfonic acid. *Int. J. Mol. Sci.* **2021**, *22*, 4325. [CrossRef]
50. Kalkan, B.; Orakdogen, N. Anionically modified N-(alkyl) acrylamide-based semi-IPN hybrid gels reinforced with $SiO_2$ for enhanced on–off switching and responsive properties. *Soft Matter.* **2022**, *18*, 4582–4603. [CrossRef]

*Review*

# Recent Advances in the Investigation of Poly(lactic acid) (PLA) Nanocomposites: Incorporation of Various Nanofillers and their Properties and Applications

**Nikolaos D. Bikiaris [1], Ioanna Koumentakou [1], Christina Samiotaki [1], Despoina Meimaroglou [1], Despoina Varytimidou [1], Anastasia Karatza [1], Zisimos Kalantzis [1], Magdalini Roussou [1], Rizos D. Bikiaris [1] and George Z. Papageorgiou [2,*]**

1 Laboratory of Polymer Chemistry and Technology, Department of Chemistry, Aristotle University of Thessaloniki, GR-54124 Thessaloniki, Greece
2 Department of Chemistry, University of Ioannina, P.O. Box 1186, GR-45110 Ioannina, Greece
* Correspondence: gzpap@uoi.gr

**Abstract:** Poly(lactic acid) (PLA) is considered the most promising biobased substitute for fossil-derived polymers due to its compostability, biocompatibility, renewability, and good thermomechanical properties. However, PLA suffers from several shortcomings, such as low heat distortion temperature, thermal resistance, and rate of crystallization, whereas some other specific properties, i.e., flame retardancy, anti-UV, antibacterial or barrier properties, antistatic to conductive electrical characteristics, etc., are required by different end-use sectors. The addition of different nanofillers represents an attractive way to develop and enhance the properties of neat PLA. Numerous nanofillers with different architectures and properties have been investigated, with satisfactory achievements, in the design of PLA nanocomposites. This review paper overviews the current advances in the synthetic routes of PLA nanocomposites, the imparted properties of each nano-additive, as well as the numerous applications of PLA nanocomposites in various industrial fields.

**Keywords:** poly(lactic acid) (PLA); nano-additives; nanocomposites; synthesis; properties; applications



**Citation:** Bikiaris, N.D.; Koumentakou, I.; Samiotaki, C.; Meimaroglou, D.; Varytimidou, D.; Karatza, A.; Kalantzis, Z.; Roussou, M.; Bikiaris, R.D.; Papageorgiou, G.Z. Recent Advances in the Investigation of Poly(lactic acid) (PLA) Nanocomposites: Incorporation of Various Nanofillers and their Properties and Applications. *Polymers* **2023**, *15*, 1196. https://doi.org/10.3390/polym15051196

Academic Editors: José Miguel Ferri, Vicent Fombuena Borràs and Miguel Fernando Aldás Carrasco

Received: 19 February 2023
Revised: 23 February 2023
Accepted: 24 February 2023
Published: 27 February 2023



## 1. Introduction

Industrial plastic production was started at the beginning of 1950, and today it is estimated that about 8.3–9.1 million metric tons (Mt) of plastics have been globally produced [1]. Most of these are used in packaging materials (>50%), and are light and very difficult to be recycled; thus, the 79% of produced plastics are discarded in landfills or accumulate in the environment [2]. Most commercially produced plastics are fossil-based materials, are very stable, and degrade slowly in the environment, mainly owing to UV irradiation from the sun, forming fragments of different sizes. With sizes ranging from 1 μm to 5 mm, called microplastics (MPs), they have been recognized in the last 10–20 years as the main threat to living microorganisms [3–5] and human health, and can be found in oceans, sediments, rivers, and sewages [6,7]. Fears concerning the effect of MPs in living organisms has lead scientists and companies working with polymers to focus their research activities on the production of eco-friendly plastics that could be degraded in the environment without harmful by-products by substituting them with nondegradable fossil-based plastics. Thus, biodegradable and biobased polymers have garnered huge interest in recent decades as an alternative solution to the growing demand for single-use petroleum-based conventional polymers [8].

The ever-growing global concern regarding the environmental impacts and raising awareness towards issues such as global climate change and plastic waste management has led consumers, and in turn fuelled industries and governments, to adopt a more sustainable approach [9]. Many waste management solutions, such as reusing, reducing, composting, and recycling, have contributed to a decrease in white pollution (i.e., single-use plastic contaminating the environment, reaching nature, oceans, and landfills), and the development of biodegradable polymers has been a favourable change, since they can be recovered through recycling and composting. In this regard, PLA has been the frontrunner in (bio)polymers since it presents a potential solution to the waste disposal problem [10,11]. Lactic acid (LA) is the precursor to PLA and can be obtained by the fermentation of renewable resources such as corn, cassava starch, and potato [12].

PLA, a linear aliphatic polyester, exhibits a lot of attractive properties and thus is facilitated in numerous commercial sectors [13,14]. Its excellent barrier to flavor and good heat sealability have increased its usage in the food packaging field, while its appealing mechanical properties, high clarity, and ease of processing have expanded its utility in medical, automotive, textile, and agriculture areas [15,16]. In addition, it is renewable, compostable, and carries a relatively low cost [11]. PLA can be also degraded slowly in the environment since is not a biodegradable polyester but a compostable, leading to the production of microplastics. From several toxicity studies it was found that PLA MPs did not show any significant effects on plants and soils [17], while in the case of the human body, PLA is a safe biopolymer, since it is used extensively in medical and biomedical applications [18]. It can be slowly degraded in the human body, producing non-toxic fragments such as lactic acid [19].

In spite of these advantages, PLA also presents some drawbacks such as fragility, high permeability to gas and vapor, low melt strength, and thermal stability, while its degradation rate is slower compared to other common natural organic wastes, such as food and yard waste. This results in a significant limitation in its acceptance into industrial food and yard composting facilities and generally in its widespread use [20]. Due to the significant improvement observed in hybrid PLA nanocomposites, even at very low nanofiller loadings, many investigations have been carried out concerning this matter [21,22]. Hybrid nanotechnology has created a new revolution in the area of material science, developing the most high-tech advanced composites for future applications. At least one particle in the nanoscale dimension is included in the nanocomposites. The addition of nanoparticles (NPs) demonstrates remarkable enhancement in thermal, mechanical, physical thermo-mechanical properties due to better distribution, a highly specific ratio, and effective polymer–filler interaction [8].

Various nanoparticles have been used for the development of PLA-based nanocomposites. Among them, clays, carbon, bioglasses, metals, and natural nanomaterials are the most common [23–25]. This paper provides a comprehensive overview of the state-of-the-art PLA nanocomposites, including past and recent achievements that could influence future research and industrial applications. It presents information about the synthesis of PLA nanocomposites and the effect of nanomaterials on the morphology and physicochemical properties (mechanical, thermal, biological, electrical, biodegradable properties, etc.) of the final materials [26].

## 2. Synthesis of Nanocomposites

PLA nanocomposites are generally produced by means of three main techniques, namely in situ polymerization, solution casting, and melt mixing processing, and to a lower extent by electrospinning and additive manufacturing. Figure 1 overviews the different preparation methods for the incorporation of various nano-additives into the PLA matrix.

### *2.1. In-Situ Polymerization*

In-situ polymerization involves the dispersion of nanoparticles in a liquid monomer or monomer solution followed by polymerization, accomplished with the help of heat, radiation, or suitable initiators [27]. This process can be used to synthesis polymer nanocomposites under controlled conditions in order to produce materials with known and homogeneous compositions as well as with covalently bond formation between nanoparticles and the polymer matrix [28,29]. It has been proven by several studies that the in-situ polymerization assists in raising the interparticle spacing, achieving uniform dispersion of nano-reinforcements in the selected polymer matrix. Moreover, it controls the nanoparticle exfoliation, even at higher concentrations with a lesser chance of agglomeration [11]. Additionally, during the synthesis of PLA nanocomposites via in situ polymerization, covalent bonds between the growing polymer chains and the active groups of nanoparticles are formed, enhancing the PLA properties [15,27]. Manju et al. [30] reinforced PLA with sepiolite nanoclay via in-situ polymerization and obtained agglomerate-freed bio-nanocomposite even at higher loading of the silicate (7 wt.%). The melting point of the PLA was increased on incorporation of the clay due to homogenous dispersion attained by the excellent interaction of C-O of PLA with the clay's Si-OH groups [15]. Yang et al. [31] synthesized poly(L-lactide) (PLLA)-thermally reduced graphene oxide (TRG) via the in-situ ring-opening polymerization of lactide, with TRG as the initiator. They proved that the chemical interaction between PLLA and the TRG sheets was strong enough to increase the nucleation rate and the overall crystallization rate of the PLLA. Song et al. [32] grafted PLA covalently with the convex surfaces and tips of the multi-walled carbon nanotubes with carboxylic functional groups (MWCNTs–COOH) via in situ polycondensation. PLA was grafted on MWCNTs–COOH, which acted as an initiator, and the TGA results showed that the grafted PLA content could be controlled by the reaction time.

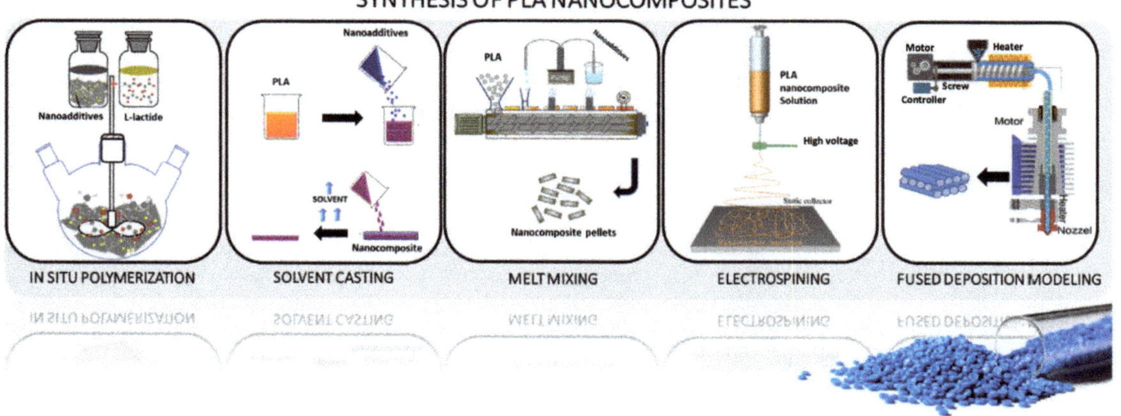


**Figure 1.** Synthetic routes employed for the preparation of PLA nanocomposites.

## 2.2. Solution Casting

In this technique, a solution of the polymer in a suitable solvent is mixed with a dispersion of the nanoparticles in the same or a different solvent. In the production of PLA nanocomposites, the solvent is usually used to pre-swell the nanofillers and the dispersion of the swollen nanofillers is mixed with the PLA solution. The sonication method is often practiced, aiding in the dispersion of nanofillers [33]. When the solvent is evaporated, the adsorbed polymer on the swelled nanofillers will reassemble, sandwiching the polymer to form nanocomposites. The solution method is an effective fabrication technique for nanocomposites due to the ease of processing nanoparticles in solvent, and it can be mainly used for nanoparticles that are thermally sensitive and degrade at high temperatures. On the other hand, in a practical sense, the solution intercalation approach is unattractive as a result of the use of organic solvents, and the polymer may be less serviceable as a result of the traces of solvent that could contaminate it [9,15,27]. The solution casting method is the conventional method for manufacturing PLA/$TiO_2$ nanocomposites. Feng et al. [34] fabricated PLA/$TiO_2$ nanocomposites, using different organic solvents and dissolution conditions. Specifically, PLA was dissolved in a chloroform solution, and magnetically stirred at 40 °C until complete. Then, 2 wt.% $TiO_2$ were dissolved in absolute ethanol to obtain five different nanocomposites. After ultrasonic dispersion for 60 min, the mixture was slowly added dropwise to the stirred PLA solution. The results showed that the maximum tensile strengths of the nanocomposites were achieved. Moreover, the incorporation of $TiO_2$ significantly decreased the water vapor transmittance rate of the nanocomposites while improving the water solubility. Fu et al. [35] employed citric acid-modified nanoclay to enhance PLA through solution casting to fabricate a shape memory nanocomposite. They showed a considerable restriction in the PLA chain segment, which supported an anisotropic force required in shape recovery. The grafted citric acid on the clay and ester groups of the PLA provided good affinity, being responsible for the shape memory property.

### 2.3. Melt Mixing

Melt mixing is a technique wherein a high temperature, higher than the melting point of the PLA, under the shear force is applied to the mixture of PLA matrix with nanofillers to prepare nanocomposites with the objective of obtaining well-dispersed nano reinforcement. However, it requires a pre-drying process of PLA granules as PLA has a high sensitivity to moisture [36].

This process presents significant advantages over solution casting and in situ intercalative polymerization methods. It supports a high volume of bulk polymer, requires no solvent/chemical, and carries a relatively low cost, faster than the solution and in situ polymerization and with the least environmental impact. Moreover, this method is compatible with processing techniques such as injection molding and extrusion, which is used in the polymer industry. Nevertheless, melt mixing subjects PLA to some level of degradation due to applying elevated temperature and mechanical shear during the process [37]. Melt blending has been widely used to prepare PLA/nanoclay composites. Ray et al. [38] prepared PLA nanocomposites containing 3, 5, and 7 wt.% OMMT (MMT modified with octadecyl ammonium cation), in a twin screw extruder (TSE), which was then converted into 0.7–2 mm thick sheets for characterization studies. On the other hand, this method is unsuitable for PLA/graphene nanocomposites since the pristine graphene has a tendency to agglomerate in the polymer matrix, unlike the other preparation routes of PLA/graphene nanomaterials that include in situ intercalative polymerization and solution intercalation [39]. Stacking occurs due to the large ratio of surfaces of graphene sheets in relation to their thickness, which yields significant van der Waals forces and a strong interaction between single sheets of graphene. The physical and chemical properties of such graphene aggregates are similar to the properties of graphite with relatively small surface areas [40]. The preparation of polymer-based nanocomposites has become a great challenge in obtaining a good distribution of the nano-reinforcement. Nonetheless, as for graphene or graphene oxide, the formation of bundles is not a problem, although the inclination of incomplete exfoliation and restacking can occur. The graphene functionalized through modification is preferred for melt intercalation [27]. This has been supported by Murariu et al. [41], who produced PLA with commercially available expanded graphite. The expanded graphite nanofiller yielded PLA composites with competitive functional properties such as high rigidity, with the Young's modulus and storage modulus increasing with the nano-additives' content.

### 2.4. Electrospinning

Recently, PLA fibers have been developed to be applied in the textile industry, for medical applications such as tissue engineering, and in food packaging applications. PLA nanofibers fabricated with the electrospinning method present a higher surface area and porosity than regular fibers and are applied to covering filtration systems, sensor assembly, tissue scaffolds, and protective clothing. In this method, the choice of solvent and the concentration of polymer and incorporation of nanofillers in the process affect the melt or solution spinning and thus the morphologies of the fabricated fibers and pores [42]. Melt spinning and solution spinning are traditional processes of PLA fiber fabrication but are mostly limited to fiber diameter on a micro-scale, which is a constraint to further PLA applications. Another drawback of these processes is that some active components such as anti-bacterial and antioxidant ingredients are thermolabile, which makes conventional melt processing unsuitable for preparing such nanocomposites [43]. On the other hand, electrospinning is a promising and versatile technique for producing multifunctional nanosized PLA nanocomposite fibers through the application of high voltage to viscoelastic polymer solution/melt with varying nano-reinforcements. The electrospinning machine set-up can be a single-nozzle or multi-nozzle type [44,45]. This technique can produce fibers with diameters ranging from the nanoscale to the submicron scale, which are typically one to two orders of magnitude lower than those of traditional solution-spun fibers. Specifically, Touny et al. [46] fabricated PLA/HNT nanocomposite fibers via electrospinning and the average diameter of the fibers spun was 230–280 nm. Shi et al. [47] produced PLA/cellulose nanocrystals (CNC) electrospun fibers and they found that PLA nanocomposite fibers with 1 wt.% CNC-loading was 1 µm while they decreased to 642 and 405 nm at 5 and 10 wt.% CNC-loading levels, respectively. In another work, Liu et al. [48] fabricated electrospun PLA/GO and PLA/nHAp fibers and they proved that the addition of 15 wt.% nHA in the PLA matrix produced fibers with a 563 ± 196 nm diameter size, while the incorporation of 3 wt.% GO decreased the diameter of the composite fibers to 412 ± 240 nm. The PLA nanofibers fabricated with the electrospinning method also present a higher surface area and porosity than regular fibers and are applied to covering filtration systems, sensor assembly, tissue scaffolds, and protective clothing. In this method, the choice of solvent and the concentration of polymer and incorporation of nanofillers in the process affect the melt or solution spinning, and thus the morphologies of the fabricated fibers and pores. PLA nanocomposites have also been used to fabricate nanofibers via electrospinning; however, some researchers have experienced difficulty in the homogeneous generation of nanocomposite fibers due to the difficulty of nano-additives' dispersion in the PLA matrix as a result PLA nanocomposites not being able to be electrospun.

Kim et al. [49] used an extremely fine hydroxyapatite powder, which was prepared using the sol–gel process, to improve the dispersion of the ceramic powder within the PLA matrix during electrospinning. Zhang et al. [50] ultrasonically dispersed ZnO nanoparticles in a PLA matrix and developed electrospun membranes with enhanced antimicrobial activity and UV blocking for food packaging applications. Xiang et al. [51] fabricated PLA/cellulose nanocomposite fibers electrospun at elevated temperatures from solutions of PLA in N,N-dimethylformamide (DMF) containing suspended cellulose nanocrystals. They optimized the electrospinning conditions to form PLA/cellulose nanocomposite fibers with uniform diameters; however, they found that the average fiber diameter decreased with the addition of cellulose nanocrystals. They also confirmed the presence of cellulose nanocrystals at the surface of PLA by surface elemental composition analysis.

### 2.5. Additive Manufacturing

Additive manufacturing (AM) is a very promising and rapid processing method permitting the manufacture of 3D-printed models or functional parts with complex geometries [52]. Fused deposition modeling (FDM) is a widely adopted AM technique due to its simplicity and low material wastage and cost. This process begins by electronically designing and "slicing" the 3D model of the item in the CAD to be loaded into the FlashPrint software, which generates the printing path. FDM is an extrusion-based process wherein material is selectively dispensed via a shaped needle or orifice [53]. The nozzle temperature has to be set above the melting temperature of the thermoplastic for it to flow. In a recent optimization study, it was found that the best properties of pure PLA prints were achieved when the printing temperature was 215 °C. Additionally, the content of water in the PLA composite filament should be less than 50–250 ppm, otherwise the presence of larger amounts of water can cause swelling of the filament, which can block the hot nozzle during the 3D process. However, these characteristics of PLA are affected by the increasing of different amounts and types of nanofillers. Moreover, the morphology of the PLA nanocomposites and the interactions between polymer—polymer and polymer—nanoparticles affect the final properties of the 3D-printed product [54].

Wang et al. [55] researched the capacity of PLA with nHAp (at a content of 0%, 10%, 20%, 30%, 40%, and 50%) to be 3D printed for medical applications. The results showed that the PLA/nHAp composite had a stable structure, and the processing of this material was highly controllable. When the proportion of nHAp was greater than 50%, the composite material could not pass through the printing nozzle coherently and stably due to its high brittleness. When the nHAp ratio was less than or equal to 50%, the composite filament material could be printed by FDM. After applying the 3D printing process of PLA/nHAp composites, the printed samples with 0% to 30% nHAp could still maintain the integrity of the structure after the pressure test, which proved that these samples maintained their elasticity. However, when the n-HA ratio reached more than 40%, the printed sample became brittle and could not maintain complete shape after compression. In a recent study by Yang et al. [56], a filament based on PLA/CNTs composites was produced by the FDM process. The effects of the CNT content on the crystallization-melting behavior and melt flow rate were tested to investigate the printability of the PLA/CNT. The results demonstrate that the CNT content has a significant influence on the mechanical properties and conductivity properties. The addition of 6 wt.% CNT resulted in a 64.12% increase in tensile strength and a 29.29% increase in flexural strength. The electrical resistivity varied from approximately $1 \times 10^{12}$ $\Omega$/sq to $1 \times 10^{2}$ $\Omega$/sq for CNT contents ranging from 0 wt.% to 8 wt.%. Three-dimensional PLA-G electrode properties allowed for graphene surface alteration and electrochemical enhancement in the sensing of molecular targets. Bustillos et al. [57] fabricated hierarchical nanocomposites fused by deposition modeling printing using filaments of PLA and PLA-graphene (Figure 2). Enhanced creep resistance in the PLA-graphene was attributed to the restriction of the polymeric chains by graphene, caused by low strain rates identified during secondary creep. Three-dimensional-printed PLA-graphene exhibited a 20.5% improvement in microscale creep resistance as compared to PLA at loads of 90 mN. Wear resistance was increased by a 14% in PLA-graphene as compared to PLA.

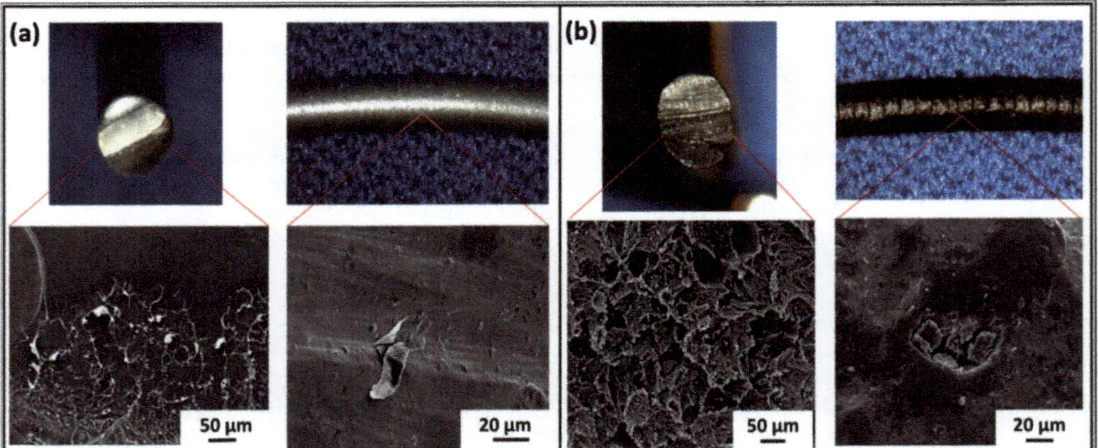


**Figure 2.** Top and cross-sectional microstructure of (**a**) neat PLA and (**b**) PLA-graphene filaments [57].

## 3. Effect of Nano-Additives on PLA/Nanocomposite Properties

### 3.1. PLA/Metal Oxides

Interest in the investigation of PLA/metal oxides containing nanoparticles has rapidly grown in recent years because of their numerous properties. Particularly, the design of metal-containing nanocomposites is a current trend in scientific research due to their high-potential practical application of antimicrobial and antiviral materials [58]. A series of metal oxide nanoparticles have been synthesized including silica dioxide ($SiO_2$), titania dioxide ($TiO_2$), zinc oxide (ZnO), magnesium oxide (MgO), boron oxide (BO), and aluminum oxide ($Al_2O_3$). They exhibit a variety of morphologies such as sphere, triangular, star, nanotubes, nanorods, nanowires, etc. The preparation of polymer-metal oxide nanocomposites with a nanophase-separated structure requires the homogeneous dispersion of the metal oxide NPs as well as a reduction in the size of the polymer–metal oxide interface, since it may alter the physical property of the nanocomposites [59]. Due to a high density and limited size, metal oxide NPs present interesting results in terms of chemical and physical properties. In general, properties including thermal stability, toughness, glass transition temperature, optical, tensile strength, etc. are found to be improved by the formation of such nanocomposites with PLA. Therefore, these nanocomposites are widely used in numerous applications, from active surface coating to biomedical applications including structural materials [13,60]. This chapter provides a general overview of the recent studies involved in the fabrication of PLA/metal oxide nanocomposites.

#### 3.1.1. Green Synthesis of Nanosized Metal-Oxides

Over the last decade, novel preparation routes/methods for nanomaterials (such as metal nanoparticles, quantum dots (QDs), and nanofibers and their composites) have been an interesting field in nanoscience [61]. To obtain nanomaterials of desirable shape, size, and properties, two different fundamental principles of synthesis are required. Figure 3 illustrates the top-down and bottom-up preparation routes investigated in the existing literature.

Nanoparticles are prepared through size reduction, disintegrating from the bulk material into fine particles in the top-down approach. This route can be accomplished through physical and chemical methods via lithographic, mechanical milling, sputtering, chemical etching, pulsed laser ablation, and photoreduction techniques. Still, the major downside in the top-down approach is the imperfection of the surface morphology. In the bottom-up approach, also known as the self-assembly approach, the nanoparticle synthesis relies on wet chemical methods (e.g., chemical reduction/oxidation of metal ions) and others, such as sol-gel chemistry, chemical vapor deposition (CVD) co-precipitation, microemulsion, laser and spray pyrolysis, hydrothermal, aerosol, and electrodeposition methods. In the bottom-up synthesis, the nanoparticles are built from smaller units, for example, by atomic and molecular condensation. However, they still suffer from the involvement of potentially hazardous and toxic substances, high-cost investments, environmental toxicity, high energy requirements, lengthy reaction times, and non-eco-friendly by-products.

Interestingly, the morphological parameters of NPs may be moderated by varying the concentrations of chemicals and reaction conditions (e.g., temperature and pH). Nonetheless, if these synthesized nanomaterials are subject to the actual/specific applications, then they can suffer from the following challenges: (i) stability in hostile environments, (ii) bioaccumulation/toxicity features, (iii) expansive analysis requirements, and (iv) recycle/reuse/regeneration. Nanoparticles prepared using heavy radiations and toxic substances such as reducing and stabilizing agents are toxic in nature, thus leading to harmful effects on humans and the environment.

The majority of chemical methods often exploits hazardous reducing agents such as sodium borohydride ($NaBH_4$), ethylene glycol, hydrazine, and sodium citrate, but they result in poor control over particle size and size distribution; thus, extra-stabilizing agents such as ascorbic acid, polymers, ligands, dendrimers, surfactants, phospholipids, and surfactants are necessary to avoid the agglomeration of nanoparticles. A high thermal decomposition treatment then takes place to eliminate the reducing and stabilizing agents adsorbed on the nanoparticle surface, which may block the active sites of particles, consequently resulting in a decreased surface area and deteriorated reactivity of the products. Finally, this procedure produces a low yield of the targeted products.

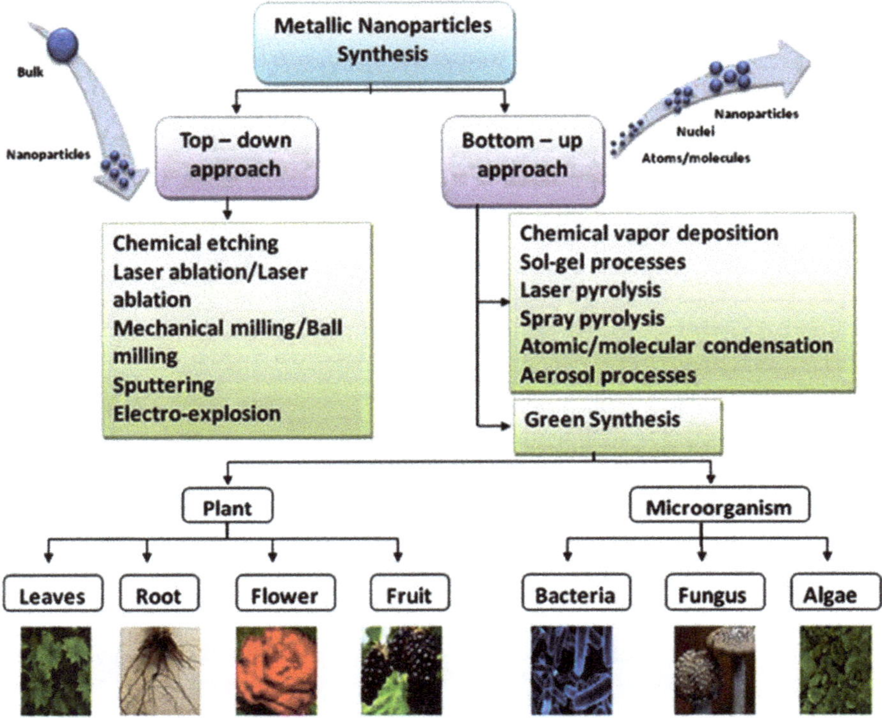


**Figure 3.** Top-down and bottom-up approaches for the fabrication of metal-based nanoparticles [62].

Green nanotechnology has developed recently with various processes to restrain environmental damage by replacing the usage of harmful compounds. More and more interest has been gained by green extraction of metal oxide NPs in the last decade, owing to the use of eco-friendly reagents such as microorganisms, fungi, enzymes, leaves, plants, etc. NPs are synthesized by this eco-friendly synthesis method, which involves a single-step pollution free method that requires less energy to initiate the reaction and the preparation time is less when compared to other synthetic routes. The utilization of ideal solvent systems and natural supplies (such as organic systems) is vital to achieve this aim.

Green synthesis methodologies based on biological precursors depend on various reaction parameters such as temperature, pressure, solvent/solvents system, and pH environment (basic, acidic, or neutral). Biosynthesis is a green synthetic approach that can be categorized as a bottom-up approach where the metal atoms are arranged into clusters and eventually into NPs. Amongst the accessible green methods of synthesis for metal/metal oxide NPs, the exploitation of plant extracts is a rather simple and easy process to produce nanoparticles on a large scale relative to bacteria- and/or fungi-mediated synthesis (Figure 4). Moreover, the plentifulness of effective phytochemicals in various plant extracts, especially in leaves, such as ketones, aldehydes, flavones, amides, terpenoids, carboxylic acids, phenols, and ascorbic acids, is the main reason for their wide extraction. These products are most commonly known as biogenic nanoparticles and have the capability of reducing metal salts into metal NPs.

Ramanujam et al. [63] extracted MgO NPs using Emblica officinalis fruit extract. The mean size of MgO NPs was found to be 27 nm. Ashwini and his group [64] conducted a study upon extracting MgO NPs using aloe vera. The average size of obtained MgO nanoparticles was found to be 50 nm.

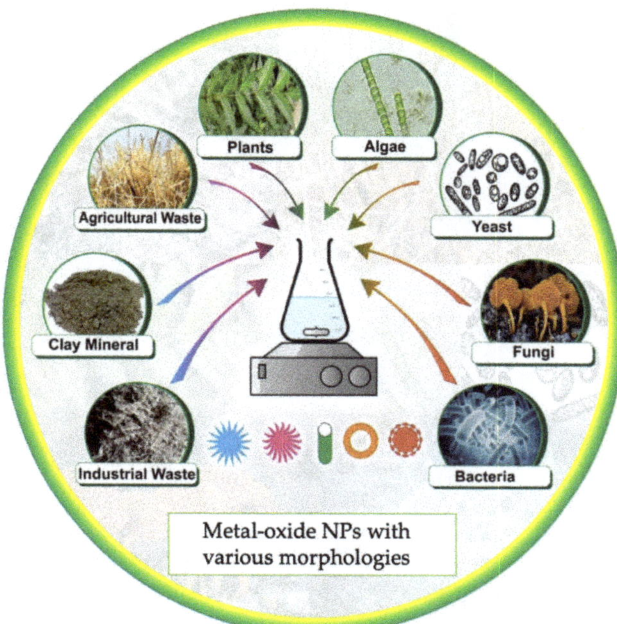


**Figure 4.** Biological resources for the green synthesis of metal-oxide NPs [65].

Sankar et al. [66] effectively synthesized $TiO_2$ nanoparticles e from the aqueous leaf extract of Azadirachta indica with an average size of 124 nm. $TiO_2$ NPs using extracts of Echinacea purpurea Herba were biosynthesized using an aquatic solution of Echinacea purpurea herba extract as a bioreductant by Dobrucka [67]. The size of the $TiO_2$ nanoparticles was found to be in the range of 120 nm with a spherical size and the presence of agglomerates. The eco-friendly synthesis of $TiO_2$ NPs using different biological sources are summarized in Table 1.

**Table 1.** Varying sizes of spherical $TiO_2$ NPs from different biological sources.

| Biological Source | Size of NPs | References |
|---|---|---|
| Azadirachta indica, leaf extract | 124 (average) nm | [66] |
| Aeromonas hydrophila, bacterium | 28–54 nm | [68] |
| Annona squamosal, fruit peel extract | 23 nm | [69] |
| Bacillus amyloliquefaciens, bacterium | 15–86 nm | [70] |
| Euphorbia prostrata, leaf extract | 83.22 nm | [71] |

Saraswathi et al. [72] biosynthesized ZnO NPs mediated by the leaves of Lagerstroemia speciosa with an average particle size of 40 nm (Figure 5). Interestingly, the obtained NPs had a spherical shape at 200 °C and were rod-shaped at 800 °C (Figure 6).

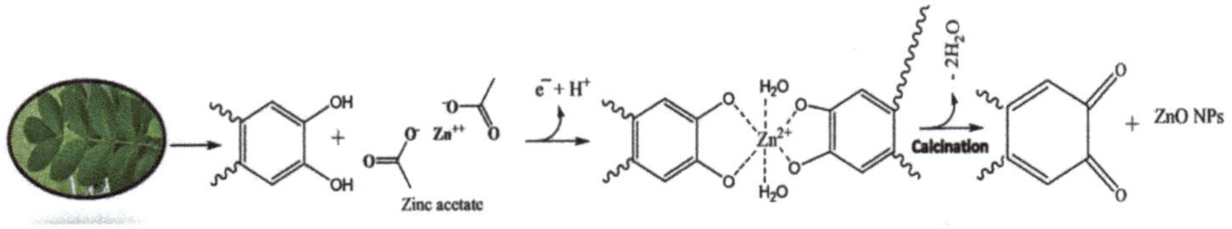

Figure 5. Biosynthesis of ZnO NPs using L.specioca leaves [72].

**Figure 6.** SEM analysis of ZnO-NPs: (**a**) 200 °C (spherical shape) and (**b**) 800 °C (nanorods) [72].

Yuvakkumar et al. [73] reported for the first time the sustainable biosynthesis of ZnO nanocrystals employing *Nephelium lappaceum* L. peel extract as a natural ligation agent. Green synthesis of zinc oxide nanocrystals was carried out via zinc-ellagate complex formation using rambutan peel wastes. The green syntheses of ZnO nanoparticles using various biological supplies are summarized in Table 2.

**Table 2.** Green synthetic sources of varying sizes of spherical ZnONPs.

| Biological Source | Size | References |
|---|---|---|
| Jacarandamimosifolia, flow erextract | 2–4 nm | [74] |
| Ruta graveolens, stem extract | ~28 nm | [75] |
| Moringa oleifera, leaf extract | ~6–10 nm | [76] |
| Polygala tenuifolia, root extract | 33.03–73.48 nm | [77] |
| Sechium edule, leaf extract | 36.2 nm (mean) | [78] |

Rahimzadeh et al. [79] described a green synthetic process to obtain $SiO_2$ NPs utilizing *Rhus coriaria* L. extract and sodium metasilicate ($Na_2SiO_3 \cdot 5H_2O$) under reflux conditions. The FESEM images revealed that $SiO_2$ NPs are spherical in shape with a minimum degree of agglomeration, and the NP sizes were between 10~15 nm. Rezaeian and his group prepared ZnO NPs from olive leaves with an average particle size between 30–40 nm and spherical in shape [80]. Rahimzadeh et al. [79] in their work synthesized silica NPs via green synthesis by utilizing leaf biomasses from sugarcane (Saccharum ravannae), weed plants (Saccharum officinarum), and rice plants (Oryza sativa) via a simple chemical method. NPs from sugarcane were hexagonal in shape with a 29.13 mean size, particles extracted from weed plants were spherical in shape with average size of 39.47, and nanoparticles from rice plants had a spherical shape with an average size of 30.56 nm.

### 3.1.2. MgO

MgO is a widely available metal-based oxide used for the reinforcement of PLA. It is a non-toxic alkaline earth metal, reproducible on a large scale, is environmentally friendly, low-cost, and is available in the form of natural periclase. Presently, MgO is listed as a generally recognized as safe (GRAS) material by the US Food and Drug Administration (US-FDA) [81]. Medically, MgO is used for antibacterial purposes, bone tissue regeneration, as well as cancer treatment [82]. Leung et al. [83] reported strong antibacterial activity of the MgO NPs even during the absence of any ROS production. Moreover, the chemical properties and the high surface area of nanosized MgO expanded its usage in numerous modern fields. In brief, MgO is an insulator exhibiting a band gap of $E_g$ = 7.8 eV and a high dielectric constant. The MgO NPs that are fabricated using sol-gel and flame spray pyrolysis techniques presented high efficiencies in textiles and methylene blue dye removal. In addition, very fine MgO NPs of $E_g$ = 4.45 eV were prepared via a combustion route using urea-formaldehyde as fuel [84]. The transition from insulators to semiconductors is due to the quantum size effects and surface effects in MgO NPs. These effects lead to discrete energy levels in the band gap, which result in its reduction [85].

#### 3.1.3. ZnO

ZnO is a naturally occurring oxidic chemical found in the rare mineral zincite that crystallizes in the hexagonal wurtzite structure P63mc. Metallic zinc is plentiful in the earth's crust and creates pure powder. Before the invention of nanotechnology, ZnO was used in bulk, but later it was also facilitated as a nanosized material for its intended properties [86]. Nanometric ZnO can appear in various structures. A dimensional structure consists most extensively of needles, hexagons, nano-rods, ribbons, belts, wires, and brushes. ZnO can be produced in two-dimensional structures, such as nanoparticles, nanosheets, and nanoplates. The variety of nanostructures given for ZnO have created more possibilities in the field of nanotechnology [87].

Amongst various semiconducting nano-scale materials, ZnO is a distinctive electronic and photonic wurtzite n-type semiconductor with a wide direct band gap of 3.37 eV and a high exciton binding energy (60 mV) at room temperature [88]. The high exciton binding energy of ZnO allows for excitonic transitions even at room temperature, which could mean high radiative recombination efficiency for spontaneous emission as well as a lower threshold voltage for laser emission. The lack of a centre of symmetry in wurtzite, combined with large electromechanical coupling, results in strong piezoelectric and pyroelectric properties, and thus the use of ZnO in mechanical actuators and piezoelectric sensors [89].

Furthermore, ZnO is biosafe and biocompatible and hence may be employed in biomedical applications without being coated. Statistics showed that approximately 99,000 scientific papers in the period of 2018–2020 were related to the search for new antimicrobial compounds. About 6% of these papers addressed ZnO-related compounds. ZnO is one of the most promising nanomaterials, owing to its excellent antimicrobial properties, good chemical stability, biocompatibility, optical and UV shielding abilities, low price, and widespread availability. ZnO NPs exhibit antimicrobial capability against a wide range of microorganisms such as *E. coli*, *Pseudomonas aeruginosa*, and *S. aureus*, becoming appealing for various applications such as packaging, textiles, and medicine [90].

#### 3.1.4. $TiO_2$

$TiO_2$ or titania is a renowned and extensively studied material due to its excellent photostability, photocatalytic activity, non-toxicity, safety, chemical stability, outstanding antimicrobial properties, and significant antibacterial properties [91]. Three distinct phases or "crystalline polymorphs" of $TiO_2$ NPs are found in nature, namely anatase, rutile, and brookite. Amongst these three, rutile is the steadiest phase, while in contrast, brookite and anatase are the labile phases at all temperatures. The anatase phase of $TiO_2$ NPs is the most responsive one, in a chemical manner. Both anatase and brookite exhibit the capability to convert into a rutile phase when heated [92]. Regarding the manufacturing methods of PLA/$TiO_2$ nanocomposites, in the literature, a great variety of methods and techniques is reported. The conventional one is the "solution casting method", in which the nanocomposites are produced by constant dispersion of the NPs in a polymer matrix. Also frequently employed is the electrospinning technique, which is important for the surface coating of in situ-produced NPs, particularly on the exterior or in the majority of a polymer matrix [34] and the "solvent casting method", followed by a or some hot-pressing steps [93]. The "Sol-gel method" serves as a doable method to increase the dispersibility of $TiO_2$ during the film-forming process and to additionally alter the thermal/mechanical properties and antimicrobial action of biopolymer films. Improving the distribution degree of metal oxides in a polymer matrix essentially affects the photocatalytic ability as well as the antibacterial efficacy and mechanical properties [91]. The "Thermally induced phase separation (TIPS) method" [94], "a spin coating process" [95], melt blending through a corotating twin-screw extruder [96], and the "breath figure (BF) method" [97] are also mentioned as potential production methods.

$TiO_2$ is the most extensively investigated photocatalyst due to its strong oxidizing abilities, super hydrophilicity, long durability, chemical stability, low cost, nontoxicity, and transparency under visible light [98]. In the last 30 years, the photocatalytic efficiency of nanosized $TiO_2$ at degrading organic contaminants in the air and water has resulted in their utilization in several environmental applications. These include photocatalytic hydrogen production via water splitting, degradation of organic pollutants in wastewater photocatalytic self-cleaning, bacterial disinfection photo-induced super hydrophilicity, as well as in photovoltaics and photosynthesis. Currently, water contamination by organic compounds, viruses, bacteria, and metals has been highlighted as a major worldwide problem. In terms of antimicrobial activity, $TiO_2$ NPs appeared to be able to eliminate a broad range of micro-organisms, such as fungi, protozoa, Gram-positive and Gram-negative bacteria, viruses, and bacteriophages. Numerous techniques have been explored to remove these pollutants from water, including adsorption and photocatalysis, which comprise eco-friendly routes [99]. Studies have shown that one of the best ways to utilize photocatalytic $TiO_2$ NPs is to immobilize them onto the surface of polymeric materials.

This excellent photocatalytic ability of $TiO_2$ has also enabled its use in food safety applications, such as packaging materials, cutting boards, conveyor belts, or non-contact surfaces such as floor, walls, and curtains. In the presence of a UV-A light source, titania has the capability to produce the transition of an electron towards the conductive band, supporting the oxidative capacity of other species by generating reactive oxygen species (ROS) such as superoxide radicals ($O\cdot^{-2}$) and hydroxyl radicals ($\cdot OH$) [98,100].

#### 3.1.5. n-$SiO_2$

Advancement in nanotechnology has led to the production of a nanosized silica, $SiO_2$, which has been widely used as filler in engineering composites. It is often added as a cementitious material since it can reduce the use of cement due to its reinforcing effect, as well as accelerate the hydration and alter the microstructure evolution of cementitious composition [101]. Many scientific works have been conducted regarding the incorporation of n-$SiO_2$ in the PLA matrix due to its numerous properties. Moreover, nano $SiO_2$ provides environmental protection, safe and non-toxic functions, abundant surface groups, a huge specific area, and a high rigidity and modulus, leading to its wide usage as a cheap but effective nanofiller for polymers. Moreover, the nontoxic nature, the high thermal and chemical stability, water resistance, and mechanical properties as well as the important role in the biomineralization of $SiO_2$ render it a biocompatible nanofiller that retains the biocompatibility of the PLA matrix. In addition, nano $SiO_2$ has a white color, so the composite filament is bright and can be colored with almost any desirable color. Statistics show that approximately 99,000 scientific papers in the period from 2018 to 2020 were related to the search for new antimicrobial compounds. About 6% of these papers addressed metal-related compounds [90].

### *3.2. Carbon-Based PLA/Nanocomposites*

#### 3.2.1. Carbon Nanotubes (CNTs)

In spite of PLA's advantages, its mechanical as well as electrical and thermal properties could be further improved so as to expand the application fields. The most effective way to overtake this disadvantage is the filling of a PLA matrix with the introduction of nanoscopic dimensions of carbon-based fillers (Figure 7) with a high aspect ratio, such as carbon nanotubes (CNTs), carbon nanofibers (CNFs), graphene nanoplatelets, and spherical nanoparticles. Upon good dispersion in the PLA matrix, such nano-additives have been found to generate an increase in both the degree and the rate of crystallization, since they provide additional sites for crystallization; in other words, they serve as additional crystallization nuclei [102–104]. Their excellent mechanical properties, such as high modulus in the direction of the nanotube's axis and excellent electrical conductivity that varies from insulating to metallic and the hollow structures of CNTs, have expanded their usage as fillers [105]. The growing interest towards carbon-based nanomaterials, including CNTs, mainly relates to their intriguing properties as conductive fillers for the fabrication of electric/electronic devices, for which production volumes have become dramatically elevated in recent years [106]. CNTs exhibit a highly specific surface area that allows for low loadings to tune the polymer key properties concerning their mechanical, thermal, electrical, and biological performance. CNTs that possess a wall structure consisting of a single graphite sheet closed in a tubular shape are called single-walled carbon nanotubes (SWCNTs), while those consisting of a plurality of graphite sheets each arranged into a tubular shape and nested one within the other are named multi-walled carbon nanotubes (MWCNTs). MWCNTs consist of smaller diameter single-walled tubes inside larger diameter tubes and may vary from a double-walled nanotube to as many as fifty concentric tubes, exhibiting diameters varying between 2 and 100 nm [107]. They have exceptional mechanical properties, aspect ratio, electrical and thermal conductivities, and chemical stability, and hence are considered excellent candidates for the creation of multifunctional materials [108,109].

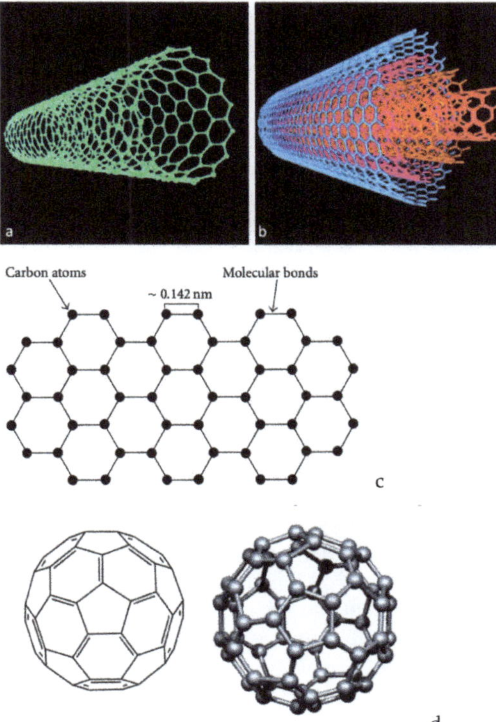


**Figure 7.** Carbon-based nano-additives. Three-dimensional presentation of (**a**) SWCNTs and (**b**) MWCNTs [110]; (**c**) schematic representation of a graphene sheet [111] and (**d**) 2D and 3D illustration of the fullerene C60 structure [112].

### 3.2.2. Graphene

In the case of graphene, it is naturally derived from graphite, which is composed of a layer of less than 100 nm and can be divided into sheets (Figure 7c) around 1–2 nm thick. The physical structure of graphene with its large specific surface area makes it have a better reinforcing effect than other nanofillers [113,114]. Graphene has high mechanical strength and electron mobility, and thus may be facilitated as a filler, even at low amounts in a polymer matrix due to its highly specific surface area and chemical interaction with the matrix by the formation of strong bonds [115,116]. Due to its two-dimensional arrangement of $sp^2$-bonded carbon atoms, graphene has been shown to enhance the wear resistance and reduction of friction. Graphene's incorporation of polymer matrices has resulted in composites exhibiting superior mechanical strength while retaining their flexibility, as well as tailorable thermal and electrical conductivity as a consequence of the generated graphene network in the matrix [57]. Another form of graphene that can be used as nanofillers are graphite nanoplatelets (GNPs) or graphite nanosheets (GNS), which also possess great mechanical, thermal, and electrical properties [117]. The aforementioned fillers of graphene or its derivatives (graphene oxide, reduced graphene oxide) are utilized in varying concentrations as fillers in a PLA matrix in order to prepare nanocomposites with enhanced properties via different techniques [116,118]. Based on the unique properties of graphene, graphene-based polymer composites are expected to offer enhanced electrical and thermal conductivity, improved dimensional stability, higher resistance to microcracking, and increased barrier properties above the matrix polymer [108].

### 3.2.3. Carbon Nanofibers (CNFs)

Nanosized carbon-based reinforcements such as carbon nanofibers (CNF) into pure thermoplastic matrices have proven to be valuable for manufacturing polymer matrix composites with enhanced mechanical performance and functionality. Carbon nanofibers have a highly specific area, elasticity, and great strength due to their nano-sized diameter. Concerning CNFs, they are characterized by diameters ranging between 50 and 200 nm, being different from the conventional carbon fibers that have diameters in the order of micrometers. Due to the preparation process employed, CNFs possess improved properties, while they can offer property enhancements similar to CNTs in a more cost-effective way [103]. Therefore, they are utilized as fillers in order to enhance mechanical and electrical properties as well as the thermal conductivity of polymers [119]. Critical parameters for enhancing the mechanical and electrical properties are weight fraction of the filler, filler length, and its orientation. Overall, carbon-based nanomaterials offer the possibility to combine PLA properties with several of their unique features [108].

### 3.2.4. Fullerene

Fullerene is a unique nano-allotropic form of carbon. Similar to carbon nanoparticles (graphene, carbon nanotubes, nanodiamonds), fullerene has been reinforced in polymeric matrices [120–123]. Fullerene and derived nanofillers affect the structural, electrical, thermal, mechanical, and physical properties of polymeric matrices. However, the main encounter in the formation of the polymer/fullerene nanocomposite is the dispersion and miscibility with the polymeric matrices. Studies have shown that the incorporation of carbon-based nanofillers with various dimensionalities such as zero-dimensional (0D) fullerenes offers an effective approach to PLA nanocomposites with synergistic enhancements in the electrical and mechanical properties when exposed to external stimuli [124]. For that reason, polymer/fullerene nanocomposites have been applied in olar cell, supercapacitor, electronic, and biomedical devices and systems [122].

### 3.3. PLA Nanocomposites with Natural Nano-Additives

Currently, amongst the key-priorities of the industry and academic sector is the replacement of petroleum-derived and unsafe complexes by the increasing inclusion of natural and green compounds that can be obtained from diverse renewable resources. To achieve promising results, cost-efficient and eco-friendly extraction methods have been designed over the years. Once these green alternatives have been isolated, they are successfully applied to many fields with very assorted aims of utilization such as coagulants, adhesives, dyes, additives, or biomolecules [125]. The main challenge is to develop high-performance polyphenol-reinforced thermoplastic composites, where the use of natural fillers replaces the usual chemical additives with non-toxic ones, not only to improve the final performance but also to increase the desired multifunctionalities (structural, antioxidant, and antibacterial) [126]. Polyphenols comprise an enormous family of secondary metabolites that are stored in vacuoles of vegetal cells such as esters or glycosides. While this family of compounds is vast, they share some mutual properties, such as the formation of coloured complexes with iron salts, oxidation by potassium permanganate in alkaline media, and easy electrophilic aromatic substitution-coupling with diazonium salts and aldehydes [127,128]. Lignin and tannins are considered polyphenols with high molecular weights. Thus, not only do they possess typical features of the polyphenols group, but the presence of a large number of hydroxyls provides them with the ability to create bonds to reach a stable cross-linked association within several molecules, such as carbohydrates or proteins. This unique characteristic differentiates them from the common group of polyphenols.

#### 3.3.1. Lignin

Lignin (Lgn) (Figure 8) is the second most abundant natural polymer after cellulose [129] and is found in every vascular plant on earth [130]. It is a biodegradable, nontoxic, and low-cost macromolecule [131,132] that has generated great interest due to its multi-functionalities, such as UV resistance, bioactive antibacterial, and anti-oxidation activities. Commercially, it is easily obtained as a byproduct in the paper and pulp industry. Its use is also regularly reported in biorefineries and in carbon-fiber manufacturing. Lignin nanoparticles (Lgn-NPs) have been proven to be beneficial reinforcing materials in polymer nanocomposites, as they enhance their thermal and mechanical properties because of the aromatic rings, while [129,131] in the PLA case specifically, lignin nanoparticles' antibacterial and antioxidant activity as well as their UV protection and reduced water sorption capacity can enhance the final nanocomposites' performance. Moreover, lignin's highly branched polyphenolic structure filled with plentiful functional groups allows for its chemical modification and polarity adjustment in such a way that lignin derivatives can be employed in copolymers, composites, and blends for a variety of applications such as innovative phenolic resins, epoxies, adhesives, and in the packaging industry [130]. Presently, the majority of food packaging materials used worldwide are non-biodegradable petrochemical-based plastics, which are responsible for polluting the environment. Approximately 380 million tons of plastic are produced globally every year and around 40% are used in the packaging industry [133]. This statement has forced researchers to widely study the incorporation of natural agents such as Lgn-NPs into biodegradable natural polymers for active packaging. Active food packaging systems involve the direct incorporation of natural bioactive compounds with antioxidant and antimicrobial activity in food packaging materials, which may possibly improve shelf life, ensure the safety of food, and control undesirable quality variations in the food during storage, transportation, and distribution [134]. Bioactive PLA-based composites could potentially replace the currently used plastic films to reduce the menace of environmental pollution and also help in the valorization of wastes from the food industry.

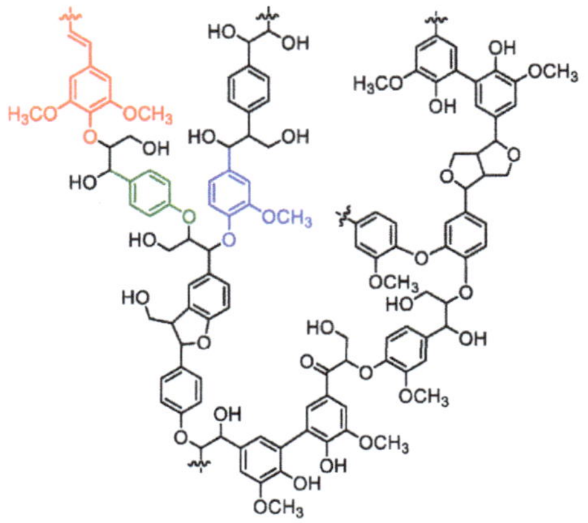

Figure 8. Chemical structure of lignin (Major monolignol units are colored: sinapyl alcohol—red, guaiacyl alcohol—blue, and p-coumaryl alcohol—green) [135].

#### 3.3.2. Tannin

The term tannin (TANN) broadly refers to a large complex of biomolecules of a polyphenolic nature (Figure 9). Tannins are found in most species throughout flora, where their functions are to protect the plants or vegetables against predation and might help in regulating plant growth. They are also used for iron gall ink production, adhesive production in wood-based industries, anti-corrosive chemical production, as a uranium-recovering chemical from seawater, and in the removal of mercury and methylmercury from solutions. There are two major groups of tannins, i.e., hydrolyzable and condensed tannins. While hydrolyzable tannins are present in few dicotyledons species, the natural presence of condensed ones is much more abundant; thus, they represent a major and more valuable source of commercial tannins [136].

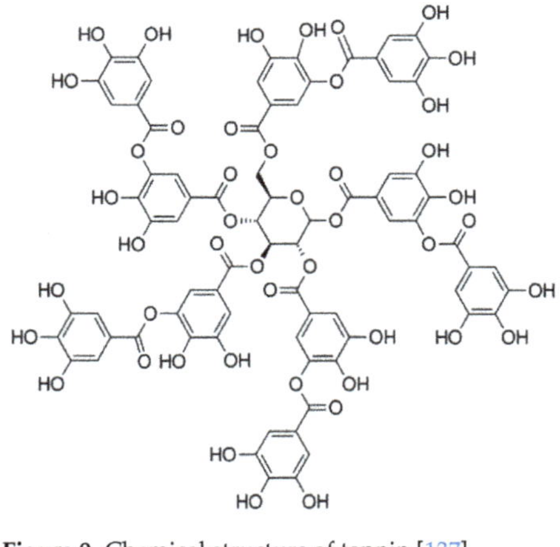

Figure 9. Chemical structure of tannin [137].

Generally, tannins are obtained from natural renewable resources, i.e., plants that are the secondary phenolic compounds of plants. More specifically, tannins are either galloyl esters or oligomeric and polymeric pro-anthocyanidins produced by the secondary metabolism of plants, i.e., synthesized by biogenetic pathways [138]. Condensed tannins, composed of flavonoid units, are one of the most abundant and sustainable biopolymers in plants. Their characteristics include antioxidant, antimicrobial, and stabilizing properties and are attractive for use in polymer materials [139]. Studies conducted in the past have shown that tannins exhibit excellent antioxidant and UV-protective properties when incorporated in polyethylene [140], poly(vinyl chloride) (PVC) [141], polypropylene [142], and poly(vinyl alcohol) (PVA) [143]. Until now, the combination of PLA and bio-fillers has been claimed to be an efficient and beneficial method to produce low-cost biocomposites with superior characteristics such as flame retardancy, mechanical performance, thermal stability, and gas-barrier properties. Considering the natural renewable reinforcements based on polyphenolic materials, lignin and tannins have great potential for the preparation of PLA nanocomposites to enhance functionality, especially toward bioactivity, whilst both phenolics act as free radical scavengers and thus as natural antioxidants that are both UV-resistant and bioactive.

In a recent work, Cresnar et al. [144] prepared two series of PLA-KL and PLA-TANN at various contents (0.5%, 1.0%, and 2.5% (*w*/*w*)) by hot melt extrusion to study their antioxidant and antibacterial properties. The results showed the accelerated antioxidant behavior of all PLA-KL and PLA-TANN composites, which increases with the filler content (Figure 10). Furthermore, the KL- and PLA-based TANN showed resistance to *E. coli*, but without a correlation trend between polyphenol filler content and structure. The water contact angle showed that neither KL nor TANN caused a significant change in the wettability, but only a minor improvement in the hydrophilicity of the PLA composites.

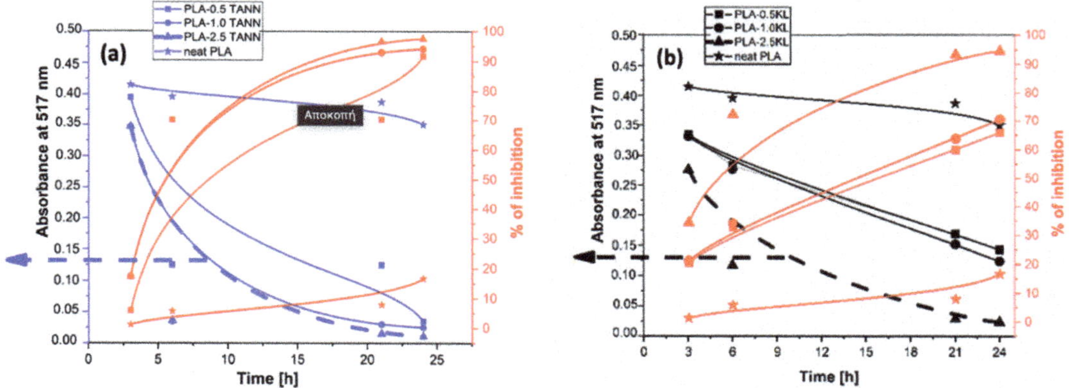


**Figure 10.** Reaction kinetics and % of inhibition of the PLA-based (**a**) TANN and (**b**) KL composites film evaluated with DPPH radical scavenging in the methanol solution indicated after 24 h [144].

Ainali et al. [145] studied the decomposition mechanism and the thermal stability of PLA nanocomposites filled with both biobased kraft-lignin (KL) and TANN in different contents using the melt extrusion method. It was established that the PLA/KL nanocomposites exhibit better thermostability compared to pristine PLA, while the addition tannin had a minor catalytic effect that can decrease the thermal stability of PLA. The calculated E$\alpha$ value of the PLA-TANN nanocomposite was lower than that of PLA-KL, leading to a considerably higher decomposition rate constant, which accelerated the thermal degradation. Cresnar et al. [146] studied the crystallization and molecular mobility of these composites. It was found that the extremely slow and weak crystallization rate of PLA was improved by the fillers, via the combined effects of providing sites for crystallization and moderately enhancing polymer mobility (chains diffusion). Remarkably, the crystal structure and semicrystalline morphology were found altered in the composites with respect to the neat PLA. Regarding the segmental mobility of PLA, the addition of both tannin and lignin was found to impose moderate changes, explicitly, slight acceleration, narrowing of relaxation times range, and suppression in fragility; however, no systematic changes in the calorimetric or dielectric strengths were detected. Despite the weak effects, it was concluded that TANN is more effective, and this seems to correlate with their smaller size and moderately better distribution as compared to lignin.

Several of these composites have been successfully applied in the field of food packaging, medical devices, textile sectors, and others, owing to the enhanced compatibility between hydrophilic biopolymers and the hydrophobic PLA polymer matrix.

#### 3.3.3. Nanocellulose (NC)

Cellulose is a polysaccharide that exists as a linear chain, consisting of repeating anhydro-D-glucose units covalently linked by β-1,4-glycosidic bonds [147]. Plants create around 75 billion tons of cellulose every year, making this an extremely abundant material. Nanocellulose (NC), prepared by breaking down cellulose fibers, is one such biodegradable, renewable nanofiller (biopolymer) that yields a low carbon footprint and generally refers to nanosized cellulose with a diameter of less than 100 nm and a length of up to several micrometers. It is considered by experts as a highly promising alternative material for petroleum-based products due to its biodegradability, renewability, eco-friendliness, and nontoxicity. It has been extensively used in diverse fields, such as fibers and clothes, the paper industry, optical sensors, the food industry, and pharmaceutics, among many others [148]. The unique characteristics of nanocellulose, such as superior mechanical properties (strength 2–3 GPa), low density (1.6 g $cm^{-3}$), highly specific surface area (200–300 $m^2g^{-1}$), and low thermal expansion coefficient (1 ppm $K^{-1}$), make it an ideal building block for flexible functional compounds and in outdoor and engineering applications [149]. Except for improvement in electrochemical performance, nanocellulose-based composites with green and abundant raw materials are also beneficial for the reduction of production costs and the construction of environmentally friendly processes.

Lately, NC has been increasingly facilitated in active packaging applications. For cellulose-derived packaging, three types of cellulose are employed, namely, cellulose nanocrystals (CNC), cellulose nanofibrils (CNFBs), and bacterial nanocellulose (BNC). Currently, CNC and CNF are incorporated as reinforcing agents into various biopolymers, such as PLA, for the preparation of green nanocomposites. NC can synergistically work with other materials to improve the barrier and the thermo-mechanical and rheological properties of the nanocomposites. Intramolecular and hydrogen bonding makes cellulose insoluble in almost all solvents as these bonds produce a great strengthening quality [150]. It has been reported that the films produced from NC can be voluntarily reprocessed and recycled into a packaging film without drastically degrading the properties of the same [151]. The most significant reason behind this would be its carbon neutrality, non-toxic nature, recyclability, and sustainability. Attributed to the distinctive properties of NC, tunable surface chemistry, barrier properties, mechanical strength, crystallinity, biodegradability, non-toxicity, and high aspect ratio, it is a growing credible renewable green substrate in food packaging applications [152].

#### 3.3.4. Nano-Biochar (n-BC)

Of the state-of-art biofuels, biochar (BC) has exhibited the potential to complement solid fuel for harvesting bioenergy, while addressing the vital issues associated with the environment. BC belongs to the category of richly carbonaceous items, and it is derived from biomasses such as agricultural and forestry wastes, crops, wood, leaves, municipal sludge, manure, woodchips, and other C-rich materials. Sludge is generated during the wastewater treatment procedure, which produces a solid waste that needs to be treated and disposed. Nevertheless, it is a promising feedstock to produce BC due to its rich carbon content and nutrients such as ammonia [153]. It can be formed through a variety of treatments such as pyrolysis, torrefaction, and gasification at a wide range of temperatures (200–700 °C) and times. Depending on the process conditions, biochar can be produced as either a primary or side product, accompanied by a variable amount of liquid and gaseous streams (such as bio-oil and syngas). Prior to thermal combustion, biomass is dried, the particles are further heated, and volatile substances are released from the solid. The volatile compounds can form permanent gases (such as $CO_2$, CO, $CH_4$, and $H_2$) or condensable organic compounds (e.g., acetic acid and methanol). Subsequent reactions in the gas phase include cracking and polymerization and can therefore alter the entire product spectrum. Three products can be distinguished from the resulting material: permanent gases, one or more liquid phase(s) (water and tar), and a solid residue [154].

The easy availability of feedstock and the inexpensive production of biochar has made it a material of significance for environmental remediation in recent years [155]. BC presents some advantages over other carbon items (single or multi carbon nanotubes, graphene, and activated carbon), such as its highly specific surface area, good stability, abundant functional groups, great carbon stability, and highly porous structure, and thus it has been used in a wide variety of fields, for example, as a pollutant absorbent and for biosensors, fuel cell, and supercapacitors [156]. These properties have expanded its usage in various sectors, ranging from heat and power production, flue gas cleaning, metallurgical applications, use in agriculture and animal husbandry, as a building material, to medical use. Attempting to decrease greenhouse gas emissions, BC has gained increasing popularity lately as a replacement for fossil-carbon carriers in several of these applications [157].

Nano-biochar is naturally formed during the processing of bulk biochar, while its yield is low (e.g., only ~2.0% in peanut shell-derived biochar). A size reduction process is required to increase the content of NPs in biochar. This process can be easily operated via grinding or milling [158]. Figure 11 presents the most common approaches for the preparation of nano-BC from biomasses.

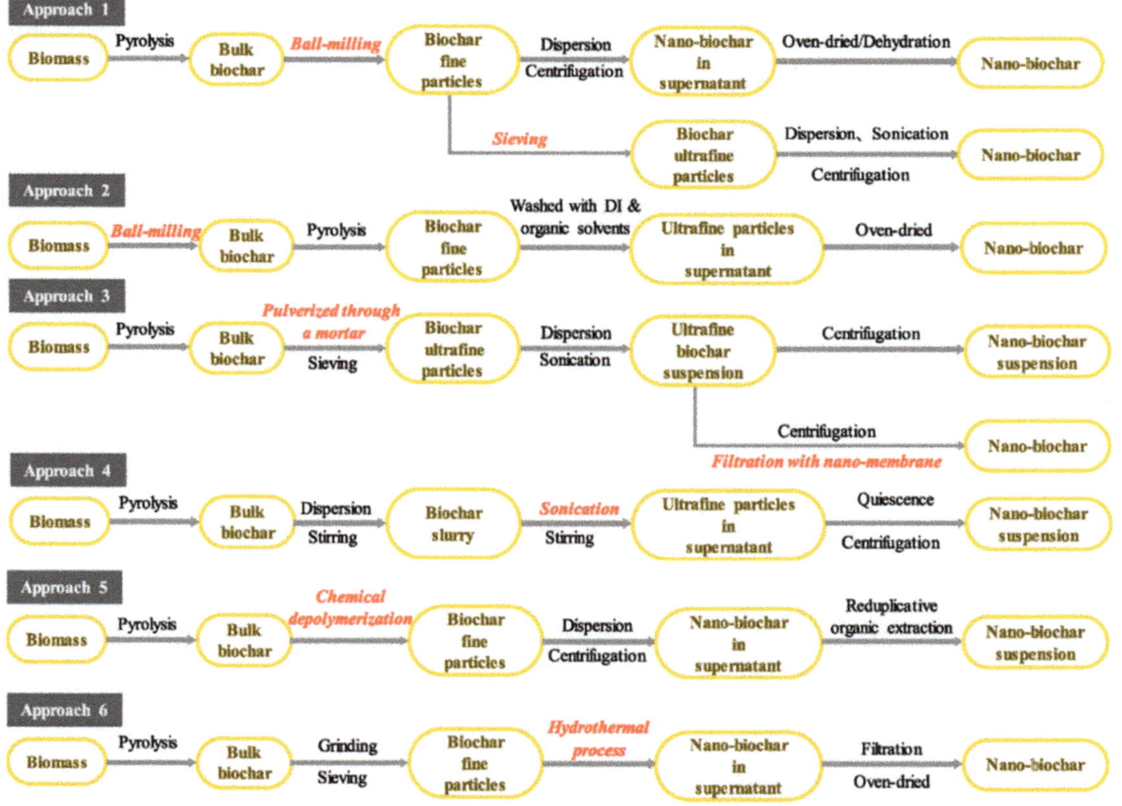


**Figure 11.** Commonly employed preparation routes for nano-biochar from biomass feedstock [159].

In comparison with bulk BC that is normally measured on a micron or millimetre scale, nano-biochar is defined as biochar with a particle size of <100 nm. However, a wide range of particle sizes has been considered as nano-biochar in the literature, with particle sizes from 0 to 600 nm being considered as nano-biochar. The particle size of nano-BC is highly dependent on the various feedstocks and preparation methods. For the same feedstock, increasing the pyrolysis temperature can lead to a decrease in particle sizes. For instance, increasing the pyrolysis temperature for rice straw from 400 to 700 °C caused a decrease in particle size from 403 to 234 nm using the centrifuging separation method, as reported by Lian et al. [160]. Likewise, when using the same nano-BC preparation route, increasing the pyrolysis temperature from 300 to 600 °C for rice hull feedstock yielded particles with sizes ranging between 190 and 59 nm.

In addition, conventional BC does not bear electrical conductivity. Instead, the biochar with its size reduced on a nano scale with its activation can be a new research trend for formulating easy and stable biosensors and nanocomposites, owing to its greater surface area and its ability to facilitate rapid electron transmission during sensing and biosensing and offer more adsorption sites for target binding [161]. As a carrier, n-BC could facilitate the migration of natural solutes and contaminants, in contrast with the positive effects of bulk BC, such as holding nutrients and immobilizing hazardous chemicals [162].

### 3.4. PLA/Ceramic Nanocomposites

The development of ceramic nanoparticles with improved properties has been studied with much success in several areas such as in synthesis and surface science. Ceramics are defined as solid compounds that are formed by the application of heat and sometimes pressure, comprising at least two elements, provided one of them is a non-metal or a metalloid. The other substance(s) may be a metal or another metalloid.

The properties that these versatile materials exhibit include high mechanical strength and hardness, good thermal and chemical stability, and viable thermal, optical, electrical, and magnetic performance. In general, ceramic components are formed as desired shapes starting from a mixture of powder with or without binders and other additives, using conventional technologies, including injection molding, die pressing, tape casting, gel casting, etc. The sintering of the green parts at elevated temperatures is furthermore essential to reaching densification. Ceramics are a class of biomaterials extensively employed in biomedical devices [163]. Owing to their ability to be fabricated into a variety of shapes, along with their high compressive strength, variable porosity, and bioactive properties in the body, ceramics are widely facilitated as implant materials. The high similarity in the chemical composition of some ceramics such as calcium phosphate with human bone minerals makes them suitable for use as orthopaedic implants (human skeleton, bones, and joints) and dental materials. These materials show excellent bioactivity, high biocompatibility, and excellent osteoconduction characteristics [164].

#### 3.4.1. Bioglass

Of late, bioactive glasses (bioglass, BG) have emerged as potential biomaterials demonstrating interesting applications as bone-cementing materials in prosthetic medical implants and drug delivery systems. Bioglass represents a subgroup of ceramic materials (i.e., based on $SiO_2$, $Na_2O$, CaO, and $P_2O_5$) that are comparable to native bone mineral components that have osteoconductivity and good mechanical properties. Owing to their lower silica content (<60%), BG materials have interesting bioactivity properties, thereby producing different forms of hydroxyapatite that ultimately stimulate the natural adhesion by means of biological and morphological fixation [165]. BG has been recognized as among the most important bioceramics for bone regeneration, exhibiting high biocompatibility and positive biological effects after implantation. This material can form a carbonated hydroxyapatite surface layer, and is accountable for the strong bonding between BG and human bone.

Nanobioactive glasses (NBGs) are used in combination with biodegradable polymers to enhance their mechanical performance. NBGs exhibit appropriate properties including biocompatibility, controlled biodegradability, ability to release the cells, ability to connect to both hard and soft tissues, and releasing of ions during the degradation process. This group of materials has the potential to be used in tissue repairs and wound healing, and their positive effects on the angiogenesis process have been demonstrated in some reports. The fidelity of NBGs in tissue repair might be attributed to their mineral structure. Accordingly, ions releasing from NBGs can stimulate tissue repair within the human body and therefore accelerate the process of recovery [166]. Furthermore, morphology and particle size influence the bioactivity of the BG. Studies concluded that an increase of the surface area and porosity of BG can dramatically enhance the bioactivity. Nanoscale BG can also increase in vitro bioactivity (HA deposition) in higher quantities than micrometric particles [167].

#### 3.4.2. n-Hydroxyapatite (HAp)

Hydroxyapatite (HAp) ($Ca_{10}(PO_4)_6(OH)_2$) is a bioactive ceramic material, thermodynamically stable in its crystalline state in body fluid and is the principal inorganic constituent of human bone, making it biocompatible with excellent bone healing properties due to its osteoconductive and osteoinductive capacities [168,169]. Natural HAp is typically extracted from biological wastes or feedstocks such as mammalian bone (e.g., bovine, camel, and horse), marine or aquatic sources (e.g., fish bone and fish scales), shell sources (e.g., cockle, clam, eggshell, and seashell), certain plants, and also from mineral sources (e.g., limestone). Synthetic HAp can be fabricated through various methods, including dry methods (solid-state and mechanochemical), wet methods (chemical precipitation, hydrolysis, sol-gel, hydrothermal, emulsion, and sonochemical), and high temperature processes (combustion and pyrolysis) [170]. Nano-hydroxyapatite (nHA) has the small-size effect of a small crystal grain diameter, large interface, high surface free energy and binding energy, and the unique physical and chemical properties of nanomaterials such as a macro quantum tunnelling effect [171].

The brittle nature of HAp makes it unsuitable for most applications when used alone since it does not meet mechanical requirements. The reinforcement of PLA with ceramics such as nanohydroxyapatite (nHAp) is heavily investigated for tissue engineering and bone regeneration applications and has effectively solved many problems such as high brittleness of hydroxyapatite itself and uncontrollable degradation rate. Moreover, HAp in its nanophase has been found to improve osteoblast cell adhesion and long-term functions both in vitro and in vivo [172].

### *3.5. Nanoclays*

Among the variety of nanofillers, nanoclays are the oldest and potentially one of the most interesting and versatile ones [173]. Clays are divided into several classes, such as kaolinite, montmorillonite, sepiolite, smectite, chlorite, illite, and halloysite based on their particle morphology as well as chemical and mineralogical composition. Due to their wide availability, relatively low cost, and relatively low environmental impact, nanoclays have been studied and developed for numerous usages. Approximately 30 different types of nanoclays can be found, which depending on their properties are used in different applications [174].

With the rapid growth of nanotechnology, clay minerals are increasingly used as natural nanomaterials. Nanoclays are nanoparticles of layered mineral silicates with layered structural units that can form complex clay crystallites by stacking these layers [175]. An individual layer unit is composed of octahedral and/or tetrahedral sheets. The different structures of nanoclays are basically composed of alternating tetrahedral silica sheets "$SiO_2$" and alumina octahedral layers "$AlO_6$" in ratios of 1:1 when one octahedral sheet is linked to one tetrahedral sheet as kaolinite or halloysite. In ratios of 2:1, this structure created from two tetrahedral sheets sandwiching an octahedral sheet such as montmorillonite and sepiolite yields the proportion of 2:1:1 (chlorite). It has been reported that the different types of nanoclays affect the properties of PLA/clays nanocomposites [176].

Solvent casting [177], melt blending [178], and in situ polymerization techniques [179] have been extensively used to synthesize nanoclays containing PLA composites. In recent years, nanoclays have been given great consideration due to being able to improve and significantly enhance mechanical and thermal properties, barrier, and flame resistance properties, as well as in their use in the accelerated biodegradation of polymers [173]. However, it has been reported that the different types of nanoclays affect the properties of PLA/clays nanocomposites [176]. Although organic montmorillonite (MMT) [180], bentonite [181], and halloysites nanotubes (HNTs) [182] are the most widely used nanoclays in the synthesis of PLA nanocomposites, the use of other clays with different morphologies, such as sepiolite (a fibrous silicate that has microporous channels running along the length of the fibers) [183] and palygorskite (a fibrous silicate with a needle-like morphology) [184] has also been reported in the literature.

Several studies have shown significant improvement in the properties of PLA because of the addition of nanoclays [185]. At present, nanoclays such as organic montmorillonite (MMT), bentonite, and halloysites nanotubes (HNTs) are being greatly considered as they possess the potential tendency to extensively improve the thermal, mechanical, and functional properties of polymers. However, it has been observed that the properties of synthesized nanocomposites are affected by the amount of nanoclays that is added to the PLA matrix due to the incompatibility between the hydrophobic polymer and the hydrophilic natural nanoclays [186,187]. Specifically, this was observed in the case of MMT nanoclays that present a hydrophilic nature, thus hindering their uniform dispersion in the hydrophobic organic PLA matrix [20]. Studies have reported that the mechanical properties of bio-nanocomposite films improved if a small content of nanoclays was added into the packaging materials. Nevertheless, the mechanical properties of the films declined with a further increase in nanoclay concentration [188,189]. Table 3 overviews the mechanical properties of PLA nanocomposites containing different nanoclays.

**Table 3.** PLA/nanoclays composites and their mechanical properties as found in the literature.

| Sample | Content of Nanoclay | Tensile Strength | Young's Modulus | Elongation at Break | Flexural Strength | Impact Strength | Bibliography |
|---|---|---|---|---|---|---|---|
| PLA/Halloysite | 3% wt. | Increase of 14% compared to neat PLA | Increase of 50% compared to neat PLA | Increase of 3% compared to neat PLA | - | - | [180] |
| PLA/Kenaf fiber (30%)/MMT | 1% wt. | Increase of 5.7% compared to PLA/Kenaf | Increase of 39.61% compared to neat PLA | - | Increase of 46.4% compared to PLA/Kenaf | Increase of 10.6% compared to PLA/Kenaf | [190] |
| PLA/Aloe vera fiber (30%)/MMT | 1% wt. | Increase of 5.72% compared to PLA/Aloe vera | Increase of 18.84% compared to neat PLA | - | Increase of 6.08% compared to PLA/Aloe Vera | Increase of 10.43% compared to PLA/Aloe vera | [186] |
| PLA/Kenaf/Aloe vera/MMT | 1% wt. | Increase of 23.2% compared to PLA/Kenaf Increase of 11.46% compared to PLA/Aloe vera | Tensile Modulus Increase of 24.61% compared to neat PLA | - | Increase of 56.43% compared to PLA/Kenaf Increase of 12.63% compared to PLA/Aloe vera | Increase of 57.5% compared to PLA/Kenaf Increase of 54.27% compared to PLA/Aloe vera | [187] |
| PLA/PCL/MMT | 4% wt. | Increase of 15% compared to the blend | Increase of 26% compared to the blend | - | - | Decrease of 33% compared to the blend | [191] |
| PLA/Halloysite nanotubes (HNTs) | 9% wt. | Decrease of 10.7% compared to neat PLA | Increase of 10.8% compared to neat PLA | Decrease of 46% compared to neat PLA | Decrease of 7.3% compared to neat PLA | Decrease of 51.4% compared to neat PLA | [192] |

## 4. Applications of PLA Nanocomposites

As mentioned above, nanofillers represent an interesting way to extend and improve the properties of PLA in order to prepare high-performance PLA-based nanocomposites. Due to the enhanced properties of PLA nanocomposites, they present the potential as promising applications in the packaging, construction, medical, textile, electronics, solar panel, and agricultural sectors (Figure 12).

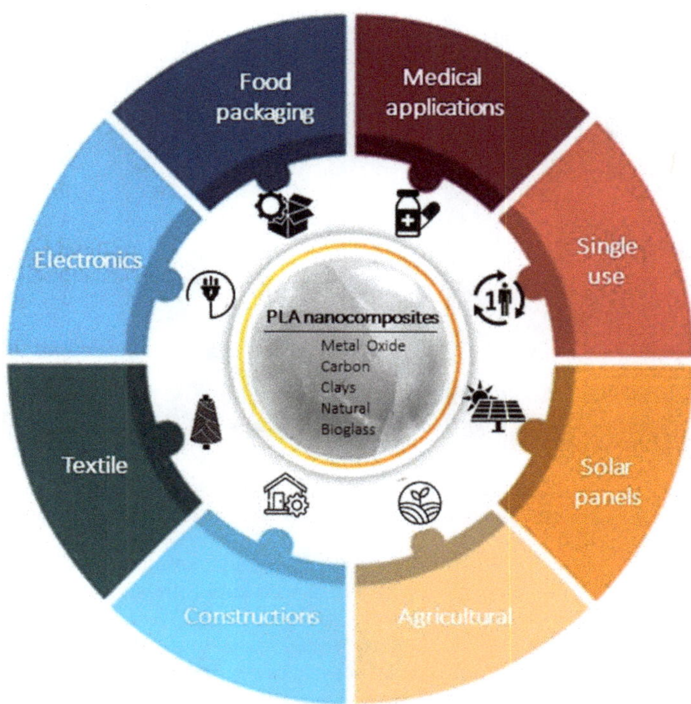


**Figure 12.** Schematic overview of the numerous applications involving PLA nanocomposites.

### 4.1. *Applications of PLA/Metal Oxides*

#### 4.1.1. Food Packaging

Metallic-based particles such as MgO, ZnO, $SiO_2$ and $TiO_2$ have been mainly used in food packaging due to their enhanced antimicrobial properties. In brief, Swaroop et al. [193] produced PLA biofilms reinforced with MgO nanoparticles (up to 4 wt.%) using the solvent casting method on a lab scale for the food packaging sector. Among the prepared biocomposite films, the 2 wt.% reinforced PLA films exhibited the maximum improvement in tensile strength and oxygen barrier properties (up to 29% and 25%, respectively) in comparison to neat PLA films. In general, the studied biofilms were transparent, capable of screening UV radiations and presented superior antibacterial efficacy against the *E. coli* bacterial culture. Taking it a step further, the same group developed for the first time an industrial level melt-processing setup for the preparation of blown PLA/MgO nanocomposite (NC) films for food packaging [194]. In that work, up to 3 wt.% PLA/MgO nanocomposite films were prepared for the investigation of key mechanical, barrier, optical, thermal, and anti-bacterial performance (Figure 13). The tensile strength and plasticity improved by nearly 22% and 146%, respectively, for the 2 wt.% MgO-reinforced films. The oxygen and water vapor barrier properties improved by nearly 65% and 57%, respectively, for the 1% material. For the 1 wt.% NC films, around 44% of the *E. coli* bacteria were killed after a 24 h treatment, indicating that these nanocomposites could be used as a sustainable alternative for petroleum-based packaging film materials.

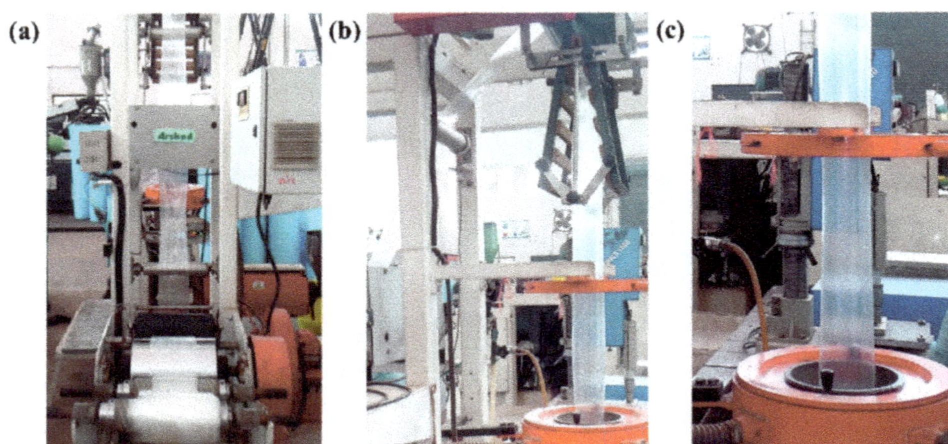

**Figure 13.** Industrial scale setup of the film blowing process: (**a**) Calendaring of the neat PLA film and (**b**,**c**) the blown film of the 1% PLA/MgO nanocomposite [194].

Chong et al. [195] conducted a literature report regarding the fabrication of PLA-ZnO nanocomposites with enhanced antibacterial properties. Although ZnO's antibacterial mechanisms are still under thorough investigation, ZnO NPs have been found to be active against biofilm formation due to the generation of hydroxyl radicals, thus increasing the antibacterial activity of commonly used antibiotics. They highlighted that their multi-functionality and versatility enable their use in antibacterial, UV-absorption, and photocatalytic applications such as food packaging products.

Zhang et al. [50] developed PLA/ZnO NPs electrospun membranes with enhanced antimicrobial activity and UV blocking for food packaging applications. The optimal performance in terms of antimicrobial activity was achieved with 0.5 wt.% ZnO NPs. Specifically, the inhibition zones of *Escherichia coli* and *Staphylococcus aureus* strains were 15.02 ± 0.18 mm and 14.74 ± 0.06 mm, respectively, concluding that this composite is a suitable alternative as a food packaging material that provides both an antibacterial effect and serves as a UV light barrier.

$TiO_2$ is a material with bactericide activity that is improved in the presence of light [196]. Baek et al. [197] modified the surface of $TiO_2$ with oleic acid (OA) to make it nonpolar and thus improve its compatibility with PLA and PLA films that were prepared using solvent casting. The surface modification with oleic acid improved the dispersion of the nanoparticles in PLA matrices more so than unmodified $TiO_2$, as verified by SEM micrographs. Oxygen permeability and water vapor permeability of 1% OT-PLA were reduced by 29% and 26%, respectively, when compared to pure PLA. OT-PLA had higher transparency than T-PLA but provided better light blocking in UVA and UVB regions than PLA. Feng et al. [34] investigated the antibacterial activity of PLA/$TiO_2$ nanocomposite films and nanofibers against *E. coli* and *S. aureus* using two different lighting conditions: a UV-A (360 nm) irradiation and a fluorescent lamp source. The pristine PLA presented almost no antibacterial properties, while with increasing amounts of $TiO_2$ NPs, and the antibacterial strength of the nanocomposite films rose evenly. The maximum inhibition ratio was observed at a $TiO_2$ content of 0.75 wt.% for both NPs and films. Segura- González et al. [93] also examined the antibacterial activity of PLA/$TiO_2$ nanocomposites against *E. coli* (DH5$\alpha$). The basic conclusion was that the presence of $TiO_2$ NPs led to a decline in the biofilm development and in the bacterial growth of *E. coli*. These results may be attributed to an immediate impedance on bacterial metabolisms. Furthermore, bacteria size on the nanocomposites incorporating nanoparticles with a diameter of 21 nm was even smaller than one grown on the nanocomposites incorporating the <100 nm nanoparticles. This observation may relate to the surface-to-volume ratio of the NPs, which might alter the oxidative catalytic performance of $TiO_2$ in the direction of the organic material related to both the EPS (extracellular polymeric substance) and the bacteria themselves.

An alternative material in the field of nano-food packaging was also proposed by Batool et al [198]. In their work, ZnO NPs hexagonal in shape and 36.5 nm in size at concentrations of 0.4% and 4% ($w/w$) were incorporated into a PLA matrix using the solution casting technique. ZnO-NPs were prepared by *aloe barbadensis'* leaves extract and their addition in PLA enhanced stability, elongation, and film thickness. The antimicrobial assay of fruit and biofilm against *E. coli* and *S. aureus* bacterial strain results confirmed that nano-ZnO-based packaging film improved the shelf life and food quality of Vitis vinifera fruit up to two weeks at 40 °C.

#### 4.1.2. Medical Applications

Except for food packaging, PLA/Metal oxide nanocomposites can be also used in medical applications such as in tissue engineering, wound healing, and drug delivery systems [199–201].

In a very recent study, Grande-Tovar et al. [199], prepared membranes based on polycaprolactone (PCL) and PLA incorporated with ZnO-NPs, glycerol (GLY), and tea tree essential oil (TTEO) for tissue engineering applications. It was found that the incorporation of the ZnO-NPs and TTEO accelerated the degradation process of the PLA/PCL matrix. All membranes exhibited biodegradability and biocompatibility after 60 days of study, allowing a healing procedure to occur with the recovery of tissue architecture and hair without the occurrence of an aggressive immune response. The membranes are fragmented and reabsorbed by inflammatory cells during the resolution process (Figure 14). Simultaneously, the implantation zone was replaced by connective tissue with type III collagen fibers, blood vessels, and some inflammatory cells that continue the reabsorption process.

In an attempt to prepare alternative materials for bone implant applications, Nonato et al. [202] molded PLA/ZnO nanofibers (1 wt.%) using the solvent-cast three-dimensional (3D) technique. Conditions were adapted to imitate a mechanical fatigue test at human body temperature–cyclic stress in an isotherm at 36.5 °C. The obtained results indicated that for temperatures above 30 °C, the storage module of PLA/ZnO nanocomposites was higher when compared to pure PLA, and in the fatigue test, PLA/ZnO withstood more than 3600 cycles, while pure PLA failed after an average of 1768 cycles. Once again, the excellent antimicrobial activity of ZnO nanocomposites was verified against numerous bacterial strains such as *Staphylococcus aureus*, *Salmonella*, *E. coli*, as well as *Candida albicans* yeast.

In another work, Rashedi et al. [203] prepared PLA/ZnO nanofibrous nanocomposites loaded with tranexamic acid (TXA), which was prepared using the electrospinning process for wound healing patches. The obtained nanofibers exhibited a small pore size that enables the appropriate permeation of atmospheric oxygen to the wound. An in vitro cytotoxicity assay of the nanofibrous mats showed the wound dressing material did not cause any harmful effect upon the human dermal fibroblast cells. Antibacterial studies against Gram-negative *Escherichia coli* and Gram-positive *Staphylococcus aureus* revealed a 75% and 98% reduction in colonies of the bacterial strains, respectively. In vivo tests on mice models evidently demonstrated that the PLA/ZnO/TXA nanofibrous nanocomposites dressing enhanced the wound healing process (Figure 15). These results encourage the use of the suggested nanocomposite as a helpful wound dressing where rapid wound healing and proliferation of skin cells are mandatory.

Monadi et al. [204] incorporated $SiO_2$ in the PLA matrix at different concentrations (1, 3, and 5 wt.%) to produce a PLA nanocomposite film with enhanced antimicrobial properties for food packaging applications. They found high antimicrobial activity against Escherichia coli and Staphylococcus aureus, whereas the water vapor permeability and oxygen transfer rate presented the lowest values, decreasing by 56.9% and 55.2%, respectively, compared to neat PLA.

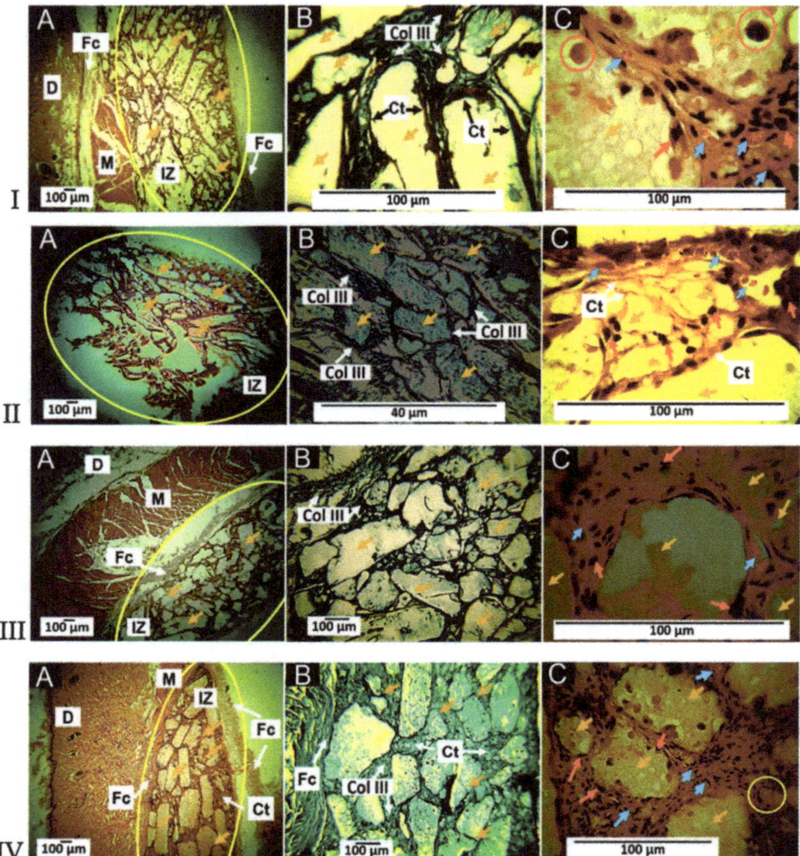


**Figure 14.** Subdermal implants of F1, F2, F3, and F4 at 60 days: (**A**) Formulations from the 4 X HE technique, (**B**) formulations from the 10 X GT technique, and (**C**) formulations from the 100 X HE technique. Yellow oval: Implantation zone. D: Dermis. M: Muscle. Fc: Fibrous capsule. IZ: Implantation zone. Col III: type III collagen. Yellow arrows: Fragments of materials. Red arrows: Inflammatory cells. Blue arrows: Blood vessels. HE: Hematoxylin and Eosin technique. GT: Gomori technique [199].

Moreover, Wang et al. [205] grafted PLA chains with a small amount of functionalized $SiO_2$ (f-$SiO_2$). Because of the improvement in dispersion and interfacial interaction in the PLA matrix, the f-$SiO_2$ illustrated an effective reinforcing and toughening effect for PLA, where the elongation at break, tensile strength, and impact toughness of PLA nanocomposite improved by 47.8, 14.9, and 30.3%, respectively, compared to PLA neat. Additionally, the degree of PLA crystallinity was significantly enhanced by the added f-$SiO_2$.

The objective of a recent study [206] was to prepare a series of poly(DL-lactic acid) (PDLLA) nanocomposites with four different amounts of silica ($SiO_2$) nanoparticles (2.5, 5, 10 and 20 wt.%), following a new two-step synthesis route: ring opening polymerization of DL-lactide and polycondensation. According to intrinsic viscosity measurements, the average molecular weight of PDLLA (Mn $\approx$ 38,097 g mol$^{-1}$) decreased with increasing $SiO_2$ content. SEM images confirmed the fine dispersion of $SiO_2$ nanoparticles inside the polymer matrix due to the interactions between ester groups of PDLLA and surface silanol groups of $SiO_2$. In addition, they confirmed that thermal stability was increased by the addition of $SiO_2$ in the PLA matrix.

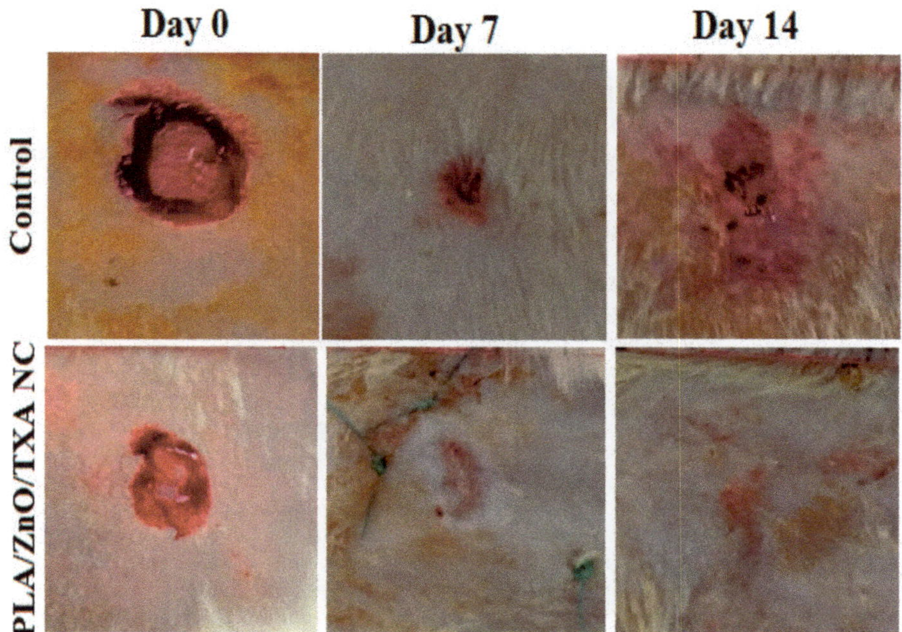


**Figure 15.** Images of skin wound healing process in mice after 0, 7, and 14 days of treatment with the PLA/ZnO/TXA nanocomposites [203].

4.1.3. Environmental Applications

In the last 30 years, the photocatalytic efficiency of nanosized metal oxides at degrading organic contaminants in the air and water has resulted in their utilization in several environmental applications [207]. These include photocatalytic hydrogen production via water splitting, degradation of organic pollutants in wastewater photocatalytic self-cleaning, bacterial disinfection photo-induced super hydrophilicity, as well as in photovoltaics and photosynthesis [99].

Bobirică et al. [208] prepared novel photocatalytic membranes based on PLA/$TiO_2$ hybrid nanofibers deposited on fiberglass supports for the removal of ampicillin from aqueous solutions via the electrospinning technique. The fiberglass fabric plain woven-type membrane showed the highest efficiency of ampicillin removal from the aquatic solutions. However, the degree of mineralization of the aqueous solution is kept low even after two hours of photocatalysis, owing to the degradation of PLA from the photocatalytic membrane. Li et al. [91] synthesized via the sol-gel method $TiO_2$ nanoparticles (6.3–11.1 nm) using titanium tera-isopropoxide (TTIP) as the hydrolysis material for the reinforcement of PLA films by casting. The photocatalytic activity of films was verified by methyl orange (MO) solution degradation under UV irradiation. Photodegradation of PLA due to $TiO_2$ NPs was associated with the development of carbon-centered radicals and acceleration of interfacial polymer chains gap by active oxygen species. For neat PLA, after 12 h of UV radiation, the degradation of MO observed was 55%, while the film containing 0.6 wt.% $TiO_2$ reached a photocatalytic degradation efficiency of 99% after 12 h of UV irradiation.

By using mild and environmentally friendly synthetic conditions, Gupta et al. [209] developed photocatalytic nanocomposites composed of porous PLA microparticles via oil/water emulsion with incorporated anatase $TiO_2$ NPs for the sorption and UV-triggered degradation of organic compounds. The authors demonstrated that the sorption capacity, dye degradability, and composite disintegration can be controlled by altering the PLA microparticles' porosity and the distribution of incorporated titania NPs. Both types of PLA/$TiO_2$ composites removed rhodamine 6G from water (up to 60% of the initial amount in six hours of UV exposure time) unlike negligible dye removal observed for $TiO_2$-absent PLA particles. The authors suggested that these composite microsponges may be facilitated as nontoxic photocatalytic materials for the efficient environmental clean-up of contaminated water.

The photocatalytic degradation of a mixture of cytostatic drugs has been studied using immobilized $TiO_2$ on polymer films (PET and PLA) [207]. Both studied nanocomposite materials have been proven to be effective. However, higher degradation rates were achieved in the presence of PET-$TiO_2$.

#### 4.1.4. Other Applications

It has also been reported that PLA/metal oxide nanocomposites are alternative candidates for air filters applications as well as in electronic and automotive applications (Table 4).

Petousis et al. [210] reported the effect of $Al_2O_3$ NPs at four different filler loadings, as a reinforcing agent of PLA using FFF for a series of potential applications. A positive reinforcement mechanism was observed at all filler loadings, while the mechanical percolation threshold with the maximum increase of performance was found between 1.0–2.0 wt.% filler loading (2.0 wt.% in PLA, and a 40.2% and 27.1% increase in strength and modulus, respectively). Sukhanova et al. [211] investigated the incorporation of epoxidized aluminum oxide nanofibers (AONF) nanofiller in improving the characteristics of composite polymer films based on biodegradable PLA by solution casting. It was shown that the incorporation of AONF results in an enhancement in the mechanical properties of neat PLA. Kangali et al. [212] prepared PLA/boron oxide nanocomposites by the solution casting method as alternative candidates for electronic, packaging, and automotives industries. The particle size of boron oxide was reduced to a nanosize using SPEX ball mill and boron oxide nanoparticles (BONPs) that were functionalized with oleic acid, [3-(trimethoxylsilyl) propylmethacrylate] (MPTMS) and triethoxyvinylsilane (VS). Mechanical and flammability tests showed that the nanosized and surface functionalized BONPs enhanced the crystallinity and mechanical performance of PLA/BONPs. The crystallinity value increased from 13.68% to 32.55% with the increasing functionalized BONPs in the PLA matrix. Likewise, the tensile strength of the nanocomposites increased from 41.25 $\pm$ 0.80 MPa to 51.12 $\pm$ 2.54 MPa with the increasing functionalized BONPs.

Wang et al. [213] prepared hybrid PLA/$TiO_2$ NPs fibrous membranes presenting good antibacterial activity via the one-step electrospinning technique for air filter applications. Filtration performance tests (Figure 16) indicated that fibers with a high surface roughness, large specific surface area, and large nanopore volume greatly improved the particle capture efficiency and facilitated the penetration of airflow. From the overall tests conducted, it was concluded by the authors that the PLA/$TiO_2$ fibrous membrane loaded with 1.75 wt.% $TiO_2$ NPs prepared at a relative humidity of 45% was the optimal material, since it showed excellent filtration efficiency (99.996%) and a relatively low pressure drop (128.7 Pa), as well as a high antibacterial activity of 99.5% against *Staphylococcus aureus*.

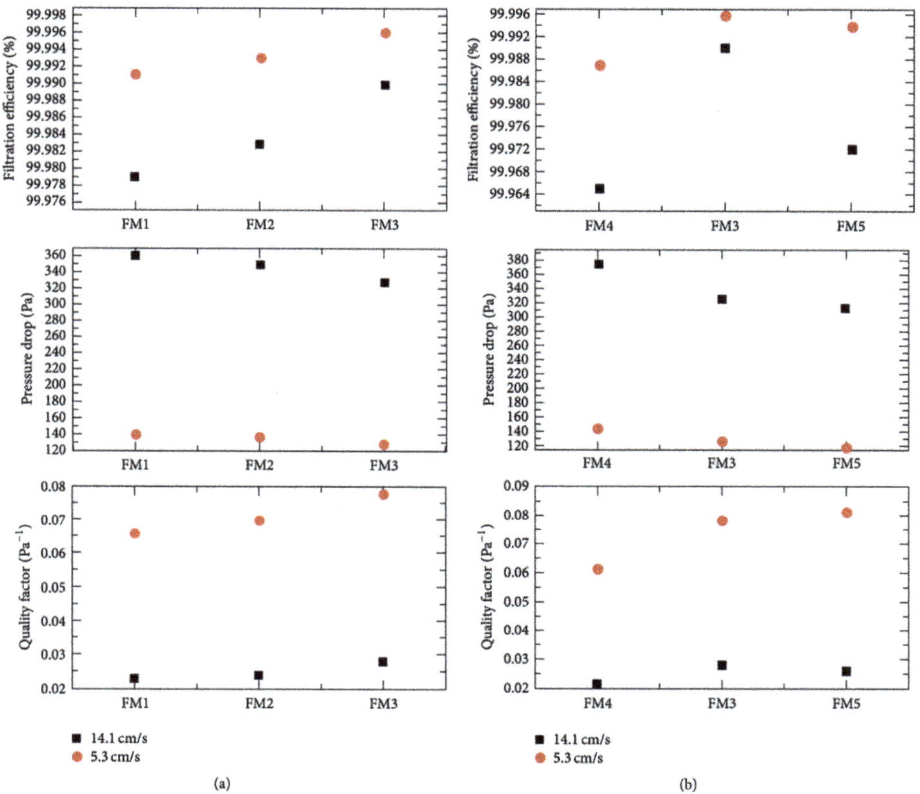


**Figure 16.** Filtration performance of FM1 (containing 0 wt.% $TiO_2$ prepared at a relative humidity of 45%), FM2 (containing 1 wt.% $TiO_2$ prepared at a relative humidity of 45%), FM3 (containing 1.75 wt.% $TiO_2$ prepared at a relative humidity of 45%), FM4 (containing 1.75 wt.% $TiO_2$ prepared at a relative humidity of 15%), and FM5 (containing 1.75 wt.% $TiO_2$ prepared at a relative humidity of 60%) at various face velocities: (**a**) filtration efficiency, pressure drop, and quality factor of FM1, FM2, and FM3 at a face velocity of 5.3 cm/s and 14.1 cm/s, respectively and (**b**) filtration efficiency, pressure drop, and quality factor of FM3, FM4, and FM5 at a face velocity of 5.3 cm/s and 14.1 cm/s, respectively [213].

**Table 4.** Applications of PLA/metal-oxide nanocomposites in various sectors.

| Material | Application | Properties | References |
|---|---|---|---|
| PLA-ZnO | Packaging, tissue engineering, wound healing, drug delivery, disposable electronics | Dielectric properties, nti-inflammatory and antibacterial activity, biocompatibility | [195,199,203] |
| PLA-$TiO_2$ | Packaging, air filters, tissue engineering, wound healing, electronics | Total anti-UV protection, optical and antibacterial properties, nanocomposites with higher kinetics of crystallization, photodegradability, etc. | [94,214,215] |
| PLA/$Al_2O_3$ and PLA/BO | Electronics, packaging and automotives industries | Anti-inflammatory properties and non-toxicity against fibroblast L929 cells | [210,211] |

### *4.2. PLA/Carbon-Based Nanocomposites*

#### 4.2.1. Thermal and Electrical Applications

It has been demonstrated that carbon-based materials including carbon black, activated carbon, graphite, CNFs, and CNTs are very effective fillers in improving the thermal insulation performance of polymer foams by reducing the radiative heat transfer [132]. Wang et al. [216] reported an eco-friendly and versatile process to fabricate ultra-low-threshold and lightweight biodegradable PLA/MWCNTs foams with segregated conductive networks for high-performance thermal insulation and electromagnetic interference shielding to meet the rising request for high-performance multifunctional materials in sustainable development. Owing to the unique structure of the microporous PLA matrix embedded by conductive 3D MWCNTs networks, the lightweight porous PLA/MWCNTs with a density of 0.045 $g/cm^3$ possess a percolation threshold of 0.00094 vol%, which was the minimum value reported at the time. Moreover, the developed material showed excellent thermal insulation performance with a thermal conductivity of 27.5 $mW{\cdot}m^{-1}{\cdot}K^{-1}$, significantly lower than the best value of common thermal insulation materials.

A simple, $CO_2$-based and eco-friendly yet effective foaming methodology for fabricating ultra-low-density PLA/CNTs nanocomposite foam for thermal insulation application was reported by Li et al. [217]. With the gradual incorporation of CNTs, three kinds of networks were generated in PLA/CNTs nanocomposites and had a distinct reinforcement influence on their melt viscoelasticity. The storage modulus of PLA/CNTs nanocomposites were three orders of magnitude higher in contrast to neat PLA. Interestingly, relative to abnormal DSC, a double melting peak phenomenon appeared in the high-pressure DSC curves of various PLA specimens. Biodegradable PLA/CNTs nanocomposite foam was successfully fabricated at a foaming temperature ($T_f$) of 121 °C with a super-high volume expansion ratio (VER) of 49.6 times. The pore size, pore density, and VER of diverse PLA specimens were tuned and regulated efficiently by the content of CNTs and foaming temperature. Finally, the authors underlined that these results offer a promising strategy for developing other thermoplastic polyester foams with ultra-high VER to obtain some unique functional attributes.

Sanusi et al. [218–220] incorporated hybrid nanofillers containing MWCNTs and montmorillonite (Mt) at concentrations of 0.5, 1 and 2 wt.% through a two-step procedure of solution and melt mixing, to create reinforcing fillers in the PLA matrix. The physicochemical studies indicated a strong hybrid–polymer interaction. The structural analyses confirmed the synergy between the nanoclay and the carbon nanotubes in reaching a homogeneous dispersion of Mt/MWCNT NPs throughout the PLA. Furthermore, the addition of a low mass fraction of the studied hybrid nanofillers led to a more than 24% and 45% enhancement in the tensile strength and elastic modulus of PLA, respectively. Finally, at low loading (0.5 wt.%) of the Mt/MWCNTs nanofiller, the performance of the thermal degradation was improved, while an increased amount of added nanohybrid implies a well-established percolation network in the nanocomposite, which might serve as a good thermal conductive material.

Graphene and its different types also have excellent electrical and thermal properties. Kalinke et al. [221] reported the comparison of the electrochemical properties of 3D-printed PLA-graphene electrodes (PLA-G) under different activation conditions and through different processes. The sensor showed good repeatability and reproducibility and the electrodes were successfully applied to DA determination in synthetic urine and human serum, showing good recovery, from 88.8 to 98.4%. Thus, the activation methods were vital for the improvement in the 3D PLA-G electrode properties, allowing graphene surface alteration and electrochemical enhancement in the sensing of molecular targets.

In another work, the introduction of exfoliated graphene into the PLA matrix produced conductive PLA/graphene nanocomposites via the solution casting method [222]. Among the concentrations of the filler (0.5, 1.5, 1.7, 2, 2.5, and 3 wt.%) added to PLA, the content of 1.7 wt.% graphene was the optimal choice since a significant drop in impedance ($10^5$ $\Omega$) compared to neat PLA ($10^{11}$ $\Omega$) was noticed. As the graphene loading was increased, the impedance gradually reached the reduced value of $10^4$ Ω. In addition, the specific nanocomposites suggest the reusability of the films for at least five cycles and the capability to reproduce these results in nature.

Finally, Spinelli et al. [223] evaluated the temperature effect on the thermophysical properties of PLA nanocomposites reinforced with two different weight percentages (3 and 6 wt.%) of GNPs. At the lowest temperature (298.15 K) measured, an enhancement of 171% was observed for the thermal conductivity compared to the unreinforced matrix due to the addition of 6 wt.% GNPs, whereas at the highest temperature (372.15 K) such an improvement was about 155%. Moreover, similar to the glass transition (expected at ~333 K), the thermal diffusivity decreased with increasing temperatures.

Masarra et al. [224] printed electrically conductive PLA/PCL composites using the FFF that were mixed with different contents of GNPs. The electrical resistivity results using the four-probes method revealed that the 3D-printed samples featuring the same graphene content are semiconductors. Varying the printing raster angles also exerted an influence on the electrical conductivity results. The electrical percolation threshold was found to be lower than 15 wt.%, whereas the rheological percolation threshold was found to be lower than 10 wt.%. Furthermore, the 20 wt.% and 25 wt.% GNP composites were able to connect to an electrical circuit. Lastly, an increase in the Young's modulus was shown with the percentage of graphene. In a PLA/poly(methyl methacrylate) (PMMA) blend (ratio of 40:60 *w*/*w*), two types of graphene were added to prepare nanocomposites via solution-mixing blending, graphene nanoplatelets (GNPs), and acid-modified graphene (FG) [225]. It was observed that the addition of FG to the PLA/PMMA blend indicates a lower percolation threshold and higher electrical conductivity compared with the blend filled with GNPs due to FG's better homogenous dispersion and uniform distribution in the polymer phases.

Silva et al. [226] synthesized biodegradable blends of PLA with poly(3-hydroxybutyrate-*co*-3-hydroxyvalerate) (PHBV), enriched with CNTs for electrical, electromagnetic, and military applications. More specifically, the PLA/PHBV blend (80/20) and PLA/PHBV blend-based nanocomposites with 0.5 and 1.0 wt.% CNT were produced. SEM analysis showed that PHBV was homogeneously dispersed in the PLA matrix and that CNTs are preferably dispersed in the PHBV phase. The CNT acted as a nucleating agent for the crystallization of PHBV and had no effect on the thermal stability of the nanocomposites. The addition of 1.0 wt.% CNT resulted in better electrical properties ($2.79 \times 10^{-2}$ S/m) and an excellent result as an electromagnetic interference shielding material (attenuation of approximately 96.9% of the radiation in the X-band).

The hydrolytic degradation of PLA/poly(ethylene oxide) (PEO) blends and their CNTs nanocomposites were also investigated by Zare et al. [227]. The fine dispersion of CNTs in the nanocomposites also demonstrated the strong interfacial interactions between the polymer and CNTs. The samples showed different melting peaks, suggesting that PLA and PEO are immiscible. The thermal decomposition of PEO was accelerated by the addition of CNTs in all samples because of the better heat transfer to the dispersed phase in the nanocomposites.

The capability of fullerenes to form supramolecular complexes with different types of molecules has been featured in applications mainly in the development of organic photovoltaic cells. The topic of supramolecular complexes of fullerenes with macromolecules such as PLA was explored by Cataldo et al. [228]. Spectrophotometric tests revealed that a new band in the UV, i.e., at 230 nm, was assigned to the charge-transfer interaction between PLLA and C60, whereas the former acts as an electron donor through the ketones of the ester group and the latter as an electron acceptor being an electron-deficient olefin. From this study, it was confirmed that PLLA and $C_{60}$ are in close contact with each other, suggesting a kind of host–guest interaction where the helical conformation of PLLA is able to induce the optical activity to the achiral $C_{60}$ electronic transitions.

#### 4.2.2. Biomedical Applications

CNTs have been also viewed as an interesting additive to improve the performance of 3D-printed PLA parts for biomedical applications, owing to the biocompatibility and biodegradability of PLA [229]. Magiera et al. [230] compared the degradation behavior of electrospun-nanofibers destined for biomedical applications, obtained from neat PLA and PLA/CNTs composites in aquatic environments. PLA and PLA/CNT composite nanofibers underwent swelling and partial degradation during incubation due to the penetration of water into the polymer matrix. An increase in the fibers' diameters occurred, although the overall porosity of the composites remained unaltered. Changes in the mechanical properties of the composite mats were higher than those observed for pure PLA mats. After 14 days of incubation, the samples that retained 47 to 78% of their initial tensile strength were higher than pure PLA samples. Morphological changes in pure PLA nanofibers were more dynamic than in composite nanofibers, which indicates their higher stability under the experiment conditions, while no significant changes in the crystallinity, wettability, and porosity of the samples were observed.

Also employing the electrospinning method, Zhang et al. [231] incorporated MWCNTs and doxorubicin (DOX) into the PLLA nanofibers for localized cancer treatment by the combination of both chemo- and thermotherapy. The multifunctional fibers presented increased cytotoxicity both in vitro and in vivo by the combination of photothermal-induced hyperthermia and chemotherapy with DOX. In total, photothermal treatment and chemotherapy were successfully integrated into one single system using the electrospun DOX/MWCNTs-loaded PLLA nanofiber mats. The fiber mats realized local drug delivery and hyperthermia and at the same time and showed significantly enhanced anti-tumor efficacy for this combination therapy strategy. The low energy of irradiation did not only induce cancer cell death via the hyperthermia effect, initiating the burst release of DOX from the fibers due to the relatively low Tg of PLLA, but also significantly increased the temperature of the fiber-covered tumor site. The latter resulted in an enhanced restraining effect on tumor growth and minor side effects on other normal organs.

Lee et al. [232] prepared micro needle patterns (Figure 17) of PLA/MWCNTs using the injection molding method and studied the effects of MWCNTs on crystallization, thermal behavior, and the replication and surface properties. It was concluded by the authors that the processing parameters greatly affect the replication and surface properties of the micro injection-molded PLA/CNT nanocomposites. Specifically, an analysis of the thermal behavior and crystallinity indicated that the MWCNTs promoted the unique $\alpha'$ to $\alpha$ crystal transition of PLA, resulting in an enhancement of surface modulus and hardness. In addition, the MWCNTs increased the activation energy for thermal degradation of PLA due to the physical barrier effect. The replication quality (ratio higher than 96%) of the micro-features in the PLA/MWCNTs nanocomposites was accomplished by elevating the injection speed (120 mm/s) and holding pressure (100 MPa), which enhances the polymer filling ability within the micro cavity.

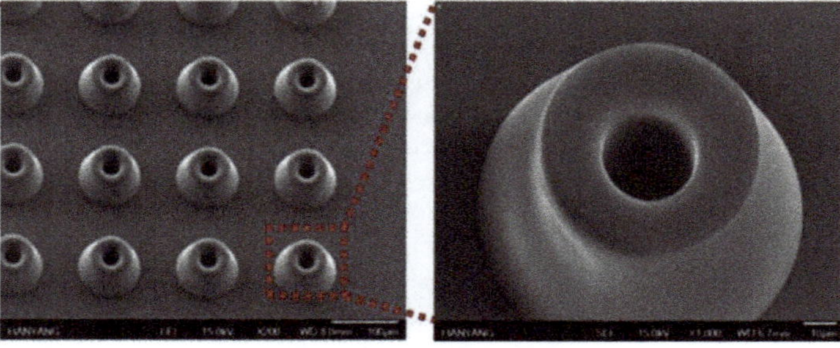

**Figure 17.** SEM micrographs of the micro needle surfaces of PLA/MWCNTs nanocomposites [232].

In the work conducted by Vidakis et al. [233], an industrially scalable method was developed for the preparation of multifunctional nanocomposite filaments suitable for multiple industrial applications, such as sensors fabrication, health monitoring devices, medicine etc. Briefly, the PLA polymeric matrix was enriched with MWCNTs at 0.5, 1.0, 2.5, and 5.0 wt.% filler loadings to fabricate novel materials using 3D-printing technology. The performed tests showed that the addition of MWCNTs at loadings higher than 1 wt.% significantly improved the mechanical properties and rendered the nanocomposites electrically conductive. Furthermore, the 5 wt.% loading showed mild antibacterial activity against *E. coli* and *S. aureus* colonies.

The combination of C60 and PLA could have possible medical application in the preparation of a new type of suture or stent with various medical functions. Keeping that in mind, Thummarungsan et al. [234] reported the preparation of electroactive samples based on the dibutyl phthalate (DBP)-plasticized PLA and fullerene (C60) produced via the solution casting method for biomedical applications. All PLA composites presented fast and reversible responses when subjected to electrical stimulus. The C60/PLA/DBP composite at a concentration of 1.0% $v/v$ showed the highest storage modulus response up to $23.51 \times 10^5$ Pa under the 1.5 kV $mm^{-1}$ electric field. Thus, the authors claimed that the electrically responsive PLA composites prepared in their work that have a short response time and a high bending deformation were demonstrated to be promising biobased materials for actuator applications.

Chen et al. [235] synthesized for the first time PLLA composites with a C60 tetragonal single crystal (C60TSC) through facile evaporation of the $CHCl_3$ solution containing PLLA and C60. TGA tests showed that in the second step of mass loss, the thermal decomposition peak temperature of PLLA in the PLLA/C60TSC composite rises to 352.6 °C, 29.7 °C higher than that of pure PLLA, indicating that the resultant C60TSC efficiently enhanced the thermal stability of PLLA. Overall, the incorporation of C60TSC also led to the formation of PLLA composites that have a higher melting temperature and higher cold crystallization temperature, as well as a higher glass transition temperature and lower relative crystallinity (Xc) than pure PLLA.

In another study, inorganic fullerene (IF)-like tungsten disulphide ($WS_2$) nanoparticles, ranging between 0.1 and 1 wt.%, from layered transition metal dichalcogenides (TMDCs), were successfully introduced into a PLLA polymer matrix by Naffakh et al. [236]. The main objective of this research was to generate novel bio-nanocomposite materials through an advantageous melt-processing route. It was discovered that the incorporation of increasing IF-$WS_2$ contents led to a progressive acceleration of the crystallization rate of PLLA. The morphology and kinetic data demonstrated the high performance of these novel nanocomposites for industrial applications.

Li et al. [237] developed a C60 L-phenylalanine (phe) derivative attached with PLA (C60-phe-PLA) for the preparation of injectable Mitoxantrone (MTX) antitumor drug multifunctional implants. C60-phe-PLA was self-assembled to form microspheres (Figure 18a) consisting of a hydrophilic antitumor drug (MTX) and a hydrophobic block (C60) using the dispersion–solvent diffusion method. The results obtained from the performed investigation were very promising for the future treatment of solid cancer tumors. Specifically, the self-assembled microspheres showed a sustained release pattern within 15 days of in vitro release studies, as presented in Figure 18b. According to the tissue distribution of C57BL mice after intratumoral administration of the microspheres, the MTX was mostly distributed amongst the tumors, and was hardly detected in the heart, liver, spleen, lungs, and kidneys. Microspheres provided high antitumor efficacy with negligible toxic effects on normal organs, owing to their significantly increased MTX tumor retention time, low MTX levels in normal organs, and strong photodynamic activity of PLA-phe-C60. In summary, the authors stated that the prepared C60-phe-PLA microsphere could serve not only as a powerful photo dynamic therapy but also as a sustained-release drug delivery vehicle, suggesting that there is great potential for this formulation for local cancer treatment.

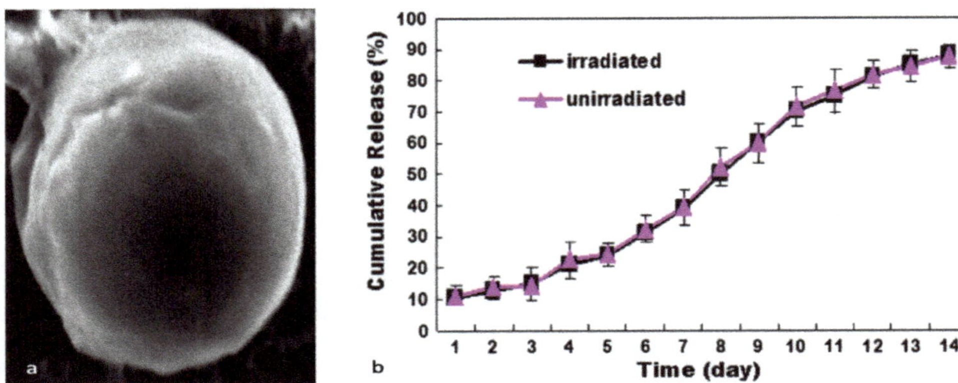


**Figure 18.** (**a**) SEM micrograph of the developed MTX microspheres and (**b**) in vitro release profiles of MTX from microspheres within 15 days of study [237].

4.2.3. Structural Applications (Mechanical-Thermal Properties Enhancement)

CNTs can enhance the crystallization rate and mechanical properties of PLA. With that knowledge, Bortoli et al. [238] functionalized CNTs with an $HNO_3$ solution to create defects and incorporate oxygen functional groups on the CNTs surface for the assessment of the thermal and mechanical properties of 3D-printed PLA/CNT nanocomposites. The results suggested that the functionalized CNTs (f-CNTs) displayed an improved dispersion in the matrix and acted as effective nucleating agents for PLA crystallization, when compared to the use of commercial CNTs (c-CNTs). The addition of just 0.5 wt.% f-CNT was adequate to yield a significant increase in the mechanical strength of the 3D- printed parts (from 29.4 ± 0.7 MPa for PLA/c-CNT to 41.6 ± 1.4 MPa for PLA/f-CNT) and provided better interfacial adhesion between 3D-printed layers, maintaining the thermal stability of the nanocomposites (Figure 19). DMA analysis indicated a significant increase in the storage modulus (~43% at 37 °C) when f-CNTs were used as a reinforcement (Figure 20). Consequently, both mechanical and thermal properties of 3D-printed PLA/CNT nanocomposites were significantly enhanced when f-CNTs were employed, also indicating that percentages of less than 1 wt.% of this additive are required.

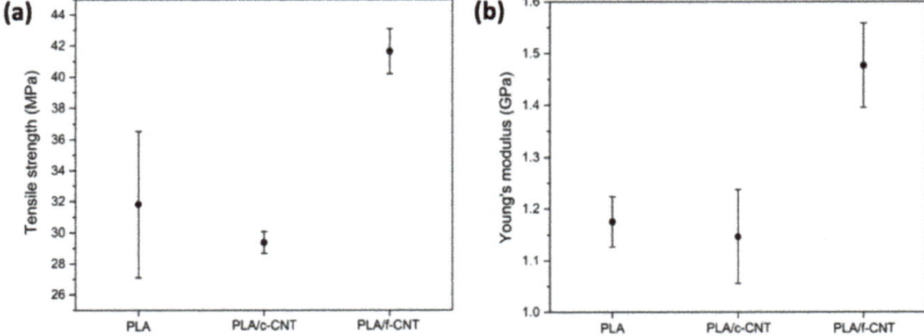


**Figure 19.** Effects of the c-CNT and f-CNT addition on the (**a**) tensile strength, and (**b**) Young's modulus [238].

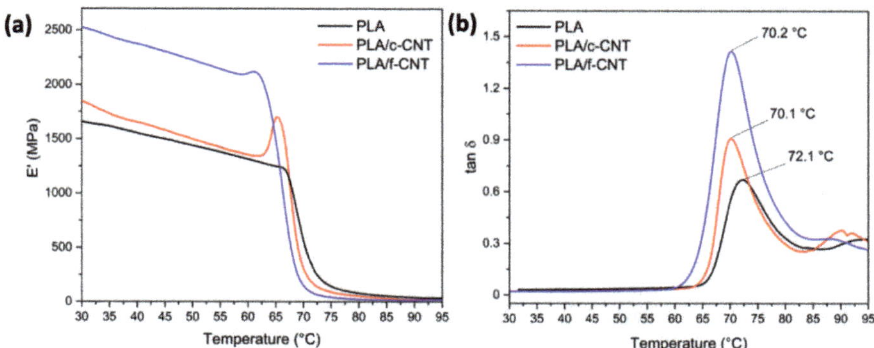


**Figure 20.** (**a**) Storage modulus (E′) and (**b**) Tan δ of pure PLA, PLA/c-CNT, and PLA/f-CNT obtained from DMA [238].

In a recent study by Yang et al. [56] a filament based on PLA/CNTs composites was prepared for the fused deposition modeling (FDM) process. The effects of the CNT content on the crystallization-melting behavior and melt flow rate were tested to investigate the printability of the PLA/CNT. The results demonstrate that the CNT content has a significant influence on the mechanical properties and conductivity properties. The addition of 6 wt.% CNT resulted in a 64.12% increase in tensile strength and a 29.29% increase in flexural strength. The electrical resistivity varied from approximately $1 \times 10^{12}$ Ω/sq to $1 \times 10^{2}$ Ω/sq for CNT contents ranging from 0 wt.% to 8 wt.%.

Lately, PLA-graphene nanocomposites have been extensively fabricated using additive manufacturing such as 3D printing due to the attractive properties offered by both materials. However, in composites, the main challenge is to understand the graphene properties' transfer from the nanoscale to the macroscale. Camargo et al. [239] investigated the effect of the variation of the infill and layer thickness parameters on the mechanical behavior of 3D-printed materials. Due to the layered production process, 3D-printed parts exhibit anisotropic behavior. The data obtained showed that the mechanical performance improved with the enhancement of the layer thickness and infill density parameters, while impact energy decreased as the infill increased. Caminero et al. [240] reported that the printed PLA-graphene composite samples showed the best performance in terms of surface texture, tensile strength, and flexural stress in comparison to pristine PLA. Nevertheless, the impact strength of the PLA-graphene composites was reduced by 1.2–1.3 times compared to that of the un-reinforced PLA matrix. Furthermore, the Plaza group examined the coupled effect of the cyclic loading amplitude and frequency on the fatigue behavior of materials (especially composites) and the induced self-heating phenomenon [241]. It was shown that at high values of frequency and applied stress during the first stage of low cycles fatigue, composites exhibit an overall fatigue response mainly governed by induced thermal fatigue (ITF). Accordingly, the mechanical fatigue (MF) nature becomes predominant during the second stage before the failure. For low frequency and applied amplitude, no significant self-heating phenomenon has been observed.

In the study conducted by El Magri et al. [242], the combined effect of process parameters, loading amplitude, and frequency on the fatigue behavior of the 3D-printed PLA-graphene specimens were analyzed. The obtained experimental results highlighted that fatigue lifetime clearly depends on the process parameters as well as the loading amplitude and frequency. In addition, when the frequency was 80 Hz, the coupling effect of thermal and mechanical fatigue caused self-heating, which decreases the fatigue lifetime.

Garcia et al. [243] compared geometric properties such as the dimensional accuracy, flatness error, surface texture, and surface roughness of a 3D-FFF-printed PLA-GNP reinforced matrix. The results showed that the dimensional accuracy was mostly affected by the build orientation, which showed an increase in the layer area on the X–Y plane and the highest dimensional deviation owing to the longer displacements of the extruder accumulating positioning errors. Therefore, it is clear that PLA-graphene filaments improved the mechanical, electrical, and thermal properties without losing their geometric quality.

Ghani et al. [244] evaluated the mechanical properties of PLA reinforced with GNPs at a concentration range of 0.1 wt.% to 1 wt.% via the meld-blending technique and then injection molding. At 0.3 wt.% filler, the tensile strength increased, reaching the maximum value of 50.3 MPa, owing to improved dispersion inside the PLA matrix. The elongation at break is enhanced with the further addition of the filler, resulting in the highest value of 2.32% with 0.5 wt.%, although with 1 wt.% graphene, this property decreased because the excess of the filler limits the motion of the polymer chains and makes the nanocomposite more brittle.

Nanocomposites of PLA reinforced with graphene oxide nanosheets using the electrospinning method also enhanced the mechanical properties [245]. The addition of GO nanosheets strengthened the PLA fiber mat matrix with the ultimate tensile strength, reaching the value of 4.4 $\pm$ 0.46 MPa from 1.7 $\pm$ 0.43 MPa of neat PLA fiber mats. The modulus was also improved from 0.53 $\pm$ 0.07 MPa of neat PLA to 0.95 $\pm$ 0.24 MPa due to the GO nanosheets, which demonstrates a strong interfacial interaction with the matrix. Nevertheless, elongation at break was decreased from 31% of neat PLA to 17.7% of PLA/GO fiber mats.

The majority of the current research on additive manufacturing has been concentrated on improving the mechanical properties, such as strength and stiffness, of polymer composites. Papon et al. [246] reinforced PLA with CNF and investigated the effect of filler concentrations (0.5 wt.% and 1 wt.%) on the composites' properties. Mechanical studies indicated an increase of 12% in the Young's modulus at the concentration of 1 wt.% of the filler. Although the failure strength demonstrated an increase at 0.5 wt.% CNF, by increasing the content up to 1 wt.%, their value decreased due to agglomeration of the filler, low CNF aspect ratio, and poor interfacial bonding between the polymer matrix and the filler.

Malafeev et al. [247] investigated the mechanical properties of PLA using vapor grown carbon nanofibers (VGCF) as a filler via a twin-screw micro-extruder. An improvement of 20% in strength was observed at high-temperature orientational stretching of the nanocomposite by four times, regardless of the concentration of the filler, while the elastic modulus and deformation at break were at the initial level of the pristine fibers. In contrast, stretching by six times caused a decrease of 15% in strength and elastic properties and a possibility of existing microcracks in polymer fibrils was suggested. Dave et al. [248] reinforced PLA using a carbon nanofiller (0.1 wt.%) in the form of carbon quantum dots to create composite scaffolds through melt blending and underwent a 7 min plasma treatment for the improvement of cell adhesion and growth, obtaining superhydrophilicity and cell proliferation, which had increased by 70% compared to untreated ones. An increase of 24.1% in tensile strength was noticed compared with pure PLA due to the enhanced surface area of intercalated nanoparticles and effective load transfer and dissipation at the interface of carbon dots with PLA.

Nanocomposites of PLA reinforced with carbon NPs via compression molding technique exhibited an improved tensile strength due to good dispersion of NPs in the PLA matrix, the formation of hydrogen bonds and the increasing composition of the filler, ranging from 2 wt.% to 8 wt.% [249]. Regarding the Young's modulus, even though there was a 25% increase, only with the addition of 8 wt.% of the filler, a decrease was observed due to additional quantity of nanocarbon that hinders the bonding of some particles with the PLA matrix.

Improved electrical resistivity, compared to a solid PLA filament, in the range of 64 $\pm$ 25 to 1.4 $\pm$ 0.48 $\Omega$ m for 6–15 wt.% PLA/carbon NPs filaments, was reported by Potnuru et al. [250]. In a further step, 3D samples using the abovementioned filaments with 25 MPa mechanical strength and 20 $\pm$ 10 $\Omega$ m electrical resistivity were produced. The authors highlighted that this method can be facilitated to build humanoid robot structures with electrical circuitry. Enhanced electrical conductivity (3.76 S/m) was also claimed by Jain et al. [251], owing to chemical bonding between PLA and the nanofiller, during the fabrication of PLA/carbon nanopowder filaments for 3D printing. Solution blending was used for PLA-NC nanocomposite fabrication and melt extrusion was employed to make cylindrical filaments.

Carbon-based nano-additives can be used also to enhance the miscibility of PLA blends. The group of Amran et al. [252,253] compared the mechanical properties of a PLA/liquid natural rubber (LNR) blend (weight ratio of 90:10 wt/wt) reinforced with varying contents (0.25 wt.%, 0.50 wt.%, 0.75 wt.%, and 1.00 wt.%) of GNPs and functionalized GNPs. PLA/LNR exhibited a tensile strength 19.5 MPa, while the maximum tensile strength was demonstrated by the blend with 0.75 wt.% at 24.4 MPa. The same pattern was observed after the incorporation of the GNP-A and GNP-T fillers. In total, the 0.75 wt.% samples yielded the highest tensile strength at 45.47 MPa and 35.70 Mpa, respectively.

Wang et al. [254] exploited vinyl functionalized graphene (VGN) as an efficient compatibilizer for the preparation of mechanically strong biodegradable nanocomposites from PLA/PCL blends. After being reactively compatibilized using 0.5, 1.0, and 2.0 wt.% VGN, the phase size of co-continuous PLA/PCL blend was remarkably decreased, and the tensile strength was increased by 200%, 280%, and 253%, respectively. The strong interfacial interactions in reactively compatibilized PLA/PCL/VGN blend nanocomposites were evidenced by the linear rheological and atomic force microscopical modulus measuring results.

Different contents of functionalized graphene nanoplatelet (FGNP) were incorporated into a PLA/chitosan (PLA/CS) (75/25 wt/wt) blend and the effect of the filler on the mechanical performance, biodegradability, and electrical properties of the blend was investigated [255]. The nanocomposite filled with 3 phr FGNP indicated an improvement of approximately 142% and 261% in tensile strength and Young's modulus, respectively, compared to the neat blend. Furthermore, with the addition of 3 phr FGNP to the PLA/CS blend, the thermal stability significantly improved. In addition, the study of biodegradability behavior demonstrated that the weight loss rate improved over time.

Wang et al. [256] prepared a PLA/poly(butylene succinate) (PBS) blend (weight ratio 70/30 wt/wt) and reduced graphene oxide (rGO) (in different contents of 0, 0.5, 1.0, 2.0, and 3.0 wt.%) as a filler via melt blending. The study showed an improvement of the mechanical properties, especially with the addition of 3.0 wt.% rGO. Elongation at break also increased to 9.2% by adding 3 wt.% rGO from 5.9% of the neat blend. Measurement of the tensile strength indicated an increase of around 65% compared to the neat PLA/PBS blend and the impact strength exhibited a value 2.9 times higher than the initial blend, owing to the good dispersion of the filler and the compatibilization of the components of the composites.

### 4.3. *Applications of PLA/Nanocomposites with Natural Nano-Additives*

#### 4.3.1. Food Packaging

Briefly, for food packaging applications, Montes et al. [257] extruded and successfully layered through thermo-compression PLA monolayer films containing Lgn-NPs and Umbelliferone (UMB). In vitro antioxidant studies using the DPPH method produced a radical scavenging activity (RSA) value of 80%, which corroborates with the results obtained by Yang et al. [258]. Moreover, the incorporation of Lgn-NPs decreased the transmittance in the visible region, reaching rather null values at wavelengths lower than 350 nm in the UV region. A food packaging material is expected to be transparent in visible light and opaque in the UV region, in order to protect the food from the oxidative deterioration, discoloration, and flavors losses causes by the UV radiation [259].

The UV, antioxidant, antibacterial, and compostability of PLA films containing 1 wt.% and 3 wt.% lignin nanoparticles (pristine (Lgn-NPs), chemically modified with citric acid (caLNP) and acetylated (aLNP)), were assessed by Cavallo et al. [130]. In general, and irrespective of whether Lgn-NPs were chemically modified or not, films containing Lgn-NPs restrained the growth of *Escherichia coli* and *Micrococcus luteus* when compared to neat

PLA. After 17 days of disintegration studies in compost conditions, no evident amount of sample was observed, indicating the complete compostable nature of the composites. As expected, due to the presence of phenolic groups (-OH), ketones, and other chromophores, unmodified Lgn-NPs were capable of trapping DPPH radicals and exhibiting UV-blocking behavior, while for caLNP and aLNP, chemical modification involving their phenolic hydroxyl groups reduced both their antioxidant properties and UV protection. Finally, migration tests yielded significantly lower values than the migration limits allowed for food contact materials, thus rendering the nanocomposites suitable for the packaging sector.

PLA composite films were produced using unmodified soda micro- or nano-lignin as a green filler at four different contents, between 0.5 wt.% and 5 wt.% [260]. It was found that the tensile strength and the Young's modulus were improved by the addition of lignin (L) and especially nanolignin (NL) (Figure 21a). This is due to the finer dispersion of NL in the PLA matrix, as verified by the TEM micrographs (Figure 21b). The UV-blocking and antioxidant properties of the composite films were also enhanced, especially at higher filler contents. As can be seen in Figure 21c, the residual DPPH content over time for different PLA–L (c) and PLA–NL (d) composites immersed in a DPPH/ethanol solution is lower in both L and NL composites, compared with neat PLA sample, which shows negligible activity. By contrast, the addition of either L or NL in the PLA matrix enhanced the antioxidant activity, as could be deduced from the higher reduction rate of DPPH• over time. The composites with higher L/NL content showed better antioxidant activity.

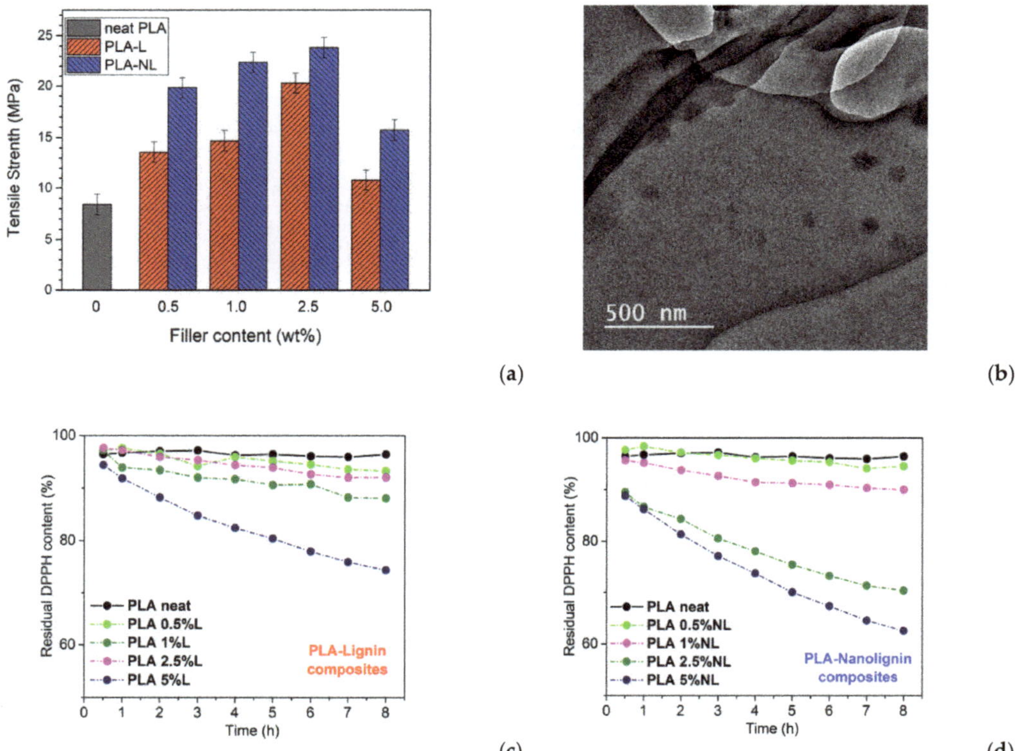

(a) (b) (c) (d)

**Figure 21.** (**a**) Tensile strength variation pf PLA/lignin and nanolignin composites, (**b**) TEM micrographs of PLA/nanolignin containing 1 wt.% nanolignin and Reaction kinetics of the free radical DPPH during immersion of PLA–Lignin, (**c**) and PLA-Nanolignin (**d**) films in ethanol solution [260].

Similarly for food packaging purposes, Cerro et al. [129] developed PLA/Lgn-NPs nanocomposite films containing cinnamaldehyde (Ci) through a combination of melt extrusion and the supercritical impregnation process. The incorporation of Lgn-NPs and Ci affected the thermal, mechanical, and colorimetric properties of the developed films resulting in biodegradable plastic materials with a solid UV-light barrier performance compared to neat PLA films. Toxicity studies conducted upon rats presented normal blood parameters after a single dose of the nanocomposites. Disintegrability tests (Figure 22) under composting conditions verified the biodegradable character of developed materials. Moreover, the Ci active agent accelerated the disintegration rate, while the PLA films with Lgn-NPs presented a slightly reduced rate of disintegration. At Day 23, a disintegration degree higher than 90% was determined for all bio-nanocomposites.

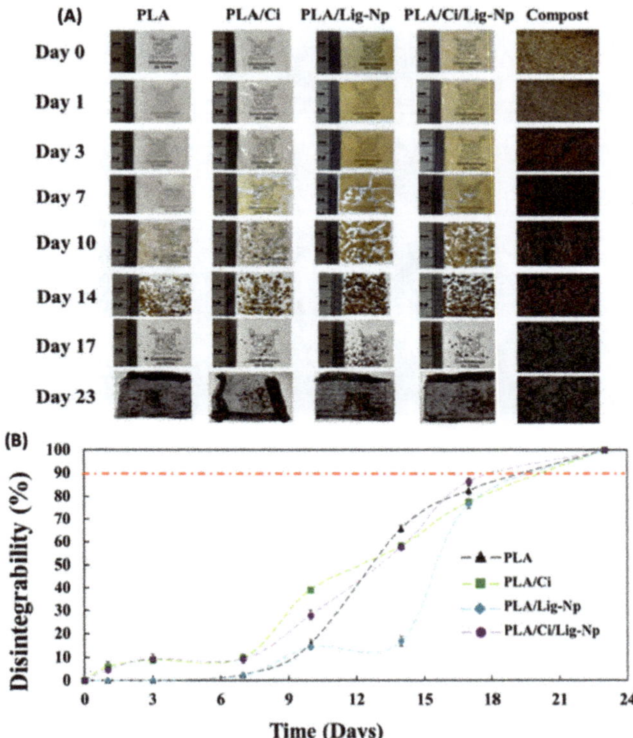


**Figure 22.** (**A**) Optical observation of prepared films before and after different incubation days under composting conditions and (**B**) % disintegrability degree under composting conditions [129].

The disintegration of PLA under composting conditions is one of the most appealing properties for food packaging purposes. Its degradation typically starts with the hydrolysis of the PLA chains induced by the diffusion of water into the composites. The influence over the degradation process strongly depends on the hydrophilicity/hydrophobicity and dispersion of the added NPs. Yang and his group [261] claimed in their work that the hydrophobic nature of Lgn-NPs (1 wt.%) delayed the disintegration process of the PLA matrix in a composting environment. Furthermore, when they added a 3 wt.% Lgn-NPs in the PLA matrix, nanoparticles aggregation and rougher film surface structures were detected, causing higher degradation.

Lgn-NPs have also been incorporated into PLA-based copolymers. In their study, Yang et al. [262] produced PLLA-PCL-lignin nanocomposites by a melt processing method, where Lgn-NPs grafted with PLLA and PCL copolymer served as interfacial compatibilizers.

Chihaoui et al. [263] examined the feasibility of incorporating lignocellulosic nanofibers (LCNFs) into a PLA/PEG blend (weight ratio of 80:20). Tensile tests and DMA indicated an enhancement in the tensile strength and Young's modulus by about 250% and 1100%, respectively, at 8% LCNFs content while maintaining a high toughness (around 16 $MJ/m^3$). The latter was attributed to the ability of the LCNFs to create a homogeneously distributed and entangled network within the PLA/PEG. The crystallization process of the blend was not affected by the addition of LCNFs, as no nucleation effects were observed. Finally, from the disintegration tests in composting conditions, both PEG and LCNFs were found to have a positive contribution to the disintegration of the PLA matrix, mostly owing to their hydrophilic nature favoring water diffusion inside the plasticized film (Figure 23).

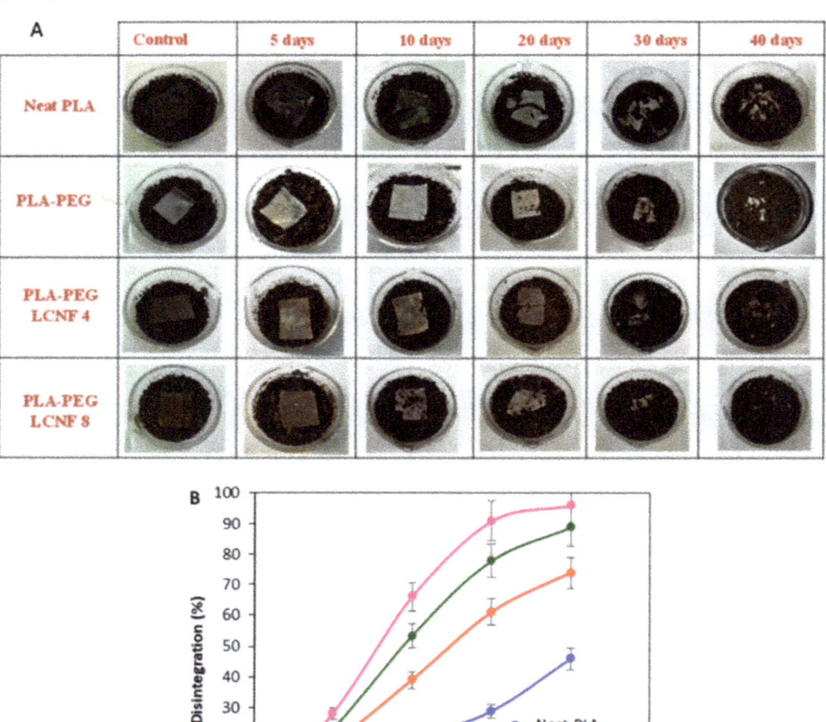


**Figure 23.** (**A**) Visual observation of PLA and PLA/PEG/LCNF nanocomposites after different days under composting environment and (**B**) disintegration degree under composting conditions as a function of time [263].

Gulzar et al. [264] discovered electrospungelatin/chitosan solutions incorporated with tannic acid (TA) and chito-oligosaccharides (COS) at varying levels on PLA films. Bead-free and smooth nanofibers were formed at higher TA and COS levels. Both TA and COS provided antioxidative properties to the film, whereas a high amount of TA rendered higher antibacterial activity against both Gram-positive and Gram-negative bacteria on the surface of the films. Water vapor permeability (WVP), tensile strength, and elongation at break of GC-NF coated PLA films were found to be improved when compared to PLA film. Moreover, light transmission of GC-NF coated PLA films was reduced and the films were cloudy in appearance.

In another study, PLA was combined with poly(2-ethyl-2-oxazoline) (PEOx) and filled with nanocellulose extracted from corn cob (CCNC) [265]. Synthesized nanocomposites were prepared using the solvent casting method and they found that PEOx induces β-crystals formation while CCNC promotes α-crystals. In addition, the percentage of the improvement in the tensile strength and tensile modulus of PLA/PEOx/CCNC was higher than that of PLA/CCNC. This was attributed to PEOx, which facilitated the dispersion and distribution of CCNC in the PLA matrix.

Sangeetha et al. [266], investigated the effect of nanocellulose on properties of ethylene vinyl alcohol (EVOH)/ethylene vinyl acetate (EVA)-toughened PLA. Bio-nanocomposites were prepared via the melt mixing technique using a twin screw extruder followed by injection molding. The addition of nanofibrillated cellulose up to 2 wt.% retained the tensile strength and increased the tensile modulus of PLA/EVA/EVOH ternary blend systems, whereas further weight loadings decreased the tensile strength significantly. Furthermore, the addition of nanofibrillated cellulose increased the stiffness of the composite, raising the elongation at break up to 214% and the impact strength up to 89%. Additionally, they found that nanocellulose acted as a nucleating agent such that it initiated a crystallization phenomenon at a lower temperature.

Rasheed et al. [267] studied the effect of cellulose nanocrystals (CNC) from bamboo fiber on the properties of PLA/poly(butylene succinate) (PBS) composites fabricated by melt mixing. Results showed uniform distribution of CNC particles in the nanocomposites, improving their thermal stability, tensile strength, and tensile modulus up to 1 wt.%, making the composite films stiffer. The highest values of tensile modulus and tensile strength were obtained at 7600 MPa and 93 MPa for the composite with 1 wt.% of CNC. However, the elongation at break decreased insignificantly due to the reduction in flexibility of composites upon the addition of CNCs. In a thermal analysis, it was found that CNCs restricted the crystallization of PLA-PBS blends.

Rigotti et al. [150] investigated PLA nanocomposites containing various amounts (from 1 to 20 wt.%) of nanocellulose esterified with lauryl chains (LNC). Results showed that the low content of LNC (up to LNC content of 6.5 wt.%) in the PLA matrix was well dispersed and formed small, sub-micrometric clusters. In contrast, higher filler contents presented oval aggregates in the micrometric range. The addition of LNC did not affect the thermal properties (glass transition temperature and melting temperature) of the PLA matrix. Concurrently, as LNC content increased, both storage and elastic moduli presented a sharp decrease of up to 5 wt.% filler, and a lower reduction for LCN concentration of 10–20 wt.%. Nanocomposites with 3 and 5 wt.% filler exhibited the highest strain at break and a large amount of plastic deformation due to a strong interfacial adhesion between the filler particles and the PLA matrix. The addition of LNC fillers improved the gas barrier properties of the PLA film to a critical LNC concentration of 6.5 wt.%, where the gas permeability of the nanocomposite resulted in being 70% lower than that of the PLA matrix. The addition of higher LNC content increased the gas permeability of the nanocomposites due to the presence of large LNC aggregates.

Jin et al. [268] compounded PLA and NC at various ratios to prepare biocomposite films via the solution casting process. To enhance the compatibility of PLA/NCC blends, the NCC was intentionally subjected to graft modification by 3- aminopropyltriethoxysilane (KH-550). The increased silanized NCC (SNCC) was found to improve the light resistance, air permeability, thermal stability, and mechanical performance of the PLA- based composite films. Especially, compared to the pristine PLA sample, the obtained PLA-based composite films with 0.5 wt.% SNCC showed increases of 53.87% and 61.46% in the tensile strength and elongation at break, respectively. Additionally, PLA/SNCC composite films displayed a decrease of 87.9% in air permeability compared with neat PLA film.

Finally, in order to improve the compatibility of PLA/NCC blends, the NCC was deliberately subjected to graft modification by 3-aminopropyltriethoxysilane (KH-550). The obtained PLA nanocomposite with 0.5 wt.% NCC increased by 53.87%, and by 61.46% for the tensile strength and elongation at break, respectively, compared with PLA neat films. Moreover, PLA/NCC composites presented a decrease of 87.9% in air permeability in comparison to pure PLA film. In another case, nanocellulose was oxidated (TOBC) using TEMPO and Pickering emulsion in order to prevent the presence of aggregates of hydrophilic nanocellulose in the PLA matrix. It was found that the Pickering emulsion improved the dispersion of nanocellulose in the PLA matrix. TOBC improved the crystallization rate of PLA as the crystallinity of the composite materials containing 1.5% TOBC was 2.16 times higher than that of pure PLA. The addition of 1.5% TOBC in PLA also enhanced the mechanical strength and toughness of the materials. Specifically, the tensile strength, elongation at break, maximum bending strength, and elastic modulus of the PLA were raised by 9.2%, 202%, 45%, and 49%, respectively [269].

Another group reinforced PLA with oxidized nanocellulose by TEMPO-mediated and polyethylene glycol (PEG) to enhance the compatibility and strength of the PLA. The tensile strength of neat PLA decreased with the addition of oxidized nanocellulose due to incompatibility between the PLA matrix and oxidized nanocellulose. However, interestingly, the tensile modulus, which represents the resistance to deformation of material, of PLA/1PEG and PLA/1PEG/oxidized nanocellulose, were higher than PLA and PLA/oxidized nanocellulose, respectively, because the heterogeneous nucleation may occur so the PLA chains can form in an orderly fashion in the crystalline regions, which results in higher tensile modulus [270].

Sobhan et al. [271] prepared biochar nanoparticle BCNP/PLA nanocomposites by the solvent casting method. They studied the electrical conductivity of the BCNP/PLA nanocomposites and they found that increasing the BCNP content in nanocomposites from 50% to 85% increased the cyclic voltammetry (CV) and differential plus voltammetry (DPV) from 5 to 22 mA and 1.9 to 12 mA, respectively. Correspondingly, they proved that the addition of BCNP improved the thermal stability of nanocomposites, increasing the content of BCNP as 85% BCNP/PLA nanocomposite had the less thermal degradation at 400 °C. These results were also confirmed by Jasim et al. [272], who developed PLA/(0.5–10) wt.% BCNP nanocomposites using thymol as a plasticizer. Results showed that electrical conductivity increased with increased BCNP. Additionally, they noticed that BCNP increased the elongation and tear resistance but reduced the tensile strength and tensile modulus and hardness. Specifically, the Young modulus of PLA neat presented a value of 2.83 GPa, while the nanocomposites' values ranged from 1.94 GPa with 0.5% biochar to 1.09 GPa with 10% biochar.

#### 4.3.2. Flame Retardants

Chollet and his group [273] examined for the first time the use of Lgn-NPs as a flame-retardant additive for PLA. Nanoparticles were grafted using diethyl chlorophosphate (LNP-diEtP) and diethyl (2-(triethoxysilyl)ethyl) phosphonate (LNP-SiP) to enhance their flame-retardant effect in the PLA matrix. In brief, they found that phosphorylated lignin nanoparticles restrained the PLA degradation during melt processing and the nanocomposites were shown to be relatively thermally stable. Even at a low concentration (5 wt.%), the grafting of phosphorus on the surface of Lgn-NPs enables a significant increase of ignition time as well as a reduction of peak of heat release rate (pHRR).

#### 4.3.3. Pesticides

In an attempt to reduce the groundwater contamination induced by traditional pesticides, leading to ecosystem destruction and food pollution, the group of Yu et al. [274] developed abamectin nanopesticide (Abam-PLA-Tannin-NS) and azoxystrobin nanopesticide (Azox-PLA-Tannin-NS) with strong adhesion to foliage via chemical modification. Hydrogen bonding was mainly responsible for the interaction between TA-coated NPs and the foliage. Abam-PLA-Tannin-NS and Azox-PLA-Tannin-NS exhibited better photostability and excellent continuous sustained release. The retention rates of Abam-PLA-Tannin-NS and Azox-PLA-Tannin-NS on the foliage surface were outstandingly improved by more than 50%, compared with untreated nanopesticides due to affinitive binding. Resultantly, the indoor toxicity of Abam-PLA-Tannin-NS and antifungal activity of Azox-PLA-Tannin-NS were enhanced.

#### 4.3.4. Implantable Medical Devices

Liao et al. [275] prepared filaments from PLA and acetylated tannin (AT) via a twin-screw extruder for 3D printing. The acetylation of tannin contributed to its well dispersion within the PLA matrix, guaranteeing the successful fabrication of PLA/AT composite filaments. Experimental results suggest that PLA can be compounded with up to 20 wt.% AT without any visible deterioration in the tensile property. The resulting composites acquired a better degradation rate in aquatic systems compared with neat PLA, especially in an alkaline medium. Moreover, the incorporation of various AT contents induced no noteworthy effects on the $T_m$ and $T_g$ of PLA because it has little influence on the intermolecular interactions or the chain flexibility of PLA polymer chains. A decrease of crystallinity was found in PLA/AT composites, leading to degradation rates in aquatic systems, particularly under alkaline environments, which would be beneficial in short-term applications such as implantable devices in the biomedical field.

The same group successfully prepared an entire biocomposite based on PLA and tannin acetate via in situ reactive extrusion using dicumyl peroxide (DCP) to enhance the interfacial adhesion [276]. The developed composites exhibited higher molecular weight as compared with reactive extruded PLA, together with a higher Young's modulus and tensile strength. A rheological study showed that reactive extruded PLA/tannin acetate composites exhibited higher complex viscosity and modulus storage, indicating the strong interfacial adhesion of PLA matrix and tannin acetate. Moreover, $T_g$, thermal stability, and crystallinity degree were all enhanced due to better interaction by radical initiated polymerization. Contact angle data revealed a more hydrophobic characteristic of the free radical extruded composites compared with pure PLA, which might be interesting for food packaging applications.

#### 4.3.5. Heavy Metals Removal

Kian et al. [277] also prepared PLA/PBS dual-layer membranes filled with a 0–3 wt.% cellulose nanowhisker (CNWs) with aim to remove metal ions from wastewater. The increase in CNWs from 0 to 3% loadings improved the membrane porosity (43–74%) but reduced pore size (2.45–0.54 μm). The thermal stability of neat membrane was enhanced by 1% CNW but decreased with loadings of 2 and 3% CNWs due to the flaming behavior of nanocellulose. Nanocomposites with 3% CNW displayed the highest tensile strength (23.5 MPa), Young's modulus (0.75 GPa), and elongation at break (7.1%) as compared to other samples. Moreover, it exhibited the highest removal efficiency for both nickel and cobalt metal ions reaching 84% and 83%, respectively.

#### 4.3.6. Optical Applications

Geng et al. [278] synthesized PLA nanocomposites with poly(ethylene glycol) (PEG)-grafted cellulose nanofibers through a uniaxial drawing method to improve the dispersion ability of nanocellulose in the PLA matrix. With the incorporation of the PEG-grafted nanocellulose in PLA, the ultimate strength and toughness were enhanced by 39% and 70%, respectively, as compared to the nanocomposite containing unmodified cellulose nanofibers. Moreover, the aforementioned nanocomposites were highly transparent and possessed an anisotropic light scattering effect, revealing its significant potential for optical applications such as solar cells and displays.

#### 4.3.7. Outdoor Usages

Ghasemi et al. [279] synthesized PLA/NC composites through melt mixing using malleated PLA (PLA-g-MA) as a compatibilizer to facilitate the interaction between NC and the matrix. The morphology studies showed that a relatively good dispersion of NC was achieved within the PLA matrix in the presence of malleated PLA. However, mechanical studies pointed out that adding NC did not improve the impact strength of the nanocomposites compared to neat PLA, while using PLA-g-MA improved the nanocomposites' impact strength significantly. In particular, with regard to samples containing 5% wt of PLA-g-MA, this increment would be 131% [280].

### 4.4. Nanoceramics

#### 4.4.1. Medical Applications

As previously stated, nanoceramics are extensively studied for medical applications because they are comparable to native bone mineral, exhibit high biocompatibility and have positive biological effects after implantation.

Specifically, Canales et al. [167] prepared PLA dense films with incorporated bioglass NPs via the melting process, at concentrations 5, 10, and 25 wt.%, obtaining nanocomposites with mainly a uniform distribution of nanometric particle agglomerates, although isolated particles could also be detected. This high level of particle dispersion demonstrates a good interaction between the polymer and nBGs, partially compensating the particle/particle interaction. The presence of nBGs increased the stiffness of the polymer matrix since the Young's modulus increased with the increase of the nBGs content, and the 25 wt.% loading exhibiting the maximum value, about 52.6% higher than neat PLA. It was observed by thermal studies that although Tg did not show any significant change with the nanoparticle incorporation, the crystallization temperature decreased by 12.9 °C with the 25 wt.% nBGs loading, indicating that nanoparticles disrupted both the regularity of the PLA and therefore the crystallization process. The presence of nanoparticles also decreased the thermal stability of the PLA matrix, as nanocomposites presented up to about 20 °C lower degradation temperatures in a nitrogen atmosphere. The bioactive capacity of the PLA/nBGs composites increased with the increase of nanoparticles content, from 5 to 10 wt.%, inducing the formation of an apatite layer on the sample's surface with a Ca/P ratio of 1.47, confirming the development of nBGs. However, the 25 wt.% loading PLA composite barely presented antibacterial behavior but exhibited higher viability on HeLa cells (cervical uterine adenocarcinoma cell line) compared to neat PLA.

Castro et al. [281] prepared fibrous scaffolds from PLA and ceramic nanobioglass (PLA/nBGs) at 5 and 10 wt.%. They observed a uniform distribution of the nBGs within the polymeric matrix and the increase in the concentration of the nBGs did not affect its internal sponge structure. Additionally, they noted that the incorporation of nBGs improved the thermal stability of the PLA due to the hydrogen bonds between the carbonyl group of the PLA and the hydroxyl groups of the nBGs and van der Waals interactions. Nanocomposites with 5 wt.% and 10 wt.% nBGs content exhibited a decrease in crystallinity by 71.8 and 71.1%, respectively. This decrease took place because nanocomposites do not have compact crystal lattices, with nucleation predominating on the material's surface. The results from the biological analysis concluded that the incorporation of the nBGs did not affect the cell viability of HeLa cells. In the case of subdermal implantation analyses in biomodels, reabsorption was very slow, showing a significant presence in the material even three months after implantation, with an inflammatory response and an inflammatory capsule. Those changes helped the nanocomposites have higher reabsorption in the subdermal tissues than neat PLA, without affecting their biocompatibility. In addition, the nBGs induced the antimicrobial activity of Gram-positive and Gram-negative bacteria (e.g., Escherichia coli, Vibrio parahaemolyticus, and Bacillus cereus) at high concentrations of 20 $w/v$%, using the TTC method.

Esmaeilzadeh et al. [282] investigated the in vivo properties of PLLA/PCL/nBGs composites for a period six months for application to bioscrews. The in vivo results from the implants inserted on canine models indicated that the weight losses of the PLLA/PCL/n-BG composite and the PDLLA/PCL blend were approximately 60 and 70%, respectively. Moreover, the obtained histological images of the animal model after six months of implantation distinguished the formation of the new bone within the implanted area, while no osteitis, osteomyelitis, or structural abnormality were observed. They attributed the lower degradation rate of the nanocomposite compared to that of the neat blend implants to the presence of nBGs and their appropriate distribution throughout the PDLLA/PCL matrix. Further, it was proved that nBGs prevented the migration of the products resulting from PDLLA and PCL degradation. The acidity of the implants was caused by the acid release from products that were neutralized by the releasing of Ca-P ions of nBGs.

Macha et al. [283] developed PLA/3-aminopropyltriethoxysilane (APTES) nanosurface-modified nBGs composites via the solution casting method with different nBGs loadings (0.1, 0.5, and 1%). The modified nBGs provided better bonding between the amine groups of the APTES and carbonyl groups in the PLA matrix resulting in improved interfacial adhesion compared to composites with untreated particles. In addition, an improvement in elongation at break was detected, which indicated that nBGs modified by 1% APTES significantly influenced the percentage elongation of the PLA/BG nanocomposite at fracture by the effect of surface treating.

Hydroxyapatite (HAp) is the most emerging bioceramic, which is widely used in various biomedical applications, mainly in orthopedics and dentistry due to HAp having close similarities with the inorganic mineral component of bone and teeth.

Morsi et al. [284] filled the PLA matrix with iron-doped nHAp (FeHA NPs) in different mass fractions (PLA/FeHA NPs wt.%: 90/10, 80/20, and 70/30) via the solution casting technique for tissue engineering applications. The FT-IR results demonstrated that FeHA NPs were linked with PLA via hydrogen bond formation between OH of $Ca(OH)_2$ and oxygen atoms in the ester group in PLA, which contributed to a better interface behavior. The inclusion of 10 wt.% FeHA NPs displayed the highest value of elastic modulus at 330 MPa, 28 MPa higher than the neat polymer.

Toong et al. [285] evaluated the nanocomposite formulations of nHAp and L-lactide functionalized nHAp and LA-nHAp with PLLA as potential thin-strut bioresorbable scaffolds. Functionalized nanofillers/PLLA exhibited improved nanofiller dispersion, and issues such as agglomeration and poor matrix/filler integration were mitigated. The uniform dispersion of nHAp improved the mechanical properties. Specifically, it was observed that the ultimate tensile strength of 10 wt.% nHAp/PLLA and 10 wt.% LA-nHAp/PLLA improved by 31.96 and 75.02 MPa, respectively, compared to neat PLLA, while 10 wt.% and 15 wt.% loadings of LA-nHAp in the matrix resulted in the highest values of elongation at break.

Michael et al. [169] modified nHAp using three different surface modifiers namely, 3-aminopropyl triethoxysilane (APTES), sodium n-dodecyl sulfate (SDS), and polyethylenimine (PEI) to improve the dispersion capacity of nHAp in the PLA matrix. The addition of 5 wt.% nHAp treated with APTES and SDS to the PLA matrix eliminated the presence of voids and agglomeration but increased the content of modified nHAp to 30 wt.%, which caused the nanofillers' agglomeration. Similar behavior was observed in the modified nHAp with PEI. In addition, after the nHAp treatment with APTES and SDS, the tensile strength of PLA-5 wt.%-modified nHAp improved by 9 and 5 MPa, respectively, compared to the unmodified nHAp composite. On the other hand, treating nHAp with PEI caused a decrease of the tensile strength by 9 MPa compared to neat PLA. Similarly, PLA composites with APTES-nHAp and SDS-nHAp attained an improved tensile modulus of 8.5 and 2.4% in comparison with PLA neat, while PLA with PEI-nHAp recorded a decrease of 17.6%. This general improvement in tensile properties can be attributed to the improved interaction and homogenous dispersion of the modified nHAp in the PLA matrix. However, increasing modified nHAp loading to 30 wt.% caused the tensile strength to drop by 34% (APTES), 29.6% (SDS), and 40.5% (PEI), respectively, due to the presence of agglomeration.

Ziaee et al. [286] investigated the effect of HA and yttria-stabilized zirconia (YSZ), an inert and biocompatible ceramic with high bending and compressive strength and toughness, as bone tissue engineering scaffolds. An SEM analysis of the PLLA-HA-YSZ nanocomposites revealed a highly interconnected porous structure for all the scaffolds. The introduction of YSZ to the nanocomposite led to a significant increase in compressive strength, modulus, and densification strain due to toughening mechanisms and zirconia phase transformation from the tetragonal to the monoclinic. In addition, flexural strength and modulus showed an upward trend by the addition of YSZ particles to the scaffolds. It should be noted that PLA-20%HA-20%YSZ indicated the highest strength (0.47 MPa) and modulus (5.1 MPa) in both the compression and bending tests, although it did not demonstrate the proper strain compared to other scaffolds. Thus, PLA-15%HA-15%YSZ has been reported as the best candidate due to its appropriate strength and strain.

Ghassemi et al. [287] investigated the effect of adding different contents of nHAp on the mechanical properties of PLA/thermoplastic polyurethane (PLA/TPU) blends. They found that the addition of nHAp improved the interfacial properties, homogeneity, and compatibility of the blends. The 5wt.% nHAp-loaded nanocomposite indicated the highest tensile strength and modulus, i.e., an improvement of approximately 14 and 36%, respectively, compared to the PLA/TPU neat blend. On the other hand, the elongation at break of the blend diminished by the enhancement of the content of nHAp. They attribute this reduction to the fact that polymer chains were entrapped into HAp nanoparticles, which diminished the mobility of polymer chains because of the interactions. They also noted that another reason behind the decrease in the elongation-at-break was the introduction of stiff mineral nanoparticles, leading to stiffening of the blend nanocomposite.

Rahman et al. [288] examined the inclusion of chitosan and gelatin in a PLA/nanohydroxyapatite composite (GEL-CS-PLA/nHA), and discovered that the 5% nHAp-containing composite is the one most compatible with osteoconduction since it displayed a pore diameter of 125 μm, while the composites with 10%, 15%, and 20% nHAp revealed a smaller pore size in the range of 15–28 μm. Additionally, the addition of 10% nHAp in the GEL-CS-PLA blend showed the best compressive strength results as it was increased by 15.3% compared to the virgin blend. These results were due to the good dispersion of nHAp into the GEL-CHT-PLA blend and the compact morphology and less porous structure of nanocomposites. Regarding the thermal properties, the 10% nHAp-loading scaffold presented the greatest thermal stability (90–94% degradation at a temperature of 600 °C), while antibacterial and cytotoxicity test results revealed that the composite is resistant towards microbial attacks and has low sensitivity in cytotoxicity.

Yan et al. [289] prepared PLA/PBAT (poly(butylene adipat-co-terephthalte))/n-HA/MCC (microcrystalline cellulose) composites through physical blending. Good nanoparticle dispersion was observed when the nHAp content was lower than 6 wt.%, while obvious agglomeration of nHAp appeared by the addition of 10 wt.%. The tensile test demonstrated that by increasing the nHAp content, the tensile strength decreased gradually. However, the addition of 2 wt.% nHAp displayed the highest tensile strength compared with the other composites but was still lower than neat PLA/PBAT. This was because the addition of nHAp gradually increased the distance between the continuous phases of the material and weakened the interaction between the macromolecular chains. However, an improvement in thermal stability was observed along with the increase of the nHAp content from 2 wt.% to 10 wt.%.

The effect of integrating small amounts of nHAp (0.5, 1 and 3 wt.%) in a PLA/PCL (70:30 wt.%) blend was studied by Peponi et al. [290]. They found that the neat polymer blend reached elongations at break up to 200% due to the plasticizing effect of PCL on the PLA/PCL matrix, and these values were maintained for the nanocomposites up to 1 wt.% of nHAp, while the stretchability strongly decreased for the highest amount of nHAp (3 wt.%). Regarding the thermal properties of nanocomposites, they proved that the lowest amount of nHA (0.5 wt.%) did not produce significant changes on the thermal degradation process of nanocomposites. Higher amounts of nHA (1 wt.%) improved the thermal degradation of the PLA/PCL blend. Nevertheless, it should be highlighted that there was no degradation at temperatures below 200 °C which is the highest temperature used during processing.

Zhang et al. [168] reported uniform, high-quality oil-soluble HAp nanorods prepared by an innovative oleic acid (OA)-assisted mixed-solvent thermal method. The HAp nanorods were dispersed in PLA by a solution method in four concentrations (1, 3, 5, and 7 wt.%), followed by hot pressing to form nanocomposites films. As it turned out, the HA nanorods dispersed homogeneously as single particles without aggregation in the PLA matrix, which was attributable to the presence of an OA monolayer on the HA nanorods' surface. Adding 3 wt.% HA nanorods elongation at break of the composites showed an increase of 4.55% compared to pure PLA, while 5 and 7 wt.% displayed a similar behavior to the neat polymer. This decrease was attributed to the higher crystallinity and crystal density. In addition, the 1 wt.% HA nanocomposite presented the highest tensile strength value, while the 3 wt.% displayed the highest Young's modulus. However, the improvement of mechanical properties of PLA by the HA nanorods was not excellent, due to the low filler content and low length to diameter ratio o (L/D = 67.48).

Mondal et al. [291] used the 3D printing process to prepare scaffolds based on PLA and nHAp-modified PLA for medical applications. They concluded that the HAp nanoparticle accumulation on PLA scaffold surface enhanced its surface roughness and controlled the porosity. Specifically, the porosity of 3D-printed scaffolds with nHAp was decreased by 5–9%, compared to the neat PLA scaffold and SEM results (Figure 24), which confirmed the good distribution of nHAp on the PLA surface. The maximum compressive stress was detected in printed PLA-nHAp scaffolds at approximately 25 MPa, which was higher than the neat PLA scaffold. A thermogravimetric analysis revealed a significant thermal stability improvement after the addition of nHAp, considering that the maximum weigh loss was approximately 6.75% for PLA-nHAp and around 97% for PLA scaffolds. Additionally, it was found that cell attachment was highly influenced by the interaction of nHAp on the PLA scaffold surface, adsorbing proteins and facilitating cellular activity. On the contrary, the fabricated PLA scaffolds showed very poor cell attachment and proliferation on their surface.

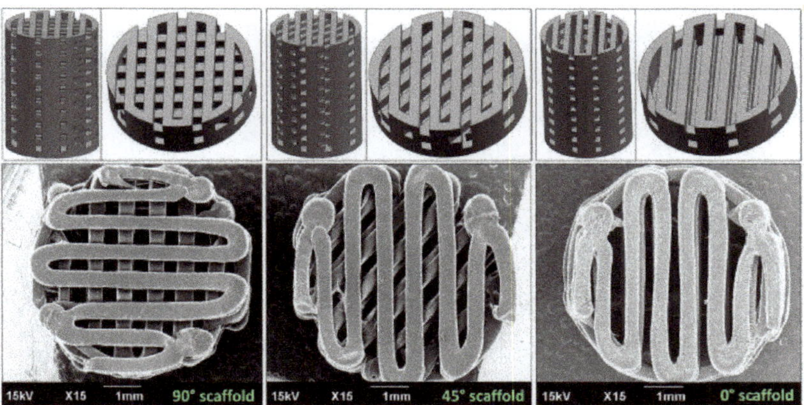


**Figure 24.** Three-dimensional-printed scaffolds destined for tissue engineering applications. The top view shows the schematic drawing from Solidworks 2016 (Dassault Systèmes, Vélizy-Villacoublay, France). The bottom view shows the original SEM image of the 3D-printed scaffold (MakerBot Z18 3D printer, New York City, New York, USA) [291].

In a similar case, Swetha et al. [292] fabricated 3D-printed PLA/Gel scaffolds and their surface was treated with Mg-nHAp in varying concentrations. Three-dimensional-PLA/Gel/Mg-nHAp scaffolds presented a uniform interconnected porous architecture. However, the porosity of the prepared scaffolds was decreased by the increasing of Mg-nHAp content (from 0.5 wt.% to 1.5 wt.%). The biological analysis showed that PLA/Gel/Mg-nHAp scaffolds exhibited improved protein adsorption in 24 h when compared with three-dimensional-PLA, which was attributable to their suitable porous architecture, as well as the presence of organic and inorganic components mimicking the biological bone matrix and apatite. It was also revealed that the PLA/Gel/Mg-nHAp three-dimensional-scaffolds were non-toxic, proving that the addition of Mg-nHAp had no influence on cell growth.

Wang et al. [55] discovered the capacity of PLA with nHAp (at a content of 0%, 10%, 20%, 30%, 40%, and 50%) to be 3D-printed in medical applications (Figure 25). The results showed that the PLA/nHAp composite had a stable structure, and the processing of this material was highly controllable. When the proportion of nHAp was greater than 50%, the composite material could not pass through the printing nozzle coherently and stably due to its high brittleness. When the nHAp ratio was less than or equal to 50%, the composite filament material could be printed by FDM. They also noticed a gradual change on the 3D-printed PLA/n-HA scaffolds' surface from smooth to rough, as the nHAp ratio increased from 0% to 50% due to the appearance of a large number of convex and concave structures on the surface, which increased the surface area of the scaffold. After applying the 3D printing process to PLA/nHAp composites, the printed samples with 0% to 30% nHAp could still maintain the integrity of the structure after the pressure test, which proved that these samples maintained their elasticity. However, when the n-HA ratio reached more than 40%, the printed sample became brittle and could not maintain a complete its shape after compression. Nevertheless, the mechanical properties of all the PLA-printed scaffolds with the integration of nHAp were much higher than that of the pure HA porous ceramic and natural human cancellous bone. The biocompatibility and osteogenic induction properties were proven to be better than that of the pure PLA scaffold.

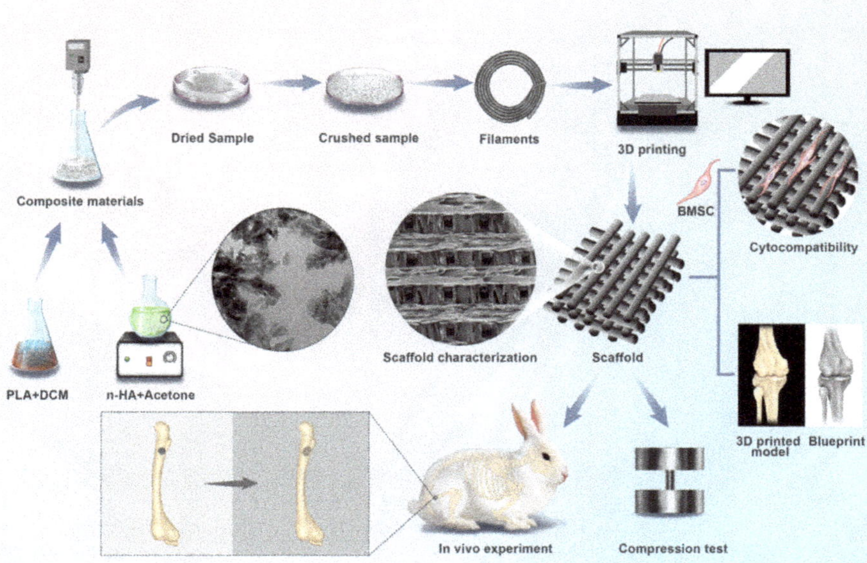


**Figure 25.** Schematic illustration of preparation and application of PLA/n-HA scaffolds for bone regeneration. The PLA/n-HA composite material was prepared by the wet mixing method, and the 3D-printed composite scaffold was made by PLA/n-HA filament. The material characterization and in vivo and in vitro experiments of the composite materials successively verified the sufficient mechanical strength and non-toxicity suitable for the promotion of the repair of bone defects [55].

#### 4.4.2. Agricultural

Hajibeygi and Shafiei-Navid [293], taking advantage of nHAp's surface functional groups such as P-OH and Ca-OH, used a new synthesized organic additive containing a-amino phosphonate and a naphthalene ring (PDA) as an organic modifier of HA NPs in order to increase the compatibility between PLA and nHAp. The presence of PDA led to improved dispersion of nHAp in the PLA matrix, while with the incorporation of only

2 wt.% nHAp and 6 wt.% PDA, the nanocomposite's tensile strength was demonstrated to be 51 MPa higher than neat PLA. The initial decomposition temperature and char residue of PLA containing 6 wt.% PDA along with 2 wt.% nHAp were increased by 20 °C and 12%, respectively.

### 4.5. PLA/Nanoclays

#### 4.5.1. Food Packaging

Ramesh et al. [186] prepared PLA nanocomposites with 30% treated aloe vera fibers and 0, 1, 2, and 3 wt.% nanoclay fillers (PLA-TAF-MMT). They noted that the addition of 1 wt.% MMT in the PLA matrix acted as a fine bonding agent between the fibers and the matrix by the bonding of nano-matrix. On the other hand, the increased content of MMT clay (2–3% wt) showed an uneven surface, small voids, and agglomerations. The morphology of nanocomposites effected their mechanical properties. Specifically, the addition of 1 wt.% MMT in PLA improved the tensile and flexural modules by 18.84% and 60.95%, respectively, more than virgin-PLA. They attributed the increase in mechanical properties to the fact that the incorporation of MMT clay in PLA reduced micro-voids, presented uniform dispersion, and improved interaction, compatibility, and adhesion between the nano-matrix. On the contrary, the high amount of nanoclays (2–3% wt.) diminished the mechanical properties of PLA nanocomposites. The decomposition temperature increased by increasing the amount of MMT in the PLA matrix; this enhancement to the MMT clay acted as a barrier to the oxygen, constrained the mobility to the chain, and slowed down the decomposition process.

Similar results were presented by Othman et al. [180], who synthesized PLA nanocomposites incorporating different types (montmorillonite (MMT) and halloysite) and concentrations (0–9 wt.%) of nanoclays. PLA with 3 wt.% concentration of nanoclays resulted in the optimum mechanical and oxygen barrier properties due to the strong interaction between nanoclays and the PLA matrix. The incorporation of 3 wt.% nanoclays MMT and Halloysite in the PLA matrix raised the Young Modulus values to around 77% for PLA/MMT (2554 MPa) and 50% for PLA/Halloysite (2158 MPa) in comparison to the Young Modulus of neat PLA (1439 MPa). Nevertheless, these properties were decreased as more nanoclays ($\geq$5 wt.%) were added into the films due to the aggregation of nanoclays. PLA/MMT nanocomposites showed better properties than PLA with halloysite nanoclays due to the nanoclay structure in nature. The addition of 3 wt.% nanoclays into virtually transparent PLA film had only minor effects on the transparency of the material as the reduction in light transmittance was only around 10%.

Alikarami et al. [26] researched the gas permeability of PLA/Cloisite-30B nanoclays composites for food packaging applications. They found that the addition of Cloisite-30B reduced the oxygen permeability by 55% compared to the pristine blend due to the tortuosity effect of nanosheets that were appropriately dispersed in the matrix. These results were confirmed by Salehi et al. [294], who synthesized poly(propylene) (PP)/PLA/nanoclay (Cloisite-30B) nanocomposite films and investigated their barrier properties against $CO_2$, $O_2$, and $N_2$.

#### 4.5.2. Engineering Applications

Montava-Jorda et al. [192] prepared PLA composites with different halloysite nanotube (HNTs) loadings (3, 6, and 9 wt.%) with the presence of PVA compatibilizer for further uses as carriers for active compounds in medicine, packaging, and other sectors. The obtained results indicated a slight decrease in elongation at break as well as in tensile and flexural strength, with both properties being related to material cohesion; however, the reduction in mechanical properties was less than 7%. On the contrary, the stiffness increased with the HNTs content. Additionally, they noted that HNTs did not affect the glass transition temperature with invariable values of about 64 °C or the melt peak temperature. On the contrary, nanoclays moved the cold crystallization process towards lower values, from 112.4 °C for neat PLA down to 105.4 °C for the composite containing 9 wt.% HNTs.

Alakrach et al. [295] illustrated a similar trend with the addition of natural HNTs, proving that a small quantity of HNTs operated as plasticizer of PLA, which dispersed the impact energy during the break. Specifically, it was found that the incorporation of 4% wt HNTs increased the tensile strength by 43% and the elongation up to 4 wt.% compared to the neat PLA. However, Eryildiz et al. [295,296] proved that a lower amount than 4 wt.% HNTs in PLA was also able to increase the mechanical properties. Specifically, 0.5 wt.% HNTs in PLA raised the tensile strength about 50% and the elongation about 70%; as well, the addition of 2 wt.% modified HNTs improved the tensile strength to 15.3% and the elongation at break to 18.6%, greater than that of neat PLA. Moreover, Nizar et al. [297] researched the thermal properties of PLA/HNTs composites and they showed that the thermal stability of nanocomposites was reduced with the increasing addition of HNTs (5–8% wt) due to the presence of HNTs agglomerates in the PLA matrix. The uneven uniformity dispersion of nanoclays increased the thermal degradation of nanocomposites as the space between PLA and HNTs allowed the volatile pyrolyzed products of PLA to accumulate, thereby accelerating their degradation. Another research group melt mixed PLA with poly(ether-ester) elastomer (TPEE) and three loadings of BTN (1, 3, and 5% wt) bentonite (BTN) on a twin screw extruder, which was then injection-molded to improve the mechanical and thermal properties of PLA [298]. The Young's modulus, impact strength, tensile strength, and thermal stability of all the prepared blends were enhanced, while the elongation at break deteriorated. Among the three PLA nanocomposites, the one with 1% wt BTN formed exfoliated structures, and therefore presented the highest elongation at break, tensile strength, and impact strength compared to the other synthesized nanocomposites.

## 5. Conclusions

The current review focuses specifically on the latest development of PLA-based nanocomposites. The demand for eco-friendly and sustainable products, exploitation of biobased feedstock, consideration of recycling routes, rising of oil price, and restrictions for the use of polymers with a high "carbon footprint" has paved the way for the swift expansion of biobased polymers. In this respect, PLA has been the most popular and promising of these biobased polymers due to its attractive renewability, biodegradability, and relatively low cost. Moreover, very recent studies underlined the huge potential of PLA in the case of higher added value for engineering applications, i.e., in food packaging, electronic and electrical devices, transportation, and mechanical and medical devices. Nevertheless, the PLA matrix must be improved in terms of thermal resistance, heat distortion temperature, and rate of crystallization while exhibiting some other specific properties (flame-retardancy, antistatic to conductive electrical characteristics, anti-UV, mechanical robustness antibacterial or barrier properties, etc.) to meet with end-user high demands. As presented in the current review paper, numerous nanofillers have been investigated towards that goal in order to prepare high-performance PLA-based nanocomposites using a variety of preparation methods.

Natural nanocompounds such as lignin and tannin NPs are especially suitable for active packaging applications since they accelerate the disintegration of PLA films and provide antioxidant and antimicrobial properties to prepared films. Incorporation of nano MgO imparts oxygen scavenger properties, limits oxidation, and enhances the shelf-life of foods, while ZnO NPs increase the UV shielding of food packaging films. Fabrication of PLA/$TiO_2$ nanocomposites can be successfully employed for the photocatalytic removal of organic compounds from water streams. On the other hand, the addition of carbon-based fillers, such as CNTs and graphene, could boost the mechanical performance and electric properties of PLA-based materials in sensors and EMI shielding. Moreover, the capability of PLA-fullerenes nanocomposites to form supramolecular complexes with different types of molecules can be exploited in the development of organic photovoltaic cells.

Although a significant amount of research has been conducted on PLA nanocomposites, much work in certain fields is still required. The dispersion within the PLA matrix and the application of totally "green" pathways and techniques pose a great challenge to academic societies and industries. Taking into consideration the above, it can be concluded that, undoubtedly, PLA-nanocomposites are currently under extensive investigation and their commercial utilization is rapidly catching up. Lastly, in-depth knowledge of these PLA nanocomposites will result in a new dimension in sustainable material development for numerous applications.

**Author Contributions:** Conceptualization, G.Z.P.; writing—original draft preparation, N.D.B. and I.K.; writing—review and editing, C.S., Z.K., D.M., M.R., A.K., R.D.B. and D.V.; supervision, G.Z.P.; project administration, G.Z.P. All authors have read and agreed to the published version of the manuscript.

**Funding:** This research received no external funding.

**Institutional Review Board Statement:** Not applicable.

**Informed Consent Statement:** Not applicable.

**Data Availability Statement:** The data presented in this study are available on request from the corresponding author.

**Conflicts of Interest:** The authors declare no conflict of interest.

## References

1. Pappa, A.; Papadimitriou-Tsantarliotou, A.; Kaloyianni, M.; Kastrinaki, G.; Dailianis, S.; Lambropoulou, D.A.; Christodoulou, E.; Kyzas, G.Z.; Bikiaris, D.N. Insights into the Toxicity of Biomaterials Microparticles with a Combination of Cellular and Oxidative Biomarkers. *J. Hazard. Mater.* **2021**, *413*, 125335. [CrossRef] [PubMed]
2. Geyer, R.; Jambeck, J.R.; Law, K.L. Production, Use, and Fate of All Plastics Ever Made—Supplementary Information. *Sci. Adv.* **2017**, *3*, 19–24. [CrossRef] [PubMed]
3. Kaloyianni, M.; Bobori, D.C.; Xanthopoulou, D.; Malioufa, G.; Sampsonidis, I.; Kalogiannis, S.; Feidantsis, K.; Kastrinaki, G.; Dimitriadi, A.; Koumoundouros, G.; et al. Toxicity and Functional Tissue Responses of Two Freshwater Fish after Exposure to Polystyrene Microplastics. *Toxics* **2021**, *9*, 289. [CrossRef] [PubMed]
4. Bobori, D.C.; Dimitriadi, A.; Feidantsis, K.; Samiotaki, A.; Fafouti, D.; Sampsonidis, I.; Kalogiannis, S.; Kastrinaki, G.; Lambropoulou, D.A.; Kyzas, G.Z.; et al. Differentiation in the Expression of Toxic Effects of Polyethylene-Microplastics on Two Freshwater Fish Species: Size Matters. *Sci. Total Environ.* **2022**, *830*, 154603. [CrossRef]
5. Bobori, D.C.; Feidantsis, K.; Dimitriadi, A.; Datsi, N.; Ripis, P.; Kalogiannis, S.; Sampsonidis, I.; Kastrinaki, G.; Ainali, N.M.; Lambropoulou, D.A.; et al. Dose-Dependent Cytotoxicity of Polypropylene Microplastics (PP-MPs) in Two Freshwater Fishes. *Int. J. Mol. Sci.* **2022**, *23*, 13878. [CrossRef]
6. Jiang, J.Q. Occurrence of Microplastics and Its Pollution in the Environment: A Review. *Sustain. Prod. Consum.* **2018**, *13*, 16–23. [CrossRef]
7. Huang, D.; Tao, J.; Cheng, M.; Deng, R.; Chen, S.; Yin, L.; Li, R. Microplastics and Nanoplastics in the Environment: Macroscopic Transport and Effects on Creatures. *J. Hazard. Mater.* **2021**, *407*, 124399. [CrossRef]
8. Kian, L.K.; Saba, N.; Jawaid, M.; Sultan, M.T.H. A Review on Processing Techniques of Bast Fibers Nanocellulose and Its Polylactic Acid (PLA) Nanocomposites. *Int. J. Biol. Macromol.* **2019**, *121*, 1314–1328. [CrossRef]
9. Banerjee, R.; Ray, S.S. An Overview of the Recent Advances in Polylactide-Based Sustainable Nanocomposites. *Polym. Eng. Sci.* **2021**, *61*, 617–649. [CrossRef]
10. Balla, E.; Daniilidis, V.; Karlioti, G.; Kalamas, T.; Stefanidou, M.; Bikiaris, N.D.; Vlachopoulos, A.; Koumentakou, I.; Bikiaris, D.N. Poly(Lactic Acid): A Versatile Biobased Polymer for the Future with Multifunctional Properties-from Monomer Synthesis, Polymerization Techniques and Molecular Weight Increase to PLA Applications. *Polymers* **2021**, *13*, 1822. [CrossRef]
11. Raquez, J.M.; Habibi, Y.; Murariu, M.; Dubois, P. Polylactide (PLA)-Based Nanocomposites. *Prog. Polym. Sci.* **2013**, *38*, 1504–1542. [CrossRef]
12. Kervran, M.; Vagner, C.; Cochez, M.; Ponçot, M.; Saeb, M.R.; Vahabi, H. Thermal Degradation of Polylactic Acid (PLA)/Polyhydroxybutyrate (PHB) Blends: A Systematic Review. *Polym. Degrad. Stab.* **2022**, *201*, 109995. [CrossRef]
13. John, A.; Črešnar, K.P.; Bikiaris, D.N.; Zemljič, L.F. Colloidal Solutions as Advanced Coatings for Active Packaging Development: Focus on PLA Systems. *Polymers* **2023**, *15*, 273. [CrossRef] [PubMed]
14. Papageorgiou, G.Z.; Achilias, D.S.; Nanaki, S.; Beslikas, T.; Bikiaris, D. PLA Nanocomposites: Effect of Filler Type on Non-Isothermal Crystallization. *Thermochim. Acta* **2010**, *511*, 129–139. [CrossRef]
15. Sanusi, O.M.; Benelfellah, A.; Bikiaris, D.N.; Aït Hocine, N. Effect of Rigid Nanoparticles and Preparation Techniques on the Performances of Poly(Lactic Acid) Nanocomposites: A Review. *Polym. Adv. Technol.* **2021**, *32*, 444–460. [CrossRef]
16. Klonos, P.A.; Lazaridou, M.; Samiotaki, C.; Kyritsis, A.; Bikiaris, D.N. Dielectric and Calorimetric Study in Renewable Polymer Blends Based on Poly(Ethylene Adipate) and Poly(Lactic Acid) with Microphase Separation. *Polymer* **2022**, *259*, 125329. [CrossRef]

17. Ainali, N.M.; Kalaronis, D.; Evgenidou, E.; Kyzas, G.Z.; Bobori, D.C.; Kaloyianni, M.; Yang, X.; Bikiaris, D.N.; Lambropoulou, D.A. Do Poly(Lactic Acid) Microplastics Instigate a Threat? A Perception for Their Dynamic towards Environmental Pollution and Toxicity. *Sci. Total Environ.* **2022**, *832*, 155014. [CrossRef]
18. Vlachopoulos, A.; Karlioti, G.; Balla, E.; Daniilidis, V.; Kalamas, T.; Stefanidou, M.; Bikiaris, N.D.; Christodoulou, E.; Koumentakou, I.; Karavas, E.; et al. Poly(Lactic Acid)-Based Microparticles for Drug Delivery Applications: An Overview of Recent Advances. *Pharmaceutics* **2022**, *14*, 359. [CrossRef]
19. Beslikas, T.; Gigis, I.; Goulios, V.; Christoforides, J.; Papageorgiou, G.Z.; Bikiaris, D.N. Crystallization Study and Comparative in Vitro-in Vivo Hydrolysis of PLA Reinforcement Ligament. *Int. J. Mol. Sci.* **2011**, *12*, 6597–6618. [CrossRef]
20. Mayekar, P.C.; Castro-Aguirre, E.; Auras, R.; Selke, S.; Narayan, R. Effect of Nano-Clay and Surfactant on the Biodegradation of Poly(Lactic Acid) Films. *Polymers* **2020**, *12*, 311. [CrossRef]
21. Sharif, A.; Mondal, S.; Hoque, M.E. Polylactic Acid (PLA)-Based Nanocomposites: Processing and Properties. *Bio-Based Polym. Nanocompos. Prep. Process. Prop. Perform.* **2019**, 233–254. [CrossRef]
22. Sha, L.; Chen, Z.; Chen, Z.; Zhang, A.; Yang, Z. Polylactic Acid Based Nanocomposites: Promising Safe and Biodegradable Materials in Biomedical Field. *Int. J. Polym. Sci.* **2016**, *2016*, 6869154. [CrossRef]
23. Najafi, N.; Heuzey, M.C.; Carreau, P.J. Polylactide (PLA)-Clay Nanocomposites Prepared by Melt Compounding in the Presence of a Chain Extender. *Compos. Sci. Technol.* **2012**, *72*, 608–615. [CrossRef]
24. Črešnar, K.P.; Aulova, A.; Bikiaris, D.N.; Lambropoulou, D.; Kuzmič, K.; Zemljič, L.F. Incorporation of Metal-based Nanoadditives into the Pla Matrix: Effect of Surface Properties on Antibacterial Activity and Mechanical Performance of Pla Nanoadditive Films. *Molecules* **2021**, *26*, 4161. [CrossRef] [PubMed]
25. Chen, B.K.; Shih, C.C.; Chen, A.F. Ductile PLA Nanocomposites with Improved Thermal Stability. *Compos. Part A Appl. Sci. Manuf.* **2012**, *43*, 2289–2295. [CrossRef]
26. Alikarami, N.; Abrisham, M.; Huang, X.; Panahi-Sarmad, M.; Zhang, K.; Dong, K.; Xiao, X. Compatibilization of PLA Grafted Maleic Anhydrate through Blending of Thermoplastic Starch (TPS) and Nanoclay Nanocomposites for the Reduction of Gas Permeability. *Int. J. Smart Nano Mater.* **2022**, *13*, 130–151. [CrossRef]
27. Norazlina, H.; Kamal, Y. Graphene Modifications in Polylactic Acid Nanocomposites: A Review. *Polym. Bull.* **2015**, *72*, 931–961. [CrossRef]
28. Vassiliou, A.A.; Bikiaris, D.; El Mabrouk, K.; Kontopoulou, M. Effect of Evolved Interactions in Poly(Butylene Succinate)/Fumed Silica Biodegradable In Situ Prepared Nanocomposites on Molecular Weight, Material Properties, and Biodegradability. *J. Appl. Polym. Sci.* **2010**, *116*, 2658–2667. [CrossRef]
29. Bikiaris, D.; Karavelidis, V.; Karayannidis, G. A New Approach to Prepare Poly(Ethylene Terephthalate)/Silica Nanocomposites with Increased Molecular Weight and Fully Adjustable Branching or Crosslinking by SSP. *Macromol. Rapid Commun.* **2006**, *27*, 1199–1205. [CrossRef]
30. Manju, P.; Krishnan, P.S.G.; Nayak, S.K. In-Situ Polymerised PLA-SEP Bionanocomposites: Effect of Silanol Groups on the Properties of PLA. *J. Polym. Res.* **2020**, *27*, 134. [CrossRef]
31. Yang, J.H.; Lin, S.H.; Lee, Y.D. Preparation and Characterization of Poly(l-Lactide)-Graphene Composites Using the in Situ Ring-Opening Polymerization of PLLA with Graphene as the Initiator. *J. Mater. Chem.* **2012**, *22*, 10805–10815. [CrossRef]
32. Song, W.; Zheng, Z.; Tang, W.; Wang, X. A Facile Approach to Covalently Functionalized Carbon Nanotubes with Biocompatible Polymer. *Polymer* **2007**, *48*, 3658–3663. [CrossRef]
33. Acevedo, J.; Escobar, G.; López, D. Functionalized Multiwalled Carbon Nanotubes for the Improvement of Dispersion Stability in Diesel Fuel-Type for Potential Application as Nano Additives. *ChemistrySelect* **2022**, *7*, e202202626. [CrossRef]
34. Feng, S.; Zhang, F.; Ahmed, S.; Liu, Y. Physico-Mechanical and Antibacterial Properties of PLA/$TiO_2$ Composite Materials Synthesized via Electrospinning and Solution Casting Processes. *Coatings* **2019**, *9*, 525. [CrossRef]
35. Fu, L.; Wu, F.; Xu, C.; Cao, T.; Wang, R.; Guo, S. Anisotropic Shape Memory Behaviors of Polylactic Acid/Citric Acid– Bentonite Composite with a Gradient Filler Concentration in Thickness Direction. *Ind. Eng. Chem. Res.* **2018**, *57*, 6265–6274. [CrossRef]
36. Lawal, A.T. Recent Progress in Graphene Based Polymer Nanocomposites. *Cogent Chem.* **2020**, *6*, 1833476. [CrossRef]
37. Tsioptsias, C.; Leontiadis, K.; Tzimpilis, E.; Tsivintzelis, I. Polypropylene Nanocomposite Fibers: A Review of Current Trends and New Developments. *J. Plast. Film Sheeting* **2021**, *37*, 283–311. [CrossRef]
38. Sinha Ray, S.; Yamada, K.; Okamoto, M.; Ueda, K. Control of Biodegradability of Polylactide via Nanocomposite Technology. *Macromol. Mater. Eng.* **2003**, *288*, 203–208. [CrossRef]
39. Botta, L.; Scaffaro, R.; Sutera, F.; Mistretta, M.C. Reprocessing of PLA/Graphene Nanoplatelets Nanocomposites. *Polymers* **2018**, *10*, 18. [CrossRef]
40. Yu, W.; Sisi, L.; Haiyan, Y.; Jie, L. Progress in the Functional Modification of Graphene/Graphene Oxide: A Review. *RSC Adv.* **2020**, *10*, 15328–15345. [CrossRef]
41. Murariu, M.; Dechief, A.L.; Bonnaud, L.; Paint, Y.; Gallos, A.; Fontaine, G.; Bourbigot, S.; Dubois, P. The Production and Properties of Polylactide Composites Filled with Expanded Graphite. *Polym. Degrad. Stab.* **2010**, *95*, 889–900. [CrossRef]
42. Bikiaris, N.D.; Koumentakou, I.; Michailidou, G.; Kostoglou, M.; Vlachou, M.; Barmpalexis, P.; Karavas, E.; Papageorgiou, G.Z. Investigation of Molecular Weight, Polymer Concentration and Process Parameters Factors on the Sustained Release of the Anti-Multiple-Sclerosis Agent Teriflunomide from Poly(ε-Caprolactone) Electrospun Nanofibrous Matrices. *Pharmaceutics* **2022**, *14*, 1693. [CrossRef] [PubMed]

43. Barkoula, N.M.; Alcock, B.; Cabrera, N.O.; Peijs, T. Flame-Retardancy Properties of Intumescent Ammonium Poly(Phosphate) and Mineral Filler Magnesium Hydroxide in Combination with Graphene. *Polym. Polym. Compos.* **2008**, *16*, 101–113.
44. Kim, G.H.; Cho, Y.S.; Kim, W.D. Stability Analysis for Multi-Jets Electrospinning Process Modified with a Cylindrical Electrode. *Eur. Polym. J.* **2006**, *42*, 2031–2038. [CrossRef]
45. SalehHudin, H.S.; Mohamad, E.N.; Mahadi, W.N.L.; Afifi, A.M. Simulation and Experimental Study of Parameters in Multiple-Nozzle Electrospinning: Effects of Nozzle Arrangement on Jet Paths and Fiber Formation. *J. Manuf. Process.* **2021**, *62*, 440–449. [CrossRef]
46. Touny, A.H.; Lawrence, J.G.; Jones, A.D.; Bhaduri, S.B. Effect of Electrospinning Parameters on the Characterization of PLA/HNT Nanocomposite Fibers. *J. Mater. Res.* **2010**, *25*, 857–865. [CrossRef]
47. Shi, Q.; Zhou, C.; Yue, Y.; Guo, W.; Wu, Y.; Wu, Q. Mechanical Properties and in Vitro Degradation of Electrospun Bio-Nanocomposite Mats from PLA and Cellulose Nanocrystals. *Carbohydr. Polym.* **2012**, *90*, 301–308. [CrossRef]
48. Liu, C.; Wong, H.M.; Yeung, K.W.K.; Tjong, S.C. Novel Electrospun Polylactic Acid Nanocomposite Fiber Mats with Hybrid Graphene Oxide and Nanohydroxyapatite Reinforcements Having Enhanced Biocompatibility. *Polymers* **2016**, *8*, 287. [CrossRef]
49. Li, M.; Mondrinos, M.J.; Chen, X.; Gandhi, M.R.; Ko, F.K.; Lelkes, P.I. Elastin Blends for Tissue Engineering Scaffolds. *J. Biomed. Mater. Res. Part A* **2006**, *79*, 963–973. [CrossRef]
50. Zhang, R.; Lan, W.; Ji, T.; Sameen, D.E.; Ahmed, S.; Qin, W.; Liu, Y. Development of Polylactic Acid/ZnO Composite Membranes Prepared by Ultrasonication and Electrospinning for Food Packaging. *Lwt* **2021**, *135*, 110072. [CrossRef]
51. Xiang, C.; Joo, Y.L.; Frey, M.W. Nanocomposite Fibers Electrospun from Poly(Lactic Acid)/Cellulose Nanocrystals. *J. Biobased Mater. Bioenergy* **2009**, *3*, 147–155. [CrossRef]
52. Alam, F.; Varadarajan, K.M.; Kumar, S. 3D Printed Polylactic Acid Nanocomposite Scaffolds for Tissue Engineering Applications. *Polym. Test.* **2020**, *81*, 106203. [CrossRef]
53. Alam, F.; Shukla, V.R.; Varadarajan, K.M.; Kumar, S. Microarchitected 3D Printed Polylactic Acid (PLA) Nanocomposite Scaffolds for Biomedical Applications. *J. Mech. Behav. Biomed. Mater.* **2020**, *103*, 103576. [CrossRef] [PubMed]
54. Bhagia, S.; Bornani, K.; Agarwal, R.; Satlewal, A.; Ďurkovič, J.; Lagaňa, R.; Bhagia, M.; Yoo, C.G.; Zhao, X.; Kunc, V.; et al. Critical Review of FDM 3D Printing of PLA Biocomposites Filled with Biomass Resources, Characterization, Biodegradability, Upcycling and Opportunities for Biorefineries. *Appl. Mater. Today* **2021**, *24*, 101078. [CrossRef]
55. Wang, W.; Zhang, B.; Li, M.; Li, J.; Zhang, C.; Han, Y.; Wang, L.; Wang, K.; Zhou, C.; Liu, L.; et al. 3D Printing of PLA/n-HA Composite Scaffolds with Customized Mechanical Properties and Biological Functions for Bone Tissue Engineering. *Compos. Part B Eng.* **2021**, *224*, 109192. [CrossRef]
56. Yang, L.; Li, S.; Zhou, X.; Liu, J.; Li, Y.; Yang, M.; Yuan, Q.; Zhang, W. Effects of Carbon Nanotube on the Thermal, Mechanical, and Electrical Properties of PLA/CNT Printed Parts in the FDM Process. *Synth. Met.* **2019**, *253*, 122–130. [CrossRef]
57. Bustillos, J.; Montero, D.; Nautiyal, P.; Loganathan, A.; Boesl, B.; Agarwal, A. Integration of Graphene in Poly(Lactic) Acid by 3D Printing to Develop Creep and Wear-Resistant Hierarchical Nanocomposites. *Polym. Polym. Compos.* **2008**, *16*, 101–113. [CrossRef]
58. Demchenko, V.; Kobylinskyi, S.; Iurzhenko, M.; Riabov, S.; Vashchuk, A.; Rybalchenko, N.; Zahorodnia, S.; Naumenko, K.; Demchenko, O.; Adamus, G.; et al. Nanocomposites Based on Polylactide and Silver Nanoparticles and Their Antimicrobial and Antiviral Applications. *React. Funct. Polym.* **2022**, *170*, 105096. [CrossRef]
59. Shifrina, Z.B.; Matveeva, V.G.; Bronstein, L.M. Role of Polymer Structures in Catalysis by Transition Metal and Metal Oxide Nanoparticle Composites. *Chem. Rev.* **2020**, *120*, 1350–1396. [CrossRef]
60. Prasanna, S.R.V.S.; Pandey, S.; Balaji, K.; Rana, S. Metal Oxide Based Nanomaterials and Their Polymer Nanocomposites. In *Nanomaterials and Polymer Nanocomposites*; Elsevier Inc.: Amsterdam, The Netherlands, 2019. [CrossRef]
61. Kim, J.S.; Kuk, E.; Yu, K.N.; Kim, J.H.; Park, S.J.; Lee, H.J.; Kim, S.H.; Park, Y.K.; Park, Y.H.; Hwang, C.Y.; et al. Antimicrobial Effects of Silver Nanoparticles. *Nanomed. Nanotechnol. Biol. Med.* **2007**, *3*, 95–101. [CrossRef]
62. Singh, J.; Dutta, T.; Kim, K.H.; Rawat, M.; Samddar, P.; Kumar, P. "Green" Synthesis of Metals and Their Oxide Nanoparticles: Applications for Environmental Remediation. *J. Nanobiotechnol.* **2018**, *16*, 84. [CrossRef] [PubMed]
63. Ramanujam, K.; Sundrarajan, M. Antibacterial Effects of Biosynthesized MgO Nanoparticles Using Ethanolic Fruit Extract of Emblica Officinalis. *J. Photochem. Photobiol. B Biol.* **2014**, *141*, 296–300. [CrossRef]
64. Anantharaman, A.; Sathyabhama, S.; George, M. Green Synthesis of Magnesium Oxide Nanoparticles Using Aloe Vera and Its Applications. *Int. J. Sci. Res. Dev.* **2016**, *4*, 2321–2613.
65. Karande, S.D.; Jadhav, S.A.; Garud, H.B.; Kalantre, V.A.; Burungale, S.H.; Patil, P.S. Green and Sustainable Synthesis of Silica Nanoparticles. *Nanotechnol. Environ. Eng.* **2021**, *6*, 29. [CrossRef]
66. Sankar, R.; Rizwana, K.; Shivashangari, K.S.; Ravikumar, V. Ultra-Rapid Photocatalytic Activity of Azadirachta Indica Engineered Colloidal Titanium Dioxide Nanoparticles. *Appl. Nanosci.* **2015**, *5*, 731–736. [CrossRef]
67. Dobrucka, R. Synthesis of Titanium Dioxide Nanoparticles Using Echinacea Purpurea Herba. *Iran. J. Pharm. Res.* **2017**, *16*, 753–759.
68. Jayaseelan, C.; Rahuman, A.A.; Roopan, S.M.; Kirthi, A.V.; Venkatesan, J.; Kim, S.K.; Iyappan, M.; Siva, C. Biological Approach to Synthesize $TiO_2$ Nanoparticles Using Aeromonas Hydrophila and Its Antibacterial Activity. *Spectrochim. Acta Part A Mol. Biomol. Spectrosc.* **2013**, *107*, 82–89. [CrossRef]

69. Roopan, S.M.; Bharathi, A.; Prabhakarn, A.; Abdul Rahuman, A.; Velayutham, K.; Rajakumar, G.; Padmaja, R.D.; Lekshmi, M.; Madhumitha, G. Efficient Phyto-Synthesis and Structural Characterization of Rutile $TiO_2$ Nanoparticles Using Annona Squamosa Peel Extract. *Spectrochim. Acta Part A Mol. Biomol. Spectrosc.* **2012**, *98*, 86–90. [CrossRef]
70. Khan, R.; Fulekar, M.H. Biosynthesis of Titanium Dioxide Nanoparticles Using Bacillus Amyloliquefaciens Culture and Enhancement of Its Photocatalytic Activity for the Degradation of a Sulfonated Textile Dye Reactive Red 31. *J. Colloid Interface Sci.* **2016**, *475*, 184–191. [CrossRef]
71. Zahir, A.A.; Chauhan, I.S.; Bagavan, A.; Kamaraj, C.; Elango, G.; Shankar, J.; Arjaria, N.; Roopan, S.M.; Rahuman, A.A.; Singh, N. Green Synthesis of Silver and Titanium Dioxide Nanoparticles Using Euphorbia Prostrata Extract Shows Shift from Apoptosis to G0/G1 Arrest Followed by Necrotic Cell Death in Leishmania Donovani. *Antimicrob. Agents Chemother.* **2015**, *59*, 4782–4799. [CrossRef]
72. Sai Saraswathi, V.; Tatsugi, J.; Shin, P.K.; Santhakumar, K. Facile Biosynthesis, Characterization, and Solar Assisted Photocatalytic Effect of ZnO Nanoparticles Mediated by Leaves of *L. speciosa*. *J. Photochem. Photobiol. B Biol.* **2017**, *167*, 89–98. [CrossRef] [PubMed]
73. Yuvakkumar, R.; Suresh, J.; Nathanael, A.J.; Sundrarajan, M.; Hong, S.I. Novel Green Synthetic Strategy to Prepare ZnO Nanocrystals Using Rambutan (*Nephelium lappaceum* L.) Peel Extract and Its Antibacterial Applications. *Mater. Sci. Eng. C* **2014**, *41*, 17–27. [CrossRef] [PubMed]
74. Sharma, D.; Sabela, M.I.; Kanchi, S.; Mdluli, P.S.; Singh, G.; Stenström, T.A.; Bisetty, K. Biosynthesis of ZnO Nanoparticles Using Jacaranda Mimosifolia Flowers Extract: Synergistic Antibacterial Activity and Molecular Simulated Facet Specific Adsorption Studies. *J. Photochem. Photobiol. B Biol.* **2016**, *162*, 199–207. [CrossRef] [PubMed]
75. Lingaraju, K.; Raja Naika, H.; Manjunath, K.; Basavaraj, R.B.; Nagabhushana, H.; Nagaraju, G.; Suresh, D. Biogenic Synthesis of Zinc Oxide Nanoparticles Using *Ruta graveolens* (L.) and Their Antibacterial and Antioxidant Activities. *Appl. Nanosci.* **2016**, *6*, 703–710. [CrossRef]
76. Matinise, N.; Fuku, X.G.; Kaviyarasu, K.; Mayedwa, N.; Maaza, M. ZnO Nanoparticles via Moringa Oleifera Green Synthesis: Physical Properties & Mechanism of Formation. *Appl. Surf. Sci.* **2017**, *406*, 339–347. [CrossRef]
77. Nagajyothi, P.C.; Cha, S.J.; Yang, I.J.; Sreekanth, T.V.M.; Kim, K.J.; Shin, H.M. Antioxidant and Anti-Inflammatory Activities of Zinc Oxide Nanoparticles Synthesized Using Polygala Tenuifolia Root Extract. *J. Photochem. Photobiol. B Biol.* **2015**, *146*, 10–17. [CrossRef] [PubMed]
78. Elavarasan, N.; Kokila, K.; Inbasekar, G.; Sujatha, V. Evaluation of Photocatalytic Activity, Antibacterial and Cytotoxic Effects of Green Synthesized ZnO Nanoparticles by Sechium Edule Leaf Extract. *Res. Chem. Intermed.* **2017**, *43*, 3361–3376. [CrossRef]
79. Rahimzadeh, C.Y.; Barzinjy, A.A.; Mohammed, A.S.; Hamad, S.M. Green Synthesis of $SiO_2$ Nanoparticles from Rhus Coriaria L. Extract: Comparison with Chemically Synthesized $SiO_2$ Nanoparticles. *PLoS ONE* **2022**, *17*, e0268184. [CrossRef]
80. Rezaeian, M.; Afjoul, H.; Shamloo, A.; Maleki, A.; Afjoul, N. Green Synthesis of Silica Nanoparticles from Olive Residue and Investigation of Their Anticancer Potential. *Nanomedicine* **2021**, *16*, 1581–1593. [CrossRef]
81. Arshad, A.N.; Wahid, M.H.M.; Rusop, M.; Majid, W.H.A.; Subban, R.H.Y.; Rozana, M.D. Dielectric and Structural Properties of Poly(Vinylidene Fluoride) (PVDF) and Poly(Vinylidene Fluoride-Trifluoroethylene) (PVDF-TrFE) Filled with Magnesium Oxide Nanofillers. *J. Nanomater.* **2019**, *2019*, 5961563. [CrossRef]
82. Das, B.; Moumita, S.; Ghosh, S.; Khan, M.I.; Indira, D.; Jayabalan, R.; Tripathy, S.K.; Mishra, A.; Balasubramanian, P. Biosynthesis of Magnesium Oxide (MgO) Nanoflakes by Using Leaf Extract of Bauhinia Purpurea and Evaluation of Its Antibacterial Property against Staphylococcus Aureus. *Mater. Sci. Eng. C* **2018**, *91*, 436–444. [CrossRef] [PubMed]
83. Leung, Y.H.; Ng, A.M.C.; Xu, X.; Shen, Z.; Gethings, L.A.; Wong, M.T.; Chan, C.M.N.; Guo, M.Y.; Ng, Y.H.; Djurišić, A.B.; et al. Mechanisms of Antibacterial Activity of Mgo: Non-Ros Mediated Toxicity of Mgo Nanoparticles towards *Escherichia coli*. *Small* **2014**, *10*, 1171–1183. [CrossRef] [PubMed]
84. Nassar, M.Y.; Mohamed, T.Y.; Ahmed, I.S.; Samir, I. MgO Nanostructure via a Sol-Gel Combustion Synthesis Method Using Different Fuels: An Efficient Nano-Adsorbent for the Removal of Some Anionic Textile Dyes. *J. Mol. Liq.* **2017**, *225*, 730–740. [CrossRef]
85. Halder, R.; Bandyopadhyay, S. Synthesis and Optical Properties of Anion Deficient Nano MgO. *J. Alloys Compd.* **2017**, *693*, 534–542. [CrossRef]
86. Look, D.C. Recent Advances in ZnO Materials and Devices. *Mater. Sci. Eng.* **2001**, *80*, 383–387. [CrossRef]
87. Sharma, P.; Hasan, M.R.; Mehto, N.K.; Deepak; Bishoyi, A.; Narang, J. 92 Years of Zinc Oxide: Has Been Studied by the Scientific Community since the 1930s- An Overview. *Sensors Int.* **2022**, *3*, 100182. [CrossRef]
88. Talam, S.; Karumuri, S.R.; Gunnam, N. Synthesis, Characterization, and Spectroscopic Properties of ZnO Nanoparticles. *ISRN Nanotechnol.* **2012**, *2012*, 372505. [CrossRef]
89. Arya, S.; Mahajan, P.; Mahajan, S.; Khosla, A.; Datt, R.; Gupta, V.; Young, S.-J.; Oruganti, S.K. Review—Influence of Processing Parameters to Control Morphology and Optical Properties of Sol-Gel Synthesized ZnO Nanoparticles. *ECS J. Solid State Sci. Technol.* **2021**, *10*, 023002. [CrossRef]
90. Gudkov, S.V.; Burmistrov, D.E.; Serov, D.A.; Rebezov, M.B.; Semenova, A.A.; Lisitsyn, A.B. A Mini Review of Antibacterial Properties of ZnO Nanoparticles. *Front. Phys.* **2021**, *9*, 641481. [CrossRef]
91. Li, S.; Chen, G.; Qiang, S.; Yin, Z.; Zhang, Z.; Chen, Y. Synthesis and Evaluation of Highly Dispersible and Efficient Photocatalytic $TiO_2$/Poly Lactic Acid Nanocomposite Films via Sol-Gel and Casting Processes. *Int. J. Food Microbiol.* **2020**, *331*, 108763. [CrossRef]

92. Zorah, M.; Mustapa, I.R.; Daud, N.; Nahida, J.H.; Sudin, N.A.S.; Majhool, A.A.; Mahmoudi, E. Improvement Thermomechanical Properties of Polylactic Acid via Titania Nanofillers Reinforcement. *J. Adv. Res. Fluid Mech. Therm. Sci.* **2020**, *70*, 97–111. [CrossRef]
93. González, E.A.S.; Olmos, D.; Lorente, M.A.; Vélaz, I.; González-Benito, J. Preparation and Characterization of Polymer Composite Materials Based on PLA/$TiO_2$ for Antibacterial Packaging. *Polymers* **2018**, *10*, 1365. [CrossRef] [PubMed]
94. Buzarovska, A.; Dinescu, S.; Chitoiu, L.; Costache, M. Porous Poly(l-Lactic Acid) Nanocomposite Scaffolds with Functionalized $TiO_2$ Nanoparticles: Properties, Cytocompatibility and Drug Release Capability. *J. Mater. Sci.* **2018**, *53*, 11151–11166. [CrossRef]
95. Mallick, S.; Ahmad, Z.; Touati, F.; Bhadra, J.; Shakoor, R.A.; Al-Thani, N.J. PLA-$TiO_2$ Nanocomposites: Thermal, Morphological, Structural, and Humidity Sensing Properties. *Ceram. Int.* **2018**, *44*, 16507–16513. [CrossRef]
96. Luo, Y.; Lin, Z.; Guo, G. Biodegradation Assessment of Poly (Lactic Acid) Filled with Functionalized Titania Nanoparticles (PLA/$TiO_2$) under Compost Conditions. *Nanoscale Res. Lett.* **2019**, *14*, 56. [CrossRef] [PubMed]
97. Shebi, A.; Lisa, S. Evaluation of Biocompatibility and Bactericidal Activity of Hierarchically Porous PLA-$TiO_2$ Nanocomposite Films Fabricated by Breath-Figure Method. *Mater. Chem. Phys.* **2019**, *230*, 308–318. [CrossRef]
98. Xie, J.; Hung, Y.C. UV-A Activated $TiO_2$ Embedded Biodegradable Polymer Film for Antimicrobial Food Packaging Application. *Lwt* **2018**, *96*, 307–314. [CrossRef]
99. Ismael, M. Latest Progress on the Key Operating Parameters Affecting the Photocatalytic Activity of $TiO_2$-Based Photocatalysts for Hydrogen Fuel Production: A Comprehensive Review. *Fuel* **2021**, *303*, 121207. [CrossRef]
100. Foster, H.A.; Ditta, I.B.; Varghese, S.; Steele, A. Photocatalytic Disinfection Using Titanium Dioxide: Spectrum and Mechanism of Antimicrobial Activity. *Appl. Microbiol. Biotechnol.* **2011**, *90*, 1847–1868. [CrossRef]
101. Ling, Y.; Zhang, P.; Wang, J.; Chen, Y. Effect of PVA Fiber on Mechanical Properties of Cementitious Composite with and without Nano-$SiO_2$. *Constr. Build. Mater.* **2019**, *229*, 117068. [CrossRef]
102. Papadopoulos, L.; Klonos, P.A.; Terzopoulou, Z.; Psochia, E.; Sanusi, O.M.; Hocine, N.A.; Benelfellah, A.; Giliopoulos, D.; Triantafyllidis, K.; Kyritsis, A.; et al. Comparative Study of Crystallization, Semicrystalline Morphology, and Molecular Mobility in Nanocomposites Based on Polylactide and Various Inclusions at Low Filler Loadings. *Polymer* **2021**, *217*, 123457. [CrossRef]
103. Charitos, I.; Klonos, P.A.; Kyritsis, A.; Koralli, P.; Kontos, A.G.; Kontou, E. Thermomechanical Performance of Biodegradable Poly (Lactic Acid)/Carbonaceous Hybrid Nanocomposites: Comparative Study. *Polym. Compos.* **2022**, *43*, 1900–1915. [CrossRef]
104. Klonos, P.; Terzopoulou, Z.; Koutsoumpis, S.; Zidropoulos, S.; Kripotou, S.; Papageorgiou, G.Z.; Bikiaris, D.N.; Kyritsis, A.; Pissis, P. Rigid Amorphous Fraction and Segmental Dynamics in Nanocomposites Based on Poly(L–Lactic Acid) and Nano-Inclusions of 1–3D Geometry Studied by Thermal and Dielectric Techniques. *Eur. Polym. J.* **2016**, *82*, 16–34. [CrossRef]
105. Bhattacharya, M. Polymer Nanocomposites-A Comparison between Carbon Nanotubes, Graphene, and Clay as Nanofillers. *Materials* **2016**, *9*, 262. [CrossRef] [PubMed]
106. Scaffaro, R.; Maio, A. Integrated Ternary Bionanocomposites with Superior Mechanical Performance via the Synergistic Role of Graphene and Plasma Treated Carbon Nanotubes. *Compos. Part B Eng.* **2019**, *168*, 550–559. [CrossRef]
107. Valdes, A.; Qu, Z.W.; Kroes, G.J.; Rossmeisl, J.; Nørskov, J.K. Oxidation and Photo-Oxidation of Water on $TiO_2$ Surface. *J. Phys. Chem. C* **2008**, *112*, 9872–9879. [CrossRef]
108. Ivanov, E.; Kotsilkova, R.; Xia, H.; Chen, Y.; Donato, R.K.; Donato, K.; Godoy, A.P.; Di Maio, R.; Silvestre, C.; Cimmino, S.; et al. PLA/Graphene/MWCNT Composites with Improved Electrical and Thermal Properties Suitable for FDM 3D Printing Applications. *Appl. Sci.* **2019**, *9*, 1209. [CrossRef]
109. Norizan, M.N.; Moklis, M.H.; Ngah Demon, S.Z.; Halim, N.A.; Samsuri, A.; Mohamad, I.S.; Knight, V.F.; Abdullah, N. Carbon Nanotubes: Functionalisation and Their Application in Chemical Sensors. *RSC Adv.* **2020**, *10*, 43704–43732. [CrossRef]
110. Tang, R.; Shi, Y.; Hou, Z.; Wei, L. Carbon Nanotube-Based Chemiresistive Sensors. *Sensors* **2017**, *17*, 882. [CrossRef]
111. Clemons, C.B.; Roberts, M.W.; Wilber, J.P.; Young, G.W.; Buldum, A.; Quinn, D.D. Continuum Plate Theory and Atomistic Modeling to Find the Flexural Rigidity of a Graphene Sheet Interacting with a Substrate. *J. Nanotechnol.* **2010**, *2010*, 868492. [CrossRef]
112. Porto, L.S.; Silva, D.N.; de Oliveira, A.E.F.; Pereira, A.C.; Borges, K.B. Carbon Nanomaterials: Synthesis and Applications to Development of Electrochemical Sensors in Determination of Drugs and Compounds of Clinical Interest. *Rev. Anal. Chem.* **2020**, *38*, 1–16. [CrossRef]
113. Milani, M.A.; Quijada, R.; Basso, N.R.S.; Graebin, A.P.; Galland, G.B. Influence of the Graphite Type on the Synthesis of Polypropylene/Graphene Nanocomposites. *J. Polym. Sci. Part A Polym. Chem.* **2012**, *50*, 3598–3605. [CrossRef]
114. Papageorgiou, G.Z.; Terzopoulou, Z.; Bikiaris, D.; Triantafyllidis, K.S.; Diamanti, E.; Gournis, D.; Klonos, P.; Giannoulidis, E.; Pissis, P. Evaluation of the Formed Interface in Biodegradable Poly(l-Lactic Acid)/Graphene Oxide Nanocomposites and the Effect of Nanofillers on Mechanical and Thermal Properties. *Thermochim. Acta* **2014**, *597*, 48–57. [CrossRef]
115. Klonos, P.; Kripotou, S.; Kyritsis, A.; Papageorgiou, G.Z.; Bikiaris, D.; Gournis, D.; Pissis, P. Glass Transition and Segmental Dynamics in Poly(l-Lactic Acid)/Graphene Oxide Nanocomposites. *Thermochim. Acta* **2015**, *617*, 44–53. [CrossRef]
116. Klonos, P.A.; Peoglos, V.; Bikiaris, D.N.; Kyritsis, A. Rigid Amorphous Fraction and Thermal Diffusivity in Nanocomposites Based on Poly(l -Lactic Acid) Filled with Carbon Nanotubes and Graphene Oxide. *J. Phys. Chem. C* **2020**, *124*, 5469–5479. [CrossRef]
117. Trivedi, D.N.; Rachchh, N.V. Graphene and Its Application in Thermoplastic Polymers as Nano-Filler- A Review. *Polymer* **2022**, *240*, 124486. [CrossRef]
118. Angelova, P. Mechanical and Thermal Properties of PLA Based Nanocomposites with Graphene and Carbon Nanotubes. *J. Theor. Appl. Mech.* **2019**, *49*, 241–256. [CrossRef]

119. Msibi, M.; Mashamba, A.; Mashinini, P.M.; Hashe, V.T. 3D Printing of Carbon Nanofiber-PLA Composite. In Proceedings of the 11th South African Conference Computational and Applied Mechanics, SACAM 2018, Vanderbijlpark, South Africa, 17–19 September 2018; pp. 670–679.
120. Díez-Pascual, A.M. State of the Art in the Antibacterial and Antiviral Applications of Carbon-Based Polymeric Nanocomposites. *Int. J. Mol. Sci.* **2021**, *22*, 10511. [CrossRef]
121. Giannopoulos, G.I. Linking MD and FEM to Predict the Mechanical Behaviour of Fullerene Reinforced Nylon-12. *Compos. Part B Eng.* **2019**, *161*, 455–463. [CrossRef]
122. Kausar, A. Polymer/Fullerene Nanocomposite Coatings—Front Line Potential. *Emergent Mater.* **2022**, *5*, 29–40. [CrossRef]
123. Pan, Y.; Guo, Z.; Ran, S.; Fang, Z. Influence of Fullerenes on the Thermal and Flame-Retardant Properties of Polymeric Materials. *J. Appl. Polym. Sci.* **2020**, *137*, 47538. [CrossRef]
124. Dhanaraj, A.; Das, K.; Keller, J.M. A Study of the Optical Band Gap Energy and Urbach Energy of Fullerene (C60) Doped PMMA Nanocomposites. In *AIP Conference Proceedings*; AIP Publishing LLC: Woodbury, NY, USA, 2020; p. 110040. [CrossRef]
125. Fraga-Corral, M.; García-Oliveira, P.; Pereira, A.G.; Lourenço-Lopes, C.; Jimenez-Lopez, C.; Prieto, M.A.; Simal-Gandara, J. Technological Application of Tannin-Based Extracts. *Molecules* **2020**, *25*, 614. [CrossRef] [PubMed]
126. Chinaglia, S.; Tosin, M.; Degli-Innocenti, F. Biodegradation Rate of Biodegradable Plastics at Molecular Level. *Polym. Degrad. Stab.* **2018**, *147*, 237–244. [CrossRef]
127. Haslam, E. Vegetable Tannins. In *Biochemistry of Plant Phenolics*; Plenum Press: New York, NY, USA, 1979.
128. Smeriglio, A.; Barreca, D.; Bellocco, E.; Trombetta, D. Proanthocyanidins and Hydrolysable Tannins: Occurrence, Dietary Intake and Pharmacological Effects. *Br. J. Pharmacol.* **2017**, *174*, 1244–1262. [CrossRef]
129. Cerro, D.; Bustos, G.; Villegas, C.; Buendia, N.; Truffa, G.; Godoy, M.P.; Rodríguez, F.; Rojas, A.; Galotto, M.J.; Constandil, L.; et al. Effect of Supercritical Incorporation of Cinnamaldehyde on Physical-Chemical Properties, Disintegration and Toxicity Studies of PLA/Lignin Nanocomposites. *Int. J. Biol. Macromol.* **2021**, *167*, 255–266. [CrossRef]
130. Cavallo, E.; He, X.; Luzi, F.; Dominici, F.; Cerrutti, P.; Bernal, C.; Foresti, M.L.; Torre, L.; Puglia, D. UV Protective, Antioxidant, Antibacterial and Compostable Polylactic Acid Composites Containing Pristine and Chemically Modified Lignin Nanoparticles. *Molecules* **2021**, *26*, 126. [CrossRef]
131. Patanair, B.; Saiter-Fourcin, A.; Thomas, S.; Thomas, M.G.; Pundarikashan, P.P.; Nair, K.G.; Kumar, V.K.; Maria, H.J.; Delpouve, N. Promoting Interfacial Interactions with the Addition of Lignin in Poly(Lactic Acid) Hybrid Nanocomposites. *Polymers* **2021**, *13*, 272. [CrossRef]
132. Bikiaris, N.D.; Koumentakou, I.; Lykidou, S.; Nikolaidis, N. Innovative Skin Product O/W Emulsions Containing Lignin, Multiwall Carbon Nanotubes and Graphene Oxide Nanoadditives with Enhanced Sun Protection Factor and UV Stability Properties. *Appl. Nano* **2022**, *3*, 1–15. [CrossRef]
133. Groh, K.J.; Backhaus, T.; Carney-Almroth, B.; Geueke, B.; Inostroza, P.A.; Lennquist, A.; Leslie, H.A.; Maffini, M.; Slunge, D.; Trasande, L.; et al. Overview of Known Plastic Packaging-Associated Chemicals and Their Hazards. *Sci. Total Environ.* **2019**, *651*, 3253–3268. [CrossRef]
134. Suppakul, P.; Miltz, J.; Sonneveld, K.; Bigger, S.W. Active Packaging Technologies with an Emphasis on Antimicrobial Packaging and Its Applications. *J. Food Sci.* **2003**, *68*, 408–420. [CrossRef]
135. Karunarathna, M.S.; Smith, R.C. Valorization of Lignin as a Sustainable Component of Structural Materials and Composites: Advances from 2011 to 2019. *Sustainability* **2020**, *12*, 734. [CrossRef]
136. Das, A.K.; Islam, M.N.; Faruk, M.O.; Ashaduzzaman, M.; Dungani, R. Review on Tannins: Extraction Processes, Applications and Possibilities. *S. Afr. J. Bot.* **2020**, *135*, 58–70. [CrossRef]
137. Chang, Z.; Zhang, Q.; Liang, W.; Zhou, K.; Jian, P.; She, G.; Zhang, L. A Comprehensive Review of the Structure Elucidation of Tannins from Terminalia Linn. *Evid. Based Complement. Altern. Med.* **2019**, *2019*, 8623909. [CrossRef] [PubMed]
138. Pizzi, A. Tannins: Prospectives and Actual Industrial Applications. *Biomolecules* **2019**, *9*, 344. [CrossRef] [PubMed]
139. Liao, J.; Brosse, N.; Hoppe, S.; Du, G.; Zhou, X.; Pizzi, A. One-Step Compatibilization of Poly(Lactic Acid) and Tannin via Reactive Extrusion. *Mater. Des.* **2020**, *191*, 108603. [CrossRef]
140. Bridson, J.H.; Kaur, J.; Zhang, Z.; Donaldson, L.; Fernyhough, A. Polymeric Flavonoids Processed with Co-Polymers as UV and Thermal Stabilisers for Polyethylene Films. *Polym. Degrad. Stab.* **2015**, *122*, 18–24. [CrossRef]
141. Shnawa, H.A.; Khaleel, M.I.; Muhamed, F.J. Oxidation of HDPE in the Presence of PVC Grafted with Natural Polyphenols (Tannins) as Antioxidant. *Open J. Polym. Chem.* **2015**, *05*, 9–16. [CrossRef]
142. Liao, J.; Brosse, N.; Pizzi, A.; Hoppe, S. Dynamically Cross-Linked Tannin as a Reinforcement of Polypropylene and UV Protection Properties. *Polymers* **2019**, *11*, 102. [CrossRef]
143. Zhai, Y.; Wang, J.; Wang, H.; Song, T.; Hu, W.; Li, S. Preparation and Characterization of Antioxidative and UV-Protective Larch Bark Tannin/PVA Composite Membranes. *Molecules* **2018**, *23*, 2073. [CrossRef]
144. Cresnar, K.P.; Zamboulis, A.; Bikiaris, D.N.; Aulova, A.; Zemljic, L.F. Kraft Lignin/Tannin as a Potential Accelerator of Antioxidant and Antibacterial Properties in an Active Thermoplastic Polyester-Based Multifunctional Material. *Polymers* **2022**, *14*, 1532. [CrossRef]
145. Ainali, N.M.; Tarani, E.; Zamboulis, A.; Črešnar, K.P.; Zemljič, L.F.; Chrissafis, K.; Lambropoulou, D.A.; Bikiaris, D.N. Thermal Stability and Decomposition Mechanism of Pla Nanocomposites with Kraft Lignin and Tannin. *Polymers* **2021**, *13*, 2818. [CrossRef] [PubMed]

146. Črešnar, K.P.; Klonos, P.A.; Zamboulis, A.; Terzopoulou, Z.; Xanthopoulou, E.; Papadopoulos, L.; Kyritsis, A.; Kuzmič, K.; Zemljič, L.F.; Bikiaris, D.N. Structure-Properties Relationships in Renewable Composites Based on Polylactide Filled with Tannin and Kraft Lignin—Crystallization and Molecular Mobility. *Thermochim. Acta* **2021**, *703*, 178998. [CrossRef]
147. Li, T.; Chen, C.; Brozena, A.H.; Zhu, J.Y.; Xu, L.; Driemeier, C.; Dai, J.; Rojas, O.J.; Isogai, A.; Wågberg, L.; et al. Developing Fibrillated Cellulose as a Sustainable Technological Material. *Nature* **2021**, *590*, 47–56. [CrossRef] [PubMed]
148. Trache, D.; Tarchoun, A.F.; Derradji, M.; Hamidon, T.S.; Masruchin, N.; Brosse, N.; Hussin, M.H. Nanocellulose: From Fundamentals to Advanced Applications. *Front. Chem.* **2020**, *8*, 392. [CrossRef]
149. Xu, T.; Du, H.; Liu, H.; Liu, W.; Zhang, X.; Si, C.; Liu, P.; Zhang, K. Advanced Nanocellulose-Based Composites for Flexible Functional Energy Storage Devices. *Adv. Mater.* **2021**, *33*, 2101368. [CrossRef]
150. Rigotti, D.; Checchetto, R.; Tarter, S.; Caretti, D.; Rizzuto, M.; Fambri, L.; Pegoretti, A. Polylactic Acid-Lauryl Functionalized Nanocellulose Nanocomposites: Microstructural, Thermo-Mechanical and Gas Transport Properties. *Express Polym. Lett.* **2019**, *13*, 858–876. [CrossRef]
151. Shanmugam, K.; Doosthosseini, H.; Varanasi, S.; Garnier, G.; Batchelor, W. Nanocellulose Films as Air and Water Vapour Barriers: A Recyclable and Biodegradable Alternative to Polyolefin Packaging. *Sustain. Mater. Technol.* **2019**, *22*, e00115. [CrossRef]
152. Ahankari, S.S.; Subhedar, A.R.; Bhadauria, S.S.; Dufresne, A. Nanocellulose in Food Packaging: A Review. *Carbohydr. Polym.* **2021**, *255*, 117479. [CrossRef]
153. Wang, J.; Wang, S. Preparation, Modification and Environmental Application of Biochar: A Review. *J. Clean. Prod.* **2019**, *227*, 1002–1022. [CrossRef]
154. Weber, K.; Quicker, P. Properties of Biochar. *Fuel* **2018**, *217*, 240–261. [CrossRef]
155. Zhang, M.; Song, G.; Gelardi, D.L.; Huang, L.; Khan, E.; Mašek, O.; Parikh, S.J.; Ok, Y.S. Evaluating Biochar and Its Modifications for the Removal of Ammonium, Nitrate, and Phosphate in Water. *Water Res.* **2020**, *186*, 116303. [CrossRef] [PubMed]
156. Wang, L.; Ok, Y.S.; Tsang, D.C.W.; Alessi, D.S.; Rinklebe, J.; Mašek, O.; Bolan, N.S.; Hou, D. Biochar Composites: Emerging Trends, Field Successes and Sustainability Implications. *Soil Use Manag.* **2022**, *38*, 14–38. [CrossRef]
157. Lehmann, J.; Cowie, A.; Masiello, C.A.; Kammann, C.; Woolf, D.; Amonette, J.E.; Cazuela, M.L.; Camprs-Arbestain, M.; Whitman, T. Biochar in Climate Change Mitigation. *Nat. Geosci.* **2021**, *13*, 883–892. [CrossRef]
158. Naghdi, M.; Taheran, M.; Brar, S.K.; Rouissi, T.; Verma, M.; Surampalli, R.Y.; Valero, J.R. A Green Method for Production of Nanobiochar by Ball Milling- Optimization and Characterization. *J. Clean. Prod.* **2017**, *164*, 1394–1405. [CrossRef]
159. Song, B.; Cao, X.; Gao, W.; Aziz, S.; Gao, S.; Lam, C.H.; Lin, R. Preparation of Nano-Biochar from Conventional Biorefineries for High-Value Applications. *Renew. Sustain. Energy Rev.* **2022**, *157*, 112057. [CrossRef]
160. Lian, F.; Yu, W.; Zhou, Q.; Gu, S.; Wang, Z.; Xing, B.; Xing, B. Size Matters: Nano-Biochar Triggers Decomposition and Transformation Inhibition of Antibiotic Resistance Genes in Aqueous Environments. *Environ. Sci. Technol.* **2020**, *54*, 8821–8829. [CrossRef]
161. Anupama; Khare, P. A Comprehensive Evaluation of Inherent Properties and Applications of Nano-Biochar Prepared from Different Methods and Feedstocks. *J. Clean. Prod.* **2021**, *320*, 128759. [CrossRef]
162. Liu, G.; Zheng, H.; Jiang, Z.; Zhao, J.; Wang, Z.; Pan, B.; Xing, B. Formation and Physicochemical Characteristics of Nano Biochar: Insight into Chemical and Colloidal Stability. *Environ. Sci. Technol.* **2018**, *52*, 10369–10379. [CrossRef]
163. Barsoum, M. *Fundamental of Ceramics*; CRC Press: Boca Raton, FL, USA, 2019.
164. Chen, Z.; Li, Z.; Li, J.; Liu, C.; Lao, C.; Fu, Y.; Liu, C.; Li, Y.; Wang, P.; He, Y. 3D Printing of Ceramics: A Review. *J. Eur. Ceram. Soc.* **2019**, *39*, 661–687. [CrossRef]
165. Gupta, N.; Singh, A.; Dey, N.; Chattopadhyay, S.; Joseph, J.P.; Gupta, D.; Ganguli, M.; Pal, A. Pathway-Driven Peptide-Bioglass Nanocomposites as the Dynamic and Self-Healable Matrix. *Chem. Mater.* **2021**, *33*, 589–599. [CrossRef]
166. Dehnavi, N.; Parivar, K.; Goodarzi, V.; Salimi, A.; Nourani, M.R. Systematically Engineered Electrospun Conduit Based on PGA/Collagen/Bioglass Nanocomposites: The Evaluation of Morphological, Mechanical, and Bio-Properties. *Polym. Adv. Technol.* **2019**, *30*, 2192–2206. [CrossRef]
167. Canales, D.; Saavedra, M.; Flores, M.T.; Bejarano, J.; Ortiz, J.A.; Orihuela, P.; Alfaro, A.; Pabón, E.; Palza, H.; Zapata, P.A. Effect of Bioglass Nanoparticles on the Properties and Bioactivity of Poly(Lactic Acid) Films. *J. Biomed. Mater. Res. Part A* **2020**, *108*, 2032–2043. [CrossRef] [PubMed]
168. Zhang, R.; Hu, H.; Liu, Y.; Tan, J.; Chen, W.; Ying, C.; Liu, Q.; Fu, X.; Hu, S.; Wong, C.P. Homogeneously Dispersed Composites of Hydroxyapatite Nanorods and Poly(Lactic Acid) and Their Mechanical Properties and Crystallization Behavior. *Compos. Part A Appl. Sci. Manuf.* **2020**, *132*, 105841. [CrossRef]
169. Michael, F.M.; Ratnam, C.T.; Khalid, M.; Ramarad, S.; Walvekar, R. Surface Modification of Nanohydroxyapatite and Its Loading Effect on Polylactic Acid Properties for Load Bearing Implants. *Polym. Polym. Compos.* **2017**, *16*, 101–113. [CrossRef]
170. Szcześ, A.; Hołysz, L.; Chibowski, E. Synthesis of Hydroxyapatite for Biomedical Applications. *Adv. Colloid Interface Sci.* **2017**, *249*, 321–330. [CrossRef] [PubMed]
171. Du, M.; Chen, J.; Liu, K.; Xing, H.; Song, C. Recent Advances in Biomedical Engineering of Nano-Hydroxyapatite Including Dentistry, Cancer Treatment and Bone Repair. *Compos. Part B Eng.* **2021**, *215*, 108790. [CrossRef]
172. Wetteland, C.L.; Liu, H. Optical and Biological Properties of Polymer-Based Nanocomposites with Improved Dispersion of Ceramic Nanoparticles. *J. Biomed. Mater. Res. Part A* **2018**, *106*, 2692–2707. [CrossRef]

173. Ray, S.S. Recent Trends and Future Outlooks in the Field of Clay-Containing Polymer Nanocomposites. *Macromol. Chem. Phys.* **2014**, *215*, 1162–1179. [CrossRef]
174. Guo, F.; Aryana, S.; Han, Y.; Jiao, Y. A Review of the Synthesis and Applications of Polymer-Nanoclay Composites. *Appl. Sci.* **2018**, *8*, 1696. [CrossRef]
175. Rafiee, R.; Shahzadi, R. Mechanical Properties of Nanoclay and Nanoclay Reinforced Polymers: A Review. *Polym. Compos.* **2019**, *40*, 431–445. [CrossRef]
176. Jawaid, M.; Qaiss, A.E.K.; Bouhfi, R. *Nanoclay Reinforced Polymer Composites Natural Fibre/Nanoclay Hybrid Composites*; Springer: Singapore, 2016. [CrossRef]
177. Rhim, J.W.; Hong, S.I.; Ha, C.S. Tensile, Water Vapor Barrier and Antimicrobial Properties of PLA/Nanoclay Composite Films. *Lwt* **2009**, *42*, 612–617. [CrossRef]
178. Pluta, M.; Galeski, A.; Alexandre, M.; Paul, M.A.; Dubois, P. Polylactide/Montmorillonite Nanocomposites and Microcomposites Prepared by Melt Blending: Structure and Some Physical Properties. *J. Appl. Polym. Sci.* **2002**, *86*, 1497–1506. [CrossRef]
179. Re, G.L.; Benali, S.; Habibi, Y.; Raquez, J.M.; Dubois, P. Stereocomplexed PLA Nanocomposites: From in Situ Polymerization to Materials Properties. *Eur. Polym. J.* **2014**, *54*, 138–150. [CrossRef]
180. Othman, S.H.; Ling, H.N.; Talib, R.A.; Naim, M.N.; Risyon, N.P.; Saifullah, M. PLA/MMT and PLA/Halloysite Bio-Nanocomposite Films: Mechanical, Barrier, and Transparency. *J. Nano Res.* **2019**, *59*, 77–93. [CrossRef]
181. Morais, D.D.S.; Barbosa, R.; Medeiros, K.M.; Araújo, E.M.; Melo, T.J.A. Preparation of Poly(Lactic Acid)/Bentonite Clay Bionanocomposite. *Mater. Sci. Forum* **2014**, *775–776*, 233–237. [CrossRef]
182. Guo, F.; Liao, X.; Li, S.; Yan, Z.; Tang, W.; Li, G. Heat Insulating PLA/HNTs Foams with Enhanced Compression Performance Fabricated by Supercritical Carbon Dioxide. *J. Supercrit. Fluids* **2021**, *177*, 105344. [CrossRef]
183. Moazeni, N.; Mohamad, Z.; Dehbari, N. Study of Silane Treatment on Poly-Lactic Acid(PLA)/Sepiolite Nanocomposite Thin Films. *J. Appl. Polym. Sci.* **2015**, *132*, 1–8. [CrossRef]
184. Kasprzhitskii, A.; Lazorenko, G.; Kruglikov, A.; Kuchkina, I.; Gorodov, V. Effect of Silane Functionalization on Properties of Poly(Lactic Acid)/Palygorskite Nanocomposites. *Inorganics* **2021**, *9*, 3. [CrossRef]
185. Cardoso, R.S.; Aguiar, V.O.; Marques, M.D.F.V. Masterbatches of Polypropylene/Clay Obtained by in Situ Polymerization and Melt-Blended with Commercial Polypropylene. *J. Compos. Mater.* **2017**, *51*, 3547–3556. [CrossRef]
186. Ramesh, P.; Prasad, B.D.; Narayana, K.L. Effect of MMT Clay on Mechanical, Thermal and Barrier Properties of Treated Aloevera Fiber/PLA-Hybrid Biocomposites. *Silicon* **2020**, *12*, 1751–1760. [CrossRef]
187. Ramesh, P.; Prasad, B.D.; Narayana, K.L. Effect of Fiber Hybridization and Montmorillonite Clay on Properties of Treated Kenaf/Aloe Vera Fiber Reinforced PLA Hybrid Nanobiocomposite. *Cellulose* **2020**, *27*, 6977–6993. [CrossRef]
188. Othman, S.H.; Hassan, N.; Talib, R.A.; Kadir Basha, R.; Risyon, N.P. Mechanical and Thermal Properties of PLA/Halloysite Bio-Nanocomposite Films: Effect of Halloysite Nanoclay Concentration and Addition of Glycerol. *J. Polym. Eng.* **2017**, *37*, 381–389. [CrossRef]
189. Risyon, N.P.; Othman, S.H.; Basha, R.K.; Talib, R.A. Effect of Halloysite Nanoclay Concentration and Addition of Glycerol on Mechanical Properties of Bionanocomposite Films. *Polym. Polym. Compos.* **2016**, *24*, 795–802. [CrossRef]
190. Ramesh, P.; Prasad, B.D.; Narayana, K.L. Influence of Montmorillonite Clay Content on Thermal, Mechanical, Water Absorption and Biodegradability Properties of Treated Kenaf Fiber/PLA-Hybrid Biocomposites. *Silicon* **2021**, *13*, 109–118. [CrossRef]
191. Rao, R.U.; Venkatanarayana, B.; Suman, K.N.S. Enhancement of Mechanical Properties of PLA/PCL (80/20) Blend by Reinforcing with MMT Nanoclay. *Mater. Today Proc.* **2019**, *18*, 85–97. [CrossRef]
192. Montava-Jorda, S.; Chacon, V.; Lascano, D.; Sanchez-Nacher, L.; Montanes, N. Manufacturing and Characterization of Functionalized Aliphatic Polyester from Poly(Lactic Acid) with Halloysite Nanotubes. *Polymers* **2019**, *11*, 1314. [CrossRef]
193. Swaroop, C.; Shukla, M. Nano-Magnesium Oxide Reinforced Polylactic Acid Biofilms for Food Packaging Applications. *Int. J. Biol. Macromol.* **2018**, *113*, 729–736. [CrossRef]
194. Swaroop, C.; Shukla, M. Development of Blown Polylactic Acid-MgO Nanocomposite Films for Food Packaging. *Compos. Part A Appl. Sci. Manuf.* **2019**, *124*, 105482. [CrossRef]
195. Chong, W.J.; Shen, S.; Li, Y.; Trinchi, A.; Pejak, D.; (Louis) Kyratzis, I.; Sola, A.; Wen, C. Additive Manufacturing of Antibacterial PLA-ZnO Nanocomposites: Benefits, Limitations and Open Challenges. *J. Mater. Sci. Technol.* **2022**, *111*, 120–151. [CrossRef]
196. Dash, K.K.; Ali, N.A.; Das, D.; Mohanta, D. Thorough Evaluation of Sweet Potato Starch and Lemon-Waste Pectin Based-Edible Films with Nano-Titania Inclusions for Food Packaging Applications. *Int. J. Biol. Macromol.* **2019**, *139*, 449–458. [CrossRef]
197. Baek, N.; Kim, Y.T.; Marcy, J.E.; Duncan, S.E.; O'Keefe, S.F. Physical Properties of Nanocomposite Polylactic Acid Films Prepared with Oleic Acid Modified Titanium Dioxide. *Food Packag. Shelf Life* **2018**, *17*, 30–38. [CrossRef]
198. Batool, M.; Abid, A.; Khurshid, S.; Bashir, T.; Ismail, M.A.; Razaq, M.A.; Jamil, M. Quality Control of Nano-Food Packing Material for Grapes (Vitis Vinifera) Based on ZnO and Polylactic Acid (PLA) Biofilm. *Arab. J. Sci. Eng.* **2022**, *47*, 319–331. [CrossRef]
199. Grande-Tovar, C.D.; Castro, J.I.; Valencia Llano, C.H.; Tenorio, D.L.; Saavedra, M.; Zapata, P.A.; Chaur, M.N. Polycaprolactone (PCL)-Polylactic Acid (PLA)-Glycerol (Gly) Composites Incorporated with Zinc Oxide Nanoparticles (ZnO-NPs) and Tea Tree Essential Oil (TTEO) for Tissue Engineering Applications. *Pharmaceutics* **2022**, *15*, 43. [CrossRef] [PubMed]
200. Roeinfard, M.; Bahari, A. Nanostructural Characterization of the Fe3O4/ZnO Magnetic Nanocomposite as an Application in Medicine. *J. Supercond. Nov. Magn.* **2017**, *30*, 3541–3548. [CrossRef]

201. Kim, I.; Viswanathan, K.; Kasi, G.; Thanakkasaranee, S.; Sadeghi, K.; Seo, J. ZnO Nanostructures in Active Antibacterial Food Packaging: Preparation Methods, Antimicrobial Mechanisms, Safety Issues, Future Prospects, and Challenges. *Food Rev. Int.* **2022**, *38*, 537–565. [CrossRef]
202. Nonato, R.C.; Mei, L.H.I.; Bonse, B.C.; Leal, C.V.; Levy, C.E.; Oliveira, F.A.; Delarmelina, C.; Duarte, M.C.T.; Morales, A.R. Nanocomposites of PLA/ZnO Nanofibers for Medical Applications: Antimicrobial Effect, Thermal, and Mechanical Behavior under Cyclic Stress. *Polym. Eng. Sci.* **2022**, *62*, 1147–1155. [CrossRef]
203. Rashedi, S.M.; Khajavi, R.; Rashidi, A.; Rahimi, M.K.; Bahador, A. Novel PLA/ZnO Nanofibrous Nanocomposite Loaded with Tranexamic Acid as an Effective Wound Dressing: In Vitro and in Vivo Assessment. *Iran. J. Biotechnol.* **2021**, *19*, 38–47. [CrossRef]
204. Sefidan, A.M. Novel Silicon Dioxide -Based Nanocomposites as an Antimicrobial in Poly (Lactic Acid) Nanocomposites Films. *Nanomed. Res. J.* **2018**, *3*, 65–70. [CrossRef]
205. Wang, B.; Zhang, X.; Zhang, L.; Feng, Y.; Liu, C.; Shen, C. Simultaneously Reinforcing and Toughening Poly(Lactic Acid) by Incorporating Reactive Melt-Functionalized Silica Nanoparticles. *J. Appl. Polym. Sci.* **2020**, *137*, 48834. [CrossRef]
206. Nerantzaki, M.; Prokopiou, L.; Bikiaris, D.N.; Patsiaoura, D.; Chrissafis, K.; Klonos, P.; Kyritsis, A.; Pissis, P. In Situ Prepared Poly(DL-Lactic Acid)/Silica Nanocomposites: Study of Molecular Composition, Thermal Stability, Glass Transition and Molecular Dynamics. *Thermochim. Acta* **2018**, *669*, 16–29. [CrossRef]
207. Ainali, N.M.; Kalaronis, D.; Evgenidou, E.; Bikiaris, D.N.; Lambropoulou, D.A. Insights into Biodegradable Polymer-Supported Titanium Dioxide Photocatalysts for Environmental Remediation. *Macromol* **2021**, *1*, 201–233. [CrossRef]
208. Bobirică, C.; Bobirică, L.; Râpă, M.; Matei, E.; Predescu, A.M.; Orbeci, C. Photocatalytic Degradation of Ampicillin Using PLA/$TiO_2$ Hybrid Nanofibers Coated on Different Types of Fiberglass. *Water* **2020**, *12*, 176. [CrossRef]
209. Gupta, N.; Kozlovskaya, V.; Dolmat, M.; Yancey, B.; Oh, J.; Lungu, C.T.; Kharlampieva, E. Photocatalytic Nanocomposite Microsponges of Polylactide-Titania for Chemical Remediation in Water. *ACS Appl. Polym. Mater.* **2020**, *2*, 5188–5197. [CrossRef]
210. Petousis, M.; Vidakis, N.; Mountakis, N.; Papadakis, V.; Tzounis, L. Three-Dimensional Printed Polyamide 12 (PA12) and Polylactic Acid (PLA) Alumina (Al2O3) Nanocomposites with Significantly Enhanced Tensile, Flexural, and Impact Properties. *Nanomaterials* **2022**, *12*, 4292. [CrossRef]
211. Sukhanova, A.; Boyandin, A.; Ertiletskaya, N.; Simunin, M.; Shalygina, T.; Voronin, A.; Vasiliev, A.; Nemtsev, I.; Volochaev, M.; Pyatina, S. Study of the Effect of Modified Aluminum Oxide Nanofibers on the Properties of PLA-Based Films. *Materials* **2022**, *15*, 6097. [CrossRef]
212. Kangallı, E.; Bayraktar, E. Preparation and Characterization of Poly(Lactic Acid)/Boron Oxide Nanocomposites: Thermal, Mechanical, Crystallization, and Flammability Properties. *J. Appl. Polym. Sci.* **2022**, *139*, e52521. [CrossRef]
213. Wang, Z.; Pan, Z.; Wang, J.; Zhao, R. A Novel Hierarchical Structured Poly(Lactic Acid)/Titania Fibrous Membrane with Excellent Antibacterial Activity and Air Filtration Performance. *J. Nanomater.* **2016**, *2016*, 39. [CrossRef]
214. Salahuddin, N.; Abdelwahab, M.; Gaber, M.; Elneanaey, S. Synthesis and Design of Norfloxacin Drug Delivery System Based on PLA/$TiO_2$ Nanocomposites: Antibacterial and Antitumor Activities. *Mater. Sci. Eng. C* **2020**, *108*, 110337. [CrossRef]
215. Marsi, T.C.O.; Ricci, R.; Toniato, T.V.; Vasconcellos, L.M.R.; Elias, C.D.M.V.; Silva, A.D.R.; Furtado, A.S.A.; Magalhães, L.S.S.M.; Silva-Filho, E.C.; Marciano, F.R.; et al. Electrospun Nanofibrous Poly (Lactic Acid)/Titanium Dioxide Nanocomposite Membranes for Cutaneous Scar Minimization. *Front. Bioeng. Biotechnol.* **2019**, *7*, 421. [CrossRef]
216. Wang, G.; Wang, L.; Mark, L.H.; Shaayegan, V.; Wang, G.; Li, H.; Zhao, G.; Park, C.B. Ultralow-Threshold and Lightweight Biodegradable Porous PLA/MWCNT with Segregated Conductive Networks for High-Performance Thermal Insulation and Electromagnetic Interference Shielding Applications. *ACS Appl. Mater. Interfaces* **2018**, *10*, 1195–1203. [CrossRef]
217. Li, Y.; Yin, D.; Liu, W.; Zhou, H.; Zhang, Y.; Wang, X. Fabrication of Biodegradable Poly (Lactic Acid)/Carbon Nanotube Nanocomposite Foams: Significant Improvement on Rheological Property and Foamability. *Int. J. Biol. Macromol.* **2020**, *163*, 1175–1186. [CrossRef] [PubMed]
218. Sanusi, O.M.; Benelfellah, A.; Papadopoulos, L.; Terzopoulou, Z.; Malletzidou, L.; Vasileiadis, I.G.; Chrissafis, K.; Bikiaris, D.N.; Aït Hocine, N. Influence of Montmorillonite/Carbon Nanotube Hybrid Nanofillers on the Properties of Poly(Lactic Acid). *Appl. Clay Sci.* **2021**, *201*, 105925. [CrossRef]
219. Sanusi, O.M.; Benelfellah, A.; Terzopoulou, Z.; Bikiaris, D.N.; Aït Hocine, N. Properties of Polylactide Reinforced with Montmorillonite/Multiwalled Carbon Nanotube Hybrid. *Mater. Today Proc.* **2021**, *47*, 3247–3250. [CrossRef]
220. Sanusi, O.M.; Benelfellah, A.; Papadopoulos, L.; Terzopoulou, Z.; Bikiaris, D.N.; Aït Hocine, N. Properties of Poly(Lactic Acid)/Montmorillonite/Carbon Nanotubes Nanocomposites: Determination of Percolation Threshold. *J. Mater. Sci.* **2021**, *56*, 16887–16901. [CrossRef]
221. Kalinke, C.; Neumsteir, N.V.; Aparecido, G.D.O.; Ferraz, T.V.D.B.; Dos Santos, P.L.; Janegitz, B.C.; Bonacin, J.A. Comparison of Activation Processes for 3D Printed PLA-Graphene Electrodes: Electrochemical Properties and Application for Sensing of Dopamine. *Analyst* **2020**, *145*, 1207–1218. [CrossRef]
222. Chakraborty, G.; Pugazhenthi, G.; Katiyar, V. Exfoliated Graphene-Dispersed Poly (Lactic Acid)-Based Nanocomposite Sensors for Ethanol Detection. *Polym. Bull.* **2019**, *76*, 2367–2386. [CrossRef]
223. Spinelli, G.; Guarini, R.; Kotsilkova, R.; Batakliev, T.; Ivanov, E.; Romano, V. Experimental and Simulation Studies of Temperature Effect on Thermophysical Properties of Graphene-Based Polylactic Acid. *Materials* **2022**, *15*, 2340. [CrossRef]

224. Masarra, N.A.; Batistella, M.; Quantin, J.C.; Regazzi, A.; Pucci, M.F.; El Hage, R.; Lopez-Cuesta, J.M. Fabrication of PLA/PCL/Graphene Nanoplatelet (GNP) Electrically Conductive Circuit Using the Fused Filament Fabrication (FFF) 3D Printing Technique. *Materials* **2022**, *15*, 762. [CrossRef]
225. Paydayesh, A.; Pashaei Soorbaghi, F.; Aref Azar, A.; Jalali-Arani, A. Electrical Conductivity of Graphene Filled PLA/PMMA Blends: Experimental Investigation and Modeling. *Polym. Compos.* **2019**, *40*, 704–715. [CrossRef]
226. Silva, A.P.B.; Montagna, L.S.; Passador, F.R.; Rezende, M.C.; Lemes, A.P. Biodegradable Nanocomposites Based on PLA/PHBV Blend Reinforced with Carbon Nanotubes with Potential for Electrical and Electromagnetic Applications. *Express Polym. Lett.* **2021**, *15*, 987–1003. [CrossRef]
227. Zare, Y.; Rhee, K.Y. Following the Morphological and Thermal Properties of PLA/PEO Blends Containing Carbon Nanotubes (CNTs) during Hydrolytic Degradation. *Compos. Part B Eng.* **2019**, *175*, 107132. [CrossRef]
228. Cataldo, F. On the Interaction of C60 Fullerene with Poly(L-Lactic Acid) or Poly(Lactide). *Fullerenes Nanotub. Carbon Nanostruct.* **2020**, *28*, 621–626. [CrossRef]
229. Shi, S.; Chen, Y.; Jing, J.; Yang, L. Preparation and 3D-Printing of Highly Conductive Polylactic Acid/Carbon Nanotube Nanocomposites: Via Local Enrichment Strategy. *RSC Adv.* **2019**, *9*, 29980–29986. [CrossRef] [PubMed]
230. Magiera, A.; Markowski, J.; Pilch, J.; Blazewicz, S. Degradation Behavior of Electrospun PLA and PLA/CNT Nanofibres in Aqueous Environment. *J. Nanomater.* **2018**, *2018*, 8796583. [CrossRef]
231. Zhang, Z.; Liu, S.; Xiong, H.; Jing, X.; Xie, Z.; Chen, X.; Huang, Y. Electrospun PLA/MWCNTs Composite Nanofibers for Combined Chemo- and Photothermal Therapy. *Acta Biomater.* **2015**, *26*, 115–123. [CrossRef] [PubMed]
232. Lee, J.H.; Park, S.H.; Kim, S.H.; Ito, H. Replication and Surface Properties of Micro Injection Molded PLA/MWCNT Nanocomposites. *Polym. Test.* **2020**, *83*, 106321. [CrossRef]
233. Vidakis, N.; Petousis, M.; Kourinou, M.; Velidakis, E.; Mountakis, N.; Fischer-Griffiths, P.E.; Grammatikos, S.; Tzounis, L. Additive Manufacturing of Multifunctional Polylactic Acid (PLA)—Multiwalled Carbon Nanotubes (MWCNTs) Nanocomposites. *Nanocomposites* **2021**, *7*, 184–199. [CrossRef]
234. Thummarungsan, N.; Pattavarakorn, D.; Sirivat, A. Electrically Responsive Materials Based on Dibutyl Phathalate Plasticized Poly(Lactic Acid) and Spherical Fullerene. *Smart Mater. Struct.* **2022**, *31*, 035029. [CrossRef]
235. Chen, L.; Pang, X.; Zhao, J.; Piao, G. Facile Method to Fabricate Poly(L-Lactic Acid)/C60 Tetragonal Single Crystal Composites with Enhanced Thermal Stability. *Chem. Phys. Lett.* **2019**, *723*, 16–21. [CrossRef]
236. Naffakh, M. Biopolymer Nanocomposite Materials Based on Poly(L-lactic Acid) and Inorganic Fullerene-like Ws2 Nanoparticles. *Polymers* **2021**, *13*, 2947. [CrossRef]
237. Li, Z.; Zhang, F.L.; Pan, L.L.; Zhu, X.L.; Zhang, Z.Z. Preparation and Characterization of Injectable Mitoxantrone Poly (Lactic Acid)/Fullerene Implants for in Vivo Chemo-Photodynamic Therapy. *J. Photochem. Photobiol. B Biol.* **2015**, *149*, 51–57. [CrossRef] [PubMed]
238. De Bortoli, L.S.; de Farias, R.; Mezalira, D.Z.; Schabbach, L.M.; Fredel, M.C. Functionalized Carbon Nanotubes for 3D-Printed PLA-Nanocomposites: Effects on Thermal and Mechanical Properties. *Mater. Today Commun.* **2022**, *31*, 103402. [CrossRef]
239. Camargo, J.C.; Machado, Á.R.; Almeida, E.C.; Silva, E.F.M.S. Mechanical Properties of PLA-Graphene Filament for FDM 3D Printing. *Int. J. Adv. Manuf. Technol.* **2019**, *103*, 2423–2443. [CrossRef]
240. Caminero, M.Á.; Chacón, J.M.; García-Plaza, E.; Núñez, P.J.; Reverte, J.M.; Becar, J.P. Additive Manufacturing of PLA-Based Composites Using Fused Filament Fabrication: Effect of Graphene Nanoplatelet Reinforcement on Mechanical Properties, Dimensional Accuracy and Texture. *Polymers* **2019**, *11*, 799. [CrossRef]
241. Garc, E.; Jos, P.; Miguel, Á.; Torija, C. Analysis of PLA Geometric Properties Processed by FFF Additive Manufacturing: E Ff Ects of Process Parameters and Plate-Extruder Precision Motion. *Polymers* **2019**, *11*, 1581.
242. El Magri, A.; Vanaei, S.; Shirinbayan, M.; Vaudreuil, S.; Tcharkhtchi, A. An Investigation to Study the Effect of Process Parameters on the Strength and Fatigue Behavior of 3d-Printed Pla-Graphene. *Polymers* **2021**, *13*, 3218. [CrossRef]
243. García, E.; Núñez, P.J.; Chacón, J.M.; Caminero, M.A.; Kamarthi, S. Comparative Study of Geometric Properties of Unreinforced PLA and PLA-Graphene Composite Materials Applied to Additive Manufacturing Using FFF Technology. *Polym. Test.* **2020**, *91*, 106860. [CrossRef]
244. Ab Ghani, N.F.; Mat Desa, M.S.Z.; Bijarimi, M. The Evaluation of Mechanical Properties Graphene Nanoplatelets Reinforced Polylactic Acid Nanocomposites. *Mater. Today Proc.* **2019**, *42*, 283–287. [CrossRef]
245. Davoodi, A.H.; Mazinani, S.; Sharif, F.; Ranaei-Siadat, S.O. GO Nanosheets Localization by Morphological Study on PLA-GO Electrospun Nanocomposite Nanofibers. *J. Polym. Res.* **2018**, *25*, 16–19. [CrossRef]
246. Papon, E.A.; Haque, A. Tensile Properties, Void Contents, Dispersion and Fracture Behaviour of 3D Printed Carbon Nanofiber Reinforced Composites. *J. Reinf. Plast. Compos.* **2018**, *37*, 381–395. [CrossRef]
247. Malafeev, K.; Ivankova, E.M.; Kamalov, A.; Yudin, V.; Popova, E.N.; Moskalyuk, O. Influence of VGCF Concentration on Properties of Biodegradable Fibers Based on Poly (Lactide Acid). In Proceedings of the 2021 IEEE Conference of Russian Young Researchers in Electrical and Electronic Engineering (ElConRus), St. Petersburg, Moscow, Russia, 26–29 January 2021; pp. 1786–1789. [CrossRef]
248. Dave, K.; Mahmud, Z.; Gomes, V.G. Superhydrophilic 3D-Printed Scaffolds Using Conjugated Bioresorbable Nanocomposites for Enhanced Bone Regeneration. *Chem. Eng. J.* **2022**, *445*, 136639. [CrossRef]

249. Segun, A.; Adewuyi, B.O.; Ojo, D.O.; Gideon, O.N. Mechanical and Structural Properties of Nanocarbon Particles Reinforced in Plasticised Polylactic Acid for High Strength Application. *J. Phys. Sci.* **2021**, *32*, 41–56. [CrossRef]
250. Potnuru, A.; Tadesse, Y. Investigation of Polylactide and Carbon Nanocomposite Filament for 3D Printing. *Prog. Addit. Manuf.* **2019**, *4*, 23–41. [CrossRef]
251. Jain, S.K.; Tadesse, Y. Fabrication of Polylactide/Carbon Nanopowder Filament Using Melt Extrusion and Filament Characterization for 3D Printing. *Int. J. Nanosci.* **2019**, *18*, 1850026. [CrossRef]
252. Amran, N.A.M.; Ahmad, S.; Chen, R.S.; Shahdan, D. Tensile Properties and Thermal Stability of Nanocomposite Poly-Lactic Acid/Liquid Natural Rubber Filled Graphene Nanoplates. In *AIP Conference Proceedings*; AIP Publishing LLC: Woodbury, NY, USA, 2019; p. 030006. [CrossRef]
253. Mohd Amran, N.A.; Ahmad, S.; Chen, R.S.; Shahdan, D.; Flaifel, M.H.; Omar, A. Assessment of Mechanical and Electrical Performances of Polylactic Acid/Liquid Natural Rubber/Graphene Platelets Nanocomposites in the Light of Different Graphene Platelets Functionalization Routes. *Macromol. Chem. Phys.* **2021**, *222*, 2100185. [CrossRef]
254. Wang, B.; Ye, X.; Wang, B.; Li, X.; Xiao, S.; Liu, H. Reactive Graphene as Highly Efficient Compatibilizer for Cocontinuous Poly(Lactic Acid)/Poly(ε-Caprolactone) Blends toward Robust Biodegradable Nanocomposites. *Compos. Sci. Technol.* **2022**, *221*, 109326. [CrossRef]
255. Paydayesh, A.; Mousavi, S.R.; Estaji, S.; Khonakdar, H.A.; Nozarinya, M.A. Functionalized Graphene Nanoplatelets/Poly (Lactic Acid)/Chitosan Nanocomposites: Mechanical, Biodegradability, and Electrical Conductivity Properties. *Polym. Compos.* **2022**, *43*, 411–421. [CrossRef]
256. Wang, D.; Lu, X.; Qu, J. Role of In Situ Thermal-Reduced Graphene Oxide on the Morphology and Properties of Biodegradable Poly(Lactic Acid)/Poly(Butylene Succinate) Blends. *Polym. Compos.* **2017**, *16*, 101–113. [CrossRef]
257. Iglesias Montes, M.L.; Luzi, F.; Dominici, F.; Torre, L.; Cyras, V.P.; Manfredi, L.B.; Puglia, D. Design and Characterization of PLA Bilayer Films Containing Lignin and Cellulose Nanostructures in Combination with Umbelliferone as Active Ingredient. *Front. Chem.* **2019**, *7*, 157. [CrossRef]
258. Yang, W.; Fortunati, E.; Bertoglio, F.; Owczarek, J.S.; Bruni, G.; Kozanecki, M.; Kenny, J.M.; Torre, L.; Visai, L.; Puglia, D. Polyvinyl Alcohol/Chitosan Hydrogels with Enhanced Antioxidant and Antibacterial Properties Induced by Lignin Nanoparticles. *Carbohydr. Polym.* **2018**, *181*, 275–284. [CrossRef]
259. Li, J.H.; Miao, J.; Wu, J.L.; Chen, S.F.; Zhang, Q.Q. Preparation and Characterization of Active Gelatin-Based Films Incorporated with Natural Antioxidants. *Food Hydrocoll.* **2014**, *37*, 166–173. [CrossRef]
260. Makri, S.P.; Xanthopoulou, E.; Klonos, P.A.; Grigoropoulos, A.; Kyritsis, A.; Tsachouridis, K.; Anastasiou, A.; Deligkiozi, I.; Nikolaidis, N.; Bikiaris, D.N. Effect of Micro- and Nano-Lignin on the Thermal, Mechanical, and Antioxidant Properties of Biobased PLA–Lignin Composite Films. *Polymers* **2022**, *14*, 5274. [CrossRef] [PubMed]
261. Yang, W.; Fortunati, E.; Dominici, F.; Kenny, J.M.; Puglia, D. Effect of Processing Conditions and Lignin Content on Thermal, Mechanical and Degradative Behavior of Lignin Nanoparticles/Polylactic (Acid) Bionanocomposites Prepared by Melt Extrusion and Solvent Casting. *Eur. Polym. J.* **2015**, *71*, 126–139. [CrossRef]
262. Yang, W.; Zhu, Y.; He, Y.; Xiao, L.; Xu, P.; Puglia, D.; Ma, P. Preparation of Toughened Poly(Lactic Acid)-Poly(ε-Caprolactone)-Lignin Nanocomposites with Good Heat- and UV-Resistance. *Ind. Crops Prod.* **2022**, *183*, 114965. [CrossRef]
263. Chihaoui, B.; Tarrés, Q.; Delgado-Aguilar, M.; Mutjé, P.; Boufi, S. Lignin-Containing Cellulose Fibrils as Reinforcement of Plasticized PLA Biocomposites Produced by Melt Processing Using PEG as a Carrier. *Ind. Crops Prod.* **2022**, *175*, 114287. [CrossRef]
264. Gulzar, S.; Tagrida, M.; Nilsuwan, K.; Prodpran, T.; Benjakul, S. Electrospinning of Gelatin/Chitosan Nanofibers Incorporated with Tannic Acid and Chitooligosaccharides on Polylactic Acid Film: Characteristics and Bioactivities. *Food Hydrocoll.* **2022**, *133*, 107916. [CrossRef]
265. Ng, W.K.; Chow, W.S.; Ismail, H. Tensile, Thermal and Optical Properties of Poly(Lactic Acid)/Poly(2-Ethyl-2-Oxazoline)/Corn Cob Nanocellulose Nanocomposite Film. *Polym. Technol. Mater.* **2022**, *61*, 207–219. [CrossRef]
266. Sangeetha, V.H.; Varghese, T.O.; Nayak, S.K. Value Addition of Waste Cotton: Effect of Nanofibrillated Cellulose on EVA/EVOH Toughened Polylactic Acid System. *Waste Biomass Valorization* **2020**, *11*, 4119–4128. [CrossRef]
267. Rasheed, M.; Jawaid, M.; Parveez, B. Bamboo Fiber Based Cellulose Nanocrystals/Poly(Lactic Acid)/Poly(Butylene Succinate) Nanocomposites: Morphological, Mechanical and Thermal Properties. *Polymers* **2021**, *13*, 1076. [CrossRef]
268. Jin, K.; Tang, Y.; Zhu, X.; Zhou, Y. Polylactic Acid Based Biocomposite Films Reinforced with Silanized Nanocrystalline Cellulose. *Int. J. Biol. Macromol.* **2020**, *162*, 1109–1117. [CrossRef]
269. Li, L.; Chen, Y.; Yu, T.; Wang, N.; Wang, C.; Wang, H. Preparation of Polylactic Acid/TEMPO-Oxidized Bacterial Cellulose Nanocomposites for 3D Printing via Pickering Emulsion Approach. *Compos. Commun.* **2019**, *16*, 162–167. [CrossRef]
270. Somphol, W.; Lampang, T.N.; Prapainainar, P.; Sae-Oui, P.; Loykulnant, S.; Suebsai, A.; Dittanet, P. Effect of Polyethylene Glycol in Nanocellulose/PLA Composites. *Key Eng. Mater.* **2019**, *821*, 89–95. [CrossRef]
271. Sobhan, A.; Muthukumarappan, K.; Wei, L. Development of Bio-Nanocomposite Films by Combination of PLA and Biochar for Smart Food Packaging. In Proceedings of the 2020 ASABE Annual International Virtual Meeting, Online, 13–15 July 2020; American Society of Agricultural and Biological Engineers: St. Joseph, MI, USA; pp. 1–9. [CrossRef]
272. Jasim, S.M.; Ali, N.A. Properties Characterization of Plasticized Polylactic Acid/Biochar (Bio Carbon) Nano-Composites for Antistatic Packaging. *Iraqi J. Phys.* **2019**, *17*, 13–26. [CrossRef]

273. Chollet, B.; Lopez-Cuesta, J.M.; Laoutid, F.; Ferry, L. Lignin Nanoparticles as a Promising Way for Enhancing Lignin Flame Retardant Effect in Polylactide. *Materials* **2019**, *12*, 2132. [CrossRef] [PubMed]
274. Yu, M.; Sun, C.; Xue, Y.; Liu, C.; Qiu, D.; Cui, B.; Zhang, Y.; Cui, H.; Zeng, Z. Tannic Acid-Based Nanopesticides Coating with Highly Improved Foliage Adhesion to Enhance Foliar Retention. *RSC Adv.* **2019**, *9*, 27096–27104. [CrossRef] [PubMed]
275. Liao, J.; Brosse, N.; Pizzi, A.; Hoppe, S.; Zhou, X.; Du, G. Characterization and 3D Printability of Poly (Lactic Acid)/Acetylated Tannin Composites. *Ind. Crops Prod.* **2020**, *149*, 112320. [CrossRef]
276. Liao, J.; Brosse, N.; Hoppe, S.; Zhou, X.; Xi, X.; Du, G.; Pizzi, A. Interfacial Improvement of Poly (Lactic Acid)/Tannin Acetate Composites via Radical Initiated Polymerization. *Ind. Crops Prod.* **2021**, *159*, 113068. [CrossRef]
277. Kian, L.K.; Jawaid, M.; Nasef, M.M.; Fouad, H.; Karim, Z. Poly(Lactic Acid)/Poly(Butylene Succinate) Dual-Layer Membranes with Cellulose Nanowhisker for Heavy Metal Ion Separation. *Int. J. Biol. Macromol.* **2021**, *192*, 654–664. [CrossRef]
278. Geng, S.; Yao, K.; Zhou, Q.; Oksman, K. High-Strength, High-Toughness Aligned Polymer-Based Nanocomposite Reinforced with Ultralow Weight Fraction of Functionalized Nanocellulos. *Biomacromolecules* **2018**, *19*, 4075–4083. [CrossRef]
279. Ghasemi, S.; Behrooz, R.; Ghasemi, I.; Yassar, R.S.; Long, F. Development of Nanocellulose-Reinforced PLA Nanocomposite by Using Maleated PLA (PLA-g-MA). *J. Thermoplast. Compos. Mater.* **2018**, *31*, 1090–1101. [CrossRef]
280. Shazleen, S.S.; Foong Ng, L.Y.; Ibrahim, N.A.; Hassan, M.A.; Ariffin, H. Combined Effects of Cellulose Nanofiber Nucleation and Maleated Polylactic Acid Compatibilization on the Crystallization Kinetic and Mechanical Properties of Polylactic Acid Nanocomposite. *Polymers* **2021**, *13*, 3226. [CrossRef] [PubMed]
281. Castro, J.I.; Valencia Llano, C.H.; Tenorio, D.L.; Saavedra, M.; Zapata, P.; Navia-Porras, D.P.; Delgado-Ospina, J.; Chaur, M.N.; Hernández, J.H.M.; Grande-Tovar, C.D. Biocompatibility Assessment of Polylactic Acid (PLA) and Nanobioglass (n-BG) Nanocomposites for Biomedical Applications. *Molecules* **2022**, *27*, 3640. [CrossRef]
282. Esmaeilzadeh, J. In Vivo Assessments of the Poly (d/l) Lactide/Polycaprolactone/Bioactive Glass Nanocomposites for Bioscrews Application. *Adv. Ceram. Prog.* **2021**, *7*, 17–22.
283. Macha, I.J.; Ben-Nissan, B.; Santos, J.; Cazalbou, S.; Stamboulis, A.; Grossin, D.; Giordano, G. Biocompatibility of a New Biodegradable Polymer-Hydroxyapatite Composite. *J. Drug Deliv. Sci. Technol.* **2017**, *38*, 72–77. [CrossRef]
284. Morsi, M.A.; Hezma, A.E.M. Effect of Iron Doped Hydroxyapatite Nanoparticles on the Structural, Morphological, Mechanical and Magnetic Properties of Polylactic Acid Polymer. *J. Mater. Res. Technol.* **2019**, *8*, 2098–2106. [CrossRef]
285. Toong, D.W.Y.; Ng, J.C.K.; Cui, F.; Leo, H.L.; Zhong, L.; Lian, S.S.; Venkatraman, S.; Tan, L.P.; Huang, Y.Y.; Ang, H.Y. Nanoparticles-Reinforced Poly-l-Lactic Acid Composite Materials as Bioresorbable Scaffold Candidates for Coronary Stents: Insights from Mechanical and Finite Element Analysis. *J. Mech. Behav. Biomed. Mater.* **2022**, *125*, 104977. [CrossRef] [PubMed]
286. Ziaee, F.; Zebarjad, S.M.; Javadpour, S. Compressive and Flexural Properties of Novel Polylactic Acid/Hydroxyapatite/Yttria-Stabilized Zirconia Hybrid Nanocomposite Scaffold. *Int. J. Polym. Mater. Polym. Biomater.* **2018**, *67*, 229–238. [CrossRef]
287. Ghassemi, B.; Estaji, S.; Mousavi, S.R.; Nemati Mahand, S.; Shojaei, S.; Mostafaiyan, M.; Arjmand, M.; Khonakdar, H.A. In-Depth Study of Mechanical Properties of Poly(Lactic Acid)/Thermoplastic Polyurethane/Hydroxyapatite Blend Nanocomposites. *J. Mater. Sci.* **2022**, *57*, 7250–7264. [CrossRef]
288. Rahman, M.M.; Shahruzzaman, M.; Islam, M.S.; Khan, M.N.; Haque, P. Preparation and Properties of Biodegradable Polymer/Nano-Hydroxyapatite Bioceramic Scaffold for Spongy Bone Regeneration. *J. Polym. Eng.* **2019**, *39*, 134–142. [CrossRef]
289. Yan, D.; Wang, Z.; Guo, Z.; Ma, Y.; Wang, C.; Tan, H.; Zhang, Y. Study on the Properties of PLA/PBAT Composite Modified by Nanohydroxyapatite. *J. Mater. Res. Technol.* **2020**, *9*, 11895–11904. [CrossRef]
290. Peponi, L.; Sessini, V.; Arrieta, M.P.; Navarro-Baena, I.; Sonseca, A.; Dominici, F.; Gimenez, E.; Torre, L.; Tercjak, A.; López, D.; et al. Thermally-Activated Shape Memory Effect on Biodegradable Nanocomposites Based on PLA/PCL Blend Reinforced with Hydroxyapatite. *Polym. Degrad. Stab.* **2018**, *151*, 36–51. [CrossRef]
291. Mondal, S.; Nguyen, T.P.; Pham, V.H.; Hoang, G.; Manivasagan, P.; Kim, M.H.; Nam, S.Y.; Oh, J. Hydroxyapatite Nano Bioceramics Optimized 3D Printed Poly Lactic Acid Scaffold for Bone Tissue Engineering Application. *Ceram. Int.* **2020**, *46*, 3443–3455. [CrossRef]
292. Swetha, S.; Balagangadharan, K.; Lavanya, K.; Selvamurugan, N. Three-Dimensional-Poly(Lactic Acid) Scaffolds Coated with Gelatin/Magnesium-Doped Nano-Hydroxyapatite for Bone Tissue Engineering. *Biotechnol. J.* **2021**, *16*, 2100282. [CrossRef] [PubMed]
293. Hajibeygi, M.; Shafiei-Navid, S. Design and Preparation of Poly(Lactic Acid) Hydroxyapatite Nanocomposites Reinforced with Phosphorus-Based Organic Additive: Thermal, Combustion, and Mechanical Properties Studies. *Polym. Adv. Technol.* **2019**, *30*, 2233–2249. [CrossRef]
294. Salehi, A.; Jafari, S.H.; Khonakdar, H.A.; Ebadi-Dehaghani, H. Temperature Dependency of Gas Barrier Properties of Biodegradable PP/PLA/Nanoclay Films: Experimental Analyses with a Molecular Dynamics Simulation Approach. *J. Appl. Polym. Sci.* **2018**, *135*, 46665. [CrossRef]
295. Alakrach, A.M.; Noriman, N.Z.; Dahham, O.S.; Hamzah, R.; Alsaadi, M.A.; Shayfull, Z.; Idrus, S.S. The Effects of Tensile Properties of PLA/HNTs-ZrO2 Bionanocomposites. *J. Phys. Conf. Ser.* **2018**, *1019*, 012066. [CrossRef]
296. Li, Y.; Li, P.; Wu, M.; Yu, X.; Naito, K.; Zhang, Q. Halloysite Nanotubes Grafted Polylactic Acid and Its Composites with Enhanced Interfacial Compatibility. *J. Appl. Polym. Sci.* **2021**, *138*, 49668. [CrossRef]

297. Mohd Nizar, M.; Hamzah, M.S.A.; Abd Razak, S.I.; Mat Nayan, N.H. Thermal Stability and Surface Wettability Studies of Polylactic Acid/Halloysite Nanotube Nanocomposite Scaffold for Tissue Engineering Studies. *IOP Conf. Ser. Mater. Sci. Eng.* **2018**, *318*, 012006. [CrossRef]
298. Chaiwutthinan, P.; Phutfak, N.; Larpkasemsuk, A. Effects of Thermoplastic Poly(Ether-Ester) Elastomer and Bentonite Nanoclay on Properties of Poly(Lactic Acid). *J. Appl. Polym. Sci.* **2021**, *138*, 8–11. [CrossRef]

MDPI

*Review*

# Therapeutic Efficacy of Polymeric Biomaterials in Treating Diabetic Wounds—An Upcoming Wound Healing Technology

**Weslen Vedakumari Sathyaraj [1,*], Lokesh Prabakaran [1], Jayavardhini Bhoopathy [1], Sankari Dharmalingam [2], Ramadoss Karthikeyan [3] and Raji Atchudan [4,5,*]**

1 Faculty of Allied Health Sciences, Chettinad Hospital and Research Institute, Chettinad Academy of Research and Education, Kelambakkam 603103, Tamil Nadu, India
2 Department of Biotechnology, College of Science and Humanities, SRM Institute of Science and Technology, Kattankulathur 603203, Tamil Nadu, India
3 School of Pharmacy, Sri Balaji Vidyapeeth, SBV Campus, Pillaiyarkuppam, Puducherry 607402, Tamil Nadu, India
4 School of Chemical Engineering, Yeungnam University, Gyeongsan 38541, Republic of Korea
5 Department of Chemistry, Saveetha School of Engineering, Saveetha Institute of Medical and Technical Sciences, Chennai 602105, Tamil Nadu, India
* Correspondence: drweslenv@care.edu.in (W.V.S.); atchudanr@yu.ac.kr (R.A.)

**Abstract:** Diabetic wounds are one of the serious, non-healing, chronic health issues faced by individuals suffering from diabetic mellitus. The distinct phases of wound healing are either prolonged or obstructed, resulting in the improper healing of diabetic wounds. These injuries require persistent wound care and appropriate treatment to prevent deleterious effects such as lower limb amputation. Although there are several treatment strategies, diabetic wounds continue to be a major threat for healthcare professionals and patients. The different types of diabetic wound dressings that are currently used differ in their properties of absorbing wound exudates and may also cause maceration to surrounding tissues. Current research is focused on developing novel wound dressings incorporated with biological agents that aid in a faster rate of wound closure. An ideal wound dressing material must absorb wound exudates, aid in the appropriate exchange of gas, and protect from microbial infections. It must support the synthesis of biochemical mediators such as cytokines, and growth factors that are crucial for faster healing of wounds. This review highlights the recent advances in polymeric biomaterial-based wound dressings, novel therapeutic regimes, and their efficacy in treating diabetic wounds. The role of polymeric wound dressings loaded with bioactive compounds, and their in vitro and in vivo performance in diabetic wound treatment are also reviewed.

**Keywords:** diabetes; polymers; biomaterials; scaffolds; wound dressings

**Citation:** Sathyaraj, W.V.; Prabakaran, L.; Bhoopathy, J.; Dharmalingam, S.; Karthikeyan, R.; Atchudan, R. Therapeutic Efficacy of Polymeric Biomaterials in Treating Diabetic Wounds—An Upcoming Wound Healing Technology. *Polymers* **2023**, *15*, 1205. https://doi.org/10.3390/polym15051205

Academic Editors: José Miguel Ferri, Vicent Fombuena Borràs and Miguel Fernando Aldás Carrasco

Received: 30 December 2022
Revised: 15 February 2023
Accepted: 22 February 2023
Published: 27 February 2023



## 1. Introduction

Diabetes is a metabolic disorder that ranks as one of the top ten reasons for death among the global population. The International Diabetes Federation (IDF) has reported 463 million diabetic cases in 2019, and this count is suspected to grow to 578 million in 2030 [1,2]. Diabetes mellitus (DM) occurs when the pancreas fails to secrete the necessary amount of insulin required to maintain a normal blood sugar level in the human body. A drastic rise in blood sugar level impairs the process of wound healing, and results in chronic non-healing wounds, which may lead to hospitalization or lower extremity amputation [3]. In diabetic patients, the different phases of wound healing are hindered by various factors such as stalled expression of growth factors, metabolic insufficiency, and reduced physiological response, which prolong the time required for wound recovery [1,3] Diabetes is also connected with different types of illnesses such as chronic kidney failure, cardiovascular disease, stroke, and peripheral neuropathy [4]. Moreover, changes in motor and sympathetic functions may result in physical deformation of the feet due to extreme skin dehydration and wound formation [1]. A recent report states that almost 50 to 70% of all limb amputations are due to diabetes, and one leg is removed every thirty seconds among patients suffering from diabetes [3,5]. The management of diabetic wounds using polymer-based dressing materials has gained a lot of attention among clinicians due to their beneficial properties such as significant antibacterial, mechanical, and wound healing properties [3]. This review highlights the recent advancements in natural and synthetic polymer-based biomaterials for treating diabetic wounds.

## 2. Wound Healing—Physiology

Wound healing is an intricate biological process that occurs when there is a loss of integrity in skin or body tissues [6]. Wound healing requires the involvement of different types of cells, growth factors, enzymes, and various components of the extracellular matrix for repairing and restoring damaged tissues and organs [3]. It occurs in four distinct stages: haemostasis, inflammation, proliferation, and remodelling (Figure 1).

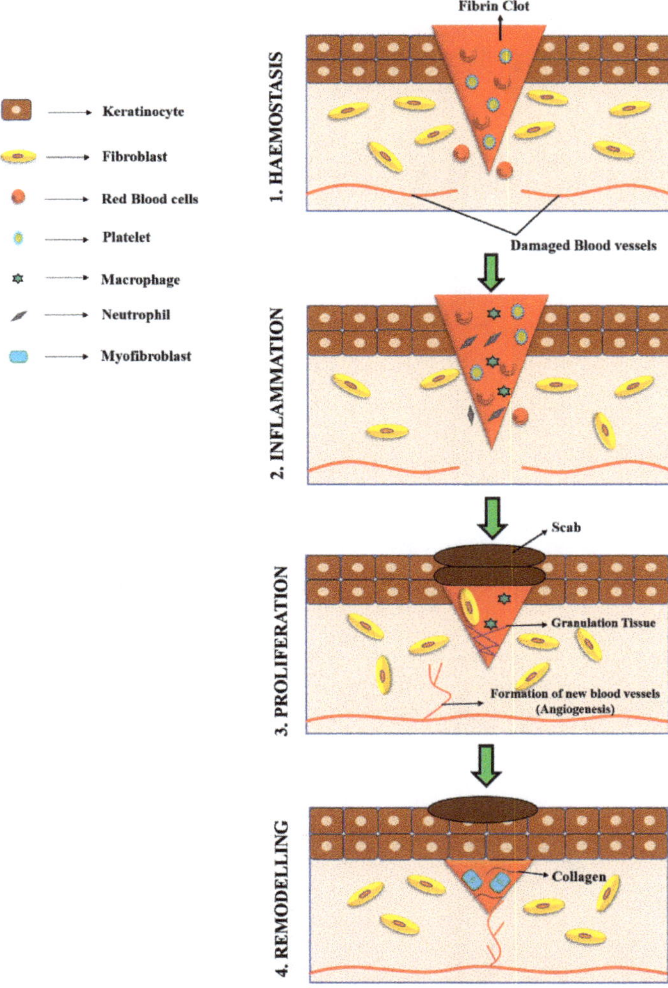


**Figure 1.** Four different phases of wound healing.

These four phases must proceed in an ordered fashion to avoid interruptions or delays in wound closure [7]. Haemostasis is a process that begins immediately after an injury to stop bleeding, and results in the formation of blood clots. During this event, platelets aggregate at the wounded site owing to the interaction with proteins such as collagen and fibronectin. Soluble fibrinogen is converted into insoluble fibrin to arrest bleeding. The area surrounding the clot and damaged tissue produces growth factors and pro-inflammatory cytokines that aid in efficient wound healing. When the bleeding has stopped, the inflammatory phase is initiated, and this involves the migration of leucocytes to the injured site to eliminate debris and infectious microorganisms [8]. This phase is characterized by the sequential infiltration of distinctive kinds of cells such as macrophages, neutrophils, and lymphocytes, which protect the wounded site from infections [9–11]. Macrophages play a vital role in all the stages of wound healing [12,13] as they release cytokines that are responsible for inflammation, activation of leukocytes, and clearance of apoptotic cells. As soon as apoptotic cells are cleared, macrophages are transformed into a pro-regenerative state that activates fibroblasts and keratinocytes, resulting in the regeneration of tissues. The proliferative phase is overlapped with the inflammatory phase, which leads to the proliferation and migration of epithelial cells. Fibroblasts and epithelial cells perform a vital task in the formation of collagen and granulation tissue at the site of the wound. The main components of the extracellular matrix—collagen, glycosaminoglycans, and proteoglycans—are synthesized by fibroblasts, and they play a vital role in the healing of the wound. At the end of the proliferative phase, the wound healing process moves into the final remodelling phase, which is characterized by the formation of granulation tissue [9,14].

## 3. Wound Healing in Diabetes

The normal phases of wound healing are disrupted due to diabetes (Figure 2). Diabetic wounds (Figure 3) continue to persist in the inflammatory phase, and the development of matured granulation tissues is inhibited by the hindering of the initiation of the proliferative phase in wound healing [3,15]. Intrinsic and extrinsic factors are involved in impairing the healing of diabetic wounds. Continuous mechanical stress and recurrent trauma can further deteriorate the healing process and result in ulcer formation [16,17]. Diabetic wound healing is delayed due to various causes such as neuropathy, poor immunity, microbial infection, oxygen deficit, and minimal activity of growth factors [3,7,18]. Numerous cells such as macrophages, neutrophils, fibroblasts, lymphocytes, keratinocytes, mast cells, and endothelial cells are actively involved in the normal wound healing process. Several growth factors and cytokines are secreted by these cells, which perform a key role in accelerating wound healing. Increased blood sugar level alters macrophage polarization, which serves as one of the chief causes for impaired diabetic wound healing. Events such as continuous secretion of pro-inflammatory cytokines, reduced angiogenic response [7], decreased activity of neutrophils, macrophages, and fibroblasts, were observed in diabetic wounds [19,20]. Diabetic wounds may also result in sensory disability towards temperature, pressure, and lesions. Lack of pain and abnormal vasodilator autoregulation together aggravate the process of wound healing [3]. Diabetic wounds may limit physical movement and cause psychiatric stress and depression [15].

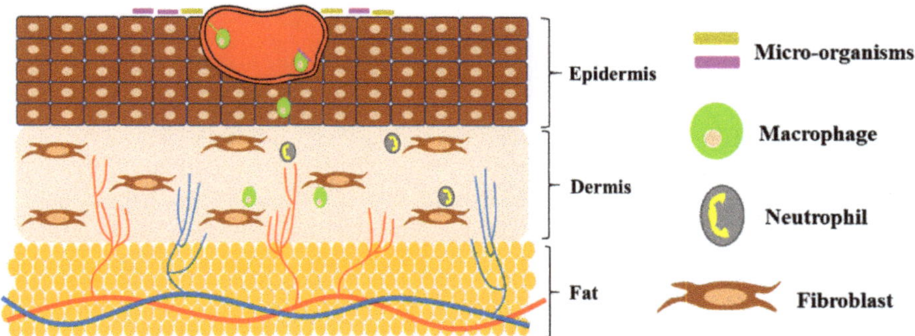


**Figure 2.** Diagrammatic representation of normal wound healing.

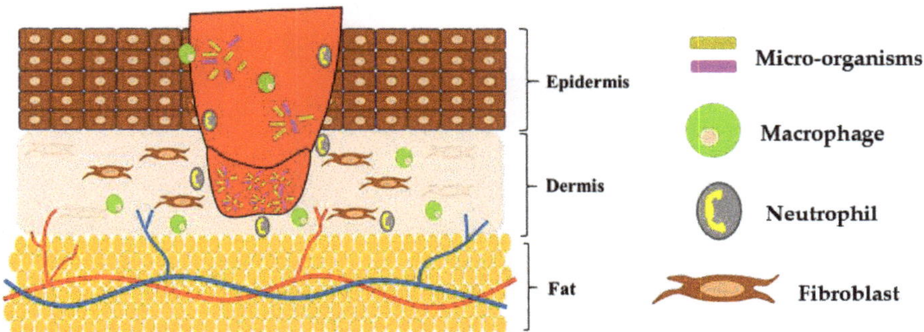


**Figure 3.** Impaired or delayed wound healing under diabetic conditions.

## 4. Types of Diabetic Wound Dressings

Wound dressings quicken the process of wound healing by allowing water transmission, providing a moist atmosphere, and aiding in improved granulation and re-epithelialization. They can be incorporated with therapeutic molecules or anti-microbial agents for efficient treatment of wounds [3]. The most commonly used diabetic wound care products available on the market are Comfeel, Granuflex, and Duoderm. However, serious concerns are raised about their use in treating infected wounds, as they may cause maceration to the surrounding tissues that are present around the wound. Intrasite Gel and Aquaform are two types of hydrogels that are used in wound treatment, but their use in treating diabetic foot lesions are restricted in individuals with limb ischemia [21]. Even though there are different types of commercially available diabetic wound dressings, the percentage of exudate absorption varies between them, which demands the development of new materials for treating different types of diabetic wounds. The new materials developed must hold a perfect balance between therapeutic molecules and antibiotics that are used to reduce healing time and the chance of formation of new ulcers [22].

The different forms of wound dressings (Figure 4) are as follows.

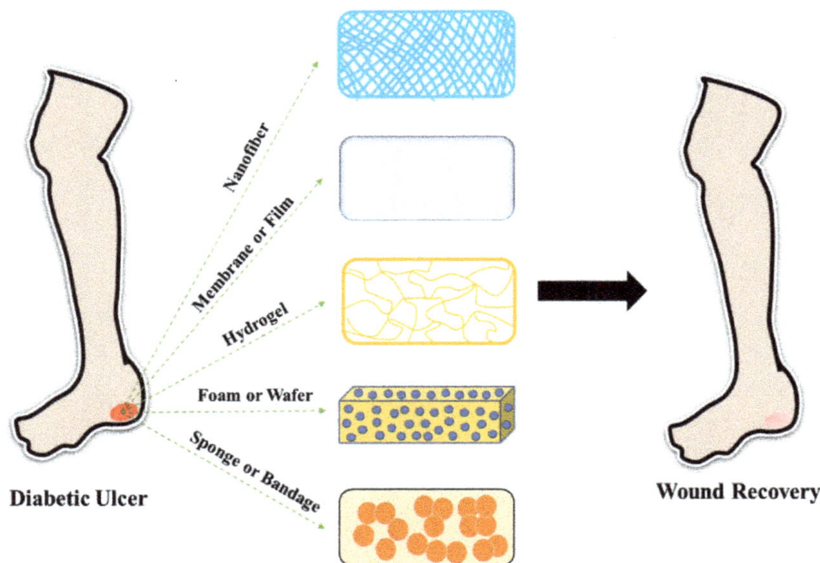


**Figure 4.** Different types of wound dressings used for diabetic wound treatment.

(i) Films are transparent, sticky materials that are widely used in the field of wound treatment. Their transparent nature assists in monitoring the healing of wounds without disturbing the injured site and dressing material [23]. They allow the permeation of gases including oxygen, water vapor, and carbon dioxide between the wounded site and the environment [24]. Film-based dressing materials have various benefits such as high flexibility and elasticity, and ease in fabricating them in the desired size and shape based on the application [25].

(ii) Hydrogels are commonly used in tissue engineering and wound recovery. They are formulated by the physical or chemical cross-linking of natural or synthetic polymers. Due to their three-dimensional polymeric network, they have the ability to absorb a high quantity of water molecules when compared with their dry weight. This property makes them superior among all wound dressing materials as they can retain excessive moisture content at the wounded site [26]. They can be fabricated in various forms and sizes, and loaded with anti-microbial substances, cells, and growth factors for reducing the time required for wound closure [27]. The ability of hydrogels to maintain a moist milieu helps in promoting granulation and re-epithelization, which, in turn, results in the regeneration of tissues.

(iii) Nanofibrous dressings are a group of nanofibers with sizes ranging from nanometres (nm) to micrometers (μm) [28]. Various strategies are employed to fabricate nanofibers, but the electrospinning technique is one of the most extensively utilized methods due to its enormous merits such as cost-effectiveness, ease, versatility, control of porosity, and tuning of mechanical properties of nanofibers. Once applied on the wound, the nanofibers can be removed easily without causing any damage to the applied site [29–31]. They can be loaded with various bioactive molecules for treating non-healing wounds [32]. Nanofibers have the ability to imitate the native extracellular matrix. They also offer an appropriate environment for cell proliferation and adhesion for rapid healing of wounds [33].

(iv) Foam is a type of wound dressing material that is composed of both hydrophilic and hydrophobic foam with bioadhesive boundaries [34]. The hydrophobic portion prevents unnecessary entry of liquids into the wound bed but permits gaseous exchange and water vapor permeation. The advantages of using a foam-based wound dressing are that they can maintain appropriate moisture content and absorb excess volume of wound exudates [35]. Based on the thickness of the wound, foam has the ability to absorb different amounts of wound exudates [36]. However, foams are inappropriate for dry wounds with fewer exudates [37].

(v) Wafer-based wound dressings are extremely porous freeze-dried polymers that have similar characteristics to those of foams. Wafers absorb the exudate of a wound and transform it into a gel or viscous solution that provides a moist atmosphere [38]. A few polymers including xanthan gum and sodium alginate have been used for the fabrication of wafers for biomedical applications [39]. The wound exudate-absorbing property of wafers helps in reducing fluid collection and microbial infection, which, in turn, aids in the quick recovery of wounds.

(vi) Sponges are soft and flexible with interconnected pores. Due to their porous nature, they have excellent swelling ability, which is an ideal feature of a wound dressing material [40]. Different kinds of sponges have been fabricated with different types of polymers for delivering therapeutic molecules for the efficient treatment of diabetic wounds. Sponges have been proven to help in cell migration and prevent microbial infection at the wound site [3]. Due to the presence of interconnected pores, sponges enhance the migration of fibroblasts, which results in faster closure of wounds.

## 5. Polymeric Biomaterials

Although there have been numerous advancements in the area of wound healing, the treatment of chronic wounds continues to be a major problem in patients suffering from diabetic foot ulcers and other major injuries [41]. An ideal wound dressing material must possess special features such as absorbing wound exudates, aiding in appropriate exchange of gas, and protecting from microbial infections. They must also support in the synthesis of biochemical mediators such as cytokines and growth factors that are essential for the proper healing of wounds [42]. Bioactive compounds extracted from natural sources have been investigated to understand their role in accelerating the process of healing diabetic wounds. The study of natural materials in the form of wound dressing has gained special attention because of their potency in inducing the formation of new tissues. Synthetic polymers are also extensively used in treating diabetic wounds as they exhibit excellent mechanical, bioinert, and biocompatible characteristics. Both natural and synthetic polymeric biomaterials are considered as satisfactory wound dressings due to their exceptional properties such as increased wound healing efficacy, less/no immunogenicity, good mechanical strength, and biocompatibility. Table 1 shows the different types of wound dressing materials prepared using both natural and synthetic polymers. Herein, we discuss the various wound dressing polymeric biomaterials that are used for treating diabetic wounds.

**Table 1.** Different types of polymeric biomaterial-based dressings.

| Authors | Material/Dressings | Therapeutic Compounds | Applications | Ref. |
|---|---|---|---|---|
| Ahmed et al. | Polyvinyl alcohol—Chitosan nanofiber mats | Zinc oxide NP | Microbial-Infected DW Care | [43] |
| Cam et al. | Polyvinylpyrrolidone-Polycaprolactone nanofibrous mats | Pioglitazone | DW Healing | [44] |
| Almasian et al. | Polyurethane—Carboxymethylcellulose nanofibers | Plant extract of *Malva sylvestris* | DW Treatment | [45] |
| Chen et al. | Poly-N-acetylglucosamine nanofibers | Polydeoxyribonucleotide | Diabetic Skin Ulcer | [46] |

**Table 1.** *Cont.*

| Authors | Material/Dressings | Therapeutic Compounds | Applications | Ref. |
| --- | --- | --- | --- | --- |
| Choi et al. | Polyethylene glycol—Polycaprolactone hybrid nanofibers | Human Epidermal Growth Factor | Diabetic Ulcer Treatment | [47] |
| Cui et al. | Polylactide-based nanofibers | Doxycycline | Chronic Wound Management | [48] |
| Grip et al. | Hydroxypropyl Methyl-cellulose/Polyethylene oxide nanofibers | β-Glucan | DW Care | [49] |
| Kanji et al. | Polyethersulfone nanofibers | Human umbilical cord blood-derived CD34+ cells | DW Management | [50] |
| Lee et al. | PLGA nanofibers | Platelet-derived growth factor, Vancomycin, and Gentamicin | Diabetic Infected Wound Care | [51] |
| Lee et al. | PLGA nanofibers | Insulin | DW Recovery | [52] |
| Merrel et al. | Polycaprolactone nanofibers | Curcumin | DW Management | [53] |
| Pinzón-García et al. | Polycaprolactone nanofibers | Bixin | DW Healing | [54] |
| Ranjbar-Mohammadi et al. | Polycaprolactone—Gum Tragacanth nanofibers | Curcumin | DW Care | [55] |
| Shalaby et al. | Cellulose acetate nanofibers | Silver NP | Microbial-Infected Diabetic Lesion Treatment | [56] |
| Zehra et al. | Polycaprolactone nanofibers | Sodium Percarbonate | DW Management | [57] |
| Lee et al. | PLGA—Collagen scaffold membranes | Glucophage | DW Management | [58] |
| Zheng et al. | PLGA—Cellulose nanocrystals nanofiber membranes | Neurotensin | DW Care | [59] |
| Liu et al. | Cellulose acetate—Zein composite nanofiber membranes | Sesamol | DW Treatment | [60] |
| Lee et al. | PLGA membranes | Metformin | DW Healing | [61] |
| Ren et al. | Poly-L-lactic acid fibrous membranes | Dimethyloxalylglycine-loaded mesoporous silica NP | DW Treatment | [62] |
| Lobmann et al. | Hyaluronic acid membranes | Human keratinocytes | Diabetic Foot Wounds | [63] |
| Augustine et al. | Poly(3-hydroxybutyrate-co-3-hydroxyvalerate) membranes | Cerium oxide NP/gelatin | DW Treatment | [64] |
| Augustine et al. | Polyvinyl alcohol—Polylactic acid hybrid membranes | Connective tissue growth factor | Wound Dressing Membranes For Diabetic Lesions And Chronic Ulcers | [65] |

**Table 1.** *Cont.*

| Authors | Material/Dressings | Therapeutic Compounds | Applications | Ref. |
|---|---|---|---|---|
| Arantes et al. | Chitosan films | Retinoic acid / solid lipid nanoparticles | DW Healing | [66] |
| Arul et al. | Collagen films | Biotinylated GHK peptide | DW Dressing | [67] |
| Inpanya et al. | Fibroin films | Aloe gel | DW Management | [68] |
| Kim et al. | Polyvinylpyrrolidone—Polyvinyl alcohol films | Sodium fusidate | Wound Healing | [69] |
| Mizuno et al. | Chitosan films | Fibroblast growth factors | DW Healing | [70] |
| Song et al. | Cellulose films | Selenium | Cutaneous DW Healing | [71] |
| Tan et al. | Sodium alginate hydrocolloid films | Vicenin-2 | DW Management | [72] |
| Tong et al. | Polyvinyl alcohol—Cellulose anocrystal films | Curcumin | DW Care | [73] |
| Voss et al. | Cellulose—Polyvinyl alcohol films | Propolis and/or Vitamin C | DW Management | [74] |
| Wu et al. | Silk Fibroin—Chitosan films | Adipose-derived stem cells | DW Care | [75] |
| Da Silva et al. | Hyaluronic acid spongy hydrogels | Human adipose stem cells | Diabetic Foot Ulcer | [76] |
| Lai et al. | Sodium carboxymethyl-cellulose hydrogels | Fern extracts *(Blechnum orientale Linn.)* | Diabetic Ulcer Treatment | [77] |
| Li et al. | Hydroxyapatite/Chitosan composite hydrogels | Exosomes (SMSCs-126) | DW Treatment | [78] |
| Masood et al. | Chitosan—Polyethylene glycol hybrid hydrogels | Silver NP | DW Healing | [79] |
| Shi et al. | Chitosan—Dextran hydrogels | Silver NP | DW Treatment | [80] |
| Thangavel et al. | Chitosan hydrogels | L-glutamic acid | DW Healing | [81] |
| Zhang et al. | Poly (γ-glutamic acid)—Heparin—Chitosan composite hydrogels | Superoxide dismutase | DW Treatment | [82] |
| Choi et al. | Polyurethane foams | Silver nanoparticles and Recombinant Human Epidermal Growth Factor | Bacteria-Infected DW Management | [83] |
| Pyun et al. | Polyurethane foams | Recombinant Human Epidermal Growth Factor | DW Treatment | [84] |
| Atia et al. | Sodium alginate—Gelatin wafers | Diosmin nanocrystals | DW Healing | [85] |
| Anisha et al. | Hyaluronic acid—Chitosan sponges | Silver nanoparticles | Wound Dressing for Diabetic Foot Ulcer | [86] |

**Table 1.** *Cont.*

| Authors | Material/Dressings | Therapeutic Compounds | Applications | Ref. |
| --- | --- | --- | --- | --- |
| Lipsky et al. | Collagen sponges | Gentamicin | Diabetic Foot Ulcer | [87] |
| Mohandas et al. | Chitosan—Hyaluronic acid composite sponges | Fibrin nanoparticles incorporated with vascular endothelial growth factors | Wound Dressing For DW | [88] |
| Shi et al. | Chitosan—Silk hybrid sponges | Gingival mesenchymal stem cell-derived exosomes | DW Healing | [89] |
| Wang et al. | Chitosan—Collagen sponges | Recombinant Human Acidic Fibroblast Growth Factors | DW Healing | [90] |
| Xia et al. | Chitosan composite sponges | Quaternary ammonium chitosan nanoparticles | Wound DressingMaterial for Diabetic Chronic Injury | [91] |
| Kondo et al. | Hyaluronic acid—Collagen sponges | Epidermal growth factors | DW Healing | [92] |
| Raveendran et al. | Chitosan bandages | Ciprofloxacin and Fluconazole-containing Fibrin nanoparticles | DW Management | [93] |
| Mohanty et al. | Sodium alginate—Chitosan bandages | Epidermal growth factor, curcumin, and mesenchymal stem cells | DW Healing | [94] |
| Kumar et al. | Chitosan hydrogel composite bandages | Zinc oxide nanoparticles | Wound Dressing Material | [95] |

NP—nanoparticles; PLGA—polylactic-co-glycolic acid; DW—diabetic wound.

### 5.1. Natural Polymers

Natural biomaterials are considered to be suitable candidates for preparing wound dressing material due to their exceptional properties such as less/no immunogenicity and good biocompatibility. They also serve as satisfactory matrices for cells that play imperative roles in the process of wound healing. Some of the widely accepted natural products extracted from natural sources that are widely used as wound dressing material are collagen, gelatin, fibrin and silk proteins.

#### 5.1.1. Collagen

Collagen provides integrity to human skin and serves as a principal component of the extracellular matrix (ECM) [96]. It is abundantly present in bones, ligaments, and tendons. It has distinctive properties such as excellent biocompatibility, thermal stability, mechanical strength, and low immunogenicity [97,98]. Collagen plays a vital role in haemostasis as it interacts with the platelets that are deposited at the site of wound through chemotaxis [99]. It mediates various pro-regenerative physiological interactions that are responsible for wound healing. Collagen is extensively used as a matrix for wound treatment and tissue regeneration. Collagen is isolated from different types of sources such as bovine, equine and, porcine tissues [100,101]. Although there are 29 types of collagens, type 1 collagen is widely available and can be extracted easily from mammalian connective tissues [102,103]. The most commonly used type 1 collagen is isolated from the tendons of rat tails [104,105]. Bovine collagen is extracted from several tissues such as bone, skin, and the Achilles tendon [106,107]. Collagen is formulated in the form of scaffolds with varying concentrations and pore sizes. These scaffolds can absorb wound exudates, attach onto the wound bed, and provide moist environment [108].

Collagen-based scaffolds (Figure 5) are commonly used as a wound dressing material for treating skin burns, foot ulcers, and pressure sores [109]. In certain cases, collagen is combined with other sources such as fibronectin or elastin to improve the fluid-binding property of the scaffolds [110,111]. Collagen is also fabricated in the form of implants that can be used as a support for delivering keratinocytes for skin regeneration [112–114]. After implantation, collagen scaffolds are infiltrated by connective tissues that are composed of glycosaminoglycan, new collagen, fibroblasts, and macrophages. Based on the cross-linking percentage, collagen scaffolds are degraded into small peptides within a few weeks of implantation, and replaced by native collagen that is synthesized by fibroblasts [115]. Apligraf®—the first tissue-engineered wound dressing material that was approved for treating diabetic ulcers—was made up of two-layered collagen hydrogels loaded with human keratinocytes and fibroblasts [116]. Subsequently, several modifications were carried out by altering the concentration of collagen to improve the mechanical strength of Apligraf®. When collagen concentration was increased, there was a significant rise in the proliferation of fibroblasts and stimulation of keratinocyte growth factor, which lead to the faster healing of wounds [117]. At present, collagen-based wound dressing materials are widely accepted in managing full-thickness wounds and skin burns. It is also possible to enhance the activity of collagen by combining it with bioactive therapeutic agents and antimicrobial compounds substances that accelerate the rate of wound closure [118]. Type 1 collagen has the ability to draw growth factors toward the wounded area and quicken the healing and regeneration of damaged tissues [98]. However, in the case of diabetic wounds or diabetic foot ulcers, the epidermis becomes ulcerous, resulting in the deficiency of type 1 collagen. This further delays the proliferation and migration of fibroblasts, which, in turn, prolongs the time required for wound healing [119]. Lee et al. determined the ability of collagen (colladerm) wound dressing in treating diabetic foot ulcers. When patients were treated with collagen dressing every 2–3 days for up to 3 weeks, a significant decrease in the wound area with 73.7% healing of the diabetic wound ulcer was observed. The results demonstrated the safety and efficacy of collagen dressing in faster healing of diabetic foot ulcers [120]. Hauck et al. demonstrated the possible use of hyaluronan/collagen hydrogels loaded with high-sulphated hyaluronan in treating dermal wounds in diabetic mice models. The hydrogel enhanced the healing rate of damaged tissues with decreased inflammation, improved vascularization, and increased pro-regenerative macrophage activation, and hastened the formation of new tissues for wound closure [121]. Shagdarova et al. prepared hydrogels using chitosan, collagen, and silver nanoparticles for treating diabetic injuries/wounds. The hydrogels had a fibrous porous structure with a better swelling ratio. When applied onto diabetic wounds, the hydrogels elevated the expression of genes such as vascular endothelial growth factor, Interleukin 1b, tissue inhibitor of metalloproteinases-1, and transforming growth factor beta 1 [122].

#### 5.1.2. Gelatin

Gelatin is a natural polymer that is obtained from the partial hydrolysis of collagen [123]. Due to its salient features such as availability, biodegradability, biocompatibility, cell-interactivity, and non-toxicity, gelatin is commonly used in the field of biomedicine [124]. When used as a scaffold, gelatin has the ability to absorb water molecules, making it an appropriate candidate for wound dressing material. The main drawback associated with gelatin is its poor stability and mechanical strength. Therefore, to increase its mechanical stability, gelatin is cross-linked with agents such as glutaraldehyde, fructose, dextran, genipin, formaldehyde, and carbodiimides [125]. Samadian et al. developed berberine-loaded cellulose acetate/gelatin electrospun mats as wound dressing for treating diabetic foot ulcers. The fibres had an average diameter of $502 \pm 150$ nm, and demonstrated antibacterial behaviour against *Staphylococcus aureus* and *Pseudomonas aeruginosa*. The electrospun mats exhibited suitable tensile strength and water uptake potency required for a wound dressing material. A haemolytic assay performed using red blood cells showed that the percentage of haemolysis was significantly low for berberine-loaded cellulose acetate/gelatin electrospun mats when compared with the positive control—water [126]. Yu et al. prepared a paeoniflorin-sodium alginate (SA)-gelatin skin scaffold with a mesh-like structure with uniform pore distribution for treating diabetic wounds. Animal models showed improved deposition of collagen with microvascular regeneration when treated with the skin scaffold, thereby proving their possible use in the field of diabetic wound treatment [127]. Sadeghi et al. prepared biodegradable scaffolds using gelatin and sulphated alginate as skin replacements to accelerate the healing of diabetic wounds. The carbodiimide mode of cross-linking followed by lyophilization was carried out to prepare the scaffolds. Cell culture analysis proved the non-toxicity of the scaffolds, with enhanced cell growth when the quantity of sulphated alginate was increased in the scaffold. Diabetic animal models proved the ability of the scaffold to cure wounds by providing the required environment for faster healing of wounds [128].

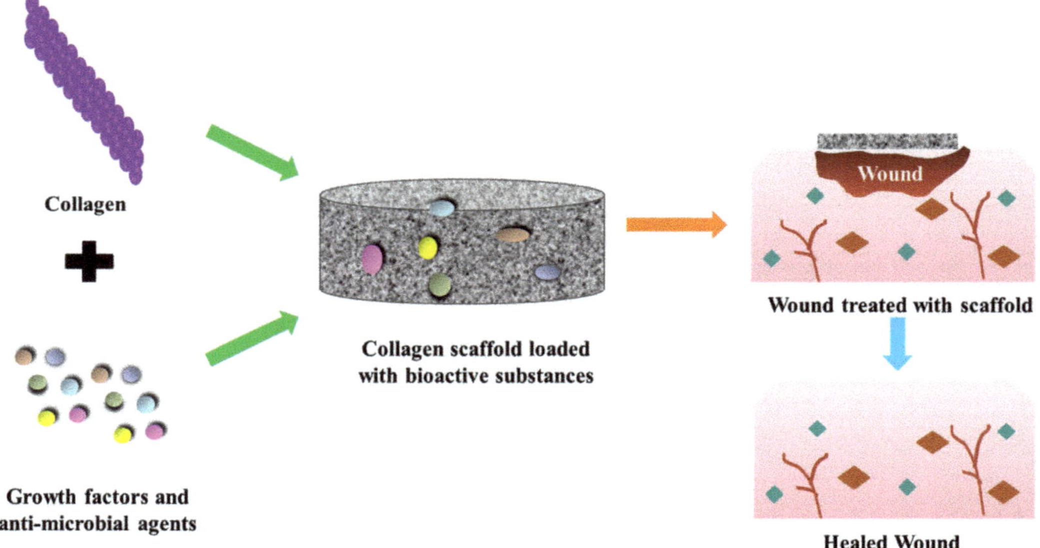


**Figure 5.** Collagen scaffold loaded with therapeutic molecules for diabetic wound treatment.

#### 5.1.3. Fibrin

Fibrin is obtained from fibrinogen, which is converted in response to tissue injury. It acts as a mesh and forms blood clots to prevent bleeding. Fibrin is extensively used for clinical applications, in the form of sealants and haemostatic agents [129]. When used in the form of scaffolds, it has the ability to deliver inflammatory cells and growth factors that are necessary for wound repair and tissue regeneration [130].

Fibrin serves as a substrate for different types of cells such as platelets, fibroblasts, endothelial cells, and macrophages. It triggers the process of cellular proliferation and new blood vessel formation, thereby leading to efficient healing of wounds. Fibrin can be formulated in different structures such as nanoparticles, hydrogels, scaffolds and films. Fibrin-based cell delivery is widely accepted for treating dermal wounds, as it stimulates neovascularization and the rejuvenation of skin cells [131]. For treating skin burns, fibrin-based hydrogels/films are utilized for the transplantation of keratinocytes to induce fibroblast formation and re-epithelization. Fibrin scaffolds loaded with vascular endothelial factor and fibroblast growth factor enhanced re-epithelization, collagen deposition, and accelerated wound closure in mice with diabetic wounds [132]. Fibrin as a scaffold has the ability to mimic the extracellular matrix and enhance the interaction of cells responsible for tissue regeneration [133,134]. Geer et al. studied the re-epithelializing performance of fibrin using human keratinocytes under in vitro conditions [135]. Falanga et al. demonstrated that when bone marrow-derived mesenchymal stem cells (Figure 6) were entrapped within fibrin, they reduced the time taken for wound closure in acute and chronic wounds [136]. When bone marrow nuclear cells were mixed and injected along with fibrin gel to treat infarcted myocardium, it resulted in neovascularization and increased tissue regeneration. Fibrin-based formulations are highly efficient in curing diabetic wounds. Crisci et al. investigated the efficacy of fibrin rich in leukocytes and platelets (FLP) in treating osteomyelitis ulcers in diabetic feet. FLP was collected from diabetic individuals suffering from osteomyelitis and skin lesions for a minimum period of 180 days. Surgical debridement was carried out to deliver FLP directly into skin lesions of patients, and the development of lesions was assessed periodically. The study report (Figure 7) stated that FLP treatment settled down skin lesions with no indication of microbial infection [137].

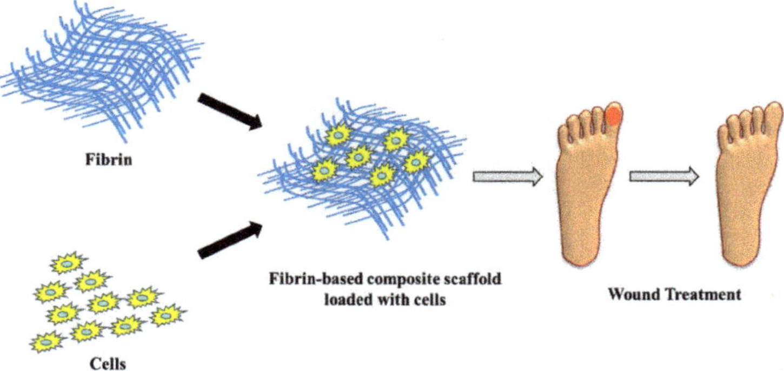


**Figure 6.** Fibrin-based therapy for wound treatment.

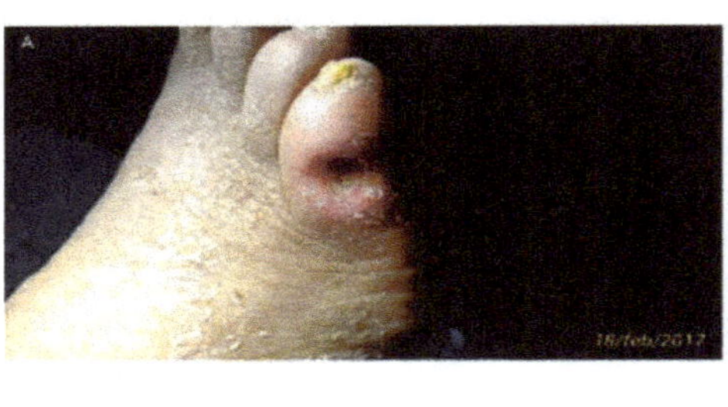


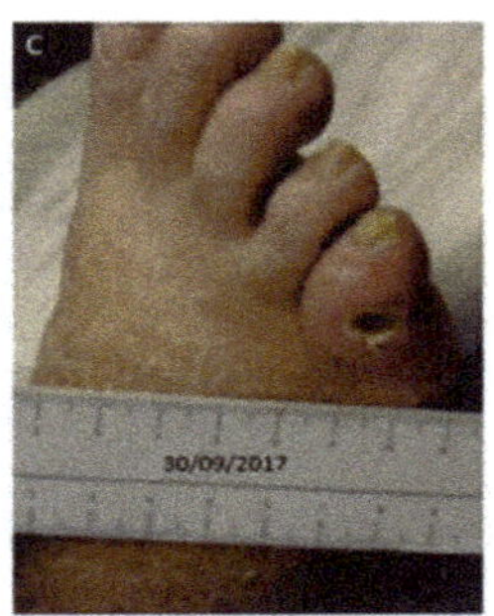


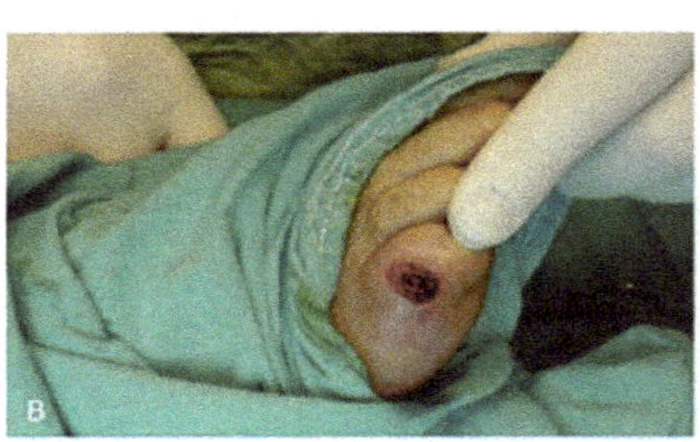


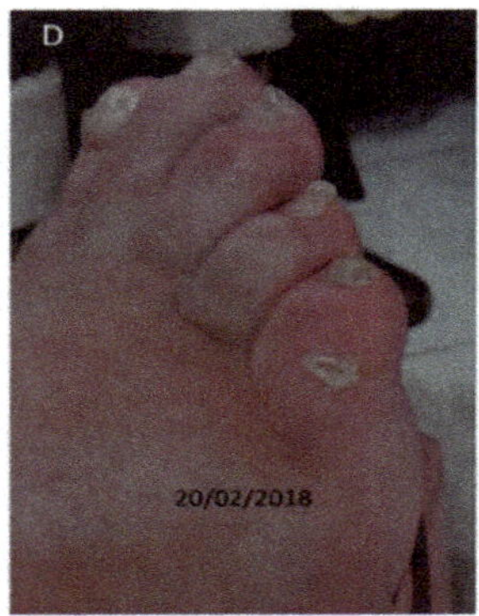


**Figure 7.** Efficacy of fibrin rich in leukocytes and platelets (FLP) in treating osteomyelitis ulcers in diabetic feet. Reprinted from Ref. [137].

Losi et al. employed electrospinning and spray phase-inversion procedures to synthesize bilayered fibrin/poly(ether)urethane scaffolds rich in platelet lysate for treating diabetic wounds. Cell culture experiments performed using L929 mouse fibroblasts proved that the efficacy of two important growth factors—platelet-derived growth factor (PDGF) and vascular endothelial growth factor (VEGF), which play a crucial role in healing chronic wounds—was retained in the electrospun scaffolds. An in vitro release study of PDGF and VEGF from the synthesized scaffolds demonstrated an initial burst release of growth factors within 24 h of study, followed by sustained release for one week. When applied onto full-thickness wounds in diabetic animal models, a significant improvement in wound closure within 2 weeks of treatment was observed. Moreover, improved re-epithelialization and collagen deposition were also witnessed in wounds treated with the scaffolds, thereby proving their potency in healing diabetic wounds/ulcers [138]. Poly(ether)urethane-polydimethylsiloxane/fibrin-based scaffolds containing poly(lactic-co-glycolic acid) nanoparticles loaded with recombinant human vascular endothelial growth factor and basic fibroblast growth factor were fabricated by Losi et al. with the intention of triggering cellular proliferation and accelerating the process of wound healing in genetically diabetic mice. The presence of growth factors in the scaffolds quickened the rate of closure of full-thickness skin wounds on day 15 in diabetic mice. Histological analysis showed extensive re-epithelialization, with increased granulation tissue formation/maturity and collagen deposition, thereby elucidating the efficiency of the prepared scaffolds in treating diabetic wounds [133].

#### 5.1.4. Silk Proteins

Silk cocoons are discarded as waste by the silk industry, but they can be used as valuable resources for fabricating wound dressings that can aid in faster healing of wounds.

Silk-based biomaterials are extensively used in the field of medicine due to their excellent biocompatibility and biodegradability. The FDA approved product silk voice is a type of scaffold, prepared using reconstituted or solubilized silk protein [139]. Silk protein has the capability to induce cell migration and proliferation, and attract cells such as keratinocytes to the wounded site, thereby accelerating the process of wound healing [140]. Two different types of proteins—silk fibroin and silk sericin—are isolated from the cocoons of silkworms (Figure 8). They are widely used in biomedical applications due to their lower immunogenicity, biodegradability, biocompatibility, moisture absorption, UV resistance, and antibacterial properties [141].

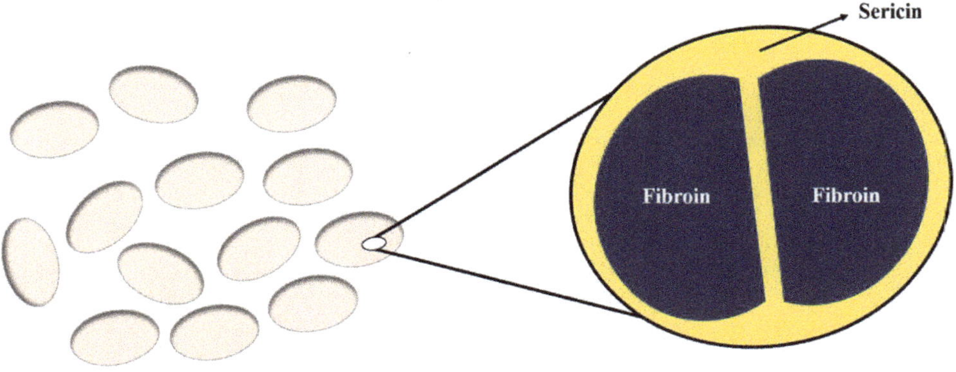


**Figure 8.** Two different types of silk proteins—silk fibroin and silk sericin.

Silk proteins are fabricated in various forms such as films, nanofibers, and sponges for biomedical applications. When used as a wound dressing material, silk proteins have enhanced fibroblast adhesion and lead to faster healing of wounds [142]. The middle and posterior silk glands of silkworm *Bombyx mori* secretes two important proteins, silk fibroin and silk sericin, which are extensively used to accelerate the healing of wounds [143]. Silk fibroin has extraordinary thermal stability and mechanical strength when compared with polymers such as collagen and polylactic acid. In addition, the presence of an RGD peptide sequence promotes the attachment, movement, and proliferation of cells such as keratinocytes, endothelial, epithelial, and glial cells, and osteoblasts for the effective healing of wounds. Porous silk fibroin scaffolds/sponges can be prepared using different types of techniques such as lyophilization, gas forming, and freeze-drying/foaming methods [144]. Liu et al. prepared silk fibroin scaffolds incorporated with neurotensin-loaded gelatin microspheres as a novel therapeutic regime for healing diabetic foot ulcers in diabetic rat models. Macroscopic evaluation of wounds showed significant reduction in wound size on day 14 in the experimental group. In addition, histological and immunofluorescence analyses demonstrated the accumulation of fibroblasts with a substantial expression of collagen at the site of the wound. The prepared scaffolds had a good porosity of approximately 85%, with an average pore size of 40–80 μm [145]. Guan et al. fabricated microneedle patches with multiple features such as anti-microbial, anti-oxidant, and pro-angiogenic properties for targeting diabetic lesions/wounds. The microneedles were constructed using silk fibroin methacryloyl with tremendous biocompatibility and mechanical stability. Two different bioactive molecules—VEGF and Prussian blue nanozymes—were loaded on the tip of microneedles with polymyxin—an anti-bacterial agent—at the base layer of the microneedle patches. The patches played a significant role in treating diabetic skin wounds [146]. Xu et al. prepared electrospun Huangbai liniment-loaded silk fibroin/poly-(L-lactide-co-caprolactone) nanofibers for treating diabetic wounds. The fibres were smooth, without any bead formation when viewed under a scanning electron microscope. The nanofibers exhibited antibacterial activity against *Escherichia coli* and *Staphylococcus aureus*. Cell culture experiments showed enhanced adhesion and proliferation of NIH-3T3 cells when cultured on the nanofibers. Animal experiments using a diabetic mice model proved that the nanofibers had the ability to elevate the synthesis of collagen and expression of the TGF-β signalling pathway, which, in turn, promoted efficient healing of diabetic wounds [147]. Silk sericin is a globular protein with significant features such as non-immunogenicity, good biocompatibility, biodegradability, anti-oxidant properties, and regenerative potency. It serves as a promising biomaterial in the field of medicine. Silk sericin-based hydrogels were prepared using horseradish peroxidase (HRP) and hydrogen peroxide cross-linking methods. A chorioallantoic membrane assay performed using chick embryos showed a minimal rise in the number of new blood vessels in the test group when compared with that of the control. The hydrogel showed collagen deposition and mild induction of superoxide dismutase and catalase in diabetic wounds treated with hydrogel in the mouse model [148]. Samad et al. formulated carboxymethyl cellulose/sericin hydrogels for diabetic wound treatment. The hydrogels had porous morphology, excessive swelling efficacy, and antimicrobial properties. When applied on to full-thickness excision wounds in diabetic rats, the upregulation of collagen deposition and downregulation of pro-inflammatory markers was witnessed, which led to the healing of wounds without any insulin treatment [149].

### 5.2. *Synthetic Polymers for Diabetic Wounds*

Synthetic polymers are used in combination with natural polymers for treating diabetic wounds because they exhibit excellent mechanical properties. They are used for tissue engineering applications owing to their inert and biocompatible characteristics. Synthetic polymers such as polycaprolactone (PCL), poly(vinyl alcohol) (PVA), Poly(2-hydroxyethyl methacrylate) (pHEMA), polylactide (PLA), and polyglycolic acid (PGA) have been used as scaffolds in tissue engineering and wound healing applications along with natural polymers [3,150,151].

#### 5.2.1. Polycaprolactone (PCL)

PCL is a hydrophobic polymer that has a great degradation rate and excellent bioactivity. It is a linear aliphatic semicrystalline polymer. PCL polymer can be modified by changing the molecular weight, crystallinity, or structure using polyethylene glycol and hydrophobic ceramics, or by creating copolymers with PLA and PGA. PCL exhibits reduced cellular attachment owing to its hydrophobicity, which can be altered by modifying its surface with other biomaterials [152]. An organic and inorganic composite scaffold containing two-dimensional nanovermiculite and PCL electrospun fibres for treating diabetic wounds were prepared by Huang et al. The results show that polycaprolactone electrospun fibres with two-dimensional vermiculite nanosheets could significantly improve neo-vascularization, re-epithelialization, and collagen formation in the diabetic wound bed [153]. Amine-terminated block copolymers containing PCL and polyethylene glycol and PCL were electrospun using electrospinning technique by Choi et al. The human epidermal growth factors (EGF) were immobilized on the surface of the nanofibers. Dorsal wounds were created in diabetic animals in order to study the wound healing efficacy of the prepared wound dressing material. Immunohistochemical studies showed that the EGF receptor were highly expressed in the nanofiber-treated groups. The results showed that the prepared nanofibers could be a potential material for treating diabetic wounds [47]. Merrell et al. used PCL nanofibers as drug delivery vehicles. He prepared PCL nanofibers loaded with curcumin and used them as a diabetic wound dressing material. In total, 70% of human foreskin fibroblast cells (HFF-1) cells were viable when treated with the prepared nanofibers. A streptozocin-induced diabetic mouse model were used for the in vivo study, which showed an increased rate of wound closure in animals treated with the nanofibers. The study proved that the prepared nanofibers are bioactive, with anti-inflammatory and antioxidant properties [53]. Lv et al. prepared a PCL/gelatin nanofiber composite scaffold containing silicate-based ceramic particles (Nagelschmidtite, NAGEL, $Ca_7P_2Si_2O_{16}$)

through the co-electrospinning technique for diabetic wound healing (Figure 9). In vivo studies revealed that these nanofiber composite scaffolds promoted angiogenesis, the deposition of collagen, and re-epithelialization at the wounded site in the diabetic mice. The results suggested that the release of Si ions and the structure of nanofibrous scaffolds have the potential for diabetic wound healing, and pave the way for biomaterials used in the field of both wound healing and tissue engineering applications [154].

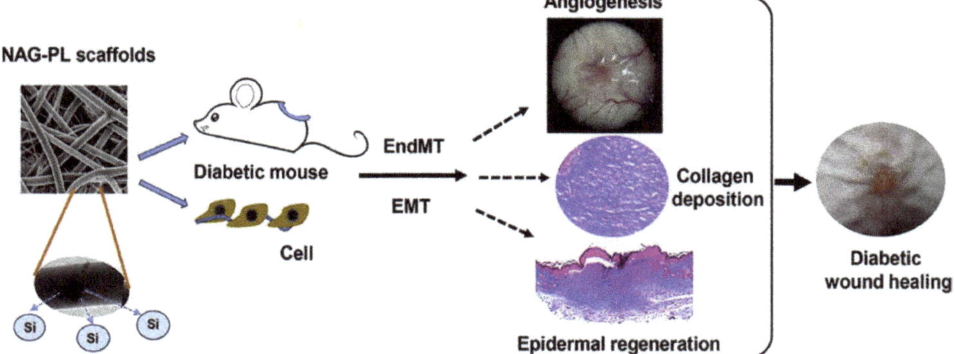


**Figure 9.** The role of poly (caprolactone)/gelatin nanofibrous scaffolds in treating diabetic wound healing. Si—Silicon ions; NAG-PL—Nagelschmidtite-Poly(caprolactone); EndMT—Endothelial mesenchymal transformation; EMT—Epithelial-to-mesenchymal transition. Reprinted from Acta Biomaterialia, Vol. Number 60; Lv F., Wang J., Xu P., Han Y., Ma H., Xu H., Chen S., Chang J., Ke Q., Liu M., Yi Z.; A conducive bioceramic/polymer composite biomaterial for diabetic wound healing, Pages No. 128–143, 2017 with permission from Elsevier [154].

#### 5.2.2. Poly(vinyl alcohol) (PVA)

PVA is an excellent biocompatible synthetic polymer produced by the hydrolysis of vinyl acetate. It is one of the US Food and Drug Administration (FDA)-approved synthetic polymeric materials. The value of PVA is growing at an enormous rate, as it is utilized for biomedical applications owing to desirable characteristics such as non-carcinogenic, non-toxic, bio-adhesive, and swelling behaviours. PVA has been potentially used in soft eye lenses, cartilages, and eye drops, etc. PVA can be fabricated in different forms such as fibres, gel, and film that support in aiding the adhesion and proliferation of cells [155]. Huang et al. fabricated electrospun brown alga-derived polysaccharide and PVA nanofibers for skin repair in diabetic mice. Brown alga-derived polysaccharide is a sea mustard found in marine areas. The result suggested that the prepared nanofibers decreased inflammation and stimulated angiogenesis at the wound site of diabetic mice [156]. Lin et al. synthesized PVA/cobalt-substituted hydroxyapatite nanocomposites as a wound dressing material for diabetic foot ulcer treatment. The nanocomposites were prepared using solvent casting method. The result showed that the prepared nanocomposites had high mechanical properties and excellent bioactivity. The nanocomposites discharged a small number of cobalt ions into the cell-cultured medium, which showed better cell growth. The prepared nanocomposites could be a potential wound dressing material for diabetic foot ulcer treatment [157]. Zhu et al. fabricated PVA hydrogel loaded with fibroblast growth factor 21 and metformin for diabetic wound healing. The fabricated hydrogel were injectable, adhesive and ROS scavenging abilities. In vivo results showed the formation of blood vessels with faster healing of diabetic wounds [158]. Wang et al. prepared PVA/chitosan nanocomposite hydrogels incorporated with Tibetan eighteen-flavour dangshen pills (TEP) for treating chronic diabetic wounds. TEP is a traditional Tibetan medicine used to treat skin diseases with analgesic, anti-inflammatory, and healing properties. This hydrogel were treated with L939 cells, which showed no cytotoxic effect and demonstrated that the new formulation can be used for treating diabetic wounds with the help of traditional medicine [159]. Cellulose and PVA-based films incorporated with Vitamin C and/or propolis for faster diabetic wound healing were created by Voss et al. Cellulose-PVA/Vitamin C and Cellulose-PVA/Vitamin C/Propolis films were prepared in order to analyse the release of Vitamin C in a precise manner. When Cellulose-PVA/Vitamin C/Propolis were used in an STZ-induced diabetic animal model, it showed faster wound closure. Histopathological analysis showed better results when treated with Cellulose-PVA/Vitamin-C/Propolis. The results suggest that the prepared PVA-based film could be a potential treatment procedure for faster wound healing (Figure 10) [74].

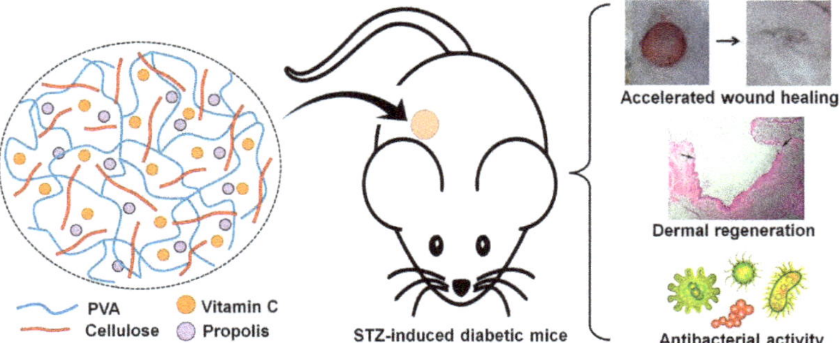


**Figure 10.** PVA-based films for diabetic wound healing. Reprinted from International Journal of Pharmaceutics, Vol. Number 552 (1–2), Voss G.T., Gularte M.S., Vogt A.G., Giongo J.L., Vaucher R.A., Echenique J.V., Soares M.P., Luchese C., Wilhelm E.A., Fajardo A.R., Polysaccharide-based film loaded with vitamin C and propolis: A promising device to accelerate diabetic wound healing, Pages No 340–351, 2018 with permission from Elseiver [74].

Ahmed et al. fabricated chitosan, PVA, and zinc oxide nanofibrous mats using electrospinning technique for faster healing of diabetic wounds. The compounds chitosan and PVA have wound-healing properties, and zinc oxide has excellent antibacterial activity. The result showed that these nanofiber mats exhibit efficient antibacterial and antioxidant properties. In vivo analysis showed that there was faster healing of diabetic wounds in groups treated with nanofiber mats [43]. Kim et al. developed a new film-forming hydrogel including PVA, polyvinylpyrrolidone, and propylene glycol incorporated with sodium fusidate for wound healing applications. The film showed excellent elasticity and flexibility, and could be an effective pharmaceutical product for wound treatment [69].

#### 5.2.3. Poly(2-Hydroxyethyl Methacrylate) (pHEMA)

pHEMA is a hydrophilic, non-biodegradable, and biocompatible polymer that is widely used for different types of wound healing, bone regeneration, and cancer treatment. pHEMA-based biomaterial gained lot of attention in wound treatment and ocular therapy due to its excellent biocompatible and minimal thrombogenic properties. Due to its transparent nature, it facilitates the tracking of wound recovery when used as wound dressing material. Bacterial cellulose pHEMA and silver were combined as a multifunctional wound dressing material with efficient antimicrobial properties [160].

#### 5.2.4. Polylactide (PLA) and Polyglycolic Acid (PGA)

PGA and PLA are extensively suitable for the fabrication of scaffolds. These two synthetic polymers serve as a suitable platform for tissue construction. Moreover, these polymers have been used as implantable materials in the field of medicine. Polyglycolic acid is the first synthetic polymer utilized in the form of suture under the name of "Dexon". At the site of the wound, PLGA and PLA stimulate the supply of lactic acid, and help in inducing angiogenesis and quickening the process of wound healing [155]. Khazaeli et al. prepared PLA/chitosan nanoscaffolds using the microwave-assisted electrospinning technique loaded with cod liver oil for diabetic wound healing. The results showed that the groups treated with nanoscaffolds exhibited wound recovery within 14 days of treatment [161]. Zheng et al. fabricated polylactic co-glycolic acid/cellulose nanocrystal nanofibers loaded with neurotensin to study their therapeutic potency in treating diabetic wounds. The prepared nanofibers were applied on wounds in diabetic mice, which showed a slow release of neurotensin for 2 weeks. The results suggested that the prepared nanofibers effectively stimulate the regeneration of tissues for diabetic foot ulcer treatment [59]. Zha et al. prepared polyglycolic acid/silk fibroin nanofibrous scaffolds incorporated with deferoxamine for diabetic wound healing application. The prepared nanofibrous scaffold had a porous three-dimensional nanofibrous structure and exhibited good mechanical strength, biodegradability, and biocompatibility, which could promote cell adhesion, growth, and migration. The in vivo results showed that the prepared nanofibrous scaffold accelerated the wound healing rate in treated groups [162].

## 6. Future Perspectives and Conclusions

To date, there are numerous polymer-based wound dressings for treating and managing diabetic wounds/foot ulcers. These dressings differ in their porosity, mechanical strength, swelling, and moisture absorption properties. They aid in cellular adhesion, proliferation, and migration without causing any cytotoxicity or immunotoxicity. Wound dressings protect wounds from microbial infections and physical damage. The efficacy of the polymeric biomaterials can be improved by loading therapeutic molecules, growth factors, and anti-microbial agents that could accelerate the process of wound closure by triggering collagen deposition and vascularization. The implantation of the dressing material exactly at the site of the wound is a challenging procedure, as it requires appropriate coordination between the scientist and physician. The method employed to synthesize scaffolds is also crucial as it plays a vital role in determining the quality and performance of the wound dressings. In addition, the fabrication technique may involve costly sophisticated instruments, which may further increase the cost of treatment. Therefore, it is important to recognize the issues that affect clinical translation, and pursue alternatives that can overcome the current problems.

Further, research in the field of 3D printing and tissue engineering can improve the potency of polymeric wound dressings for efficient diabetic wound treatment. Wound dressing materials that are 3D printed can serve as a unique platform and can be incorporated with different types of bioactive compounds and antimicrobial agents that can speed up the rate of wound healing. The use of 3D printing can overcome the disadvantages associated with the conventional techniques, and is highly reliable and low-cost. Additionally, the combination of biomarkers and nanoparticles that can be used to monitor wound recovery can be loaded with the wound dressing materials for efficient diabetic wound treatment.

**Author Contributions:** Conceptualization, writing—original draft and supervision, W.V.S.; Writing—original draft, L.P. and J.B.; Investigation, S.D. and R.K.; Investigation and visualization, R.A. All authors have read and agreed to the published version of the manuscript.

**Funding:** The authors Weslen Vedakumari Sathyaraj and Lokesh Prabakaran thank the Science and Engineering Research Board (SERB), Government of India for supporting this work under the Promoting Opportunities for Women in Exploratory Research (SERB-POWER) Scheme [File Number: SPG/2021/003353].

**Institutional Review Board Statement:** Not applicable.

**Informed Consent Statement:** Not applicable.

**Data Availability Statement:** Not applicable.

**Conflicts of Interest:** The authors declare no conflict of interest.

## References


1. Patel, S.; Srivastava, S.; Singh, M.R.; Singh, D. Mechanistic insight into diabetic wounds: Pathogenesis, molecular targets and treatment strategies to pace wound healing. *Biomed. Pharmacother.* **2019**, *112*, 108615. [CrossRef] [PubMed]
2. Selle, J.J.; Aminuddin, A.; Chellappan, K. Benefit of Foot Thermogram Analysis in the Treatment of Diabetic Foot Ulcer: A Systematic Review. *Biomed. Res. Ther.* **2022**, *9*, 5029–5042. [CrossRef]
3. Alven, S.; Peter, S.; Mbese, Z.; Aderibigbe, B.A. Polymer-Based Wound Dressing Materials Loaded with Bioactive Agents: Potential Materials for the Treatment of Diabetic Wounds. *Polymers* **2022**, *14*, 724. [CrossRef] [PubMed]
4. Okonkwo, U.A.; DiPietro, L.A. Diabetes and Wound Angiogenesis. *Int. J. Mol. Sci.* **2017**, *18*, 1419. [CrossRef] [PubMed]
5. Spampinato, S.F.; Caruso, G.I.; De Pasquale, R.; Sortino, M.A.; Merlo, S. The Treatment of Impaired Wound Healing in Diabetes: Looking among Old Drugs. *Pharmaceuticals* **2020**, *13*, 60. [CrossRef]
6. Singh, S.; Young, A.; McNaught, C.-E. The physiology of wound healing. *Surgery* **2017**, *35*, 473–477. [CrossRef]
7. Guo, S.; DiPietro, L.A. Factors Affecting Wound Healing. *J. Dent. Res.* **2010**, *89*, 219–229. [CrossRef]
8. Subramaniam, T.; Fauzi, M.; Lokanathan, Y.; Law, J. The Role of Calcium in Wound Healing. *Int. J. Mol. Sci.* **2021**, *22*, 6486. [CrossRef]
9. Gosain, A.; DiPietro, L.A. Aging and Wound Healing. *World J. Surg.* **2004**, *28*, 321–326. [CrossRef]
10. Broughton, G.; Janis, J.E.; Attinger, C.E. The Basic Science of Wound Healing. *Plast. Reconstr. Surg.* **2006**, *117*, 12S–34S. [CrossRef]
11. Keylock, K.T.; Vieira-Potter, V.; Wallig, M.A.; DiPietro, L.A.; Schrementi, M.; Woods, J.A. Exercise accelerates cutaneous wound healing and decreases wound inflammation in aged mice. *Am. J. Physiol. Integr. Comp. Physiol.* **2008**, *294*, R179–R184. [CrossRef]
12. Krzyszczyk, P.; Schloss, R.; Palmer, A.; Berthiaume, F. The Role of Macrophages in Acute and Chronic Wound Healing and Interventions to Promote Pro-wound Healing Phenotypes. *Front. Physiol.* **2018**, *9*, 419. [CrossRef] [PubMed]
13. Mosser, D.M.; Edwards, J.P. Exploring the full spectrum of macrophage activation. *Nat. Rev. Immunol.* **2008**, *8*, 958–969. [CrossRef]
14. Campos, A.C.; Groth, A.K.; Branco, A.B. Assessment and nutritional aspects of wound healing. *Curr. Opin. Clin. Nutr. Metab. Care* **2008**, *11*, 281–288. [CrossRef]
15. Oro, F.B.; Sikka, R.S.; Wolters, B.; Graver, R.; Boyd, J.L.; Nelson, B.J.; Swiontkowski, M. Autograft Versus Allograft: An Economic Cost Comparison of Anterior Cruciate Ligament Reconstruction. *Arthrosc. J. Arthrosc. Relat. Surg.* **2011**, *27*, 1219–1225. [CrossRef] [PubMed]
16. Flynn, M.; Tooke, J. Aetiology of Diabetic Foot Ulceration: A Role for the Microcirculation? *Diabet. Med.* **1992**, *9*, 320–329. [CrossRef] [PubMed]
17. Tsourdi, E.; Barthel, A.; Rietzsch, H.; Reichel, A.; Bornstein, S.R. Current Aspects in the Pathophysiology and Treatment of Chronic Wounds in Diabetes Mellitus. *BioMed Res. Int.* **2013**, *2013*, 1–6. [CrossRef] [PubMed]
18. Burgess, J.L.; Wyant, W.A.; Abdo Abujamra, B.; Kirsner, R.S.; Jozic, I. Diabetic Wound-Healing Science. *Medicina* **2021**, *57*, 1072. [CrossRef] [PubMed]
19. Okizaki, S.-I.; Ito, Y.; Hosono, K.; Oba, K.; Ohkubo, H.; Amano, H.; Shichiri, M.; Majima, M. Suppressed recruitment of alternatively activated macrophages reduces TGF-β1 and impairs wound healing in streptozotocin-induced diabetic mice. *Biomed. Pharmacother.* **2015**, *70*, 317–325. [CrossRef]
20. Loots, M.A.; Kenter, S.B.; Au, F.L.; van Galen, W.; Middelkoop, E.; Bos, J.D.; Mekkes, J.R. Fibroblasts derived from chronic diabetic ulcers differ in their response to stimulation with EGF, IGF-I, bFGF and PDGF-AB compared to controls. *Eur. J. Cell Biol.* **2002**, *81*, 153–160. [CrossRef]
21. Hilton, J.R.; Williams, D.T.; Beuker, B.; Miller, D.R.; Harding, K.G. Wound Dressings in Diabetic Foot Disease. *Clin. Infect. Dis.* **2004**, *39*, S100–S103. [CrossRef] [PubMed]
22. Gianino, E.; Miller, C.; Gilmore, J. Smart Wound Dressings for Diabetic Chronic Wounds. *Bioengineering* **2018**, *5*, 51. [CrossRef] [PubMed]
23. Sood, A.; Granick, M.S.; Tomaselli, N.L. Wound Dressings and Comparative Effectiveness Data. *Adv. Wound Care* **2014**, *3*, 511–529. [CrossRef] [PubMed]
24. Negut, I.; Grumezescu, V.; Grumezescu, A.M. Treatment Strategies for Infected Wounds. *Molecules* **2018**, *23*, 2392. [CrossRef] [PubMed]
25. Gupta, B.; Agarwal, R.; Alam, M. Textile-Based Smart Wound Dressings. *Indian J. Fibre Text. Res.* **2010**, *35*, 174–184.
26. Tavakoli, S.; Klar, A.S. Advanced Hydrogels as Wound Dressings. *Biomolecules* **2020**, *10*, 1169. [CrossRef]
27. Gupta, P.; Vermani, K.; Garg, S. Hydrogels: From controlled release to pH-responsive drug delivery. *Drug Discov. Today* **2002**, *7*, 569–579. [CrossRef] [PubMed]
28. Xu, S.-C.; Qin, C.-C.; Yu, M.; Dong, R.-H.; Yan, X.; Zhao, H.; Han, W.-P.; Zhang, H.-D.; Long, Y.-Z. A battery-operated portable handheld electrospinning apparatus. *Nanoscale* **2015**, *7*, 12351–12355. [CrossRef]
29. Hajilou, H.; Farahpour, M.R.; Hamishehkar, H. Polycaprolactone nanofiber coated with chitosan and Gamma oryzanol functionalized as a novel wound dressing for healing infected wounds. *Int. J. Biol. Macromol.* **2020**, *164*, 2358–2369. [CrossRef]
30. Mulholland, E.J.; Ali, A.; Robson, T.; Dunne, N.J.; McCarthy, H.O. Delivery of RALA/siFKBPL nanoparticles via electrospun bilayer nanofibers: An innovative angiogenic therapy for wound repair. *J. Control. Release* **2019**, *316*, 53–65. [CrossRef]
31. Zhao, Y.; Qiu, Y.; Wang, H.; Chen, Y.; Jin, S.; Chen, S. Preparation of Nanofibers with Renewable Polymers and Their Application in Wound Dressing. *Int. J. Polym. Sci.* **2016**, *2016*, 1–17. [CrossRef]

32. Liu, M.; Duan, X.P.; Li, Y.M.; Yang, D.P.; Long, Y.Z. Electrospun nanofibers for wound healing. *Mater. Sci. Eng. C* **2017**, *76*, 1413–1423. [CrossRef] [PubMed]
33. Heydari, P.; Varshosaz, J.; Kharazi, A.Z.; Karbasi, S. Preparation and evaluation of poly glycerol sebacate/poly hydroxy butyrate core-shell electrospun nanofibers with sequentially release of ciprofloxacin and simvastatin in wound dressings. *Polym. Adv. Technol.* **2018**, *29*, 1795–1803. [CrossRef]
34. Dhivya, S.; Padma, V.V.; Santhini, E. Wound dressings—A review. *BioMedicine* **2015**, *5*, 22. [CrossRef]
35. Morgan, D. Wounds—What Should a Dressing Formulary Include? *Hosp. Pharm.* **2002**, *9*, 216–261.
36. Vijayan, A.; Nanditha, C.K.; Kumar, G.S.V. ECM-mimicking nanofibrous scaffold enriched with dual growth factor carrying nanoparticles for diabetic wound healing. *Nanoscale Adv.* **2021**, *3*, 3085–3092. [CrossRef]
37. Ramos-E-Silva, M.; de Castro, M.C.R. New dressings, including tissue-engineered living skin. *Clin. Dermatol.* **2002**, *20*, 715–723. [CrossRef]
38. Lipsky, B.A.; Hoey, C. Topical Antimicrobial Therapy for Treating Chronic Wounds. *Clin. Infect. Dis.* **2009**, *49*, 1541–1549. [CrossRef]
39. Matthews, K.; Stevens, H.; Auffret, A.; Humphrey, M.; Eccleston, G. Formulation, stability and thermal analysis of lyophilised wound healing wafers containing an insoluble MMP-3 inhibitor and a non-ionic surfactant. *Int. J. Pharm.* **2008**, *356*, 110–120. [CrossRef]
40. Yang, X.; Liu, W.; Xi, G.; Wang, M.; Liang, B.; Shi, Y.; Feng, Y.; Ren, X.; Shi, C. Fabricating antimicrobial peptide-immobilized starch sponges for hemorrhage control and antibacterial treatment. *Carbohydr. Polym.* **2019**, *222*, 115012. [CrossRef]
41. Toleubayev, M.; Dmitriyeva, M.; Kozhakhmetov, S.; Sabitova, A. Efficacy of erythropoietin for wound healing: A systematic review of the literature. *Ann. Med. Surg.* **2021**, *65*, 102287. [CrossRef] [PubMed]
42. Aramwit, P. Introduction to Biomaterials for Wound Healing. In *Wound Healing Biomaterials*; Elsevier: Amsterdam, The Netherlands, 2016; pp. 3–38.
43. Ahmed, R.; Tariq, M.; Ali, I.; Asghar, R.; Khanam, P.N.; Augustine, R.; Hasan, A. Novel electrospun chitosan/polyvinyl alcohol/zinc oxide nanofibrous mats with antibacterial and antioxidant properties for diabetic wound healing. *Int. J. Biol. Macromol.* **2018**, *120*, 385–393. [CrossRef]
44. Cam, M.E.; Yildiz, S.; Alenezi, H.; Cesur, S.; Ozcan, G.S.; Erdemir, G.; Edirisinghe, U.; Akakin, D.; Kuruca, D.S.; Kabasakal, L.; et al. Evaluation of burst release and sustained release of pioglitazone-loaded fibrous mats on diabetic wound healing: An in vitro and in vivo comparison study. *J. R. Soc. Interface* **2020**, *17*, 20190712. [CrossRef] [PubMed]
45. Almasian, A.; Najafi, F.; Eftekhari, M.; Ardekani, M.R.S.; Sharifzadeh, M.; Khanavi, M. Polyurethane/carboxymethylcellulose nanofibers containing Malva sylvestris extract for healing diabetic wounds: Preparation, characterization, in vitro and in vivo studies. *Mater. Sci. Eng. C* **2020**, *114*, 111039. [CrossRef]
46. Chen, X.; Zhou, W.; Zha, K.; Liu, G.; Yang, S.; Ye, S.; Liu, Y.; Xiong, Y.; Wu, Y.; Cao, F. Treatment of chronic ulcer in diabetic rats with self assembling nanofiber gel encapsulated-polydeoxyribonucleotide. *Am. J. Transl. Res.* **2016**, *8*, 3067–3076. [PubMed]
47. Choi, J.S.; Leong, K.W.; Yoo, H.S. In vivo wound healing of diabetic ulcers using electrospun nanofibers immobilized with human epidermal growth factor (EGF). *Biomaterials* **2008**, *29*, 587–596. [CrossRef] [PubMed]
48. Cui, S.; Sun, X.; Li, K.; Gou, D.; Zhou, Y.; Hu, J.; Liu, Y. Polylactide nanofibers delivering doxycycline for chronic wound treatment. *Mater. Sci. Eng. C* **2019**, *104*, 109745. [CrossRef]
49. Grip, J.; Engstad, R.E.; Skjæveland, I.; Škalko-Basnet, N.; Isaksson, J.; Basnet, P.; Holsæter, A.M. Beta-glucan-loaded nanofiber dressing improves wound healing in diabetic mice. *Eur. J. Pharm. Sci.* **2018**, *121*, 269–280. [CrossRef]
50. Kanji, S.; Das, M.; Joseph, M.; Aggarwal, R.; Sharma, S.M.; Ostrowski, M.; Pompili, V.J.; Mao, H.-Q.; Das, H. Nanofiber-expanded human CD34+ cells heal cutaneous wounds in streptozotocin-induced diabetic mice. *Sci. Rep.* **2019**, *9*, 8415. [CrossRef]
51. Lee, C.-H.; Liu, K.-S.; Cheng, C.-W.; Chan, E.-C.; Hung, K.-C.; Hsieh, M.-J.; Chang, S.-H.; Fu, X.; Juang, J.-H.; Hsieh, I.-C.; et al. Codelivery of Sustainable Antimicrobial Agents and Platelet-Derived Growth Factor via Biodegradable Nanofibers for Repair of Diabetic Infectious Wounds. *ACS Infect. Dis.* **2020**, *6*, 2688–2697. [CrossRef]
52. Lee, C.-H.; Hung, K.-C.; Hsieh, M.-J.; Chang, S.-H.; Juang, J.-H.; Hsieh, I.-C.; Wen, M.-S.; Liu, S.-J. Core-shell insulin-loaded nanofibrous scaffolds for repairing diabetic wounds. *Nanomedicine* **2020**, *24*, 102123. [CrossRef]
53. Merrell, J.G.; McLaughlin, S.W.; Tie, L.; Laurencin, C.T.; Chen, A.F.; Nair, L.S. Curcumin loaded poly (ε-caprolactone) nanofibers: Diabetic wound dressing with antioxidant and anti-inflammatory properties. *Clin. Exp. Pharmacol. Physiol.* **2009**, *36*, 1149–1156. [CrossRef] [PubMed]
54. García, A.D.P.; Cassini-Vieira, P.; Ribeiro, C.C.; Jensen, C.E.D.M.; Barcelos, L.S.; Cortes, M.E.; Sinisterra, R.D. Efficient cutaneous wound healing using bixin-loaded PCL nanofibers in diabetic mice. *J. Biomed. Mater. Res. Part B Appl. Biomater.* **2017**, *105*, 1938–1949. [CrossRef] [PubMed]
55. Ranjbar-Mohammadi, M.; Rabbani, S.; Bahrami, S.H.; Joghataei, M.; Moayer, F. Antibacterial performance and in vivo diabetic wound healing of curcumin loaded gum tragacanth/poly(ε-caprolactone) electrospun nanofibers. *Mater. Sci. Eng. C* **2016**, *69*, 1183–1191. [CrossRef] [PubMed]
56. Shalaby, T.; Fekry, N.M.; El Sodfy, A.S.; Elsheredy, A.; Moustafa, M.E.S.S.A. Preparation and characterisation of antibacterial silver-containing nanofibers for wound healing in diabetic mice. *Int. J. Nanoparticles* **2015**, *8*, 82. [CrossRef]

57. Zehra, M.; Zubairi, W.; Hasan, A.; Butt, H.; Ramzan, A.; Azam, M.; Mehmood, A.; Falahati, M.; Chaudhry, A.A.; Rehman, I.U.; et al. Oxygen Generating Polymeric Nano Fibers That Stimulate Angiogenesis and Show Efficient Wound Healing in a Diabetic Wound Model. *Int. J. Nanomed.* **2020**, *15*, 3511–3522. [CrossRef] [PubMed]
58. Lee, C.-H.; Chang, S.-H.; Chen, W.-J.; Hung, K.-C.; Lin, Y.-H.; Liu, S.-J.; Hsieh, M.-J.; Pang, J.-H.S.; Juang, J.-H. Augmentation of diabetic wound healing and enhancement of collagen content using nanofibrous glucophage-loaded collagen/PLGA scaffold membranes. *J. Colloid Interface Sci.* **2015**, *439*, 88–97. [CrossRef]
59. Zheng, Z.; Liu, Y.; Huang, W.; Mo, Y.; Lan, Y.; Guo, R.; Cheng, B. Neurotensin-loaded PLGA/CNC composite nanofiber membranes accelerate diabetic wound healing. *Artif. Cells, Nanomed. Biotechnol.* **2018**, *46*, 493–501. [CrossRef]
60. Liu, F.; Li, X.; Wang, L.; Yan, X.; Ma, D.; Liu, Z.; Liu, X. Sesamol incorporated cellulose acetate-zein composite nanofiber membrane: An efficient strategy to accelerate diabetic wound healing. *Int. J. Biol. Macromol.* **2020**, *149*, 627–638. [CrossRef]
61. Lee, C.-H.; Hsieh, M.-J.; Chang, S.-H.; Lin, Y.-H.; Liu, S.-J.; Lin, T.-Y.; Hung, K.-C.; Pang, J.-H.S.; Juang, J.-H. Enhancement of Diabetic Wound Repair Using Biodegradable Nanofibrous Metformin-Eluting Membranes: In Vitro and in Vivo. *ACS Appl. Mater. Interfaces* **2014**, *6*, 3979–3986. [CrossRef]
62. Ren, X.; Han, Y.; Wang, J.; Jiang, Y.; Yi, Z.; Xu, H.; Ke, Q. An aligned porous electrospun fibrous membrane with controlled drug delivery—An efficient strategy to accelerate diabetic wound healing with improved angiogenesis. *Acta Biomater.* **2018**, *70*, 140–153. [CrossRef] [PubMed]
63. Lobmann, R.; Pittasch, D.; Mühlen, I.; Lehnert, H. Autologous human keratinocytes cultured on membranes composed of benzyl ester of hyaluronic acid for grafting in nonhealing diabetic foot lesions. *J. Diabetes Its Complicat.* **2003**, *17*, 199–204. [CrossRef] [PubMed]
64. Augustine, R.; Hasan, A.; Patan, N.K.; Dalvi, Y.B.; Varghese, R.; Antony, A.; Unni, R.N.; Sandhyarani, N.; Al Moustafa, A.-E. Cerium Oxide Nanoparticle Incorporated Electrospun Poly(3-hydroxybutyrate-co-3-hydroxyvalerate) Membranes for Diabetic Wound Healing Applications. *ACS Biomater. Sci. Eng.* **2020**, *6*, 58–70. [CrossRef] [PubMed]
65. Augustine, R.; Zahid, A.A.; Hasan, A.; Wang, M.; Webster, T.J. CTGF Loaded Electrospun Dual Porous Core-Shell Membrane for Diabetic Wound Healing. *Int. J. Nanomed.* **2019**, *14*, 8573–8588. [CrossRef] [PubMed]
66. Arantes, V.T.; Faraco, A.A.; Ferreira, F.B.; Oliveira, C.A.; Martins-Santos, E.; Cassini-Vieira, P.; Barcelos, L.S.; Ferreira, L.A.; Goulart, G.A. Retinoic acid-loaded solid lipid nanoparticles surrounded by chitosan film support diabetic wound healing in in vivo study. *Colloids Surf. B Biointerfaces* **2020**, *188*, 110749. [CrossRef]
67. Arul, V.; Kartha, R.; Jayakumar, R. A therapeutic approach for diabetic wound healing using biotinylated GHK incorporated collagen matrices. *Life Sci.* **2007**, *80*, 275–284. [CrossRef]
68. Inpanya, P.; Faikrua, A.; Ounaroon, A.; Sittichokechaiwut, A.; Viyoch, J. Effects of the blended fibroin/aloe gel film on wound healing in streptozotocin-induced diabetic rats. *Biomed. Mater.* **2012**, *7*, 035008. [CrossRef]
69. Kim, D.W.; Kim, K.S.; Seo, Y.G.; Lee, B.-J.; Park, Y.J.; Youn, Y.S.; Kim, J.O.; Yong, C.S.; Jin, S.G.; Choi, H.-G. Novel sodium fusidate-loaded film-forming hydrogel with easy application and excellent wound healing. *Int. J. Pharm.* **2015**, *495*, 67–74. [CrossRef]
70. Mizuno, K.; Yamamura, K.; Yano, K.; Osada, T.; Saeki, S.; Takimoto, N.; Sakurai, T.; Nimura, Y. Effect of chitosan film containing basic fibroblast growth factor on wound healing in genetically diabetic mice. *J. Biomed. Mater. Res.* **2003**, *64*, 177–181. [CrossRef]
71. Song, S.H.; Kim, J.E.; Koh, E.K.; Sung, J.E.; Lee, H.A.; Yun, W.B.; Hong, J.T.; Hwang, D.Y. Selenium-loaded cellulose film derived from Styela clava tunic accelerates the healing process of cutaneous wounds in streptozotocin-induced diabetic Sprague–Dawley rats. *J. Dermatol. Treat.* **2018**, *29*, 606–616. [CrossRef]
72. Tan, W.S.; Arulselvan, P.; Ng, S.-F.; Taib, C.N.M.; Sarian, M.N.; Fakurazi, S. Improvement of diabetic wound healing by topical application of Vicenin-2 hydrocolloid film on Sprague Dawley rats. *BMC Complement. Altern. Med.* **2019**, *19*, 20. [CrossRef] [PubMed]
73. Tong, W.Y.; bin Abdullah, A.Y.K.; binti Rozman, N.A.S.; bin Wahid, M.I.A.; Hossain, M.; Ring, L.C.; Lazim, Y.; Tan, W.-N. Antimicrobial wound dressing film utilizing cellulose nanocrystal as drug delivery system for curcumin. *Cellulose* **2018**, *25*, 631–638. [CrossRef]
74. Voss, G.T.; Gularte, M.S.; Vogt, A.G.; Giongo, J.L.; Vaucher, R.A.; Echenique, J.V.; Soares, M.P.; Luchese, C.; Wilhelm, E.A.; Fajardo, A.R. Polysaccharide-based film loaded with vitamin C and propolis: A promising device to accelerate diabetic wound healing. *Int. J. Pharm.* **2018**, *552*, 340–351. [CrossRef] [PubMed]
75. Wu, Y.-Y.; Jiao, Y.-P.; Xiao, L.-L.; Li, M.-M.; Liu, H.-W.; Li, S.-H.; Liao, X.; Chen, Y.-T.; Li, J.-X.; Zhang, Y. Experimental Study on Effects of Adipose-Derived Stem Cell–Seeded Silk Fibroin Chitosan Film on Wound Healing of a Diabetic Rat Model. *Ann. Plast. Surg.* **2018**, *80*, 572–580. [CrossRef]
76. da Silva, L.P.; Santos, T.C.; Rodrigues, D.B.; Pirraco, R.P.; Cerqueira, M.T.; Reis, R.L.; Correlo, V.M.; Marques, A.P. Stem Cell-Containing Hyaluronic Acid-Based Spongy Hydrogels for Integrated Diabetic Wound Healing. *J. Investig. Dermatol.* **2017**, *137*, 1541–1551. [CrossRef]
77. Lai, J.C.-Y.; Lai, H.-Y.; Rao, N.K.; Ng, S.-F. Treatment for diabetic ulcer wounds using a fern tannin optimized hydrogel formulation with antibacterial and antioxidative properties. *J. Ethnopharmacol.* **2016**, *189*, 277–289. [CrossRef]
78. Li, M.; Ke, Q.-F.; Tao, S.-C.; Guo, S.-C.; Rui, B.-Y.; Guo, Y.-P. Fabrication of hydroxyapatite/chitosan composite hydrogels loaded with exosomes derived from miR-126-3p overexpressed synovial mesenchymal stem cells for diabetic chronic wound healing. *J. Mater. Chem. B* **2016**, *4*, 6830–6841. [CrossRef]

79. Masood, N.; Ahmed, R.; Tariq, M.; Ahmed, Z.; Masoud, M.S.; Ali, I.; Asghar, R.; Andleeb, A.; Hasan, A. Silver nanoparticle impregnated chitosan-PEG hydrogel enhances wound healing in diabetes induced rabbits. *Int. J. Pharm.* **2019**, *559*, 23–36. [CrossRef]
80. Shi, G.; Chen, W.; Zhang, Y.; Dai, X.; Zhang, X.; Wu, Z. An Antifouling Hydrogel Containing Silver Nanoparticles for Modulating the Therapeutic Immune Response in Chronic Wound Healing. *Langmuir* **2019**, *35*, 1837–1845. [CrossRef]
81. Thangavel, P.; Ramachandran, B.; Chakraborty, S.; Kannan, R.; Lonchin, S.; Muthuvijayan, V. Accelerated Healing of Diabetic Wounds Treated with L-Glutamic acid Loaded Hydrogels Through Enhanced Collagen Deposition and Angiogenesis: An In Vivo Study. *Sci. Rep.* **2017**, *7*, 10701. [CrossRef]
82. Zhang, L.; Ma, Y.; Pan, X.; Chen, S.; Zhuang, H.; Wang, S. A composite hydrogel of chitosan/heparin/poly (γ-glutamic acid) loaded with superoxide dismutase for wound healing. *Carbohydr. Polym.* **2018**, *180*, 168–174. [CrossRef] [PubMed]
83. Choi, H.J.; Thambi, T.; Yang, Y.H.; Bang, S.I.; Kim, B.S.; Pyun, D.G.; Lee, D.S. AgNP and rhEGF-incorporating synergistic polyurethane foam as a dressing material for scar-free healing of diabetic wounds. *RSC Adv.* **2017**, *7*, 13714–13725. [CrossRef]
84. Pyun, D.G.; Choi, H.J.; Yoon, H.S.; Thambi, T.; Lee, D.S. Polyurethane foam containing rhEGF as a dressing material for healing diabetic wounds: Synthesis, characterization, in vitro and in vivo studies. *Colloids Surfaces B Biointerfaces* **2015**, *135*, 699–706. [CrossRef] [PubMed]
85. Atia, N.M.; Hazzah, H.A.; Gaafar, P.M.; Abdallah, O.Y. Diosmin Nanocrystal–Loaded Wafers for Treatment of Diabetic Ulcer: In Vitro and In Vivo Evaluation. *J. Pharm. Sci.* **2019**, *108*, 1857–1871. [CrossRef] [PubMed]
86. Anisha, B.; Biswas, R.; Chennazhi, K.; Jayakumar, R. Chitosan–hyaluronic acid/nano silver composite sponges for drug resistant bacteria infected diabetic wounds. *Int. J. Biol. Macromol.* **2013**, *62*, 310–320. [CrossRef]
87. Lipsky, B.A.; Kuss, M.; Edmonds, M.; Reyzelman, A.; Sigal, F. Topical Application of a Gentamicin-Collagen Sponge Combined with Systemic Antibiotic Therapy for the Treatment of Diabetic Foot Infections of Moderate Severity. *J. Am. Podiatr. Med. Assoc.* **2012**, *102*, 223–232. [CrossRef]
88. Mohandas, A.; Anisha, B.; Chennazhi, K.; Jayakumar, R. Chitosan–hyaluronic acid/VEGF loaded fibrin nanoparticles composite sponges for enhancing angiogenesis in wounds. *Colloids Surfaces B Biointerfaces* **2015**, *127*, 105–113. [CrossRef]
89. Shi, Q.; Qian, Z.; Liu, D.; Sun, J.; Wang, X.; Liu, H.; Xu, J.; Guo, X. GMSC-Derived Exosomes Combined with a Chitosan/Silk Hydrogel Sponge Accelerates Wound Healing in a Diabetic Rat Skin Defect Model. *Front. Physiol.* **2017**, *8*, 904. [CrossRef]
90. Wang, W.; Lin, S.; Xiao, Y.; Huang, Y.; Tan, Y.; Cai, L.; Li, X. Acceleration of diabetic wound healing with chitosan-crosslinked collagen sponge containing recombinant human acidic fibroblast growth factor in healing-impaired STZ diabetic rats. *Life Sci.* **2008**, *82*, 190–204. [CrossRef]
91. Xia, G.; Zhai, D.; Sun, Y.; Hou, L.; Guo, X.; Wang, L.; Li, Z.; Wang, F. Preparation of a novel asymmetric wettable chitosan-based sponge and its role in promoting chronic wound healing. *Carbohydr. Polym.* **2020**, *227*, 115296. [CrossRef]
92. Kondo, S.; Niiyama, H.; Yu, A.; Kuroyanagi, Y. Evaluation of a Wound Dressing Composed of Hyaluronic Acid and Collagen Sponge Containing Epidermal Growth Factor in Diabetic Mice. *J. Biomater. Sci. Polym. Ed.* **2012**, *23*, 1729–1740. [CrossRef]
93. Thattaruparambil Raveendran, N.; Mohandas, A.; Ramachandran Menon, R.; Somasekharan Menon, A.; Biswas, R.; Jayakumar, R. Ciprofloxacin- and Fluconazole-Containing Fibrin-Nanoparticle-Incorporated Chitosan Bandages for the Treatment of Polymicrobial Wound Infections. *ACS Appl. Bio Mater.* **2019**, *2*, 243–254. [CrossRef] [PubMed]
94. Mohanty, C.; Pradhan, J. A human epidermal growth factor-curcumin bandage bioconjugate loaded with mesenchymal stem cell for in vivo diabetic wound healing. *Mater. Sci. Eng. C* **2020**, *111*, 110751. [CrossRef] [PubMed]
95. Kumar, P.T.S.; Lakshmanan, V.-K.; Anilkumar, T.; Ramya, C.; Reshmi, P.; Unnikrishnan, A.; Nair, S.V.; Jayakumar, R. Flexible and Microporous Chitosan Hydrogel/Nano ZnO Composite Bandages for Wound Dressing: In Vitro and In Vivo Evaluation. *ACS Appl. Mater. Interfaces* **2012**, *4*, 2618–2629. [CrossRef]
96. Kahan, V.; Andersen, M.; Tomimori, J.; Tufik, S. Stress, immunity and skin collagen integrity: Evidence from animal models and clinical conditions. *Brain Behav. Immun.* **2009**, *23*, 1089–1095. [CrossRef] [PubMed]
97. Shoulders, M.D.; Raines, R.T. Collagen Structure and Stability. *Annu. Rev. Biochem.* **2009**, *78*, 929–958. [CrossRef]
98. Naomi, R.; Fauzi, M.B. Cellulose/Collagen Dressings for Diabetic Foot Ulcer: A Review. *Pharmaceutics* **2020**, *12*, 881. [CrossRef]
99. Davison-Kotler, E.; Marshall, W.S.; García-Gareta, E. Sources of Collagen for Biomaterials in Skin Wound Healing. *Bioengineering* **2019**, *6*, 56. [CrossRef]
100. Ruszczak, Z. Effect of collagen matrices on dermal wound healing. *Adv. Drug Deliv. Rev.* **2003**, *55*, 1595–1611. [CrossRef]
101. Chen, F.-M.; Liu, X. Advancing biomaterials of human origin for tissue engineering. *Prog. Polym. Sci.* **2016**, *53*, 86–168. [CrossRef]
102. Sorushanova, A.; Delgado, L.M.; Wu, Z.; Shologu, N.; Kshirsagar, A.; Raghunath, R.; Mullen, A.M.; Bayon, Y.; Pandit, A.; Raghunath, M.; et al. The Collagen Suprafamily: From Biosynthesis to Advanced Biomaterial Development. *Adv. Mater.* **2019**, *31*, 1801651. [CrossRef] [PubMed]
103. Ricard-Blum, S. The Collagen Family. *Cold Spring Harb. Perspect. Biol.* **2011**, *3*, a004978. [CrossRef] [PubMed]
104. Parenteau-Bareil, R.; Gauvin, R.; Berthod, F. Collagen-Based Biomaterials for Tissue Engineering Applications. *Materials* **2010**, *3*, 1863–1887. [CrossRef]
105. Brodsky, B.; Eikenberry, E.F. [5] Characterization of Fibrous Forms of Collagen. In *Methods in Enzymology*; Academic Press: Cambridge, MA, USA, 1982; pp. 127–174.
106. Haut, R.C. Age-Dependent Influence of Strain Rate on the Tensile Failure of Rat-Tail Tendon. *J. Biomech. Eng.* **1983**, *105*, 296–299. [CrossRef]

107. Ferraro, V.; Gaillard-Martinie, B.; Sayd, T.; Chambon, C.; Anton, M.; Santé-Lhoutellier, V. Collagen type I from bovine bone. Effect of animal age, bone anatomy and drying methodology on extraction yield, self-assembly, thermal behaviour and electrokinetic potential. *Int. J. Biol. Macromol.* **2017**, *97*, 55–66. [CrossRef]
108. Asghar, A.; Henrickson, R.L. Chemical, Biochemical, Functional, and Nutritional Characteristics of Collagen in Food Systems. In *Advances in Food Research*; Elsevier: Amsterdam, The Netherlands, 1982; pp. 231–372.
109. Nehrer, S. Chondrocyte-seeded collagen matrices implanted in a chondral defect in a canine model. *Biomaterials* **1998**, *19*, 2313–2328. [CrossRef]
110. Geesin, J.C.; Brown, L.J.; Liu, Z.; Berg, R.A. Development of a Skin Model Based on Insoluble Fibrillar Collagen. *J. Biomed. Mater. Res.* **1996**, *33*, 1–8. [CrossRef]
111. Doillon, C.J.; Silver, F.H.; Berg, R.A. Fibroblast growth on a porous collagen sponge containing hyaluronic acid and fibronectin. *Biomaterials* **1987**, *8*, 195–200. [CrossRef]
112. Rovira, A.; Amedee, J.; Bareille, R.; Rabaud, M. Colonization of a calcium phosphate/ elastin-solubilized peptide-collagen composite material by human osteoblasts. *Biomaterials* **1996**, *17*, 1535–1540. [CrossRef]
113. Leipziger, L.S.; Glushko, V.; DiBernardo, B.; Shafaie, F.; Noble, J.; Nichols, J.; Alvarez, O.M. Dermal wound repair: Role of collagen matrix implants and synthetic polymer dressings. *J. Am. Acad. Dermatol.* **1985**, *12*, 409–419. [CrossRef]
114. McPherson, J.M.; Sawamura, S.; Armstrong, R. An examination of the biologic response to injectable, glutaraldehyde cross-linked collagen implants. *J. Biomed. Mater. Res.* **1986**, *20*, 93–107. [CrossRef] [PubMed]
115. Gerrits, L.; Hammink, R.; Kouwer, P.H.J. Semiflexible polymer scaffolds: An overview of conjugation strategies. *Polym. Chem.* **2021**, *12*, 1362–1392. [CrossRef]
116. Chattopadhyay, S.; Raines, R.T. Collagen-based biomaterials for wound healing. *Biopolymers* **2014**, *101*, 821–833. [CrossRef] [PubMed]
117. Zaulyanov, L.; Kirsner, R.S. A review of a bi-layered living cell treatment (Apligraf) in the treatment of venous leg ulcers and diabetic foot ulcers. *Clin. Interv. Aging* **2007**, *2*, 93–98. [CrossRef] [PubMed]
118. Rennert, R.C.; Rodrigues, M.; Wong, V.W.; Duscher, D.; Hu, M.; Maan, Z.; Sorkin, M.; Gurtner, G.C.; Longaker, M.T. Biological therapies for the treatment of cutaneous wounds: Phase III and launched therapies. *Expert Opin. Biol. Ther.* **2013**, *13*, 1523–1541. [CrossRef]
119. Munish, T.; Ramneesh, G.; Sanjeev, S.; Jasdeep, S.; Jaspal, S.; Nikhil, G. Comparative Study of Collagen Based Dressing and Standard Dressing in Diabetic Foot Ulcer. *J. Evol. Med. Dent. Sci.* **2015**, *4*, 3614–3621. [CrossRef]
120. Lee, D.S.; Lee, Y.N.; Han, S.K.; Namgoong, S. Effect of Collagen Dressing on Diabetic Wound Healing-A Pilot Study. *J. Korean Wound Manag. Soc.* **2015**, *11*, 1–10.
121. Hauck, S.; Zager, P.; Halfter, N.; Wandel, E.; Torregrossa, M.; Kakpenova, A.; Rother, S.; Ordieres, M.; Räthel, S.; Berg, A.; et al. Collagen/hyaluronan based hydrogels releasing sulfated hyaluronan improve dermal wound healing in diabetic mice via reducing inflammatory macrophage activity. *Bioact. Mater.* **2021**, *6*, 4342–4359. [CrossRef]
122. Shagdarova, B.; Konovalova, M.; Zhuikova, Y.; Lunkov, A.; Zhuikov, V.; Khaydapova, D.; Il'Ina, A.; Svirshchevskaya, E.; Varlamov, V. Collagen/Chitosan Gels Cross-Linked with Genipin for Wound Healing in Mice with Induced Diabetes. *Materials* **2021**, *15*, 15. [CrossRef]
123. Michelacci, Y.M. Collagens and proteoglycans of the corneal extracellular matrix. *Braz. J. Med. Biol. Res.* **2003**, *36*, 1037–1046. [CrossRef]
124. Duconseille, A.; Astruc, T.; Quintana, N.; Meersman, F.; Sante-Lhoutellier, V. Gelatin structure and composition linked to hard capsule dissolution: A review. *Food Hydrocoll.* **2015**, *43*, 360–376. [CrossRef]
125. Rodríguez-Rodríguez, R.; Espinosa-Andrews, H.; Velasquillo-Martínez, C.; García-Carvajal, Z.Y. Composite Hydrogels based on gelatin, chitosan and polyvinyl alcohol to biomedical applications: A review. *Int. J. Polym. Mater. Polym. Biomater.* **2020**, *69*, 1–20. [CrossRef]
126. Samadian, H.; Zamiri, S.; Ehterami, A.; Farzamfar, S.; Vaez, A.; Khastar, H.; Alam, M.; Ai, A.; Derakhshankhah, H.; Allahyari, Z.; et al. Electrospun cellulose acetate/gelatin nanofibrous wound dressing containing berberine for diabetic foot ulcer healing: In vitro and in vivo studies. *Sci. Rep.* **2020**, *10*, 8312. [CrossRef]
127. Yu, H.; Gong, W.; Mei, J.; Qin, L.; Piao, Z.; You, D.; Gu, W.; Jia, Z. The efficacy of a paeoniflorin-sodium alginate-gelatin skin scaffold for the treatment of diabetic wound: An in vivo study in a rat model. *Biomed. Pharmacother.* **2022**, *151*, 113165. [CrossRef] [PubMed]
128. Sadeghi, A.; Zare-Gachi, M.; Najjar-Asl, M.; Rajabi, S.; Fatemi, M.J.; Forghani, S.F.; Daemi, H.; Pezeshki-Modaress, M. Hybrid gelatin-sulfated alginate scaffolds as dermal substitutes can dramatically accelerate healing of full-thickness diabetic wounds. *Carbohydr. Polym.* **2023**, *302*, 120404. [CrossRef] [PubMed]
129. Zheng, Y.; Yuan, W.; Liu, H.; Huang, S.; Bian, L.; Guo, R. Injectable supramolecular gelatin hydrogel loading of resveratrol and histatin-1 for burn wound therapy. *Biomater. Sci.* **2020**, *8*, 4810–4820. [CrossRef] [PubMed]
130. Wolberg, A.S. Thrombin generation and fibrin clot structure. *Blood Rev.* **2007**, *21*, 131–142. [CrossRef]
131. Uzunalli, G. Nanomaterials for Wound Healing. In *Nanomaterials for Regenerative Medicine*; Humana: Cham, Switzerland, 2019; pp. 81–117.
132. Heher, P.; Mühleder, S.; Mittermayr, R.; Redl, H.; Slezak, P. Fibrin-based delivery strategies for acute and chronic wound healing. *Adv. Drug Deliv. Rev.* **2018**, *129*, 134–147. [CrossRef]

133. Losi, P.; Briganti, E.; Errico, C.; Lisella, A.; Sanguinetti, E.; Chiellini, F.; Soldani, G. Fibrin-based scaffold incorporating VEGF- and bFGF-loaded nanoparticles stimulates wound healing in diabetic mice. *Acta Biomater.* **2013**, *9*, 7814–7821. [CrossRef]
134. Al Kayal, T.; Losi, P.; Pierozzi, S.; Soldani, G. A New Method for Fibrin-Based Electrospun/Sprayed Scaffold Fabrication. *Sci. Rep.* **2020**, *10*, 5111. [CrossRef]
135. Geer, D.J.; Swartz, D.D.; Andreadis, S.T. Fibrin Promotes Migration in a Three-Dimensional *in Vitro* Model of Wound Regeneration. *Tissue Eng.* **2002**, *8*, 787–798. [CrossRef] [PubMed]
136. Falanga, V.; Iwamoto, S.; Chartier, M.; Yufit, T.; Butmarc, J.; Kouttab, N.; Shrayer, D.; Carson, P. Autologous Bone Marrow–Derived Cultured Mesenchymal Stem Cells Delivered in a Fibrin Spray Accelerate Healing in Murine and Human Cutaneous Wounds. *Tissue Eng.* **2007**, *13*, 1299–1312. [CrossRef]
137. Crisci, A.; Marotta, G.; Licito, A.; Serra, E.; Benincasa, G.; Crisci, M. Use of Leukocyte Platelet (L-PRF) Rich Fibrin in Diabetic Foot Ulcer with Osteomyelitis (Three Clinical Cases Report). *Diseases* **2018**, *6*, 30. [CrossRef] [PubMed]
138. Losi, P.; Al Kayal, T.; Buscemi, M.; Foffa, I.; Cavallo, A.; Soldani, G. Bilayered Fibrin-Based Electrospun-Sprayed Scaffold Loaded with Platelet Lysate Enhances Wound Healing in a Diabetic Mouse Model. *Nanomaterials* **2020**, *10*, 2128. [CrossRef] [PubMed]
139. Chouhan, D.; Mandal, B.B. Silk biomaterials in wound healing and skin regeneration therapeutics: From bench to bedside. *Acta Biomater.* **2020**, *103*, 24–51. [CrossRef]
140. Altman, G.H.; Diaz, F.; Jakuba, C.; Calabro, T.; Horan, R.L.; Chen, J.; Lu, H.; Richmond, J.; Kaplan, D.L. Silk-based biomaterials. *Biomaterials* **2003**, *24*, 401–416. [CrossRef] [PubMed]
141. Park, Y.R.; Sultan, T.; Park, H.J.; Lee, J.M.; Ju, H.W.; Lee, O.J.; Lee, D.J.; Kaplan, D.L.; Park, C.H. NF-κB signaling is key in the wound healing processes of silk fibroin. *Acta Biomater.* **2018**, *67*, 183–195. [CrossRef]
142. Capar, G.; Aygun, S.S.; Gecit, M.R. Separation of sericin from fatty acids towards its recovery from silk degumming wastewaters. *J. Membr. Sci.* **2009**, *342*, 179–189. [CrossRef]
143. Tariq, M.; Tahir, H.M.; Butt, S.A.; Ali, S.; Ahmad, A.B.; Raza, C.; Summer, M.; Hassan, A.; Nadeem, J. Silk derived formulations for accelerated wound healing in diabetic mice. *Peerj* **2021**, *9*, e10232. [CrossRef]
144. Sultan, M.T.; Lee, O.J.; Kim, S.H.; Ju, H.W.; Park, C.H. Silk Fibroin in Wound Healing Process. In *Novel Biomaterials for Regenerative Medicine*; Springer: Singapore, 2018; pp. 115–126.
145. Liu, J.; Yan, L.; Yang, W.; Lan, Y.; Zhu, Q.; Xu, H.; Zheng, C.; Guo, R. Controlled-release neurotensin-loaded silk fibroin dressings improve wound healing in diabetic rat model. *Bioact. Mater.* **2019**, *4*, 151–159. [CrossRef]
146. Guan, G.; Zhang, Q.; Jiang, Z.; Liu, J.; Wan, J.; Jin, P.; Lv, Q. Multifunctional Silk Fibroin Methacryloyl Microneedle for Diabetic Wound Healing. *Small* **2022**, *18*, 2203064. [CrossRef] [PubMed]
147. Xu, X.; Wang, X.; Qin, C.; Khan, A.U.R.; Zhang, W.; Mo, X. Silk fibroin/poly-(L-lactide-co-caprolactone) nanofiber scaffolds loaded with Huangbai Liniment to accelerate diabetic wound healing. *Colloids Surfaces B Biointerfaces* **2021**, *199*, 111557. [CrossRef] [PubMed]
148. Baptista-Silva, S.; Bernardes, B.G.; Borges, S.; Rodrigues, I.; Fernandes, R.; Gomes-Guerreiro, S.; Pinto, M.T.; Pintado, M.; Soares, R.; Costa, R.; et al. Exploring Silk Sericin for Diabetic Wounds: An In Situ-Forming Hydrogel to Protect against Oxidative Stress and Improve Tissue Healing and Regeneration. *Biomolecules* **2022**, *12*, 801. [CrossRef] [PubMed]
149. El-Samad, L.M.; Hassan, M.A.; Basha, A.A.; El-Ashram, S.; Radwan, E.H.; Aziz, K.K.A.; Tamer, T.M.; Augustyniak, M.; El Wakil, A. Carboxymethyl cellulose/sericin-based hydrogels with intrinsic antibacterial, antioxidant, and anti-inflammatory properties promote re-epithelization of diabetic wounds in rats. *Int. J. Pharm.* **2022**, *629*, 122328. [CrossRef]
150. Alven, S.; Nqoro, X.; Aderibigbe, B.A. Polymer-Based Materials Loaded with Curcumin for Wound Healing Applications. *Polymers* **2020**, *12*, 2286. [CrossRef]
151. Mir, M.; Ali, M.N.; Barakullah, A.; Gulzar, A.; Arshad, M.; Fatima, S.; Asad, M. Synthetic polymeric biomaterials for wound healing: A review. *Prog. Biomater.* **2018**, *7*, 1–21. [CrossRef]
152. Mondal, D.; Griffith, M.; Venkatraman, S.S. Polycaprolactone-based biomaterials for tissue engineering and drug delivery: Current scenario and challenges. *Int. J. Polym. Mater. Polym. Biomater.* **2016**, *65*, 255–265. [CrossRef]
153. Huang, X.; Wang, Q.; Mao, R.; Wang, Z.; Shen, S.G.; Mou, J.; Dai, J. Two-dimensional nanovermiculite and polycaprolactone electrospun fibers composite scaffolds promoting diabetic wound healing. *J. Nanobiotechnology* **2022**, *20*, 343. [CrossRef]
154. Lv, F.; Wang, J.; Xu, P.; Han, Y.; Ma, H.; Xu, H.; Chen, S.; Chang, J.; Ke, Q.; Liu, M.; et al. A conducive bioceramic/polymer composite biomaterial for diabetic wound healing. *Acta Biomater.* **2017**, *60*, 128–143. [CrossRef]
155. Awasthi, A.; Gulati, M.; Kumar, B.; Kaur, J.; Vishwas, S.; Khursheed, R.; Porwal, O.; Alam, A.; Kr, A.; Corrie, L.; et al. Recent Progress in Development of Dressings Used for Diabetic Wounds with Special Emphasis on Scaffolds. *BioMed Res. Int.* **2022**, *2022*, 1–43. [CrossRef]
156. Huang, X.; Guan, N.; Li, Q. A Marine-Derived Anti-Inflammatory Scaffold for Accelerating Skin Repair in Diabetic Mice. *Mar. Drugs* **2021**, *19*, 496. [CrossRef]
157. Lin, W.-C.; Tang, C.-M. Evaluation of Polyvinyl Alcohol/Cobalt Substituted Hydroxyapatite Nanocomposite as a Potential Wound Dressing for Diabetic Foot Ulcers. *Int. J. Mol. Sci.* **2020**, *21*, 8831. [CrossRef]
158. Zhu, H.; Xu, J.; Zhao, M.; Luo, H.; Lin, M.; Luo, Y.; Li, Y.; He, H.; Wu, J. Adhesive, injectable, and ROS-responsive hybrid polyvinyl alcohol (PVA) hydrogel co-delivers metformin and fibroblast growth factor 21 (FGF21) for enhanced diabetic wound repair. *Front. Bioeng. Biotechnol.* **2022**, *10*, 968078. [CrossRef] [PubMed]

159. Wang, Z.; Gao, S.; Zhang, W.; Gong, H.; Xu, K.; Luo, C.; Zhi, W.; Chen, X.; Li, J.; Weng, J. Polyvinyl alcohol/chitosan composite hydrogels with sustained release of traditional Tibetan medicine for promoting chronic diabetic wound healing. *Biomater. Sci.* **2021**, *9*, 3821–3829. [CrossRef] [PubMed]
160. Zare, M.; Bigham, A.; Zare, M.; Luo, H.; Ghomi, E.R.; Ramakrishna, S. pHEMA: An Overview for Biomedical Applications. *Int. J. Mol. Sci.* **2021**, *22*, 6376. [CrossRef] [PubMed]
161. Khazaeli, P.; Alaei, M.; Khaksarihadad, M.; Ranjbar, M. Preparation of PLA/chitosan nanoscaffolds containing cod liver oil and experimental diabetic wound healing in male rats study. *J. Nanobiotechnology* **2020**, *18*, 176. [CrossRef] [PubMed]
162. Zha, S.; Utomo, Y.K.S.; Yang, L.; Liang, G.; Liu, W. Mechanic-Driven Biodegradable Polyglycolic Acid/Silk Fibroin Nanofibrous Scaffolds Containing Deferoxamine Accelerate Diabetic Wound Healing. *Pharmaceutics* **2022**, *14*, 601. [CrossRef] [PubMed]

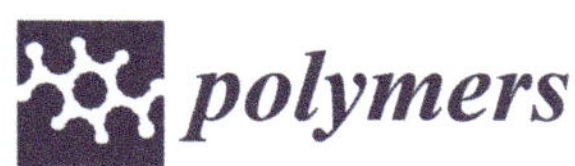

*Article*

# Hyaluronic Acid-Coated Chitosan Nanoparticles as an Active Targeted Carrier of Alpha Mangostin for Breast Cancer Cells

**Lisna Meylina [1,2], Muchtaridi Muchtaridi [3], I Made Joni [4,5], Khaled M. Elamin [6] and Nasrul Wathoni [1,*]**

1 Department of Pharmaceutics and Pharmaceutical Technology, Faculty of Pharmacy, Universitas Padjadjaran, Bandung 45363, Indonesia
2 Department of Pharmaceutics and Pharmaceutical Technology, Faculty of Pharmacy, Universitas Mulawarman, Samarinda 75119, Indonesia
3 Department of Pharmaceutical Analysis and Medicinal Chemistry, Faculty of Pharmacy, Universitas Padjadjaran, Bandung 45363, Indonesia
4 Department of Physics, Faculty of Mathematics and Natural Sciences, Universitas Padjadjaran, Bandung 45363, Indonesia
5 Functional Nano Powder University Center of Excellence, Universitas Padjadjaran, Bandung 45363, Indonesia
6 Graduate School of Pharmaceutical Sciences, Kumamoto University, 5-1 Oe-honmachi, Chuo-ku, Kumamoto 862-0973, Japan
* Correspondence: nasrul@unpad.ac.id; Tel.: +62-2-842-888888 (ext. 3510)

**Abstract:** Alpha mangostin (AM) has potential anticancer properties for breast cancer. This study aims to assess the potential of chitosan nanoparticles coated with hyaluronic acid for the targeted delivery of AM (AM-CS/HA) against MCF-7 breast cancer cells. AM-CS/HA showed a spherical shape with an average diameter of 304 nm, a polydispersity index of 0.3, and a negative charge of 24.43 mV. High encapsulation efficiency (90%) and drug loading (8.5%) were achieved. AM released from AM-CS/HA at an acidic pH of 5.5 was higher than the physiological pH of 7.4 and showed sustained release. The cytotoxic effect of AM-CS/HA ($IC_{50}$ 4.37 µg/mL) on MCF-7 was significantly higher than AM nanoparticles without HA coating (AM-CS) ($IC_{50}$ 4.48 µg/mL) and AM ($IC_{50}$ 5.27 µg/mL). These findings suggest that AM-CS/HA enhances AM cytotoxicity and has potential applications for breast cancer therapy.

**Keywords:** alpha mangostin; chitosan; hyaluronic acid; polymeric nanoparticle; cytotoxic



**Citation:** Meylina, L.; Muchtaridi, M.; Joni, I.M.; Elamin, K.M.; Wathoni, N. Hyaluronic Acid-Coated Chitosan Nanoparticles as an Active Targeted Carrier of Alpha Mangostin for Breast Cancer Cells. *Polymers* **2023**, *15*, 1025. https://doi.org/10.3390/polym15041025

Academic Editors: José Miguel Ferri, Vicent Fombuena Borràs and Miguel Fernando Aldás Carrasco

Received: 30 December 2022
Revised: 15 February 2023
Accepted: 15 February 2023
Published: 18 February 2023



## 1. Introduction

Breast cancer is the most common type of malignancy and the second leading cause of cancer-related death worldwide [1,2]. In general, the treatment of breast cancer involves various combinations of surgery, radiation therapy, chemotherapy, and hormone therapy that have many drawbacks, such as limited effectiveness and unwanted side effects. In addition, chemotherapy shows low efficacy due to multidrug resistance and is highly toxic to healthy cells due to its non-specific targeting [3–5].

Alpha mangostin is a derivative of xanthone compounds isolated from the rind of the mangosteen fruit (*Garcinia mangostan*). AM has antiproliferative activity and apoptotic effects on different types of cancer, one of which is breast cancer, among the mechanisms of inducing apoptosis in breast cancer cells through the downregulation of B-cell lymphoma 2 (Bcl2) and the upregulation of Bcl-2-associated X protein (Bax) against breast cancer cells [6–10]. In addition to its anticancer activity, AM has limitations due to its poor solubility [11], the first fast metabolism reaction, efflux reactions caused by intercellular transporters, rapid drug release, and low selectivity for cancer cells [6,12–14].

The advancement of nanoparticle delivery system technology has the potential to improve delivery efficiency while minimizing side effects by directly targeting cancer cells [15–19]. Polymeric nanoparticles are a drug delivery system approach that utilizes polymers as carriers in the form of nanoparticles [20–22]. Polymeric nanoparticles are frequently produced from biopolymers such as chitosan in their formulation because they offer several advantages over other synthetic polymers, such as being green economy-friendly, eco-friendly, easy to make, bio-compatible, biodegradable, and low in their toxicity [23,24]. They have also been investigated for their ability to increase drug macromolecular permeation on epithelial membranes through the reversible opening of the transmembrane gap (tight junction) [25,26]. Chitosan nanoparticles have previously been used as AM carriers for breast cancer cells. In this study, AM was successfully encapsulated in chitosan nanoparticles, significantly increasing the cytotoxicity of AM ($IC_{50}$ 6.7 μg/mL) compared to that which was not prepared in the polymeric nanoparticle formulation ($IC_{50}$ 8.2 μg/mL) against the MCF-7 cell line [27,28].

Modification of nanoparticles aims to improve drug targeting through passive or active targeting [29,30]. Passive targeting can enhance the penetration of nanoparticles to the tumor tissue site through an enhanced permeability and retention (EPR) effect [31–35]. Meanwhile, active targeting contains structural modifications and surface functionalization of nanoparticles that lead to more specific targeting capabilities [36,37]. The limited selectivity of nanoparticles against cancer excludes the benefits of nanoparticle drug delivery for effective chemotherapy. It is critical to improve the selectivity of nanoparticles for cancer cells so that they can deliver more therapeutic agents to targeted cells than healthy cells, boosting therapeutic efficacy and minimizing adverse effects [38]. Therapeutic targeting can be accomplished by decorating the surface of nanoparticles with specific ligands to target the appropriate receptor cells, which are overexpressed on cancer cell membranes [39,40]. There are many candidates for ligand targeting, such as folate, antibodies, and hyaluronic acid, which have shown efficacy in breast cancer targeting [41,42]. Several studies have found that hyaluronic acid (HA) is one of the most often utilized ligands for coating chitosan nanoparticles for targeting breast cancer. HA is a natural polysaccharide made up of D-glucuronic acid and *N*-acetyl-D-glucosamine, which shows a high affinity for the integral membrane glycoprotein cluster differentiation-44 (CD44) on the cell surface in breast cancer [43,44]. CD44 is a cell surface receptor that is overexpressed in breast cancer, and targeting this receptor could facilitate intracellular uptake of nanoparticles, thereby increasing drug concentrations in cancer cells through CD44 receptor-mediated endocytosis [45–51].

In this study, HA-coated AM nanoparticles were developed and applied to actively target MCF-7 breast cancer cells that express CD44. For this purpose, the cytotoxic effect of HA-coated AM nanoparticles will be compared with that of AM and AM nanoparticles without HA.

## 2. Materials and Methods

### 2.1. Material

AM was obtained from Chengdu Biopurify Phytochemicals (Chengdu, Sichuan, China). Chitosan (CS), with MW: 1526.5 g/mol and DD: 81.38%, was isolated with a purity of 70%. HA (MW = 60 KDa) was purchased from Kangcare Bioindustry (Nanjing, China), and sodium tripolyphosphate (TPP) from Kristata (Bandung, West Java, Indonesia). The MCF-7 breast cancer cell line was obtained from the American Type Culture Collection (Manassas, VA, USA).

### 2.2. Method

#### 2.2.1. Fabrication of AM-CS

The ionic gelation technique was used to produce AM-CS. Briefly, AM (1 mg/mL) was dissolved in ethanol, and CS (1 mg/mL) was dissolved in acetic acid, then stirred overnight at room temperature with a magnetic stirrer, respectively. TPP (1 mg/mL) was dissolved into demineralized water. AM and CS solutions were mixed and transferred drop-by-drop to TPP solutions while being constantly magnetically stirred. The mixture was kept on a magnetic stirrer overnight at room temperature, then sonicated for 30 min.

Finally, nanoparticles were separated from the mixture by centrifugation at 13,552× $g$ for 30 min [27,28].

2.2.2. Fabrication of Surface Functionalization of AM-CS

For the coating process, AM-CS and HA were dispersed in an acetate buffer at pH 5. Then, AM-CS was added dropwise to various concentrations of HA (Table 1) with constant vigorous stirring (30 min, 1200 rpm). The nanoparticles were then purified by centrifugation at 13,552× $g$ for 30 min [42,52,53].

**Table 1.** AM nanoparticles' formulation.

| Formulation | AM (mg/mL) | CS (mg/mL) | TPP (mg/mL) | HA (mg/mL) |
|---|---|---|---|---|
| AM-CS | 1 | 10 | 2 | - |
| AM-CS/HA1 | 1 | 10 | 2 | 20 |
| AM-CS/HA2 | 1 | 10 | 2 | 40 |
| AM-CS/HA3 | 1 | 10 | 2 | 60 |

2.2.3. Particle Size, Polydispersity Index (PDI), and Zeta Potential

The particle size, PDI, and zeta potential of the AM nanoparticles' formulation were evaluated using the dynamic light scattering (DLS) analyzer (SZ 100 Horiba, Kyoto, Japan) [54,55].

2.2.4. Morphology Studies

The morphology of AM-CS and AM-CS/HA was examined by scanning electron microscopy (SEM) (Model SU3500 SEM; Hitachi, Tokyo, Japan). The samples were placed into the stub and coated with platinum (30 s, 10 mA). AM-CS and AM-CS/HA photomicrographs were taken at 10 kV with 20,000 magnifications [27,28].

2.2.5. Determination of Entrapment Efficiency and Drug Loading

The entrapment efficiency (EE) and drug loading (DL) of nanoparticles were calculated by spectroscopy. Briefly, AM-CS/HA was mixed with ethyl acetate and centrifuged (6000× $g$ rpm, 5 min). After collecting the supernatant, the absorbance at 245 nm was measured with a spectrophotometer. The supernatant was then resuspended in sufficient ethanol to determine the amount of AM encapsulated and the total amount of AM. Serial concentrations of AM (2–12 µg/mL) were measured at 245 nm to generate the standard curve. EE and DL of AM in AM-CS/HA were calculated by Equations (1) and (2) [27,28]:

$$EE\ (\%) = \frac{mass\ of\ the\ AM\ in\ AM-CS/HA}{mass\ of\ AM\ used} \times 100\% \tag{1}$$

$$DL\ (\%) = \frac{mass\ of\ the\ AM\ in\ AM-CS/HA}{mass\ of\ AM-CS/HA} \times 100\% \tag{2}$$

2.2.6. Fourier-Transform Infrared Spectroscopy Analysis

The chemical interaction of raw materials and nanoparticles was investigated using a Fourier-transform infrared spectrophotometer (FTIR) (Thermo Fisher, Waltham, MA, USA) and measured at 4000–400 cm$^{-1}$ [27,56].

2.2.7. X-ray Diffraction Analysis

X-ray diffraction (XRD) (X-pert MPD diffractometer type, Rigaku International, Tokyo, Japan) was used to examine the crystallinity of AM-CS/HA. The samples were scanned throughout an angular range (2 theta) of 5–60° [27,57].

2.2.8. Differential Scanning Calorimetry Analysis

Differential scanning calorimetry (DSC) (Perkin Elmer DSC-6, MA, USA) was used to study the thermal properties of AM-CS/HA. The samples were carried out at a heating rate of 10 °C/min from 30 to 300 °C, with a stream of flowing nitrogen at 50 mL/min [27].

2.2.9. In Vitro Release Studies

The release profile in phosphate-buffered saline (PBS) solution was investigated at pH 7.4 and 5.5. Typically, 5 mg of nanoparticles were dispersed in PBS and transferred to a dialysis tube (molecular weight cut-off 12,000 Da). The dialysis tube was immersed in PBS medium before being put into a beaker containing 50 mL of release medium at 37 °C and 100 rpm. At determined time intervals, 5 mL of dissolution medium was taken and replaced with an equal quantity of fresh medium. The collected samples were then measured using a spectrophotometer at a wavelength of 245 nm [55].

2.2.10. Cytotoxicity Studies

The MTT assay was used to assess the cytotoxic activity of AM and nanoparticles on MCF-7 cells. Here, 5000 cells/well of MCF-7 cells (ATCC) were seeded on 96-well plates in the presence of RPMI culture media containing 10% FCS for 24 h. Then, the media was aspirated and replaced with cell culture media containing various amounts of AM (2–6 µg/mL). Next, 0.5 mg/mL of MTT solution was added and incubated for 4 h at 37 °C. The formed formazan crystals were treated with 100 µL of SDS in 0.01% HCl, and then the absorbance was measured at 450 nm using an ELISA plate reader (EpochTM Microplate Spectrophotometer, VT, USA). Cell viability was represented as a percentage of the treated cells compared to the control cells, as stated in Equation (3), and $IC_{50}$ was calculated from the dose–response curves [58]:

$$Cell\ viability\ (\%) = \frac{absorbance\ of\ treated\ sample}{absorbance\ of\ control\ sample} \times 100\% \tag{3}$$

2.2.11. Statistical Analysis

The quantitative data were expressed as the mean ± standard error of the mean (S.E.M.). The two-way ANOVA was used for statistical analysis. $p$-values < 0.05 were considered significant.

## 3. Results

### 3.1. Characterization of AM Nanoparticles

3.1.1. Particle Size, PDI, Zeta Potential, Morphology, EE, and DL

The mean particle size, PDI, and zeta potential of various AM nanoparticle formulas are shown in Table 2. The data show that the nanoparticle size is in the range of 200–400 nm. The zeta potential of AM-CS showed a positive value, then the AM-CS/HA showed a negative value. In addition, the PDI of all formulas was <1. In this study, AM-CS/HA1 was selected for further characterization and cytotoxicity evaluation on MCF-7 cells because this formula produced the smallest particle size.

**Table 2.** Particle size, distribution, and zeta potential of AM nanoparticles.

| Formulation | Particle Size (nm) | PDI | Zeta Potential (mV) |
|---|---|---|---|
| AM-CS | 229.133 ± 5.685 | 0.382 ± 0.015 | 33.83 ± 1.92 |
| AM-CS/HA1 | 304.833 ± 6.288 | 0.362 ± 0.038 | −24.43 ± 1.76 |
| AM-CS/HA2 | 369.300 ± 2.467 | 0.360 ± 0.028 | −28.44 ± 2.26 |
| AM-CS/HA3 | 412.767 ± 6.001 | 0.346 ± 0.034 | −33.31 ± 1.85 |

The morphologies of nanoparticles were examined by SEM (Figure 1). As shown in Figure 1, the nanoparticles were approximately spherical. The EE and DL are displayed in Table 3. The average entrapment efficiency of the AM-CS and AM-CS/HA1 was 85.32% ± 0.40% and 90.40% ± 0.161%, respectively, indicating that AM did not escape from the nanoparticles during the HA coating process.

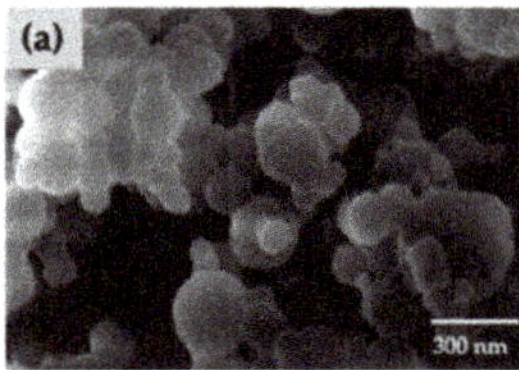

**Figure 1.** SEM photomicrographs of (**a**) AM-CS and (**b**) AM-CS/HA1.

**Table 3.** The average of EE and DL of the nanoparticles.

| Formulation | EE (%) | DL (%) |
|---|---|---|
| AM-CS | 88.325 ± 3.340 | 8.674 ± 0.018 |
| AM-CS/HA1 | 90.404 ± 2.161 | 8.514 ± 0.007 |

3.1.2. FTIR Analysis

The results of the FTIR analysis are displayed in Figure 2. The AM spectrum showed the presence of O–H stretch at 3411.15 and 3234.88 $cm^{-1}$, stretching vibrations of C–H at 2988.11, 2961.04, and 2910.40 $cm^{-1}$, C=O at 1638.20 $cm^{-1}$, C–C at 1448.33 $cm^{-1}$, orto–$OCH_3$ stretch at 1197.83 $cm^{-1}$, and C–O–C stretch at 1073.82 $cm^{-1}$ [59,60]. The CS spectrum displayed broad peaks around 3332.14 $cm^{-1}$ corresponding to the amide (N-H) and O-H groups, C–H stretch at 2871.62 $cm^{-1}$, C=O stretch at 1637.26 $cm^{-1}$, N–H bend at 1582.86 $cm^{-1}$, C–H bend at 1422.38 cm-1, C–N at 1375.93 $cm^{-1}$, C–O–C stretch at 1149.98 $cm^{-1}$, and C–O at 1022.44 $cm^{-1}$ [61,62]. The characteristic absorption peaks of HA were 3409 $cm^{-1}$ corresponding to the N–H and O–H groups, amide II and III at 1557 and 1337 $cm^{-1}$, C–C stretching of the COONa group was observed at 1404 $cm^{-1}$, and C–O stretch at 1042 $cm^{-1}$ [63]. The spectra of AM-CS/HA1 presented absorption bands at 1515.21 and 1735.45 $cm^{-1}$ due to –$NH_3$ of CS and –COOH of HA, respectively [64].

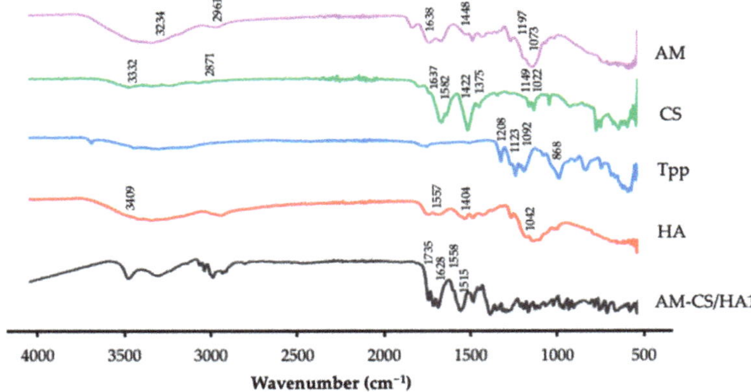


**Figure 2.** FTIR spectrum of raw material and AM-CS/HA1.

3.1.3. XRD Analysis

The XRD patterns are displayed in Figure 3. The AM showed sharp multiple peaks at 2θ of 5.4°, 11.6°, and 13.3°, which indicated a crystalline pattern [65,66], and the CS showed peaks at 10.4°, 19.7°, and 29.3° that exhibited semi-crystalline patterns [67–69]. The XRD spectrum of HA showed no specific diffraction pattern, indicating the amorphous nature of HA [64]. The peaks exhibited by the AM-CS/HA1 resembled those of HA and showed an amorphous nature.

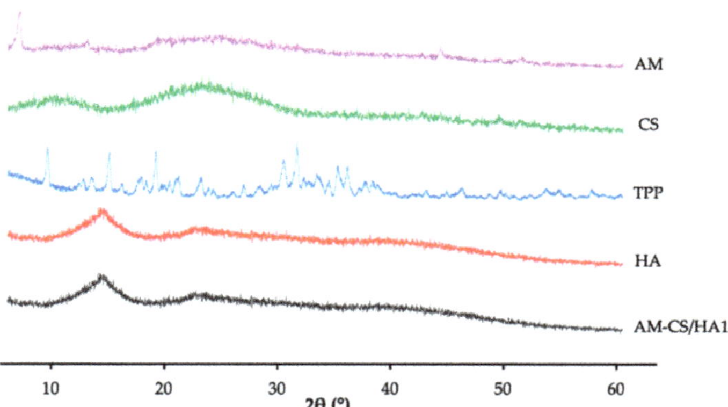


**Figure 3.** XRD patterns of raw material and AM-CS/HA1.

#### 3.1.4. DSC Analysis

DSC thermograms for AM exhibited the endothermic phase at 177 °C, and HA had an obvious glass transition peak at 85 °C and an exothermic peak at 241 °C. The DSC thermogram of chitosan showed an endothermic peak between 95.1 and 102.3 °C and an exothermic peak between 303.77 and 304.28 °C. The AM-CS/HA1 displayed patterns that corresponded to the glass transition (103.1 °C). The results of the DSC analysis are shown in Figure 4.

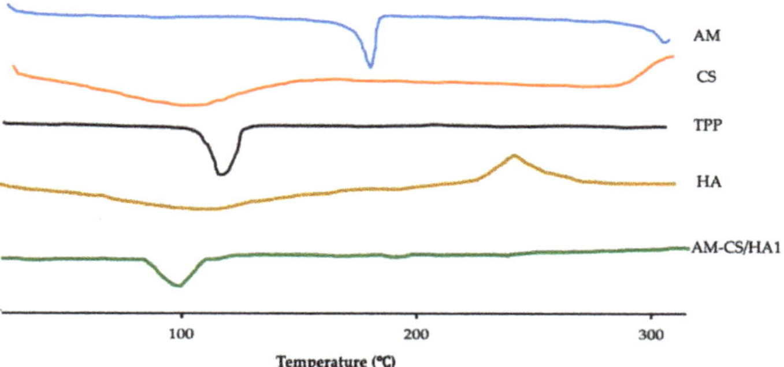


**Figure 4.** DSC thermographs of raw material and AM-CS/HA1.

### 3.2. *In Vitro Release Studies*

The profile of the in vitro release of AM from nanoparticles in PBS (pH 7.4 and 5.5) within 96 h is shown in Figure 5, and the Higuchi parameters for release kinetics are summarized in Table 4. The release of AM from nanoparticles demonstrated an initial burst of up to 11% during the first hour, followed by a sustained release for 96 h.

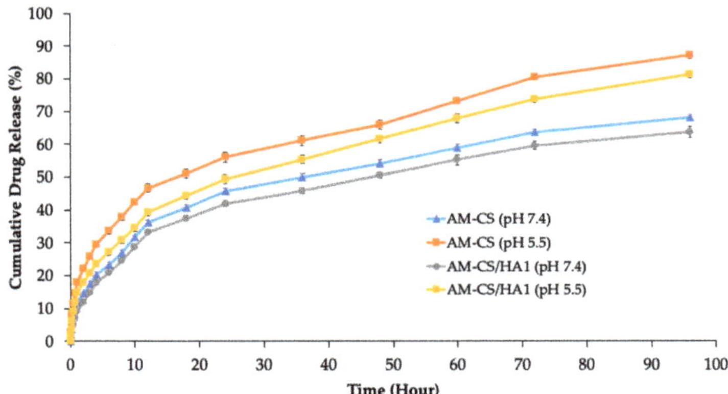


**Figure 5.** AM release profiles from AM-loaded nanoparticles at pH 7.4 and 5.5. Each value represents the mean ± S.E.M.

**Table 4.** Higuchi regression parameter for AM release from AM-CS and AM-CS/HA1.

| Parameter | AM-CS | | AM-CS/HA1 | |
|---|---|---|---|---|
| | pH 7.4 | pH 5.5 | pH 7.4 | pH 5.5 |
| Intercept | 0.852 ± 0.477 | 3.265 ± 0.129 | 0.889 ± 0.169 | 1.723 ± 0.3428 |
| Slope | 9.281 ± 0.489 | 11.821 ± 0.377 | 9.097 ± 0.160 | 10.289 ± 0.340 |
| Correlation coefficient (*r*) | 0.994 ± 0.002 | 0.981 ± 0.005 | 0.995 ± 0.001 | 0.987 ± 0.002 |

### 3.3. Cytotoxicity Studies

The cytotoxic activity of AM, -CS/HA1, AM, AM-CS, and AM-CS/HA1 was evaluated on MCF-7 cells, as shown in Figure 6. For -CS/HA, no cytotoxic activity was observed in MCF-7 cells. On the other hand, the cytotoxicity of AM, AM-CS, and AM-CS/HA significantly differed. AM, AM-CS, and AM-CS/HA had $IC_{50}$ of 5.27, 4.48, and 4.37 μg/mL, respectively. Thus, these results demonstrated that AM-CS/HA has higher cytotoxicity compared to AM and AM-CS.

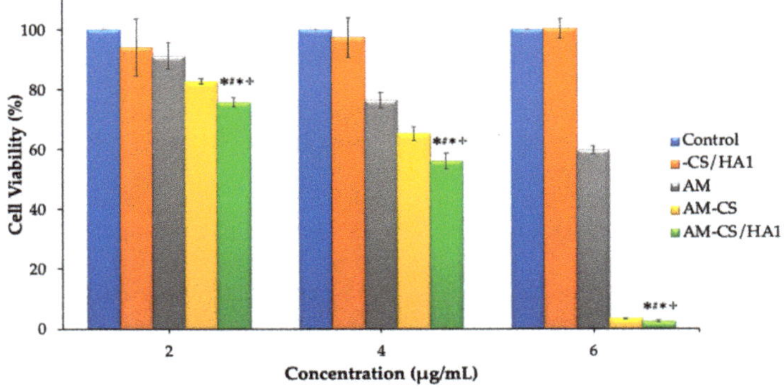


**Figure 6.** Cytotoxic activity of AM, -CS/HA1, AM-CS, and AM-CS/HA1 in MCF-7 cells. Each value represents the mean ± S.E.M. ✻ *p* < compared to the control; # *p* < compared to -CS/HA1; ✶ *p* < compared to AM; ✢ *p* < compared to AM-CS.

## 4. Discussion

AM has shown potentiality in the treatment of various types of cancer. Previous research has demonstrated that AM nanoparticles have a remarkable therapeutic impact on breast cancer [6,7]. In this study, AM-loaded nanoparticles were coated with HA for breast cancer targeting. Nanocarriers with tumor-targeting moiety attachments, such as hyaluronic acid, have the potential to increase tumor-targeted delivery, while minimizing pharmacological adverse effects [42,70–73].

In this study, AM-CS was fabricated with HA (AM-CS/HA) via an electrostatic deposition technique. As an experimental variable, three different concentrations of HA (20, 40, and 60 mg) were used. As shown in Table 2, there was a correlation between the variations in HA concentrations and particle size, or zeta potential. The higher concentration of HA resulted in larger particle sizes and a lower zeta potential value. At low concentrations, HA will enter more readily and deeply through the pores in AM-CS, resulting in denser particles, which will further increase the particle size when the HA concentration increases owing to the accumulation of the coated polymer chains on the exterior of the nanoparticles. The coating of AM nanoparticles resulted in a conversion of the nanoparticles' surface charge. HA has been found to exert a negative charge on the nanoparticles' surface because HA molecules are mostly located in the outermost shell of nanoparticles. Positively charged nanocarriers promote membrane attachment, uptake, and release of endosomes, while nanocarriers with a negative zeta potential exhibit more selective and efficient absorption, particularly when coated with targeting ligands [53,54,74].

The particle size of nanoparticles plays an important role in chemotherapeutic drug delivery systems because it can affect cellular uptake via endocytosis and determine their fate during systemic circulation [27]. Studies have shown that the nanoparticle size range between 40 and 400 nm is suitable for extending the circulation time and increasing drug accumulation in tumors [75]. This is a further reason for selecting AM-CS/HA1 for further investigation to evaluate its characteristics and cytotoxicity.

The spectrum of the AM-CS/HA1 exhibited some characteristic vibrations of HA and CS, then shifted to a higher wave number. The signal shift demonstrated that both macromolecular chains were involved in the production of the nanoparticles. The absorption band at 1735.45 $cm^{-1}$ showed the protonation that occurred in the formation of the polyelectrolyte complex [63,76]. Moreover, the amplification of the peak corresponding to the amide I and II bands, with a small shift to wave numbers 1628.39 and 1558.62 $cm^{-1}$, showed effective amide bonding between the amino and carboxylic groups on the HA and the surface of the nanoparticles [42,63,77].

The AM-CS/HA1 diffractogram data demonstrated the transformation of the crystalline or semi-crystalline phase of the material component into an amorphous form. The termination of the amine and hydroxy groups is thought to be the origin of CS's semi-crystalline transition, resulting in the development of an amorphous complex with the coated polymer (HA). Furthermore, the AM crystal lattice no longer appeared, suggesting that AM has been uniformly dispersed and encapsulated in the system [27,78].

The DSC thermogram of AM nanoparticles coated with HA exhibited a loss of the peak from CS accompanied by shifting of the HA peak to 94 °C and a loss of the exothermic peak at 236 °C from HA, which is thought to be due to the structural modification of HA after electrostatic interaction with CS [77,79]. Furthermore, AM exhibited a significant endothermal peak around 178 °C due to the melting of AM crystals. However, the AM-CS/HA diffractogram did not show an endothermic peak of AM. It can be explained that the crystallization of AM is inhibited by the nanoparticle matrix, and AM may be in a molecular or amorphous state in the nanoparticle system [78].

The release profile of AM from the nanoparticle system exhibited biphasic behavior, with early and fast release phases, followed by sustained release. Both coated and uncoated nanoparticles showed an initial burst of AM release, which was related to the quick diffusion of free drug adsorbed on the particles [42,76]. The delayed release rate of HA-coated nanoparticles compared to uncoated nanoparticles indicated that the HA coating on the surface of the nanoparticles inhibits the diffusion of drugs trapped in the nanoparticle system to be released. It happens because the coating of HA on the surface of the nanoparticles increases their density and structural hardness due to increased crosslinking interactions between the constituent components, and reduces the release of the active substance [80]. Furthermore, because chitosan and hyaluronic acid are pH-sensitive polymers, pH influences the release of AM from nanoparticles [81]. Due to the breakdown in the electrostatic balance between CS and TPP in AM-CS or between CS, TPP, and HA in -CS/HA in an acidic environment, AM release was larger at pH 5.5 than at pH 7.4. This pH-dependent release mechanism reduces the drug's systemic toxicity due to decreased bioavailability in healthy organs at physiological pH, which could reduce drug side effects for patients [82,83]. Subsequently, pH-sensitive drug delivery systems result in higher bioavailability for drugs at tumor sites at acidic pH and increase their efficiency in malignant tissues [54,76,84,85]. To estimate the kinetic profiles of AM release from the nanoparticle system, we carried out an analysis using the Higuchi model. Based on the value of the correlation coefficient ($r$), this indicated that the type of release of AM from nanoparticles was a matrix type, based on Fickian diffusion [80,86]. It is known that Higuchi's kinetic model involves drug release from the polymer matrix system, which releases drugs in a controlled and sustainable manner [87,88]. This is very important for the release of chemotherapy drugs to reduce their toxicity [89,90].

The cytotoxic study on drug-free nanoparticles (-CS/HA1) showed activity on cell viability > 90% at all tested concentrations. These results indicate that the nanoparticle carrier exhibited good biocompatibility and was less toxic to the tested MCF-7 cells. In contrast, cells that were treated with AM, AM-CS, or AM-CS/HA1 demonstrated a dose-dependent response to the drug. Moreover, the cells utilized were more sensitive to AM-CS and HA than to AM and AM-CS. The cytotoxic activity of AM and AM-CS/HA1 at the same doses was significantly different ($p < 0.05$). In conclusion, AM has lower cytotoxicity than AM-CS/HA because the HA coating of nanoparticles interacts with the CD44 receptor and is then internalized via receptor-mediated endocytosis.

## 5. Conclusions

The development of targeted drug delivery systems is necessary for the delivery of anticancer drugs to reduce systemic side effects and increase the effectiveness of therapy. Surface-modified nanoparticle delivery systems using specific ligands, such as hyaluronic acid, to target cell receptors that are overexpressed on breast cancer cell membranes, such as the CD44 receptor, have the potential to increase the efficiency of anticancer drug delivery to breast cancer cells [42]. This research succeeded in developing a targeted delivery system of hyaluronic acid-coated chitosan nanoparticles for the targeted delivery of alpha mangostin for breast cancer. Our findings showed that alpha mangostin loaded in our delivery system had a significant impact on MCF-7 cancer cells at a lower dose ($IC_{50}$ 4.37 μg/mL) compared to free alpha mangostin ($IC_{50}$ 5.27 μg/mL) or nanoparticles of alpha mangostin with chitosan carriers without a hyaluronic acid coating ($IC_{50}$ 4.48 μg/mL, $IC_{50}$ 6.7 μg/mL [27], $IC_{50}$ 4.90 μg/mL [91]). The most conclusive findings of this study indicated that the developed alpha mangostin targeted nanoparticle delivery system can be used as an effective treatment for breast cancer by specifically targeting cancer cells. Further research needs to be conducted in vivo to determine the bioavailability, toxicity, and anticancer activity of alpha mangostin nanoparticles coated with hyaluronic acid.

**Author Contributions:** Conceptualization, N.W. and L.M.; methodology, N.W.; software, L.M.; validation, N.W.; formal analysis, M.M.; investigation, I.M.J.; resources, M.M.; data curation, N.W.; writing—original draft preparation, L.M.; writing—review and editing, L.M., I.M.J., M.M., N.W., and K.M.E.; visualization, I.M.J. and M.M.; supervision, N.W.; project administration, N.W.; funding acquisition, M.M. and N.W. All authors have read and agreed to the published version of the manuscript.

**Funding:** The APC and Research were funded by Universitas Padjadjaran with Academic Leadership Grant 2023.

**Institutional Review Board Statement:** Not applicable.

**Data Availability Statement:** The data presented in this research are accessible upon request from the corresponding author.

**Acknowledgments:** The authors thank the Rector of Universitas Padjadjaran for funding this studies.

**Conflicts of Interest:** The authors declare no conflict of interest.

## References

1. Giaquinto, A.N.; Sung, H.; Miller, K.D.; Kramer, J.L.; Newman, L.A.; Minihan, A.; Jemal, A.; Siegel, R.L. Breast Cancer Statistics, 2022. *CA Cancer J. Clin.* **2022**, *72*, 524–541. [CrossRef] [PubMed]
2. Siegel, R.L.; Miller, K.D.; Fuchs, H.E.; Jemal, A. Cancer Statistics, 2022. *CA Cancer J. Clin.* **2022**, *72*, 7–33. [CrossRef] [PubMed]
3. Hong, R.; Xu, B. Breast Cancer: An up-to-Date Review and Future Perspectives. *Cancer Commun.* **2022**, *42*, 913–936. [CrossRef] [PubMed]
4. Luo, Z.; Dai, Y.; Gao, H. Development and Application of Hyaluronic Acid in Tumor Targeting Drug Delivery. *Acta Pharm. Sin. B* **2019**, *9*, 1099–1112. [CrossRef]
5. Feng, Y.; Spezia, M.; Huang, S.; Yuan, C.; Zeng, Z.; Zhang, L.; Ji, X.; Liu, W.; Huang, B.; Luo, W.; et al. Breast Cancer Development and Progression: Risk Factors, Cancer Stem Cells, Signaling Pathways, Genomics, and Molecular Pathogenesis. *Genes Dis.* **2018**, *5*, 77–106. [CrossRef]
6. Kritsanawong, S.; Innajak, S.; Imoto, M.; Watanapokasin, R. Antiproliferative and Apoptosis Induction of α-Mangostin in T47D Breast Cancer Cells. *Int. J. Oncol.* **2016**, *48*, 2155–2165. [CrossRef]
7. Shibata, M.A.; Iinuma, M.; Morimoto, J.; Kurose, H.; Akamatsu, K.; Okuno, Y.; Akao, Y.; Otsuki, Y. α-Mangostin Extracted from the Pericarp of the Mangosteen (*Garcinia mangostana* Linn) Reduces Tumor Growth and Lymph Node Metastasis in an Immunocompetent Xenograft Model of Metastatic Mammary Cancer Carrying a P53 Mutation. *BMC Med.* **2011**, *9*, 69. [CrossRef]
8. Benjakul, R.; Kongkaneramit, L.; Sarisuta, N.; Moongkarndi, P.; Müller-Goymann, C.C. Cytotoxic Effect and Mechanism Inducing Cell Death of α-Mangostin Liposomes in Various Human Carcinoma and Normal Cells. *Anti-Cancer Drugs* **2015**, *26*, 824–834. [CrossRef]
9. Muchtaridi, M.; Wijaya, C.A. Anticancer Potential of α-Mangostin. *Asian J. Pharm. Clin. Res.* **2017**, *10*, 440–445. [CrossRef]
10. Ibrahim, M.Y.; Hashim, N.M.; Mohan, S.; Kamalidehghan, B.; Ghaderian, M.; Dehghan, F.; Ali, L.Z.; Arbab, I.A.; Yahayu, M.; Lian, G.E.C. α-Mangostin from Cratoxylum Arborescens Demonstrates Apoptogenesis in MCF-7 with Regulation of NF-KB and Hsp70 Protein Modulation in Vitro, and Tumor Reduction in Vivo. *Drug Des. Devel. Ther.* **2014**, *8*, 1629–1647. [CrossRef]
11. Meylina, L.; Muchtaridi, M.; Joni, I.M.; Mohammed, A.F.A.; Wathoni, N. Nanoformulations of α-Mangostin for Cancer Drug Delivery System. *Pharmaceutics* **2021**, *13*, 1993. [CrossRef] [PubMed]
12. Verma, R.K.; Yu, W.; Shrivastava, A.; Shankar, S.; Srivastava, R.K. α-Mangostin-Encapsulated PLGA Nanoparticles Inhibit Pancreatic Carcinogenesis by Targeting Cancer Stem Cells in Human, and Transgenic (KrasG12D, and KrasG12D/Tp53R270H) Mice. *Sci. Rep.* **2016**, *6*, 32743. [CrossRef]
13. Gutierrez-Orozco, F.; Failla, M.L. Biological Activities and Bioavailability of Mangosteen Xanthones: A Critical Review of the Current Evidence. *Nutrients* **2013**, *5*, 3163–3183. [CrossRef]
14. Li, L.; Brunner, I.; Han, A.R.; Hamburger, M.; Kinghorn, A.D.; Frye, R.; Butterweck, V. Pharmacokinetics of α-Mangostin in Rats after Intravenous and Oral Application. *Mol. Nutr. Food Res.* **2011**, *55* (Suppl. S1), 67–74. [CrossRef]
15. Aydin, R.S.T. Herceptin-Decorated Salinomycin-Loaded Nanoparticles for Breast Tumor Targeting. *J. Biomed Mater. Res.—Part A* **2013**, *101*, 1405–1415. [CrossRef] [PubMed]
16. Masood, F. Polymeric Nanoparticles for Targeted Drug Delivery System for Cancer Therapy. *Mater. Sci. Eng. C* **2016**, *60*, 569–578. [CrossRef] [PubMed]
17. Aghebati-Maleki, A.; Dolati, S.; Ahmadi, M.; Baghbanzhadeh, A.; Asadi, M.; Fotouhi, A.; Yousefi, M.; Aghebati-Maleki, L. Nanoparticles and Cancer Therapy: Perspectives for Application of Nanoparticles in the Treatment of Cancers. *J. Cell. Physiol.* **2020**, *235*, 1962–1972. [CrossRef]
18. Ashfaq, U.A.; Riaz, M.; Yasmeen, E.; Yousaf, M. Recent Advances in Nanoparticle-Based Targeted Drug-Delivery Systems against Cancer and Role of Tumor Microenvironment. *Crit. Rev. Ther. Drug Carr. Syst.* **2017**, *34*, 317–353. [CrossRef]
19. Pérez-Herrero, E.; Fernández-Medarde, A. Advanced Targeted Therapies in Cancer: Drug Nanocarriers, the Future of Chemotherapy. *Eur. J. Pharm. Biopharm.* **2015**, *93*, 52–79. [CrossRef]
20. Madej, M.; Kurowska, N.; Strzalka-Mrozik, B. Polymeric Nanoparticles—Tools in a Drug Delivery System in Selected Cancer Therapies. *Appl. Sci.* **2022**, *12*, 9479. [CrossRef]
21. Perumal, S. Polymer Nanoparticles: Synthesis and Applications. *Polymers* **2022**, *14*, 5449. [CrossRef] [PubMed]
22. Xiao, X.; Teng, F.; Shi, C.; Chen, J.; Wu, S.; Wang, B.; Meng, X.; Essiet Imeh, A.; Li, W. Polymeric Nanoparticles—Promising Carriers for Cancer Therapy. *Front. Bioeng. Biotechnol.* **2022**, *10*, 1024143. [CrossRef] [PubMed]
23. Khan, M.U.A.; Razak, S.I.A.; Haider, S.; Mannan, H.A.; Hussain, J.; Hasan, A. Sodium Alginate-f-GO Composite Hydrogels for Tissue Regeneration and Antitumor Applications. *Int. J. Biol. Macromol.* **2022**, *208*, 475–485. [CrossRef] [PubMed]
24. Khan, M.U.A.; Al-Arjan, W.S.; Ashammakhi, N.; Haider, S.; Amin, R.; Hasan, A. Multifunctional Bioactive Scaffolds from ARX-g-(Zn@rGO)-HAp for Bone Tissue Engineering: In Vitro Antibacterial, Antitumor, and Biocompatibility Evaluations. *ACS Appl. Bio Mater.* **2022**, *5*, 5445–5456. [CrossRef] [PubMed]

25. Hassani, S.; Laouini, A.; Fessi, H.; Charcosset, C. Preparation of Chitosan-TPP Nanoparticles Using Microengineered Membranes—Effect of Parameters and Encapsulation of Tacrine. *Colloids Surfaces A Physicochem. Eng. Asp.* **2015**, *482*, 34–43. [CrossRef]
26. Palanikumar, L.; Al-Hosani, S.; Kalmouni, M.; Nguyen, V.P.; Ali, L.; Pasricha, R.; Barrera, F.N.; Magzoub, M. PH-Responsive High Stability Polymeric Nanoparticles for Targeted Delivery of Anticancer Therapeutics. *Commun. Biol.* **2020**, *3*, 95. [CrossRef]
27. Wathoni, N.; Meylina, L.; Rusdin, A.; Abdelwahab Mohammed, A.F.; Tirtamie, D.; Herdiana, Y.; Motoyama, K.; Panatarani, C.; Joni, I.M.; Lesmana, R.; et al. The Potential Cytotoxic Activity Enhancement of α-Mangostin in Chitosan-Kappa Carrageenan-Loaded Nanoparticle against Mcf-7 Cell Line. *Polymers* **2021**, *13*, 1681. [CrossRef]
28. Wathoni, N.; Rusdin, A.; Febriani, E.; Purnama, D.; Daulay, W.; Azhary, S.Y.; Panatarani, C.; Joni, I.M.; Lesmana, R.; Keiichi, M.; et al. Formulation and Characterization of α-Mangostin in Chitosan Nanoparticles Coated by Sodium Alginate, Sodium Silicate, and Polyethylene Glycol. *J. Pharm. Bioallied Sci.* **2019**, *20*, S619. [CrossRef]
29. Alavi, M.; Hamidi, M. Passive and Active Targeting in Cancer Therapy by Liposomes and Lipid Nanoparticles. *Drug Metab. Pers. Ther.* **2019**, *34*, 1–8. [CrossRef]
30. Tu, Y.; Yao, Z.; Yang, W.; Tao, S.; Li, B.; Wang, Y.; Su, Z.; Li, S. Application of Nanoparticles in Tumour Targeted Drug Delivery and Vaccine. *Front. Nanotechnol.* **2022**, *4*, 948705. [CrossRef]
31. Kher, C.; Kumar, S. The Application of Nanotechnology and Nanomaterials in Cancer Diagnosis and Treatment: A Review. *Cureus* **2022**, *14*, e29059. [CrossRef] [PubMed]
32. Choi, K.A.; Kim, J.H.; Ryu, K.; Kaushik, N. Current Nanomedicine for Targeted Vascular Disease Treatment: Trends and Perspectives. *Int. J. Mol. Sci.* **2022**, *23*, 2397. [CrossRef] [PubMed]
33. Yang, T.; Zhai, J.; Hu, D.; Yang, R.; Wang, G.; Li, Y.; Liang, G. "Targeting Design" of Nanoparticles in Tumor Therapy. *Pharmaceutics* **2022**, *14*, 1919. [CrossRef] [PubMed]
34. Sun, Y.; Yang, Q.; Xia, X.; Li, X.; Ruan, W.; Zheng, M.; Zou, Y.; Shi, B. Polymeric Nanoparticles for Mitochondria Targeting Mediated Robust Cancer Therapy. *Front. Bioeng. Biotechnol.* **2021**, *9*, 755727. [CrossRef]
35. Drozdov, A.S.; Nikitin, P.I.; Rozenberg, J.M. Systematic Review of Cancer Targeting by Nanoparticles Revealed a Global Association between Accumulation in Tumors and Spleen. *Int. J. Mol. Sci.* **2021**, *22*, 13011. [CrossRef]
36. Peltonen, L.; Singhal, M.; Hirvonen, J. *Principles of Nanosized Drug Delivery Systems*; Elsevier Ltd.: Amsterdam, The Netherlands, 2020. [CrossRef]
37. Dadwal, A.; Baldi, A.; Kumar Narang, R. Nanoparticles as Carriers for Drug Delivery in Cancer. *Artif. Cells Nanomed. Biotechnol.* **2018**, *46* (Suppl. S2), 295–305. [CrossRef]
38. Zhu, R.; Zhang, F.; Peng, Y.; Xie, T.; Wang, Y.; Lan, Y. Current Progress in Cancer Treatment Using Nanomaterials. *Front. Oncol.* **2022**, *12*, 930125. [CrossRef]
39. Zhou, Z.; Badkas, A.; Stevenson, M.; Lee, J.Y.; Leung, Y.K. Herceptin Conjugated PLGA-PHis-PEG PH Sensitive Nanoparticles for Targeted and Controlled Drug Delivery. *Int. J. Pharm.* **2015**, *487*, 81–90. [CrossRef]
40. Jurj, A.; Braicu, C.; Pop, L.A.; Tomuleasa, C.; Gherman, C.D.; Berindan-Neagoe, I. The New Era of Nanotechnology, an Alternative to Change Cancer Treatment. *Drug Des. Devel. Ther.* **2017**, *11*, 2871–2890. [CrossRef]
41. Dosio, F.; Arpicco, S.; Stella, B.; Fattal, E. Hyaluronic Acid for Anticancer Drug and Nucleic Acid Delivery. *Adv. Drug Deliv. Rev.* **2016**, *97*, 204–236. [CrossRef]
42. Hashad, R.A.; Ishak, R.A.H.; Geneidi, A.S.; Mansour, S. Surface Functionalization of Methotrexate-Loaded Chitosan Nanoparticles with Hyaluronic Acid/Human Serum Albumin: Comparative Characterization and in Vitro Cytotoxicity. *Int. J. Pharm.* **2017**, *522*, 128–136. [CrossRef]
43. Louderbough, J.M.V.; Schroeder, J.A. Understanding the Dual Nature of CD44 in Breast Cancer Progression. *Mol. Cancer Res.* **2011**, *9*, 1573–1586. [CrossRef] [PubMed]
44. Al-Othman, N.; Alhendi, A.; Ihbaisha, M.; Barahmeh, M.; Alqaraleh, M.; Al-Momany, B.Z. Role of CD44 in Breast Cancer. *Breast Dis.* **2020**, *39*, 1–13. [CrossRef] [PubMed]
45. Dey, A.; Koli, U.; Dandekar, P.; Jain, R. Investigating Behaviour of Polymers in Nanoparticles of Chitosan Oligosaccharides Coated with Hyaluronic Acid. *Polymer* **2016**, *93*, 44–52. [CrossRef]
46. Park, J.H.; Cho, H.J.; Yoon, H.Y.; Yoon, I.S.; Ko, S.H.; Shim, J.S.; Cho, J.H.; Park, J.H.; Kim, K.; Kwon, I.C.; et al. Hyaluronic Acid Derivative-Coated Nanohybrid Liposomes for Cancer Imaging and Drug Delivery. *J. Control. Release* **2014**, *174*, 98–108. [CrossRef] [PubMed]
47. Gupta, G.; Asati, P.; Jain, P.; Mishra, P.; Mishra, A.; Singour, P. Recent Advancements in Cancer Targeting Therapy with the Hyaluronic Acid as a Potential Adjuvant Avances Recientes En La Terapia Dirigida Al Cáncer Con El Ácido Hialurónico Como Adyuvante Potencial. *Ars Pharm.* **2022**, *63*, 387–409. [CrossRef]
48. Elamin, K.M.; Yamashita, Y.; Higashi, T.; Motoyama, K.; Arima, H. Supramolecular Complex of Methyl-β-Cyclodextrin with Adamantane-Grafted Hyaluronic Acid as a Novel Antitumor Agent. *Chem. Pharm. Bull.* **2018**, *66*, 277–285. [CrossRef] [PubMed]
49. Elamin, K.M.; Motoyama, K.; Higashi, T.; Yamashita, Y.; Tokuda, A.; Arima, H. Dual Targeting System by Supramolecular Complex of Folate-Conjugated Methyl-β-Cyclodextrin with Adamantane-Grafted Hyaluronic Acid for the Treatment of Colorectal Cancer. *Int. J. Biol. Macromol.* **2018**, *113*, 386–394. [CrossRef]
50. Bhattacharya, D.; Svechkrev, D.; Soucheck, J.; Hill, T.; Taylor, M.; Natarajan, A.; Mohs, A. Impact of Structurally Modifying Hyaluronic Acid on CD44 Interaction. *J. Mater. Chem. B* **2017**, *5*, 8183–8192. [CrossRef] [PubMed]

51. Jia, Y.; Chen, S.; Wang, C.; Sun, T.; Yang, L. Hyaluronic Acid-Based Nano Drug Delivery Systems for Breast Cancer Treatment: Recent Advances. *Front. Bioeng. Biotechnol.* **2022**, *10*, 990145. [CrossRef]
52. Nasti, A.; Zaki, N.M.; De Leonardis, P.; Ungphaiboon, S.; Sansongsak, P.; Rimoli, M.G.; Tirelli, N. Chitosan/TPP and Chitosan/TPP-Hyaluronic Acid Nanoparticles: Systematic Optimisation of the Preparative Process and Preliminary Biological Evaluation. *Pharm. Res.* **2009**, *26*, 1918–1930. [CrossRef] [PubMed]
53. Almalik, A.; Donno, R.; Cadman, C.J.; Cellesi, F.; Day, P.J.; Tirelli, N. Hyaluronic Acid-Coated Chitosan Nanoparticles: Molecular Weight-Dependent Effects on Morphology and Hyaluronic Acid Presentation. *J. Control. Release* **2013**, *172*, 1142–1150. [CrossRef] [PubMed]
54. Nokhodi, F.; Nekoei, M.; Goodarzi, M.T. Hyaluronic Acid-Coated Chitosan Nanoparticles as Targeted-Carrier of Tamoxifen against MCF7 and TMX-Resistant MCF7 Cells. *J. Mater. Sci. Mater. Med.* **2022**, *33*, 24. [CrossRef]
55. Aisha, A.F.A.; Abdulmajid, A.M.S.; Ismail, Z.; Alrokayan, S.A.; Abu-Salah, K.M. Development of Polymeric Nanoparticles of Garcinia Mangostana Xanthones in Eudragit RL100/RS100 for Anti-Colon Cancer Drug Delivery. *J. Nanomater.* **2015**, *2015*, 4–7. [CrossRef]
56. Smith, B.C. *Fundamentals of Fourier Transform Infrared Spectroscopy*, 2nd ed.; CRC Press: Boca Raton, FL, USA, 2011. [CrossRef]
57. Zak, A.; Majid, W.; ME, A.; Yousefi, R. X-ray Analysis of ZnO Nanoparticles by Williamson–Hall and Size–Strain Plot Methods. *Solid State Sci.* **2011**, *13*, 251. [CrossRef]
58. CCRC, U. Standard Procedure of the Cytotoxic Test MTT Method. *Cancer Chemoprevention Res. Cent.* **2012**, *1*, 1–7.
59. Guo, M.; Wang, X.; Lu, X.; Wang, H.; Brodelius, P.E. α-Mangostin Extraction from the Native Mangosteen (*Garcinia mangostana* L.) and the Binding Mechanisms of α-Mangostin to HSAorTRF. *PLoS ONE* **2016**, *11*, e0161566. [CrossRef] [PubMed]
60. Rohman, A.; Arifah, F.H.; Irnawati; Alam, G.; Muchtaridi, M. The Application of FTIR Spectroscopy and Chemometrics for Classification of Mangosteen Extract and Its Correlation with Alpha-Mangostin. *J. Appl. Pharm. Sci.* **2020**, *10*, 149–154. [CrossRef]
61. Zaman, M.; Butt, M.H.; Siddique, W.; Iqbal, M.O.; Nisar, N.; Mumtaz, A.; Nazeer, H.Y.; Alshammari, A.; Riaz, M.S. Fabrication of PEGylated Chitosan Nanoparticles Containing Tenofovir Alafenamide: Synthesis and Characterization. *Molecules* **2022**, *27*, 8401. [CrossRef] [PubMed]
62. Oh, J.W.; Chun, S.C.; Chandrasekaran, M. Preparation and in Vitro Characterization of Chitosan Nanoparticles and Their Broad-Spectrum Antifungal Action Compared to Antibacterial Activities against Phytopathogens of Tomato. *Agronomy* **2019**, *9*, 21. [CrossRef]
63. Carneiro, J.; Döll-Boscardin, P.M.; Fiorin, B.C.; Nadal, J.M.; Farago, P.V.; De Paula, J.P. Development and Characterization of Hyaluronic Acid-Lysine Nanoparticles with Potential as Innovative Dermal Filling. *Braz. J. Pharm. Sci.* **2016**, *52*, 645–651. [CrossRef]
64. Hussain, Z.; Pandey, M.; Choudhury, H.; Ying, P.C.; Xian, T.M.; Kaur, T.; Jia, G.W.; Gorain, B. Hyaluronic Acid Functionalized Nanoparticles for Simultaneous Delivery of Curcumin and Resveratrol for Management of Chronic Diabetic Wounds: Fabrication, Characterization, Stability and in Vitro Release Kinetics. *J. Drug Deliv. Sci. Technol.* **2020**, *57*, 101747. [CrossRef]
65. Iqbal, A.; Muhammad Shuib, N.A.; Darnis, D.S.; Miskam, M.; Abdul Rahman, N.R.; Adam, F. Synthesis and Characterisation of Rice Husk Ash Silica Drug Carrier for α-Mangostin. *J. Phys. Sci.* **2018**, *29*, 95–107. [CrossRef]
66. Sriyanti, I.; Edikresnha, D.; Rahma, A.; Munir, M.M.; Rachmawati, H.; Khairurrijal, K. Mangosteen Pericarp Extract Embedded in Electrospun PVP Nanofiber Mats: Physicochemical Properties and Release Mechanism of α-Mangostin. *Int. J. Nanomed.* **2018**, *13*, 4927–4941. [CrossRef] [PubMed]
67. Mulia, K.; Rachman, D.; Krisanti, E.A. Preparation, Characterization and Release Profile of Chitosan Alginate Freeze Dried Matrices Loaded with Mangostins. In *Journal of Physics: Conference Series*; IOP Publishing: Bristol, UK, 2019; Volume 1295. [CrossRef]
68. Eddya, M.; Tbib, B.; EL-Hami, K. A Comparison of Chitosan Properties after Extraction from Shrimp Shells by Diluted and Concentrated Acids. *Heliyon* **2020**, *6*, e03486. [CrossRef] [PubMed]
69. Podgorbunskikh, E.; Kuskov, T.; Rychkov, D.; Lomovskii, O.; Bychkov, A. Mechanical Amorphization of Chitosan with Different Molecular Weights. *Polymers* **2022**, *14*, 4438. [CrossRef]
70. Jain, A.; Jain, S.K.; Ganesh, N.; Barve, J.; Beg, A.M. Design and Development of Ligand-Appended Polysaccharidic Nanoparticles for the Delivery of Oxaliplatin in Colorectal Cancer. *Nanomed. Nanotechnol. Biol. Med.* **2010**, *6*, 179–190. [CrossRef]
71. Muhamad, N.; Plengsuriyakarn, T.; Na-Bangchang, K. Application of Active Targeting Nanoparticle Delivery System for Chemotherapeutic Drugs and Traditional/Herbal Medicines in Cancer Therapy: A Systematic Review. *Int. J. Nanomed.* **2018**, *13*, 3921–3935. [CrossRef] [PubMed]
72. Campos, J.; Varas-Godoy, M.; Haidar, Z.S. Physicochemical Characterization of Chitosan-Hyaluronan-Coated Solid Lipid Nanoparticles for the Targeted Delivery of Paclitaxel: A Proof-of-Concept Study in Breast Cancer Cells. *Nanomedicine* **2017**, *12*, 473–490. [CrossRef]
73. Almeida, P.V.; Shahbazi, M.A.; Mäkilä, E.; Kaasalainen, M.; Salonen, J.; Hirvonen, J.; Santos, H.A. Amine-Modified Hyaluronic Acid-Functionalized Porous Silicon Nanoparticles for Targeting Breast Cancer Tumors. *Nanoscale* **2014**, *6*, 10377–10387. [CrossRef]
74. Almalik, A.; Karimi, S.; Ouasti, S.; Donno, R.; Wandrey, C.; Day, P.J.; Tirelli, N. Hyaluronic Acid (HA) Presentation as a Tool to Modulate and Control the Receptor-Mediated Uptake of HA-Coated Nanoparticles. *Biomaterials* **2013**, *34*, 5369–5380. [CrossRef] [PubMed]

75. Subhan, M.A.; Yalamarty, S.S.K.; Filipczak, N.; Parveen, F.; Torchilin, V.P. Recent Advances in Tumor Targeting via Epr Effect for Cancer Treatment. *J. Pers. Med.* **2021**, *11*, 571. [CrossRef]
76. Chiesa, E.; Dorati, R.; Conti, B.; Modena, T.; Cova, E.; Meloni, F.; Genta, I. Hyaluronic Acid-Decorated Chitosan Nanoparticles for CD44-Targeted Delivery of Everolimus. *Int. J. Mol. Sci.* **2018**, *19*, 2310. [CrossRef]
77. Pereira, F.M.; Melo, M.N.; Santos, Á.K.M.; Oliveira, K.V.; Diz, F.M.; Ligabue, R.A.; Morrone, F.B.; Severino, P.; Fricks, A.T. Hyaluronic Acid-Coated Chitosan Nanoparticles as Carrier for the Enzyme/Prodrug Complex Based on Horseradish Peroxidase/Indole-3-Acetic Acid: Characterization and Potential Therapeutic for Bladder Cancer Cells. *Enzym. Microb. Technol.* **2021**, *150*, 109889. [CrossRef] [PubMed]
78. Samprasit, W.; Akkaramongkolporn, P.; Jaewjira, S.; Opanasopit, P. Design of Alpha Mangostin-Loaded Chitosan/Alginate Controlled-Release Nanoparticles Using Genipin as Crosslinker. *J. Drug Deliv. Sci. Technol.* **2018**, *46*, 312–321. [CrossRef]
79. Polexe, R.C.; Delair, T. Elaboration of Stable and Antibody Functionalized Positively Charged Colloids by Polyelectrolyte Complexation between Chitosan and Hyaluronic Acid. *Molecules* **2013**, *18*, 8563–8578. [CrossRef]
80. Herdiana, Y.; Wathoni, N.; Shamsuddin, S.; Muchtaridi, M. Drug Release Study of the Chitosan-Based Nanoparticles. *Heliyon* **2022**, *8*, e08674. [CrossRef] [PubMed]
81. Ofridam, F.; Tarhini, M.; Lebaz, N.; Gagnière, E.; Ofridam, F.; Tarhini, M.; Lebaz, N.; Gagnière, E.; Mangin, D.; Ofridam, F.; et al. PH-Sensitive Polymers: Classification and Some Fine Potential Applications To Cite This Version: HAL Id: Hal-03132353 PH-Sensitive Polymers: Classification and Some Fine Potential Applications. *Polym. Adv. Technol.* **2021**, *32*, 1455–1484. [CrossRef]
82. Bai, X.; Smith, Z.L.; Wang, Y.; Butterworth, S.; Tirella, A. Sustained Drug Release from Smart Nanoparticles in Cancer Therapy: A Comprehensive Review. *Micromachines* **2022**, *13*, 1623. [CrossRef]
83. Nunes, D.; Andrade, S.; Ramalho, M.J.; Loureiro, J.A.; Pereira, M.C. Polymeric Nanoparticles-Loaded Hydrogels for Biomedical Applications: A Systematic Review on In Vivo Findings. *Polymers* **2022**, *14*, 1010. [CrossRef]
84. Azadi, A.; Hamidi, M.; Khoshayand, M.R.; Amini, M.; Rouini, M.R. Preparation and Optimization of Surface-Treated Methotrexate-Loaded Nanogels Intended for Brain Delivery. *Carbohydr. Polym.* **2012**, *90*, 462–471. [CrossRef] [PubMed]
85. Kim, S.W.; Oh, K.T.; Youn, Y.S.; Lee, E.S. Hyaluronated Nanoparticles with PH- and Enzyme-Responsive Drug Release Properties. *Colloids Surfaces B Biointerfaces* **2014**, *116*, 359–364. [CrossRef]
86. Ullah, F.; Shah, K.U.; Shah, S.U.; Nawaz, A.; Nawaz, T.; Khan, K.A.; Alserihi, R.F.; Tayeb, H.H.; Tabrez, S.; Alfatama, M. Synthesis, Characterization and In Vitro Evaluation of Chitosan Nanoparticles Physically Admixed with Lactose Microspheres for Pulmonary Delivery of Montelukast. *Polymers* **2022**, *14*, 3564. [CrossRef]
87. Saleh, A.; Akkuş-Dağdeviren, Z.B.; Friedl, J.D.; Knoll, P.; Bernkop-Schnürch, A. Chitosan—Polyphosphate Nanoparticles for a Targeted Drug Release at the Absorption Membrane. *Heliyon* **2022**, *8*, e10577. [CrossRef]
88. Manimaran, D.; Elangovan, N.; Mani, P.; Subramanian, K.; Ali, D.; Alarifi, S.; Palanisamy, C.P.; Zhang, H.; Rangasamy, K.; Palanisamy, V.; et al. Isolongifolene-Loaded Chitosan Nanoparticles Synthesis and Characterization for Cancer Treatment. *Sci. Rep.* **2022**, *12*, 19250. [CrossRef]
89. Waqas, M.K.; Safdar, S.; Buabeid, M.; Ashames, A.; Akhtar, M.; Murtaza, G. Alginate-Coated Chitosan Nanoparticles for PH-Dependent Release of Tamoxifen Citrate. *J. Exp. Nanosci.* **2022**, *17*, 522–534. [CrossRef]
90. Tapdiqov, S.; Taghiyev, D.; Zeynalov, N.; Safaraliyeva, S.; Fatullayeva, S.; Hummetov, A.; Raucci, M.; Mustafayev, M.; Jafarova, R.; Shirinova, K. Cumulative Release Kinetics of Levothyroxine-Na Pentahydrate from Chitosan/Arabinogalactane Based PH Sensitive Hydrogel and It's Toxicology. *React. Funct. Polym.* **2022**, *178*, 105334. [CrossRef]
91. Herdiana, Y.; Wathoni, N.; Shamsuddin, S.; Muchtaridi, M. Cytotoxicity Enhancement in MCF-7 Breast Cancer Cells with Depolymerized Chitosan Delivery of α-Mangostin. *Polymers* **2022**, *14*, 3139. [CrossRef]

Article

# Strengthening Polylactic Acid by Salification: Surface Characterization Study

**Jessica Schlosser, Michael Keller, Kamran Fouladi and Babak Eslami ***

Department of Mechanical Engineering, Widener University, Chester, PA 19013, USA
* Correspondence: beslami@widener.edu; Tel.: +1-610-499-1391

**Abstract:** Polylactic acid (PLA) is one of the market's most commonly used biodegradable polymers, with diverse applications in additive manufacturing, specifically fused deposition modeling (FDM) 3D printing. The use of PLA in complex and sophisticated FDM applications is continually growing. However, the increased range of applications requires a better understanding of the material properties of this polymer. For example, recent studies have shown that PLA has the potential to be used in artificial heart valves. Still, the durability and longevity of this material in such a harsh environment are unknown, as heart valve failures have been attributed to salification. Additionally, there is a gap in the field for in situ material characterization of PLA surfaces during stiffening. The present study aims to benchmark different dynamic atomic force microscopy (AFM) techniques available to study the salification phenomenon of PLA at micro-scales using different PLA thin films with various salt concentrations (i.e., 10%, 15%, and 20% of sodium chloride (NaCl)). The measurements are conducted by tapping mode AFM, bimodal AFM, the force spectroscopy technique, and energy quantity analysis. These measurements showed a stiffening phenomenon occurring as the salt solution is increased, but the change was not equally sensitive to material property differences. Tapping mode AFM provided accurate topographical information, while the associated phase images were not considered reliable. On the other hand, bimodal AFM was shown to be capable of providing the topographical information and material compositional mapping through the higher eigenmode's phase channel. The dissipated power energy quantities indicated that how the polymers become less dissipative as salt concentration increases can be measured. Lastly, it was shown that force spectroscopy is the most sensitive technique in detecting the differences in properties. The comparison of these techniques can provide a helpful guideline for studying the material properties of PLA polymers at micro- and nano-scales that can prove beneficial in various fields.

**Keywords:** surface characterization; scanning probe microscopy; additive manufacturing; PLA; multifrequency AFM



**Citation:** Schlosser, J.; Keller, M.; Fouladi, K.; Eslami, B. Strengthening Polylactic Acid by Salification: Surface Characterization Study. *Polymers* **2023**, *15*, 492. https://doi.org/10.3390/polym15030492

Academic Editors: José Miguel Ferri, Vicent Fombuena Borràs and Miguel Fernando Aldás Carrasco

Received: 15 November 2022
Revised: 2 January 2023
Accepted: 9 January 2023
Published: 17 January 2023

## 1. Introduction

The idea of artificial heart valves made of polylactic acid (PLA) is getting closer to reality. However, the extent of degradation of plastic valves by salification is not well understood and has not been extensively investigated. The present study is focused on investigating the impact of salification on the valves through a comprehensive material characterization effort utilizing various AFM methods.

### 1.1. Atomic Force Microscopy Techniques

Atomic force microscopy (AFM), invented in the 1980s, is from the family of scanning probe microscopies providing major advantages in material characterization. AFM is capable of extracting topographical information from a wide range of samples (conductive to nonconductive, metals to polymers) while allowing material property characterization to be performed, such as mechanical properties, chemical properties, and electrical and magnetic properties [1–5]. The main component of AFM is the micro-cantilever. The deflection of the cantilever is measured by shining a laser over the back of the cantilever and measuring the signal through the laser's reflection on the photodiode detector. The analysis and manipulation of this signal will differ based on the imaging scheme. Figure 1 depicts a schematic of how dynamic AFM imaging works. One specific mode of AFM is the tapping mode, in which the cantilever oscillates at the cantilever's resonance frequency (i.e., the first eigenmode frequency) using a piezoelectric driver. Although the excitation frequency is fixed, the distance between the tip and sample is modulated so that the oscillation amplitude remains constant. In this method, named amplitude modulation AFM (AM-AFM), the height (i.e., topography) information is gathered from the amplitude signal [6]. The phase signal, the phase lag between excitation at the base of the cantilever and the oscillation of the tip, provides information on the compositional mapping of the surface. Tapping mode can be carried out in two different regimes, namely, attractive or repulsive. Attractive mode is when the phase signal is greater than 90°, and repulsive mode is when the phase signal is less than 90° [7]. In the attractive regime, the cantilever does not physically touch the surface but interacts with long-range interaction forces such as electrostatic forces or van der Waals forces. The *bistability* phenomenon will occur if the oscillation setpoint and free oscillation amplitude are set so that the solution of the equation of motion of the cantilever can have two roots [8]. Ideally, the repulsive regime creates higher quality images, but the attractive regime can be good for soft samples, as it applies smaller forces on the sample.

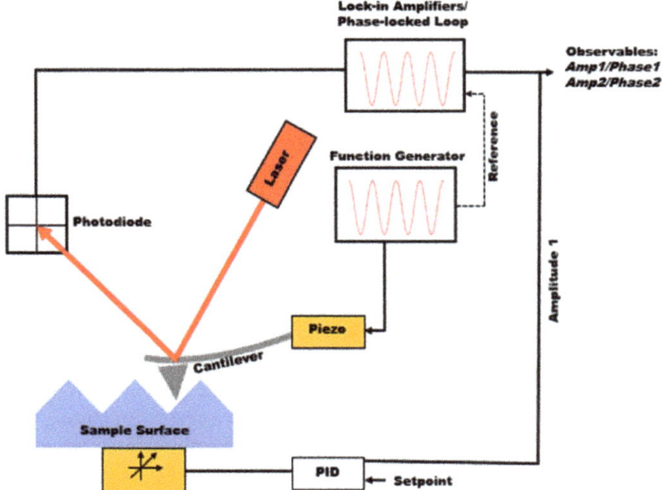


**Figure 1.** Schematic of dynamic atomic force microscopy.

Although tapping mode AFM can provide both topography and compositional mapping, there is no guarantee that the quality of height and phase channels will be good for all samples. Additionally, the information observed from phase images is not directly related to mechanical properties, rendering them not quantitatively useful. Therefore, there has been significant enthusiasm toward force spectroscopy and material models needed to fit the data during the past few years. In force spectroscopy techniques, the cantilever is approached to the sample, while its deflection is measured, as shown in Figure 1 [9–11], going through the attractive and then repulsive forces. Using the feedback loop, the cantilever is pulled out of the sample after reaching the given maximum deflection. The observed deflection can then be converted into force versus tip sample separation to extract mechanical properties such as adhesion, stiffness (modulus), and the indentation depth. Force spectroscopy only collects a single point measurement, so force mapping is useful to collect a grid of force curves over an area of a sample and allows for the mechanical properties within that area to be compared [5,12].

In the early 2000s, it was found AFM tapping mode might not provide both topography and material composition at the same time when imaging the soft matter [13,14]. Therefore, a branch of imaging techniques known as multifrequency AFM was introduced [15]. Bimodal AFM, the most common multifrequency technique, is accomplished through the simultaneous excitation of two eigenmode frequencies of the AFM cantilever. This technique simultaneously generates multiple additional properties, such as a second phase channel. It can also allow for higher resolution and more detailed information about the sample. Subsequently, it was shown that higher eigenmode AFMs could provide depth penetration through samples and enhance AFM capabilities to image subsurface features.

Many AFM techniques have been developed and used over the years. However, there is still a need to fundamentally understand the strengths and limitations of these techniques. This work aims to fill this knowledge gap when dealing with biodegradable polymers.

### 1.2. 3D-Printed Artificial Heart Valves

Cardiovascular disease (CVD) is the current leading cause of death in the United States, resulting in 696,962 deaths in 2020 [16]. Heart valve disease (HVD), which concerns any of the four heart valves, is a major contributor to CVD. Moreover, the aortic valve can thicken and become stiff (aortic valve stenosis), which is the most common cause of the need for surgery. The stiffening causes the valve to be unable to open fully, reducing the blood flow to the body [17]. Aortic valve stenosis is commonly caused by calcium and calcium phosphate buildup on the aortic valve, otherwise known as salification. Therefore, it is critical to firstly understand the stiffening phenomenon on the polymer surface of an artificial aortic valve and secondly visualize this concept to be able to design better valves.

Generally, two types of heart valves can be used for aortic valve replacement—biological or mechanical. Biological or bioprosthetic valves replace the original aortic valve with a new valve typically made from bovine (cow) or porcine (pig) tissue [17]. This animal tissue is referred to as a xenograft. Similarly, allografts are valves taken from human cadavers or brain-dead patients [17]. However, these valves are less readily available as they cannot be mass-produced or available commercially.

Mechanical valves, generally made of carbon or other sturdy materials, last longer than biological valves and do not need to be replaced. However, they do require the patients to take blood-thinning medications for the rest of their lives. Recent emerging technologies have turned aortic heart valve replacement towards 3D printing. It is believed that the interaction between the artificial heart valve and the patient's native anatomy could be significantly improved through the 3D printing of patient-specific models. Unlike mechanical or biological heart valve replacements, 3D-printed heart valves can be designed specifically for the size and anatomy of the patient. Although salification is a concern for all heart valve types, it can play a major role in the longevity and performance of 3D-printed valves. However, there is little understanding of if and how the salification process stiffens plastic valves.

### 1.3. Motivation and Objectives

Recent studies in the cardiovascular field are touting artificial heart valves made of polylactic acid (PLA), a biodegradable polymer commonly used in additive manufacturing and extensively in fused deposition modeling. However, 3D-printed heart valves can be negatively impacted by salification, similar to other artificial heart valves. Moreover, the extent of degradation of plastic valves by salification is not well understood and has not been extensively investigated.

In the present study, we undertake a comprehensive surface characterization effort focused on the salification process of biodegradable polymer-based heart valves. More specifically, our research is focused on how salification affects PLA thin films at the microscopic level. Various AFM methods are utilized to help gain a better understanding of how salification affects the material properties and quality of PLA at both quantitative and qualitative levels. The methods include tapping mode images, multifrequency AFM, energy-based AFM analysis, and force spectroscopy. The range of AFM methods in the present study can be used to provide a comprehensive guide to researchers studying salification effects on PLA to generate the most meaningful information using AFM.

The knowledge gained by the present investigation will help to understand if 3D-printed valves are viable for heart valve replacement. The findings can also help determine if salification can strengthen PLA for additive manufacturing applications. If results show that salification has caused stiffening on the surface of the PLA polymer, one can decide how to treat the polymer to reduce the chance of salt bonding to the surface.

## 2. Methods and Materials

Three PLA film films with increasing salt concentrations were produced to understand how salification affects the material properties of PLA. Although there is a difference between bulk and surface material properties, it is understood that the performance of artificial heart valves is dictated by the changes on their surfaces. Additionally, for experimental purposes, spin-coated thin films provide more controlled methods of measurement for fundamental understanding of the matter. Initially, small beads of PLA were placed in a glass beaker, and methylene chloride was added to dissolve the polymer. A magnetic stirrer was then added, and the sample was mixed until all the PLA was dissolved. Three sodium chloride (NaCl) solutions were then prepared. The first solution had a salt to di-water weight ratio of 1:10, the second solution had a ratio of 1.5:10, and the third solution had a ratio of 2:10. With a pipet, equal parts of sodium chloride solution and PLA-methylene chloride were combined in a beaker. This procedure was repeated for each sodium chloride solution. Dime-sized silicon wafers were then prepared through a cleaning process with an ultra-sonic cleaner and an isopropyl and de-ionized water rinse. Each silicon (Si) wafer was then placed on the SCK-300 digital spin-coater device. The PLA-methylene chloride-salt solution was deposited on the Si wafers. The sample was then spun at 3000 rpm for 60 s. This process was repeated for each of the three NaCl concentrations. Each sample was placed on a glass slide and set aside overnight to fully dry. The samples are referenced in Table 1.

**Table 1.** Sample NaCl concentration with given sample numbers.

| Sample Number | % Weight NaCl Concentration |
|---|---|
| 1 | 10 |
| 2 | 15 |
| 3 | 20 |

Once the samples were ready for measurement, the Asylum Research MFP-3D AFM controlled with an ARC2 controller was used to perform various AFM studies. The same type of cantilever (Multi75-G), manufactured by Budget Sensors, was used throughout the measurements for proper comparison of the AFM results. The cantilever's spring constant was measured by Sader's method [18] and was found to be 2.07 N/m with a resonance frequency of 78 kHz. These cantilevers are around 225 μm in length, 28 μm in width, and 3 μm in thickness. The tip radious is around 10 nm and made out of silicon.

The conventional tapping mode was the first method used to measure each sample's topography and phase. Simultaneously, the observables of the conventional tapping mode were analyzed from the energy-based method. Specifically, the amplitude and phase signals collected from tapping mode imaging were converted into virial ($V_{ts}$) and dissipated power ($P_{ts}$), which are convolutions of the tip–sample interactions with position and velocity, respectively. Equations (1) and (2) describe the conversions, where index $i$ specifies the corresponding eigenmode under study, $k$ is the stiffness, $A$ the instantaneous amplitude, $A_0$ the free amplitude, $f\_exc$ the excitation frequency, $f_0$ the free resonance frequency,

$\phi$ the phase, and $Q$ the quality factor. Based on these equations, if the imaging mode is simple tapping mode where the excitation frequency is equal to the resonance frequency, the $f_{exc}/f$ will be reduced to one and will simplify the equations [19,20].

$$V_{ts,i} = \frac{k_i A_i}{2}\left[A_i\left(1 - \frac{f_{exc,i}^2}{f_i^2}\right) - \frac{A_{0,i}}{Q_i}cos\phi_i\right] \quad (1)$$

$$P_{ts,i} = \frac{\pi f_{exc,i} k_i A_i^2}{Q_i}\left[\frac{A_{0,i}}{A_i}sin\phi_i - \frac{f_{exc,i}}{f_i}\right] \quad (2)$$

The second AFM method performed on each of the samples was a multifrequency technique called bimodal AFM. In this method, imaging is performed through the simultaneous excitation of two different eigenmodes, while the first eigenmode is modulated with the feedback loop, and the second eigenmode is in an open loop. The technique results in a series of images, including topography, the first eigenmode amplitude, the second eigenmode amplitude, the first eigenmode phase, and the second eigenmode phase. Since one can optimize the two relatively weakly coupled eigenmodes separately, both topography and compositional mapping are guaranteed through this technique. In this method, the free amplitude of the higher eigenmode is the key parameter governing the intensity of the tip–sample interactions for a given free amplitude and amplitude setpoint of the fundamental eigenmode [21]. The user can also control the drive frequency, which should be selected close to the eigenmode frequencies. By observing the governing equation of motion of the cantilever, it can be seen that the greater indentation capability of higher eigenmodes is possible with the use of higher eigenmodes:

$$\frac{d^2\underline{z}}{d\underline{t}^2} = -\underline{z} + \frac{1}{Q}\left[-\frac{d\underline{z}}{d\underline{t}} + \cos(\underline{t})\right] + \frac{F_{ts}(\underline{z}_{ts})}{kA_0} \quad (3)$$

where $A_0$ is the free amplitude, $\underline{z} = z/A_0$ is the normalized tip displacement, $\underline{z}_{ts}$ is the normalized tip–sample distance, $\underline{t} = \omega_0 t$ is the dimensionless time, $k$ is the cantilever force constant, and $F_{ts}$ is the tip–sample force interaction. The free oscillation amplitude is assumed to be equal to $\frac{F_0 Q}{k}$ based on the work done by Ricardo Garcia et al. [21]. The damping and excitation terms are combined with the $\frac{1}{Q}$ factor. As shown in Equation (1), the last term on the right-hand side is normalized by the product of the cantilever force constant and the free oscillation amplitude. Therefore, as the denominator of this fraction increases, the effect of tip–sample force interactions on the dynamics of the cantilever diminishes. For the bimodal AFM case, the force constant of higher eigenmodes is approximately 36 times the first eigenmode. Therefore, the cantilever becomes less sensitive to forces when excited with two eigenmodes. Consequently, surface penetration is observed, and if the soft matter under study can be compressed, the AFM tip will compress the film and a stiffer surface is observed [22].

The method of force spectroscopy was carried out to understand how increased salification affects the stiffness of PLA. An approach curve was captured on a stiff glass slide while the deflection in volts was measured. To collect the appropriate data, an area of $5 \times 5$ μm was imaged using tapping mode AFM, which was also for virial and dissipated power analyses. Next, we zoomed into the image twice on both a PLA area and salt contamination area and collected a smaller image around $1 \times 1$ μm. Once we ensured that the image focused on one material in particular, a force map of 20 contact mode force curves was collected over an area of approximately $200 \times 200$ nm. The same process was repeated twice per three samples.

Contact mechanics models were used to extract material properties of the samples by fitting the force curves to the appropriate model. The three primary models included Hertz, Derjaguin–Muller–Toporove (DMT), and Johnson–Kendall–Roberts (JKR) [23]. The Hertz model neglects adhesion and friction, which works well for nanoindentation and liquid imaging; however, it does not work for most AFM tip–sample interactions, since the

AFM tip does interact with the sample. Neutral atoms and molecules still experience forces between one another, so adhesion cannot be neglected. The DMT model is appropriate for stiffer, longer-range adhesion, while the JKR model is appropriate for stronger and shorter-range adhesion. Based on this information, the DMT contact model was selected for force spectroscopy analysis.

## 3. Results and Analysis

AFM tapping mode was the first method utilized to collect information from each of the three calcified PLA samples. Figure 2 compares the results of this method for both topography and phase. Although the topography images show some increases in surface roughness as we increased the salt concentration on the samples, they are not completely distinguishable. The main differences between the results were seen in the phase results. It should be noted that when the cantilever was tuned in air (not interacting with the surface), the phase value near or at the resonance frequency was about 90 degrees. Based on the equation of motion of a simple harmonic excited system, as the cantilever interacted with the surface (stiffer than air), the phase value decreased to values below 90 degrees. Therefore, a general rule in analyzing phase images states that the lower phase values (i.e., darker colors in phase images) represent stiffer surfaces. There were clear regions where the salt contamination on surfaces was shown. This was shown as islands on samples 2 and 3. However, more importantly, an overall increase in the stiffness of samples was shown going from sample 1 to sample 3. Overall lower phase values among samples represented the stiffer surface and tip–sample force interactions, while the contrast between the islands represented the material composition.

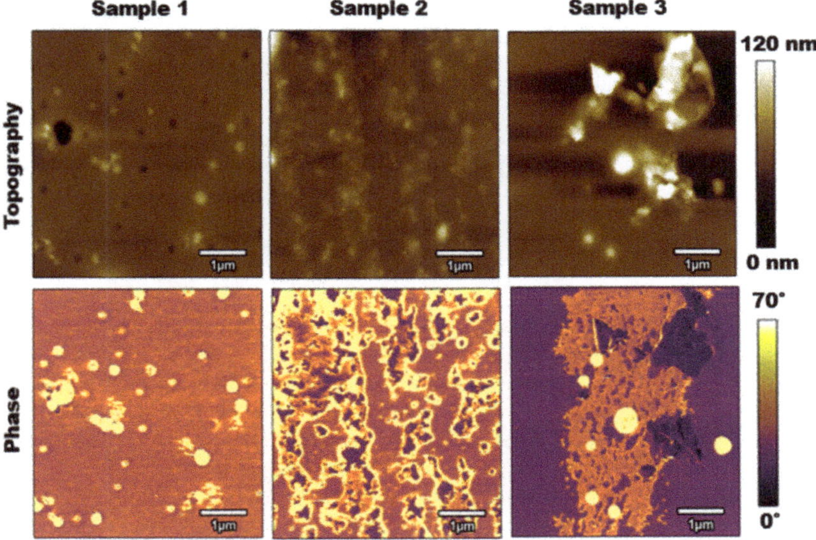


**Figure 2.** Tapping mode AFM results: **Top Row**: topography images for samples 1, 2, and 3 from left to right. **Bottom Row**: Phase images for samples 1, 2, and 3 from left to right. Scan sizes are 5 μm × 5 μm with a scan rate of 1 Hz, free oscillation amplitude of 100 nm, with 60% setpoint.

For sample 1, which had a 10% salt concentration, the salt presented itself in small, scattered circles over 6.74% of the surface. The size and shape of these salt particles appeared to be very similar to the size and shape of the humidity pores depicted by the dark circles in the topography. Even where those pores presented themselves in the topography, the phase images showed that they were the same material as where the raised surfaces occurred. Therefore, the conclusion can be drawn that at the microscale, the salt in a calcified PLA sample with a 10% salt concentration binds to both the surface of the PLA as well as the pores.

For sample 2, which had a salt concentration of 15%, the salt presented itself in large, scattered islands over 25.44% of the surface. Compared to samples 1 and 3, the salt concentrations were distributed more evenly over the surface of the PLA. Here, the salt adhered to the actual surface of the PLA rather than distributed into the pores or binding with itself.

For sample 3, which had a salt concentration of 20%, the salt presented itself in one large island with a few scattered and raised circular patches over 33.44% of the surface. The separation of the PLA and salt concentrations was likely caused by an over-concentration of the salt solution with the PLA. The solutions were likely separated even before the spin-coating process occurred, causing the salt to bind to itself and therefore adhere to the surface of the PLA as singular large islands rather than small scattered islands, as seen in sample 2.

The subsequent analysis that was performed was a virial and dissipative analysis based on the amplitude modulation channels from the tapping mode images collected from each sample. Figure 3 displays the results of this study.

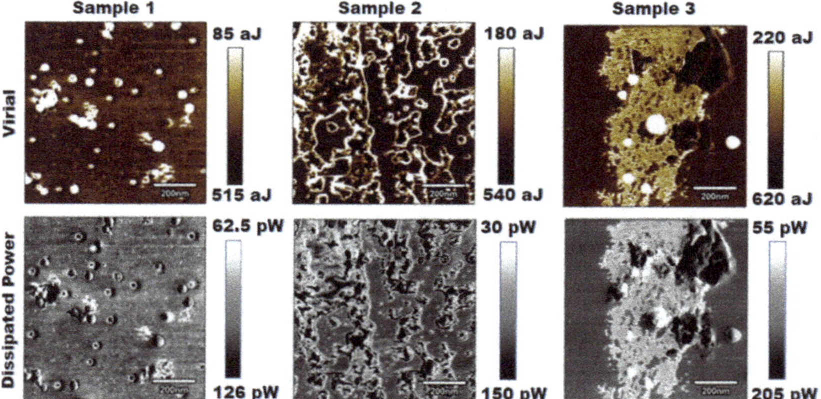


**Figure 3.** Results for virial and dissipated power for three PLA samples of increasing salt concentrations.

By visual inspection, both the virial and dissipated power images yielded similar contrasts to the tapping mode phase images. The dissipated power analysis nearly depicted the same images, while the virial depicted slightly more detail [24,25]. Again, visually, it was difficult to interpret a change in stiffness from these images, but further numerical analysis could show that the results correlated with an increase in stiffness across samples.

The next method used to analyze the calcified PLA samples was bimodal imaging, where the results were generated through the simultaneous excitation of two eigenmodes of the AFM cantilever. Figure 4 displays the results of this method, including the topography, phase 1 generated through the first eigenmode, and phase 2 generated through the second eigenmode.

The bimodal imaging results shown in Figure 4 compared well with the results from normal tapping mode presented in Figure 2, and the topographies appeared to be nearly identical. Similarly, there was some increase in surface roughness as the salt concentration increased, but the materials were not completely distinguishable. The phase 1 images were also nearly identical to the phase images from Figure 2, which clearly distinguished the salt contaminations from the PLA surface. Moreover, areas of the surface became more detailed when considering the phase 2 images based on the second eigenmode. This change was most notable for samples 2 and 3, as the gaps in the salt contaminations began to become more apparent.

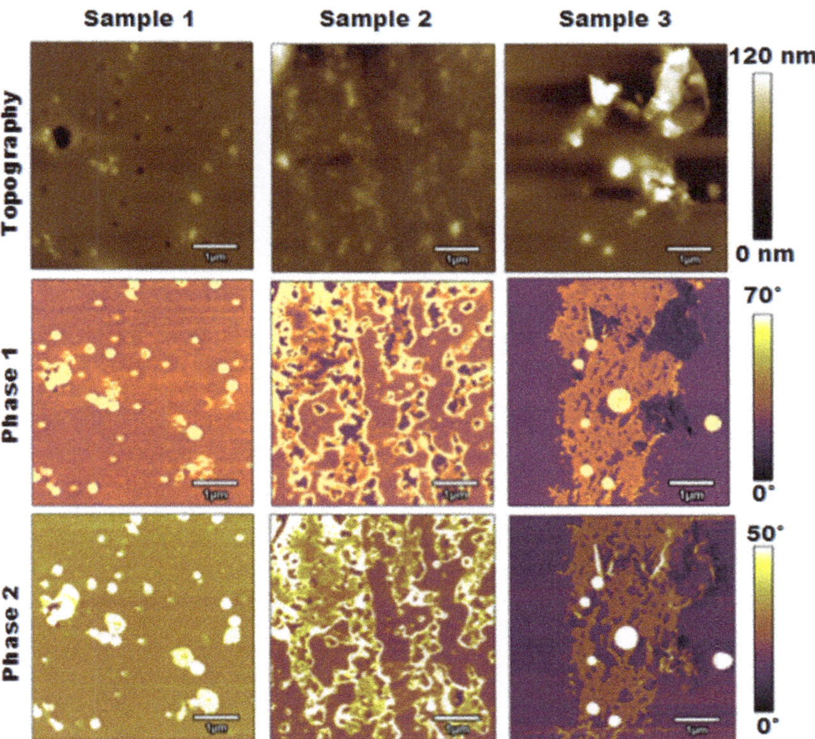


**Figure 4.** Topography, phase 1, and phase 2 results for bimodal imaging of three PLA samples of increasing salt concentration.

Additionally, the raised area in the top right corner of sample 3 differed between phase 1 and phase 2. The phase 1 image showed that area as PLA or an area with lower stiffness, while salt contaminations began to appear in that region on the phase 2 image. Overall, comparing these sets of images, bimodal imaging allows for a more detailed understanding of the materials present on the surface compared to normal tapping mode.

The final analysis performed on the three calcified PLA samples was force spectroscopy, which allowed for the Young's modulus, or stiffness, of each sample to be collected. A series of force curves was collected using force mapping over an area of high salt concentration and primarily PLA. For example, for sample 3, force maps were collected in the dark purple outer region and the pink inner region depicted in Figure 5a. Figure 5b is an example of two force versus separation curves for the different areas on sample 2, which were converted from the raw deflection versus distance curves collected through force spectroscopy. Figure 5c displays the effective Young's modulus values for each sample, which were calculated by fitting the DMT contact model to every curve and performing a particle analysis to determine the percentage of the surface covered in salt. Figure 5d is a physical representation of the DMT contact model, in which the DMT spring is represented as an infinite series of springs. Based on this representation, more springs are activated as the AFM tip goes deeper into the surface. Therefore, stiffness depends both on the position of the tip and the contact area.

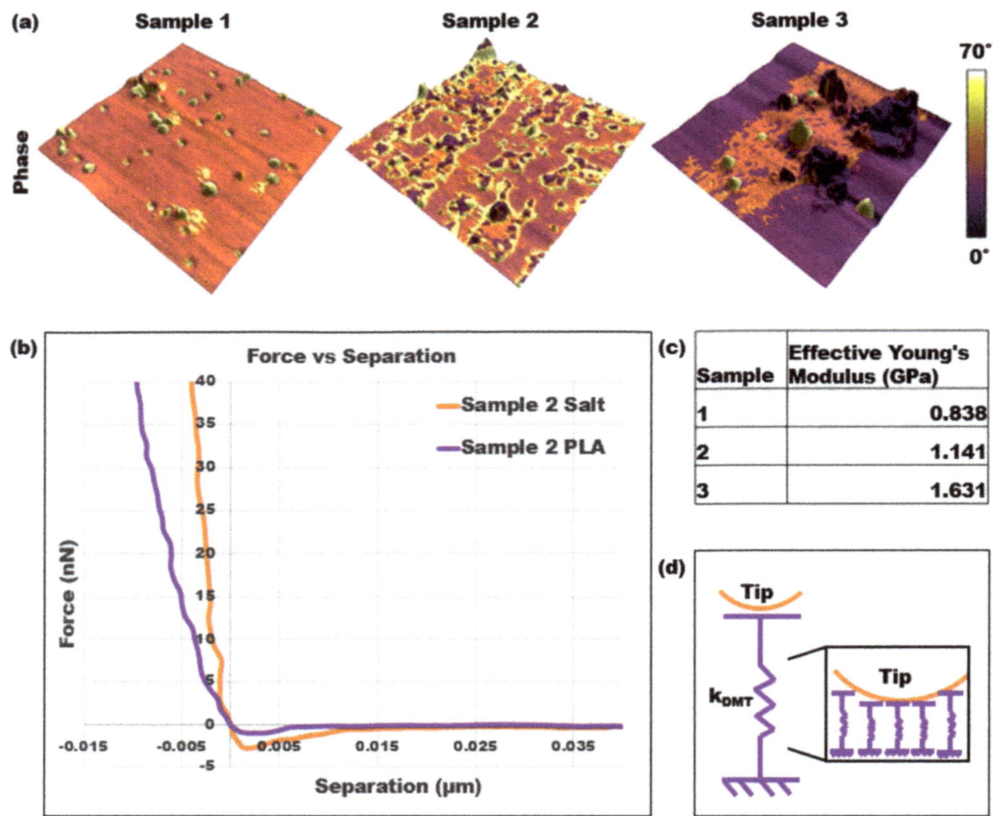


**Figure 5.** Force spectroscopy results: (**a**) three-dimensional topography of three PLA samples overlayed with a color scale representing the phases; (**b**) two force curves of sample 2 on an area of high salt concentration vs. a primarily PLA area; (**c**) effective Young's modulus results using force spectroscopy and the DMT contact model; (**d**) spring and AFM tip representation of the DMT contact model.

The 3D representation of the topography superimposed with the phase clearly distinguished the areas of high salt concentrations as well as the impact on the roughness of the samples. Additionally, the slopes of the force versus separation curves over those two different areas distinguished the stiffnesses. The orange-colored force curve over the salt contamination area was steeper than the purple curve over the PLA area. This trend was consistent over each sample, and the Young's modulus over the PLA areas also increased from sample 1 to sample 3. Therefore, as the salt contaminations covered a greater percentage of the surface and the overall NaCl concentration increased, the effective Young's modulus of the samples also increased.

It is important to note the quantitative and qualitative differences when analyzing and comparing the results from each method, including tapping mode, a virial and dissipative analysis, bimodal imaging, and force spectroscopy. First, in examining the phase change across the samples, specifically the PLA regions, it could be interpreted that the stiffness of the PLA decreases as the NaCl concentration increases. For example, as shown in Figure 5a, the phases for the PLA regions of samples 1, 2, and 3 were approximately 35°, 30°, and 15°, respectively. However, the force spectroscopy results proved otherwise. The effective Young's modulus was calculated as an accumulation of force spectroscopy results for both PLA and salt contamination regions on each sample along with their respective percent surface area. Through this analysis, the force spectroscopy results gathered that the average Young's modulus over the PLA regions increased across the samples as the NaCl concentration increased. Specifically, the average Young's modulus in these regions for samples 1, 2, and 3 were 0.774 GPa, 1.062 GPa, and 1.154 GPa, respectively. The standard deviations were 0.53 kPa, 0.79 kPa, and 0.61 kPa, respectively. A similar trend could be seen for the salt contamination areas, where phase images indicated that the pink salt areas of sample 3 had a lower stiffness than the yellow salt areas of samples 1 and 2. Once again, the force spectroscopy results proved otherwise, as sample 3 had the highest average Young's modulus value of 2.582 compared to values for samples 1 and 2 of 1.718 GPa and 1.373 GPa, respectively. It is also worth noting that these phase trends from tapping mode were consistent with the phase 1 and phase 2 images from bimodal imaging.

Theoretically, each AFM method should provide the same information about each sample. However, the results proved there were differences. While there were similarities between the methods, each method offered additional unique information and could provide guidelines for researchers when choosing different characterization techniques for samples with varying mechanical properties. Figure 6 displays the data normalization results comparing the results of each AFM method to understand the sensitivity of each AFM characterization technique performed. Sample 2 and sample 3 data were normalized by sample 1 data for the corresponding method of measurement. For example, in the tapping mode AFM column, sample 2 and sample 3 average phase values are divided by sample 1 average phase values. Since sample 1 is the untreated polymer, we used this sample as the reference point in our study. The closer the value was to one in this plot, the less of a difference was observed by the measurement technique. The dotted horizontal line represents the threshold. The average phase for simple tapping mode, average dissipated power, average phase 2 for bimodal imaging, and the effective Young's modulus from force spectroscopy were selected for comparison.

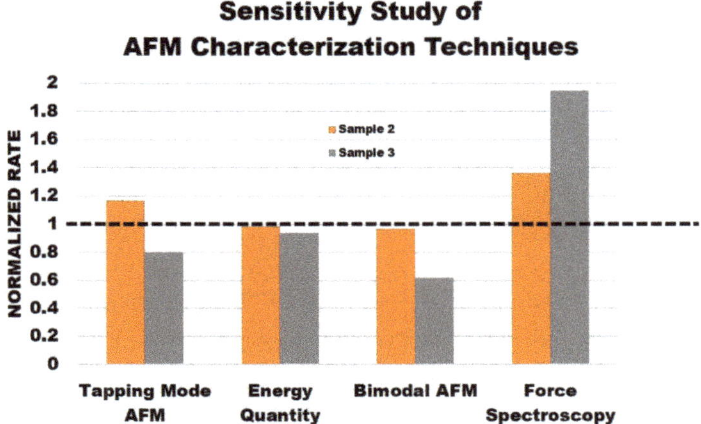


**Figure 6.** Normalized data of sensitivity study of AFM characterization techniques.

Overall, the sensitivity study shows that the phase results from simple tapping mode are not reliable. Although they provide a good sample topography, the phase images are not necessarily reliable regarding the stiffness of the samples. The energy quantity results do provide useful information as the sample surfaces become less dissipative, indicating increased stiffness as salt concentration increases. Additionally, the bimodal phase 2 images do provide useful information about the samples. The results show that the higher the salt concentration, the lower the phase values, indicating that stiffness increases across the samples. Finally, the force spectroscopy results follow the trend exactly as the Young's modulus and stiffness increase across the samples. These results also show that the force spectroscopy results are more sensitive to the sample change.

Conducting each method proved that using the same technique across samples of increasing stiffness may not be viable. It also shows that some techniques, such as bimodal phase imaging and force spectroscopy, provide more useful information than others, such as simple tapping mode phase. In other words, force spectroscopy and bimodal AFM are more sensitive to material differences over a given surface. Finally, the comparison between the normal tapping mode images and bimodal imaging proves that both methods do not need to be conducted. It would be more useful for the researcher to only consider bimodal imaging, as the addition of phase 2 and the implementation of two eigenmodes provide similar yet more accurate information about the samples. This holds true as long as the product of $kA_0$ explained in Equation (1) does not increase drastically, so the forces applied on soft matter cause damage to the surface.

## 4. Conclusions

In this study, we compared tapping mode AFM, bimodal AFM, energy analysis (dissipated power), and force spectroscopy techniques while characterizing one of the most commonly used biodegradable polymers (PLA). During this study, it was shown that as the salt concentration on PLA surfaces increases, AFM techniques are capable of detecting the material property differences. However, it was also shown that each technique has its own sensitivity to these property changes. Tapping mode AFM is not a reliable characterization technique for material properties. However, using the amplitude and phase signals of tapping mode AFM, we derived the virial and dissipated power, which verified that dissipated power (a combination of amplitude and phase information) is more sensitive to sample differences. In addition to simple tapping mode AFM, bimodal AFM was shown to be a useful technique that can detect the material changes while still providing topographical information. However, its sensitivity is not as good as force spectroscopy analysis. This study concluded that in order to detect different material properties, force spectroscopy is the most sensitive technique, although it cannot provide topographical information. Therefore, based on this work, it is recommended that investigators perform bimodal AFM imaging, followed by a force map, that can fit different material models discussed in the paper for a comprehensive analysis.

**Author Contributions:** Conceptualization, K.F. and B.E.; methodology, J.S. and M.K.; software, J.S. and B.E.; validation, J.S. and M.K.; formal analysis, J.S. and B.E.; investigation, B.E. and K.F.; resources, K.F. and B.E.; data curation, J.S. and M.K.; writing-original draft preparation, J.S. and M.K.; writing-review and editing, B.E. and K.F.; visualization, J.S. and B.E.; funding acquisition, B.E. and K.F. All authors have read and agreed to the published version of the manuscript.

**Funding:** This research was funded by Pennsylvania Department of Community and Economic Development grant number 1060170-458516.

**Institutional Review Board Statement:** Not applicable.

**Data Availability Statement:** Data will be available upon request.

**Conflicts of Interest:** The authors declare no conflict of interest.

## References

1. Lozano, J.R.; Garcia, R. Theory of phase spectroscopy in bimodal atomic force microscopy. *Phys. Rev. B* **2009**, *79*, 14110. [CrossRef]
2. Yu, J.; Ma, E.; Ma, T. Harvesting energy from low-frequency excitations through alternate contacts between water and two dielectric materials. *Sci. Rep.* **2017**, *7*, 17145. [CrossRef] [PubMed]
3. Benaglia, S.; Amo, C.A.; Garcia, R. Fast quantitative and high resolution mapping of viscoelastic properties with bimodal AFM. *Nanoscale* **2019**, *11*, 15289–15297. [CrossRef] [PubMed]
4. Martínez, N.F.; Lozano, J.R.; Herruzo, E.T.; Garcia, F.; Richter, C.; Sulzbach, T.; Garcia, R. Bimodal atomic force microscopy imaging of isolated antibodies in air and liquids. *Nanotechnology* **2008**, *19*, 384011. [CrossRef] [PubMed]
5. Katan, A.J.; van Es, M.H.; Oosterkamp, T.H. Quantitative force versus distance measurements in amplitude modulation AFM: A novel force inversion technique. *Nanotechnology* **2009**, *20*, 165703. [CrossRef]

6. Garcia, R.; Perez, R. Dynamic atomic force microscopy methods. *Surf. Sci. Rep.* **2002**, *47*, 197–301. [CrossRef]
7. Proksch, R.; Yablon, D.G. Loss tangent imaging: Theory and simulations of repulsive-mode tapping atomic force microscopy. *Appl. Phys. Lett.* **2012**, *100*, 073106. [CrossRef]
8. García, R.; San Paulo, A. Attractive and repulsive tip-sample interaction regimes in tapping-mode atomic force microscopy. *Phys. Rev. B* **1999**, *60*, 4961–4967. [CrossRef]
9. Leite, F.L.; Bueno, C.C.; Da Róz, A.L.; Ziemath, E.C.; Oliveira, O.N. Theoretical models for surface forces and adhesion and their measurement using atomic force microscopy. *Int. J. Mol. Sci.* **2012**, *13*, 12773–12856. [CrossRef]
10. Garcia, R. Nanomechanical mapping of soft materials with the atomic force microscope: Methods, theory and applications. *Chem. Soc. Rev.* **2020**, *5850*, 5850. [CrossRef]
11. Payam, A.F.; Martin-Jimenez, D.; Garcia, R. Force reconstruction from tapping mode force microscopy experiments. *Nanotechnology* **2015**, *26*, 185706. [CrossRef]
12. Hu, S.; Raman, A. Inverting amplitude and phase to reconstruct tip–sample interaction forces in tapping mode atomic force microscopy. *Nanotechnology* **2008**, *19*, 375704. [CrossRef]
13. Santos, S. Phase contrast and operation regimes in multifrequency atomic force microscopy. *Appl. Phys. Lett.* **2014**, *104*, 143109. [CrossRef]
14. Rodríguez, T.R.; García, R. Compositional mapping of surfaces in atomic force microscopy by excitation of the second normal mode of the microcantilever. *Appl. Phys. Lett.* **2004**, *84*, 449–451. [CrossRef]
15. Garcia, R.; Herruzo, E.T. The emergency of multifrequency force microscopy. *Nat. Nanotechnol.* **2012**, *7*, 217–226. [CrossRef]
16. Centers for Disease Control and Prevention. Available online: https://www.cdc.gov/nchs/fastats/heart-disease.html (accessed on 1 November 2022).
17. Brown, E. A Study in the Design of an Accelerated Wear Tester that s Compatible with a Particle image Velocimetry and High Speed Camera Setup. Master's Thesis, The Pennsylvania State University, State College, PA, USA, 2014.
18. Sader, J.E. Atomic Force Microscopy: Cantilever Calibration. *Encycl. Surf. Colloid Sci.* **2015**, 469–479.
19. Guerra, E.A.L. Analytical Developments for the Measurement of Viscoelastic Properties with the Atomic Force Micro-scope. Ph.D. Thesis, The George Washington University, Washington, DC, USA, 2018.
20. Shi, S.; Guo, D.; Luo, J. Enhanced phase and amplitude image contrasts of polymers in bimodal atomic force microscopy. *RSC Adv.* **2017**, *7*, 11768–11776. [CrossRef]
21. Ebeling, D.; Eslami, B.; Solares, S.D.J. Visualizing the subsurface of soft matter: Simultaneous topographical imaging, depth modulation, and compositional mapping with triple frequency atomic force microscopy. *ACS Nano* **2013**, *7*, 10387–10396. [CrossRef]
22. Eslami, B.; Ebeling, D.; Solares, S.D. Trade-offs in sensitivity and sampling depth in bimodal atomic force microscopy and comparison to the trimodal case. *Beilstein J. Nanotechnol.* **2014**, *5*, 1144–1151. [CrossRef]
23. López-Guerra, E.A.; Solares, S.D. Modeling viscoelasticity through spring-dashpot models in intermittent-contact atomic force microscopy. *Beilstein J. Nanotechnol.* **2014**, *5*, 2149–2163. [CrossRef]
24. Cleveland, J.P.; Anczykowski, B.; Schmid, A.E.; Elings, V.B. Energy dissipation in tapping-mode atomic force microscopy. *Appl. Phys. Lett.* **1998**, *72*, 2613–2615. [CrossRef]
25. Kiracofe, D.; Raman, A.; Yablon, D. Multiple regimes of oepration bimodal AFM: Understanding the energy of cantilever eigenmodes. *Beilstein J. Nanotechnol.* **2013**, *4*, 385–393. [CrossRef]

*Article*

# Green Composites Based on Mater-Bi® and *Solanum lycopersicum* Plant Waste for 3D Printing Applications

**Roberto Scaffaro [1,2,*], Maria Clara Citarrella [1] and Marco Morreale [3,*]**

[1] Department of Engineering, University of Palermo, Viale delle Scienze, 90128 Palermo, Italy
[2] INSTM, Consortium for Materials Science and Technology, Via Giusti 9, 50125 Florence, Italy
[3] Faculty of Engineering and Architecture, Kore University of Enna, Cittadella Universitaria, 94100 Enna, Italy
* Correspondence: roberto.scaffaro@unipa.it (R.S.); marco.morreale@unikore.it (M.M.)

**Abstract:** 3D printability of green composites is currently experiencing a boost in importance and interest, envisaging a way to valorise agricultural waste, in order to obtain affordable fillers for the preparation of biodegradable polymer-based composites with reduced cost and environmental impact, without undermining processability and mechanical performance. In this work, an innovative green composite was prepared by combining a starch-based biodegradable polymer (Mater-Bi®, MB) and a filler obtained from the lignocellulosic waste coming from *Solanum lycopersicum* (i.e., tomato plant) harvesting. Different processing parameters and different filler amounts were investigated, and the obtained samples were subjected to rheological, morphological, and mechanical characterizations. Regarding the adopted filler amounts, processability was found to be good, with adequate dispersion of the filler in the matrix. Mechanical performance was satisfactory, and it was found that this is significantly affected by specific process parameters such as the raster angle. The mechanical properties were compared to those predictable from the Halpin–Tsai model, finding that the prepared systems exceed the expected values.

**Keywords:** green composites; 3D printing; FDM; biopolymers; solanum lycopersicum

**Citation:** Scaffaro, R.; Citarrella, M.C.; Morreale, M. Green Composites Based on Mater-Bi® and *Solanum lycopersicum* Plant Waste for 3D Printing Applications. *Polymers* **2023**, *15*, 325. https://doi.org/10.3390/polym15020325

Academic Editors: José Miguel Ferri, Vicent Fombuena Borràs and Miguel Fernando Aldás Carrasco

Received: 22 November 2022
Revised: 23 December 2022
Accepted: 5 January 2023
Published: 8 January 2023



## 1. Introduction

Over the last few decades, increasing attention has been focused on the ways to improve the cost-effectiveness of the production of polymer-related items, by possibly replacing part of the polymer needed to manufacture a certain product with waste materials and/or by-products coming from other industrial, or agricultural, operations; at the same time, the need and interest in reducing the environmental impact related to the entire life cycle of polymer-based goods have grown exponentially, suggesting to replace (at least) part of the polymer itself with materials coming from renewable sources and/or biodegradable [1]. Furthermore, it is obvious that a more significant reduction in the environmental impacts requires replacing traditional polymers (coming from non-renewable sources) with bio-based and, preferably, also biodegradable polymers. Among the waste materials which can be conveniently used as fillers for polymers systems, agricultural, marine, or industrial wastes from wood processing are particularly attractive; at the same time, it is important to use biodegradable polymers in order to reduce the environmental pollution related to plastics [1–8] and to focus on obtaining a satisfactory mechanical behaviour [1–5].

In this background, the biopolymers which are more typically used in the preparation of green composites are poly (lactic acid) (PLA), polybutylene adipate terephthalate (PBAT), polycaprolactone (PCL), cellulose and starch-based polymers [2,6]. For instance, Mater-Bi® (MB) is a family of commercial starch-based biopolymers that have been finding interesting applications in many fields, thanks to satisfactory mechanical properties, good processability, adequate thermal stability, biodegradability/compostability and suitability to be reinforced with natural-organic fillers, as already reported by many papers [4,5,7]. It is important to observe that the addition of a natural-organic filler to such polymer matrices was found to improve the biodegradability [6,7] and, often, to improve the mechanical behaviour [2,7,9–11]: therefore, plant-based biomasses should be investigated for their actual potential in achieving both of such fundamental targets and they should preferably hold the prerequisite of being easily available, cheap, and widely present on the territory.

The Mediterranean area offers a wide variety of plant species (or, in general, lignocellulosic sources), coming from either the agricultural or the marine environment, that can effectively find applications in the preparation of polymer-based biocomposites. For instance, these can include Opuntia Ficus Indica (OFI), Posidonia Oceanica (PO) and Hedysarum coronarium (HC). OFI has already been studied in combination with PLA, to produce green composites via the compression moulding technique [9]. PO and in particular PO leaves (POL) have been investigated in several studies, focused on the structure–properties relationships; finding that the mechanical behaviour can be enhanced and, quite interestingly, that the degradability can be accelerated by the presence of POL [10,12,13]. HC is very abundant in the Mediterranean area and is known for applications in the agri-food sector [14,15] but has been recently investigated also regarding the formulation and preparation of green composites [16,17].

However, the formulation and preparation of innovative and effective green composites cannot be based only on the choice of the polymer matrix and the filler, but it must also consider the choice and setup of the optimal processing technique. To this point, it should be observed that thermoplastic-based green composites are usually produced by compression moulding, extrusion, or injection moulding [18]; on the other hand, the continuous development of new and more versatile production solutions, has led to a significant interest in fused deposition modelling (FDM), a technique (often referred to as "3D printing") which is now known for its great versatility: it allows obtaining elaborated geometries while still granting significant reductions in time and costs, and thus it is already one of the most promising also with concern to green composites [19–23].

More specifically, there are some recent works where lignocellulosic wastes have been used as fillers for green composites and investigated for actual suitability to FDM manufacturing. HC was combined with Mater-Bi® (MB) [16] or PLA [17] and the green composites were prepared via two different routes, i.e., compression moulding (CM) or FDM. It was found [16] that FDM could be preferable up to 10% HC content, leading to better mechanical properties (in particular, with regard to the elastic modulus) in comparison to CM, likely due to rectilinear infill and fibres orientation; furthermore, it was possible to get more dense structures than by CM [17], obtaining quite significant improvements of the mechanical properties (especially flexural ones) in comparison to the neat polymer. OFI and/or POL were investigated in combination with PLA and processed via FDM, finding that it was possible to replace up to 20% of the polymer matrix [24], with final samples characterized by good mechanical properties and satisfactory filler dispersion as well as filler–matrix adhesion, with very interesting potential applications in the release of fertilizers [25].

As pointed out several times over this brief bibliographic overview, one of the main goals related to the development and use of green composites depends on the utilization of natural-organic wastes, coming from flora (both terrestrial and marine) or fauna [26]. From this point of view, one interesting source may come from *Solanum lycopersicum*, i.e., tomato plant. This plant, widely grown in temperate zones across the world, and also in greenhouses, is one of the most important for its edible purpose. Tomato production in 2020 was led by China with almost 65 million tons, followed by India, Turkey, the United States and Egypt [27]. During the production and transportation stages, several wastes are typically produced, accounting for an estimated 10–15% of the total volume and are commonly used for compost or animal feed [28]. These wastes basically consist of skin, seeds, and tomato pomace (a by-product of tomato processing, based on peel, seeds and small amounts of pulp) and many investigations are recently focused on how to exploit them for higher-value purposes, such as extraction of lycopene, carotenoids, bases for biofuels, etc. [28,29]. However, much less attention is focused on the lignocellulosic wastes coming from the plants after extirpation of the fruits. Such lignocellulosic wastes are usually driven to incineration or, when discarded on the ground, they can represent a significant hazard, since they may contribute to feed fires and related events. It would be therefore preferable to find alternative solutions for such wastes, and their proper incorporation into green composites may be an optimal way. To our best knowledge, there is no evidence in the literature about systematic studies on green composites based on biodegradable polymers (in particular, from the Mater-Bi® family) and fillers obtained from *Solanum lycopersicum*, let alone via a more innovative technique such as FDM.

In this paper, therefore, we prepared composites based on a Mater-Bi® polymer and wastes coming from *Solanum lycopersicum*, processing them via FDM, in order to explore the actual suitability to 3D-printing applications. The obtained samples were characterized from the rheological, mechanical, and morphological points of view.

## 2. Materials and Methods

### 2.1. Materials

The biopolymeric matrix used to prepare the green composites was a sample of Mater-Bi® EF51L (MB) supplied by Novamont SpA (Novara, Italy), a polymer based on blends of aromatic and aliphatic biodegradable co-polyesters with proprietary composition. In order to avoid hydrolytic chain scissions during the melt processing, neat MB and MB-based composites were vacuum-dried overnight at 60 °C before each process.

*Solanum lycopersicum* plant waste (SL) used in this study was kindly supplied by a local farm (Sicily, Italy) The plants were mowed after tomato harvesting. In this study, the whole plant was ground as received in order to optimize production time and costs. More in detail, the obtained plant wastes were washed and dried in a vacuum oven (NSV9035, ISCO, Milan, Italy) at T = 40 °C for 3 days, and finally ground using a laboratory grinder (Retsch, Germany).

SL dried stem showed a Young's modulus of 404 MPa. The flour, obtained by grinding the whole plant as described above, displayed an average density of 1.87 g/cm$^3$. It was further vacuum-dried, overnight at 40 °C, prior to the melt mixing process in order to reduce potential MB chain scission phenomena during processing.

### 2.2. Composites Preparations

Firstly, the dried SL plant was ground for 3 min in a grinder (Retsch, Germany). The resulting powder was then sieved to obtain particles of a size suitable for the 3D printer (Next Generation, Sharebot, Nibionno, Italy), which, therefore, do not lead to obstructions in the nozzle. To this aim, and based on previous studies [11,20], the sieving fraction under 150 µm was selected. Prior to processing, the obtained SL flours and MB pellets were dried overnight in a vacuum oven (NSV9035, ISCO, Milan, Italy) at 40 °C and 60 °C, respectively.

In order to obtain a homogeneous dispersion of the filler, according to previous studies, the filler amounts chosen to prepare the MB-based biocomposites were 5, 10, 15 wt%. All of the composites (namely MB/SL5, MB/SL10, MB/SL15) and neat MB, for comparison, were prepared by melt compounding in an internal mixer (Plasticorder, Brabender, Duisburg, Germany; T = 160 °C, rotor speed = 64 rpm, t = 5 min).

The obtained materials were then ground into pellets and processed in a Polylab single-screw extruder (Haake Technik GmbH, Vreden, Germany; L/D = 25; D = 19.05 mm), operating at 40 rpm screw speed and 130–140–150–160 °C temperature profile. The extrudates were drawn with the help of a conveyor belt system (take-up speed = 5.5 m/min), to obtain filaments with a diameter suitable to the printer (1.75 mm).

The samples obtained for fused deposition modelling (FDM) were first designed with the help of CAD Solid Edge 2019® software (Plano, TX, USA), and the STL files produced were elaborated on Simplify3D® software (Cincinnati, OH, USA) to obtain the related gcode files. For each formulation, 60 mm × 10 mm × 1 mm samples were printed using a Sharebot Next Generation (Nibionno, Italy) 3D printer. FDM operating parameters are reported in Table 1. Nozzle temperature was chosen after some trials, aiming to avoid nozzle obstructions and to obtain good printability performance. The other parameters were chosen based on the scientific literature [16,17,24–26]. In particular, a 100% infill rate and a rectilinear infill pattern with a 0° or ±45° raster angle were chosen in order to evaluate its influence on the tensile properties of the composites; 45 mm/s printing speed was chosen to maximize the production rate without compromising the mechanical performance.

**Table 1.** FDM process parameters.

| FDM Operating Parameter | Value |
| --- | --- |
| Nozzle temperature | 160 °C |
| Bed temperature | 60 °C |
| Infill rate | 100% |
| Infill pattern | Rectilinear |
| Raster angle | 0° or ±45° |
| Layer thickness | 0.1 mm |
| Extrusion width | 0.4 mm |
| Printing speed | 50 mm/s |
| Perimeter shells | 1 |
| Sample Orientation | flat |

Sample formulations and sample codes are reported in Table 2.

**Table 2.** Formulation of investigated samples.

| Sample Code | MB Content (wt%) | SL Content (wt%) | SL Mesh Size (μm) | Raster Angle |
| --- | --- | --- | --- | --- |
| MB 0° | 100 | 0 | - | 0° |
| MB/SL5 0° | 95 | 5 | <150 | 0° |
| MB/SL10 0° | 90 | 10 | <150 | 0° |
| MB/SL15 0° | 85 | 15 | <150 | 0° |
| MB 45° | 100 | 0 | - | ±45° |
| MB/SL5 45° | 95 | 5 | <150 | ±45° |
| MB/SL10 45° | 90 | 10 | <150 | ±45° |
| MB/SL15 45° | 85 | 15 | <150 | ±45° |

### 2.3. Characterizations

#### Rheological Characterization

Rheological properties of the samples were analysed, using a rotational rheometer (ARES-G2, TA Instruments, New Castle, DE, USA) equipped with a 25 mm parallel-plate geometry. All the tests were performed at 160 °C, in frequency sweep mode in the range 1–100 rad/s, by imposing a constant stress of 1 Pa.

### 2.4. Morphological Analysis

The morphology of SL powder, composites filaments and FDM samples was observed by using a scanning electron microscope (Phenom ProX, Phenom-World, Eindhoven, The Netherlands) with an optical magnification range of 20–135×, electron magnification range of 80–1.3 × $10^5$, maximal digital zoom of 12×, and acceleration voltages of 15 kV. The microscope is equipped with a temperature controlled (25 °C) sample holder. The samples were fixed on an aluminium stub (pin stub 25 mm, Phenom-World, Eindhoven, The Netherlands) using a glued carbon tape.

### 2.5. Mechanical Characterization

The mechanical behaviour of SL plant, composites filaments and FDM-printed samples was investigated by tensile tests, carried out using a laboratory dynamometer (mod.3365, Instron, Norwood, MA, USA) equipped with a 1 kN load cell. The tests were performed on rectangular-shaped specimens (60 mm × 10 mm) according to ASTM D638. In particular, the measurements were performed by using a double crosshead speed: 1 mm min$^{-1}$ for 2 min and 50 mm min$^{-1}$ until fracture occurred. The grip distance was 30 mm, whereas the sample thickness was measured before each test. Eight specimens were tested for each sample, and the results for elastic modulus (E), tensile strength (TS) and elongation at break (EB) have been reported as the average values ± standard deviations.

### 2.6. X-ray Diffraction

X-ray diffraction patterns were collected by using a RIGAKU diffractometer (D-MAX 25600 HK, Rigaku, Tokyo, Japan). All diffraction patterns were obtained in the 2θ range from 5° to 80° by means of copper Kα radiation (λ = 1.54 Å) with the following setup conditions: tube voltage and current of 40 kV and 30 mA, respectively, scan speed of 4°/min with a sampling of 0.004°.

### 2.7. Differential Scanning Calorimetry Analysis

Differential scanning calorimetry (DSC) analysis was carried out on a Chip-DSC 10 (Linseis Messgeraete GmbH, Selb, Germany) by heating the samples to 200 °C at a heating rate of 40 °C/min.

### 2.8. Density Measurements

Density measurements were performed by a Thermo Pycnomatic Helium Pycnometer (Pycnomatic ATC, Thermofisher, Waltham, MA, USA), using 99.99% pure helium. Measures were repeated at least six times for each sample, at 25 °C.

### 2.9. Theoretical Modelling

The outcomes of the tensile tests were compared with those predicted by the Halpin–Tsai model, which allows esteeming the modulus of composites once are known volume fractions and elastic moduli of the starting components, and the filler aspect ratio. According to the Halpin–Tsai model, for composites reinforced with fibres randomly oriented, the composite modulus $E_{C,HT}$ is determined by the following equation:

$$E_{C,\,HT} = \frac{3}{8}E_L + \frac{5}{8}E_T \tag{1}$$

where $E_L$ and $E_T$ are, respectively, the longitudinal and transverse moduli of the composite.

In this case, $E_L$ and $E_T$ are given by:

$$E_L = E_m\left[\frac{1+(2l/d)\eta_L v_f}{1-\eta_L v_f}\right] \qquad E_T = E_m\left[\frac{1+2\eta_T v_f}{1-\eta_L v_f}\right]$$

where $v_f$ and $v_m$ are the volume fractions of EE fillers and MB, respectively, $l/d$ is the aspect ratio of the fillers while $\eta_L$ and $\eta_T$ are constants given by:

$$\eta_L = \frac{\left(E_f/E_m\right)-1}{\left(E_f/E_m\right)+(2l/d)} \qquad \eta_T = \frac{\left(E_f/E_m\right)-1}{\left(E_f/E_m\right)+2}$$

where, $E_f$ and $E_m$ are, respectively, the Young's moduli of filler and MB.

Volume fractions are determined from the weight fractions and the densities of each component (i.e., SL and MB) measured experimentally by a helium pycnometer.

## 3. Results and Discussion

The samples loaded at 5% (MB/SL5), 10% (MB/SL10) and 25% (MB/SL25) filler where properly extruded into the related filaments, to be subjected to FDM thereafter.

Filament printability (i.e., processability in FDM mode) is directly correlated to the morphological properties. More in detail, not only the diameter of the filament must be suitable for the specific 3D printer used, but its surface must be as even and homogeneous as possible [26,30]. In addition, printability depends also on the rheological and mechanical properties of the filaments, which were thus investigated as well. The obtained results are discussed in the following.

### 3.1. SL Powder and Filament Characterization

Morphological characterization was carried out first. The main results are shown in the SEM micrographs reported in Figure 1 for SL powder and in Figure 2 for MB/SL5, MB/SL10 and MB/SL15, respectively. From the SEM micrograph of the powder (Figure 1), it is possible to notice that SL powder contains elements with different morphology, reasonably belonging to different parts of the plant: stem and leaf.

**Figure 1.** SEM images of SL powder.

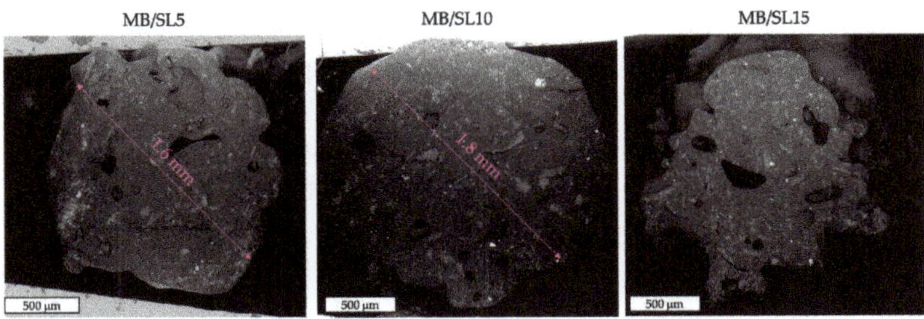


**Figure 2.** SEM images of MB/SL5, MB/SL10 and MB/SL15 filaments.

From the samples' cross-section micrographs (Figure 2), it can be observed that the SL particles are homogeneously dispersed in the MB matrix, only a few voids are present, and the general adhesion between the matrix and the particles is good. Furthermore, the diameters of the MB/SL5 and the MB/SL10 filaments are even and homogeneous, in the range 1.6–1.8 mm (respectively) which is suitable for the actual printer used. On the other hand, the MB/SL15 filaments showed uneven diameters.

Rheological measurements were performed on specimens obtained from the filaments, in order to evaluate the actual processability for FDM purposes.

Figure 3 reports the rheological values of MB and the composite filaments, on increasing the filler content.

As predictable, MB shows a clear non-Newtonian behaviour. The addition of 5% SL leads to an increase of viscosity over the entire frequency range, as well as a more marked non-Newtonian behaviour. This tendency further increases by adding 10% SL. When 15% SL is added to the MB matrix, there is a much more drastic increase in the viscosity and the onset of yield stress phenomena. Such results suggest that only the rheological behaviour of MB/SL5 and MB/SL10 appears compatible with the 3D printing process, whereas MB/SL15 may be not adequately printable, due to the excessively high viscosity which may lead to nozzle clogging during the process [26].

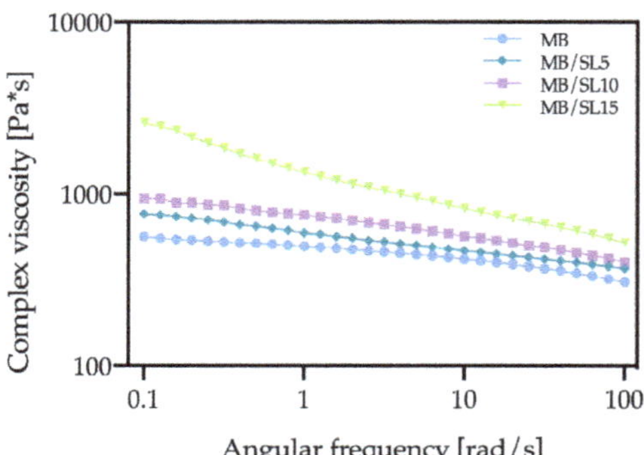


**Figure 3.** Rheological curves of MB/SL5, MB/SL10 and MB/SL15 filaments.

Anyway, optimal printability depends also on the tensile properties of the filaments. Figure 4 reports the values of elastic modulus (E), tensile strength (TS) and elongation at break (EB), on increasing the SL content. It can be noticed that the filaments become stiffer on increasing the SL content, although the effect is much more significant only in the case of MB/SL15; the tensile strength is similar to that of the neat MB, or even higher, and this is a satisfactory result since it suggests that the filament should not undergo rupture too easily, during the process; on the other hand, the deformability drops even at just 5% SL content.

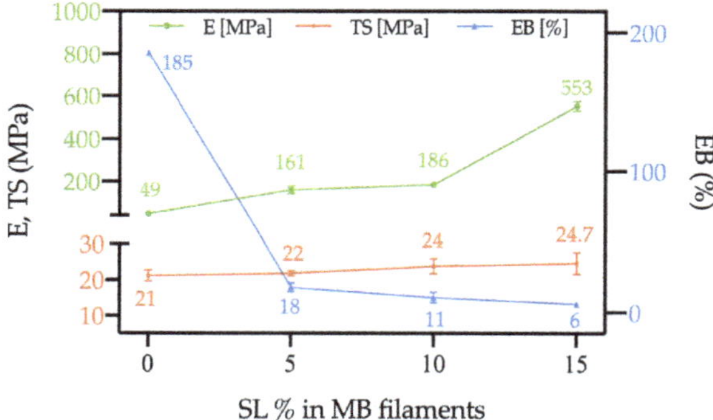


**Figure 4.** Tensile properties of MB/SL5, MB/SL10 and MB/SL15 filaments as a function of the SL amount.

In Figure 5, photos of MB/SL5, MB/SL10 and MB/SL15 filaments before (Figure 5a–c, respectively), during (Figure 5d–f, respectively) and after (Figure 5g–i, respectively) tensile test are reported. MB/SL5 and MB/SL10 filaments present a homogenous shape, and their fracture occurs a few seconds after the 50 mm min$^{-1}$ speed was applied. On the other hand, the MB/SL15 filament presents an irregular shape due to the high content of filler. In this latter case, the fracture occurred instantaneously when the 50 mm min$^{-1}$ speed was applied.

The results of the rheological and mechanical tests allow drawing some general considerations, propaedeutic for the FDM stage, since viscoelasticity and tensile strength measurements help to predict problems in printability and possible printing errors [31] In particular, too high viscosities are not suitable for the process, since the low deformability can lead to filament blocking at the nozzle of the 3D printer, and subsequent clogging and rupture (Figure 6a). On the other hand, if the filament is too soft (high decline of viscosity at low temperatures), it tends to flow too easily while not pulling correctly, resulting in nozzle clogging (Figure 6b); furthermore, if it is too brittle, it will break (Figure 6c) [31].

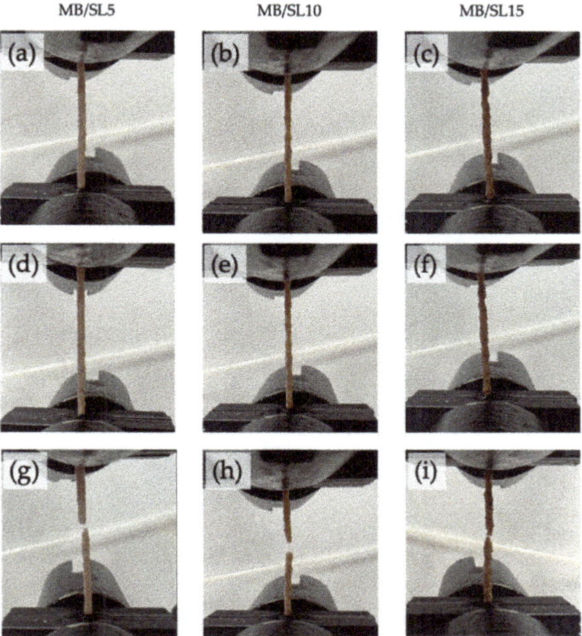


**Figure 5.** Photos of MB/SL5, MB/SL10 and MB/SL15 filaments before (**a–c**, respectively), during (**d–f**, respectively) and after (**g–i**, respectively) tensile test.

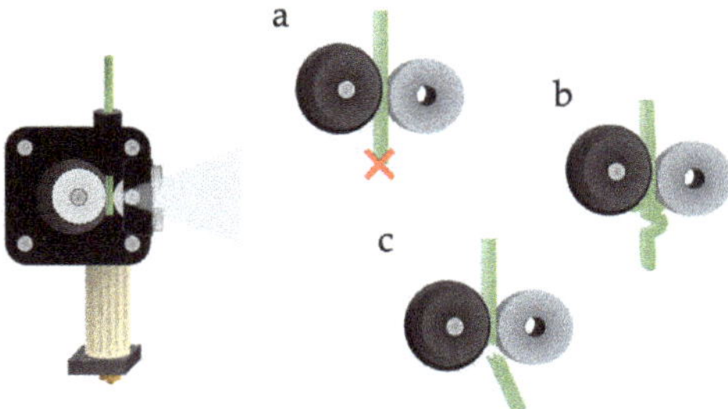


**Figure 6.** Different behaviours of the filament upon entering the melting chamber, at different viscoelastic and mechanical properties.

These considerations, therefore, suggest that MB/SL5 and MB/SL10 should be easily printable and without significant defects in the obtained samples (on the other hand, neat MB may lead to some uncertainty due to the relatively high deformability), while problems may arise with MB/SL15.

### 3.2. Printing of the Composites Filaments

Actual 3D printing was then carried out. As expected, based on the previous considerations, neat MB as well as MB/SL5 and MB/SL10 were easily processed, while the filament containing 15% SL showed to be not printable since the high viscosity caused obstruction of the nozzle and the filament broke easily.

The samples were printed with both 0° and 45° raster angles, in order to evaluate the printability and the effect of the angle on the mechanical properties of the obtained 3D specimens.

### 3.3. Characterizations of 3D Printed Samples

First, the 3D-printed samples were subjected to morphological analysis on cryofractured surfaces.

Figure 7 shows SEM images of fractured surfaces, at increasing magnification from left to right, of MB/SL10, 0° raster samples. In general, it can be stated that filler dispersion and adhesion are good, as clearly visible from the filler particle circled in green (right), where no significant voids can be found at the filler–matrix interface.

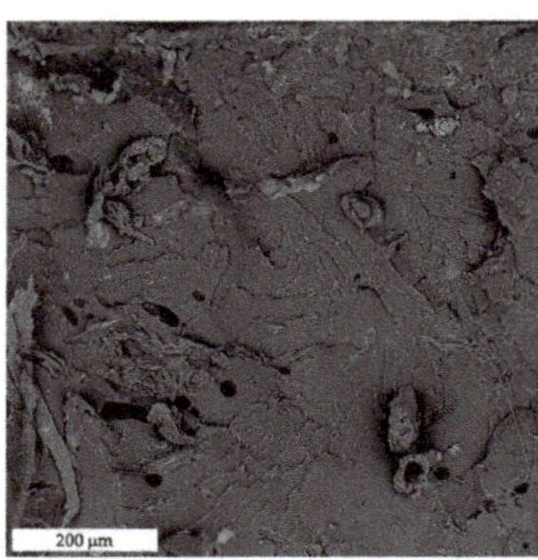


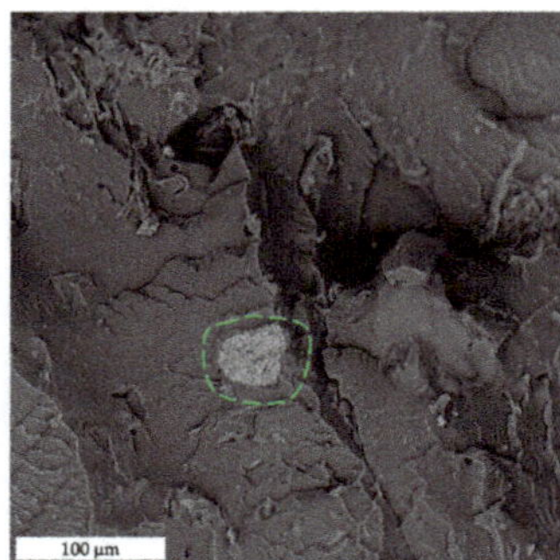


**Figure 7.** SEM images of fracture surfaces of MB/SL10 0° samples at increasing magnification (from left to right). The green circle highlights the good adhesion between the matrix and the filler.

Figures 8–11 show the fracture surfaces after tensile tests of the composite samples. Overall, it may be stated that some fibre pull-out and debonding phenomena are more visible in the SL5 rather than in the SL10 samples and, especially, in 45° samples (Figure 11) as opposed to 0° ones (Figure 10).

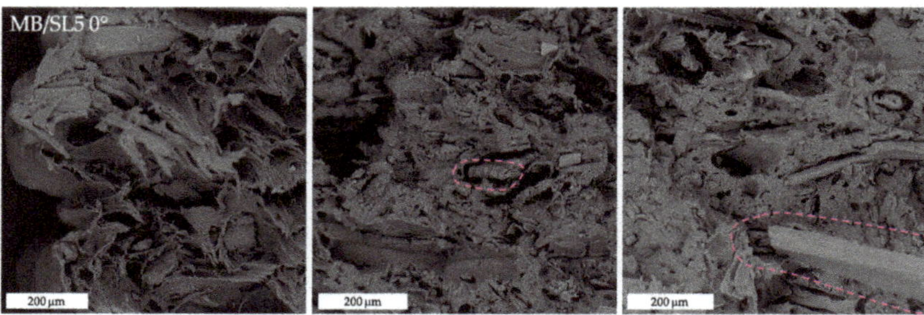


**Figure 8.** SEM images of tensile fracture surfaces of MB/SL5 0° samples. The pink circles highlight fibre pull-out and debonding phenomena.

The actual tensile properties of the 3D-printed samples are shown in Figure 12, in the case of raster angle = 0° (left) and raster angle = 45° (right). It can be observed that, in both cases, the elastic modulus and the tensile strength increase on increasing the filler content, while the deformability decreases. However, such a decrease is significantly less marked in the case of raster angle = 0°, and the overall results of all the tensile properties are better, with excellent reproducibility. This confirms the first indications from the morphological analysis, which could allow supposing higher breaking resistance of the 0° samples, in comparison to the 45° ones.

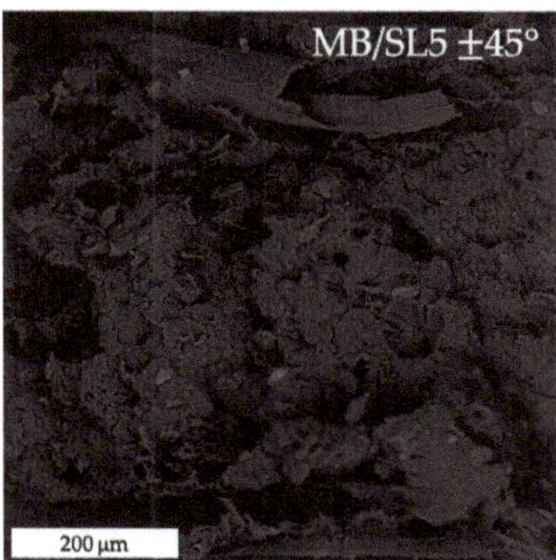


**Figure 9.** SEM image of tensile fracture surface of MB/SL5 ± 45° sample.

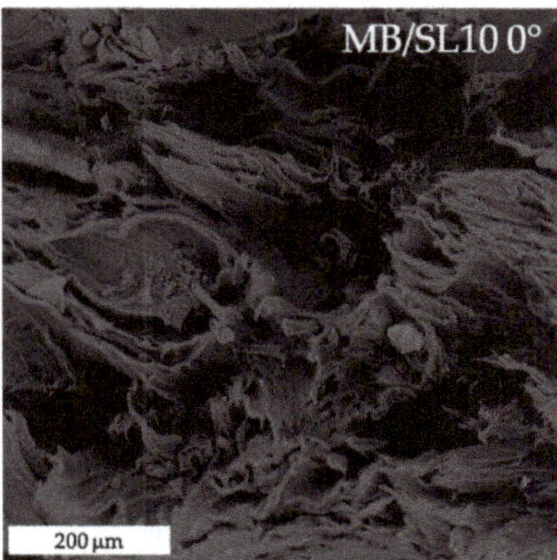


**Figure 10.** SEM image of tensile fracture surface of MB/SL10 0° sample.

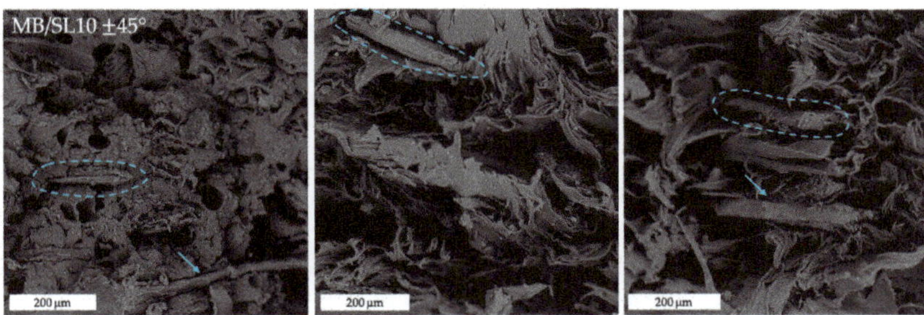


**Figure 11.** SEM images of tensile fracture surfaces of MB/SL10 ± 45° samples. The blue circles and arrows highlight fibre pull-out and debonding phenomena.

Such evidence can be further deduced from Figure 13. The 0° raster angle during printing definitively optimizes the tensile properties, confirming data from the literature, obtained on similar systems, where 0° raster angle usually optimizes tensile properties, whereas 45° leads to optimization of flexural and impact properties [10,26,32,33].

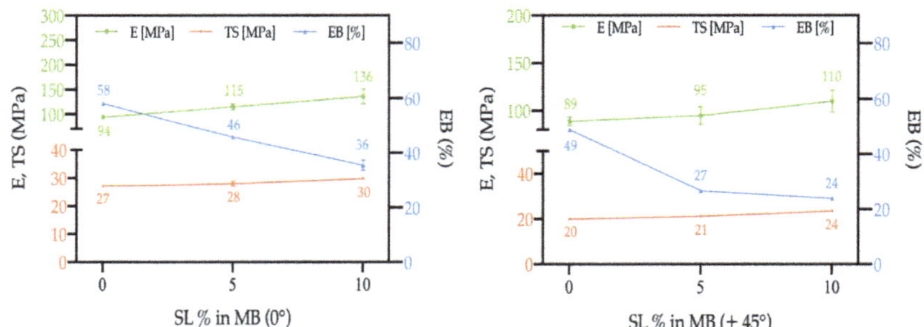


**Figure 12.** Elastic modulus (E), tensile strength (TS) and elongation at break (EB) of 3D-printed samples as a function of the filler content; raster angle = 0° (left) and raster angle = 45° (right).

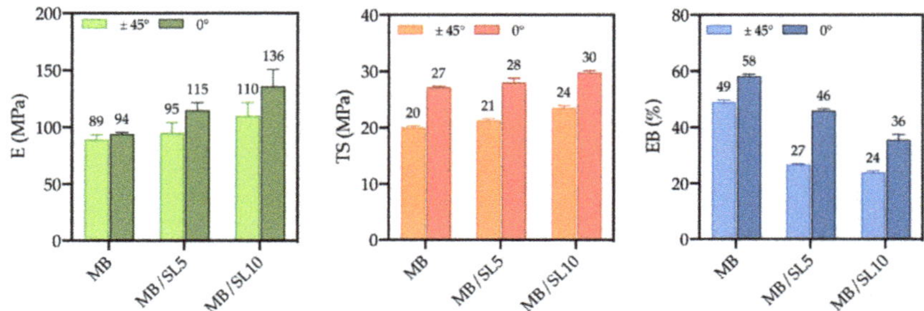


**Figure 13.** Elastic modulus (E), tensile strength (TS) and elongation at break (EB) of 3D-printed samples with different filler content and raster angle.

### 3.4. XRD and DSC Characterizations

In order to verify if the addition of SL filler leads to some crystallinity variation in the polymeric matrix, XRD and DSC analysis were performed on neat MB and MB/SL printed composites and the related outcomes are reported in Figure 14a,b, respectively. No differences can be noted in XRD curves (Figure 14a) when 5 or 10% of SL is added to the polymeric matrix. Moreover, the addition of SL powder to MB does not lead to any significant change in its melting temperature or melting enthalpy (see Figure 14b and Table 3).

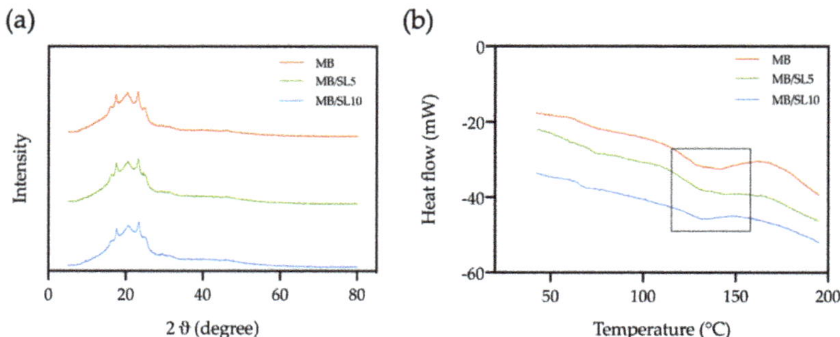


**Figure 14.** XRD spectra (**a**) and DSC analysis (**b**) of neat MB and MB/SL−printed composites.

**Table 3.** Melting temperature and melting enthalpy of 3D-printed samples obtained by DSC analysis.

| Sample | Weight (mg) | Melting Temperature (°C) | Melting Enthalpy (mJ/mg) |
|---|---|---|---|
| MB | 10.9 | 132.3 | 15.8 |
| MB/SL5 | 8.8 | 132.7 | 16.3 |
| MB/SL10 | 3.3 | 131.7 | 15.9 |

These outcomes confirm that the increase in the tensile property of SL composites, if compared to the pure matrix, can be effectively attributed to the reinforcing effect given by the filler.

### 3.5. Halpin–Tsai Model

Figure 15 shows the trends of Ec/Em (ratio between the elastic modulus of the composite and that of the matrix) on increasing the SL content, both from the experimental (Exp) results (at 0- and 45-degree raster angles) and the theoretical trend calculated according to the Halpin–Tsai model (HT). This semiempirical model allows assessing the composite modulus (Ec), once five parameters are known, i.e., the elastic modulus of matrix (Em) and filler (Ef), their volume fractions and the filler aspect ratio [34].

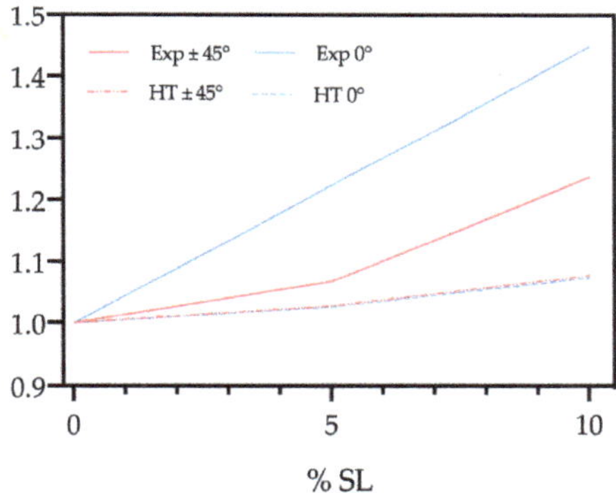


**Figure 15.** Ratio between elastic modulus of the composite and the polymer matrix, as a function of the SL content, according to the Halpin–Tsai model (HT) and the experimental results (Exp).

The trends clearly outline that the model significantly underestimates the values of Ec, especially at higher filler contents. This may be due to the filler particles coming from different parts (wastes) of the tomato plant, thus presenting some natural differences in terms of morphology and/or mechanical properties. An additional likely explanation may involve the capability of the polymer matrix to, at least partially, enter the void channels of SL particles, as presumable on the basis of the SEM images and of the results from our previous studies on similar (i.e., biodegradable polymer/natural-organic plant waste filler) systems [9].

## 4. Conclusions

In this paper, composites based on a Mater-Bi® polymer and wastes coming from *Solanum lycopersicum* were prepared and processed via FDM, in order to explore the actual suitability to 3D-printing applications. Different processing parameters and different filler amounts were investigated, and the obtained samples were characterized from the rheological, mechanical and morphological point of view. The adopted processing parameters allowed optimal processability up to 10% filler content, with satisfactory dispersion of the filler in the matrix; the same holds for the interfacial adhesion. Mechanical characterization showed that the tensile strength was kept or even improved upon increasing the filler content, the elastic modulus was enhanced and only a "physiological" reduction in the elongation at break was found; moreover, the processing parameters and, in particular, the raster angle significantly affected the tensile resistance, with 0° being preferable to ±45°. The experimental mechanical behaviour was compared to the Halpin–Tsai model, finding positive deviations for the prepared systems. Moreover, the addition of a natural waste would allow lowering the final cost of the product. Actually, the cost of *Solanum Lycopersicum* plant waste used in this work is virtually zero, since these are residues from the harvesting, and they would not find many significantly valuable alternative uses. Overall, these green composites have great potential for the development of sustainable bio-based materials aimed at several applications.

**Author Contributions:** Conceptualization, R.S., M.C.C. and M.M.; methodology, R.S., M.C.C. and M.M.; software, M.C.C.; validation, R.S., M.C.C. and M.M.; formal analysis, R.S. and M.C.C.; investigation, M.C.C.; resources, R.S.; data curation, R.S., M.C.C. and M.M.; writing—original draft preparation, R.S., M.C.C. and M.M.; writing—review and editing, M.M., M.C.C. and R.S.; visualization, M.C.C.; supervision, R.S. and M.M.; project administration, R.S.; funding acquisition, R.S. All authors have read and agreed to the published version of the manuscript.

**Funding:** This research was funded by PNRR PE_11 3A-ITALY CUP: B73C22001270006.

**Institutional Review Board Statement:** Not applicable.

**Data Availability Statement:** The rough/processed data that support our study are available from the corresponding author on reasonable request.

**Conflicts of Interest:** The authors declare no conflict of interest.

## References

1. La Mantia, F.P.; Morreale, M. Green Composites: A brief review. *Compos. Part A Appl. Sci. Manuf.* **2011**, *42*, 579–588. [CrossRef]
2. Tejyan, S.; Baliyan, N.K.; Patel, V.K.; Patnaik, A.; Singh, T. Polymer Green Composites Reinforced with Natural Fibers: A Comparative Study. *Mater. Today Proc.* **2020**, *44*, 4767–4769. [CrossRef]
3. Scaffaro, R.; Maio, A.; Sutera, F.; Gulino, E.F.; Morreale, M. Degradation and Recycling of Films Based on Biodegradable Polymers: A Short Review. *Polymers* **2019**, *11*, 651. [CrossRef] [PubMed]
4. Re, G.L.; Morreale, M.; Scaffaro, R.; la Mantia, F.P. Biodegradation Paths of Mater-Bi®/Kenaf Biodegradable Composites. *J. Appl. Polym. Sci.* **2013**, *129*, 3198–3208. [CrossRef]
5. Morreale, M.; Scaffaro, R.; Maio, A.; la Mantia, F.P. Mechanical Behaviour of Mater-Bi®/Wood Flour Composites: A Statistical Approach. *Compos. Part A Appl. Sci. Manuf.* **2008**, *39*, 1537–1546. [CrossRef]
6. Rafiee, K.; Schritt, H.; Pleissner, D.; Kaur, G.; Brar, S.K. Biodegradable Green Composites: It's Never Too Late to Mend. *Curr. Opin. Green Sustain. Chem.* **2021**, *30*, 100482. [CrossRef]
7. Bordón, P.; Paz, R.; Peñalva, C.; Vega, G.; Monzón, M.; García, L. Biodegradable Polymer Compounds Reinforced with Banana Fiber for the Production of Protective Bags for Banana Fruits in the Context of Circular Economy. *Agronomy* **2021**, *11*, 242. [CrossRef]
8. Lo Re, G.; Morreale, M.; Scaffaro, R.; la Mantia, F.P. Kenaf-Filled Biodegradable Composites: Rheological and Mechanical Behaviour. *Polym. Int.* **2012**, *61*, 1542–1548. [CrossRef]
9. Scaffaro, R.; Maio, A.; Gulino, E.F.; Megna, B. Structure-Property Relationship of PLA-Opuntia Ficus Indica Biocomposites. *Compos. Part B Eng.* **2019**, *167*, 199–206. [CrossRef]
10. Benito-González, I.; López-Rubio, A.; Martínez-Sanz, M. Potential of Lignocellulosic Fractions from Posidonia Oceanica to Improve Barrier and Mechanical Properties of Bio-Based Packaging Materials. *Int. J. Biol. Macromol.* **2018**, *118*, 542–551. [CrossRef]
11. Boudjema, H.L.; Bendaikha, H.; Maschke, U. Green Composites Based on Atriplex Halimus Fibers and PLA Matrix. *J. Polym. Eng.* **2020**, *40*, 693–702. [CrossRef]
12. Scaffaro, R.; Maio, A.; Gulino, E.F. Hydrolytic Degradation of PLA/Posidonia Oceanica Green Composites: A Simple Model Based on Starting Morpho-Chemical Properties. *Compos. Sci. Technol.* **2021**, *213*, 108930. [CrossRef]
13. Seggiani, M.; Cinelli, P.; Mallegni, N.; Balestri, E.; Puccini, M.; Vitolo, S.; Lardicci, C.; Lazzeri, A. New Bio-Composites Based on Polyhydroxyalkanoates and Posidonia Oceanica Fibres for Applications in a Marine Environment. *Materials* **2017**, *10*, 326. [CrossRef]
14. Collected, L.; Candido, V.; Avato, P. Chemical Identification of Specialized Metabolites from Sulla. *Molecules* **2021**, *26*, 4606.
15. Squartini, A.; Struffi, P.; Do, H.; Selenska-pobell, S.; Tola, E.; Giacomini, A.; Vendramin, E.; Mateos, P.F.; Martı, E. Rhizobium Sullae, the Root-Nodule Microsymbiont of *Hedysarum Coronarium* L. *Int. J. Syst. Evol. Microbiol.* **2002**, *52*, 1267–1276. [PubMed]
16. Scaffaro, R.; Citarrella, M.C.; Gulino, E.F.; Morreale, M. *Hedysarum coronarium-Based* Green Composites Prepared by Compression Molding and Fused Deposition Modeling. *Materials* **2022**, *15*, 465. [CrossRef] [PubMed]

17. Scaffaro, R.; Gulino, E.F.; Citarrella, M.C.; Maio, A. Green Composites Based on *Hedysarum coronarium* with Outstanding FDM Printability and Mechanical Performance. *Polymers* **2022**, *14*, 1198. [CrossRef] [PubMed]
18. Gholampour, A.; Ozbakkaloglu, T. *A Review of Natural Fiber Composites: Properties, Modification and Processing Techniques, Characterization, Applications*; Springer: New York, NY, USA, 2020; Volume 55. ISBN 1085301903990.
19. Liu, Z.; Lei, Q.; Xing, S. Mechanical Characteristics of Wood, Ceramic, Metal and Carbon Fiber-Based PLA Composites Fabricated by FDM. *J. Mater. Res. Technol.* **2019**, *8*, 3743–3753. [CrossRef]
20. Prabhu, R.; Devaraju, A. Recent Review of Tribology, Rheology of Biodegradable and FDM Compatible Polymers. *Mater. Today: Proc.* **2020**, *39*, 781–788. [CrossRef]
21. Kopparthy, S.D.S.; Netravali, A.N. Review: Green Composites for Structural Applications. *Compos. Part C Open Access* **2021**, *6*, 100169. [CrossRef]
22. Rajendran Royan, N.R.; Leong, J.S.; Chan, W.N.; Tan, J.R.; Shamsuddin, Z.S.B. Current State and Challenges of Natural Fibre-Reinforced Polymer Composites as Feeder in Fdm-Based 3d Printing. *Polymers* **2021**, *13*, 2289. [CrossRef]
23. Mazzanti, V.; Malagutti, L.; Mollica, F. FDM 3D Printing of Polymers Containing Natural Fillers: A Review of Their Mechanical Properties. *Polymers* **2019**, *11*, 1094. [CrossRef] [PubMed]
24. Scaffaro, R.; Maio, A.; Gulino, E.F.; Alaimo, G.; Morreale, M. Green Composites Based on PLA and Agricultural or Marine Waste Prepared by FDM. *Polymers* **2021**, *13*, 1361. [CrossRef] [PubMed]
25. Scaffaro, R.; Citarrella, M.C.; Gulino, E.F. Opuntia Ficus Indica based green composites for NPK fertilizer controlled release produced by compression molding and fused deposition modeling. *Compos. Part A Appl. Sci. Manuf.* **2022**, *159*, 107030. [CrossRef]
26. Scaffaro, R.; Citarrella, M.C.; Catania, A.; Settanni, L. Green composites based on biodegradable polymers and anchovy (Engraulis Encrasicolus) waste suitable for 3D printing applications. *Compos. Sci. Technol.* **2022**, *230*, 109768. [CrossRef]
27. FAO. Agricultural production statistics. 2000–2020. FAOSTAT Analytical Brief Series No. 41. 2022. Available online: https://www.fao.org/3/cb9180en/cb9180en.pdf (accessed on 4 January 2023).
28. Campos-Lozada, G.; Pérez-Marroquín, X.A.; Callejas-Quijada, G.; Campos-Montiel, R.G.; Morales-Peñaloza, A.; León-López, A.; Aguirre-Álvarez, G. The Effect of High-Intensity Ultrasound and Natural Oils on the Extraction and Antioxidant Activity of Lycopene from Tomato (*Solanum lycopersicum*) Waste. *Antioxidants* **2022**, *11*, 1404. [CrossRef]
29. Kiralan, M.; Ketenoglu, O. Utilization of Tomato (*Solanum lycopersicum*) by-Products: An Overview. In *Mediterranean Fruits Bio-Wastes*; Ramadan, M.F., Farag, M.A., Eds.; Springer: Cham, Switzerland, 2022. [CrossRef]
30. Tao, Y.; Liu, M.; Han, W.; Li, P. Waste office paper filled polylactic acid composite filaments for 3D printing. *Compos. Part B Eng.* **2021**, *221*, 108998. [CrossRef]
31. Lima, A.L.; Pires, F.Q.; Leandro, A.H.; Sa-Barreto, L.L.; Gratieri, T.; Gelfuso, G.M.; Cunha-Filho, M. Oscillatory shear rheology as an in-process control tool for 3D printing medicines production by fused deposition modeling. *J. Manif. Proc.* **2022**, *76*, 850–862. [CrossRef]
32. Shahabaz, S.M.; Sharma, S.; Shetty, N.; Shetty, S.D.; Gowrishankar, M.C. Influence of temperature on mechanical properties and machining of fibre reinforced polymer composites: A review. *Eng. Sci.* **2021**, *16*, 26–46. [CrossRef]
33. Singh, H.; Batra, N.K.; Dikshit, I. Development of new hybrid jute/carbon/fishbone reinforced polymer composite. *Mater. Today Proc.* **2021**, *38*, 29–33. [CrossRef]
34. Halpin, A.J.C.; Kardos, J.L. The Halpin-Tsai equations: A review. *Polym. Eng. Sci.* **1976**, *16*, 344–352. [CrossRef]

Article

# Enhanced Synaptic Properties in Biocompatible Casein Electrolyte via Microwave-Assisted Efficient Solution Synthesis

**Hwi-Su Kim [1], Hamin Park [2] and Won-Ju Cho [1,*]**

1 Department of Electronic Materials Engineering, Kwangwoon University, Gwangun-ro 20, Nowon-gu, Seoul 01897, Republic of Korea

2 Department of Electronic Engineering, Kwangwoon University, Gwangun-ro 20, Nowon-gu, Seoul 01897, Republic of Korea

* Correspondence: chowj@kw.ac.kr; Tel.: +82-2-940-5163

**Abstract:** In this study, we fabricated an electric double-layer transistor (EDLT), a synaptic device, by preparing a casein biopolymer electrolyte solution using an efficient microwave-assisted synthesis to replace the conventional heating (heat stirrer) synthesis. Microwave irradiation (MWI) is more efficient in transferring energy to materials than heat stirrer, which significantly reduces the preparation time for casein electrolytes. The capacitance–frequency characteristics of metal–insulator–metal configurations applying the casein electrolyte prepared through MWI or a heat stirrer were measured. The capacitance of the MWI synthetic casein was 3.58 $\mu F/cm^2$ at 1 Hz, which was higher than that of the heat stirrer (1.78 $\mu F/cm^2$), confirming a stronger EDL gating effect. Electrolyte-gated EDLTs using two different casein electrolytes as gate-insulating films were fabricated. The MWI synthetic casein exhibited superior EDLT electrical characteristics compared to the heat stirrer. Meanwhile, essential synaptic functions, including excitatory post-synaptic current, paired-pulse facilitation, signal filtering, and potentiation/depression, were successfully demonstrated in both EDLTs. However, MWI synthetic casein electrolyte-gated EDLT showed higher synaptic facilitation than the heat stirrer. Furthermore, we performed an MNIST handwritten-digit-recognition task using a multilayer artificial neural network and MWI synthetic casein EDLT achieved a higher recognition rate of 91.24%. The results suggest that microwave-assisted casein solution synthesis is an effective method for realizing biocompatible neuromorphic systems.

**Keywords:** microwave; biocompatible polymers; casein; electric double layer; synaptic transistors; artificial synapses; neuromorphic computing



**Citation:** Kim, H.-S.; Park, H.; Cho, W.-J. Enhanced Synaptic Properties in Biocompatible Casein Electrolyte via Microwave-Assisted Efficient Solution Synthesis. *Polymers* **2023**, *15*, 293. https://doi.org/10.3390/polym15020293

Academic Editors: José Miguel Ferri, Vicent Fombuena Borràs and Miguel Fernando Aldás Carrasco

Received: 5 December 2022
Revised: 26 December 2022
Accepted: 3 January 2023
Published: 6 January 2023



## 1. Introduction

The human brain has the ability to simultaneously calculate and memorize complex information with an ultra-low energy consumption of ~20 W [1]. This high-efficiency biological computing is accomplished through massive parallel processing, fault tolerance, and self-learning through a nervous system with ~$10^{11}$ neurons and ~$10^{15}$ synapses [2,3]. Classical computing systems based on the von Neumann architecture, unlike the human brain, faced increasingly higher power consumption and low computing speed limitations in terms of information processing due to bottlenecks between physically separated processors and memory units [4–6]. To overcome these limitations, neuromorphic systems, an innovative computing architecture inspired by the exceptional performance of the human brain in hardware, have received significant attention [7,8]. For implementing an efficient neuromorphic computing system, analog electronics that mimic synaptic functions, including two- and three-terminal devices, are key. This is because analog synaptic devices can analogically modulate the conductance of devices such as biological synaptic plasticity [9,10].

Two-terminal synaptic devices, such as memristor [11–13], spintronics device [14], and phase-change device [15–17], have been proposed through neural operation and geometrical advantages [8,18]. However, unlike the human-brain system, the two-terminal devices have limitations in that learning and signaling are performed separately. This is because the terminal receives feedback from post-neurons in the learning process and outputs signals from pre-neurons for signal transmission [19]. This limitation represents an incomplete implementation of biological synapses in two-terminal devices. In contrast, a three-terminal synaptic device can simultaneously learn and transmit signals through different parts, the gate terminal and the channel, respectively [20]. As a result, synaptic behavior can be fully emulated by a three-terminal device. In recent years, three-terminal synaptic transistors using ion-conducting electrolytes as gate insulators have been studied. Ion-conducting electrolytes contribute to electric double-layer (EDL) formation, allowing the synaptic function of electric double-layer transistors (EDLTs) to be achieved [21,22]. EDLs, which act as nanogap capacitors, have large capacitances (>1 $\mu F/cm^2$), resulting in very strong gate coupling between interfacial electrons/ions in EDLTs [22–24]. Thus, EDLTs can control synaptic plasticity with low power through channel conductance modulated by ion migration inside the electrolyte in response to gate bias [25,26]. In addition, as interest in eco-friendly and biocompatible electronic devices increases, biomaterials based on polymers, such as chitosan [27,28], starch [29,30], gelatin [31,32], and pectin [33], have been applied to EDLTs. Recently, it has been reported that casein, which is a biopolymer accounting for 80% of milk protein, has been utilized for EDLTs mimicking synaptic behavior based on its rich internal protons [34]. Moreover, these electrolytes are abundant in nature and inexpensive, especially in solution-based processes [28].

Meanwhile, chemical synthesis through microwave irradiation (MWI) was first reported in 1986 [35]. The polar molecules in the reaction mixture react quickly to the rapidly changing electric field of the MWI. As a result, rotation and friction of the polar molecules are induced, enabling direct and homogeneous internal heating in the reaction mixture [36,37]. Therefore, until now, MWI has been spotlighted and applied to various organic synthesis and material preparation due to its feature of being an environmentally friendly high-efficiency heating method [38,39]. Furthermore, MWI has several advantages over conventional heating methods, such as faster heating speed, faster processing time, volumetric heating, controllable heating, higher heat transfer efficiency, low energy consumption, and cost-effective processing [40–42].

In this study, we prepared a casein biopolymer electrolyte solution using efficient microwave heating by replacing the conventional heat stirrer and implemented improved synaptic properties through casein electrolyte-based EDLTs prepared by microwave-assisted solution synthesis. Prior to the fabrication of the MWI synthetic casein electrolyte-gated EDLTs, the frequency-dependent capacitance of metal–insulator–metal (MIM) capacitors with indium tin oxide (ITO) electrodes and casein electrolyte (synthesized through MWI) was evaluated to verify the effect of the EDL gating effect in the MWI casein electrolyte. By constructing our proposed EDLTs, essential functions related to synaptic plasticity, including excitatory post-synaptic current (EPSC), paired-pulse facilitation (PPF), signal filtering, and potentiation/depression, were evaluated. In addition, we performed recognition simulations using the Modified National Institute of Standards and Technology (MNIST) dataset via a multilayer artificial neural network (ANN) to demonstrate the application of the proposed EDLTs in neuromorphic systems. The outcomes were compared with those of EDLTs gated by the heat stirrer synthetic casein electrolyte. The notable findings of this study are, firstly, a significant reduction in casein electrolyte synthesis time through MWI heating; secondly, enhanced synaptic properties based on higher facilitation; and, thirdly, the improved recognition rate of MWI casein EDLTs.

## 2. Materials and Methods

### 2.1. Fabrication of Devices

We fabricated casein electrolyte-gated EDLTs on glass substrates (Corning Inc., New York, NY, USA) using either MWI synthesis or heat-stirrer synthesis approaches. To form the bottom-gate electrode, a 300 nm-thick ITO film ($In_2O_3$:$SnO_2$ = 9:1 mol%, THIFINE Co.,

Ltd., Incheon, Republic of Korea) was deposited on the glass substrate via radio frequency (RF) magnetron sputtering. The RF power, chamber pressure, and Ar flow of ITO sputtering were 100 W, 3 mTorr, and 20 sccm, respectively. To prepare two types of casein solutions, 3 wt.% of casein powder (technical grade, Sigma-Aldrich, St. Louis, MO, USA) and 3 wt.% of acetic acid (purity > 99%, Sigma-Aldrich) were dissolved in 94 wt.% of deionized water using the MWI or conventional heat-stirrer synthesis method. For the MWI synthesis process, a microwave of frequency 2.45 GHz and power 250 W was irradiated for 5 min. Meanwhile, a magnetic heat stirring of 800 rpm was performed at 130 °C for 6 h for the heat-stirrer synthesis process. These synthesized precursor solutions were filtered through a 5 µm-pore-size polytetrafluoroethylene syringe filter to remove contaminants. Then, each casein solution was spin-coated on an ITO/glass substrate at 3000 rpm for 30 s and dried in air for 24 h. An indium gallium zinc oxide (IGZO) film ($In_2O_3$:$Ga_2O_3$:ZnO = 4:2:4.1 mol%, THIFINE Co., Ltd.) with a thickness of 50 nm was deposited through a shadow mask using RF magnetron sputtering to form a transistor channel layer with dimensions (width × length) 1000 µm × 80 µm. Finally, 150 nm-thick ITO source/drain electrodes with dimensions 1000 µm × 200 µm were deposited through a shadow mask using RF magnetron sputtering.

### 2.2. Characterizations of the Devices

The fabricated casein electrolyte-gated EDLTs were placed in a dark box to block external light and electrical noise. The capacitance versus frequency (C–*f*) characteristic curves of ITO/casein electrolyte/ITO capacitors were analyzed using the Agilent 4284A precision LCR meter (Hewlett-Packard Corp., Palo Alto, CA, USA). The electrical properties and synaptic functions of the casein electrolyte-gated EDLTs were characterized using an Agilent 4156B precision semiconductor parameter analyzer (Hewlett-Packard Corp., USA). In addition, the Agilent 8110A pulse generator (Hewlett-Packard Corp., USA) was used to apply electrical pre-synaptic stimuli to validate the synaptic behavior of the fabricated EDLTs.

## 3. Results

### 3.1. Preparation of a Casein Electrolyte Solution

Figure 1a,b show diagrams of the synthesis equipment and internal temperature profiles of casein electrolytes for heat-stirrer and MWI processes, respectively. In the conventional heat-stirrer synthesis, the solution temperature has a non-uniform temperature distribution where the bottom of the vessel is the highest while the top surface is the lowest. Therefore, a considerable period of time is required until the temperature difference decreases due to conduction and convection phenomena. In contrast, since MWI transfers electromagnetic energy directly to the solution, the uniform temperature distribution is achieved by volumetric heating that occurs simultaneously with the initiation of the synthesis process. The temperature profiles of the casein precursor solution during the heat-stirrer and MWI synthesis processes are shown in Figure 1c. The temperatures of the solution measured using an infrared (IR) thermometer were 135 °C for the heat stirrer and 150 °C for the MWI. The thermal budgets for both processes were obtained by integrating the corresponding temperature profiles with respect to the processing times. The conventional heat-stirrer synthesis process was performed at 130 °C for 6 h and had a thermal budget of $2.3 \times 10^6$ °C·s. The MWI synthesis process was performed at 250 W for 5 min and had a thermal budget of $3.8 \times 10^4$ °C·s. Consequently, the MWI process is more efficient in casein electrolyte synthesis because it takes less time and has 82-times lower thermal budget than a heat-stirrer process.

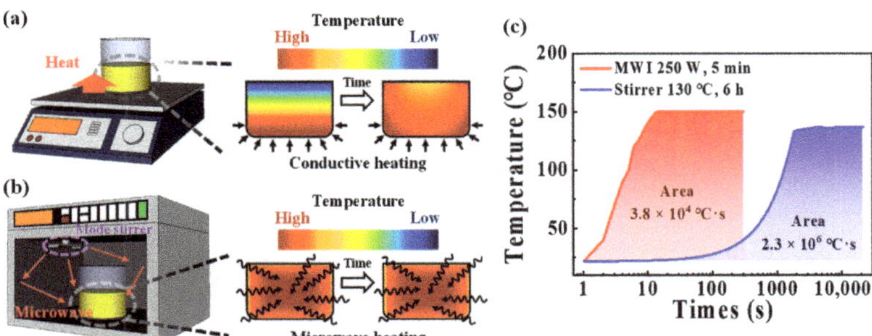


**Figure 1.** Schematic illustration of solution synthesis equipment and internal temperature profile of casein electrolyte solution for (**a**) heat-stirrer and (**b**) microwave irradiation (MWI) synthetic processes. (**c**) Temperature profiles and thermal budgets of conventional heat stirrer (130 °C, 6 h) and MWI (250 W, 5 min) processes for preparing casein electrolyte solution.

### 3.2. EDL Operation of MWI Synthetic Casein Electrolytes

To identify the difference in the chemical properties of the spin-coated casein electrolyte film prepared through MWI or heat-stirrer synthesis, Fourier-transform infrared spectroscopy (FT-IR) analysis was executed in a range of 900 to 4000 cm$^{-1}$, as shown in Figure 2a,b. Broad peaks at approximately 3273 and 3381 cm$^{-1}$ in MWI and heat-stirrer synthetic casein electrolytes, respectively, are attributed to O–H stretching [43]. In particular, casein, which is abundant in mobile protons, is a protein substance that accounts for about 80% of the total protein in milk, and the main component of the protein is an amide group [44]. Therefore, peaks related to the amide group were observed from 1700 to 1500 cm$^{-1}$ in both casein electrolyte films, in which the peaks from 1700 to 1600 cm$^{-1}$ and 1600 to 1500 cm$^{-1}$ corresponded to the amide I and amide II groups, respectively. The peaks at 1630 and 1626 cm$^{-1}$ in MWI and heat-stirrer synthetic casein electrolytes, respectively, are caused by C=O stretching. The peaks at 1526 and 1527 cm$^{-1}$ in the MWI and heat-stirrer synthetic casein electrolytes, respectively, are due to N–H bending and C–N stretching. In addition, peaks near 1106 and 1103 cm$^{-1}$ in MWI and heat-stirrer synthetic casein electrolytes, respectively, are associated with C–O stretching [45,46]. For quantitative analysis of the peaks, peak deconvolution was performed and the peak area was divided by the total spectra area to obtain a percentage of the peak area [47,48]. The area percentages of the O–H, amide I, amide II, and C–O related peaks were 26.92, 19.67, 12.94, and 2.63%, respectively, for the MWI synthetic casein electrolyte, while they were 16.38, 24.45, 7.93, and 6.05% in the heat-stirrer synthetic casein electrolyte, respectively. The presence of O–H groups could help with proton conduction, which facilitates the formation of EDL by promoting proton migration within the casein electrolyte [29]. Therefore, the MWI synthetic casein electrolyte has more O–H groups than heat-stirrer synthetic casein electrolyte, leading to higher ionic conduction. MWI is a way to facilitate chemical reactions [49], and microwaves loosen the structure of protein molecules, exposing more active groups to the synthesis process [50]. Therefore, more chemical reactions occur for the MWI synthesis process of casein solution, resulting in a higher O–H group ratio in the MWI casein electrolyte. Subsequently, MIM capacitors with a vertical sandwich structure were fabricated using casein electrolytes as an insulating layer and ITO as a metal layer to verify the EDL gating effect in synthetic casein electrolytes. Figure 2c shows the frequency-dependent capacitance characteristics over a wide frequency range from 1 Hz to 1 MHz. The maximum capacitance appeared at 1 Hz and the capacitance decreased with increasing frequency. This change is due to the difference in the response time of the mobile protons in the casein electrolyte according to the frequency. At low frequencies, protons accumulate at the two interfaces (electrolyte/channel and electrolyte/gate) to form an EDL because protons have a sufficient response time to reach the interface. The result is a huge capacitance due to the formed EDL, which is a nanometer-thick parallel capacitor [51]. It is noteworthy that the maximum capacitance of the MWI synthetic casein electrolyte was 3.58 μF/cm$^2$, which was higher than that of the heat-stirrer synthetic casein electrolyte (1.78 μF/cm$^2$). The large capacitance of the EDL has a strong gating effect between the channel and the gate, allowing for tuning of the conductivity, even at low voltages [22]. Therefore, more energy-efficient artificial synapses can be implemented using the MWI synthetic casein electrolyte-gated EDLT.

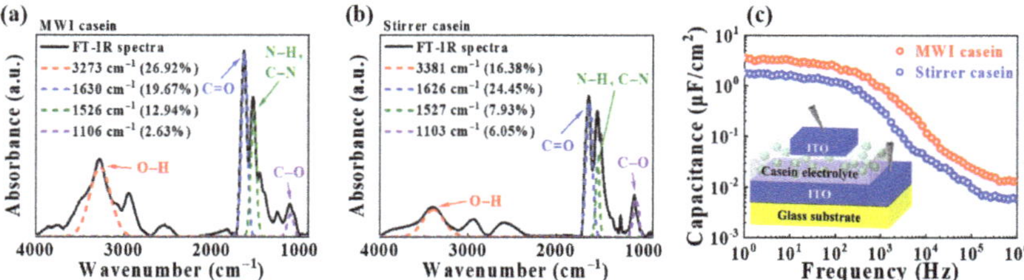


**Figure 2.** FT-IR spectra of casein electrolyte films produced through (**a**) MWI (250 W, 5 min) and (**b**) heat-stirrer (130 °C, 6 h) processes. (**c**) Frequency-dependent specific capacitance of capacitors using casein electrolyte films synthesized by MWI or heat-stirrer processes. Inset: schematic illustration of the measured ITO/casein electrolyte/ITO vertical sandwich structure.

### 3.3. MWI Synthetic Casein Electrolyte-Gated EDLT

Figure 3a shows the structure of the proposed microwave-assisted synthetic casein electrolyte-gated EDLT. A structural comparison between the functional mechanisms of the proposed EDLT and biological synapses is shown in Figure 3b. The gate electrode and channel layer of the EDLT are considered pre- and post-synapse, respectively. The casein electrolyte between the gate and the channel is considered a synaptic cleft. In biological synapses, changes in synaptic weight are caused by the transmission of neurotransmitters, including $K^+$ and $Na^+$ ions, through the synaptic cleft, which is analogous to the channel conductance changes in the EDLT induced by proton transport of casein electrolytes. Therefore, the drain current associated with the channel conductance is indicative of EPSC and its change is indicative of synaptic plasticity.

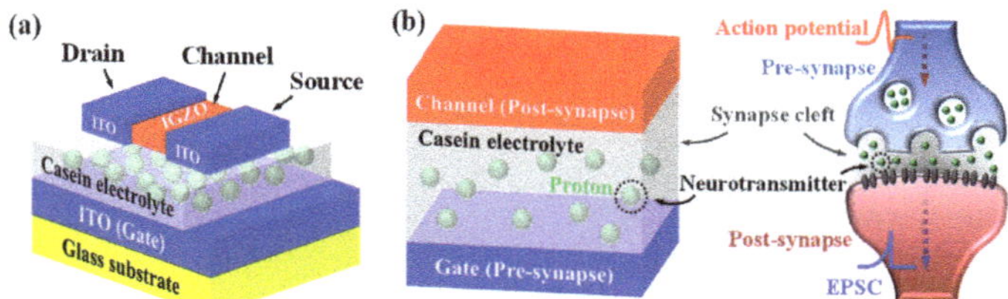


**Figure 3.** (**a**) Schematic of the proposed microwave-assisted synthetic casein electrolyte-gated electric double-layer transistor (EDLT). (**b**) Comparison between the functional mechanism of the proposed synaptic transistor and biological synapses.

### 3.4. Electrical Properties of Devices

Figure 4a shows the transfer characteristic curves ($I_D$–$V_G$) of MWI and heat-stirrer synthetic casein electrolyte-gate EDLTs measured in double-sweep mode of gate voltage ($V_G$). The $V_G$ double-sweep as the backward after the forward sweeps was performed at a drain voltage ($V_D$) of 1 V, increasing the maximum $V_G$ from 0 to 5 V in 0.5 V increments. In the double-sweep transfer curve, the direction of the hysteresis window exhibits a counterclockwise shape and the hysteresis window increases with increasing maximum $V_G$ sweep range. This hysteresis characteristic is due to the slow polarization of mobile protons in the casein electrolyte. In the $V_G$ forward sweep, the larger the maximum $V_G$, the more protons are induced at the interface of the casein electrolyte/IGZO channel. Then, in the $V_G$ backward sweep, the protons progressively diffuse back in the opposite direction, increasing the counterclockwise hysteresis window. Figure 4b shows the hysteresis window ($\Delta V$) according to the maximum $V_G$ obtained from the double-sweep transfer characteristic curves. The $\Delta V$ expanded from 0.22 V to 2.64 V with a 0.5 V/V slope in the MWI synthetic casein EDLT, indicating a greater increase in the $\Delta V$ than that of the heat stirrer (from 0.17 V to 1.75 V with a 0.3 V/V slope). Figure 4c shows the output characteristic curve ($I_D$–$V_D$), where the drain current ($I_D$) increases linearly as $V_D$ increases, followed by pinch-off and saturation characteristics. Since the MWI synthetic casein electrolyte has a stronger EDL gate effect, the MWI synthetic casein EDLT exhibits larger drain current and $\Delta V$ slope than that of the heat stirrer.

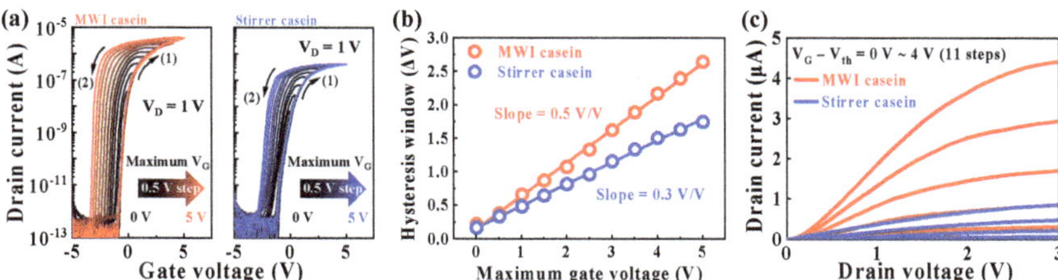


**Figure 4.** (**a**) Double-sweep transfer characteristic curves of MWI and heat-stirrer synthetic casein electrolyte-gate EDLTs as a function of gate voltage sweep range (maximum $V_G$ from 0 V to 5 V in 0.5 V increments) at a $V_D$ of 1 V. (**b**) Extracted hysteresis window as a function of maximum $V_G$. (**c**) Output characteristic curves of MWI and heat-stirrer synthetic casein electrolyte-gated EDLTs versus gate voltage—threshold voltage ($V_G - V_{th}$ = 0–4 V in 0.4 V increments).

Table 1 summarizes the electrical parameters of MWI and heat-stirrer synthetic casein EDLTs extracted from transfer characteristic curves at a maximum $V_G$ of 5 V. The MWI synthetic casein EDLTs exhibited improved transfer properties compared to the heat-stirrer synthetic casein electrolyte-gated EDLTs, with higher on/off current ratio ($I_{on}/I_{off}$), field-effect mobility ($\mu_{FE}$), and hysteresis window ($\Delta V$), but lower threshold voltage ($V_{th}$), subthreshold swing (SS), and interface trap density ($D_{it}$). In particular, the higher $\mu_{FE}$ and lower SS are attributed to the better interfacial state, lower root-mean-square surface roughness, and $D_{it}$ values of MWI synthetic casein. The thickness and surface roughness of the coated films are shown in Figure S1 (Supplementary Materials).

**Table 1.** Electrical parameters of MWI and heat-stirrer synthetic casein electrolyte-gated EDLTs.

| EDL Type | $I_{on}/I_{off}$ (A/A) | $V_{th}$ (V) | $\Delta V$ (V) | SS (mV/dec) | $\mu_{FE}$ (cm$^2$/V·s) | $D_{it}$ (cm$^2$·eV$^{-1}$) |
|---|---|---|---|---|---|---|
| MWI casein | $4.31 \times 10^7$ | 0.61 | 2.51 | 131.59 | 18.62 | $1.88 \times 10^{11}$ |
| Heat stirrer casein | $4.12 \times 10^6$ | 0.99 | 1.75 | 332.87 | 2.24 | $2.03 \times 10^{11}$ |

### 3.5. Synaptic Properties of Devices

In the human brain, biological signals are transmitted by diffusing the neurotransmitters between pre- and post-synapses, and in this process, EPSCs, which are transient current, are generated. The EPSC represents the synaptic weight regulated by the ion flux [28,52], which increases as the duration of stimulation increases, resulting in an increase in the EPSC [9]. Similarly, in EDLT, the channel conductivity, and EPSC increase as more mobile protons inside the casein electrolyte accumulate at the electrolyte/channel interface during longer spike widths [32].

Figure 5a shows EPSCs of MWI and heat-stirrer synthetic casein electrolyte-gated EDLTs triggered by a single pre-synaptic spike with a spike amplitude of 1 V for a spike width of 50 ms at a $V_D$ of 1 V. EPSCs for differences in pre-synaptic spike widths are shown in Figure S2 (Supplementary Materials). Figure 5b shows the maximum EPSC values with various pre-synaptic spike widths. The maximum EPSC value of MWI synthetic casein EDLT, which increased with the increase in the pre-synaptic spike width, was higher than that of the heat stirrer. PPF is the essential characteristic of short-term synaptic plasticity and represents the neuronal facilitation event as a function of spike interval ($\Delta t_{inter}$), in which the post-synaptic potentials evoked by the second pre-synaptic spike increase further when the second closely follows the previous one [27,53,54]. The first pre-synaptic spike induced the gathering of mobile protons at the electrolyte/channel interface. Then, a second spike is applied, where when $\Delta t_{inter}$ is short, in addition to incomplete relaxation protons, mobile protons are continuously accumulated at the interface, increasing the channel conductivity. Figure 5c,d show EPSCs facilitated by two consecutive pre-synaptic spikes (1 V, 100 ms) with a $\Delta t_{inter}$ of 50 ms for the MWI and heat-stirrer synthetic case EDLTs. EPSCs triggered by various paired pre-synaptic spikes with $\Delta t_{inter}$ of 50–1500 ms are shown in Figures S3 and S4 (Supplementary Materials). Figure 5e shows the PPF index as a function of $\Delta t_{inter}$ calculated as the ratio of the first ($A_1$) and second ($A_2$) EPSCs, where the PPF index increases with shorter $\Delta t_{inter}$ but decreases with longer $\Delta t_{inter}$. In particular, MWI synthetic casein EDLT exhibits a higher facilitation, with a PPF index of 180%, than that of the heat stirrer (PPF index of 141%) at a $\Delta t_{inter}$ of 50 ms based on a stronger EDL gating effect. The calculated PPF index was fitted with the following double exponential decay relationship [55]:

$$PPF\ index = A + C_1 \exp(-\Delta t_{inter}/\tau_1) + C_2 \exp(-\Delta t_{inver}/\tau_2) \quad (1)$$

where $A$ is a constant value, $C_1$ and $C_2$ are the magnitudes of the initial facilitation, and $\tau_1$ and $\tau_2$ are the characteristic relaxation times. The values of $\tau_1$ and $\tau_2$ in the MWI synthetic casein were 102.8 and 1401 ms, respectively, and 78.4 and 1039.2 ms in the heat-stirrer synthetic casein, respectively. The results are nearly identical to biological synapses and indicate that the proposed device allows for subdivisions of the synaptic temporal scale into rapid and slow stages lasting tens and hundreds of milliseconds [53].

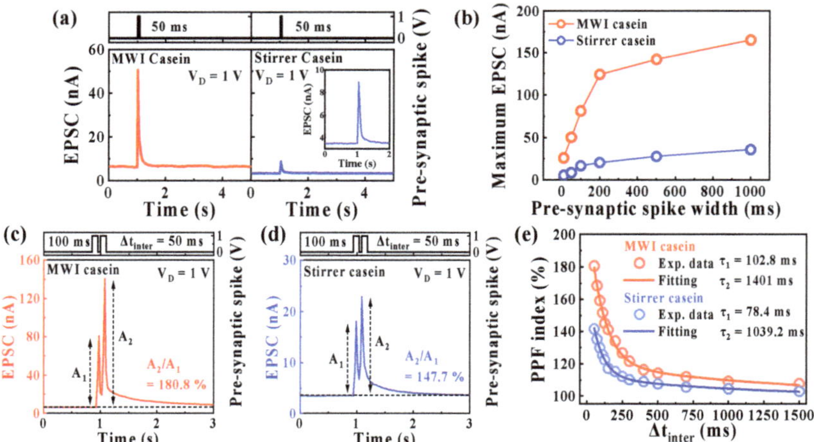


**Figure 5.** (**a**) Excitatory post-synaptic current (EPSC) of MWI and heat-stirrer synthetic casein electrolyte-gated EDLTs triggered by the pre-synaptic spike (1 V, 50 ms) at $V_D$ = 1 V. (**b**) Maximum EPSCs with different pre-synaptic spike widths (10, 50, 100, 200, 500, and 1000 ms). EPSC facilitated by a paired pre-synaptic spike (1 V, 100 ms) with a 50 ms spike interval ($\Delta t_{inter}$) of (**c**) MWI and (**d**) heat-stirrer synthetic casein electrolyte-gated EDLT. (**e**) Paired pulse facilitation index of MWI and heat-stirrer synthetic casein electrolyte-gated EDLTs as a function of $\Delta t_{inter}$ from 50 to 1500 ms of the pre-synaptic spikes.

Table 2 shows the performance comparisons between the proposed MWI synthetic casein electrolyte-gated EDLT and various biopolymer-based EDLTs. In our MWI synthetic casein EDLT, the time of EDL synthesis is considerably reduced and the performances are superior ($I_{on}/I_{off}$ and PPF index) or comparable (Capacitance and SS).

**Table 2.** Performance comparison of this EDLT with other biopolymer EDL-based EDLTs.

| Biopolymer EDL | Synthesis Condition | Capacitance at 1 Hz | $I_{on}/I_{off}$ | SS | PPF Index | Ref. |
|---|---|---|---|---|---|---|
| Starch | Heat-stirrer (90 °C, 30 min) | ~1.5 μF/cm$^2$ | ~1.6 × 10$^6$ A/A | ~140 mV/dec | ~162% ($\Delta t_{inter}$ = 10 ms) | [30] |
| Pectin | Heat-stirrer (70 °C, 30 min) | ~3.7 μF/cm$^2$ | ~6.6 × 10$^5$ A/A | ~196 mV/dec | ~150% ($\Delta t_{inter}$ = 10 ms) | [33] |
| PVA | Heat-stirrer (100 °C, 4 h) | ~1.63 μF/cm$^2$ | ~7.8 × 10$^6$ A/A | ~129.7 mV/dec | ~123% ($\Delta t_{inter}$ = 50 ms) | [54] |
| Casein | MWI 250 W (≈150 °C, 5 min) | ~3.58 μF/cm$^2$ | ~4.31 × 10$^7$ A/A | ~131.59 mV/dec | ~180% ($\Delta t_{inter}$ = 50 ms) | This work |

In addition, synapses perform the function of dynamically filtering information transactions based on short-term synaptic plasticity, as shown in Figure 6a [56]. Figure 6b,c show the EPSCs of MWI and heat-stirrer synthetic casein EDLTs in response to 10 successive pre-synaptic spikes with various frequencies from 1 to 9.8 Hz, respectively. Here, the rising and falling edge time of each pre-synaptic spike is 10 ns. EPSCs generated by sequential spikes remain almost constant at 1 Hz, increasing with progressively increasing frequency, indicating short-term facilitation. Thus, the high-pass temporal filtering functions of the EDLTs were successfully demonstrated [57]. Figure 6d shows the EPSC gain as a function of frequency, calculated from the ratio of dividing the EPSC-triggered 10th pre-synaptic spike ($A_{10}$) by the EPSC-triggered first spike ($A_1$). At a frequency of 1 Hz, the EPSC gain was 1 for both EDLTs; however, when the frequency was gradually increased to 9.8 Hz, the EPSC gain of the MWI synthetic casein EDLT rose to 2.13, which is greater than that of the heat stirrer (1.79). The MWI synthetic casein EDLT has a larger EPSC gain value based on higher facilitation than the heat stirrer, indicating that the high-pass filter can be operated over a wider gain range.

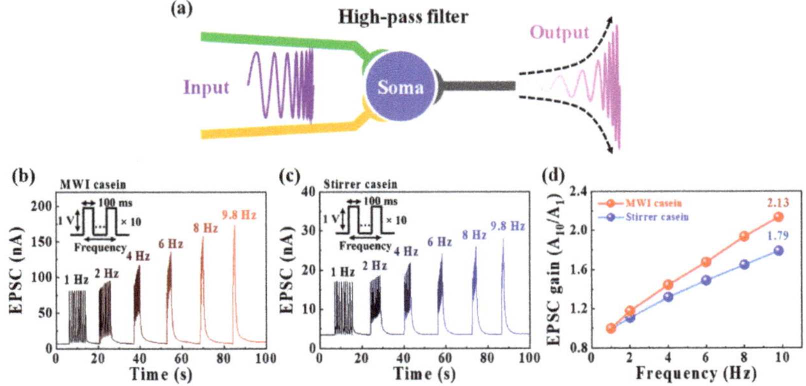


**Figure 6.** (**a**) Diagram of high-pass filtering function in biological synapses. EPSCs for 10 consecutive pre-synaptic spikes (1 V, 100 ms) at different frequencies (1 to 9.8 Hz) of (**b**) MWI and (**c**) heat-stirrer synthetic casein electrolyte-gated EDLT. (**d**) EPSC gain plotted against pre-synaptic spike frequency defined as $A_{10}/A_1$.

Synaptic weights are reinforced by repetitive stimuli to indicate long-term changes, which is long-term plasticity as opposed to short-term plasticity. Long-lasting strengthening or weakening of synaptic weights is considered long-term potentiation (LTP) or long-term depression (LTD) [58,59].

Figure 7a shows the modulation of synaptic weights in potentiation and depression by repetitive pre-synaptic stimuli, which consequently characterizes LTP and LTD. In both EDLTs, the channel conductance was changed by a potentiation pulse (5 V, 100 ms) and a depression pulse (−2.5 V, 100 ms). In MWI synthetic casein EDLT, the conductance increased from 30.7 to 105.4 nS for the potentiation stimulus, whereas it decreased from 99 to 28.7 nS for the depression stimulus. This indicates that the stimulation-dependent conductivity modulation of MWI is larger than that of the heat stirrer, which is 20 to 39.2 nS for the potentiation stimulus and 37.6 to 19.2 nS for the depression stimulus. The endurance characteristics of potentiation/depression in five cycles of potentiation and depression pulses are shown in Figure 7b, where the conductivity modulation was maintained as almost constant.

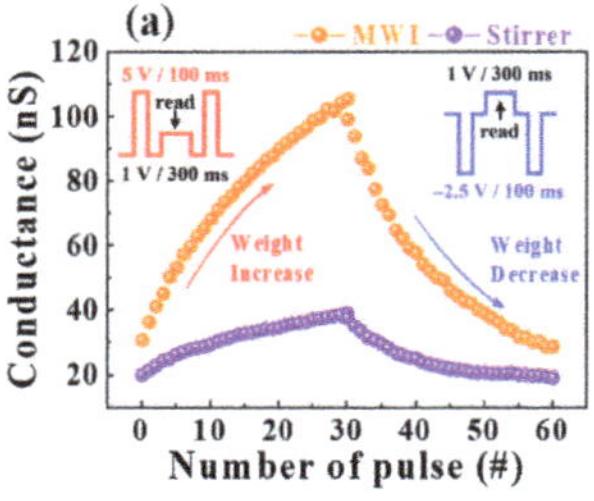


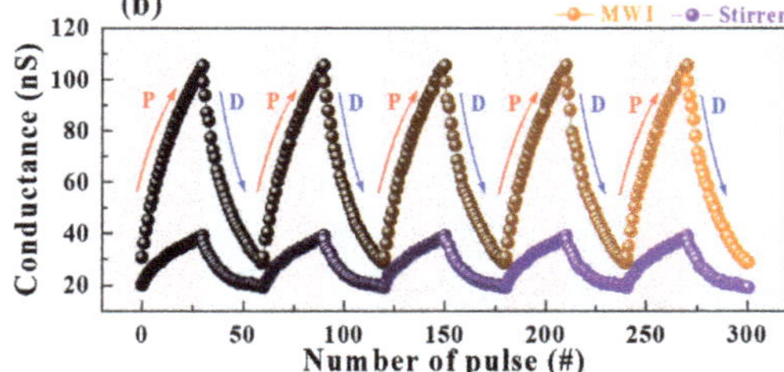


**Figure 7.** (**a**) Modulation of synaptic weights in potentiation and depression properties of MWI and heat-stirrer synthetic casein electrolyte-gated EDLTs by the number of the applied pre-synaptic spikes (#). Insets show schematic diagrams of reinforcement and depressive spikes. (**b**) Endurance test for five cycles of MWI and heat-stirrer synthetic casein electrolyte-gated EDLTs.

### *3.6. Recognition Simulation*

ANN is a computing architecture that performs complicated calculations, such as recognition and perception, with a complex network structure in which neurons are interconnected, similar to the human brain. To implement neuromorphic systems, it is essential to build ANNs based on synaptic devices in terms of hardware [60,61]. Figure 8a shows a designed multilayer ANN model in which the input, hidden, and output layers are fully connected via synaptic weights. The input and output layers were made up of 784 input neurons corresponding to 28 by 28 MNIST data and 10 output neurons related to the ten varieties of digits (0–9), respectively. This model was used for the handwritten MNIST learning test to evaluate the capability of neuromorphic computing. Figure 8b,c show the normalized potentiation and depression conductance of MWI and heat-stirrer synthetic casein EDLTs, respectively. The normalized conductance ($G_{\#}/G_1$) was determined by dividing the conductance of each step ($G_{\#}$) by the first conductance ($G_1$) and was used as a synaptic weight related to the synaptic strength connecting each neuron in the developed model. Through nonlinearity analysis of the normalized conductance, important characteristics, including dynamic range (DR), asymmetric ratio (AR), and linearity, which are highly correlated with the accuracy of learning and recognition, were obtained. The DR represents the range of conductance modulation defined by dividing $G_{max}$ by $G_{min}$. Higher DR values are required for better accuracy and performance in simulations [62]. The DR value of the MWI synthetic casein EDLT was 3.42, which was 1.75-times greater than that of the heat stirrer (1.95). The AR indicates the asymmetry of the changes in the potentiation and depression conductance. The more symmetrical the conductance modulation is, the closer the AR is to the ideal value of 0 and the higher the learning accuracy. The AR can be extracted using the equation [62]:

$$\text{AR} = \frac{max|G_p(n) - G_d(n)|}{G_{p(30)} - G_{d(30)}} \text{ for } n = 1 \text{ to } 30 \quad (2)$$

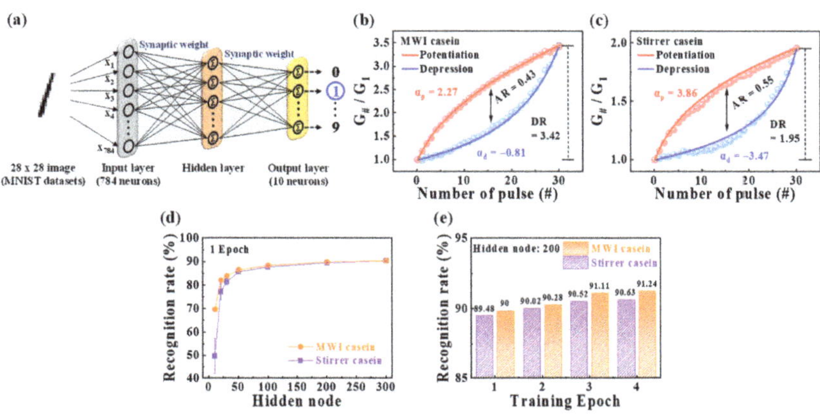


**Figure 8.** (**a**) Schematic diagram of an artificial neural network model with input, hidden, and output layers fully connected via synaptic weights for simulating MNIST recognition. Normalized potentiation and depression by nonlinearity analysis of (**b**) MWI and (**c**) heat-stirrer synthetic casein electrolyte-gated EDLTs. (**d**) Simulated MNIST recognition rates with varying numbers of hidden neurons. (**e**) Recognition rate by training epoch.

The obtained ARs are 0.43 and 0.55 for MWI and heat-stirrer synthetic casein EDLT, respectively, indicating that the MWI is closer to the ideal value and the conductance change is symmetric. Moreover, the linearity of the increase and decrease in the conductance is critical for recognition simulation. The nonlinearity factors can be defined as [63]:

$$G = \begin{cases} G_{min} \times \left(\frac{G_{max}}{G_{min}}\right)^{w} & , \ if\ \alpha = 0 \\ \left((G_{max}^{\alpha} - G_{min}^{\alpha}) \times w + G_{min}^{\alpha}\right)^{1/\alpha} & , \ if\ \alpha \neq 0 \end{cases} \quad (3)$$

where $G_{max}$ and $G_{min}$ are the maximum and minimum conductance, respectively, and w represents an internal variable ranging from 0 to 1. Further, $\alpha_p$ and $\alpha_d$ donate the nonlinearity factors for potentiation and depression, with an ideal value of 1. The calculated $\alpha_p$ and $\alpha_d$ of MWI synthetic casein EDLT were 2.27 and −0.81, which exhibited higher linearity in the increase and decrease in conductivity than that of the heat stirrer ($\alpha_p$ = 3.86, $\alpha_d$ = −3.47). Subsequently, an ANN model was designed using the normalized conductance and the obtained factors, trained with 60,000 MNIST images from the training dataset, and tested for recognition with 10,000 handwritten images from the test dataset in each training epoch. Figure 8d,e show the increase in the recognition rate with the increasing number of hidden nodes and training epochs, respectively. Based on the improved potentiation and depression characteristics and factors in the MWI synthetic casein, a higher recognition rate was achieved than that of heat-stirrer synthetic casein. The recognition rates of MWI and heat-stirrer synthetic casein EDLTs were 90% and 89.48%, with 200 hidden nodes, and 91.24% and 90.63% after four epochs, respectively. Therefore, the MWI synthetic casein EDLT represents potential as a promising artificial synapse compared to the heat stirrer.

## 4. Conclusions

We fabricated casein biopolymer electrolyte-gated EDLTs by introducing microwave-assisted synthesis to casein electrolyte solution preparation. Through MWI assistance, the synthesis time of the casein electrolyte was significantly reduced from 6 h to 5 min. Further, FT-IR analysis and C–*f* characterization on synthetic casein electrolytes demonstrated that the MWI processing had a higher EDL gating effect and ionic conductivity between the channel and gate than the conventional heat-stirrer processing. The MWI synthetic casein electrolyte-based EDLT exhibited superior double-swept transfer and output properties compared to the heat-stirrer synthetic casein electrolyte-based EDLT. Furthermore, the MWI synthetic casein EDLT successfully implemented improved plasticity functions of synaptic devices, such as EPSC, PPF, signal filter, and potentiation/depression, based on higher synaptic facilitation compared to the heat stirrer. Additionally, in the handwritten MNIST learning test, the MWI synthetic casein EDLT achieved higher recognition rates than the heat stirrer under the same learning conditions. Therefore, these results indicated that the casein electrolyte-gated EDLTs based on the microwave-assisted solution synthesis process with advanced characteristics are expected to play a key role in building bio-friendly neuromorphic computing systems, replacing conventional heat-stirrer synthesis.

**Supplementary Materials:** The following supporting information can be downloaded at: https://www.mdpi.com/article/10.3390/polym15020293/s1, Figure S1: Thickness and surface roughness of spin-coated casein electrolyte films prepared through the MWI and heat stirrer; Figure S2: EPSCs triggered by single pre-synaptic spikes with different spike widths (10, 50, 100, 200, 500, 1000 ms) of MWI and heat-stirrer synthetic casein electrolyte-gated EDLTs; Figure S3: EPSCs facilitated by paired pre-synaptic spikes with various spike intervals from 50 to 1500 ms of MWI synthetic casein electrolyte-gated EDLT; Figure S4: EPSCs facilitated by paired pre-synaptic spikes with various spike intervals from 50 to 1500 ms of heat-stirrer synthetic casein electrolyte-gated EDLT.

**Author Contributions:** Conceptualization, H.-S.K. and W.-J.C.; investigation, H.-S.K. and W.-J.C.; writing—original draft preparation, H.-S.K., H.P. and W.-J.C.; MNIST simulation, H.P.; writing—review and editing, H.-S.K., H.P. and W.-J.C.; supervision, W.-J.C.; project administration, W.-J.C.; funding acquisition, W.-J.C. All authors have read and agreed to the published version of the manuscript.

**Funding:** This research was funded by the National Research Foundation of Korea (NRF), grant funded by the Korean government (MSIT). (No. 2020R1A2C1007586). This study was also supported by the Korea Institute for Advancement of Technology (KIAT), grant funded by the Korean Government (MOTIE). (P0002397, The Competency Development Program for Industry Specialist).

**Institutional Review Board Statement:** Not applicable.

**Data Availability Statement:** Not applicable.

**Acknowledgments:** The present research has been conducted by the Research Grant of Kwangwoon University in 2022 and the Excellent research support project of Kwangwoon University in 2022.

**Conflicts of Interest:** The authors declare no conflict of interest.

## References


1. Machens, C.K. Neuroscience. Building the human brain. *Science* **2012**, *338*, 1156. [CrossRef] [PubMed]
2. Cole, M.W.; Bassett, D.S.; Power, J.D.; Braver, T.S.; Petersen, S.E. Intrinsic and task-evoked network architectures of the human brain. *Neuron* **2014**, *83*, 238. [CrossRef] [PubMed]
3. Ebbinghaus, H. Memory: A contribution to experimental psychology. *Ann. Neurosci.* **2013**, *20*, 155. [CrossRef] [PubMed]
4. Von Neumann, J. The principles of large-scale computing machines. *IEEE Ann. Hist. Comput.* **1981**, *3*, 263. [CrossRef]
5. Kuzum, D.; Yu, S.; Philip Wong, H.P. Synaptic electronics: Materials, devices and applications. *Nanotechnology* **2013**, *24*, 382001. [CrossRef] [PubMed]
6. Zidan, M.A.; Strachan, J.P.; Lu, W.D. The future of electronics based on memristive systems. *Nat. Electron.* **2018**, *1*, 22. [CrossRef]
7. Jain, A.K.; Mao, J.; Mohiuddin, K.M. Artificial neural networks: A tutorial. *Computer* **1996**, *29*, 31. [CrossRef]
8. Furber, S. Large-scale neuromorphic computing systems. *J. Neural Eng.* **2016**, *13*, 051001. [CrossRef]
9. Jo, S.H.; Chang, T.; Ebong, I.; Bhadviya, B.B.; Mazumder, P.; Lu, W. Nanoscale memristor device as synapse in neuromorphic systems. *Nano Lett.* **2010**, *10*, 1297. [CrossRef]
10. Abbas, H.; Abbas, Y.; Hassan, G.; Sokolov, A.S.; Jeon, Y.R.; Ku, B.; Kang, C.J.; Choi, C. The coexistence of threshold and memory switching characteristics of ALD HfO2 memristor synaptic arrays for energy-efficient neuromorphic computing. *Nanoscale* **2020**, *12*, 14120. [CrossRef]

11. Liu, L.; Xiong, W.; Liu, Y.; Chen, K.; Xu, Z.; Zhou, Y.; Han, J.; Ye, C.; Chen, X.; Song, Z.; et al. Designing high-performance storage in $HfO_2/BiFeO_3$ memristor for artificial synapse applications. *Adv. Electron. Mater.* **2020**, *6*, 1901012. [CrossRef]
12. Rajasekaran, S.; Simanjuntak, F.M.; Chandrasekaran, S.; Panda, D.; Saleem, A.; Tseng, T.Y. Flexible ta 2 O 5/WO 3-based memristor synapse for wearable and neuromorphic applications. *IEEE Electron Dev. Lett.* **2021**, *43*, 9. [CrossRef]
13. Covi, E.; Brivio, S.; Serb, A.; Prodromakis, T.; Fanciulli, M.; Spiga, S. HfO2-Based Memristors for Neuromorphic Applications. In Proceedings of the IEEE International Symposium on Circuits and Systems (ISCAS), Montreal, QC, Canada, 22–25 May 2016; Volume 2016, pp. 393–396. [CrossRef]
14. Fukami, S.; Ohno, H. Perspective: Spintronic synapse for artificial neural network. *J. Appl. Phys.* **2018**, *124*, 151904. [CrossRef]
15. Kuzum, D.; Jeyasingh, R.G.; Lee, B.; Wong, H.S.P. Nanoelectronic programmable synapses based on phase change materials for brain-inspired computing. *Nano Lett.* **2012**, *12*, 2179. [CrossRef]
16. La Barbera, S.; Ly, D.R.B.; Navarro, G.; Castellani, N.; Cueto, O.; Bourgeois, G.; De Salvo, B.; Nowak, E.; Querlioz, D.; Vianello, E. Narrow heater bottom electrode-based phase change memory as a bidirectional artificial synapse. *Adv. Electron. Mater.* **2018**, *4*, 1800223. [CrossRef]
17. Suri, M.; Bichler, O.; Querlioz, D.; Cueto, O.; Perniola, L.; Sousa, V.; Vuillaume, D.; Gamrat, C.; DeSalvo, B. Phase change memory as synapse for ultra-dense neuromorphic systems: Application to complex visual pattern extraction. In Proceedings of the International Electron Devices Meeting, Washington, DC, USA, 5–7 December 2011; Volume 2011, p. 4.4.1. [CrossRef]
18. Chang, T.C.; Chang, K.C.; Tsai, T.M.; Chu, T.J.; Sze, S.M. Resistance random access memory. *Mater. Today* **2016**, *19*, 254. [CrossRef]
19. Nishitani, Y.; Kaneko, Y.; Ueda, M.; Morie, T.; Fujii, E. Three-terminal ferroelectric synapse device with concurrent learning function for artificial neural networks. *J. Appl. Phys.* **2012**, *111*, 124108. [CrossRef]
20. Monalisha, P.; Kumar, A.P.S.; Wang, X.R.; Piramanayagam, S.N. Emulation of synaptic plasticity on a cobalt-based synaptic transistor for neuromorphic computing. *ACS Appl. Mater. Interfaces* **2022**, *14*, 11864. [CrossRef]
21. Panzer, M.J.; Frisbie, C.D. Exploiting ionic coupling in electronic devices: Electrolyte-gated organic field-effect transistors. *Adv. Mater.* **2008**, *20*, 3177. [CrossRef]
22. Kim, S.H.; Hong, K.; Xie, W.; Lee, K.H.; Zhang, S.; Lodge, T.P.; Frisbie, C.D. Electrolyte-gated transistors for organic and printed electronics. *Adv. Mater.* **2013**, *25*, 1822. [CrossRef]
23. Yuan, H.; Shimotani, H.; Ye, J.; Yoon, S.; Aliah, H.; Tsukazaki, A.; Kawasaki, M.; Iwasa, Y. Electrostatic and electrochemical nature of liquid-gated electric-double-layer transistors based on oxide semiconductors. *J. Am. Chem. Soc.* **2010**, *132*, 18402. [CrossRef] [PubMed]
24. Liu, Y.; Wan, X.; Zhu, L.Q.; Shi, Y.; Wan, Q. Laterally coupled dual-gate oxide-based transistors on sodium alginate electrolytes. *IEEE Electron Dev. Lett.* **2014**, *35*, 1257. [CrossRef]
25. Wan, C.J.; Zhu, L.Q.; Wan, X.; Shi, Y.; Wan, Q. Organic/inorganic hybrid synaptic transistors gated by proton conducting methylcellulose films. *Appl. Phys. Lett.* **2016**, *108*, 043508. [CrossRef]
26. Liu, Y.H.; Qiang Zhu, L.; Shi, Y.; Wan, Q. Proton conducting sodium alginate electrolyte laterally coupled low-voltage oxide-based transistors. *Appl. Phys. Lett.* **2014**, *104*, 133504. [CrossRef]
27. Liu, Y.H.; Zhu, L.Q.; Feng, P.; Shi, Y.; Wan, Q. Freestanding artificial synapses based on laterally proton-coupled transistors on chitosan membranes. *Adv. Mater.* **2015**, *27*, 5599. [CrossRef] [PubMed]
28. Liu, R.; Zhu, L.Q.; Wang, W.; Hui, X.; Liu, Z.P.; Wan, Q. Biodegradable oxide synaptic transistors gated by a biopolymer electrolyte. *J. Mater. Chem. C* **2016**, *4*, 7744. [CrossRef]
29. Gao, W.T.; Zhu, L.Q.; Tao, J.; Wan, D.Y.; Xiao, H.; Yu, F. Dendrite integration mimicked on starch-based electrolyte-gated oxide dendrite transistors. *ACS Appl. Mater. Interfaces* **2018**, *10*, 40008. [CrossRef]
30. Guo, Y.B.; Zhu, L.Q.; Long, T.Y.; Wan, D.Y.; Ren, Z.Y. Bio-polysaccharide electrolyte gated photoelectric synergic coupled oxide neuromorphic transistor with Pavlovian activities. *J. Mater. Chem. C* **2020**, *8*, 2780–2789. [CrossRef]
31. He, Y.; Sun, J.; Qian, C.; Kong, L.A.; Jiang, J.; Yang, J.; Li, H.; Gao, Y. Solution-processed natural gelatin was used as a gate dielectric for the fabrication of oxide field-effect transistors. *Org. Electron.* **2016**, *38*, 357. [CrossRef]
32. Lai, D.; Li, E.; Yan, Y.; Liu, Y.; Zhong, J.; Lv, D.; Ke, Y.; Chen, H.; Guo, T. Gelatin-hydrogel based organic synaptic transistor. *Org. Electron.* **2019**, *75*, 105409. [CrossRef]
33. Wen, J.; Zhu, L.Q.; Qi, H.F.; Ren, Z.Y.; Wang, F.; Xiao, H. Brain-inspired biodegradable pectin based proton conductor gated electronic synapse. *Org. Electron.* **2020**, *82*, 105782. [CrossRef]
34. Kim, H.S.; Park, H.; Cho, W.J. Biocompatible casein electrolyte-based electric-double-layer for artificial synaptic transistors. *Nanomaterials* **2022**, *12*, 2596. [CrossRef] [PubMed]
35. Gedye, R.; Smith, F.; Westaway, K.; Ali, H.; Baldisera, L.; Laberge, L.; Rousell, J. The use of microwave ovens for rapid organic synthesis. *Tetrahedron Lett.* **1986**, *27*, 279–282. [CrossRef]
36. Surati, M.A.; Jauhari, S.; Desai, K.R. A brief review: Microwave assisted organic reaction. *Arch. Appl. Sci. Res.* **2012**, *4*, 645–661.
37. Nüchter, M.; Ondruschka, B.; Bonrath, W.; Gum, A. Microwave assisted synthesis–a critical technology overview. *Green Chem.* **2004**, *6*, 128–141. [CrossRef]
38. Kappe, C.O. High-speed combinatorial synthesis utilizing microwave irradiation. *Curr. Opin. Chem. Biol.* **2002**, *6*, 314. [CrossRef]
39. Ge, H.C.; Luo, D.K. Preparation of carboxymethyl chitosan in aqueous solution under microwave irradiation. *Carbohydr. Res.* **2005**, *340*, 1351. [CrossRef] [PubMed]

40. Yu, H.P.; Zhu, Y.J.; Lu, B.Q. Highly efficient and environmentally friendly microwave-assisted hydrothermal rapid synthesis of ultralong hydroxyapatite nanowires. *Ceram. Int.* **2018**, *44*, 12352. [CrossRef]
41. Nomanbhay, S.; Ong, M.Y. A review of microwave-assisted reactions for biodiesel production. *Bioengineering* **2017**, *4*, 57. [CrossRef]
42. Akbarian-Tefaghi, S.; Wiley, J.B. Microwave-assisted routes for rapid and efficient modification of layered perovskites. *Dalton Trans.* **2018**, *47*, 2917. [CrossRef]
43. Long, T.Y.; Zhu, L.Q.; Ren, Z.Y.; Guo, Y.B. Global modulatory heterosynaptic mechanisms in bio-polymer electrolyte gated oxide neuron transistors. *J. Phys. D Appl. Phys.* **2020**, *53*, 435105. [CrossRef]
44. Cordeschi, M.; Di Paola, L.; Marrelli, L.; Maschietti, M. Net proton charge of β- and κ-casein in concentrated aqueous electrolyte solutions. *Biophys. Chem.* **2003**, *103*, 77. [CrossRef] [PubMed]
45. Szyk-Warszyńska, L.; Raszka, K.; Warszyński, P. Interactions of casein and polypeptides in multilayer films studied by FTIR and molecular dynamics. *Polymers* **2019**, *11*, 920. [CrossRef] [PubMed]
46. Curley, D.M.; Kumosinski, T.F.; Unruh, J.J.; Farrell, H.M., Jr. Changes in the secondary structure of bovine casein by Fourier transform infrared spectroscopy: Effects of calcium and temperature. *J. Dairy Sci.* **1998**, *81*, 3154. [CrossRef] [PubMed]
47. Fadzallah, I.A.; Majid, S.R.; Careem, M.A.; Arof, A.K. A study on ionic interactions in chitosan–oxalic acid polymer electrolyte membranes. *J. Membr. Sci.* **2014**, *463*, 65. [CrossRef]
48. Qin, L.; Bi, J.R.; Li, D.M.; Dong, M.; Zhao, Z.Y.; Dong, X.P.; Zhou, D.Y.; Zhu, B.W. Unfolding/refolding study on collagen from sea cucumber based on 2D fourier transform infrared spectroscopy. *Molecules* **2016**, *21*, 1546. [CrossRef]
49. Dallinger, D.; Kappe, C.O. Microwave-assisted synthesis in water as solvent. *Chem. Rev.* **2007**, *107*, 2563–2591. [CrossRef]
50. Li, Y.; Cai, L.; Chen, H.; Liu, Z.; Zhang, X.; Li, J.; Shi, S.Q.; Li, J.; Gao, Q. Preparation of a high bonding performance soybean protein-based adhesive with low crosslinker addition via microwave chemistry. *Int. J. Biol. Macromol.* **2022**, *208*, 45–55. [CrossRef]
51. Fujimoto, T.; Awaga, K. Electric-double-layer field-effect transistors with ionic liquids. *Phys. Chem. Chem. Phys.* **2013**, *15*, 8983. [CrossRef]
52. Drachman, D.A. Do we have brain to spare? *Neurology* **2005**, *64*, 2004. [CrossRef]
53. Zucker, R.S.; Regehr, W.G. Short-term synaptic plasticity. *Annu. Rev. Physiol.* **2002**, *64*, 355. [CrossRef]
54. Guo, L.; Zhang, G.; Han, H.; Hu, Y.; Cheng, G. Light/electric modulated approach for logic functions and artificial synapse behaviors by flexible IGZO TFTs with low power consumption. *J. Phys. D Appl. Phys.* **2022**, *55*, 195108. [CrossRef]
55. Zhao, S.; Ni, Z.; Tan, H.; Wang, Y.; Jin, H.; Nie, T.; Xu, M.; Pi, X.; Yang, D. Electroluminescent synaptic devices with logic functions. *Nano Energy* **2018**, *54*, 383. [CrossRef]
56. Huang, J.; Chen, J.; Yu, R.; Zhou, Y.; Yang, Q.; Li, E.; Chen, Q.; Chen, H.; Guo, T. Tuning the synaptic behaviors of biocompatible synaptic transistor through ion-doping. *Org. Electron.* **2021**, *89*, 106019. [CrossRef]
57. Feng, P.; Xu, W.; Yang, Y.; Wan, X.; Shi, Y.; Wan, Q.; Zhao, J.; Cui, Z. Printed neuromorphic devices based on printed carbon nanotube thin-film transistors. *Adv. Funct. Mater.* **2017**, *27*, 1604447. [CrossRef]
58. Ohno, T.; Hasegawa, T.; Tsuruoka, T.; Terabe, K.; Gimzewski, J.K.; Aono, M. Short-term plasticity and long-term potentiation mimicked in single inorganic synapses. *Nat. Mater.* **2011**, *10*, 591. [CrossRef] [PubMed]
59. Bliss, T.V.; Collingridge, G.L. A synaptic model of memory: Long-term potentiation in the hippocampus. *Nature* **1993**, *361*, 31. [CrossRef]
60. Jang, J.W.; Park, S.; Burr, G.W.; Hwang, H.; Jeong, Y.H. Optimization of conductance change in Pr 1–x Ca x MnO 3-based synaptic devices for neuromorphic systems. *IEEE Electron Device Lett.* **2015**, *36*, 457–459. [CrossRef]
61. Chen, X.; Li, E.; Zhang, X.; Chen, Q.; Yu, R.; Ye, Y.; Chen, H.; Guo, T. Printed Organic Synaptic Transistor Array for One-to-Many Neural Response. *IEEE Electron Device Lett.* **2022**, *43*, 394–397. [CrossRef]
62. Wang, C.; Li, Y.; Wang, Y.; Xu, X.; Fu, M.; Liu, Y.; Lin, Z.; Ling, H.; Gkoupidenis, P.; Yi, M.; et al. Thin-film transistors for emerging neuromorphic electronics: Fundamentals, materials, and pattern recognition. *J. Mater. Chem. C* **2021**, *9*, 11464. [CrossRef]
63. Jang, J.W.; Park, S.; Jeong, Y.H.; Hwang, H. ReRAM-based synaptic device for neuromorphic computing. In Proceedings of the IEEE International Symposium on Circuits and Systems (ISCAS), Melbourne, VIC, Australia, 1–5 June 2014; Volume 2014, pp. 1054–1057. [CrossRef]

*Article*

# Composite Cement Materials Based on β-Tricalcium Phosphate, Calcium Sulfate, and a Mixture of Polyvinyl Alcohol and Polyvinylpyrrolidone Intended for Osteanagenesis

**Kseniya Stepanova [1], Daria Lytkina [1], Rustam Sadykov [1], Kseniya Shalygina [1], Toir Khojazoda [2], Rashidjon Mahmadbegov [2] and Irina Kurzina [1,*]**

[1] Faculty of Chemistry, Tomsk State University, 634050 Tomsk, Russia
[2] Faculty of Natural Science, Russian-Tajik (Slavonic) University, Dushanbe 734000, Tajikistan
* Correspondence: kurzina99@mail.ru; Tel.: +7-913-882-1028

**Abstract:** The primary purpose of the study, presented in this article, was to obtain a composite cement material intended for osteanagenesis. The β-tricalcium phosphate powder (β-TCP, β-$Ca_3(PO_4)_2$) was obtained by the liquid-phase method. Setting and hardening of the cement system were achieved by adding calcium sulfate hemihydrate (CSH, $CaSO_4 \cdot 1/2H_2O$). An aqueous solution of polyvinyl alcohol (PVA), polyvinylpyrrolidone (PVP), and a PVA/PVP mixture were used as a polymer component. The methods of capillary viscometry and Fourier-transform infrared spectroscopy (FTIR) revealed the formation of intermolecular hydrogen bonds between polymer components, which determines the good miscibility of polymers. The physicochemical properties of the synthesized materials were characterized by X-ray diffraction (XRD) and FTIR methods, and the added amount of polymers does not significantly influence the processes of phase formation and crystallization of the system. The size of crystallites CSD remained in the range of 32–36 nm, regardless of the ratio of polymer components. The influence of the composition of composites on their solubility was investigated. In view of the lower solubility of pure β-TCP, as compared to calcium sulfate dihydrate (CSD, $CaSO_4 \cdot 2H_2O$), the solubility of composite materials is determined to a greater degree by the CSD solubility. Complexometric titration showed that the interaction between PVA and PVP impeded the diffusion of calcium ions, and at a ratio of PVA to PVP of 1/1, the smallest exit of calcium ions from the system is observed. The cytotoxicity analysis results allowed us to establish the fact that the viability of human macrophages in the presence of the samples varied from 80% to 125% as compared to the control.

**Keywords:** bone cement; composite; β-TCP; PVA; PVP



**Citation:** Stepanova, K.; Lytkina, D.; Sadykov, R.; Shalygina, K.; Khojazoda, T.; Mahmadbegov, R.; Kurzina, I. Composite Cement Materials Based on β-Tricalcium Phosphate, Calcium Sulfate, and a Mixture of Polyvinyl Alcohol and Polyvinylpyrrolidone Intended for Osteanagenesis. *Polymers* **2023**, *15*, 210. https://doi.org/10.3390/polym15010210

Academic Editors: José Miguel Ferri, Vicent Fombuena Borràs and Miguel Fernando Aldás Carrasco

Received: 3 December 2022
Revised: 26 December 2022
Accepted: 29 December 2022
Published: 31 December 2022



## 1. Introduction

The development of new biomaterials for bone tissue restoration is a relevant topic in the field of regenerative medicine [1]. This trend is conditioned by the prevalence of bone tissue defects resulting from fractures, infectious and tumor processes, age-related osteoporosis, and some other pathological conditions of the bone. Bone tissue regeneration is a limited and time-consuming process; therefore, it is necessary to use bone substitutes [2]. In addition, serious disadvantages of natural bone grafts are known: An autograft requires additional surgical interventions and does not exclude the probability of donor site morbidity, and an allograft can cause tissue rejection and transmit infections [3].

Bone cement can be defined as a family of materials consisting of powder and liquid phases that, after mixing, form a plastic paste that can self-cure once implanted into the body. This means that the material is moldable, which ensures a perfect fit to the implant site and good bone-to-material contact, even in geometrically complex defects [4]. Recent advances in orthopedic surgery are associated with the use of minimally invasive surgical techniques. In this area, it is critical to have injectable materials available, and in this sense, bone cement can play a decisive role, provided that injectable cement is developed. An example is the performance of some minimally invasive surgical procedures, namely vertebroplasty and kyphoplasty, to repair vertebral compression fractures, in which bone cement is injected into the vertebral body [4].

Improving the properties of bone cement can be achieved by reinforcing the cement matrix with either particles or fibers [5–10]. In some cases, the addition of water-soluble polymers can be used to change the adhesion of the cement slurry or its rheological properties. For construction cement, there is evidence that the combined use of sulfates and polyvinyl alcohol adversely affects the strength of cement [11]. However, for bone cement, there are studies confirming that the addition of certain water-soluble polymers can be very effective in improving the adhesion of bone cement pastes [12,13].

Of the many biomaterials, synthetic calcium phosphates (CaP) have proven themselves in a number of studies because of their chemical similarity to the bone mineral component [14,15]. CaP is used in the form of ceramic products [16,17], coatings on metal implants [18], or in the form of cement pastes [19,20]. Bone cement is characterized by a paste-like consistency, set at the body temperature, which allows molding the implant at the site of the defect with a firm adherence to the surrounding tissues, as well as using a minimally invasive technique of a surgical operation by means of injection [21–26].

Among CaP, hydroxyapatite (HA, $Ca_{10}(PO_4)_6(OH)_2$) and tricalcium phosphate (TCP, $Ca_3(PO_4)_2$) have become preferable for bone tissue regeneration [3,27–30]. Based on published results [15,31], β-TCP tends to have higher osteoconductivity as compared to that of HA and α-TCP. HA has been reported to behave similarly to an inert implant [32], while β-TCP is moderately dissolved as a result of cell-mediated resorption [33–37]. The main disadvantage of α-TCP is that the resorption rate is too fast [38]. The research results in [39] show that β-TCP is an excellent material with respect to stem cell differentiation and in vivo osteoinduction compared to pure HA and an HA/TCP mixture.

We selected a β-TCP-based material for the study. The cement powder was obtained by adding calcium sulfate (CS), which has been used as an orthopedic biomaterial for many years [40–42]. In [43], the effect of the CS/β-TCP composite was studied as compared to that of β-TCP with the example of treating iliac bone defects occurring in dogs. Four months later, CS/β-TCP demonstrated a much greater contribution to the new bone formation as opposed to that of pure β-TCP. The addition of medical gypsum will allow for obtaining a material with selective resorption, where the calcium-phosphate matrix acts as a relatively stable framework and CS acts as a pore-former and setting component. The inclusion of biocompatible polymers in the cement formulation will allow the most complete imitation of the natural matrix of the bone tissue and improve the mechanical properties of the material. Polyvinyl alcohol (PVA) and polyvinylpyrrolidone (PVP) are suitable biodegradable and nontoxic polymers [44,45], and a mixture based on them improves the in vivo stability of polymers, that is, it allows for preventing the PVA suspension formation under physiological conditions [30].

## 2. Materials and Methods

β-TCP was synthesized by the liquid-phase method at a pH of 11 according to the following equations:

$$9Ca(NO_3)_2 + 6(NH_4)_2HPO_4 + 6NH_4OH \rightarrow Ca_9(HPO_4)(PO_4)_5OH + 18NH_4NO_3$$

$$Ca_9(HPO_4)(PO_4)_5OH \rightarrow 3\beta\text{-}Ca_3(PO_4)_2 + H_2O$$

Solutions of calcium nitrate Ca(NO3)2 1.2 M and ammonium hydrophosphate $(NH_4)_2HPO_4$ 0.8 M were prepared separately at room temperature (25 °C). $Ca(NO_3)_2$ and $(NH4)_2HPO_4$ were purchased from Sigma-Aldrich (MO, USA). The initial reagents were taken in equimolar volumes so that the molar ratio of Ca/P elements was 3/2. The $Ca(NO_3)_2$ solution was added dropwise to the $(NH_4)_2HPO_4$ solution while stirring on a magnetic stirrer. pH 11 was maintained by adding a concentrated ammonia solution (35%).

Next, to establish the effect of mixing time on the phase composition of the phosphate, half of the suspension continued to be stirred for 12 h; in the case of the second half of the suspension, the next synthesis step immediately began. The white precipitate was washed with distilled water and isopropyl alcohol. The suspension was subjected to vacuum filtration. The formed precipitate was dried at 80 °C for a day. The dried powder was crushed and then calcined at 900 °C for two hours to form the TCP beta phase. In the first part, after mixing, all the same operations were performed.

PVA (with an average molecular weight of 104,500 g/mol) was purchased from Sigma Aldrich (Saint Louis, MO, USA). PVP (with a molecular weight of 3500 g/mol) was provided by Acros Organics (Geel, Belgium). The PVA 5 wt% solution was prepared in distilled water at 90 °C during constant mixing with a magnetic stirrer for 1 h. The aqueous PVP 5 wt% solution was prepared at room temperature during constant stirring for 15 min. The PVA/PVP 5 wt% mixture was prepared by mixing freshly prepared solutions in the ratios of 1/3, 1/1, and 3/1.

The minimum amount of medical gypsum required for setting the β-TCP-based cement was selected experimentally. For this reason, calcium sulfate hemihydrate (CSH, $CaSO_4 \cdot 1/2H_2O$) was mixed in amounts of 50, 40, 30, 20, and 10 wt% with β-TCP and water in a powder-to-liquid ratio of 1/1. When adding 30 wt% of gypsum, the setting was noticeable, whereas, at 20 and 10 wt%, the consistency was still quite liquid. The following powder composition was chosen: 30 wt% of CSH and 70 wt% of β-TCP.

Composite cement materials were obtained by direct mechanical mixing of β-TCP/CSH powder components with the addition of pure water and the PVA, PVP, and PVA/PVP solutions in the ratios of 1/3, 1/1, and 3/1. The powder and liquid were mixed in a ratio of 1/1. The synthesis was carried out in Petri dishes. The setting time of the samples varied from 7 to 10 min. The obtained cement paste was left to harden completely.

The phase composition of the initial components and composites was determined using a MiniFlex 600 diffractometer (Rigaku). The survey photographing of the samples was performed under monochromatic CuKα (α = 1.5418 Å) radiation in the reflection angular range of 2θ = 5–100° in increments of 0.02°, with a voltage of 40 kV and a speed of 3 °/min. The phases were decoded and identified using the ICDD diffraction database (PDF-2/Release 2012 RDB). To calculate the coherent scattering region (CSR), the reflex value at its half-height was determined using the Gaussian function in Origin software.

The spectra of the surface layer of the samples were recorded by means of the Nicolet 6700 IR-Fourier spectrometer (Thermo Scientific). The spectra were recorded with a resolution of 4 cm−1 in the range of 400–4000 $cm^{-1}$, using the FTIR attachment.

The 5 wt% solution viscosity of the PVA/PVP mixture, depending on the mixing time of the solutions, was measured at room temperature using a capillary viscometer (d = 1.31 mm).

To determine the solubility, the samples were soaked in the physiological solution at 37 °C for 20 days. The total weight loss of the samples and the total concentration of calcium ions in the solution were recorded every 5 days.

The weight losses were calculated as a percentage using the formula:

$$\text{Losses, wt\%} = (mi - m)\cdot 100\%/mi$$

where mi is the initial weight of the sample and m is the sample weight after soaking the sample in distilled water.

The $Ca^{2+}$ concentration in the solution was determined by complexometric titration in the presence of Eriochrome Black T and the ammoniac buffer solution with a pH of 10.

The viability of immune system cells on the surface of the studied materials was assessed using the Alamar Blue indicator. The optical density was measured using a Tecan Infinite 200 microrider at wavelengths of 573 and 605 nm.

The data were analyzed using Student's *t*-test and presented as an average ± standard deviation. Three duplicate samples were used in all the experiments. The data probability was considered statistically significant for values of $p < 0.05$.

## 3. Results and Discussion

The X-ray diffraction (XRD) measurement results allowed us to establish that the sample (Figure 1a), left to mix for 12 h, contained two phases: A major β-TCP and a hydroxyapatite (HA) admixture. At the same time, the sample (Figure 1b) obtained without mixing contained the third phase: Beta-form calcium pyrophosphate (β-$Ca_2P_2O_7$). In this way, such mixing influences the phase formation process of calcium phosphate. The appearance of HA and β-$Ca_2P_2O_7$ admixtures may be caused by an inhomogeneous distribution of the components in the solution during the deposition of β-$Ca_3(PO_4)_2$ [43]. HA forms at a temperature of approximately 900 °C in air, most likely via HA dehydroxylation [46]. We selected sample (a) for further work.

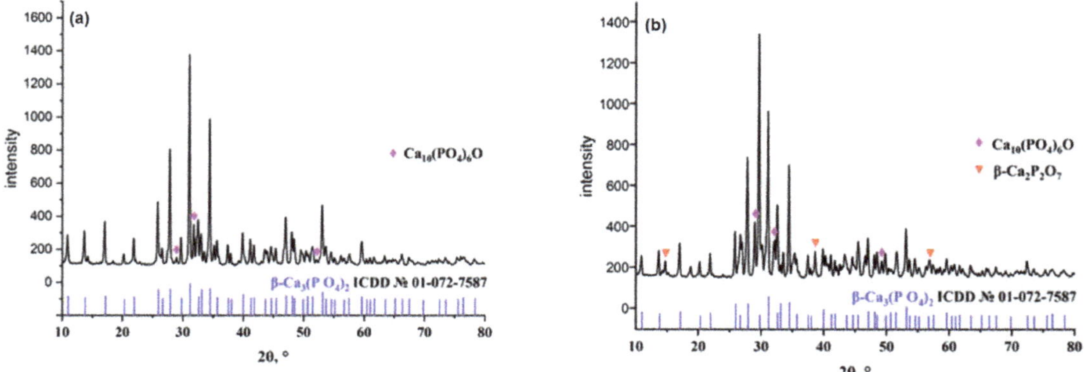


**Figure 1.** Sample left to mix for 12 h: (**a**) β-$Ca_3(PO_4)_2$, (**b**) $Ca_{10}(PO_4)_6O$.

According to the XRD data, at different ratios of the polymers, the composite samples are characterized by the same phase composition: β-$Ca_3(PO_4)_2$, $CaSO_4·2H_2O$, $Ca_{10}(PO_4)_6O$ (Figure 2).

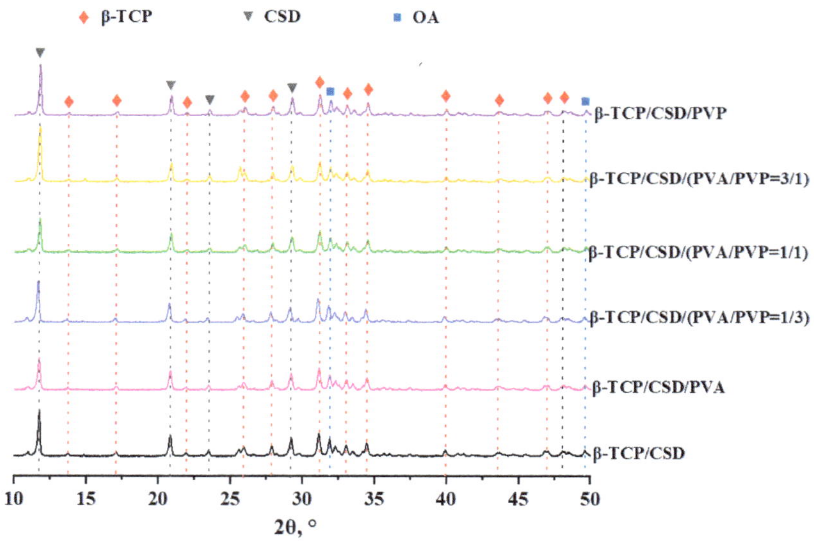


**Figure 2.** Diffraction patterns of composite cement materials.

The size of the crystallites in the composite cement samples varied in the range of 28–31 nm and 32–36 nm for the β-$Ca_3(PO_4)_2$ and $CaSO_4 \cdot 2H_2O$ phases, respectively (Table 1), i.e., there is a fairly uniform distribution of phases by the size of the crystallites. The CSR values for the β-$Ca_3(PO_4)_2$ phase, included in the initial CaP and part of the composites, essentially do not change. In the case of CSH, the size of the crystallites is somewhat smaller than that of CSD included in the composites, since hydration occurs when mixing with water. Then, calcium sulfate dihydrate microcrystals grow in size, intertwine, and grow together. As a result, the addition of polymers does not significantly influence the processes of phase formation and crystallization of the system.

**Table 1.** CSR values of cement and initial powder components.

| Sample | CSR Values for Different Phases, nm | | |
|---|---|---|---|
| | β-TCP | CSD | CSH |
| β-TCP | 30 | – | – |
| CSH | – | – | 28 |
| β-TCP/CSD | 31 | 36 | – |
| β-TCP/CSD/PVA | 30 | 34 | – |
| β-TCP/CSD/(PVA/PVP = 1/3) | 29 | 32 | – |
| β-TCP/CSD/(PVA/PVP = 1/1) | 28 | 33 | – |
| β-TCP/CSD/(PVA/PVP = 3/1) | 33 | 36 | – |
| β-TCP/CSD//PVP | 28 | 32 | – |

The Fourier-transform infrared spectroscopy (FTIR) method was used to study films of the 5% aqueous solutions of PVA, PVP, and PVA/PVP. Figure 3 shows the fundamental absorption bands of functional groups. The following specific bands were found for pure PVA [47,48]. A wide band of approximately 3343 $cm^{-1}$ corresponds to the valence vibration of the νOH hydroxyl group. Two bands are observed at 2930 $cm^{-1}$ and 2908 $cm^{-1}$, corresponding to the asymmetric and symmetric valence vibrations of $\nu CH_2$, respectively. The bands at 1418 $cm^{-1}$ and 828 $cm^{-1}$ are related to the scissor vibration of $\delta CH_2$ and stretching of the zigzag-shaped carbon base of νC-C, respectively. The absorption band at 1323 $cm^{-1}$ characterizes the deformation vibrations of δ(CH+OH). The wagging vibration of δCH is poorly observed at 1237 $cm^{-1}$. The absorption of approximately 1084 $cm^{-1}$ corresponds to the stretching of the acetyl groups of νC-O that are present in the main PVA chain. The weak band at 916 $cm^{-1}$ can be referred to as the rocking vibration of $\delta CH_2$.

The following specific bands were found for pure PVP [47–51]. The observed wide low-intensity band of approximately 3468 $cm^{-1}$ is related to the stretching of the OH-group of the water adsorbed on the polymer surface. The absorption band at 2951 $cm^{-1}$ corresponds to the asymmetrical vibration of $\nu CH_2$. The bands at 1421 $cm^{-1}$ and 840 $cm^{-1}$ correspond to the scissor vibrations of $\delta CH_2$ and the chain bending of νC-C, respectively. The bands at 1657 $cm^{-1}$ and 1284 $cm^{-1}$ can be referred to as valence vibrations of the νC=O carbonyl group and the stretching of the group of the νC-N pyrrolidone ring, respectively.

The participation of the hydroxyl group in the formation of intermolecular hydrogen bonds is manifested through the shift of the absorption band towards lower frequencies and a significant increase in its intensity [51]. The absorption band of the PVA hydroxyl group at 3343 $cm^{-1}$ shifts towards lower wave numbers at 3313 $cm^{-1}$ for the PVA/PVP mixture due to the formation of an intermolecular hydrogen bond between the PVP carbonyl group and the PVA hydroxyl group [40]. At the same time, a slight shift of the absorption band of the electron donor group by 10–20 $cm^{-1}$ is characteristic [51]; therefore, the PVP carbonyl band group is shifted from 1656 $cm^{-1}$ to 1649 $cm^{-1}$.

The IR spectra peculiarity of CaP is the presence of two doubled absorption bands. The first is detected at wave numbers such as ν = 900–1200 $cm^{-1}$ and the second at ν = 350–650 $cm^{-1}$ [52]. Figure 4 shows that the vibrations of the β-TCP phosphate groups are located at approximately 550 and 1020 $cm^{-1}$.

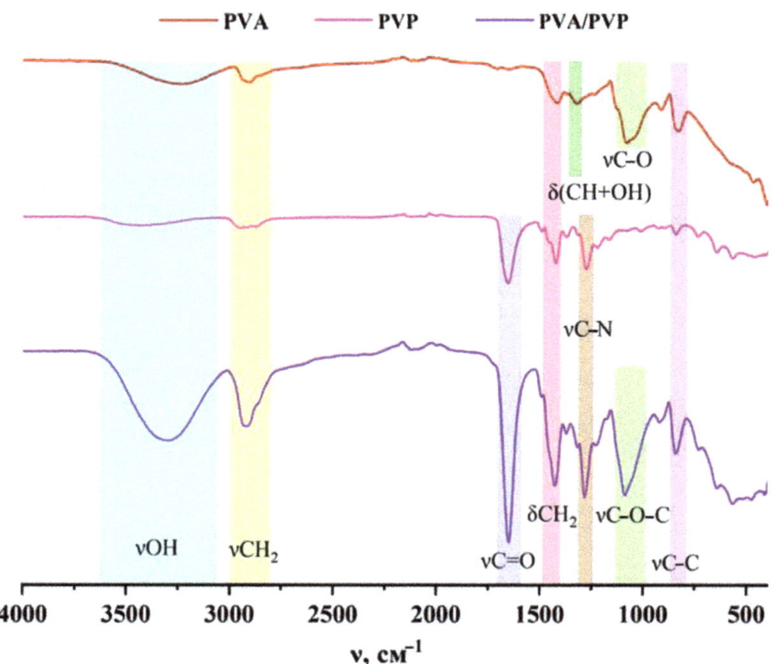


**Figure 3.** IR-spectrum of the 5% aqueous solutions of PVA, PVP, and PVA/PVP.

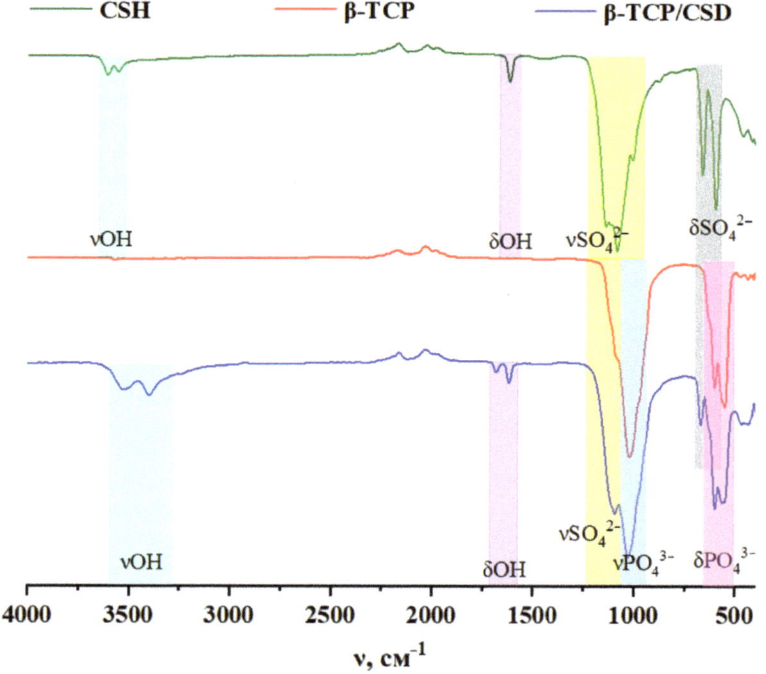


**Figure 4.** IR spectrum of CSH, β-TCP, and β-TCP/CSD water cement.

In the IR spectrum of β-TCP/CSD, there is a slight shift of the bands of valence vibrations of the $\nu PO_4^{3-}$ groups to a longer wavelength region due to the formation of a new phase on phosphate particles and anion geometry distortion as a result of nonsymmetric interactions in the crystal. Specific absorption bands are also observed at 3400 cm$^{-1}$ and 1620 cm$^{-1}$, which are related to the adsorbed water (bound OH-group), its valence, and deformation vibrations, respectively. The band at 3527 cm$^{-1}$ appears because of the valence vibrations of OH-ions of the crystallohydrate lattice (free OH-group) [30].

The sulfate ion can be identified by valence and deformation vibrations in the range of 980–1100 cm$^{-1}$ and 550–650 cm$^{-1}$, respectively [52]. In the case of $CaSO_4 \cdot 1/2H_2O$, there are fluctuations in the sulfate and hydroxide ions of the lattice, as shown in Figure 4.

The IR spectra of the composites (Figure 5) differ by the small shifts of the sulfate group band, which may be related to anion geometry distortion, inductive, and mesomeric effects as a result of the interaction with polymeric chains.

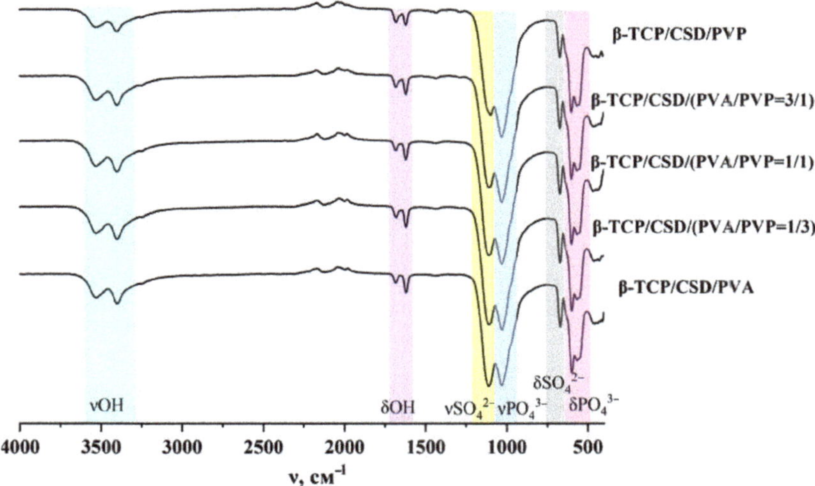


**Figure 5.** IR spectra of composite cement materials.

A graph was constructed based on the capillary viscometry results (Figure 6). In the first 30–40 min of mixing the polymers, the relative viscosity increases significantly. These results can be explained by the interaction between PVP and PVA, conditioned by the formation of intermolecular hydrogen bonds between the carbonyl group of the pyrrolidone ring and the side hydroxyl PVA group.

A series of experiments were conducted to study the solubility of the samples. The data in Table 2 demonstrate that the weight losses of cement materials increased along with an increase in the soaking time of the samples. In view of the lower solubility of pure β-TCP as compared to CSD, the solubility of the composite materials is more determined by the CSD solubility. In the case of the composites, weight losses varied in the range of 68 to 76 wt%, while in the occasion of pure β-TCP, the losses reached only 47 wt%.

In the case of the composites containing a mixture of polymers, the release of calcium ions is lower than that in the composites containing only one polymer (Figure 7). The interaction between PVA and PVP manifests itself in a solution viscosity increase, which hinders the diffusion of calcium ions.

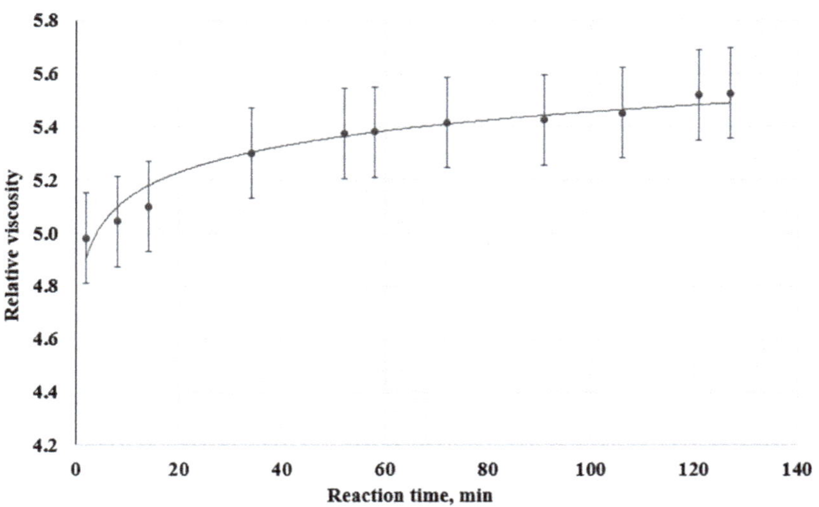


**Figure 6.** Relative viscosity measurement of the 5% solution of the PVA/PVP mixture.

**Table 2.** Weight losses of the composites and the initial β-TCP depending on the duration of sample soaking in the physiological solution.

| Number of Days | Weight Losses, wt% | | | | | | |
|---|---|---|---|---|---|---|---|
| | β-TCP | β-TCP/ CSD | β-TCP/ CSD/PVA | β-TCP/CSD/ (PVA/PVP = 1/3) | β-TCP/CSD/ (PVA/PVP = 1/1) | β-TCP/CSD/ (PVA/PVP = 3/1) | β-TCP/ CSD/PVP |
| 5 | 33 | 34 | 54 | 63 | 60 | 47 | 34 |
| 10 | 40 | 48 | 62 | 65 | 64 | 60 | 40 |
| 15 | 46 | 67 | 75 | 67 | 69 | 72 | 69 |
| 20 | 47 | 68 | 76 | 70 | 72 | 76 | 69 |

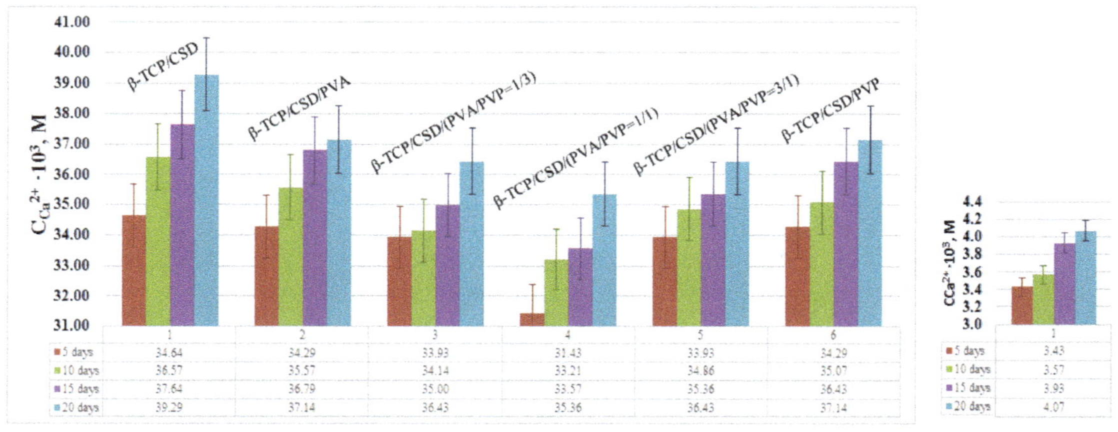


**Figure 7.** Release of $Ca^{2+}$ ions within 20 days for the composites and the initial β-TCP.

The studies conducted on human macrophages (Figure 8) have shown that, in the presence of the samples, most of the cells do not die. The viability of monocytes varies from 80% to 125% as compared to the control (100%).

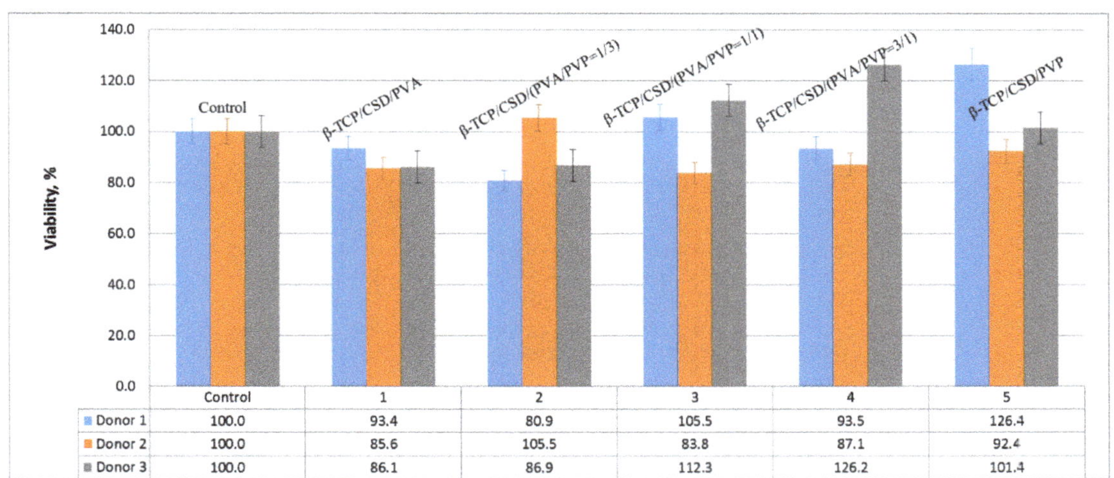


| | Control | 1 | 2 | 3 | 4 | 5 |
|---|---|---|---|---|---|---|
| Donor 1 | 100.0 | 93.4 | 80.9 | 105.5 | 93.5 | 126.4 |
| Donor 2 | 100.0 | 85.6 | 105.5 | 83.8 | 87.1 | 92.4 |
| Donor 3 | 100.0 | 86.1 | 86.9 | 112.3 | 126.2 | 101.4 |

**Figure 8.** Viability of human macrophages in the presence of composite cement.

Within the series, there is no unique dependence of cytotoxicity on different concentration ratios of polymers in the cement composition. Comparing the composites with individual polymers, it is possible to state that cell viability is higher in the presence of PVP, which confirms its hemocompatibility. It is also possible to observe that some donors have an individual reaction to the material: Donor 2 shows lower viability relative to other samples, which suggests that the viability of monocytes in the presence of materials is comparable to the control or higher. The fact that more cells survive in some donors than in controls indicates that the environment in the presence of materials is more favorable and more cells survive after incubation. From parallel studies [53] involving hydroxyapatite and polyvinyl alcohol-based gels, we know that PVA supplementation can significantly improve macrophage survival. In the case of bone cement, we observe the same effect. Thus, it can be assumed that the co-administration of polymers has a beneficial effect on the viability of macrophages, which shows the possibility of further development of these studies, which may consist of finding a narrower range of effective concentrations of the components, as well as establishing and improving mechanical characteristics.

## 4. Conclusions

1. In this work, a number of composite cement materials based on β-TCP and CSD were obtained and characterized with the addition of PVA and PVP solutions and a PVA/PVP mixture solution in the ratios of 1/3, 1/1, and 3/1.
2. The selected polymers mix well with each other and form a stable homogeneous solution. The formation of intermolecular hydrogen bonds between the polymer components was detected by capillary viscometry and IR-spectroscopy.
3. According to the XRD results, the added amount of polymers does not significantly influence the processes of phase formation and crystallization of the system.
4. It has been found that the solubility of the cement is determined to a greater extent by the CSD solubility in view of the lower solubility of pure β-TCP. The manifested interaction between the polymeric components hinders the diffusion and release of calcium ions.
5. The study of the cytotoxicity level of the composite cement materials has shown that most of the cells do not die in the presence of the samples. Consequently, these composites are very promising biomaterials for bone regeneration.

**Author Contributions:** Conceptualization, I.K.; methodology, I.K.; investigation, K.S. (Kseniya Stepanova), R.S. and K.S. (Kseniya Shalygina); writing—original draft preparation, K.S. (Kseniya Stepanova), D.L., I.K. and R.M.; writing—review and editing, D.L. and T.K. All authors have read and agreed to the published version of the manuscript.

**Funding:** Synthesis and studies of the phases and solubility of calcium phosphates were supported by grants from the President of the Russian Federation for state support of young Russian scientists-PhDs (MK-2182.2022.1.3). The study of IR spectra, degradation, and in vitro studies were supported by the Tomsk State University Development Programme (Priority-2030).

**Institutional Review Board Statement:** The study was conducted according to the guidelines of the Declaration of Helsinki and approved by the Ethics Committee of Tomsk State University (Protocol of the meeting of NR TSU Bioethics Commission, Protocol No 3 from 7 March 2022). Informed consent was obtained from all subjects involved in the study. Written informed consent has been obtained from the patient to publish this paper.

**Informed Consent Statement:** Not applicable.

**Data Availability Statement:** All data presented in this study are contained within the article.

**Conflicts of Interest:** The authors declare no conflict of interest.

## References

1. Pina, S.; Oliveira, J.M.; Reis, R.L. Natural-based nanocomposites for bone tissue engineering and regenerative medicine: A review. *Adv. Mater.* **2015**, *27*, 1143–1169. [CrossRef] [PubMed]
2. Sun, X.; Su, W.; Ma, X.; Zhang, H.; Sun, Z.; Li, X. Comparison of the osteogenic capability of rat bone mesenchymal stem cells on collagen, collagen/hydroxyapatite, hydroxyapatite and biphasic calcium phosphate. *Regen. Biomater.* **2018**, *5*, 93–103. [CrossRef]
3. Kucko, N.W.; Herber, R.P.; Leeuwenburgh, S.C.; Jansen, J.A. Calcium phosphate bioceramics and cements. In *Principles of Regenerative Medicine*, 3rd ed.; Academic Press: London, UK, 2019; pp. 591–611.
4. Ginebra, M.-P.; Montufar, E.B. Cements as bone repair materials. In *Bone Repair Biomaterials*; Woodhead Publishing: Cambridge, UK, 2019; pp. 233–271.
5. Guida, G.; Riccio, V.; Gatto, S.; Migliaresi, C.; Nicodemo, L.; Nicolais, L.; Palomba, C. *Biomaterials and Biomechanics*; Elsevier Science: Amsterdam, The Netherlands, 1984; Volume 19.
6. Henning, W.; Blencke, B.A.; Brömer, H.; Deutscher, K.K.; Gross, A.; Ege, W. Investigations with bioactivated polymethylmethacrylates. *J. Biomed. Mater. Res.* **1979**, *13*, 89–99. [CrossRef] [PubMed]
7. Murakami, A.; Behiri, J.C.; Bonfield, W. Rubber-modified bone cement. *J. Mater. Sci.* **1988**, *23*, 2029–2036. [CrossRef]
8. Liu, Y.K.; Park, J.B.; Njus, G.O.; Steinstra, D. Bone particle impregnated bone cement: In vitro studies. *J. Biomed. Mater. Res.* **1987**, *21*, 247–261. [CrossRef]
9. Henrich, D.E.; Cram, A.E.; Park, J.B.; Liu, Y.K.; Reddi, H. Inorganic bone and demineralized bone matrix impregnated bone cement: A preliminary in vivo study. *J. Biomed. Mater. Res.* **1993**, *27*, 277–280. [CrossRef]
10. Topoleski, L.D.T.; Ducheyne, P.; Cuckler, J.M. The fracture toughness of titanium fibre reinforced bone cement. *J. Biomed. Mater. Res.* **1992**, *26*, 1599–1617. [CrossRef]
11. Haider, M. Al-Baghdadi Experimental study on sulfate resistance of concrete with recycled aggregate modified with polyvinyl alcohol (PVA). *Case Stud. Constr. Mater.* **2021**, *14*, e00527.
12. Ishikawa, K.; Miyamoto, Y.; Kon, M.; Nagayama, M.; Asaoka, K. Non-decay type fast-setting calcium phosphate cement: Composite with sodium alginate. *Biomaterials* **1995**, *16*, 527–532. [CrossRef]
13. Khairoun, I.; Boltong, M.G.; Driessens, F.C.; Planell, J.A. Effect of calcium carbonate on clinical compliance of apatitic calcium phosphate bone cement. *J. Biomed. Mater. Res.* **1997**, *38*, 356–360. [CrossRef]
14. Cama, G. *Calcium Phosphate Cements for Bone Regeneration. Biomaterials for Bone Regeneration*; Woodhead Publishing: Cambridge, UK, 2014; pp. 3–25.
15. Tang, Z.; Li, X.; Tan, Y.; Fan, H.; Zhang, X. The material and biological characteristics of osteoinductive calcium phosphate ceramics. *Regen. Biomater.* **2018**, *5*, 43–59. [CrossRef] [PubMed]
16. Wang, Y.; Li, X.; Luo, Y.; Zhang, L.; Chen, H.; Min, L.; Chang, Q.; Zhou, Y.; Tu, C.; Zhu, X.; et al. Application of osteoinductive calcium phosphate ceramics in giant cell tumor of the sacrum: Report of six cases. *Regen. Biomater.* **2022**, *9*, rbac017. [CrossRef] [PubMed]
17. Tian, P.; Liu, X. Surface modification of biodegradable magnesium and its alloys for biomedical applications. *Regen. Biomater.* **2015**, *2*, 135–151. [CrossRef] [PubMed]
18. Meng, D.; Dong, L.; Yuan, Y.; Jiang, Q. In vitro and in vivo analysis of the biocompatibility of two novel and injectable calcium phosphate cements. *Regen. Biomater.* **2019**, *6*, 13–19. [CrossRef]
19. Wang, P.; Song, Y.; Weir, M.D.; Sun, J.; Zhao, L.; Simon, C.G.; Xu, H.H. A self-setting iPSMSC-alginate-calcium phosphate paste for bone tissue engineering. *Dent. Mater.* **2016**, *32*, 252–263. [CrossRef]

20. Low, K.L.; Tan, S.H.; Zein, S.H.S.; Roether, J.A.; Mourino, V.; Boccaccini, A.R. Calcium phosphate-based composites as injectable bone substitute materials. *J. Biomed. Mater. Res. Part B Appl. Biomater.* **2010**, *94*, 273–286. [CrossRef]
21. Ginebra, M.P.; Traykova, T.; Planell, J.A. Calcium phosphate cements as bone drug delivery systems: A review. *J. Control. Release* **2006**, *113*, 102–110. [CrossRef]
22. Zhang, J.; Liu, W.; Schnitzler, V.; Tancret, F.; Bouler, J.M. Calcium phosphate cements for bone substitution: Chemistry, handling and mechanical properties. *Acta Biomater.* **2014**, *10*, 1035–1049. [CrossRef]
23. O'Neill, R.; McCarthy, H.O.; Montufar, E.B.; Ginebra, M.P.; Wilson, D.I.; Lennon, A.; Dunne, N. Critical review: Injectability of calcium phosphate pastes and cements. *Acta Biomater.* **2017**, *50*, 1–19. [CrossRef]
24. Chow, L.C.; Eanes, E.D. Calcium phosphate cements. *Monogr. Oral Sci.* **2001**, *18*, 148–163.
25. Komath, M.; Varma, H.K. Development of a fully injectable calcium phosphate cement for orthopedic and dental applications. *Bull. Mater. Sci.* **2003**, *26*, 415–422. [CrossRef]
26. Ducheyne, P.; Hastings, G.W. *Metal and Ceramic Biomaterials*; CRC Press: Boca Raton, FL, USA, 1984; p. 31.
27. Sharpe, J.R.; Sammons, R.L.; Marquis, P.M. Effect of pH on protein adsorption to hydroxyapatite and tricalcium phosphate ceramics. *Biomaterials* **1997**, *18*, 471–476. [CrossRef] [PubMed]
28. Samavedi, S.; Whittington, A.R.; Goldstein, A.S. Calcium phosphate ceramics in bone tissue engineering: A review of properties and their influence on cell behavior. *Acta Biomater.* **2013**, *9*, 8037–8045. [CrossRef] [PubMed]
29. Uma Maheshwari, S.; Govindan, K.; Raja, M.; Raja, A.; Pravin, M.B.S.; Vasanth Kumar, S. Preliminary studies of PVA/PVP blends incorporated with HAp and β-TCP bone ceramic as template for hard tissue engineering. *Bio-Med. Mater. Eng.* **2017**, *28*, 401–415. [CrossRef] [PubMed]
30. Topsakal, A.; Ekren, N.; Kilic, O.; Oktar, F.N.; Mahirogullari, M.; Ozkan, O.; Sasmazel, H.T.; Turk, M.; Bogdan, L.M.; Stan, G.E.; et al. Synthesis and characterization of antibacterial drug loaded β-tricalcium phosphate powders for bone engineering applications. *J. Mater. Sci. Mater. Med.* **2020**, *31*, 1–17. [CrossRef]
31. Pan, Y.; Huang, J.L.; Shao, C.Y. Preparation of β-TCP with high thermal stability by solid reaction route. *J. Mater. Sci.* **2003**, *38*, 1049–1056. [CrossRef]
32. Bohner, M.; Santoni, B.L.G.; Döbelin, N. β-tricalcium phosphate for bone substitution: Synthesis and properties. *Acta Biomater.* **2020**, *113*, 23–41. [CrossRef]
33. Eggli, P.S.; Müller, W.; Schenk, R.K. Porous hydroxyapatite and tricalcium phosphate cylinders with two different pore size ranges implanted in the cancellous bone of rabbits. A comparative histomorphometric and histologic study of bony ingrowth and implant substitution. *Clin. Orthop. Relat. Res.* **1988**, *232*, 127–138. [CrossRef]
34. Yamada, S.; Heymann, D.; Bouler, J.M.; Daculsi, G. Osteoclastic resorption of calcium phosphate ceramics with different hydroxyapatite/β-tricalcium phosphate ratios. *Biomaterials* **1997**, *18*, 1037–1041. [CrossRef]
35. Kondo, N.; Ogose, A.; Tokunaga, K.; Ito, T.; Arai, K.; Kudo, N.; Inoue, H.; Irie, H.; Endo, N. Bone formation and resorption of highly purified β-tricalcium phosphate in the rat femoral condyle. *Biomaterials* **2005**, *26*, 5600–5608. [CrossRef]
36. Zerbo, I.R.; Bronckers, A.L.; De Lange, G.; Burger, E.H. Localization of osteogenic and osteoclastic cells in porous β-tricalcium phosphate particles used for human maxillary sinus floor elevation. *Biomaterials* **2005**, *26*, 1445–1451. [CrossRef] [PubMed]
37. Abadi, M.B.H.; Ghasemi, I.; Khavandi, A.; Shokrgozar, M.A.; Farokhi, M.; Homaeigohar, S.S.; Eslamifar, A. Synthesis of nano β-TCP and the effects on the mechanical and biological properties of β-TCP/HDPE/UHMWPE nanocomposites. *Polym. Compos.* **2010**, *31*, 1745–1753. [CrossRef]
38. Yuan, H.; Fernandes, H.; Habibovic, P.; De Boer, J.; Barradas, A.M.; De Ruiter, A.; Walsh, W.R.; Van Blitterswijk, C.A.; De Bruijn, J.D. Osteoinductive ceramics as a synthetic alternative to autologous bone grafting. *Proc. Natl. Acad. Sci. USA* **2010**, *107*, 13614–13619. [CrossRef] [PubMed]
39. Tay, B.K.; Patel, V.V.; Bradford, D.S. Calcium sulfate– and calcium phosphate–based bone substitutes: Mimicry of the mineral phase of bone. *Orthop. Clin. N. Am.* **1999**, *30*, 615–623. [CrossRef]
40. Huan, Z.; Chang, J. Self-setting properties and in vitro bioactivity of calcium sulfate hemihydrate–tricalcium silicate composite bone cements. *Acta Biomater.* **2007**, *3*, 952–960. [CrossRef]
41. Cheng, K.; Zhu, W.; Weng, X.; Zhang, L.; Liu, Y.; Han, C.; Xia, W. Injectable tricalcium phosphate/calcium sulfate granule enhances bone repair by reversible setting reaction. *Biochem. Biophys. Res. Commun.* **2021**, *557*, 151–158. [CrossRef]
42. Podaropoulos, L.; Veis, A.A.; Papadimitriou, S.; Alexandridis, C.; Kalyvas, D. Bone regeneration using b-tricalcium phosphate in a calcium sulfate matrix. *J. Oral Implantol.* **2009**, *35*, 28–36. [CrossRef]
43. Teodorescu, M.; Bercea, M. Poly(vinylpyrrolidone)—A versatile polymer for biomedical and beyond medical applications. *Polym.-Plast. Technol. Eng.* **2015**, *54*, 923–943. [CrossRef]
44. Muppalaneni, S.; Omidian, H. Polyvinyl alcohol in medicine and pharmacy: A perspective. *J. Dev. Drugs* **2013**, *2*, 1–5. [CrossRef]
45. Tonsuaadu, K.; Gross, K.A.; Plūduma, L.; Veiderma, M. A review on the thermal stability of calcium apatites. *J. Therm. Anal. Calorim.* **2012**, *110*, 647–659. [CrossRef]
46. Zidan, H.M.; Abdelrazek, E.M.; Abdelghany, A.M.; Tarabiah, A.E. Characterization and some physical studies of PVA/PVP filled with MWCNTs. *J. Mater. Res. Technol.* **2019**, *8*, 904–913. [CrossRef]
47. Omkaram, I.; Chakradhar, R.S.; Rao, J.L. EPR, optical, infrared and Raman studies of $VO^{2+}$ ions in polyvinylalcohol films. *Phys. B Condens. Matter* **2007**, *388*, 318–325. [CrossRef]

48. Abdelghany, A.M.; Abdelrazek, E.M.; Badr, S.I.; Morsi, M.A. Effect of gamma-irradiation on (PEO/PVP)/Au nanocomposite: Materials for electrochemical and optical applications. *Mater. Des.* **2016**, *97*, 532–543. [CrossRef]
49. Mondal, D.; Mollick, M.M.R.; Bhowmick, B.; Maity, D.; Bain, M.K.; Rana, D.; Mukhopadhyay, A.; Dana, K.; Chattopadhyay, D. Effect of poly (vinyl pyrrolidone) on the morphology and physical properties of poly (vinyl alcohol)/sodium montmorillonite nanocomposite films. *Prog. Nat. Sci. Mater. Int.* **2013**, *23*, 579–587. [CrossRef]
50. Elashmawi, I.S.; Baiet, H.A. Spectroscopic studies of hydroxyapatite in PVP/PVA polymeric matrix as biomaterial. *Curr. Appl. Phys.* **2012**, *12*, 141–146. [CrossRef]
51. Kupletskaya, N.B.; Kazitsyna, P.A. *Application of UV, IR, and NMR-Spectroscopy in Organic Chemistry*; High School: Moscow, Russia, 1971; pp. 214–234.
52. Nakamoto, K. *IR and Raman Spectra of Inorganic and Coordination Compounds*; High School: Moscow, Russia, 1991.
53. Sadykov, R.; Lytkina, D.; Stepanova, K.; Kurzina, I. Synthesis of Biocompatible Composite Material Based on Cryogels of Polyvinyl Alcohol and Calcium Phosphates. *Polymers* **2022**, *14*, 3420. [CrossRef]

*Review*

# Fabrication of Biodegradable and Biocompatible Functional Polymers for Anti-Infection and Augmenting Wound Repair

**Shuhua Deng [1,2], Anfu Chen [1,2,*], Weijia Chen [1], Jindi Lai [1], Yameng Pei [1], Jiahua Wen [1], Can Yang [3], Jiajun Luo [4], Jingjing Zhang [1], Caihong Lei [1], Swastina Nath Varma [2] and Chaozong Liu [2,*]**

[1] Guangdong Provincial Key Laboratory of Functional Soft Condensed Matter, School of Materials and Energy, Guangdong University of Technology, Guangzhou 510006, China
[2] Institute of Orthopaedics and Musculoskeletal Science, University College London, Royal National Orthopaedic Hospital, London HA4 4LP, UK
[3] Sino-German College of Intelligent Manufacturing, Shenzhen Technology University, Shenzhen 518118, China
[4] Centre for the Cellular Microenvironment, University of Glasgow, Glasgow G12 8LT, UK
* Correspondence: anfuchen@gdut.edu.cn (A.C.); chaozong.liu@ucl.ac.uk (C.L.)

**Abstract:** The problem of bacteria-induced infections threatens the lives of many patients. Meanwhile, the misuse of antibiotics has led to a significant increase in bacterial resistance. There are two main ways to alleviate the issue: one is to introduce antimicrobial agents to medical devices to get local drug releasing and alleviating systemic toxicity and resistance, and the other is to develop new antimicrobial methods to kill bacteria. New antimicrobial methods include cationic polymers, metal ions, hydrophobic structures to prevent bacterial adhesion, photothermal sterilization, new biocides, etc. Biodegradable biocompatible synthetic polymers have been widely used in the medical field. They are often used in tissue engineering scaffolds as well as wound dressings, where bacterial infections in these medical devices can be serious or even fatal. However, such materials usually do not have inherent antimicrobial properties. They can be used as carriers for drug delivery or compounded with other antimicrobial materials to achieve antimicrobial effects. This review focuses on the antimicrobial behavior, preparation methods, and biocompatibility testing of biodegradable biocompatible synthetic polymers. Degradable biocompatible natural polymers with antimicrobial properties are also briefly described. Finally, the medical applications of these polymeric materials are presented.

**Keywords:** biodegradable; biocompatible; antibacterial; synthetic polymers; natural polymers; medical applications

**Citation:** Deng, S.; Chen, A.; Chen, W.; Lai, J.; Pei, Y.; Wen, J.; Yang, C.; Luo, J.; Zhang, J.; Lei, C.; et al. Fabrication of Biodegradable and Biocompatible Functional Polymers for Anti-Infection and Augmenting Wound Repair. *Polymers* **2023**, *15*, 120. https://doi.org/10.3390/polym15010120

Academic Editors: José Miguel Ferri, Vicent Fombuena Borràs and Miguel Fernando Aldás Carrasco

Received: 24 November 2022
Revised: 20 December 2022
Accepted: 24 December 2022
Published: 28 December 2022

## 1. Introduction

Infections, particularly implant-associated and hospital-acquired infections, present a global medical risk and cause serious injuries and deaths every year [1]. More than 23.5 million immunocompromised patients require immunosuppressive drugs, such as acquired immunodeficiency syndrome (AIDS), rheumatoid arthritis, and organ transplant patients. These drugs suppress the body's normal immune response, making patients more susceptible to fungal and bacterial infections. For example, there are 300,000 HIV-related infections and 10,000 deaths in the United States every year [2]. In addition, exposure to other common pathogens such as *Pseudomonas aeruginosa* (*P. aeruginosa*), *Escherichia coli* (*E. coli*), *Salmonella*, and *Staphylococcus aureus* (*S. aureus*) can escalate into threatening infections that may lead to sepsis-related death [3]. It has been shown that the implant-associated infections are caused by incomplete preoperative sterilization procedures, the use of non-standard protocols during surgery, and hematogenous sources to the implant surface after surgery, resulting in financial hardship or even death for patients [4,5]. Infections caused by implants are usually attributed to bacteria, and the biofilm facilitates the bacteria adhering to the implant from the host's defense system and bactericidal agents [6].

Further, opportunistic pathogen infections will pose a serious threat in organ transplants, prosthetic grafts, and tissue-engineered structures [7]. To solve this problem, some researchers have added antibacterial substances, such as silver and antibiotics, to the scaffold [8]. The overuse of antibiotics has led to the emergence of drug-resistant strains and a growing problem of bacterial resistance [9]. The scaffold may also lose its antimicrobial activity after the complete release of the antimicrobial substances [8]. New antimicrobial treatments are therefore essential for the control of hospital-acquired infections arising as a complication of surgery and post-surgery [10].

To alleviate the problems caused by antibiotic resistance, efforts have been made to develop polymer materials with antimicrobial properties featuring the ability to prevent bacteria from adhering or even to disrupt bacterial cell membranes [11]. In addition, the structure and shape of synthetic polymers can be engineered to satisfy medical requirements [12]. Amongst these, biodegradable and biocompatible polymers have attracted a great deal of attention. Typical antibacterial polymers stand out in terms of their chemical stability. Moreover, the polymer composition needs to be biocompatible and degradable in vivo and in vitro [13]. The biodegradability of polymers can be achieved by the presence or incorporation of unstable chemical linkages (e.g., ester, amide, and carbonate bonds) [14]. Biodegradable polymers are usually divided into two categories: synthetic polymers and natural polymers [15]. Biocompatible synthetic polymers are often biodegradable, and the products of degradation can be absorbed by the human body. Natural polymers are metabolized into metabolites that are easily cleared by the kidneys or can be used by the body [16]. Common synthetic polymers include polyglycolic acid (PGA), polylactic acid (PLA), poly(lactide-co-glycolide) (PLGA) and polycaprolactone (PCL). Some natural polymers, such as cellulose, gelatin, and chitosan, are also common degradable biocompatible polymers [17].

A variety of antimicrobial methods have been developed on degradable biocompatible polymers. For example, as an environmentally friendly class of antimicrobial agents, the cationic polymer is less likely to cause resistance because it kills bacteria primarily by disrupting cell membranes. Currently, cationic polymerization has been used in polymers such as waterborne polyurethane [18,19], and natural cationic polymers such as chitosan have been extensively studied for antibacterial purposes [20]. In addition, metal ions are also common additions to polymers for antibacterial purposes, such as $Ag^+$, $Zn^+$, $Cu^+$, and so on. Degradable polymers achieve antimicrobial effect by slowly releasing metal ions. Nonionic antimicrobial polymers have also been extensively studied recently. Instead of interacting with bacteria through ions, nonionic polymers produce antimicrobial effects by adding some nonionic substances, such as curcumin, piperine, indole, aspirin, and so on [21]. These non-ionic antimicrobial agents may interact with bacteria due to hydrogen bonding, hydrophobic, dipole, etc. Although many non-antibiotics using antimicrobial methods have been investigated, the use of antibiotics is still a convenient and effective method for antimicrobial resistance. Degradable polymers not only allow the slow local release of antibiotics, but also reduce bacterial resistance compared to systemic antibiotic use. Therefore, the controlled release of drugs on degradable polymers is also a widely investigated topic.

For chronic osteomyelitis [22], bone tissue engineering [23], and wound healing [24], long-term antimicrobial therapy or avoidance of bacterial infection at the injury site is required. However, the long-term systemic use of antibiotics is very expensive and often causes systemic adverse effects [25]. Therefore, the focus of antibiotic therapy is to achieve high concentrations of antibiotics at the site of infection using a local delivery system to avoid the side effects associated with systemic administration. If a non-degradable drug delivery system is used, a secondary procedure is required for removal [26]. Therefore, efforts have been made to develop drug-delivery systems based on biodegradable biopolymers. The rate of drug release can also be regulated by adjusting the degradation rate of the biodegradable polymer.

This paper aims to review the antimicrobial behaviors and applications of biodegradable and biocompatible polymers in novel antimicrobial approaches developed for the alleviation of antibiotic resistance. These polymers hold great promise for applications, such as tissue engineering, drug delivery, and wound healing.

## 2. Approaches to Enhance Antibacterial Potential

Typically, two main methods are used to resist bacterial infection; the first method is the construction of antifouling biocompatible surfaces through hydrophilic polymers or superhydrophobic structures for resistance to bacterial adhesion [27]. Hydrophobic structures also can be inserted into the cell membrane, causing the rupturing of cell membrane, releasing cytoplasmic components (such as DNA and RNA), leading to the death of bacteria. Figure 1 shows the prevention of bacterial adhesion on a microstructured surface and the destruction of the bacterial cell membrane by surface microstructure. However, once bacteria are attached to the polymer surfaces, it will be difficult to inactivate the bacteria due to the formation of biofilms [28].

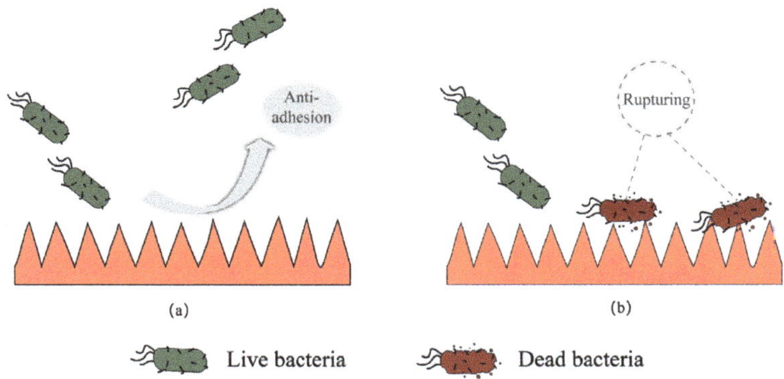


**Figure 1.** (**a**) Prevention of bacterial adhesion on microstructured surface; (**b**) destruction of bacterial cell membrane by surface microstructure.

Yi et al. [29] developed a way to sterilize ZnO nanorods. The dead bacteria attached to the nanorods can be removed in an aqueous solution, thus restoring the antimicrobial properties of the microstructure. Wang et al. [30] prepared Zn-liganded polydopamine coatings that have surface microstructures and are superhydrophilic. Fajstavr et al. [31] used a high-energy excimer laser to treat the polydimethylsiloxane (PDMS) surface and obtained ripple-like surface nanopatterns. The nanopattern had a significant bactericidal effect. Li et al. [32] synthesized structurally homogeneous small-diameter mesoporous nanospheres that can convert hydrogen peroxide ($H_2O_2$) into reactive oxygen species for bactericidal effects. Cheng et al. [33] proposed activatable nanostructures. The structure provides an acoustodynamic sterilization method and enhances the proliferation and differentiation of osteoblasts. Nanostructures often produce antimicrobial effects through acoustodynamic or photodynamic catalysis or kill bacteria by physical puncture. Nanostructures can also be combined with other antimicrobial agents to produce a synergistic antimicrobial effect.

The second method uses antibiotics to actively kill bacteria and has a high bactericidal efficiency [27]. Antibiotics are a great invention. Since the discovery of penicillin, antibiotics have saved countless lives [34]. However, with the misuse of antibiotics, bacterial resistance has caused increasingly serious problems [35]. Despite ongoing efforts to find new drugs or alternatives to antibiotics, no antibiotic or its alternative have been clinically approved in the past 30 years [36]. To address the drawbacks of these two methods, some antimicrobial substances can be added to the hydrophobic surface to enhance the antimicrobial activity. Spasova et al. [37] prepared the superhydrophobic nanofiber materials of polyvinylidene fluoride and polyvinylidene fluoride-co-hexafluoropropylene by electrospinning technique and incorporated with ZnO nanoparticles. The addition of ZnO not only increased the surface hydrophobicity, but also imparted antibacterial properties to the materials. Mirzadeh et al. [38] prepared PU/$TiO_2$ nanoparticles/graphene nanosheet composite films by the spraying technique. Non-solvent-induced phase separation increased the surface roughness and hydrophobicity of the films. The $TiO_2$ was incorporated as an antimicrobial agent. Antibiotics can also be added to biodegradable materials to achieve local drug delivery and reduce systemic toxicity as well as drug resistance. Antibiotic-containing degradable polymer nanosystems (e.g., polymer vesicles and micelles) are used to deliver antibiotics [13]. One or more antibiotics are loaded into biodegradable polymer scaffolds for implantation into the body to accomplish the local release of antibiotics [39]. Studies have shown that bacterial cells carry more negative charges than mammalian cells do [40]. Therefore, this feature can also be exploited to explore new antibacterial methods, which can be bactericidal without affecting the normal growth of the cells. Cationic polymeric antimicrobial agents, such as polyhexamethylene biguanide and chitosan take advantage of this property. The positively charged cationic polymers rely on electrostatic attraction to bacterial cell membranes, which leads to cell membrane cleavage and bacterial death. Metal ions are also common bactericides, taking $Ag^+$ as an example. The $Ag^+$ can rely on electrostatic adsorption on the cell membrane, and peptidoglycan reaction with the cell wall, further penetrating the cell, leading to cell membrane cleavage. The $Ag^+$ will react with enzymes and proteins after entering the cell, leading to the loss of normal cell function. The $Ag^+$ also produces reactive oxygen radicals, ROSs, which bind to DNA and inhibit DNA, RNA, protein synthesis, thus leading to bacterial inactivation.

## 3. Synthetic Biodegradable Polymers for Antibacterial Applications

Synthetic biodegradable polymers are considered biocompatible and highly safe and have found numerous applications in the biomedical field [41]. Due to the degradable nature of polymer implants, they can be removed at the end of the functional life of the implant without surgical intervention. In tissue engineering, synthetic biodegradable polymers usually provide suitable mechanical properties and facilitate cell proliferation and differentiation, making it an excellent material for scaffolds [42]. Synthetic polymers have the advantage of being easy to process and mold, so there are more possibilities in terms of usage properties. Synthetic biodegradable polymers usually have no inherent antimicrobial ability. Therefore, the antimicrobial activity of synthetic polymers can be effectively improved by mixing with antimicrobial substances. Some inorganic substances can also be incorporated to endow them with antimicrobial ability. Different synthetic polymers have their own characteristics, such as degradation rate and hardness. To maximize their advantages, different polymers are often added to the material system to form composites or blends in previous studies. At the same time, these polymers have great potential as drug carriers, due in large part to their biodegradability, biocompatibility and the possibility of developing sustained/controlled/pulsed release and targeted drug delivery [43].

### *3.1. Polylactic Acid (PLA)*

The PLA is one of the most widely used polymer materials for polymeric scaffolds in tissue engineering application because it has natural advantages of good mechanical properties, degradability, biocompatibility, and low cost [44]. However, the PLA has some limitations. For example, its hydrophobicity may undermine its biocompatibility. Human cells may be damaged if they are exposed to lactic acid, a degradation product of the PLA, for long periods of time [45]. Different parameters may result in the PLA exhibiting different mechanical properties, such as crystallinity, molecular weight, and processing [46]. The properties of the PLA can also be changed by tuning the material formulation, such as adding plasticizers, preparing blends, or composites with other materials. Thus, some chemical constituents are normally added into the PLA to improve various properties [47].

Llorens et al. [48] prepared the PLA nanofibers equipped with polybiguanide (PHMB) and with an average diameter between 560 and 630 nm by using the electrostatic spinning method. The PHMB-loaded PLA scaffolds have antimicrobial properties. On the one hand, the loading of PHMB increases the hydrophobicity of the scaffold, making it difficult for bacteria to adhere. On the other hand, the PHMB, as a cationic oligomer, has strong antibacterial activity. Scaffolds prepared by the electrostatic spinning method have porous structures that facilitate cell adhesion and proliferation. However, high concentrations of the PHMB are toxic to human cells, so it is important to control the concentration of the drug in the scaffolds and the rate of scaffold degradation. The experimental results showed that the concentration of PHMB did not affect the growth of cells when it was lower than 1.5 wt%. The PHMB-loaded PLA scaffolds showed biocompatibility in terms of the adhesion and proliferation of fibroblast and epithelial cell lines. Moreover, the slow and controlled drug release allowed the addition of PHMB in the scaffold at higher than safe concentrations.

Han et al. [45] proposed an innovative method for the preparation of PLA/chitosan (CS) composite films by using non-solvent induce phase separation (NIPS). The PLA/CS films prepared using the NIPS are more hydrophilic than those prepared using the casting method, which will result in better biocompatibility and a faster degradation rate of the films. The films prepared using the NIPS can also adjust the pore size of the porous structure to regulate the degradation rate. The PLA-based films were tested for their antibacterial activity against *E. coli*. The results showed that the antibacterial ability of the pure PLA film is not obvious due to the polymer structure of the PLA. The addition of CS greatly increases the antimicrobial activity, indicating that the CS is the main antimicrobial component in the film.

The porous film was experimentally demonstrated to be degradable, self-supporting, antibacterial, and transparent. The degradation of CS releases small alkaline molecules, which can neutralize the acidic degradation products of PLA and avoid some inflammation, resulting from acidic degradation to a certain extent.

Douglass et al. [49] combined the nitric oxide (NO) donor S-nitrosoglutathione (GSNO) with polyhydroxybutyrate (PHB) and the fiber-grade PLA for the manufacture of antimicrobial NO releasing nanofibers. In this study, the PLA was first mixed with the PHB in solution, stirred for 2 h, and then the GSNO was added into the solution. The fibers are formed by electrostatic spinning after 4 h. The fiber grade PLA synthesized with different length–diameter ratios overcomes some of its disadvantages, such as poor heat resistance and fragility. When the PLA was mixed with the PHB by a ratio of 3:1, the blends showed the best plasticity and maintained a certain tensile strength. The GSNO was added as a source of NO release because of the great role of NO in regulating antimicrobial behavior in blood vessels. Compared to PLA/PHB fibers, GSNO-containing fibers showed a significant reduction in colony forming units (CFU) measurements of *S. aureus* after both 2 and 24 h of exposure. After 2 h exposure, the bacterial survival rate of fibers containing GSNO was significantly lower than that of PLA/PHB fibers. The bacterial cell membrane on PLA/PHB remains intact, while that on PLA/PHB + 20 wt% GSNO is disrupted. This is due to the release of NO that disrupts the cell membrane.

Although the release of NO helps to kill bacteria, high concentrations of NO will be toxic to mammalian cells. Therefore, mouse fibroblasts were exposed to a 24 h leachate of fibers. The result shows that none of the fibers caused a significant decrease in cell viability compared to cells that were not exposed to the leachate. This demonstrates that nanofibers are not cytotoxic to mammalian cells. Therefore, NO-releasing nanofibers are a promising coating for blood-contacting medical devices.

Sharif et al. [50] synthesized composite scaffolds of the PLA/PCL blended with nano-hydroxyapatite (n(HA)) and cefixime-β cyclodextrin (Cfx-βCD) by electrospinning. In this study, the PLA and the PCL were mixed with the HA, Cfx, and Cfx-βCD to form composites film by electrospinning. Their antibacterial ability and effects on cell growth were compared. The antibacterial activity of the membrane was determined by using the *S. aureus* strain. After adding Cfx, the number of bacteria on the membrane decreased significantly within 24 h. The number of bacteria on the membrane continued to decrease when HA was added to the composite membrane. The best antibacterial activity was PLA-PCL-βCD-Cfx (PPH-βCD-Cfx) composite membrane.

The mouse pre-osteblast cell line (MC3T3) was cultured on the membrane to evaluate the cell viability on different membranes. The MC3T3 can attach and proliferate on all three membranes.

The addition of HA improved the osteoconduction of the composite. The βCD realizes the control of drug release as a carrier of antibiotics. The membranes were shown to have good antibacterial properties and to promote cell proliferation and attachment.

### 3.2. *Polycaprolactone (PCL)*

The PCL is a medical synthetic polymer with biocompatible and biodegradable properties. Due to its flexibility, low density, and easy processing, it has been widely used in tissue engineering scaffolds. In addition, it has good mechanical strength, rigidity, and heat resistance [51]. Its breakdown products form naturally occurring metabolites, which are readily metabolized by the body and eliminated without toxicity [52].

In previous research study, natural antimicrobial compounds, such as curcumin, piperine, eugenol, and rutin, were loaded into electrospun nanofibers based on the PCL [53]. A wound dressing was prepared by electrospinning. The SEM images showed that the nanofibers prepared by electrospinning were more uniform after adding curcumin and piperine. The diameter of the fiber can vary with the change in curcumin concentration. Novel three-component systems of curcumin–piperonin–eugenol (PCPiEu) and curcumin–piperonin–rutin (PCPiR) were designed and prepared. The growth of *S. aureus* in the presence of different wound dressings was studied. The growth of bacteria on pure PCL is tremendous, and so pure PCL has no antibacterial activity. With either the addition of curcumin or piperine, the number of bacteria was greatly reduced. In addition, the PC has almost no antibacterial activity against Enterococcus faecalis (Gram-negative), while both PCPiEu and PCPiR have a killing rate of more than 95%.

The MTT test was performed on different wound dressings using human fibroblasts to evaluate the cytotoxicity of wound dressings. The experiment proved that the pure PCL exhibited the highest amount of cell proliferation, which proves that the PCL has high biocompatibility. The result shows that the cell viability of piperine based on PCL samples is the lowest, which proves that piperine has cytotoxicity. The cell viability of PCR samples was higher than 100%, indicating that it was favorable for cell growth. In the three-component samples, the cell viability of PCPiEu and PCPiR was more than 90%. It seems that a better effect can be achieved by adjusting the amount of these natural compounds in the sample.

In general, the three-component wound dressing obtained good results in both the antibacterial test and the cytotoxicity test. In particular, it showed good antibacterial activity against Gram-negative bacteria. Although the mechanical properties of the system decreased compared with the PCL, the three-component wound dressing showed broad-spectrum antibacterial activity.

In another study, graphene (GP), bioglass, and zinc-doped bioglass were added to the PCL filaments, and their antimicrobial activity was analyzed comparatively. Materials for research were produced using the 3D-printing technique. The experimental results showed that the addition of a small amount of GP (0.5%) to PCL filaments resulted in a significant increase in antimicrobial activity compared to the pure PCL. This is because the structure of the GP may cause damage to the cell membranes of microbes and thus eliminate them [54].

Recently, PCL nanofibers containing Atropa belladonna were fabricated using the electrospinning technique [55]. The fruits, roots, and stems of belladonna are used in the treatment of many diseases and have strong antioxidant and anticancer properties. In this study, Atropa belladonna extract was used to encapsulate Ag nanoparticles (AgNPs).

This study concludes that both AgNPs and eAgNPs improved the antibacterial activity of PCL nanofibers against Gram-negative and Gram-positive bacteria. In addition, the cell viability of the PCL doped with eAgNPs was higher compared to the neat PCL. The main reasons for the greater cell viability of the PCL doped with eAgNPs may be that it is more hydrophilic than the pure PCL, the toxicity of AgNPs is reduced by coating on the surface of the nanoparticles, and the positive effect of free radicals in the Atropa belladonna structure on cell proliferation.

Felice et al. [56] synthesized PCL scaffolds compounded with hydroxyapatite (HA) and different concentrations of zinc oxide (ZnO) for bone tissue engineering by electrospinning techniques. The ZnO is an inorganic material with osteoinductive and osteoconductive properties. It has been reported that $Zn^{2+}$ can induce osteoblast differentiation. In addition, ZnO is considered an effective antimicrobial agent against broad-spectrum microorganisms. The addition of HA is beneficial to increase the osteoconductivity of the scaffold and can shorten the degradation time of the PCL. To assess the antimicrobial activity of the scaffolds, a series of samples were immersed in the PBS at 37 °C for 0 and 30 days for in vitro degradation, respectively. Sterile samples were immersed in broth containing *S. aureus* for 18 h at 35 ± 2 °C, and the surviving CFUs were counted. The surviving CFU of samples after 0 and 30 days of degradation were compared after 18 h of bacterial incubation, respectively. The results showed that the presence of ZnO in the samples degraded for 0 days led to a reduction in the initial bacterial load compared with the negative control group. The higher the concentration of ZnO, the higher the antibacterial activity. On the PCL scaffold containing 6% ZnO, this corresponds to a 96% reduction in bacteria. On the PCL scaffold containing the HA and 1% ZnO, an almost 99% reduction of the initial bacterial load was achieved. Notably, on the scaffold containing the HA, the antibacterial activity decreased instead with increasing the ZnO concentration.

Human fetal osteoblast cell line (HFOb) proliferation was assessed after 3, 7, and 14 days of incubation on PCL, PCL:HA, and PCL:HA:ZnO nanofiber scaffolds. On the PCL and PCL:HA:ZnO 1%, cell proliferation increased exponentially with time. In contrast, cell proliferation in the other samples was constant from day 3 or from day 7. In addition, higher concentrations of ZnO may lead to reduced cell proliferation.

These scaffolds have been shown to have an antibacterial effect against *S. aureus*. This activity rises with increasing levels of the ZnO. However, high concentrations of the ZnO may reduce cell proliferation. Therefore, low-concentration ZnO scaffolds may be promising regenerative medicine products with antibacterial ability.

### 3.3. Polyglycolic Acid (PGA)

The PGA is a semi-crystalline synthetic polymer with good biocompatibility and biodegradability. Once it degrades, the non-crystalline part first will degrade to glycolic acid which can be readily metabolized by the body; the crystalline part then will degrade to harmless water and carbon dioxide [57]. Like the PLA, it produces acidic degradation products that may trigger inflammation. The PGA has high mechanical strength and high crystallinity. However, the PGA has poor toughness and a relatively high price. Therefore, the PGA is always blended with other polymer materials.

Shuai et al. [58] prepared polymer scaffolds by laser sintering. The PGA solution and the PLLA solution were mixed, stirred, and ultrasonically dispersed. Meanwhile, the graphene oxide (GO) solution and the nano Ag solution were mixed, stirred, and ultrasonically dispersed. Finally, the two solutions were mixed together, filtered, and dried to obtain powders. The powders were sintered layer by layer to form 3D scaffolds. The SEM images show that AgNPs and GO are evenly distributed. Both the GO and nano Ag are easy to form agglomeration, so uniform dispersion is very important for the antibacterial activity of polymer scaffolds. The *E. coli* suspension was placed together with scaffolds with different GO and Ag ratios for 24 h, and the antibacterial activity of the scaffolds was evaluated by turbidimetry. Only when GO exists, the antibacterial effect is not obvious. The antibacterial effect was significantly improved only when Ag was present. When GO and Ag were simultaneously present, the antibacterial effect was further increased.

The GO-Ag nanosystem showed a synergistic antibacterial effect by combining the trapping effect of GO nanosheets with the killing effect of Ag. The GO can interact with bacterial cell membranes and adsorb on bacterial cells, resulting in an increased concentration of AgNPs around the bacteria. The antibacterial effect of AgNPs mainly depends on the release of $Ag^+$ and the promotion of reactive oxygen species (ROS) production. Figure 2 shows the synergistic antibacterial mechanism of the GO-Ag.

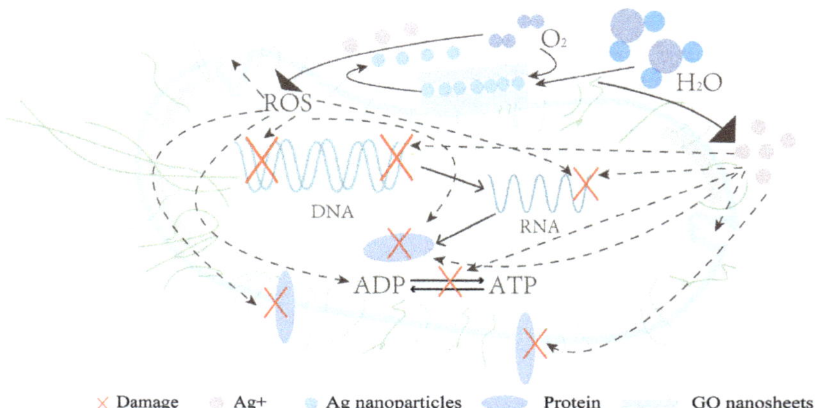


**Figure 2.** Synergistic antibacterial mechanism of GO-Ag.

Among them, the scaffolds containing 1 wt% GO and 1 wt% Ag showed good cytocompatibility without affecting MG63 cell adhesion, viability, and proliferation because the presence of the GO has a positive effect on cell adhesion, and excessive Ag has a negative effect.

Wu et al. [59] found that total alkaloids from Semen Strychnine (TASS) was loaded into polyetheretherketone (PEEK)/PGA composite scaffolds prepared by 3D printing technology to obtain long-lasting antibacterial activity. The PEEK and PGA were dissolved in ethanol solution at a mass ratio of 6:4, and then different levels of TASS were dissolved in the PEEK/PGA suspension. The TASS-PEEK/PGA suspension was stirred magnetically for 2 h, and then the suspension was dried at 50 °C until the precipitate was of constant weight. Finally, the precipitate was ground homogeneously, and the scaffolds were prepared by selective laser sintering technique. The TASS has antibacterial, anti-inflammatory, and analgesic properties. The relationship between the antimicrobial activity of the scaffolds and the TASS content was investigated. As the TASS content increases, the antibacterial activity hinders both *S. aureus* and *E. coli* from proliferating.

However, the effective TASS level is close to the toxic dose, so making the TASS localized and controllable in release is critical. The human fetal osteoblastic cell line (hFOB 1.19 cells) was used to assess the cytotoxicity of the scaffolds. Different concentrations of TASS-PEEK/PGA were incubated with hFOB 1.19 cells for 1, 3, and 5 days. Compared to PEEK/PGA, 2.5% TASS-PEEK/PGA had a slight effect on cell survival. The 7.5% TASS-PEEK/PGA had fewer cells on the third day of incubation than on the first day, but the number of cells increased again on the fifth day. This may be due to faster drug release and higher TASS levels in the first three days. However, the slow release of the drug after that promoted the cell growth.

Gao et al. [60] reported a hyperbranched poly(amide-amine)-capped Ag shell and Au core nanoparticle (Ag@Au NP)-embedded fiber membrane-structured PGA/PLGA ureteral stent. This stent showed efficient sterilization properties, taking 5 min to kill *E. coli* and 10 min to kill *S. aureus*, respectively, with a 99% bacterial inhibition rate. In a 16-day in vitro assay, the stent showed durable antibacterial activity, low release of Ag and Au and low cytotoxicity. Gradient degradation of PGA/PLGA allowed constant exposure of Ag@AuNPs to the stent surface, which acted as a bactericide and eliminated bacterial adhesion.

### 3.4. Poly(Lactic-Co-Glycolic Acid) (PLGA)

The PLGA is a copolymer of the PLA and the PGA, which biodegrades faster than the PLA due to the presence of the PGA [61]. However, the surface properties of PLGA are not ideal for cell growth [62]. PLGA has been approved by the US Food and Drug Administration (FDA) for use as an implant, and PLGA has excellent mechanical properties and degradability. Surface modification of PLGA scaffolds is therefore a promising approach, which would provide useful surface properties to the polymer without changing the native properties [63].

Jing et al. [64] prepared nanoparticles based on PLGA-CS conjugates and PLGA-alendronate (Alen) conjugates. The PLGA-CS combines the good biocompatibility and biodegradability of PLGA and CS, facilitating drug delivery. Alen, a potent anti-osteoporosis drug, was used as a model drug for modifying the PLGA in this experiment. The cytotoxicity of the nanoparticles was assessed by CCK-8 assay. The cell survival rate was higher than 95% when nanoparticles with different concentrations were placed together with MC3T3 cells for 24 h. This indicates that the nanoparticles have no significant cytotoxicity to MC3T3 cells. To assess the specific cellular uptake of nanoparticles, nanoparticles with Alen (NP4) or without Alen (NP5) were incubated with HDF cells and MC3T3 cells for 3 and 24 h, respectively. The results showed that the uptake of NP4 into MC3T3 cells was significantly greater than that of NP5 after 24 h. This demonstrated that the Alen-modified nanoparticles had specific uptake into MC3T3 cells. Although the antimicrobial activity of nanoparticles was not tested in the article, the addition of CS may give the nanoparticles some antimicrobial activity.

De Faria et al. [65] prepared PLGA and CS blend fibers by using electrostatic spinning method. The PLGA-CS mats were functionalized with GO-Ag through a chemical reaction between the carboxyl group of GO and the primary amine functional group on PLGA-CS fibers.

To evaluate the antibacterial activity of PLGA-CS after modification with GO-Ag, unmodified PLGA-CS was used as a control. The GO-Ag modified PLGA-CS and PLGA-CS were exposed to *E. coli*, *P. aeruginosa* (Gram-negative), and *S. aureus* (Gram-positive) for 3 h. The result showed that the antibacterial activity was greatly improved after GO-Ag modification. The inactivation of *E. coli* and *P. aeruginosa* reached more than 98%, while the inactivation of *S. aureus* was lower at 79.4 ± 6.1%. This may be due to the thicker peptidoglycan layer of Gram-positive bacteria, which played a protective role against *S. aureus* cells.

Azzazy et al. [66] designed PLGA nanoparticles with CS coating and loaded with harmala alkaloid-rich fraction (HARF) (H/CS/PLGA) by the emulsion–solvent evaporation method. HARF has been reported to increase collagen and fibroblasts in the microenvironment near the wound, which allows it to accelerate wound healing. The lactic acid produced by the degradation of PLGA accelerates reparative angiogenesis, while CS has antibacterial, bioadhesive, and hemostatic properties. To evaluate the cytotoxicity of these nanoparticles, human skin fibroblasts were treated with different concentrations of H/CS/PLGA NPs to test cell viability. The cell viability was higher than 85% for all concentrations and did not differ significantly from the untreated cells.

The antibacterial activities of free HARF, CS/PLGA NPs, and H/CS/PLGA NPs were evaluated as shown in Table 1. The H/CS/PLGA NPs showed the highest antibacterial activity against *S. aureus* and *E. coli*. The nanoparticles can decompose near the wound and release loaded HARF, in which the alkaloids bind to bacterial DNA to act as antibacterial agents.

**Table 1.** Antibacterial activity of optimal H/CS/PLGA NPs and free HARF against *S. aureus* and *E. coli*.

| Bacterial Strain | Minimum Inhibitory Concentration (MIC in mg/mL) | | |
|---|---|---|---|
| | **HARF** | **CS/PLGA NPs** | **H/CS/PLGA NPs** |
| *S. aureus* | 0.5 | 0.18 | 0.13 |
| *E. coli* | 0.5 | 0.18 | 0.06 |

### 3.5. Summary of Antimicrobial Strategies for Degradable Synthetic Polymers

Antimicrobial applications of biodegradable synthetic polymers and mammalian cells used for testing biocompatibility are shown in Table 2.

**Table 2.** Summary of antimicrobial applications and biocompatibility testing of degradable synthetic polymers.

| Systems | Approaches | Bacterial Strains | Mammalian Cells Used for Testing | Antibacterial Mechanism | Refs. |
|---|---|---|---|---|---|
| PLA/PHMB | Electrostatic spinning | *E. coli* *M. luteus* | fibroblast and epithelial cell lines | PHMB cationic polymer antibacterial | [48] |
| PLA/CS | Non-solvent induce phase separation | *E. coli* | - | CS cationic polymer antibacterial | [45] |
| PLA/GSNO/PHB | Electrostatic spinning | *S. aureus* | fibroblasts | GSNO releases NO to modulate the polarity shift of macrophages and produce anti-inflammatory effects. | [49] |
| PLA/PCL/n(HA)/cfx-βCD | Electrostatic spinning | *S. aureus* | Mouse pre-osteblast cell line | Antibacterial effects caused by antibiotics | [50] |
| PCL/curcumin/ piperine/eugenol/rutin | Electrostatic spinning | *S. aureus* *Enterococcus faecalis* | fibroblasts | Antibacterial effect of natural plant extracts | [53] |
| PCL/GP/Bioglass | 3D printing | *S. aureus* *E. coli* *C. albicans* | - | Structure of GP disrupts bacterial cell membrane | [54] |
| PCL/Atropa/AgNPs | Electrostatic spinning | *S. aureus* *E. coli* | HaCaT cells | The bactericidal effect of $Ag^{+}$ | [55] |
| PCL/HA/ZnO | Electrostatic spinning | *S. aureus* | Human fetal osteoblast cell line | Release of $Zn^{+}$ from ZnO to produce antibacterial effect | [56] |
| PGA/PLLA/GO/Ag | Laser sintering | *E. coli* | MG63 cells | The trapping effect of GO is synergistically bactericidal with $Ag^{+}$ bactericidal action. | [58] |
| PGA/PEEK/TASS | 3D printing | *S. aureus* *E. coli* | Human fetal osteoblast cell line | Antibacterial effects caused by antibiotics | [59] |
| PGA/PLGA/Ag@Au NPs | - | *S. aureus* *E. coli* | L929 cells | Metal ion sterilization effect | [60] |
| PLGA/CS/Alen | - | - | Mouse pre-osteblast cell line | CS cationic polymer antibacterial | [64] |
| PLGA/CS/GO/Ag | Electrostatic spinning | *S. aureus* *E. coli* *P. aeruginosa* | - | 1. CS cationic polymer antibacterial 2. The trapping effect of GO is synergistically bactericidal with $Ag^{+}$ bactericidal action. | [65] |
| PLGA/CS/HARF | Emulsion-solvent evaporation | *S. aureus* *E. coli* | Human skin fibroblasts | 1. CS cationic polymer antibacterial 2. Antibacterial effect of alkaloids | [66] |

These material systems were tested for their antimicrobial efficacy and biocompatibility. In PLA/PHMB, the complete growth inhibition of both bacteria occurred when the PHMB concentration was above 1.5 wt% (bacterial growth was lower than 1% compared to the PLA control). In contrast, when the PHMB concentration was below 0.75 wt%, bacterial growth was not significantly inhibited, but bacterial growth on the scaffold was still lower than that of the control. For the *E. coli,* it was about 30% lower than the control, and for the *M. luteus,* it was about 20% lower than the control. The antibacterial effect of the system was demonstrated to be dependent on the loading concentration of PHMB. For fibroblast and epithelial cell adhesion, it was 275% and 175% when the PHMB concentration was 1.5 wt% compared to the control, while it decreased to 175% and 100% when PHMB was increased to 2.5 wt%. This indicates that the system has good biocompatibility in a range of concentrations. Overall, low concentrations of PHMB inhibited bacterial adhesion and colonization due to the controlled and sustained release of PHMB. In contrast, significant inhibition of bacterial growth requires a concentration of PHMB higher than 0.75 wt%. In PLA/CS, the average antibacterial rate increased from 84.90% to 99.77% when PLA:CS was increased from 8:1 to 3:1. This indicates that the antimicrobial effect of the system depends on the relative content of CS. In the PLA/GSNO/PHB system, the bacterial adhesion rate of *S. aureus* was about 72.9% and 79.7% after 2 and 24 h exposure to the fiber.

Only about 20% of the bacteria remained viable after 2 h of exposure. Mouse fibroblasts showed greater than 90% viability compared to the control group without fiber exposure. In PLA/PCL/n(HA)/cfx-βCD, the growth of *S. aureus* was reduced by 90% within 24 h. The cell viability of MC3T3 cells increased continuously from day 3 to day 7 and surpassed that of the control at day 14, contributing to cell proliferation. Antibacterial testing of *S. aureus* in PCL/curcumin/piperine/eugenol/rutin showed that the dressing containing piperine killed 100% of the bacteria and the dressing containing curcumin killed more than 90% of the bacteria. Two three-component systems, PCPiEu and PCPiR, achieved bactericidal rates of approximately 80%. Antibacterial tests on *E. faecalis* showed that curcumin had no antibacterial activity against this bacterium, while the three-component systems PCPiEu and PCPiR exhibited bactericidal rates of 99.47% and 96.88%. In cellular tests performed on human fibroblasts, the cell viability of the dressing containing only piperine was only 16.5%, whereas the three-component systems PCPiEu and PCPiR showed 94.2% and 98.5% cell viability. Therefore, the two three-component systems showed good overall performance in terms of antimicrobial and biocompatibility. In PCL/Atropa/AgNPs, the antimicrobial effect was brought about by AgNPs, and the antimicrobial effect was slightly reduced by the addition of Atropa. In contrast, in the cytotoxicity test of HaCaT cells, the cell survival rate was increased by about 30% with the addition of Atropa than with the addition of AgNPs only. In PCL/HA/ZnO, PCL/HA/ZnO 1% reduced the initial *S. aureus* bacterial load by almost 99%. In PGA/PLLA/GO/Ag, antimicrobial tests were conducted using *E. coli*. GO showed good synergistic bactericidal effect with Ag. GO itself had no significant bactericidal effect, while the addition of GO increased the bactericidal rate of PGA/PLLA/GO/Ag by 17.1% over PGA/PLLA/Ag to 95.4%. With the increase of Ag content to 1.5%, the bactericidal rate reached 99.9%. However, the 1.5% Ag content made the scaffolds less cytocompatible. While 1% content of Ag still maintained good biocompatibility. In PGA/PEEK/TASS, the antimicrobial rate of the scaffold reached 55.71% for E. coil and 15.84% for *S. aureus* when the TASS content was 7.5%. In the biocompatibility test, a TASS content of 2.5% better promoted the proliferation and differentiation of human fetal osteoblasts. Among PGA/PLGA/Ag@AuNPs, Ag@AuNPs showed strong bactericidal rates, reaching over 90% and 99.9999% against *E. coli* and *S. aureus*, respectively, and the bactericidal rate continued to increase with the increase of nanoparticles. The degradation of the outer layer of the scaffold does not release large amounts of Au and AgNPs, so the scaffold has good biocompatibility. In PLGA/CS/GO/Ag, the killing rate of both *E. coli* and *P. aeru-ginosa* was over 98%, while the inactivation rate of *S. aureus* was only about 79.4%.

It seems that all these antimicrobial methods have good bactericidal effect; however, they still have some drawbacks. For example, good bactericidal methods are often accompanied by strong cytotoxicity, which requires control of its concentration and release rate. Most of the methods are effective against only one type of bacteria but are not effective against others. More antimicrobial material systems need to be investigated in order to achieve an antimicrobial effect while promoting cell proliferation.

## 4. Natural Polymers

Natural polymers are derived from plant and animal organisms, and they are usually biocompatible, biodegradable, and non-toxic [67]. Materials of natural or biological origin are the first biodegradable biomaterials used in clinical practice [68]. The application of naturally derived polymers as bionics is attractive, as it includes motifs that are recognized by cells that facilitate cell adhesion and proliferation [69]. However, these materials usually have poor mechanical properties and are difficult to process. Therefore, chemical modifications are required to be used as suitable biomedical tools [70].

### 4.1. Chitosan

Chitosan is a polysaccharide mainly derived from the exoskeleton of crustaceans and has been developed to have a variety of therapeutic functions [71]. Chitosan has been shown to have higher antimicrobial activity, higher bactericidal rates, a wider range of activity, and low toxicity to mammalian cells [72,73]. The properties of chitosan can be modified due to the presence of reactive functional groups. Moreover, chitosan derivatives can be specifically produced to enhance the desired properties while maintaining their biocompatibility and degradation properties [74,75].

The chitosan has antibacterial activity against many bacteria and fungi, and this unique property is mainly attributed to the polycationic nature of chitosan [76]. The antibacterial effect of fungal chitosan against the Gram-negative bacterium *E. coli* can be strongly attributed to its interaction with cell membrane components, as evidenced by the dramatic morphological changes in cell shape and structure following chitosan treatment [77]. Gram-negative bacteria usually have complex cell walls [78], and the antimicrobial effect of chitosan may be due to the biochemical attachment caused by the cationic charge on the chitosan particles and the anionic charge of the cell wall components [79]. In addition, proteins covalently linked to peptidoglycan in the cell membrane are largely responsible for the antigenic properties of bacteria and can interact strongly with charged chitosan particles, leading to cell death [77]. Figure 3 shows the bactericidal mechanism of chitosan against Gram-negative bacteria.

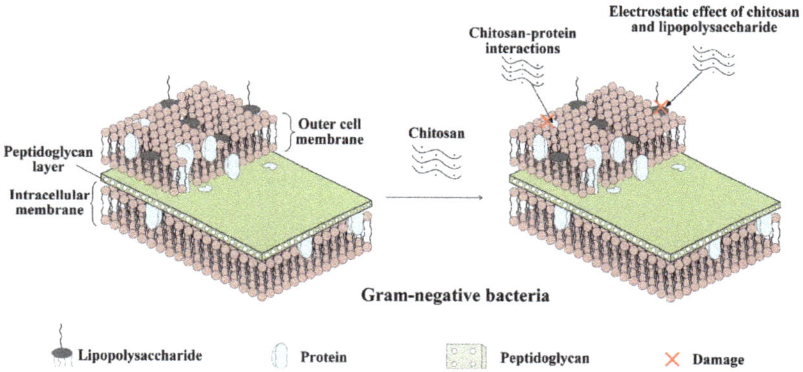


**Figure 3.** Bactericidal mechanism of chitosan against Gram-negative bacteria.

Gram-positive bacteria usually have a thick peptidoglycan layer. The chitosan has a high ability to attach peptidoglycans as an acidic polymer [80]. In addition, the phosphopeptidic acid on peptidoglycan has a polyanionic nature, and the electrostatic interaction between chitosan and phosphopeptidic acid disrupts the function of phosphopeptidic acid and causes functional disorders in bacteria. Figure 4 shows the antibacterial mechanism of chitosan against Gram-positive bacteria.

Chitosan is widely studied in medical applications because of its unique antibacterial properties and good biocompatibility. Chen et al. [81] prepared the polyelectrolyte components (PEC) with good biocompatibility by polyelectrolyte assembly, using carboxymethyl starch (CMS) and chitosan oligosaccharides (COS). The CMS/COS-PEC has controlled physicochemical properties and antibacterial activity against *S. aureus*. It can be used as a degradable hemostatic agent. Hu et al. [82] reported for the first time an innovative three-dimensional porous chitosan-vanillin-bioglass (CVB) scaffold with vanillin and bioglass as non-toxic and osteoconductive cross-linkers. The scaffold had strong antibacterial activity and improved mechanical properties. With high porosity, it significantly promoted osteoblast differentiation.

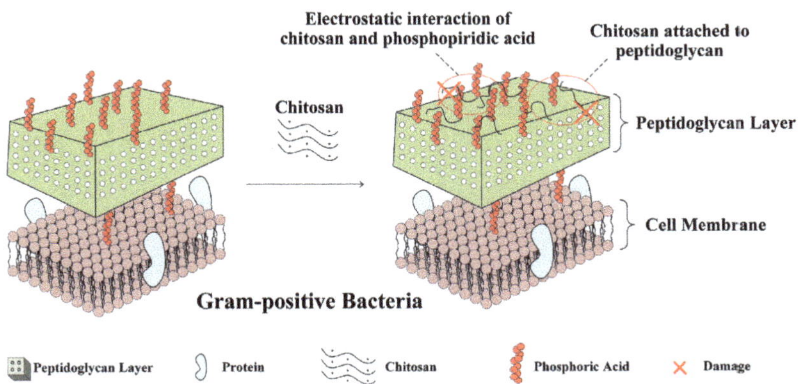


**Figure 4.** The antibacterial mechanism of chitosan against Gram-positive bacteria.

### 4.2. Gelatin

Gelatin is a biopolymer easily obtained from the partial hydrolysis of collagen and is one of the most used biopolymers approved by the FDA due to its biocompatibility, biodegradability, and ease of access [83]. In addition, the process of gelatin degradation produces little inflammatory response [84]. The degradation products are amino acids, which benefit the body, support cell adhesion, and promote cell proliferation [85,86].

Polyurethane/gelatin hybrid nanofiber scaffolds were prepared by using electrostatic spinning to promote cell growth and barrier to bacteria [86]. Cryogels cross-linked with gelatin and dopamine are effective hemostatic materials and can promote wound healing. Dopamine confers near-infrared radiation-assisted antimicrobial capacity to the cryogel [87]. Zhao et al. [85] also developed a gelatin-based physical double-network hydrogel with photothermal antibacterial activity. Photothermal sterilization holds promise as a physical antimicrobial method to kill multidrug-resistant bacteria. Metals and metal compounds, such as Au/Ag [88] and Zno [89], have also been added to gelatin to obtain antimicrobial activity.

### 4.3. Cellulose

Cellulose is a linear polysaccharide found in large quantities in several natural sources [90]. Cellulose and its derivatives are biocompatible polymers that have attracted considerable interest for biomedical applications due to their suitable physical and mechanical properties [91].

Orlando et al. [92] have suggested a green chemistry strategy to the functionalization of cellulose for the introduction of antibacterial functional groups. In the case of the functionalized material compared to the unmodified material, a reduction in the bacterial cell population by roughly half was seen within 24 h for both strains of *S. aureus* and *E. coli*. Additionally, cytotoxicity tests have shown that cellulose hydrogels do not harm keratin-forming cells when they come into touch with them directly or indirectly, even after exposure for 6 days. Fernandes et al. [93] obtained antimicrobial bacterial cellulose membranes by chemically grafting aminoalkyl groups onto the surface of their nanofiber network. These bacterial cellulose membranes showed antibacterial activity against both *S. aureus* and *E. coli*, whereas they were non-toxic to human-adipose-derived mesenchymal stem cells. Hassanpour et al. [94] chemically modified the surface of nanocellulose with a phenanthrene silica salt. The modified samples showed promising results against both Gram-positive (*S. aureus*) and Gram-negative (*E. coli*) bacteria. Moreover, it does not affect the viability of normal HDF cells at low concentrations.

The antimicrobial strategies and biocompatibility tests based on natural polymers are shown in Table 3.

**Table 3.** Summary of antimicrobial applications and biocompatibility testing of degradable natural polymers.

| Systems | Approaches | Bacterial Strains | Mammalian Cells Used for Testing | Antibacterial Mechanism | Refs. |
|---|---|---|---|---|---|
| CMS/COS | Polyelectrolyte assembly | *E. coli* *S. aureus* | MC3T3-L1 fibroblasts | Antibacterial activity of COS | [81] |
| Vanillin/bioglass/chitosan | Crosslinking | *S. gordonii* *S. sanguinis* | MC3T3-E1 | Antimicrobial effect of chitosan | [82] |
| Modified gelatin/$Fe^{+}$ | Crosslinking | *E. coli* *S. aureus* | - | Near-infrared radiation photothermal antibacterial | [85] |
| Gelatin/polydopamine | Crosslinking | *E. coli* *S. aureus* | L929 cells | Near-infrared radiation photothermal antibacterial | [87] |
| Au/Ag@gelatin | - | *P. aeruginosa* | - | Produces ROS under white light irradiation, which kills bacteria. | [88] |
| Gelatin/ZnO | Electrostatic spinning | *S. aureus* *E. coli* | MRC-5 cells | ZnO generates superoxide radicals and damages bacterial cell walls | [89] |
| Chemical modification of bacterial cellulose | Crosslinking | *S. aureus* *E. coli* | HaCaT cells | Introduction of antimicrobial groups | [92] |
| Chemical modification of bacterial cellulose | Crosslinking | *S. aureus* *E. coli* | Human adipose-derived mesenchymal stem cells | Introduction of antimicrobial groups | [93] |
| Chemical modification of nanofibrillated cellulose | - | *S. aureus* *E. coli* | Human dermal fibroblasts | Introduction of antimicrobial groups | [94] |

## 5. Application of Biodegradable Biocompatible Polymers

### 5.1. Wound Healing

Wound healing is a natural physiological process that occurs in response to any tissue injury or damage [95]. There is a high risk of infection during the healing process because damaged skin cannot form a proper barrier to bacteria. Therefore, antibiotics are often used to combat bacteria and significantly reduce the mortality rate from bacterial diseases worldwide [96]. The use of antibiotics (e.g., penicillin, vancomycin, and gentamicin) is currently the standard in the management of bacterial infections [97]. Unfortunately, mutations in bacteria have led to a resistance to antibiotics for them, and it has significantly changed the trend toward treatment [98]. Therefore, an ideal wound dressing should have proper mechanical properties, excellent hemostatic properties, and air permeation as well as antibacterial activity [99,100].

Natural or synthetic polymers with biocompatible and biodegradable properties play a crucial role in cell adhesion and proliferation because of their unique structure and excellent mechanical properties [101,102]. For example, chitosan promotes wound healing by enhancing the migration of fibroblasts and the deposition of collagen in the wound area. In addition, chitosan has hemostatic and antibacterial properties [103–105]. The degradation product of PLGA and lactic acid accelerates the repair of blood vessels and facilitates wound healing.

### 5.2. Tissue Engineering

The intricate hierarchical structure of human tissues makes them susceptible to injury, cancer, and some degenerative diseases over the course of a person's lifespan [106]. Therefore, the regeneration and repair of damaged tissues are very important for the healing process. Tissue engineering is the most promising approach, promoting cell growth by implanting biomaterials [107]. In bone tissue engineering, osteoblasts, chondrocytes, and mesenchymal stem cells are obtained from the patient's hard and soft tissues. Then, they are propagated in cultures and inoculated onto the scaffold. The scaffold is implanted into the patient, slowly degrades, and resorbs as the tissue structures grow [108,109].

Orthopedic implants exhibit good biocompatibility, biodegradability, porosity, and mechanical strength, but lack antimicrobial ability [110]. The development of aseptic surgical techniques and prophylactic systemic antibiotic therapy have reduced the incidence of infection. However, the bacterial colonization of medical devices or implants is still a serious risk [111,112]. With the increase in antibiotic-resistant bacteria, new antimicrobial approaches are being explored [113]. Most of these methods are developed on degradable biocompatible polymers.

As we have known, gelatin mixes well with natural and synthetic polymers to promote high biomechanical and bioaffinity of the scaffold. The 3D porous scaffolds and nanofiber scaffolds prepared with gelatin as the main material are mainly used for large bone defects [114]. The PCL [115,116], PGA, PLA, and PLGA copolymers [117,118] are the most used synthetic biodegradable polymers for 3D scaffolds in tissue engineering. For example, many variants of the PCL facilitate the induction of bone tissue differentiation. The PLA and the PGA are easy to process, and their degradation rates and physical and mechanical properties can be adjusted over a wide range by changing parameters [119,120].

*5.3. Drug Deliver*

Conventional antimicrobial drugs face a few difficulties, including frequent resistance, formulation-related restrictions, subpar drug targeting, and subpar drug release, all of which can result in toxicity in mammalian cells or ineffectiveness at the site of action [121]. Global human health is plainly at risk due to the growing resistance to already prescribed antibiotics and the declining availability of novel antibiotic medications. Therefore, one of the main areas of attention for the internationally acknowledged research priority is the quest for new and efficient techniques to improve medication therapy against existing antibiotics [122].

To balance the biochemical events of inflammation in chronic wounds and promote healing, chitosan-based hydrogels are well-suited for intelligent administration and can be loaded with antibacterial agents, growth factors, stem cells, and peptides [123]. Gelatin is a flexible biopolymer that has historically made it possible to develop a variety of drug delivery methods, including fibers, hydrogels, microparticles, and nanoparticles. These various methods all have certain qualities that make them particularly well-suitable for medication delivery [124]. After being effectively used to entrain antimicrobial agents, the PCL is widely used as a drug delivery method to promote bone growth and regeneration in the treatment of bone diseases [125].

## 6. Future Prospects and Challenges

Injuries and deaths caused by infections have become a global public health concern. With the emergence of multiple drug-resistant strains, however, the problem of bacterial resistance has become increasingly serious. Therefore, there is an urgent need to develop new antimicrobial methods to combat the problems caused by infections. Biodegradable biocompatible polymers have become a hot topic of research, both as scaffolds for tissue engineering and as carriers for drug delivery. Biodegradable and biocompatible polymers can be divided into synthetic polymers and natural polymers. Synthetic polymers have good mechanical properties and stability but lack the ability to grow cells. Synthetic polymers often have no antimicrobial activity. Therefore, antimicrobial agents are often added to synthetic polymers to endow them with antimicrobial activity. These antimicrobial agents include metals, antibiotics, and some antimicrobial agents derived from plants. However, all these methods have some shortcomings, such as the tendency of metal ions to cause toxicity to cells when killing bacteria. Long-term use of antibiotics may lead to drug resistance, yet new antibiotics are rarely invented. Plant-derived antimicrobial agents require strict dosage control. Otherwise, they may be toxic to cells instead of being effective. Therefore, modulating the degradation rate of degradable polymers is a promising direction, which could lead to the controlled release of antimicrobial agents. Natural polymers have good biocompatibility, but rather poor mechanical properties and are not easy to process. Several methods have been reported to improve their disadvantages, and antimicrobial features have been developed on natural polymers. Natural polymers have a variety of origins and can be chemically modified to obtain several variants. Due to their poor mechanical properties, they are mostly used to fabricate wound dressings or hydrogels. There are also many studies on blending natural polymers with synthetic polymers to exploit their respective advantages. Some physical sterilization methods have also been investigated, such as the construction of surface microstructures to puncture bacterial biofilms and near-infrared light irradiation for sterilization. Therefore, biodegradable biocompatible polymers with antimicrobial properties and the ability to promote cell growth will be a promising approach to alleviate the problem of antibiotic resistance.

Degradable polymers play a key role in the sustained release of drugs and more and more antimicrobial methods are being developed on degradable polymers. For antimicrobial substances incorporated into degradable polymers, however, the antimicrobial activity is often accompanied by cytotoxicity more or less. Therefore, controlling the sustained, stable, and quantitative release of drugs is still a subject that needs to be investigated in depth.

## 7. Conclusions

Infections caused by bacteria can be very dangerous to people's health. The most effective way to traditionally fight back against bacteria is the widespread use of antibiotics. However, due to the misuse of antibiotics, a variety of drug-resistant bacteria have emerged. Therefore, there is an urgent need to seek alternative antimicrobial methods. Biodegradable and biocompatible functional polymers are widely used in medical treatments as tissue engineering scaffolds and wound dressings. These medical devices are susceptible to bacterial infections. Moreover, a great deal of effort has been put into the research of biodegradable polymers with antimicrobial properties. Currently, effective antimicrobial methods include cationic polymer antimicrobial, metal ion antimicrobial, hydrophobic structure to prevent bacterial adhesion, photothermal sterilization, nonionic antimicrobial agents, etc. Meanwhile, biodegradable polymers can be used as carriers for drug delivery to load antibiotics, thus achieving local release of antibiotics. Biodegradable polymers with antibacterial functions have a wide range of applications, including the prevention of bacterial infections caused by medical devices and the construction of drug delivery systems.

**Author Contributions:** Conceptualization, A.C. and C.L. (Chaozong Liu); methodology, C.Y. and S.N.V.; software, J.L. (Jindi Lai) and Y.P.; validation, J.Z., C.L. (Caihong Lei) and C.L. (Chaozong Liu); investigation, J.L. (Jiajun Luo) and S.D.; data curation, S.D., W.C. and J.W.; writing—original draft preparation, S.D. and A.C.; writing—review and editing, A.C. and C.L. (Chaozong Liu); funding acquisition, A.C. and C.L. (Chaozong Liu). All authors have read and agreed to the published version of the manuscript.

**Funding:** This work was supported by the National Natural Science Foundation of China (Grant no. 52003057), the Foshan Science and Technology Innovation Project (grant number FS0AA-KJ919-4402-0145), MRC-UCL Therapeutic Acceleration Support (TAS) Fund (project no: 564022), NIHR UCLH BRC-UCL Therapeutic Acceleration Support (TAS) Fund (project no. 564021), and Wellcome Trust-Translational Partnership Award-UCL Regenerative Medicine TIN Pilot Dara Fund (Project no.: 569576), and Engineering and Physical Sciences Research Council via DTP CASE Programme (Grant no.: EP/T517793/1).

**Institutional Review Board Statement:** Not applicable.

**Data Availability Statement:** Not applicable.

**Conflicts of Interest:** The authors declare no conflict of interest.

## References


1. Wang, X.; Shan, M.; Zhang, S.; Chen, X.; Liu, W.; Chen, J.; Liu, X. Stimuli-responsive antibacterial materials: Molecular structures, design principles, and biomedical applications. *Adv. Sci.* **2022**, *9*, 2104843. [CrossRef] [PubMed]
2. Amato, D.V.; Amato, D.N.; Blancett, L.T.; Mavrodi, O.V.; Martin, W.B.; Swilley, S.N.; Sandoz, M.J.; Shearer, G.; Mavrodi, D.V.; Patton, D.L. A bio-based pro-antimicrobial polymer network via degradable acetal linkages. *Acta Biomater.* **2018**, *67*, 196–205. [CrossRef] [PubMed]
3. Ribeiro, M.; Monteiro, F.J.; Ferraz, M.P. Infection of orthopedic implants with emphasis on bacterial adhesion process and techniques used in studying bacterial-material interactions. *Biomatter* **2012**, *2*, 176–194. [CrossRef]
4. Hetrick, E.M.; Schoenfisch, M.H. Reducing implant-related infections: Active release strategies. *Chem. Soc. Rev.* **2006**, *35*, 780–789. [CrossRef] [PubMed]
5. Ma, K.; Gong, L.; Cai, X.; Huang, P.; Cai, J.; Huang, D.; Jiang, T. A green single-step procedure to synthesize Ag-containing nanocomposite coatings with low cytotoxicity and efficient antibacterial properties. *Int. J. Nanomed.* **2017**, *12*, 3665–3679. [CrossRef] [PubMed]
6. Liu, J.; Liu, J.; Attarilar, S.; Wang, C.; Tamaddon, M.; Yang, C.; Xie, K.; Yao, J.; Wang, L.; Liu, C.; et al. Nano-modified titanium implant materials: A way toward improved antibacterial properties. *Front. Bioeng. Biotechnol.* **2020**, *8*, 576969. [CrossRef]
7. Darouiche, R.O. Treatment of infections associated with surgical implants. *N. Engl. J. Med.* **2004**, *350*, 1422–1429. [CrossRef]
8. Zhao, X.; Li, P.; Guo, B.; Ma, P.X. Antibacterial and conductive injectable hydrogels based on quaternized chitosan-graft-polyaniline/oxidized dextran for tissue engineering. *Acta Biomater.* **2015**, *26*, 236–248. [CrossRef]
9. Fischbach, M.A.; Walsh, C.T. Antibiotics for emerging pathogens. *Science* **2009**, *325*, 1089–1093. [CrossRef]
10. Milazzo, M.; Gallone, G.; Marcello, E.; Mariniello, M.D.; Bruschini, L.; Roy, I.; Danti, S. Biodegradable polymeric micro/nano-structures with intrinsic antifouling/antimicrobial properties: Relevance in damaged skin and other biomedical applications. *J. Func. Biomater.* **2020**, *11*, 60. [CrossRef]
11. Park, H.H.; Sun, K.; Lee, D.; Seong, M.; Cha, C.; Jeong, H.E. Cellulose acetate nanoneedle array covered with phosphorylcholine moiety as a biocompatible and sustainable antifouling material. *Cellulose* **2019**, *26*, 8775–8788. [CrossRef]
12. Qiu, H.; Si, Z.; Luo, Y.; Feng, P.; Wu, X.; Hou, W.; Zhu, Y.; Mary, B.; Xu, L.; Huang, D. The Mechanisms and the applications of antibacterial polymers in surface modification on medical devices. *Front. Bioeng. Biotechnol.* **2020**, *8*, 910. [CrossRef] [PubMed]
13. Ding, X.; Wang, A.; Tong, W.; Xu, F. Biodegradable antibacterial polymeric nanosystems: A new hope to cope with multidrug-resistant bacteria. *Small* **2019**, *15*, 1900999. [CrossRef]
14. Cuervo-Rodríguez, R.; López-Fabal, F.; Muñoz-Bonilla, A.; Fernández-García, M. Antibacterial polymers based on poly(2-hydroxyethyl methacrylate) and thiazolium groups with hydrolytically labile linkages leading to inactive and low cytotoxic compounds. *Materials* **2021**, *14*, 7477. [CrossRef] [PubMed]
15. Liu, X.; Holzwarth, J.M.; Ma, P.X. Functionalized synthetic biodegradable polymer scaffolds for tissue engineering. *Macromol. Biosci.* **2012**, *12*, 911–919. [CrossRef] [PubMed]
16. Tian, H.; Tang, Z.; Zhuang, X.; Chen, X.; Jing, X. Biodegradable synthetic polymers: Preparation, functionalization and biomedical application. *Prog. Polym. Sci.* **2012**, *37*, 237–280. [CrossRef]
17. Asghari, F.; Samiei, M.; Adibkia, K.; Akbarzadeh, A.; Davaran, S. Biodegradable and biocompatible polymers for tissue engineering application: A review. *Artif. Cells Nanomed. Biotechnol.* **2017**, *45*, 185–192. [CrossRef]
18. Liang, H.; Liu, L.; Lu, J.; Chen, M.; Zhang, C. Castor oil-based cationic waterborne polyurethane dispersions: Storage stability, thermo-physical properties and antibacterial properties. *Ind. Crops Prod.* **2018**, *117*, 169–178. [CrossRef]
19. Liang, H.; Lu, Q.; Liu, M.; Ou, R.; Wang, Q.; Quirino, R.L.; Luo, Y.; Zhang, C. UV absorption, anticorrosion, and long-term antibacterial performance of vegetable oil based cationic waterborne polyurethanes enabled by amino acids. *Chem. Eng. J.* **2021**, *421*, 127774. [CrossRef]
20. Wang, Y.; Guo, Q.; Wang, H.; Qian, K.; Tian, L.; Yao, C.; Song, W.; Shu, W.; Chen, P.; Qi, J. Evaluation of the antibacterial activity of a cationic polymer in aqueous solution with a convenient electrochemical method. *Anal. Bioanal. Chem.* **2017**, *409*, 1627–1633. [CrossRef]
21. Li, X.; İlk, S.; Liu, Y.; Raina, D.B.; Demircan, D.; Zhang, B. Nonionic nontoxic antimicrobial polymers: Indole-grafted poly(vinyl alcohol) with pendant alkyl or ether groups. *Polym. Chem.* **2022**, *13*, 2307–2319. [CrossRef]
22. Liu, Y.; Li, X.; Liang, A. Current Research Progress of Local Drug Delivery Systems Based on Biodegradable Polymers in Treating Chronic Osteomyelitis. *Front. Bioeng. Biotechnol.* **2022**, *10*, 1042128. [CrossRef] [PubMed]
23. Du, L.; Yang, S.; Li, W.; Li, H.; Feng, S.; Zeng, R.; Yu, B.; Xiao, L.; Nie, H.Y.; Tu, M. Scaffold composed of porous vancomycin-loaded poly(lactide-co-glycolide) microspheres: A controlled-release drug delivery system with shape-memory effect. *Mater. Sci. Eng. C* **2017**, *78*, 1172–1178. [CrossRef] [PubMed]
24. Dwivedi, C.; Pandey, H.; Pandey, A.C.; Patil, S.; Ramteke, P.W.; Laux, P.; Luch, A.; Singh, A.V. In vivo biocompatibility of electrospun biodegradable dual carrier (antibiotic + growth factor) in a mouse model—Implications for rapid wound healing. *Pharmaceutics* **2019**, *11*, 180. [CrossRef]
25. Huang, C.Y.; Hsieh, R.W.; Yen, H.T.; Hsu, T.C.; Chen, C.Y.; Chen, Y.C.; Lee, C.C. Short- versus long-course antibiotics in osteomyelitis: A systematic review and meta-analysis. *Int. J. Antimicrob. Agents* **2019**, *53*, 246–260. [CrossRef]

26. Mills, D.K.; Jammalamadaka, U.; Tappa, K.; Weisman, J. Studies on the cytocompatibility, mechanical and antimicrobial properties of 3D printed poly(methyl methacrylate) beads. *Bioact. Mater.* **2018**, *3*, 157–166. [CrossRef]
27. Ding, X.; Duan, S.; Ding, X.; Liu, R.; Xu, F. Versatile antibacterial materials: An emerging arsenal for combatting bacterial pathogens. *Adv. Funct. Mater.* **2018**, *28*, 1802140. [CrossRef]
28. Jiang, R.; Hao, L.; Song, L.; Tian, L.; Fan, Y.; Zhao, J.; Liu, C.; Ming, W.; Ren, L. Lotus-leaf-inspired hierarchical structured surface with non-fouling and mechanical bactericidal performances. *Chem. Eng. J.* **2020**, *398*, 125609. [CrossRef]
29. Yi, Y.; Jiang, R.; Liu, Z.; Dou, H.; Song, L.; Tian, L.; Ming, W.; Ren, L.; Zhao, J. Bioinspired nanopillar surface for switchable mechano-bactericidal and releasing actions. *J. Hazard. Mater.* **2022**, *432*, 128685. [CrossRef]
30. Wang, P.; Zhang, Y.L.; Fu, K.L.; Liu, Z.; Zhang, L.; Liu, C.; Deng, Y.; Xie, R.; Ju, X.J.; Wang, W.; et al. Zinc-coordinated polydopamine surface with a nanostructure and superhydrophilicity for antibiofouling and antibacterial Applications. *Mater. Adv.* **2022**, *3*, 5476–5487. [CrossRef]
31. Fajstavr, D.; Frýdlová, B.; Rimpelová, S.; Kasálková, N.S.; Sajdl, P.; Švorčík, V.; Slepička, P. KrF laser and plasma exposure of pdms–carbon composite and its antibacterial properties. *Materials* **2022**, *15*, 839. [CrossRef] [PubMed]
32. Li, P.; Feng, Y.; Cheng, D.; Wei, J. Self-template synthesis of mesoporous vanadium oxide nanospheres with intrinsic peroxidase-like activity and high antibacterial performance. *J. Colloid. Interface Sci.* **2022**, *625*, 435–445. [CrossRef] [PubMed]
33. Cheng, Y.; Zhang, Y.; Zhao, Z.; Li, G.; Li, J.; Li, A.; Xue, Y.; Zhu, B.; Wu, Z.; Zhang, X. Guanidinium-decorated nanostructure for precision sonodynamic-catalytic therapy of mrsa-infected osteomyelitis. *Adv. Mater.* **2022**, *34*, e2206646. [CrossRef] [PubMed]
34. Lee, A.R.; Park, S.B.; Kim, S.W.; Jung, J.W.; Chun, J.H.; Kim, J.; Kim, Y.R.; Lazarte, J.M.S.; Jang, H.B.; Thompson, K.D.; et al. Membrane vesicles from antibiotic-resistant *Staphylococcus aureus* transfer antibiotic-resistance to antibiotic-susceptible *Escherichia coli*. *J. Appl. Microbiol.* **2022**, *132*, 2746–2759. [CrossRef] [PubMed]
35. Ferri, M.; Ranucci, E.; Romagnoli, P.; Giaccone, V. Antimicrobial resistance: A global emerging threat to public health systems. *Crit. Rev. Food Sci. Nutr.* **2017**, *57*, 2857–2876. [CrossRef] [PubMed]
36. Liu, Y.; Tong, Z.; Shi, J.; Li, R.; Upton, M.; Wang, Z. Drug repurposing for next-generation combination therapies against multidrug-resistant bacteria. *Theranostics* **2021**, *11*, 4910–4928. [CrossRef]
37. Spasova, M.; Manolova, N.; Markova, N.; Rashkov, I. Superhydrophobic PVDF and PVDF-HFP nanofibrous mats with antibacterial and anti-biofouling properties. *Appl. Surf. Sci.* **2016**, *363*, 363–371. [CrossRef]
38. Mirzadeh, S.; Asefnejad, A.; Khonakdar, H.A.; Jafari, S.H. Improved surface properties in spray-coated PU/TiO2/Graphene hybrid nanocomposites through nonsolvent-induced phase separation. *Surf. Coat. Technol.* **2021**, *405*, 126507. [CrossRef]
39. Kim, J.Y.; Kim, J.H.; Ahn, G.; An, S.H.; Ryu, R.H.; Kim, J.S.; Kim, Y.C.; Shim, J.H.; Kim, S.Y.; Yun, W.S. In vitro study of three-dimensional printed metal-polymer hybrid scaffold incorporated dual antibiotics for treatment of periprosthetic joint infection. *Mater. Lett.* **2018**, *212*, 263–266. [CrossRef]
40. Li, P.; Li, X.; Saravanan, R.; Li, C.; Susanna, S. Antimicrobial macromolecules: Synthesis methods and future applications. *RSC Adv.* **2012**, *2*, 4031–4044. [CrossRef]
41. Sumitha, M.S.; Shalumon, K.T.; Sreeja, V.N.; Jayakumar, R.; Nair, S.V.; Menon, D. Biocompatible and antibacterial nanofibrous poly (ϵ-caprolactone)-nanosilver composite scaffolds for tissue engineering applications. *J. Macromol. Sci. Part A-Pure Appl. Chem.* **2012**, *49*, 131–138. [CrossRef]
42. Costa, L.M.M.; de Olyveira, G.M.; Cherian, B.M.; Leão, A.L.; de Souza, S.F.; Ferreira, M. Bionanocomposites from electrospun PVA/pineapple nanofibers/stryphnodendron adstringens bark extract for medical applications. *Ind. Crops Prod.* **2013**, *41*, 198–202. [CrossRef]
43. Fathi, M.; Barar, J. Perspective highlights on biodegradable polymeric nanosystems for targeted therapy of solid tumors. *Bioimpacts* **2017**, *7*, 49–57. [CrossRef] [PubMed]
44. You, Y.; Min, B.M.; Lee, S.J.; Lee, T.S.; Park, W.H. In vitro degradation behavior of electrospun polyglycolide, polylactide, and poly (lactide-co-glycolide). *J. Appl. Polym. Sci.* **2005**, *95*, 193–200. [CrossRef]
45. Han, W.; Ren, J.; Xuan, H.; Ge, L. Controllable degradation rates, antibacterial, free-standing and highly transparent films based on polylactic acid and chitosan. *Colloid Surf. A-Physicochem. Eng. Asp.* **2018**, *541*, 128–136. [CrossRef]
46. Avérous, L. Chapter 21—Polylactic Acid: Synthesis, Properties and Applications. In *Monomers, Polymers and Composites from Renewable Resources*; Belgacem, M.N., Gandini, A., Eds.; Elsevier: Amsterdam, The Netherlands, 2008; pp. 433–450.
47. Škrlová, K.; Malachová, K.; Muñoz-Bonilla, A.; Měřinská, D.; Rybková, Z.; Fernández-García, M.; Plachá, D. Biocompatible polymer materials with antimicrobial properties for preparation of stents. *Nanomaterials* **2019**, *9*, 1548. [CrossRef]
48. Llorens, E.; Calderón, S.; del Valle, L.J.; Puiggalí, J. Polybiguanide (PHMB) loaded in PLA scaffolds displaying high hydrophobic, biocompatibility and antibacterial properties. *Mater. Sci. Eng. C-Mater. Biol. Appl.* **2015**, *50*, 74–84. [CrossRef]
49. Douglass, M.; Hopkins, S.; Pandey, R.; Singha, P.; Norman, M.; Handa, H. S-nitrosoglutathione-based nitric oxide-releasing nanofibers exhibit dual antimicrobial and antithrombotic activity for biomedical applications. *Macromol. Biosci.* **2021**, *21*, 2000248. [CrossRef]
50. Sharif, F.; Tabassum, S.; Mustafa, W.; Asif, A.; Zarif, F.; Tariq, M.; Gilani, M.; Rehman, I.; MacNeil, S. Bioresorbable antibacterial PCL-PLA-nHA composite membranes for oral and maxillofacial defects. *Polym. Compos.* **2019**, *40*, 1564–1575. [CrossRef]
51. Park, C.-H.; Kim, E.K.; Tijing, L.D.; Amarjargal, A.; Pant, H.R.; Kim, C.S.; Shon, H.K. Preparation and characterization of PLA/PCL composite fibers containing beta tricalcium phosphate ($\beta$-TCP) particles. *Ceram. Int.* **2014**, *40*, 5049–5054. [CrossRef]

52. Bosworth, L.A.; Downes, S. Physicochemical characterisation of degrading polycaprolactone scaffolds. *Polym. Degrad. Stabil.* **2010**, *95*, 2269–2276. [CrossRef]
53. Safdari, F.; Gholipour, M.D.; Ghadami, A.; Saeed, M.; Zandi, M. Multi-antibacterial agent-based electrospun polycaprolactone for active wound dressing. *Prog. Biomater.* **2022**, *11*, 27–41. [CrossRef] [PubMed]
54. Hajduga, M.B.; Bobiński, R.; Dutka, M.; Izabela, U.W.; Bujok, J.; Pajak, C.; Cwiertnia, M.; Kurowska, A.; Dziadek, M.; Rajzer, I. Analysis of the antibacterial properties of polycaprolactone modified with graphene, bioglass and zinc-doped bioglass. *Acta Bioeng. Biomech.* **2021**, *23*, 131–138. [CrossRef] [PubMed]
55. Avci, M.O.; Muzoglu, N.; Yilmaz, A.E.; Yarman, B.S. Antibacterial, Cytotoxicity and biodegradability studies of polycaprolactone nanofibers holding green synthesized Ag nanoparticles using Atropa Belladonna extract. *J. Biomater. Sci.-Polym. Ed.* **2022**, *33*, 1157–1180. [CrossRef]
56. Felice, B.; Sánchez, M.A.; Socci, M.C.; Sappia, L.D.; Gómez, M.I.; Cruz, M.K.; Felice, C.J.; Martí, M.; Pividori, M.I.; Simonelli, G.; et al. Controlled degradability of PCL-ZnO nanofibrous scaffolds for bone tissue engineering and their antibacterial activity. *Mater. Sci. Eng. C-Mater. Biol. Appl.* **2018**, *93*, 724–738. [CrossRef]
57. Gao, C.; Peng, S.; Feng, P.; Shuai, C. Bone biomaterials and interactions with stem cells. *Bone Res.* **2017**, *5*, 1–33. [CrossRef]
58. Shuai, C.; Guo, W.; Wu, P.; Yang, W.; Hu, S.; Xia, Y.; Feng, P. A graphene oxide-Ag co-dispersing nanosystem: Dual synergistic effects on antibacterial activities and mechanical properties of polymer scaffolds. *Chem. Eng. J.* **2018**, *347*, 322–333. [CrossRef]
59. Wu, P.; Hu, S.; Liang, Q.; Guo, W.; Xia, Y.; Shuai, C.; Li, Y. A polymer scaffold with drug-sustained release and antibacterial activity. *Int. J. Polym. Mater. Polym. Biomat.* **2020**, *69*, 398–405. [CrossRef]
60. Gao, L.; Wang, Y.; Li, Y.; Xu, M.; Sun, G.; Zou, T.; Wang, F.; Xu, S.; Da, J.; Wang, L. Biomimetic biodegradable Ag@Au nanoparticle-embedded ureteral stent with a constantly renewable contact-killing antimicrobial surface and antibiofilm and extraction-free properties. *Acta Biomater.* **2020**, *114*, 117–132. [CrossRef]
61. Zhao, W.; Li, J.; Jin, K.; Liu, W.; Qiu, X.; Li, C. Fabrication of functional PLGA-based electrospun scaffolds and their applications in biomedical engineering. *Mater. Sci. Eng. C-Mater. Biol. Appl.* **2016**, *59*, 1181–1194. [CrossRef]
62. Yang, J.; Shi, G.; Bei, J.; Wang, S.; Cao, Y.; Shang, Q.; Yang, G.; Wang, W. Fabrication and surface modification of macroporous poly(L-Lactic Acid) and poly(L-Lactic-Co-Glycolic Acid) (70/30) cell scaffolds for human skin fibroblast cell culture. *J. Biomed. Mater. Res.* **2002**, *62*, 438–446. [CrossRef] [PubMed]
63. Croll, T.I.; O'Connor, A.J.; Stevens, G.W.; Cooper-White, J.J. Controllable surface modification of poly(Lactic-Co-Glycolic Acid) (PLGA) by hydrolysis or aminolysis I: Physical, chemical, and theoretical aspects. *Biomacromolecules* **2004**, *5*, 463–473. [CrossRef] [PubMed]
64. Jing, C.; Chen, S.; Bhatia, S.S.; Li, B.; Liang, H.; Liu, C.; Liang, Z.; Liu, J.; Li, H.; Liu, Z.; et al. Bone-targeted polymeric nanoparticles as alendronate carriers for potential osteoporosis treatment. *Polym. Test.* **2022**, *110*, 107584. [CrossRef]
65. De Faria, A.F.; Perreault, F.; Shaulsky, E.; Arias Chavez, L.H.; Elimelech, M. Antimicrobial electrospun biopolymer nanofiber mats functionalized with graphene oxide–silver nanocomposites. *ACS Appl. Mater. Interfaces* **2015**, *7*, 12751–12759. [CrossRef] [PubMed]
66. Azzazy, H.M.E.-S.; Fahmy, S.A.; Mahdy, N.K.; Meselhy, M.R.; Bakowsky, U. Chitosan-coated PLGA nanoparticles loaded with peganum harmala alkaloids with promising antibacterial and wound healing activities. *Nanomaterials* **2021**, *11*, 2438. [CrossRef]
67. Li, F.; Li, S.; Liu, Y.; Zhang, Z.; Li, Z. Current advances in the roles of doped bioactive metal in biodegradable polymer composite scaffolds for bone repair: A mini review. *Adv. Eng. Mater.* **2022**, *24*, 2101510. [CrossRef]
68. Bouhlouli, M.; Pourhadi, M.; Karami, F.; Talebi, Z.; Ranjbari, J.; Khojasteh, A. Applications of bacterial cellulose as a natural polymer in tissue engineering. *Asaio J.* **2021**, *67*, 709–720. [CrossRef] [PubMed]
69. Naghieh, S.; Lindberg, G.; Tamaddon, M.; Liu, C. Biofabrication strategies for musculoskeletal disorders: Evolution towards clinical applications. *Bioengineering* **2021**, *8*, 123. [CrossRef]
70. Pina, S.; Oliveira, J.M.; Reis, R.L. Natural-based nanocomposites for bone tissue engineering and regenerative medicine: A review. *Adv. Mater.* **2015**, *27*, 1143–1169. [CrossRef]
71. Boles, L.R.; Bumgardner, J.D.; Fujiwara, T.; Haggard, W.O.; Guerra, F.D.; Jennings, J.A. Characterization of trimethyl chitosan/polyethylene glycol derivatized chitosan blend as an injectable and degradable antimicrobial delivery system. *Int. J. Biol. Macromol.* **2019**, *133*, 372–381. [CrossRef]
72. Younes, I.; Rinaudo, M. Chitin and chitosan preparation from marine sources. structure, properties and applications. *Mar. Drugs* **2015**, *13*, 1133–1174. [CrossRef] [PubMed]
73. Inzana, J.A.; Schwarz, E.M.; Kates, S.L.; Awad, H.A. Biomaterials approaches to treating implant-associated osteomyelitis. *Biomaterials* **2016**, *81*, 58–71. [CrossRef] [PubMed]
74. Ahmed, T.A.; Aljaeid, B.M. Preparation, characterization, and potential application of chitosan, chitosan derivatives, and chitosan metal nanoparticles in pharmaceutical drug delivery. *Drug Des. Dev. Ther.* **2016**, *10*, 483–507. [CrossRef] [PubMed]
75. Mourya, V.K.; Inamdar, N.N. Chitosan-modifications and applications: Opportunities galore. *React. Funct. Polym.* **2008**, *68*, 1013–1051. [CrossRef]
76. Vedula, S.S.; Yadav, G.D. Chitosan-based membranes preparation and applications: Challenges and opportunities. *J. Indian Chem. Soc.* **2021**, *98*, 100017. [CrossRef]
77. Tayel, A.A.; Gharieb, M.M.; Zaki, H.R.; Elguindy, N.M. Bio-clarification of water from heavy metals and microbial effluence using fungal chitosan. *Int. J. Biol. Macromol.* **2016**, *83*, 277–281. [CrossRef]

78. McHale, P. Biochemistry and molecular biology of antimicrobial drug action. *Br. J. Biomed. Sci.* **2000**, *57*, 183.
79. Tayel, A.A.; Moussa, S.; Opwis, K.; Knittel, D.; Schollmeyer, E.; Nickisch-Hartfiel, A. Inhibition of microbial pathogens by fungal chitosan. *Int. J. Biol. Macromol.* **2010**, *47*, 10–14. [CrossRef]
80. Helander, I.M.; Nurmiaho-Lassila, E.-L.; Ahvenainen, R.; Rhoades, J.; Roller, S. Chitosan disrupts the barrier properties of the outer membrane of Gram-negative bacteria. *Int. J. Food Microbiol.* **2001**, *71*, 235–244. [CrossRef]
81. Chen, X.; Yan, Y.; Li, H.; Wang, X.; Tang, S.; Li, Q.; Wei, J.; Su, J. Evaluation of absorbable hemostatic agents of polyelectrolyte complexes using carboxymethyl starch and chitosan oligosaccharide both in vitro and in vivo. *Biomater. Sci.* **2018**, *6*, 3332–3344. [CrossRef]
82. Hu, J.; Wang, Z.; Miszuk, J.M.; Zhu, M.; Lansakara, T.I.; Tivanski, A.V.; Banas, J.A.; Sun, H. Vanillin-bioglass cross-linked 3D porous chitosan scaffolds with strong osteopromotive and antibacterial abilities for bone tissue engineering. *Carbohydr. Polym.* **2021**, *271*, 118440. [CrossRef] [PubMed]
83. Deng, L.; Zhang, X.; Li, Y.; Que, F.; Kang, X.; Liu, Y.; Feng, F.; Zhang, H. Characterization of gelatin/zein nanofibers by hybrid electrospinning. *Food Hydrocoll.* **2018**, *75*, 72–80. [CrossRef]
84. Dong, Y.; Sigen, A.; Rodrigues, M.; Li, X.; Kwon, S.H.; Kosaric, N.; Khong, S.; Gao, Y.; Wang, W.; Gurtner, G.C. Injectable and tunable gelatin hydrogels enhance stem cell retention and improve cutaneous wound healing. *Adv. Funct. Mater.* **2017**, *27*, 1606619. [CrossRef]
85. Zhao, X.; Liang, Y.; Huang, Y.; He, J.; Han, Y.; Guo, B. Physical double-network hydrogel adhesives with rapid shape adaptability, fast self-healing, antioxidant and NIR/PH stimulus-responsiveness for multidrug-resistant bacterial infection and removable wound dressing. *Adv. Funct. Mater.* **2020**, *30*, 1910748. [CrossRef]
86. Kim, S.E.; Heo, D.N.; Lee, J.B.; Kim, J.R.; Park, S.H.; Jeon, S.H.; Kwon, I.K. Electrospun gelatin/polyurethane blended nanofibers for wound healing. *Biomed. Mater.* **2009**, *4*, 044106. [CrossRef]
87. Huang, Y.; Zhao, X.; Zhang, Z.; Liang, Y.; Yin, Z.; Chen, B.; Bai, L.; Han, Y.; Guo, B. Degradable gelatin-based IPN cryogel hemostat for rapidly stopping deep noncompressible hemorrhage and simultaneously improving wound healing. *Chem. Mater.* **2020**, *32*, 6595–6610. [CrossRef]
88. Wang, X.; Guo, J.; Zhang, Q.; Zhu, S.; Liu, L.; Jiang, X.; Wei, D.H.; Liu, R.S.; Li, L. Gelatin sponge functionalized with gold/silver clusters for antibacterial application. *Nanotechnology* **2020**, *31*, 134004. [CrossRef]
89. Chen, Y.; Lu, W.; Guo, Y.; Zhu, Y.; Song, Y. Electrospun gelatin fibers surface loaded ZnO particles as a potential biodegradable antibacterial wound dressing. *Nanomaterials* **2019**, *9*, 525. [CrossRef]
90. Janmohammadi, M.; Nazemi, Z.; Salehi, A.O.M.; Seyfoori, A.; John, J.V.; Nourbakhsh, M.S.; Akbari, M. Cellulose-based composite scaffolds for bone tissue engineering and localized drug delivery. *Bioact. Mater.* **2023**, *20*, 137–163. [CrossRef]
91. Seddiqi, H.; Oliaei, E.; Honarkar, H.; Jin, J.; Geonzon, L.C.; Bacabac, R.G.; Klein-Nulend, J. Cellulose and its derivatives: Towards biomedical applications. *Cellulose* **2021**, *28*, 1893–1931. [CrossRef]
92. Orlando, I.; Basnett, P.; Nigmatullin, R.; Wang, W.; Knowles, J.C.; Roy, I. Chemical modification of bacterial cellulose for the development of an antibacterial wound dressing. *Front. Bioeng. Biotechnol.* **2020**, *8*, 557885. [CrossRef] [PubMed]
93. Fernandes, S.C.M.; Sadocco, P.; Alonso-Varona, A.; Palomares, T.; Eceiza, A.; Silvestre, A.J.D.; Mondragon, I.; Freire, C.S.R. Bioinspired antimicrobial and biocompatible bacterial cellulose membranes obtained by surface functionalization with aminoalkyl groups. *ACS Appl. Mater. Interfaces* **2013**, *5*, 3290–3297. [CrossRef] [PubMed]
94. Hassanpour, A.; Asghari, S.; Mansour Lakouraj, M.; Mohseni, M. Preparation and Characterization of contact active antibacterial surface based on chemically modified nanofibrillated cellulose by phenanthridinium silane salt. *Int. J. Biol. Macromol.* **2018**, *115*, 528–539. [CrossRef]
95. Kumar Nethi, S.; Das, S.; Ranjan Patra, C.; Mukherjee, S. Recent advances in inorganic nanomaterials for wound-healing applications. *Biomater. Sci.* **2019**, *7*, 2652–2674. [CrossRef] [PubMed]
96. Giano, M.C.; Ibrahim, Z.; Medina, S.H.; Sarhane, K.A.; Christensen, J.M.; Yamada, Y.; Brandacher, G.; Schneider, J.P. Injectable bioadhesive hydrogels with innate antibacterial properties. *Nat. Commun.* **2014**, *5*, 4095. [CrossRef] [PubMed]
97. Rastin, H.; Ramezanpour, M.; Hassan, K.; Mazinani, A.; Tung, T.T.; Vreugde, S.; Losic, D. 3D bioprinting of a cell-laden antibacterial polysaccharide hydrogel composite. *Carbohydr. Polym.* **2021**, *264*, 117989. [CrossRef] [PubMed]
98. Solomon, S.L.; Oliver, K.B. Antibiotic resistance threats in the United States: Stepping back from the brink. *Am. Fam. Physician* **2014**, *89*, 938–941. [PubMed]
99. Saeed, S.M.; Mirzadeh, H.; Zandi, M.; Barzin, J. Designing and fabrication of curcumin loaded PCL/PVA multi-layer nanofibrous electrospun structures as active wound dressing. *Prog. Biomater.* **2017**, *6*, 39–48. [CrossRef]
100. Sadeghianmaryan, A.; Yazdanpanah, Z.; Soltani, Y.A.; Sardroud, H.A.; Nasirtabrizi, M.H.; Chen, X. Curcumin-loaded electrospun polycaprolactone/montmorillonite nanocomposite: Wound dressing application with anti-bacterial and low cell toxicity properties. *J. Biomater. Sci.-Polym. Ed.* **2020**, *31*, 169–187. [CrossRef]
101. Shende, P.; Gupta, H. Formulation and comparative characterization of nanoparticles of curcumin using natural, synthetic and semi-synthetic polymers for wound healing. *Life Sci.* **2020**, *253*, 117588. [CrossRef]
102. Karri, V.V.S.R.; Kuppusamy, G.; Talluri, S.V.; Mannemala, S.S.; Kollipara, R.; Wadhwani, A.D.; Mulukutla, S.; Raju, K.R.S.; Malayandi, R. Curcumin loaded chitosan nanoparticles impregnated into collagen-alginate scaffolds for diabetic wound healing. *Int. J. Biol. Macromol.* **2016**, *93*, 1519–1529. [CrossRef] [PubMed]

103. Shariatinia, Z.; Jalali, A.M. Chitosan-based hydrogels: Preparation, properties and applications. *Int. J. Biol. Macromol.* **2018**, *115*, 194–220. [CrossRef] [PubMed]
104. Francis Suh, J.K.; Matthew, H.W.T. Application of chitosan-based polysaccharide biomaterials in cartilage tissue engineering: A review. *Biomaterials* **2000**, *21*, 2589–2598. [CrossRef] [PubMed]
105. Ignatova, M.; Manolova, N.; Rashkov, I. Electrospun antibacterial chitosan-based fibers. *Macromol. Biosci.* **2013**, *13*, 860–872. [CrossRef]
106. Godoy-Gallardo, M.; Eckhard, U.; Delgado, L.M.; de Roo Puente, Y.J.D.; Hoyos-Nogués, M.; Gil, F.J.; Perez, R.A. Antibacterial approaches in tissue engineering using metal ions and nanoparticles: From mechanisms to applications. *Bioact. Mater.* **2021**, *6*, 4470–4490. [CrossRef]
107. Biswas, M.C.; Jony, B.; Nandy, P.K.; Chowdhury, R.A.; Halder, S.; Kumar, D.; Ramakrishna, S.; Hassan, M.; Ahsan, M.A.; Hoque, M.E.; et al. Recent advancement of biopolymers and their potential biomedical applications. *J. Polym. Environ.* **2022**, *30*, 51–74. [CrossRef]
108. Hutmacher, D.W. Scaffolds in tissue engineering bone and cartilage. *Biomaterials* **2000**, *21*, 2529–2543. [CrossRef]
109. Babensee, J.E.; Anderson, J.M.; McIntire, L.V.; Mikos, A.G. Host response to tissue engineered devices. *Adv. Drug Deliv. Rev.* **1998**, *33*, 111–139. [CrossRef]
110. Chen, Z.Y.; Gao, S.; Zhang, Y.W.; Zhou, R.B.; Zhou, F. Antibacterial biomaterials in bone tissue engineering. *J. Mat. Chem. B* **2021**, *9*, 2594–2612. [CrossRef]
111. Costa, F.; Carvalho, I.F.; Montelaro, R.C.; Gomes, P.; Martins, M.C.L. Covalent immobilization of antimicrobial peptides (AMPs) onto biomaterial surfaces. *Acta Biomater.* **2011**, *7*, 1431–1440. [CrossRef]
112. Makvandi, P.; Ali, G.W.; Della Sala, F.; Abdel-Fattah, W.I.; Borzacchiello, A. Hyaluronic acid/corn silk extract based injectable nanocomposite: A biomimetic antibacterial scaffold for bone tissue regeneration. *Mater. Sci. Eng. C-Mater. Biol. Appl.* **2020**, *107*, 110195. [CrossRef] [PubMed]
113. Ribeiro, M.; Ferraz, M.P.; Monteiro, F.J.; Fernandes, M.H.; Beppu, M.M.; Mantione, D.; Sardon, H. Antibacterial silk fibroin/nanohydroxyapatite hydrogels with silver and gold nanoparticles for bone regeneration. *Nanomed.-Nanotechnol. Biol. Med.* **2017**, *13*, 231–239. [CrossRef] [PubMed]
114. Echave, M.C.; Sánchez, P.; Pedraz, J.L.; Orive, G. Progress of gelatin-based 3D approaches for bone regeneration. *J. Drug Deliv. Sci. Technol.* **2017**, *42*, 63–74. [CrossRef]
115. Roosa, S.M.M.; Kemppainen, J.M.; Moffitt, E.N.; Krebsbach, P.H.; Hollister, S.J. The pore size of polycaprolactone scaffolds has limited influence on bone regeneration in an in vivo model. *J. Biomed. Mater. Res. Part A* **2010**, *92A*, 359–368. [CrossRef]
116. Christy, P.N.; Basha, S.K.; Kumari, V.S.; Bashir, A.K.H.; Maaza, M.; Kaviyarasu, K.; Arasu, M.V.; Al-Dhabi, N.A.; Ignacimuthu, S. Biopolymeric nanocomposite scaffolds for bone tissue engineering applications—A review. *J. Drug Deliv. Sci. Technol.* **2020**, *55*, 101452. [CrossRef]
117. Song, R.; Murphy, M.; Li, C.; Ting, K.; Soo, C.; Zheng, Z. Current development of biodegradable polymeric materials for biomedical applications. *Drug Des. Dev. Ther.* **2018**, *12*, 3117–3145. [CrossRef]
118. Bernardini, G.; Chellini, F.; Frediani, B.; Spreafico, A.; Santucci, A. Human platelet releasates combined with polyglycolic acid scaffold promote chondrocyte differentiation and phenotypic maintenance. *J. Biosci.* **2015**, *40*, 61–69. [CrossRef]
119. Seal, B.L.; Otero, T.C.; Panitch, A. Polymeric biomaterials for tissue and organ regeneration. *Mater. Sci. Eng. R-Rep.* **2001**, *34*, 147–230. [CrossRef]
120. Fournier, E.; Passirani, C.; Montero-Menei, C.N.; Benoit, J.P. Biocompatibility of implantable synthetic polymeric drug carriers: Focus on brain biocompatibility. *Biomaterials* **2003**, *24*, 3311–3331. [CrossRef]
121. Singh, S.; Hussain, A.; Shakeel, F.; Ahsan, M.J.; Alshehri, S.; Webster, T.J.; Lal, U.R. Recent insights on nanomedicine for augmented infection control. *Int. J. Nanomed.* **2019**, *14*, 2301–2325. [CrossRef]
122. Kalhapure, R.S.; Suleman, N.; Mocktar, C.; Seedat, N.; Govender, T. Nanoengineered drug delivery systems for enhancing antibiotic therapy. *J. Pharm. Sci.* **2015**, *104*, 872–905. [CrossRef] [PubMed]
123. Liu, H.; Wang, C.; Li, C.; Qin, Y.; Wang, Z.; Yang, F.; Li, Z.; Wang, J. A functional chitosan-based hydrogel as a wound dressing and drug delivery system in the treatment of wound healing. *RSC Adv.* **2018**, *8*, 7533–7549. [CrossRef] [PubMed]
124. Foox, M.; Zilberman, M. Drug delivery from gelatin-based systems. *Expert Opin. Drug Deliv.* **2015**, *12*, 1547–1563. [CrossRef] [PubMed]
125. Singh, R.K.; Jin, G.-Z.; Mahapatra, C.; Patel, K.D.; Chrzanowski, W.; Kim, H.W. Mesoporous silica-layered biopolymer hybrid nanofibrous scaffold: A novel nanobiomatrix platform for therapeutics delivery and bone regeneration. *ACS Appl. Mater. Interfaces* **2015**, *7*, 8088–8098. [CrossRef] [PubMed]

MDPI

Article

# Crystallinity Dependence of PLLA Hydrophilic Modification during Alkali Hydrolysis

Jiahui Shi [1], Jiachen Zhang [1], Yan Zhang [1], Liang Zhang [1], Yong-Biao Yang [2,*], Ofer Manor [3] and Jichun You [1,*]

[1] Key Laboratory of Organosilicon Chemistry and Material Technology, Ministry of Education, College of Material, Chemistry and Chemical Engineering, Hangzhou Normal University, Hangzhou 311121, China
[2] School of Chemistry and Chemical Engineering, Liaoning Normal University, Dalian 116029, China
[3] The Wolfson Department of Chemical Engineering, Technion-Israel Institute of Technology, Haifa 32000, Israel
* Correspondence: yongbiao@lnnu.edu.cn (Y.-B.Y.); you@hznu.edu.cn (J.Y.)

**Abstract:** Poly(L-lactic acid) (PLLA) has been extensively used in tissue engineering, in which its surface hydrophilicity plays an important role. In this work, an efficient and green strategy has been developed to tailor surface hydrophilicity via alkali hydrolysis. On one hand, the ester bond in PLLA has been cleaved and generates carboxyl and hydroxyl groups, both of which are beneficial to the improvement of hydrophilicity. On the other hand, the degradation of PLLA increases the roughness on the film surface. The resultant surface wettability of PLLA exhibits crucial dependence on its crystallinity. In the specimen with high crystallinity, the local enrichment of terminal carboxyl and hydroxyl groups in amorphous regions accelerates the degradation of ester group, producing more hydrophilic groups and slit valleys on film surface. The enhanced contact between PLLA and water in aqueous solution (i.e., the Wenzel state) contributes to the synergistic effect between generated hydrophilic groups and surface roughness, facilitating further degradation. Consequently, the hydrophilicity has been improved significantly in the high crystalline case. On the contrary, the competition effect between them leads to the failure of this strategy in the case of low crystallinity.

**Keywords:** PLLA; hydrophilicity; alkali hydrolysis; crystallinity



**Citation:** Shi, J.; Zhang, J.; Zhang, Y.; Zhang, L.; Yang, Y.-B.; Manor, O.; You, J. Crystallinity Dependence of PLLA Hydrophilic Modification during Alkali Hydrolysis. *Polymers* **2023**, *15*, 75. https://doi.org/10.3390/polym15010075

Academic Editors: José Miguel Ferri, Vicent Fombuena Borràs and Miguel Fernando Aldás Carrasco

Received: 25 November 2022
Revised: 13 December 2022
Accepted: 21 December 2022
Published: 25 December 2022



## 1. Introduction

With the increasing requirement of health care and the improved understanding of natural tissue, more and more attention is being paid to medical techniques. Tissue engineering, as a newly emerged field, aims at the development of functional substitutes for damaged tissue [1]. In tissue engineering, one of the ideal characteristics for the designed and manufactured medical material is that its surface can lead to increased affinity of biomolecules [2]. Poly(lactic acid) (or polylactide, abbreviated as PLA) is a semi-crystalline and environmentally friendly material used in tissue engineering in recent decades. Lactic acid can be produced by an industrial process from petrochemical feedstocks, or alternatively by a fermentation process from renewable sources such as corn, wheat or rice. The difference is that the petrochemical route results in the racemic mixture of D- and L-enantiomers, while the production from the natural route is mostly (99.5%) the L-isomer which leads to poly(L-lactic acid) (PLLA) in the next step [3,4].

Although PLLA has excellent biocompatibility and biodegradability properties and thus has been widely used in medical purposes such as implants, bone grafting and fracture fixation devices [5,6], it is in nature a relatively hydrophobic material with a static water contact angle in the range of 75°–85° [7–9]. For biomedical applications such hydrophobicity will lead to low cell affinity and thus affect the cell adhesion onto its surface. In order to improve the cytocompatibility of PLLA, many surface modification techniques have been proposed, including polymerization grafting [10], ozone oxidization [11,12], plasma treatment [13–16], alkaline hydrolysis [17,18], surface coating [19], layer-by-layer self-assembly [20,21], etc.

In addition to relatively high cost, low efficiency or residual toxic solvent, the reported strategies always concern either complicated chemical reaction or special requirement of equipment, which does limit their applications [22,23]. The hydrolysis of PLA as a potential method has attracted the attention of researchers. Previous results pay more attention to the evolution of PLA's matrix properties [24–26] rather than its surface properties.

The degradation of ester bonds with the help of alkaline hydrolysis treatment has been considered as a reliable, green, low labor-intensive and less time-consuming method [27–29]. It is well known that in alkaline hydrolysis ester bonds undergo the nucleophilic attack of hydroxide ions, and finally the ester bonds are cleaved to create carboxyl and hydroxyl groups. By varying the chemical composition of material surface, alkaline hydrolysis can improve PLA's hydrophilicity and reduce its water contact angle. Consequently, much effort has been made to develop the method of alkaline hydrolysis for PLLA in recent years. For instance, using ethanol to assist the hydroxide nucleophilic attack on ester bond in NaOH treatment, Yang et al. modified the surface of the microporous PLLA film to enhance its cell affinity [17]. Sabee and his co-workers investigated the dependence of surface modification for PLLA microspheres on sodium hydroxide (NaOH) concentration [30]. For the preparation of bioactivated PLLA scaffolds, the flexible surface functionalization by the specific alkaline treatment has been successfully developed by Meng et al. [18]. Tsuji et al. investigated the stability and degradation of PLLA in NaOH. Their results indicated that the hydrolysis took place preferentially in the amorphous regions. The specimen with low crystallinity, therefore, exhibited shorter degradation period.

In this work, therefore, an alkaline hydrolysis strategy, inspired by the preferential degradation in amorphous region, is proposed to establish a more convenient, effective and green approach for the surface modification of PLLA with the consideration of its crystallinity. The crystallinity dependence of PLLA hydrophilic modification during alkali hydrolysis has been investigated in detail. The random degradation on the specimen with low crystallinity and selective degradation on alternating structures including crystal lamellar and amorphous region are compared. Based on the surface chemical composition (obtained from XPS and FTIR) and surface roughness (determined by SEM and AFM), the important role of crystallinity in alkali hydrolysis has been clarified.

## 2. Experiment

### 2.1. Materials

PLLA (3001D, *Mw* = 120,000, PDI = 1.9) was purchased from Nature works, USA. Sodium hydroxide (NaOH, AR, ≥96.0%) was bought from Lingfeng Chemical Reagent Co., Ltd., Shanghai, China.

### 2.2. Preparation and Hydrolysis of PLLA Films

PLLA films were prepared by hot-pressing at 200 °C and 10 MPa (with a thickness of 500 μm). The melted PLLA films were quickly cooled to 120 °C under 10 MPa and annealed for 60 min, and were named as $PLLA_{60}$. The unannealed films were named as $PLLA_0$. After heat treatment procedure, all PLLA films were quenched in an ice-water mixture at 0 °C to stop crystallization. Hydrophilic modification of PLLA film was done as follows by surface alkaline hydrolysis. Firstly, the PLLA films were immersed in 1 mol/L NaOH solution, and alkaline hydrolysis was performed at 55 °C for a predetermined time (1 min, 2 min, 3 min, 4 min, 5 min, 10 min, 15 min, 20 min and 25 min). Secondly, the treated PLLA films were rinsed with deionized water and soaked for 2 h. Finally, the films were dried in a vacuum oven at 55 °C for 48 h.

### 2.3. Characterization

Differential scanning calorimetry (DSC Q2000, TA) measurements were performed in a nitrogen atmosphere at the heating and cooling rate of 10 °C/min. A universal tensile testing machine (Instron 5966) was used for tensile property tests at room temperature. The tensile speed was 10 mm/min. The PLLA samples (before or after alkaline hydrolysis) were cut into dumbbell shapes (18 × 0.5 × 3 mm, ASTM D 412-80). At least five specimens were tested for each sample, and the result of the tensile property test was the average value of 5 stretches. The surface structure of the PLLA samples was observed by scanning electron microscopy (SEM, Hitachi S-4800) and atomic force microscopy (AFM, Seiko E-Sweep). AFM was used to measure the surface roughness of PLLA films, in which a tip with a force constant of 2 N/m was adopted. The water contact angle of the PLLA films was obtained by drop-shape analysis (DSA, Krüss DSA100). The surface composition of the PLLA film was characterized by means of Thermo ESCALAB 250 X-ray photoelectron spectrometer (XPS) and Fourier transform infrared resonance (FTIR, Brucker VERTEX 70 V) with attenuated total reflection (ATR) mode.

## 3. Results and Discussion

### 3.1. DSC Analysis of the Crystallinities of PLLA Films

First of all, the crystallization kinetics of PLLA at 120 °C (isothermal crystallization) were investigated with the help of DSC. As shown in Figure 1A, there is an obvious exothermic peak at the beginning stage, corresponding to the crystallization of PLLA. It takes ~41 min for PLLA to fully crystallize. In this work, two specimens were adopted to clarify the crystallinity dependence of alkali hydrolysis. They were named as $PLLA_0$ and $PLLA_{60}$ since these specimens were crystallized at 120 °C for 0 min and 60 min, respectively. Then DSC measurements were performed to identify the crystallinities of them. In the black curve in Figure 1B, the glass transition temperature is located at 62.3 °C. During heating, there is a broad exothermic peak. It starts at ~90 °C and does not disappear until the occurrence of endothermic peak. The endothermic peak can be attributed to the melting of PLLA. On the contrary, the exothermic peak arises from the cold crystallization of PLLA since the temperature is between its glass transition temperature and melting temperature. It is facile to calculate the crystallinities of PLLA according to Equation (1).

$$X_w = \frac{\Delta H_m}{\Delta H_m^0} \tag{1}$$

where $X_w$, $\Delta H_m$ and $\Delta H_m^0$ are crystallinity, the measured melting enthalpy and standard melting enthalpy (93.7 J/g [31]), respectively. For $PLLA_0$, its crystallinity is only 4.1%. In the red curve, $PLLA_{60}$ has a similar glass transition temperature to $PLLA_0$. Its melting temperature, however, exhibits a slightly higher magnitude (166.2 °C) relative to $PLLA_0$ (165.4 °C). Such an improvement of melting temperature can be attributed to the thicker crystal lamellae resulting from sufficient isothermal crystallization. As for the calculated crystallinity of $PLLA_{60}$, it is 46.3%. Because of the high crystallinity in $PLLA_{60}$ during isothermal crystallization, no cold crystallization (exothermic peak) is observed in the red curve.

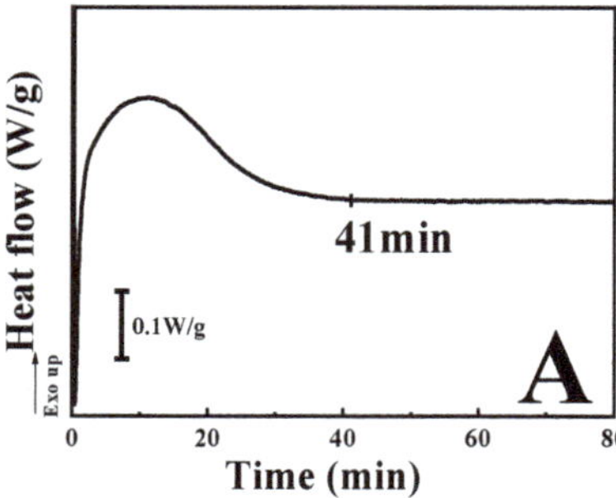


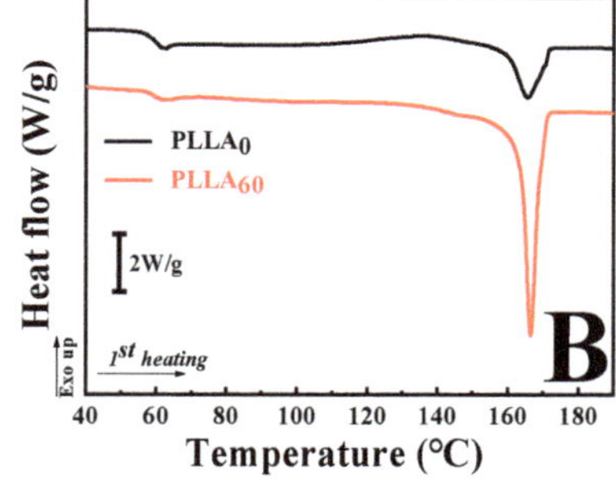


**Figure 1.** The crystallization kinetics of PLLA film at 120 °C (**A**) and the first heating DSC curve of $PLLA_0$ and $PLLA_{60}$ (**B**).

### 3.2. DSA and Tensile Test during Alkali Hydrolysis

Figure 2A gives the hydrolysis time dependence of water contact angle of PLLA films with different crystallinities based on the DSA method. In agreement with previous reports [7–9], the water contact angles of $PLLA_0$ and $PLLA_{60}$ films before alkali hydrolysis are located at ~74.0°, suggesting that they are not very hydrophilic materials. Upon alkali hydrolysis, the evolution of contact angle exhibits remarkable dependence on the crystallinities of PLLA. The pictures showing hydrophilicity through contact angle measurement have been provided in Figure S1. For $PLLA_0$ films (black line in Figure 2A), the contact angle did not change significantly within the experiment period (25 min) but only increased slightly. For the annealed PLLA films ($PLLA_{60}$), there were two stages in the evolution of the water contact angle. At the first one, the contact angle decreased quickly with alkali hydrolysis time (down to 22.4° at 15 min), while at the second stage (15 min to 25 min), the value was not much sensitive to alkali hydrolysis anymore. In other words, the surfaces of PLLA films with higher crystallinity become hydrophilic if the hydrolysis time is long enough. A facile strategy, therefore, has been developed to tailor the surface wettability of PLLA via alkali hydrolysis. In the assessment of mechanical performance, our attention will be paid only to the specimen of $PLLA_{60}$ since for $PLLA_0$ its wettability is hard to manipulate by means of hydrolysis. As shown in Figure 2B, the elongation at break of untreated $PLLA_{60}$ film is ~4.5% and the yield strength is ~60 MPa. After alkali hydrolysis, those two values decrease to 3.1% and 55 MPa, respectively. Both variations are within the experimental errors. This result indicates that the alkali hydrolysis takes place only on the film surface and has no remarkable influence on their mechanical properties in the period up to 25 min. This is significant for the modified PLLA in medical applications.

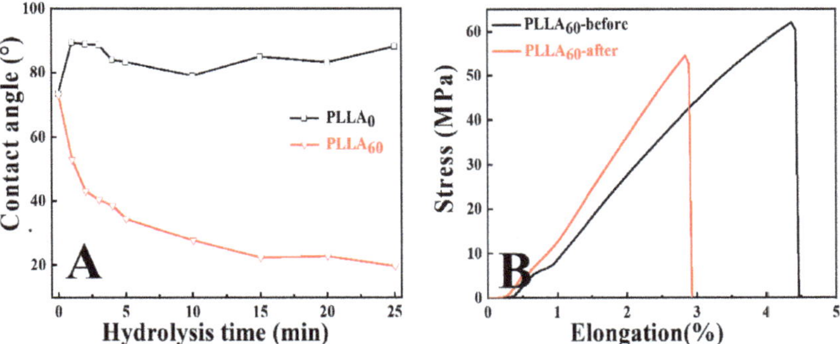


**Figure 2.** The hydrolysis time dependence of water contact angle in $PLLA_0$ and $PLLA_{60}$ (**A**) and strain–stress curves of $PLLA_{60}$ before and after hydrolysis (**B**).

In some special scenarios (e.g., fog capture), hydrophilicity or hydrophobicity gradient play key roles. According to the strategy developed as discussed for Figure 2A, it is facile to fabricate a wettability gradient with the help of home-made setup (Figure 3A). During the experiment, a specimen ($PLLA_{60}$) with certain length (2.5 cm, here) was kept in alkali aqueous solution and drawn out with the speed of 1 mm/min. The top part of the specimen corresponded to a shorter (down to 0 min) alkali hydrolysis time, contributing to higher magnitudes of water contact angles (Figure 3C). This can also be verified by the hemispherical droplets at the left part in Figure 3B. On the contrary, the bottom part stayed in alkali aqueous solution for a longer period (up to 25 min), yielding complete wetting of water on the specimen upon sufficient alkali hydrolysis (right parts of Figure 3B,C). Based on the results discussed above, the manipulation of the wettability gradient on PLLA films has been achieved successfully by varying the alkali hydrolysis period. The tunable water contact angle can meet unique requirements for the customized surface in various applications. Obviously, the adjustment of water contact angles (Figure 2A) and its wettability gradient (Figure 3B,C) are under the control of both surface composition and structures [16,21]. In the following parts, we will discuss them one by one.

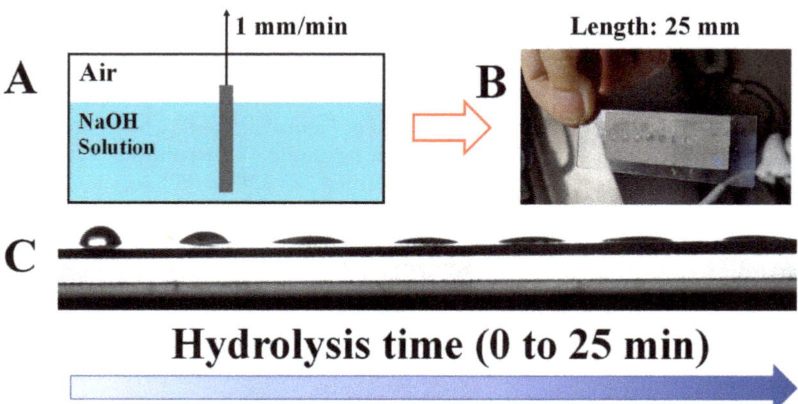


**Figure 3.** The home-made setup for the fabrication of hydrophilic gradient via alkali hydrolysis (**A**). (**B**,**C**) show the photos of PLLA films with tunable water contact angle.

### 3.3. XPS and FTIR Characterization during Alkali Hydrolysis

When the specimen is kept in alkali aqueous solution, the degradation of PLLA takes place, producing carboxyl and hydroxyl groups, which is shown in Equation (2) [32].

$$\text{R-COO-R}' + \text{NaOH} \rightarrow \text{RCOO}^- + \text{HO-R}' + \text{Na}^+ \tag{2}$$

During alkali hydrolysis, the surface composition was identified with the help of XPS and FTIR in ATR mode. Figure 4A shows the C1s spectra of specimens before and after alkali hydrolysis. There are three characteristic peaks at 284.6 eV (Peak 1), 286.8 eV (Peak 2) and 288.8 eV (Peak 3), corresponding to C-C, C-O-C and C=O groups, respectively [33,34]. In the black curve, Peak 2 exhibits higher area fraction relative to Peak 3. As for hydrolysis on the specimen of $PLLA_0$, they are comparable (red curve). While in the case of $PLLA_{60}$ after treatment, Peak 3 corresponds to higher area fraction (blue curve) compared with Peak 2. The lower magnitude of area ratio between Peak 2 and Peak 3 indicates that the hydrolysis of PLLA takes place, during which the ester bond has been broken and converted into hydroxyl and carboxyl groups (as Equation (2)). In this process, the accounts of C-O-C (Peak 2) in XPS C1s spectra decrease. This conclusion has also been supported by FTIR results as shown in Figure 4B. To compare the absorbance of specimen before and after hydrolysis, all curves have been normalized according to the characteristic peak of $-CH_3$ at 1360 $cm^{-1}$. The characteristic peak at 1180 $cm^{-1}$ is related to the stretching vibration of C-O-C bond in PLLA. It exhibits a higher intensity before hydrolysis (black curve). The treatment for $PLLA_0$ leads to the reduced absorbance of the corresponding peak (red curve). Such decrease has been further enhanced in the case of $PLLA_{60}$ upon hydrolysis (blue curve). The XPS (Figure 4A) and FTIR (Figure 4B) results reveal that some ester bonds have been cleaved during alkali hydrolysis. In Figure 4C and 4D, the peaks at 3647.3 $cm^{-1}$ and 1717.9 $cm^{-1}$ can be indexed to the hydroxyl and carboxyl groups, respectively. The absorbances of them at the indicated position exhibit much higher magnitudes relative to the specimen before treatment, suggesting the higher content of carboxyl and hydroxyl groups after alkali treatment. The combination of degradation of ester group and generation of carboxyl/hydroxyl groups makes it clear that alkali hydrolysis cleaves the ester bond and produces hydrophilic groups. Furthermore, the comparison between red and blue curves in Figure 4 reveals that there are more hydrophilic (carboxyl and hydroxyl) groups in the specimen of $PLLA_{60}$.

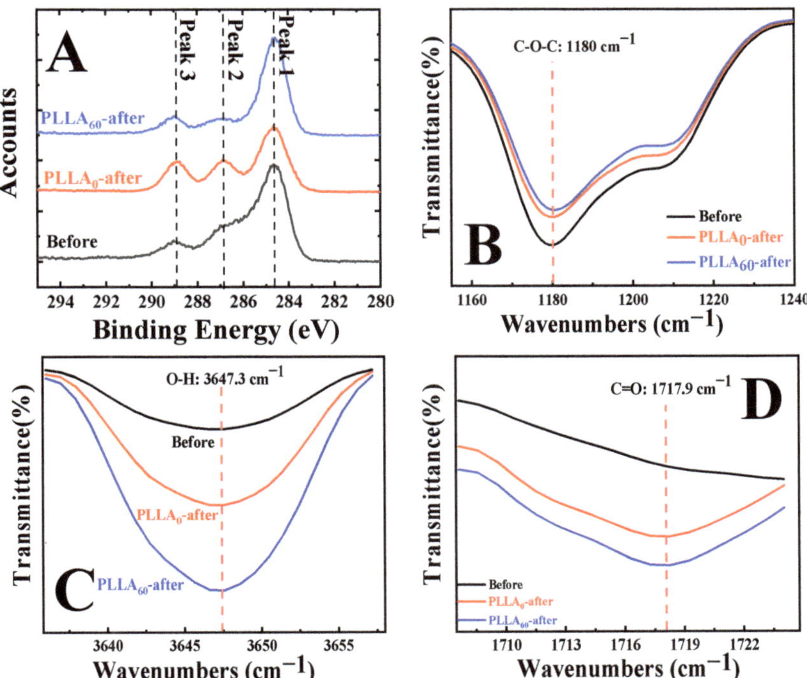


**Figure 4.** XPS C1s spectra (**A**) and FTIR spectra (**B–D**) of specimens before and after alkali hydrolysis of $PLLA_0$ and $PLLA_{60}$.

### 3.4. SEM and AFM Characterization during Alkali Hydrolysis

As mentioned in the Introduction, the surface roughness of the PLLA films may be varied during alkaline hydrolysis. In this work, SEM has been employed to characterize the surface morphology of PLLA films. From Figure 5A to Figure 5C, it can be seen that the alkaline hydrolysis happens randomly on the surface of $PLLA_0$ films, producing concave structures (marked by red arrows). As the time of alkaline hydrolysis increase, the density and diameter of the concave structures gradually increase. Finally, they connect with each other, contributing to higher magnitudes of surface roughness. In the case of $PLLA_{60}$, the resultant morphologies are of great difference, as shown in Figure 5D–F. There are some slit valleys in Figure 5D (hydrolysis for 1 min). Then their size increases while some new valleys appear in Figure 5E (hydrolysis for 5 min). Finally, these structures crowd together, accounting for the significant improvement of surface roughness. The reason for the occurrence of these slit valleys (marked by green arrows) will be discussed in the following parts. The SEM results in Figure 5 make the following points clear. On one hand, the alkali hydrolysis on $PLLA_0$ and $PLLA_{60}$ produces concave and slit structures, respectively; on the other hand, the specimen of $PLLA_{60}$ exhibits rougher surface relative to $PLLA_0$.

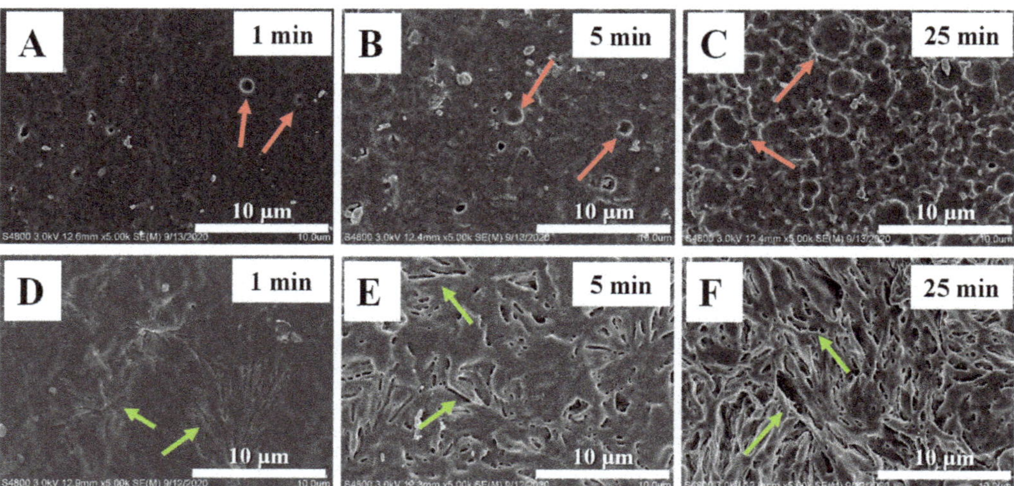


**Figure 5.** SEM images of modified PLLA membranes upon alkali hydrolysis for the indicated time in $PLLA_0$ (**A–C**) and $PLLA_{60}$ (**D–F**).

To identify the surface roughness upon hydrolysis quantitatively, AFM has been employed. The 3D AFM images are shown in Figure 6. Obviously, the surface becomes more and more rough during the alkali treatment in $PLLA_0$ (Figure 6A–C). The increase of surface roughness in $PLLA_{60}$ (Figure 6D–F) is more remarkable. Therefore, the AFM topography images have excellent agreement with SEM images. The root-mean-square (RMS) surface roughness has been calculated with the help of commercial software equipped on AFM based on Equation (3).

$$\text{Roughness} = \sqrt{\frac{1}{S}\iint \{F(x,\ y) - Z\}^2 dxdy} \tag{3}$$

where $S$ and $Z$ are the corresponding area and average height at the calculated position, respectively. The resultant surface roughness is shown in Figure 6G. At the surface of hot-pressed film, the RMS roughness of $PLLA_0$ and $PLLA_{60}$ is located at ~100 nm, suggesting a relatively flat film surface. After the treatment in alkali aqueous solution, the roughness exhibits higher magnitudes. In the case of $PLLA_0$, it increases to 156 nm (5 min), 279 nm (15 min) and 395 nm (25 min). The alkali hydrolysis on $PLLA_{60}$ results in a more significant increase in RMS roughness. It is 178 nm and 406 nm upon treatment for 5 min and 15 min, respectively. Finally, it reaches 537 nm at 25 min. Both AFM images and corresponding surface roughness are all consistent with SEM results shown in Figure 5. According to the discussion above, the alkali hydrolysis destroys the film surface, contributing to the rough surface and playing an important role in the consequent wettability. This treatment produces a much rougher surface in the specimen with higher crystallinity ($PLLA_{60}$).

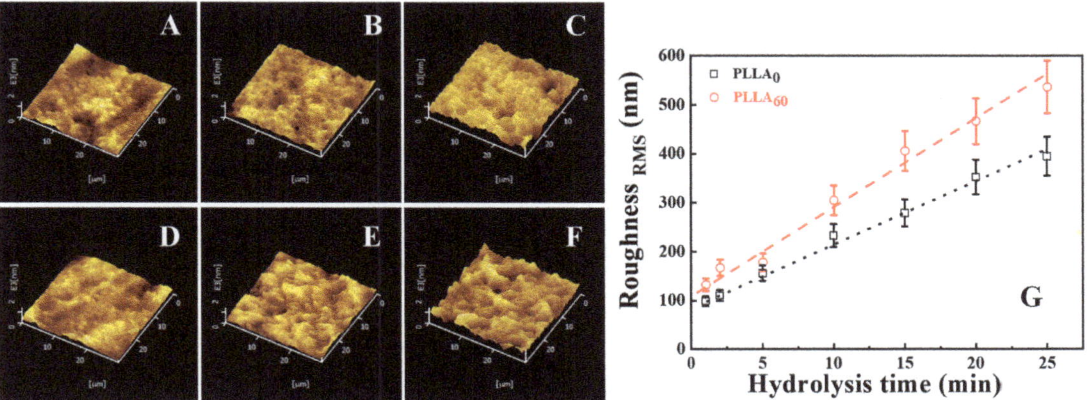


**Figure 6.** AFM images of modified PLLA membranes: $PLLA_0$ with alkaline hydrolysis time for 1 min, 5 min and 25 min (**A–C**); $PLLA_{60}$ with alkaline hydrolysis time for 1 min, 5 min and 25 min (**D–F**); (**G**) shows the roughness evolution during hydrolysis in $PLLA_0$ and $PLLA_{60}$.

### *3.5. PLLA Surface Alkali Hydrolysis Mechanism*

According to the discussion above, we can describe the hydrophilicity of PLLA in NaOH solution as well as its crystallinity/hydrolysis time dependences as follows. When the specimen with high crystallinity ($PLLA_{60}$) is kept in alkali aqueous solution, the degradation of PLLA takes place. As a result, the ester bond has been cleaved, producing carboxyl and hydroxyl groups. Both are beneficial to the improvement of PLLA hydrophilicity. At the same time, the degradation of PLLA yields rough surface. The roughness can enhance the hydrophilicity further, accounting for water contact angles down to 20° (Figures 2A and 3). However, the reason that such a hydrophilicity evolution cannot happen to the amorphous PLLA films remains obscure. It is well known that surface roughness and chemical composition are both important factors affecting the hydrophilicity of PLLA. Our attention, therefore, has been paid to the following points to understand the difference between $PLLA_0$ and $PLLA_{60}$.

(1) The densities of hydrophilic groups, defined as the number of these groups per unit area, are of great difference in specimens with low and high crystallinities. In PLLA film with very low crystallinity ($PLLA_0$), there are only terminal carboxyl and hydroxyl groups in PLLA chain. They randomly distribute on the whole surface (Figure 7A), leading to the lower density of the hydrophilic groups. However, for a PLLA film upon sufficient crystallization, the terminal carboxyl and hydroxyl groups have been expelled out from crystal lamellae and locate in the inter-lamellar regions (Figure 7B). The enrichment of carboxyl and hydroxyl groups contributes to a higher local density of the hydrophilic group.

(2) When our specimen is immersed in alkali aqueous solution, the local density of the hydrophilic group produces significant influence on the hydrolysis. In $PLLA_0$, the surface is not hydrophilic enough to enhance the contact between PLLA film and water since the water contact angle of untreated PLLA locates at 74.0° (Figure 2A). In the case of $PLLA_{60}$, the alkali solution prefers to wet the hydrophilic regions between crystal lamellae, accelerating the hydrolysis. Consequently, more carboxyl and hydroxyl groups have been generated, corresponding to the lower content of C-O-C in XPS results (Figure 4A) and enhanced absorbance in FTIR (Figure 4C, D).

(3) The surface roughness resulting from hydrolysis exhibits crucial dependence on the crystallinity. In the case of $PLLA_0$, the hydrolysis takes place randomly at the whole film surface, contributing to concave structures. The density of hydrophilic groups is not high enough to induce the sufficient contact between PLLA and water. The competition effect between chemical composition and roughness is the reason for the Cassie state in which there are air cushions, depressing the further degradation. The situation in $PLLA_{60}$,

however, is different. The area containing PLLA crystal lamellae remains stable while the hydrophilic area in inter-lamellar regions accelerates the hydrolysis since it occurs preferentially in the amorphous region of PLLA films [32]. The selective hydrolysis results in slit valleys shown in SEM (Figure 5) and AFM (Figure 6) images, and higher magnitudes of height difference in two regions, yielding higher surface roughness (Figure 6G). There are so many hydrophilic groups in the valleys that the contact between PLLA and water has been enhanced, accounting for the Wenzel state in which sufficient contact can promote further hydrolysis. This is the so-called synergistic effect.

(4) With the combination of surface roughness and chemical composition, the specimen of $PLLA_{60}$ exhibits strong hydrophilicity and lower water contact angle (down to 20°). In the case of $PLLA_0$, the hydrolysis has been slowed down, leading to lower content of carboxyl and hydroxyl groups on film surface. It is difficult to enlarge the hydrophilicity by surface roughness. The water contact angle, therefore, increases slightly upon hydrolysis. In one word, the competitive effect and synergistic effect between surface roughness and chemical composition dominate the wettability in $PLLA_0$ and $PLLA_{60}$, respectively.

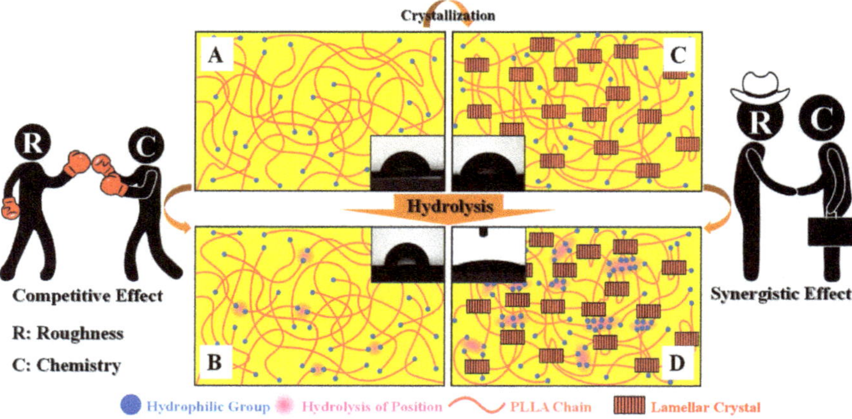


**Figure 7.** Schematic diagram of PLLA surface alkali hydrolysis mechanism: (**A**) Untreated amorphous PLLA film surface; (**B**) amorphous PLLA film surface after hydrolysis; (**C**) untreated crystallized PLLA film surface; (**D**) crystallized PLLA film surface after hydrolysis.

## 4. Conclusions

In this work, a green and convenient strategy of surface modification for PLLA film has been developed via alkaline hydrolysis during which the crystallinity of PLLA plays an important role. In the case of high crystallinity, the water contact angle decreases remarkably upon alkali hydrolysis (from 74.0° down to 18.2°). The evolutions of surface roughness (from ~100 nm up to 537 nm) and composition indicate that the local enrichment of terminal carboxyl and hydroxyl groups during isothermal crystallization contributes to sufficient contact between PLLA and water in aqueous solution, accelerating the degradation of PLLA. This is the reason for high-density carboxyl and hydroxyl groups and higher magnitudes of surface roughness in the high crystalline case. The synergistic effect between them promotes the further alkaline hydrolysis and improves the hydrophilicity of PLLA. On the contrary, terminal hydrophilic groups distribute on the whole surface randomly in the specimen with low crystallinity, which leads to limited hydrophilic modification. In that case, the surface of PLLA is still hydrophobic and the increase of surface roughness contributes to the Cassie state in which there are air cushions, depressing the further degradation. As a result, the competition effect between surface roughness and composition leads to failure of this strategy. Note that in this work, our study was performed at fixed temperature and pH levels. The conditions of extreme temperature and pH may result in a complicated Cassie state, a Wenzel state and even transition between them which will be investigated in future publication. Our results are significant for the surface modification of PLLA as well as its applications in medical fields.

**Supplementary Materials:** The following supporting information can be downloaded at: https://www.mdpi.com/article/10.3390/polym15010075/s1, Figure S1: Contact angle pictures of modified PLLA film: amorphous PLLA films with different hydrolysis time($A_1$–$A_{10}$); Different hydrolysis time of PLLA films annealed for 60 min ($B_1$–$B_{10}$).

**Author Contributions:** Conceptualization, J.Y. and Y.-B.Y.; methodology, J.Y. and Y.-B.Y.; validation, J.Y., Y.-B.Y. and Y.Z.; formal analysis, J.S. and J.Z.; investigation, J.S., J.Z. and L.Z.; resources, J.S.; data curation, J.S. and Y.Z.; writing—original draft preparation, J.S. and Y.Z.; writing—review and editing, J.S., J.Z. and O.M.; visualization, J.Y.; supervision, J.Y.; project administration, J.Y.; funding acquisition, J.Y. All authors have read and agreed to the published version of the manuscript.

**Funding:** This work was financially supported by the National Natural Science Foundation of China (51973048).

**Institutional Review Board Statement:** Not applicable.

**Informed Consent Statement:** Not applicable.

**Data Availability Statement:** The data presented in this study are available on request from the corresponding author.

**Conflicts of Interest:** The authors declare no conflict of interest.

## References

1. Langer, R.; Vacanti, J.P. Tissue engineering. *Science* **1993**, *260*, 920–926. [CrossRef] [PubMed]
2. Castner, D.G.; Ratner, B.D. Biomedical surface science: Foundations to frontiers. *Surf. Sci.* **2002**, *500*, 28–60. [CrossRef]
3. Dorgan, J.R.; Lehermeier, H.J.; Palade, L.-I.; Cicero, J. Polylactides: Properties and prospects of an environmentally benign plastic from renewable resources. *Macromol. Symp.* **2001**, *175*, 55–66. [CrossRef]
4. Cicero, J.A.; Dorgan, J.R.; Garrett, J.; Runt, J.; Lin, J.S. Effects of molecular architecture on two-step, melt-spun poly(lactic acid) fibers. *J. Appl. Polym. Sci.* **2002**, *86*, 2839–2846. [CrossRef]
5. Koons, G.L.; Diba, M.; Mikos, A.G. Materials design for bone-tissue engineering. *Nat. Rev. Mater.* **2020**, *5*, 584–603. [CrossRef]
6. Feng, Y.; Zhu, S.; Mei, D.; Li, J.; Zhang, J.; Yang, S.; Guan, S. Application of 3D Printing Technology in Bone Tissue Engineering: A Review. *Curr. Drug Deliv.* **2021**, *18*, 847–861. [CrossRef]
7. Ratner, B.D. Surface modification of polymers: Chemical, biological and surface analytical challenges. *Biosens. Bioelectron.* **1995**, *10*, 797–804. [CrossRef]
8. Burg, K.J.L.; Holder, W.D.; Culberson, C.R.; Beiler, R.J.; Greene, K.G.; Loebsack, A.B.; Roland, W.D.; Mooney, D.J.; Halberstadt, C.R. Parameters affecting cellular adhesion to polylactide films. *J. Biomater. Sci. Polym. Ed.* **1999**, *10*, 147–161. [CrossRef]
9. Jiao, Y.-P.; Cui, F.-Z. Surface modification of polyester biomaterials for tissue engineering. *Biomed. Mater.* **2007**, *2*, R24–R37. [CrossRef]
10. Ma, Z.; Gao, C.; Gong, Y.; Shen, J. Chondrocyte behaviors on poly-l-lactic acid (PLLA) membranes containing hydroxyl, amide or carboxyl groups. *Biomaterials* **2003**, *24*, 3725–3730. [CrossRef]
11. Hintz, H.; Egelhaaf, H.-J.; Peisert, H.; Chassé, T. Photo-oxidation and ozonization of poly(3-hexylthiophene) thin films as studied by UV/VIS and photoelectron spectroscopy. *Polym. Degrad. Stab.* **2010**, *95*, 818–825. [CrossRef]
12. Nagata, T.; Oh, S.; Chikyow, T.; Wakayama, Y. Effect of UV–ozone treatment on electrical properties of PEDOT:PSS film. *Org. Electron.* **2011**, *12*, 279–284. [CrossRef]
13. Desmet, T.; Morent, R.; De Geyter, N.; Leys, C.; Schacht, E.; Dubruel, P. Nonthermal Plasma Technology as a Versatile Strategy for Polymeric Biomaterials Surface Modification: A Review. *Biomacromolecules* **2009**, *10*, 2351–2378. [CrossRef] [PubMed]
14. Jacobs, T.; Declercq, H.; De Geyter, N.; Cornelissen, R.; Dubruel, P.; Leys, C.; Beaurain, A.; Payen, E.; Morent, R. Plasma surface modification of polylactic acid to promote interaction with fibroblasts. *J. Mater. Sci. Mater. Med.* **2013**, *24*, 469–478. [CrossRef] [PubMed]
15. Hergelová, B.; Zahoranová, A.; Kováčik, D.; Stupavská, M.; Černák, M. Polylactic acid surface activation by atmospheric pressure dielectric barrier discharge plasma. *Open Chem.* **2015**, *13*. [CrossRef]
16. Moraczewski, K.; Rytlewski, P.; Malinowski, R.; Żenkiewicz, M. Comparison of some effects of modification of a polylactide surface layer by chemical, plasma, and laser methods. *Appl. Surf. Sci.* **2015**, *346*, 11–17. [CrossRef]
17. Yang, J.; Wan, Y.; Tu, C.; Cai, Q.; Bei, J.; Wang, S. Enhancing the cell affinity of macroporous poly(L-lactide) cell scaffold by a convenient surface modification method. *Polym. Int.* **2003**, *52*, 1892–1899. [CrossRef]

18. Meng, J.; Boschetto, F.; Yagi, S.; Marin, E.; Adachi, T.; Chen, X.; Pezzotti, G.; Sakurai, S.; Sasaki, S.; Aoki, T.; et al. Enhancing the bioactivity of melt electrowritten PLLA scaffold by convenient, green, and effective hydrophilic surface modification. *Biomater. Adv.* **2022**, *135*, 112686. [CrossRef]
19. Bertlein, S.; Hochleitner, G.; Schmitz, M.; Tessmar, J.; Raghunath, M.; Dalton, P.D.; Groll, J. Permanent Hydrophilization and Generic Bioactivation of Melt Electrowritten Scaffolds. *Adv. Healthc. Mater.* **2019**, *8*, e1801544. [CrossRef]
20. Yu, D.-G.; Lin, W.-C.; Yang, M.-C. Surface Modification of Poly(l-lactic acid) Membrane via Layer-by-Layer Assembly of Silver Nanoparticle-Embedded Polyelectrolyte Multilayer. *Bioconjug. Chem.* **2007**, *18*, 1521–1529. [CrossRef]
21. Nemani, S.K.; Annavarapu, R.K.; Mohammadian, B.; Raiyan, A.; Heil, J.; Haque, M.A.; Abdelaal, A.; Sojoudi, H. Surface Modification of Polymers: Methods and Applications. *Adv. Mater. Interfaces* **2018**, *5*, 1801247. [CrossRef]
22. Xie, C.; Gao, Q.; Wang, P.; Shao, L.; Yuan, H.; Fu, J.; Che, W.; He, Y. Structure-induced cell growth by 3D printing of hetero-geneous scaffolds with ultrafine fibers. *Mater. Des.* **2019**, *181*, 108092–108102. [CrossRef]
23. Krylova, V.; Dukštienė, N.; Lelis, M.; Tučkutė, S. PES/PVC textile surface modification by thermochemical treatment for improving its hydrophilicity. *Surf. Interfaces* **2021**, *25*, 101184–101199. [CrossRef]
24. Elsawy, M.A.; Kim, K.-H.; Park, J.-W.; Deep, A. Hydrolytic degradation of polylactic acid (PLA) and its composites. Renew. *Sustain. Energy Rev.* **2017**, *79*, 1346–1352. [CrossRef]
25. Gorrasi, G.; Pantani, R. Hydrolysis and Biodegradation of Poly(lactic acid). In *Synthesis, Structure and Properties of Poly(Lactic Acid)*; Di Lorenzo, M.L., Androsch, R., Eds.; Springer: Cham, Switzerland, 2018; pp. 119–151. [CrossRef]
26. Vaid, R.; Yildirim, E.; Pasquinelli, M.A.; King, M.W. Hydrolytic Degradation of Polylactic Acid Fibers as a Function of pH and Exposure Time. *Molecules* **2021**, *26*, 7554. [CrossRef]
27. Baran, E.H.; Erbil, H.Y. Surface Modification of 3D Printed PLA Objects by Fused Deposition Modeling: A Review. *Colloids Interfaces* **2019**, *3*, 43. [CrossRef]
28. Schneider, M.; Fritzsche, N.; Puciul-Malinowska, A.; Baliś, A.; Mostafa, A.; Bald, I.; Zapotoczny, S.; Taubert, A. Surface Etching of 3D Printed Poly(lactic acid) with NaOH: A Systematic Approach. *Polymers* **2020**, *12*, 1711. [CrossRef]
29. Chen, W.; Nichols, L.; Brinkley, F.; Bohna, K.; Tian, W.; Priddy, M.W.; Priddy, L.B. Alkali treatment facilitates functional nano-hydroxyapatite coating of 3D printed polylactic acid scaffolds. *Mater. Sci. Eng. C* **2021**, *120*, 111686. [CrossRef]
30. Sabee, M.M.S.M.; Kamalaldin, N.A.; Yahaya, B.H.; Hamid, A.A. Characterization and in vitro study of surface modified PLA microspheres treated with NaOH. *J. Polym. Mater.* **2016**, *33*, 191.
31. Qiao, T.; Song, P.; Guo, H.; Song, X.; Zhang, B.; Chen, X. Reinforced electrospun PLLA fiber membrane via chemical cross-linking. *Eur. Polym. J.* **2016**, *74*, 101–108. [CrossRef]
32. Tsuji, H.; Mizuno, A.; Ikada, Y. Properties and Morphology of Poly(L-lactide). III. Effects of Initial Crystallinity on Long-Term In Vitro Hydrolysis of High Molecular Weight Poly(L-lactide) Film in Phosphate-Buffered Solution. *J. Appl. Polym. Sci.* **2000**, *77*, 1452–1464. [CrossRef]
33. Ton-That, C.; Shard, A.G.; Teare, D.O.H.; Bradley, R.H. XPS and AFM surface studies of solvent-cast PS/PMMA blends. *Polymer* **2001**, *42*, 1121–1129. [CrossRef]
34. You, J.; Shi, T.; Liao, Y.; Li, X.; Su, Z.; An, L. Temperature dependence of surface composition and morphology in polymer blend film. *Polymer* **2008**, *49*, 4456–4461. [CrossRef]

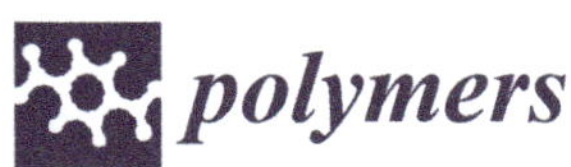

*Article*

# Effects of Accelerating the Ageing of 1D PLA Filaments after Fused Filament Fabrication

**Jaime Orellana-Barrasa, Sandra Tarancón * and José Ygnacio Pastor**

Centro de Investigación en Materiales Estructurales (CIME), Universidad Politécnica de Madrid, 28040 Madrid, Spain
* Correspondence: sandra.tarancon@upm.es

**Abstract:** The effects of post-treatment temperature-based methods for accelerating the ageing of PLA were studied on 1D single-PLA filaments after fused filament fabrication (FFF). The goal was to answer the questions whether the PLA can be safely aged—i.e., without degrading—at higher temperatures; at which temperatures, if any; how long it takes for the PLA to fully age at the chosen temperature; and which are the main differences between the material aged at room temperature and the material aged at higher temperatures. We also share other helpful information found. The use of 1D filaments allows for decoupling the variables related to the 3D structure (layer height, raster angle, infill density, and layers adhesion) from the variables solely related to the material (here, we analysed the molecular weight, the molecular orientation, and the crystallinity). 1D PLA filaments were aged at 20, 39, 42, 51, 65, 75, and 80 °C in a water-bath-inspired process in which the hydrolytic degradation of the PLA was minimised for the ageing temperatures of interest. Those temperatures were selected based on a differential scanning calorimetry (DSC) scan of the PLA right after it was printed in order to study the most effective ageing temperature, 39 °C, and highlight possible degradation mechanisms during ageing. The evolution of the thermal and mechanical properties of the PLA filaments at different temperatures was recorded and compared with those of the material aged at room temperature. A DSC scan was used to evaluate the thermal and physical properties, in which the glass transition, enthalpic relaxation, crystallisation, and melting reactions were analysed. A double glass transition was found, and its potential implications for the scientific community are discussed. Tensile tests were performed to evaluate the tensile strength and elastic modulus. The flow-induced molecular orientation, the degradation, the logistic fitting, and the so-called summer effect—the stabilisation of properties at higher values when aged at higher temperatures—are discussed to assess the safety of accelerating the ageing rate and the differences between the materials aged at different temperatures. It was found that the PLA aged at 39 °C (1) reached almost stable properties with just one day of ageing, i.e., the ageing rate accelerated by 875% for the elastic modulus and by 1635% for the yield strength; (2) the stable properties were higher than those from the PLA aged at room temperature; and (3) no signs of degradation were identified for the ageing temperature of interest.

**Keywords:** 1D PLA filaments; mechanical properties; thermal properties; temperature; ageing

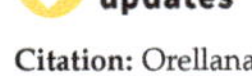

**Citation:** Orellana-Barrasa, J.; Tarancón, S.; Pastor, J.Y. Effects of Accelerating the Ageing of 1D PLA Filaments after Fused Filament Fabrication. *Polymers* **2023**, *15*, 69. https://doi.org/10.3390/polym15010069

Academic Editors: José Miguel Ferri, Vicent Fombuena Borràs and Miguel Fernando Aldás Carrasco

Received: 16 November 2022
Revised: 15 December 2022
Accepted: 21 December 2022
Published: 24 December 2022

## 1. Introduction

Additive manufacturing is a growing field that can produce 3D-printed components with few geometrical restrictions. These manufacturing processes are suitable for polymers, metals, ceramics, and composites [1–6]. Fused filament fabrication (FFF) is shared among the different additive manufacturing techniques due to its low cost and accessibility. The FFF works with materials that melt and flow (i.e., that are not suitable for thermosets), such as PLA [7–9], PETG [10], and ABS [11,12]. In this work, we study polylactic acid (PLA), a polymeric biomaterial commonly studied in many applications [13], from basic furniture and decoration [5,14] or textile and fashion [9] to advanced medical applications [7,15,16]. All of this is of great interest, as an accessible technique (FFF) is suitable for producing components with almost no restrictions with a relatively cheap material (PLA). However, before testing any PLA or PLA-based composite materials produced via FFF, it must be considered that FFF processing melts and extrudes the PLA. This creates a thermal treatment that returns the polymer to an out-of-equilibrium state from which it slowly ages, which in this work means an evolution of the material due to the thermal vibration of the macromolecules to their equilibrium position. Because of this, the properties of the printed PLA slowly evolve (age) and increase the elastic modulus, the tensile strength, the yield strength, the glass transition, the enthalpic relaxation, and any other property related to the conformation of the amorphous part of the PLA. Logically, and knowing this phenomenon, it should be mandatory to fully age the PLA—and explicitly prove that it is entirely aged—to make possible a reliable comparison of results between researchers when the effect of any variable is studied on the properties of printed PLA filaments or structures. That said, in our first work on PLA [17], we performed a comprehensive study on how the 1D filaments of printed neat PLA 2003D (one of the many PLA grades) naturally aged. We found that this high-molecular-weight PLA required up to 90 days—at room temperature, inside zip bags with a desiccant—to stabilise its mechanical and thermal properties and that these properties increased by up to 75% simply because of the ageing. An essential and argued aspect of that work was the use of 1D filaments instead of the standardised 3D-printed samples. Additive manufacturing via FFF consists in placing a continuous flow of fused material at particular locations, layer by layer, until the addition of material creates an object. The manner of producing these components layer by layer makes them highly anisotropic, and several variables need to be defined that have a significant influence on the properties of the final printed product: infill density, nozzle temperature, printing speed, layer thickness, raster angle, and bed temperature [18–21]. This inherent fact about any 3D-printed component makes understanding the material difficult using 3D-printed samples. It is almost impossible to separate the structure-related variables (the layer adhesion or the concentration of stresses on corners) and the material-related variables (the molecular weight, the crystallinity, and the orientation of the molecules). Surprisingly, no researchers, as far as we know, have studied the properties of 1D-printed PLA specimens by decoupling the material properties from the structural variables [22]. Nonetheless, studies on 3D-printed samples have highlighted the difficulties of addressing the multivariable problem of studying 3D-printed components [18].

Altogether, the facts from our previous work [18], in which the PLA 2003D required up to 90 days at room temperature to be entirely aged, raise the question of how to control the ageing rate of the printed PLA filaments and measure the real impact on the PLA. We performed a second work demonstrating that the PLA's ageing can be safely stopped (without any decrease in the mechanical properties) by freezing the PLA at −24 °C in a conventional freezer [23]. However, we only studied lower temperatures—as our goal in that study was to completely stop the ageing—and did not conduct any research on higher temperatures for accelerating its ageing. In reviewing the literature, we could not find answers to our research questions posed here. Thus, we waited three months before making any characterisation of neat PLA material afterwards, at least until now (except for composite PLA-based materials). It is not new information that the ageing rate of a polymer can be accelerated by increasing the temperature. However, increasing temperatures might be problematic for PLA materials, as this enhances any potential degradation of the PLA, especially in humid atmospheres, based on the hydrolytic degradation due to the cleavage of ester groups [24–29]. This degradation should not be a problem at temperatures below the PLA's glass transition temperature ($T_g$) [19].

As the macromolecules gain more and more kinetic energy due to an increase in the temperature, they have more energy to reach more stable locations on the microstructure of the polymer, reducing its free space and ageing the material. For effective ageing (in which the properties increase until they stabilise), it would be ideal to use temperatures as high as possible but consistently below the $T_g$. However, if temperatures above $T_g$ are used, the macromolecules have so much kinetic energy that they can no longer be in ordered and stable positions—it is as though the amorphous part were to melt—and the ageing of the PLA is no longer possible.

The $T_g$ is a second-order transition in which the polymer chains significantly increase their mobility. It can be measured using differential scanning calorimetry (DSC) or fast scanning calorimetry (FSC) as an endothermal drop related to the change in heat capacity [30], or through dynamic mechanical analysis (DMA) as a drop on the elastic modulus, as an increase in the loss modulus, and as an increase as well of the tan δ [31,32]. In this work, the DSC technique was used. The glass transition was not instantaneous. The material did not heat evenly during the DSC (depending on the mass of the material, the manner in which it is placed inside the DSC crucible, and the heating rate). The $T_g$ also depends on a molecular weight distribution which, by its essence, makes the $T_g$ into a distribution of temperatures. Even though the $T_g$ is defined at some point between a changing line in the DSC graph (a transition), there is an onset and offset temperature. There are different theories for defining the specific $T_g$ value from a DSC, but they all provide a $T_g$ value higher than the onset temperature. Thus, lower values than the usually defined $T_g$ should be used to ensure that all the PLA is ageing. More precisely, the ageing temperatures should be lower than the onset temperature measured at a slow heating rate and with as much surface of the PLA in direct contact with the DSC crucible as possible. This will ensure that all the amorphous part of the PLA (which, as a polymer, is formed through a weight distribution of macromolecules) remains restricted to low degrees of mobility, leading to an effective ageing process [24]. Otherwise, for a $T_g$ value measured at 10 °C, ageing performed at 1 degree below that measured $T_g$ might not age the polymer or, at least, not all of its structure.

Regarding the literature about ageing PLA, many researchers have studied the effect of heating the PLA after it was processed via FFF [19,33,34]. For example, some researchers have studied thermal treatments at temperatures above $T_g$ to improve the mechanical properties of PLA by increasing the crystallinity [19,34]. Alternatively, other work mentions an improvement in mechanical properties due to the enhanced bond between the layers and an increase in crystallinity [35]. However, as far as we know, no research has considered the isolated 1D-printed filament but the whole 3D-printed structure after FFF [12–14,27,29].

This paper analyses and discusses the effects solely affecting the PLA when it is accelerated by raising the ageing temperature in a low-humidity atmosphere by studying those 1D filaments. Our research questions to answer are:

1. Is it safe to effectively accelerate the ageing of the PLA by heating it? To answer this question, the degradation of the PLA was analysed with regard to the shifting of peaks in the DSC scans (decreased temperature in the crystallisation reactions, increase in the crystallisation enthalpies, a decrease in the melting point) and the decrease in both the elastic modulus and tensile strength. As the primary degradation mechanism is the hydrolytic degradation with the air humidity, the change in colour of a desiccant inside the ageing device was used to determine the low-humidity conditions during the ageing qualitatively.
2. At which temperature can the PLA be effectively and safely accelerated? How long does the PLA need to be entirely aged at that temperature? Different temperatures were studied. The temperatures were chosen based on a DSC scan on the PLA right after it was printed after finding an optimal ageing temperature and forcing the degradation mechanism. The evolution of the thermal and mechanical properties was measured at different times to ensure that stable properties were reached and the minimum ageing time was set for the central ageing temperature of interest.

3. What are the main differences between the PLA that was slowly aged at room temperature and the one aged at higher temperatures? The PLA aged at room temperature is compared with the material effectively aged at higher temperatures. This makes possible a better understanding of how future 3D-printed samples of PLA will be affected—at least for the thermal and mechanical properties of the 3D structures that are no longer slowly aged during 3 months at 20 °C but accelerated at a higher temperature.
4. Has any other phenomenon of interest been detected? As different properties were studied, the results obtained during this research were analysed to determine other phenomena related to the PLA and, if something interesting is found, to share it with the research community.

To answer these questions and solely address the variables related to the material, 1D filaments were studied, following the methodology developed in previous research [15,17,23] based on the UNE-EN ISO 527-2:2012. The degradation, the flow-induced molecular orientation, the so-called summer effect, and other phenomena of interest (a hidden $T_g$) are analysed. This paper complements our previous study on controlling the ageing of the PLA [23], in which the effects of freezing the PLA to stop its ageing were understood. The results of that paper are used in this research. Now, we are looking to understand the opposite, that is, how accelerating the ageing rate by increasing the ageing temperature impacts the mechanical and thermal properties of the filaments after FFF, the building block of any 3D-printed structure for answering those questions.

Answering our research questions will provide the following benefits:

5. Our main aim is to find and understand a reliable, safe, and effective method to accelerate the ageing of PLA 4043D-based materials (decoupled from the effects of the structure) to be sure of how the ageing method is affecting the PLA and reduce the research schedule from months (time required until the PLA is entirely aged at room temperature) to just a few days (as demonstrated in this study).
6. An emphasis on the importance of studying aged materials is essential for comparing results from different research with confidence.
7. It would allow us to understand and report how the thermal treatments affect the material decoupled from the variables related to the structure. Nonetheless, this study will have some limitations for extrapolation to 3D structures, as the thermal history of the PLA during the printing of 3D models is different depending on printed geometry, printer parameters, and ambient conditions [20]. However, studying the most fundamental building block of any 3D-printed structure, a 1D filament, will help unveil the phenomenon underlying the changes observed in a 3D-printed e-structure after thermal treatment, simplifying the multivariable problem of 3D-printed parts from a novel perspective not found in the literature.

## 2. Materials and Methods

This study analysed the first three months of the natural ageing of 1D filaments of PLA aged at room temperature and higher temperatures. Our previous study showed that a similar high-molecular-weight PLA required three months of natural ageing at 20 °C in order to stabilise its properties; this is the reason for conducting the ageing studies for up to 91 days. As the primary objective was to compare the samples at room temperature (20 °C) with the samples aged at higher temperatures (39, 51, 65, 75, and 80 °C), no longer times were considered. For a better understanding of the content of this work, a summary of the material production, storage, ageing, and experiments is provided in Figure 1, as well as the research question.

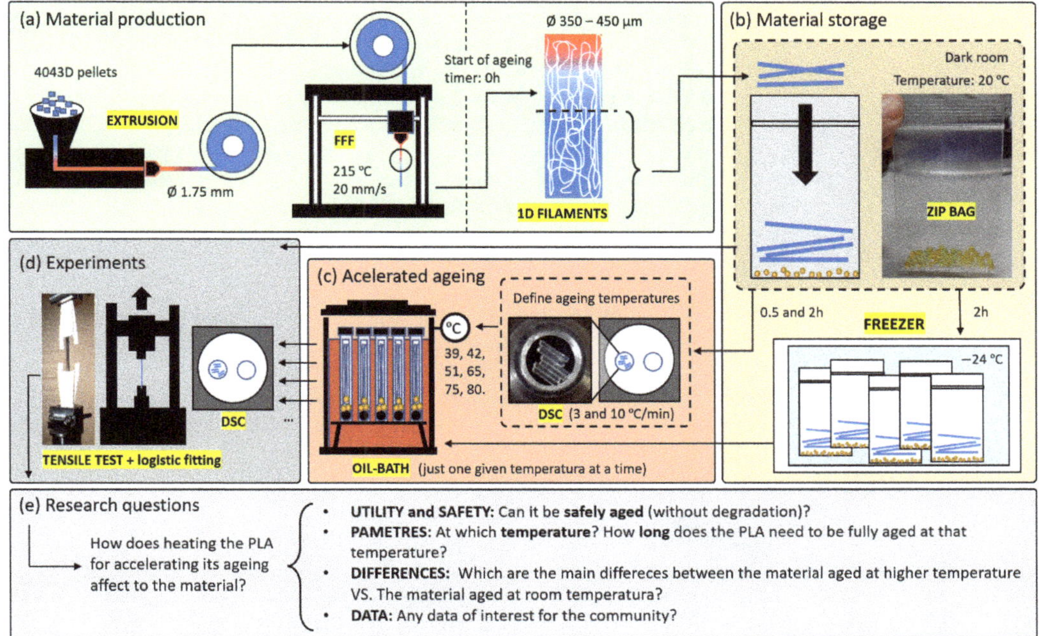


**Figure 1.** Workflow of the experiments performed: (**a**) production of PLA 1D filaments from pellets; (**b**) storage of PLA filaments; (**c**) accelerated ageing prior to definition of ageing temperatures with a DSC scan on the material right after it was printed; (**d**) experiments performed; (**e**) research questions.

### 2.1. Materials and FFF

The material studied was the 4043D (a commonly studied PLA suitable for 3D printing applications) from Nature Works and supplied by Prusa in the form of filaments with a diameter of 1.75 ± 0.02 mm. The material was printed in a Prusa i3MK3S+ with a 0.4mm nozzle at 215 ± 1 °C (a temperature between 200 and 220 °C, typical printing temperatures for neat PLA and recommended by Prusa) and at an extrusion rate of 20 mm/s, a relatively slow printing speed found in accurate printing processes and studied by other authors [36]. Filaments with diameters between 350 and 450 micrometres were obtained (Figure 1a). This 4043D material was compared with the 2003D from Nature Works, a high-molecular-weight PLA (182.000 g/mol) used in our previous studies [15,17,23]. The material 2003D has been discontinued; thus, we looked for a similar PLA currently used in the literature, and found the 4043D [36]. The PLA 4043D was printed in a room at 23 °C and ambient humidity, then frozen at −24 °C after 2 h of natural ageing at 20 °C to ensure that all the PLA studied were under the same conditions [23], as summarised in Figure 1b.

### 2.2. Ageing Procedure

Printed filaments were stored inside zip bags with a desiccant inside to ensure a low-humidity-controlled atmosphere, essential for avoiding any degradation through the hydrolysis of the PLA. Ageing at room temperature was performed inside these bags, as indicated in Figure 1b. The bags were later placed in a dark room, with no UV light sources and at a controlled temperature of 20 ± 1 °C. The ageing temperatures were 39, 42, 51, 65, 75, and 80 °C. The reason for these temperatures is explained in the discussion, but in summary, they were chosen based on the DSC scan of a 2h aged sample of PLA right after it was printed (Figure 1c) to try to force any potential degradation mechanism inside the ageing device and to probe the importance of adequately ageing the material before making conclusions about it. The ageing at those six temperatures higher than room temperature was performed inside an oil bath, using an "oil bath" device, schematised in Figure 2a. Premium Mineral Oil 10W-40 (Repsol) was used as the liquid bath in the device, as shown in Figure 2b. Then, the PLA filaments were introduced inside PET zip bags with a desiccant inside. After this, PP test tubes were prepared (Figure 2c) by placing a steel ball at the bottom—to ensure that the test tube sinks—and adding 2–3 g of desiccant. Then, the zip bags with the filaments were introduced inside PP test tubes. The PP tubes were sealed with vacuum grease applied to the caps-tube joints (Figure 2c). Finally, the tubes were immersed inside the oil, with the oil previously stabilised at the required temperature. The temperature was monitored with a precision of ±0.1 °C with the help of a thermocouple (FLUKE 50D K/J Thermometer) introduced in a reference test tube—and not directly into the oil—ensuring that the measured temperatures were the actual ageing temperatures. The oil level in the oil bath was slightly below the cap-tube joint, as schematised in Figure 2a. Different methods were tested for the ageing at elevated temperatures, involving water as the hot liquid in the water bath device; however, any method involving water instead of oil failed as the desiccant completely changed its colour in just a few minutes. The cost of the ageing device with the oil and desiccant was EUR 150, making it an economical and practical method for the rest of the researcher community.

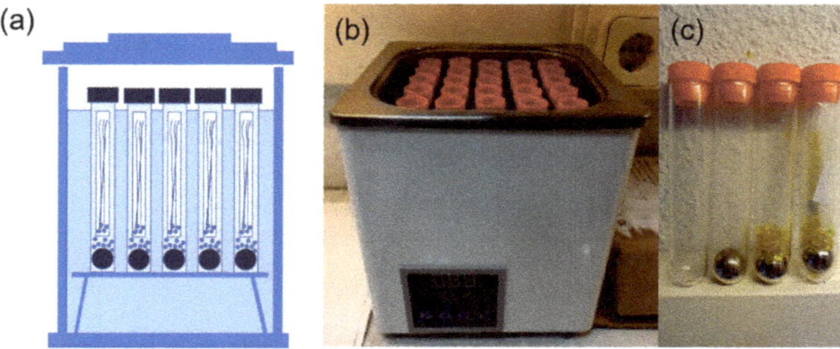


**Figure 2.** Ageing device. (**a**) Scheme of the ageing device; (**b**) actual picture of the ageing device with 25 test tubes in a 5-by-5 grid, with the middle one used for measuring the temperature; (**c**) preparation of PP test tubes, left to right: PP tube, with a ball of steel, with desiccant, with the samples inside a zip bag with more desiccant inside.

### 2.3. Differential Scanning Calorimetry

The differential scanning calorimetry (DSC) was performed on a Mettler Toledo 822e device at a heating rate of 10 °C/min from 30 to 180 °C (Figure 1d). The device was calibrated following the Indium standard. Aluminium crucibles of 40 microlitres were used, in which 4.5 to 5.5 mg of PLA filaments were cut into 2–3 mm pieces and carefully placed inside, ensuring that all the pieces were in contact with the bottom. This was important for minimising measurement errors, as it was found that differences of up to 1 °C were obtained if not all the PLA was in direct contact with the crucible [16].

### 2.4. Tensile Test

Mechanical tests were performed on an Instron 5866 universal tensile test machine with a load cell of 1 kN (Figure 1d). Before the mechanical testing, the samples were attempered at room temperature inside zip bags with a desiccant for 2 h in a room with no sources of sunlight. Then, the same procedure was followed that was described in our previous study for testing the 1D filaments [15,17,23]. The strain rate used was 1 mm/min (consistent with the UNE-EN ISO 527-2:2012) on filaments of 20 mm length, in addition to 7.5 mm extra length on each side glued with a cyanoacrylate-based adhesive to a cardboard frame (Figure 3). Then, the cardboard frame was firmly held with mechanical clamps, as shown in Figure 3b. The clamps were joined to the load cell with rotulas to avoid non-tensile stresses on the filaments during the tests. For modelling mechanical properties, the logistic model proposed in our previous study was used [1].

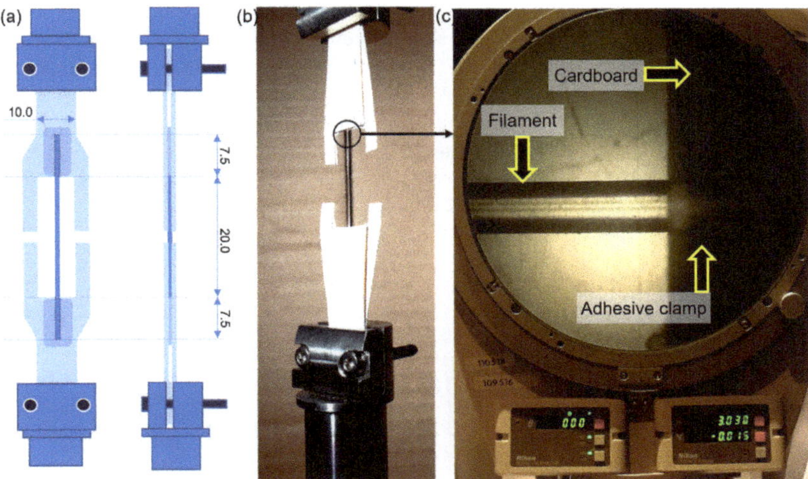


**Figure 3.** Clamps setup used for the mechanical testing of 1D-printed PLA filaments. (**a**) Scheme of the setup, (**b**) picture of an actual sample prior to testing, (**c**) profilometer detail of the filament, cardboard, and adhesive clamp.

This is the expression used for the logistic fitting:

$$S(t) = \frac{S_\infty}{1 + Ae^{-Bt}} = \frac{(1 + A)S_0}{1 + Ae^{-Bt}} \quad (1)$$

where

1. $S_\infty$, the value of the property at an infinite time of ageing. It corresponds to the stabilised value. It has the units of the property (e.g., stress units such as MPa or GPa for the tensile strength or elastic modulus).
2. A, the ageing potential (adimensional). It describes the gap between the property at zero days (on the material right after it was printed) and infinite days of ageing (stabilized).
3. B, is the ageing rate in units of (1/time), which describes the speed at which the material evolves towards the steady state. This fitting was proven to properly define the evolution of the tensile strength, yield strength and elastic modulus with the natural ageing at room temperature by using just those 3 parameters. The objective was to increase the value of the B parameter, the ageing rate, by increasing the ageing temperature. B was used to quantify the extent to which the ageing rate varied at different ageing temperatures. The logistic fitting is impossible to apply if the material is not ageing, i.e., for the temperatures higher than $T_g$ (51 °C and 65 °C). When the logistic fitting was not possible because the material did not evolve with the ageing temperature (51 °C in this study), the linear fitting y = mx + c was used. The 65 °C was not fitted to any curve as there were not enough points.

## 3. Results

### 3.1. Thermal Properties

All the results for the thermal properties measured on the DSC are summarised in Table 1.

**Table 1.** Thermal properties of the 1D-printed PLA. All samples were printed at 215 °C. Temperature measured in °C with a ± 0.5 error (except as indicated), enthalpies in J/g with a 3.4% error of total value following Mettler Toledo guidelines, time in hours (h) or days (d) with an error of ± 0.1h. All aged samples at temperatures above 20 °C were aged after 2 h of natural ageing (for example, one day at 39 °C means that the material has been aged 2 h at 20 °C and one day at 39 °C), crystallinity in percentage (%) with an error of ± 1. Positive enthalpy changes indicate that the material absorbs energy, following the IUPAC convention. $T_{g.1}$ and $T_{g2}$: first and second glass transition on the DSC during heating; $T_{ER}$: enthalpic relaxation temperature; $\Delta H_{ER}$: enthalpy of enthalpic relaxation; $T_{CC}$: cold crystallisation temperature; $\Delta H_{CC}$: enthalpy of cold crystallisation; $T_m$: melting temperature; $\Delta H_m$: enthalpy of melting; $\chi_{\%}$: crystallinity percentage. The symbol "-" is used for undetected values.

| Ageing | Ageing Temperature | $T_{g.1}$ | $T_{ER.1}$ | $\Delta H_{ER.1}$ | $T_{g.2}$ | $T_{CC}$ | $\Delta H_{CC}$ | $T_m$ | $\Delta H_m$ | $\chi_{\%}$ |
|---|---|---|---|---|---|---|---|---|---|---|
| 0.5 h | 20 ± 0.1 | - | - | - | 54.4 | 123 | −15.2 | 152 | 15.2 | <3 |
| 2 h [1] | | 45.5 | - | - | 57.5 | 123 | −15.2 | 152 | 15.2 | <3 |
| 3 h | | 46.4 | - | - | 57.6 | 123 | −14.7 | 151 | 16.4 | <3 |
| 1 d | | 51.5 | - | - | 59.9 | 124 | −17.0 | 154 | 17.2 | <3 |
| 4 d | | 54.8 | - | - | 61.1 | 121 | −16.8 | 152 | 16.4 | <3 |
| 7 d | | 56.1 | 58.6 | 0.15 | 62.6 | 122 | −15.6 | 152 | 15.9 | <3 |
| 14 d | | 56.6 | 58.9 | 0.79 | - | 121 | −15.8 | 152 | 16.3 | <3 |
| 24 d | | 56.9 | 60.0 | 1.61 | - | 122 | −16.2 | 152 | 16.5 | <3 |
| 49 d | | 57.2 | 60.6 | 2.5 | - | 122 | −15.9 | 152 | 16.3 | <3 |
| 91 d | | 57.5 | 61.0 | 3.7 | - | 123 | −15.7 | 153 | 16.5 | <3 |
| 1 d | 39 ± 0.1 | 58.4 | 61.4 | 2.0 | - | 125 | −14.4 | 153 | 15.2 | <3 |
| 2 d | | 59.0 | 61.7 | 3.3 | - | 124 | −17.8 | 152 | 18.6 | <3 |
| 4 d | | 59.4 | 62.2 | 4.0 | - | 125 | −14.5 | 152 | 15.7 | <3 |
| 7 d | | 60.7 | 63.2 | 4.4 | - | 125 | −14.7 | 152 | 15.5 | <3 |
| 14 d | | 61.0 | 63.5 | 4.7 | - | 124 | −15.0 | 152 | 15.7 | <3 |
| 24 d | | 61.2 | 63.5 | 5.0 | - | 124 | −15.9 | 152 | 17.2 | <3 |
| 1 d | 42 ± 0.1 | 59.5 | 62.1 | 3.2 | - | 124 | −14.0 | 152 | 15.4 | <3 |
| 4 d | | 63.3 | 65.6 | 5.7 | - | 124 | −15.2 | 152 | 16.9 | <3 |
| 1 d | 51 ± 0.1 | 57.5 | 61.5 | 0.6 | - | 122 | −21 | 152 | 22 | <3 |
| 4 d | | 58.2 | 62.5 | 0.9 | - | 122 | −22 | 152 | 23 | <3 |
| 7 d | | 57.7 | 62.0 | 0.8 | - | 121 | −25 | 151 | 24 | <3 |
| 14 d | | 57.1 | 63.0 | 0.7 | - | 121 | −23 | 152 | 23 | <3 |
| 24 d | | 57.5 | 62.3 | 0.8 | - | 121 | −24 | 152 | 24 | <3 |
| 42 d | | 57.2 | 62.0 | 0.8 | - | 121 | −24 | 152 | 24 | <3 |
| 1 d | 65 ± 0.1 | 57.7 | 61.9 | 0.6 | - | 118 | −27 | 150 | 27 | <3 |
| 1 d | 75 ± 0.1 | 60.8 | - | - | - | - [2] | - [2] | 154 | 28 | 20 |
| 1 d | 80 ± 0.1 | 61.9 | - | - | - | - | 0 | 153 | 27 | 28 |

[1] Accelerated ageing materials start from this reference at 2 h of natural ageing at 20 °C. [2] A precise calculation is not possible. Values detected, but highly mixed α′- and α-related reactions.

It is generally said that the ageing rate of a polymer can be accelerated only at temperatures below $T_g$. For this reason, $T_g$ was measured on PLA samples right after printing (0.5 and 2 h) to set a maximum ageing temperature. However, how this maximum ageing temperature should be obtained is not further explained. Our goal was to use the highest temperature possible. Two DSC tests were performed to obtain a picture of the $T_g$ reaction on the PLA right after it was printed. Figure 4 shows the DSC scans at 10 °C/min of the PLA aged at 20 °C at different times. A phenomenon was found, not described yet in the literature as far as we know, in which two glass transition reactions appeared. These were later sought and found on the PLA 2003D.

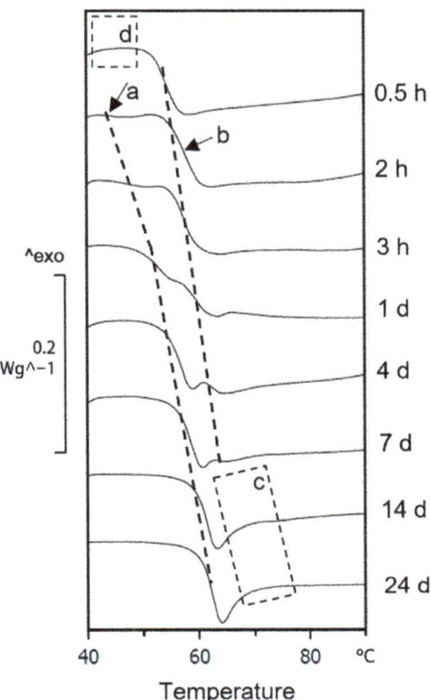


**Figure 4.** Detail of DSC scans at 10 °C/min of PLA 4043D samples aged at 20 °C, for illustrating the double $T_g$ observed: (a) hidden $T_{g1}$ on PLA right after it was printed, which was used to determine the maximum temperature at which the material could be aged and highlight the importance of studying aged materials; (b) second glass transition, $T_{g2}$; (c) missing $T_{g2}$; (d) missing $T_{g1}$.

In Figure 4, a clear endothermal transition consistent with $T_g$, called $T_{g2}$, was identified after 0.5 h of the printed material. A slight endothermal process was noticed before the $T_{g2}$ on the samples aged for 2 h. The experiment was conducted at heating rates of 10 °C/min during the following days (Figure 4). We observed: (1) that the $T_{g2}$ was disappearing, and (2) that the slight endothermal reaction was becoming larger and forming the typical shape of the glass transition; we called it $T_{g1}$ as it was the first glass transition to appear on the DSC heating scan. For the $T_{g1}$ to show up consistently, at least 2 h of natural ageing at 20 °C was required after the FFF process (Figure 4). Samples studied at 0.5 h after being printed did not show $T_{g1}$, making this double $T_g$ a potential problem for (a) any DSC in the literature on PLA samples right after printing at ageing times lower than 2 h, (b) for the typically used procedure "first heating – cooling – second heating" [37], as our findings provide information about $T_{g2}$ but not about $T_{g1}$. Notice that the PLA 4043D-based materials are commonly studied in the literature [27–32]. Moreover, this $T_{g1}$ phenomenon is very subtle during the first hours. A large mass of PLA-to-surface contact between the PLA and the aluminium crucible made it even more challenging to detect it. This result shows the importance of carefully placing the material inside the DSC crucible in the most controlled and reproducible manner and with a similar mass (not in the range of 2 to 10 mg). Both variables were responsible for shifts of up to +1 °C in the temperature peaks and a widening of the peaks. The evolution of $T_{g1}$ was found to be extremely fast (compare its value at 2 and 3 h on samples aged at 20 °C in Table 1), making it hard to define a maximum temperature limit for accelerated ageing.

After this discussion on the double glass transitions and considering that the $T_g$ is defined approximately as the halfway point of the glass transition—depending on the standardised method used—not the glass transition but the onset temperature of the glass transition was used as the maximum temperature at which the ageing could be effectively performed. The onset temperature of the first glass transition ($T_{g1}$) was 42.9 °C (Figure 5a).

This temperature was measured from a DSC at 10 °C/min after 2 h of natural ageing at 20 °C, and it is known that for obtaining precise values, slower DSC scans should be used. The heating rate of 10 °C/min is a heating rate that can be easily found in research on PLA from other authors in the literature [30,38–44] as well as in our previous works [15,17,23]. However, notice the difference between the scan at 10 °C and 3 °C in Figure 5. The onset temperature of $T_{g1}$ on the DSC scan at 3 °C/min was 41.7 °C, decreasing by 1.2 degrees the measured onset temperature.

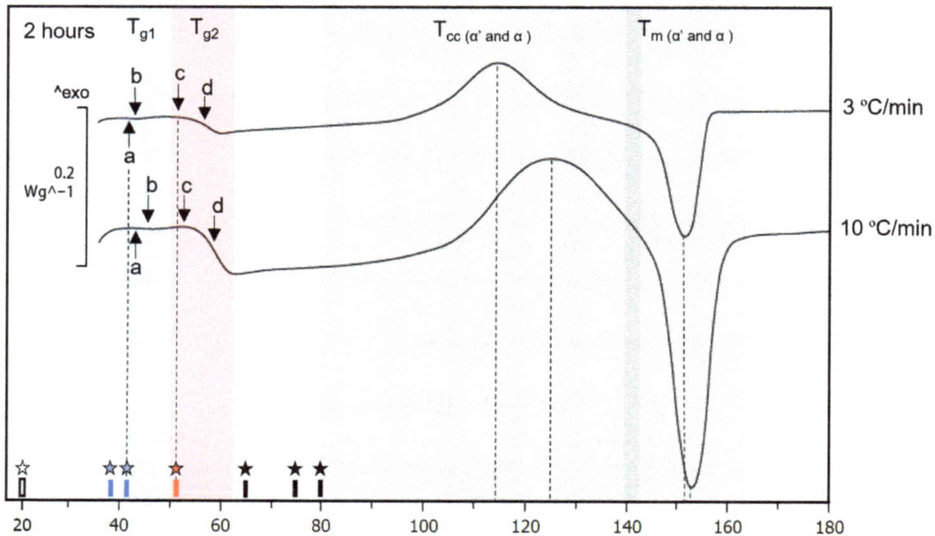


**Figure 5.** DSC of neat PLA right after printing via FFF at heating rates of 3 and 10 °C/min. Stars indicate the ageing temperatures used: 20, 39, 42, 51, 65, 75, and 80 °C. (a) T.onset of the first glass transition; (b) $T_g$ of the first glass transition; (c) T.onset of the second glass transition; (d) $T_g$ of the second glass transition. Crystallisation and melting reactions are also indicated. Dashed lines are included for reference.

The evolution of $T_{g1}$ was relatively fast, making it even harder to define a maximum temperature limit for the accelerated ageing. It was found that 42 °C worked very well for accelerating the ageing of the PLA 4043D naturally aged 2 h after FFF (results in Table 1). That temperature was above the measured 41.7 °C, but we decided to use a more conservative temperature to be on the safe side of the daily use of ageing samples 2 h right after printing. The chosen temperature was 39 °C, approximately 3 degrees below the onset temperature, 6 degrees lower than the first glass transition measured ($T_{g1}$ at 10 °C/min on samples aged 2 h at 20 °C after FFF), and 15 degrees lower than the $T_{g2}$ measured at 0.5 h after printing. The 51 °C ageing temperature was chosen as it was close to but below the second glass transition ($T_{g2}$) found on the PLA right after it was printed in order to demonstrate that the $T_{g2}$ measured cannot be directly used as an indicator of the maximum temperature for accelerating the ageing of the PLA, especially if it was measured on a sample right after printing. Temperatures of 65, 75, and 80 °C were used to provide further information on how ageing at higher temperatures could affect the PLA (inside our ageing device) and to show any potential degradation evidence together with the mechanical tests to confirm the safeness of accelerating the ageing of the PLA. The DSC results are discussed together with the mechanical properties.

### 3.2. Mechanical Properties

All the results from the tensile tests on the PLA 4043D are summarised in Figures 6 and 7. Tests at 0.5 h were not viable, as the adhesive clamp needs time to harden, and the minimum possible ageing time at room temperature was 2 h (approximately 0.1 days) from printing to a reliable tensile test.

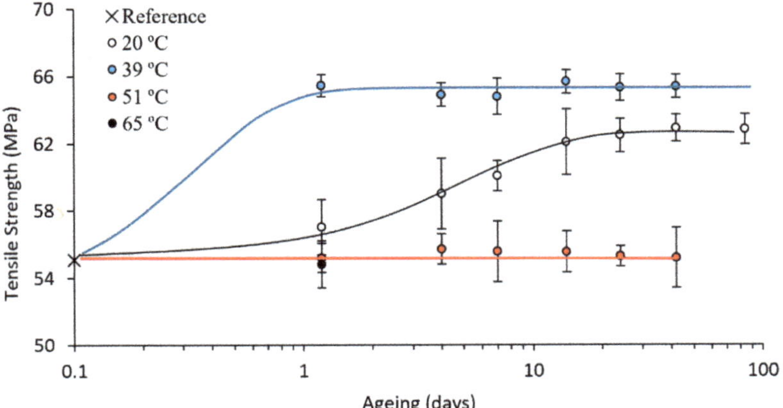


**Figure 6.** Tensile strength evolution with ageing time at different ageing temperatures. Logistic fittings for 20 and 39 °C, linear fitting for 51 °C. These fittings correspond to Equations (2), (4), and (6), respectively.

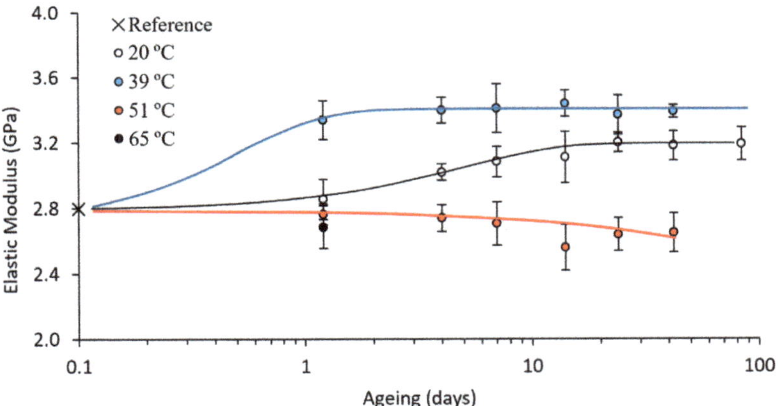


**Figure 7.** Elastic modulus evolution with ageing time at different ageing temperatures. Logistic fittings for 20 and 39 °C, linear fitting for 51 °C. These fittings correspond to Equations (3), (5), and (7), respectively.

Thermal and mechanical properties were measured, and they are highly interrelated. For an appropriate discussion, all the results must be considered together. The samples aged at 20 °C were used for setting a reference. In this way, it was possible to compare the evolution of the samples aged at higher temperatures with these controls. As expected, the ageing of these PLA 4043D samples at room temperature was consistent with those in our previous study on a similar PLA, the 2003D. The trends in the evolution of the properties remained similar: (1) the enthalpic relaxation (both temperature and enthalpy) increased until stabilisation; (2) the enthalpic relaxation enthalpy continued to increase even after the stabilisation of the glass transition; (3) the crystallisation and melting reactions remained unaltered; (4) the elastic modulus and yield strength evolved until they stabilised, requiring times similar to the stabilisation of the glass transition. This similar evolution trend was further proven by applying our proposed logistic fitting [17] to the mechanical properties—elastic modulus and yield strength—and the fitting provided an excellent description of the evolution of these two properties.

The logistic fittings in Figures 6 and 7, for both the tensile strength and the elastic modulus, correspond with the following Equations (1) and (2):

$$\sigma_y(\text{MPa}) = \frac{62.7}{1 + 0.13\text{e}^{-0.20\text{t}}} \tag{2}$$

$$\sigma_y(\text{MPa}) = \frac{3.18}{1 + 0.13\text{e}^{-0.24\text{t}}} \tag{3}$$

where $\sigma_Y$ is the yield strength in MPa, $E$ is the elastic modulus in GPa and t is the ageing time in days. This confirms that our proposed logistic fittings correctly described the evolution of the mechanical properties with the ageing at room temperature.

The samples aged at 39 °C showed almost stable properties after just one day of ageing, indicating a significant improvement in the ageing rate, the B parameter on the logistic fitting. The logistic fittings were calculated, and Equations (4) and (5) were obtained:

$$\sigma_y(\text{MPa}) = \frac{65.4}{1 + 0.26\text{e}^{-3.47\text{t}}} \tag{4}$$

$$\sigma_y(\text{MPa}) = \frac{3.41}{1 + 0.28\text{e}^{-2.34\text{t}}} \tag{5}$$

In comparing the parameters obtained in the logistic fittings on samples aged at 20 and 39 °C (Table 2), we observed not only that the ageing was accelerated (increase in B), but also that it improved the value of the stable properties (increase in $S_\infty$). This result was again consistent with results in the literature, as it was at the limit of the maximum temperature that can be used for accelerating the ageing, as explained beforehand, and it was specifically chosen for this reason. Properties close to the stable ones were obtained after just one day (ageing rates increased by, approximately, 1000 % regarding B values calculated from fitting equations (Table 2)).

**Table 2.** Comparison of logistic fitting parameters for the yield strength and elastic modulus on samples aged at 20 °C and 39 °C.

| Yield Strength | A | B | $S_\infty$ |
|---|---|---|---|
| 20 °C | 0.13 | 0.20 | 62.7 |
| 39 °C | 0.26 | 3.47 | 65.4 |
| Variation | 100% | 1635% | 4.7% |
| Elastic modulus | A | B | $S_\infty$ |
| 20 °C | 0.13 | 0.24 | 3.18 |
| 39 °C | 0.28 | 2.34 | 3.41 |
| Variation | 115% | 875% | 7.6% |

It is known, and was here observed, that the $T_g$ increases with the ageing of samples after printing. Thus, the maximum ageing temperature could be increased proportionally to the increase in $T_g$, but fully optimising the accelerated ageing process was not the objective of this work.

The evolution of the properties to higher $\mathbf{S_\infty}$ values is explained as follows. The macromolecules with a higher kinetic energy—higher ageing temperature—can surmount higher energy barriers, reach more stable locations, and provide a stronger and more rigid material as well as a higher $T_g$ and enthalpic relaxation. This effect was referred to as the summer effect, as samples which might be stable at ambient conditions, if not stored properly, could evolve to higher values due to summer temperatures. The summer effect also applies for explaining the outstanding results obtained on samples aged at 42 °C, in which even higher values were obtained on the DSC scans after just four days for the thermal properties, reaching the $T_{g1}$ 63.3 °C and its enthalpic relaxation 65.6 °C and 5.7 J/g without any signal of degradation.

Samples aged at 51 °C did not evolve, as the temperature was over the $T_{g1}$, proving the importance of studying aged materials. The samples, however, had a relatively high $T_g$, as if they were aged. However, the enthalpic relaxation value was as it would be if the material had been just printed, raising a contradiction for any direct relation between the glass transition and the enthalpic relaxation: samples which are not aged regarding the enthalpic relaxation might show $T_g$ values similar to those of an aged material. This shows that it is difficult, if not impossible, to correlate some of the thermal properties measured on the PLA. For example, a $T_{g1}$ of 60 °C, an average value of $T_g$ found on aged PLA materials, is not necessarily a sign of a high enthalpic relaxation enthalpy (the figure typically used in the literature to indicate how aged the PLA is). However, some trends can be found and modelled [17]. The desiccant remained stable during the first days of ageing, but, after 42 days, it turned into a blueish colour indicative of adsorbed humidity and highlighting a potential hydrolytic degradation of the PLA, Figure 8. No significant signs of hydrolytic degradation were found on the PLA aged at 51 °C on the DSC scans, except for a slight decrease in the crystallisation temperature and an increase in the crystallisation enthalpy. Concerning the mechanical tests, the material behaved mainly as the PLA 2 h aged at 20 °C right after it was printed. The logistic fitting was not possible for these samples, as they did not age at 51 °C. Linear fitting was used instead (y = mx + c), resulting in Equations (6) and (7).

$$\sigma_y(\mathrm{MPa}) = -0.002t + 55.5 \tag{6}$$

$$E(\mathrm{GPa}) = -0.0023t + 2.72 \tag{7}$$

where the meaning of m and x are the slope and the theoretical property at time zero, respectively; $\sigma_Y$ is the yield strength in MPa; $E$ is the elastic modulus in GPa; and t is the ageing time in days. It was noticed that the elastic modulus had a slightly decreasing negative slope in Equation (7); note that the relation of -0.0027 relative to 2.23 in Equation (7) is an order of magnitude higher than $-0.002$ compared with 55.5 in Equation (6). This could be associated with the loss of the flow-induced molecular orientation, which is a known phenomenon that occurs when a polymer is extruded (Figure 9a), due to the shear stress produced in the PLA-nozzle interface (Figure 9b,c) and mainly affects to the elastic modulus [45–47]. This phenomenon is more noticeable for small extrusion shapes, such as our filaments, which were in the range of 350 to 450 micrometres. There is almost no literature on the effect of this phenomenon on printed polymers via FFF.

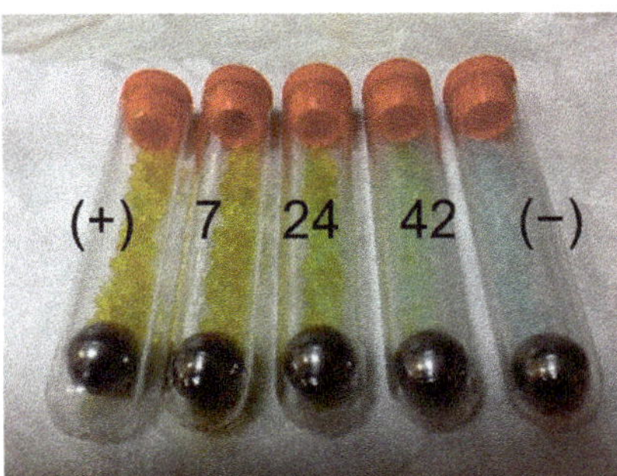


**Figure 8.** Desiccant control samples after 7, 24, and 42 days of ageing at 51 °C. (+) stands for the positive reference with a desiccant at 0 days, and (−) stands for a negative reference with the desiccant in direct contact with the air for some minutes.

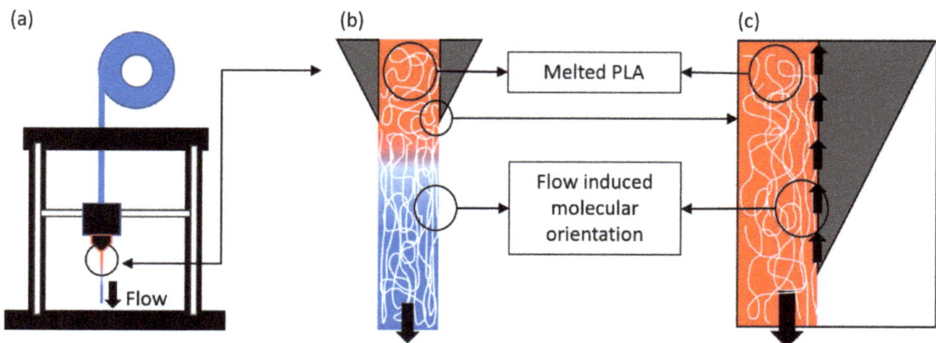


**Figure 9.** Flow-induced molecular orientation: (**a**) printing process, (**b**) detail of the nozzle and printed PLA, (**c**) detail of the shear forces between the nozzle and the PLA.

A decrease in the molecular weight—degradation—could also explain the slight decrease in the elastic modulus due to the degradation of the material. This degradation is especially important when the temperature increases over 50 degrees in a humid ambient environment [48]. Although a low-humidity atmosphere was obtained with the different barriers used during the ageing—the oil bath, the PP tube with desiccant, the vacuum grease used in the joint, and the PET zip bag with even more desiccant inside—the desiccant ended up turning blueish after 42 days at 51 °C. As a reference, the desiccant only remained yellow for a few minutes if it was placed under ambient conditions, and if it was placed inside our ageing device, it required some weeks (Figure 8).

Considering both the loss of flow-induced molecular orientation and the degradation as the two possibilities for explaining the decreases in the elastic modulus, the PLA samples were aged and tested at even higher temperatures for one day, forcing the mechanisms of both the degradation and the loss of flow-induced molecular orientation. Note that the influence of the crystallinity was discarded entirely, as the measured values were below 3% for all the samples. Three higher temperatures were analysed: 65, 75, and 80 °C. Both the 75 and the 80 °C temperatures were discarded from the mechanical test, as we observed on the DSC that they crystallised after one day of ageing (up to 28 % of crystallinity content compared with the 2–3% crystallinity contents on the samples aged at lower temperatures), introducing new variables to consider in the system but which are out of the scope of this study. Nonetheless, the desiccant remained bright yellow, as the positive in Figure 9, after one day of ageing at temperatures of 75 and 80 °C. After discarding the temperatures of 75 and 80 °C, the chosen temperature was 65 °C, which did not crystallise during the ageing.

The results from the PLA aged at 65 °C were insufficient to provide information to support any of the mechanisms: (1) it had similar mechanical values to the 42-day-aged PLA at 51 °C, and (2) the error bars of the measurements were within the error bars of the reference samples. The DSC results showed that the material was slightly degraded as both a decrease in the crystallisation temperature and an increase in the crystallisation enthalpies -related to higher mobility of the polymeric chains were detected. After all, the same analysis considering the error bars applies to the samples aged at 51 °C. The elastic modulus decreased by 5–6 % considering the average values, but it is also true that the error bars fell within the error bars of the reference sample. This makes it impossible to conclude why the elastic modulus of the PLA aged at 51 °C decreased. However, the results provide useful information: the degradation of the PLA and the loss of flow-induced mechanical orientation are almost negligible for a sample aged for 42 days at 51 °C, thus making those phenomena completely negligible for samples aged at 39 °C—potentially at 42 °C—for some days.

## 4. Conclusions

With all these data, we can ensure that accelerating the ageing rate of the PLA 4043D at 39 °C for a few days in an oil-bath device is safe, that it effectively ages the PLA, and that it will produce slightly higher properties than if it is slowly aged at room temperature. Furthermore, more critically, no variables related to the 3D structure were mixed in the discussion, making this a potentially essential reference for discussing the effect of heat treatments on 3D-printed parts produced via FFF.

Summarising the results obtained, our answers to our initial research questions are:

- Questions 1 and 2:

It is safe to accelerate the ageing rate of the PLA at 39 °C for several weeks in our proposed oil-bath ageing device, which aged the material properties close to the stable ones after just one day, but some extra days are needed to fully reach stable properties

Any degradation of the PLA during the ageing for some days at 39 °C can be discarded as it is hardly observable in a sample aged for 42 days at 51 °C.

The mechanical tests, combined with the DSC, were not suitable for decoupling the potential degradation from the loss of flow-induced molecular orientation.

Potentially, higher temperatures could be used with the increase in $T_{g1}$ during the ageing process to obtain faster ageing rates and higher properties due to the so-called summer effect.

- Question 3:

The main difference between the samples aged at 39 °C and those aged at 20 °C was the higher stability of the material aged at 39 °C. This has been defined as the summer effect, in which ageing at higher temperatures produces higher stabilisation properties (for the tensile strength, elastic modulus, glass transition, and enthalpic relaxation) due to the higher kinetic energy of the PLA macromolecules that can access more stable locations.

- Question 4:
  - Two glass transitions ($T_{g1}$ and $T_{g2}$) were found and characterised which had not been previously described in the literature. For these two glass transitions to be noticed, the printed PLA needed at least 2 h of ageing at 20 °C. This could induce possible mistakes in selecting the ageing temperature, as the 51 °C (below $T_{g2}$) was found to be too high for ageing the material.
  - Samples aged at temperatures above $T_{g1}$ did not age as expected as they increased their $T_g$ values whilst remaining at a negligible enthalpic relaxation enthalpy. This makes it impossible to correlate the $T_g$ values with those of a material that is or is not aged.
  - The PLA 2003D from our previous studies and the PLA 4043D studied here were found to age similarly and to be relatively similar regarding their thermal and mechanical properties.

This covers the objective of this research: understanding the effects of accelerating the ageing rate on the PLA. We conclude that it can be safely done and that it is worth doing. Future work should use these results to decrease the PLA's ageing time, with reference to the main difference between the samples aged at an accelerated rate and those studied until now at a standard ageing rate discussed in our previous works. Some questions remain open, such as how to properly decouple the effect of the loss of flow-induced molecular orientation from the effects of the degradation. Furthermore, it would be interesting to see to what extent the PLA properties could be increased by pushing the summer effect to its limits.

Combined with our previous research examining the effects of freezing the PLA to stop its ageing [23], researchers have the option to produce as many materials as they wish in only one day and, under the same conditions, to freeze all the produced material. They can then test the fully aged materials with just a few days of ageing at a higher temperature. This research will provide the research community with data on how the FFF-printed material is affected by these thermal treatments decoupled from all the related structural variables. We consider this a crucial step for properly understanding any 3D-printed structure, though we are also aware of its limitations.

The limitations of this work are that every different geometry that is printed will undergo a different thermal history (for example, a slower cooling rate if the PLA is deposited over a still-hot PLA layer). This might affect the crystallinity contents of the material. This variable was not considered in this study, as the crystallinity percentages were very low and materials with higher contents (samples aged at 75 and 80 °C) were discarded.

**Author Contributions:** Conceptualization, J.O.-B. and J.Y.P.; methodology, J.O.-B. and S.T.; software, J.O.-B. and S.T.; validation, J.O.-B., S.T. and J.Y.P.; formal analysis, J.O.-B. and J.Y.P.; investigation, J.O.-B. and S.T.; resources, J.O.-B. and J.Y.P.; data curation, J.O.-B.; writing—original draft preparation, J.O.-B.; writing—review and editing, J.O.-B., S.T. and J.Y.P.; visualization, J.O.-B.; supervision, S.T. and J.Y.P.; project administration, J.Y.P.; funding acquisition, J.Y.P. All authors have read and agreed to the published version of the manuscript.

**Funding:** This research was funded by the Spanish Government (PID2019-106631GB-C44, MICINN/FEDER, UE) and the Comunidad de Madrid Government (P2018/NMT-4511 NMAT2D-CM, P2018/NMT-4411 ADITIMAT-CM, FEDER-UE). J. Orellana acknowledges a scholarship provided by UPM and the Ministerio de Educación, Cultura y Deporte of Spain (FPU17/02035) and Prusa Research for facilitating the PLA-4043D filaments.

**Institutional Review Board Statement:** Not applicable.

**Informed Consent Statement:** Not applicable.

**Data Availability Statement:** Not applicable.

**Conflicts of Interest:** The authors declare no conflict of interest.

## References

1. Singh, S.; Ramakrishna, S.; Berto, F. 3D Printing of polymer composites: A short review. *Mater. Des. Process. Commun.* **2020**, *2*, e97. [CrossRef]
2. Ngo, T.D.; Kashani, A.; Imbalzano, G.; Nguyen, K.; Hui, D. Additive manufacturing (3D printing): A review of materials, methods, applications and challenges. *Compos. B Eng.* **2018**, *143*, 172–196. [CrossRef]
3. Khosravani, M.R.; Frohn-Sörensen, P.; Reuter, J.; Engel, B.; Reinicke, T. Fracture studies of 3D-printed continuous glass fiber reinforced composites. *Theor. Appl. Fract. Mech.* **2022**, *119*, 103317. [CrossRef]
4. Nugraha, A.D.; Ruli; Supriyanto, E.; Rasgianti; Prawara, B.; Martides, E.; Junianto, E.; Wibowo, A.; Sentanuhady, J.; Muflikhun, M.A. First-rate manufacturing process of primary air fan (PAF) coal power plant in Indonesia using laser powder bed fusion (LPBF) technology. *J. Mater. Res. Technol.* **2022**, *18*, 4075–4088. [CrossRef]
5. Patmonoaji, A.; Mahardika, M.A.; Nasir, M.; She, Y.; Wang, W.; Muflikhun, M.A.; Suekane, T. Stereolithography 3D Printer for Micromodel Fabrications with Comprehensive Accuracy Evaluation by Using Microtomography. *Geosciences* **2022**, *12*, 183. [CrossRef]
6. Masiuchok, O.; Iurzhenko, M.; Kolisnyk, R.; Mamunya, Y.; Godzierz, M.; Demchenko, V.; Yermolenko, D.; Shadrin, A. Polylactide/Carbon Black Segregated Composites for 3D Printing of Conductive Products. *Polymers* **2022**, *14*, 4022. [CrossRef] [PubMed]
7. Arefin, A.M.E.; Khatri, N.; Kulkarni, N.; Egan, P.F. Polymer 3D Printing Review: Materials, Process, and Design Strategies for Medical Applications. *Polymers* **2021**, *13*, 1499. [CrossRef]
8. Buj-Corral, I.; Zayas-Figueras, E.E. Comparative study about dimensional accuracy and form errors of FFF printed spur gears using PLA and Nylon. *Polym. Test.* **2023**, *117*, 107862. [CrossRef]
9. Muflikhun, M.A.; Sentanu, D.A. Characteristics and performance of carabiner remodeling using 3D printing with graded filler and different orientation methods. *Eng. Fail. Anal.* **2021**, *130*, 105795. [CrossRef]
10. Nieto, D.M.; Alonso-García, M.; Pardo-Vicente, M.; Rodríguez-Parada, L. Product Design by Additive Manufacturing for Water Environments: Study of Degradation and Absorption Behavior of PLA and PETG. *Polymers* **2021**, *13*, 1036. [CrossRef]
11. Khan, I.; Kumar, N. Fused deposition modelling process parameters influence on the mechanical properties of ABS: A review. *Mater. Today Proc.* **2021**, *44*, 4004–4008. [CrossRef]
12. Selvamani, S.K.; Samykano, M.; Subramaniam, S.R.; Ngui, W.K.; Kadirgama, K.; Kanagaraj, G.; Idris, M.S. 3D printing: Overview of ABS evolvement. *AIP Conf. Proc.* **2019**, *2059*, 020041. [CrossRef]
13. Fico, D.; Rizzo, D.; Casciaro, R.; Corcione, C.E. A Review of Polymer-Based Materials for Fused Filament Fabrication (FFF): Focus on Sustainability and Recycled Materials. *Polymers* **2022**, *14*, 465. [CrossRef] [PubMed]

14. Pringle, A.M.; Rudnicki, M.; Pearce, J.M. Wood Furniture Waste–Based Recycled 3-D Printing Filament. *For. Prod. J.* **2018**, *68*, 86–95. [CrossRef]
15. Pastor, J.Y.; Orellana-Barrasa, J.; Ferrández-Montero, A.; Ferrari, B.; Boccaccini, A.R. The Mechanical, Thermal, and Chemical Properties of PLA-Mg Filaments Produced via Colloidal Route for Fused Filament Fabrication. *Polymers* **2022**, *14*, 5414. [CrossRef]
16. Cifuentes, S.C.; Lieblich, M.; López, F.; Benavente, R.; González-Carrasco, J.L. Effect of Mg content on the thermal stability and mechanical behaviour of PLLA/Mg composites processed by hot extrusion. *Mater. Sci. Eng. C Mater. Biol. Appl.* **2017**, *72*, 18–25. [CrossRef]
17. Barrasa, J.O.; Ferrández-Montero, A.; Ferrari, B.; Pastor, J.Y. Characterisation and Modelling of PLA Filaments and Evolution with Time. *Polymers* **2021**, *13*, 2899. [CrossRef]
18. Kechagias, J.D.; Vidakis, N.; Petousis, M.; Mountakis, N. A multi-parametric process evaluation of the mechanical response of PLA in FFF 3D printing. *Mater. Manuf. Process.* **2022**, *37*, 1–13. [CrossRef]
19. Liao, Y.; Liu, C.; Coppola, B.; Barra, G.; Di Maio, L.; Incarnato, L.; Lafdi, K. Effect of Porosity and Crystallinity on 3D Printed PLA Properties. *Polymers* **2019**, *11*, 1487. [CrossRef]
20. Vanaei, H.R.; Shirinbayan, M.; Costa, S.F.; Duarte, F.M.; Covas, J.A.; Deligant, M.; Khelladi, S.; Tcharkhtchi, A. Experimental study of PLA thermal behavior during fused filament fabrication. *J. Appl. Polym. Sci.* **2021**, *138*, 49747. [CrossRef]
21. Kanakannavar, S.; Pitchaimani, J. Fracture toughness of flax braided yarn woven PLA composites. *Int. J. Polym. Anal. Charact.* **2021**, *26*, 364–379. [CrossRef]
22. Popescu, D.; Zapciu, A.; Amza, C.; Baciu, F.; Marinescu, R. FDM process parameters influence over the mechanical properties of polymer specimens: A review. *Polym. Test.* **2018**, *69*, 157–166. [CrossRef]
23. Orellana-Barrasa, J.; Ferrández-Montero, A.; Ferrari, B.; Pastor, J.Y. Natural Ageing of PLA Filaments, Can It Be Frozen? *Polymers* **2022**, *14*, 3361. [CrossRef]
24. Deroiné, M.; Le Duigou, A.; Corre, Y.-M.; Le Gac, P.-Y.; Davies, P.; César, G.; Bruzaud, S. Accelerated ageing of polylactide in aqueous environments: Comparative study between distilled water and seawater. *Polym. Degrad. Stab.* **2014**, *108*, 319–329. [CrossRef]
25. Saha, S.K.; Tsuji, H. Hydrolytic Degradation of Amorphous Films of L-Lactide Copolymers with Glycolide and D-Lactide. *Macromol. Mater. Eng.* **2006**, *291*, 357–368. [CrossRef]
26. Polidar, M.; Metzsch-Zilligen, E.; Pfaendner, R. Controlled and Accelerated Hydrolysis of Polylactide (PLA) through Pentaerythritol Phosphites with Acid Scavengers. *Polymers* **2022**, *14*, 4237. [CrossRef] [PubMed]
27. Rodriguez, E.J.; Marcos, B.; Huneault, M.A. Hydrolysis of polylactide in aqueous media. *J. Appl. Polym. Sci.* **2016**, *133*, 44152. [CrossRef]
28. Agrawal, C.M.; Huang, D.; Schmitz, J.; Athanasiou, K.A. Elevated Temperature Degradation of a 50:50 Copolymer of PLA-PGA. *Tissue Eng.* **2007**, *3*, 345–352. [CrossRef]
29. Li, S.; McCarthy, S. Further investigations on the hydrolytic degradation of poly (DL-lactide). *Biomaterials* **1999**, *20*, 35–44. [CrossRef]
30. Monnier, X.; Saiter, A.; Dargent, E. Physical aging in PLA through standard DSC and fast scanning calorimetry investigations. *Thermochim. Acta* **2017**, *648*, 13–22. [CrossRef]
31. Mofokeng, J.P.; Luyt, A.; Tábi, T.; Kovács, J. Comparison of injection moulded, natural fibre-reinforced composites with PP and PLA as matrices. *J. Thermoplast. Compos. Mater.* **2011**, *25*, 927–948. [CrossRef]
32. Cristea, M.; Ionita, D.; Iftime, M.M. Dynamic Mechanical Analysis Investigations of PLA-Based Renewable Materials: How Are They Useful? *Materials* **2020**, *13*, 5302. [CrossRef] [PubMed]
33. Chalgham, A.; Ehrmann, A.; Wickenkamp, I. Mechanical Properties of FDM Printed PLA Parts before and after Thermal Treatment. *Polymers* **2021**, *13*, 1239. [CrossRef] [PubMed]
34. Jayanth, N.; Jaswanthraj, K.; Sandeep, S.; Mallaya, N.; Siddharth, S.R. Effect of heat treatment on mechanical properties of 3D printed PLA. *J. Mech. Behav. Biomed. Mater.* **2021**, *123*, 104764. [CrossRef]
35. Akhoundi, B.; Nabipour, M.; Hajami, F.; Shakoori, D. An Experimental Study of Nozzle Temperature and Heat Treatment (Annealing) Effects on Mechanical Properties of High-Temperature Polylactic Acid in Fused Deposition Modeling. *Polym. Eng. Sci.* **2020**, *60*, 979–987. [CrossRef]
36. Backes, E.H.; Pires, L.D.N.; Costa, L.C.; Passador, F.R.; Pessan, L.A. Analysis of the Degradation During Melt Processing of PLA/Biosilicate®Composites. *J. Compos. Sci.* **2019**, *3*, 52. [CrossRef]
37. Bhagia, S.; Bornani, K.; Agrawal, R.; Satlewal, A.; Ďurkovič, J.; Lagaňa, R.; Bhagia, M.; Yoo, C.G.; Zhao, X.; Kunc, V.; et al. Critical review of FDM 3D printing of PLA biocomposites filled with biomass resources, characterization, biodegradability, upcycling and opportunities for biorefineries. *Appl. Mater. Today* **2021**, *24*, 101078. [CrossRef]
38. Cao, X.; Mohamed, A.; Gordon, S.; Willett, J.; Sessa, D.J. DSC study of biodegradable poly(lactic acid) and poly(hydroxy ester ether) blends. *Thermochim. Acta* **2003**, *406*, 115–127. [CrossRef]
39. Jia, S.; Yu, D.; Zhu, Y.; Wang, Z.; Chen, L.; Fu, L. Morphology, Crystallization and Thermal Behaviors of PLA-Based Composites: Wonderful Effects of Hybrid GO/PEG via Dynamic Impregnating. *Polymers* **2017**, *9*, 528. [CrossRef]
40. Leonés, A.; Peponi, L.; Lieblich, M.; Benavente, R.; Fiori, S. In Vitro Degradation of Plasticized PLA Electrospun Fiber Mats: Morphological, Thermal and Crystalline Evolution. *Polymers* **2020**, *12*, 2975. [CrossRef]

41. Kaavessina, M.; Ali, I.; Al-Zahrani, S.M. The Influences of Elastomer toward Crystallization of Poly(lactic acid). *Procedia Chem.* **2012**, *4*, 164–171. [CrossRef]
42. Murariu, M.; Paint, Y.; Murariu, O.; Laoutid, F.; Dubois, P. Tailoring and Long-Term Preservation of the Properties of PLA Composites with 'Green' Plasticizers. *Polymers* **2022**, *14*, 4836. [CrossRef] [PubMed]
43. De Bortoli, L.S.; De Farias, R.; Mezalira, D.; Schabbach, L.; Fredel, M.C. Functionalized carbon nanotubes for 3D-printed PLA-nanocomposites: Effects on thermal and mechanical properties. *Mater. Today Commun.* **2022**, *31*, 103402. [CrossRef]
44. Da Silva Barbosa Ferreira, E.; Luna, C.; Siqueira, D.; Araújo, E.; De França, D.; Wellen, R.M.R. Annealing Effect on Pla/Eva Blends Performance. *J. Polym. Environ.* **2022**, *30*, 541–554. [CrossRef]
45. Coxon, L.D.; White, J.R. Residual stresses and aging in injection molded polypropylene. *Polym. Eng. Sci.* **1980**, *20*, 230–236. [CrossRef]
46. Chan, T.W.D.; Lee, L.J. Analysis of molecular orientation and internal stresses in extruded plastic sheets. *Polym. Eng. Sci.* **1989**, *29*, 163–170. [CrossRef]
47. White, J.R. Origins and measurements of internal stress in plastics. *Polym. Test.* **1984**, *4*, 165–191. [CrossRef]
48. Niaounakis, M.; Kontou, E.; Xanthis, M. Effects of aging on the thermomechanical properties of poly(lactic acid). *J. Appl. Polym. Sci.* **2011**, *119*, 472–481. [CrossRef]

Article

# Thermo-Mechanical and Creep Behaviour of Polylactic Acid/Thermoplastic Polyurethane Blends

**Yi-Sheng Jhao [1], Hao Ouyang [1], Fuqian Yang [2] and Sanboh Lee [1,*]**

1 Department of Materials Science and Engineering, National Tsing Hua University, Hsinchu 300, Taiwan
2 Materials Program, Department of Chemical and Materials Engineering, University of Kentucky, Lexington, KY 40506, USA
* Correspondence: sblee@mx.nthu.edu.tw

**Abstract:** There is a great need to develop biodegradable thermoplastics for a variety of applications in a wide range of temperatures. In this work, we prepare polymer blends from polylactic acid (PLA) and thermoplastic polyurethane (TPU) via a melting blend method at 200 °C and study the creep deformation of the PLA/TPU blends in a temperature range of 10 to 40 °C with the focus on transient and steady-state creep. The stress exponent for the power law description of the steady state creep of PLA/TPU blends decreases linearly with the increase of the mass fraction of TPU from 1.73 for the PLA to 1.17 for the TPU. The activation energies of the rate processes for the steady-state creep and transient creep decrease linearly with the increase of the mass fraction of TPU from 97.7 ± 3.9 kJ/mol and 59.4 ± 2.9 kJ/mol for the PLA to 26.3 ± 1.3 kJ/mol and 25.4 ± 1.7 kJ/mol for the TPU, respectively. These linearly decreasing trends can be attributed to the weak interaction between the PLA and the TPU. The creep deformation of the PLA/TPU blends consists of the contributions of individual PLA and TPU.

**Keywords:** polylactic acid; thermoplastic polyurethane; creep; non-linear Burgers model; activation energy

**Citation:** Jhao, Y.-S.; Ouyang, H.; Yang, F.; Lee, S. Thermo-Mechanical and Creep Behaviour of Polylactic Acid/Thermoplastic Polyurethane Blends. *Polymers* **2022**, *14*, 5276. https://doi.org/10.3390/polym14235276

Academic Editors: José Miguel Ferri, Vicent Fombuena Borràs and Miguel Fernando Aldás Carrasco

Received: 27 September 2022
Accepted: 28 November 2022
Published: 2 December 2022

## 1. Introduction

Polylactic acid (PLA), a thermoplastic aliphatic polyester, has been regarded as a prospective and promising biodegradable material with the potential of replacing petrochemical plastics [1]. The biodegradable characteristics of PLA have attracted great interest for a variety of potential applications. However, the applications of PLA are limited by its brittleness, thermal stability, and impact resistance.

To increase the applications of PLA in a variety of fields, PLA-based composites and PLA-polymer blends have been developed. For example, Matta et al. [2] used melt blending to produce a biodegradable polymer blend consisting of PLA and polycaprolactone (PCL), which possesses better impact strength and toughness than respective PLA and PCL. Ho et al. [3] grafted PLA onto maleic-anhydride functionalized thermoplastic polyolefin elastomer to form thermoplastic polyolefin elastomer-grafted polylactide, which improves the toughness of PLA. Tokoro et al. [4] used three different bamboo fibers as PLA reinforcement materials to form PLA/bamboo composites, which exhibited higher bending strength and impact strength than PLA at room temperature.

Currently, most studies have been focused on tensile and impact deformation of PLA-based composites and blends [5–9]. There are few studies on the creep deformation of PLA-based composites and blends. Yang et al. [10] examined the effect of short bamboo fibers on the short-time creep of PLA-based composites and observed the increase of creep resistance with the increase of the weight fraction of bamboo fibers up to 60%. Georgiopoulos et al. [11] assessed the short-time creep deformation of PLA/PBAT (poly(butylene adipate-terephthalate)) blend reinforced with wood fibers and used Findley's and Burger's models in the creep analysis. Morreale et al. [12] studied the tensile creep of PLA-based bio-composites and found a strong dependence of the creep deformation on temperature and fabrics. Ye et al. [13] used a four-element Burger's model in analyzing the creep of PLA prepared by the fused-filament process and observed the effects of the printing angle and stress. Waseem et al. [14] used the response surface methodology in the tensile-creep analysis of the PLA produced by three-dimensional printing for additive manufacturing. Guedes et al. [15] characterized the creep deformation and stress relaxation of PLA/PCL fibers and used Burger's model and the standard solid model in the analysis. Niaza et al. [16] studied the long-term creep of PLA/HA (hydroxyapatite) composites and found that PLA/HA scaffolds under mechanical loading up to 10 MPa did not change shape and lose mechanical strength.

Thermoplastic polyurethane (TPU) has good elastic properties, transparency, and wear resistance. TPU is also biocompatible and bio-stable and is a promising material for a variety of implantable medical devices [17]. Blending TPU with PLA can alter the brittle characteristic of PLA and increase the toughness of PLA [18] due to the good elastic properties of TPU, while it is unclear if blending TPU with PLA can hinder or promote the motion of polymer chains under constant loading. This work aims to study the creep deformation of PLA/TPU blends. The effect of the mass ratio of PLA to TPU on the activation energies of the creep deformation of the blends is assessed.

It should be noted that there are extensive studies on the creep deformation of a variety of polymers, including polyurethanes [19,20]. The analysis has been based on the use of dashpot-spring models in describing the creep curves. However, there are little studies on the calculations of the activation energies of the rate processes for the creep deformation of polymers.

## 2. Experimental Details

The PLA (4032D, $T_g$ around 55–60 °C) and TPU (300-grade series, ester type) used in this work were from Natureworks LLC (Minnetonka, MN, USA) and Bayer Co., Ltd. (Leverkusen, Germany), respectively. The PLA was a semi-crystalline polymer with 98% L-isomer and 2% D-isomer.

PLA/TPU blends were prepared with different mass ratios of PLA to TPU (30/70, 50/50, and 70/30) via a melting blend method at 200 °C. Briefly, the PLA and TPU mixture of a preset mass ratio was dried in an oven at 80 °C for 4 h. The PLA/TPU mixture was heated to 200 °C to a molten state, which was then injected into a mold at 2.94 MPa to form a PLA/TPU plate. The plate was cooled down to 30 °C. Using laser cutting, specimens in a dumbbell shape, as shown in Figure 1, were prepared. The specimens were mechanically ground with CarbiMet papers of 400, 800, 1200, and 2500 grit, consecutively, and then polished with 1 μm alumina slurry. The polished specimens were annealed in the air in a furnace at 50 °C for 24 h to release the residual stresses introduced during the sample preparation and then naturally cooled down to room temperature in the furnace.

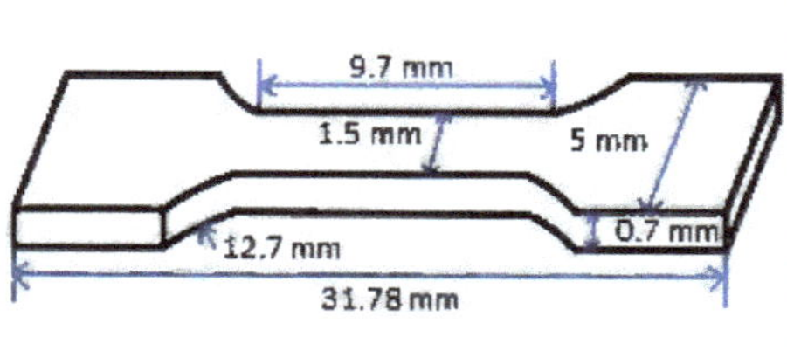


**Figure 1.** Schematic of the geometrical dimensions of specimens for creep tests.

Thermal analysis of the prepared PLA/TPU blends was performed in a temperature range of −80 °C to 220 °C on a differential scan calorimeter (DSC) (Netzsch 200F3, Erich NETZSCH GmbH & Co., Selb, Germany). Nitrogen gas was used during the DSC scan to prevent the specimens from oxidation. The flow rate of nitrogen was 100 mL/min.

Thermal cycling of the specimens of 8–10 mg in the mass loading was conducted with the heating and cooling rates of 10 °C/min and 40 °C/min, respectively, and 10 min each at 220 °C and −80 °C, respectively. The thermal cycling eliminated the thermal history of the specimens. After the thermal cycling, the specimens were reheated to 220 °C at a heating rate of 10 °C/min for thermal analysis. The glass transition temperatures, $T_g$, of PLA, TPU, and PLA/TPU blends were determined to be the temperature at the midpoint of the corresponding heat-capacity jump in the heat-flow vs temperature plots.

The creep tests of the PLA, TPU, and PLA/TPU blends were performed on a dynamic mechanical analyzer (TA Q800 DMA, TA instrument, New Castle, DE, USA) in a temperature range of 10 to 40 °C. Due to the limitation of the instrument and the differences in the properties of the materials, the stresses applied to the PLA, TPU, and PLA/TPU blends were different. Before the creep test, each specimen was maintained at the preset temperature for 5 min to reach thermal equilibrium. The time for the creep tests was 40 min. After the creep test, the crept specimen was maintained in a stress-free state for 40 min. The strain was measured as a function of time during the test.

## 3. Results

Figure 2 shows the DSC curves of the PLA, TPU, and PLA/TPU blends. For the PLA and TPU, the glass transition temperatures are 57.4 °C and −40.1 °C, respectively. For the PLA/TPU blends, there are two glass transition temperatures presented in the DSC curves with a weak one around −40 °C associated with the TPU glass transition temperature and the other one around 57 °C associated with the PLA glass transition temperature. Such a result suggests that PLA and TPU are thermodynamically immiscible, as supported by the dual melting for the PLA [17] and PLA/TPU blends [21,22]. It is interesting to note that the melting temperature of the PLA/TPU blends decreases with the increase of the mass fraction of TPU, revealing the contribution of TPU.

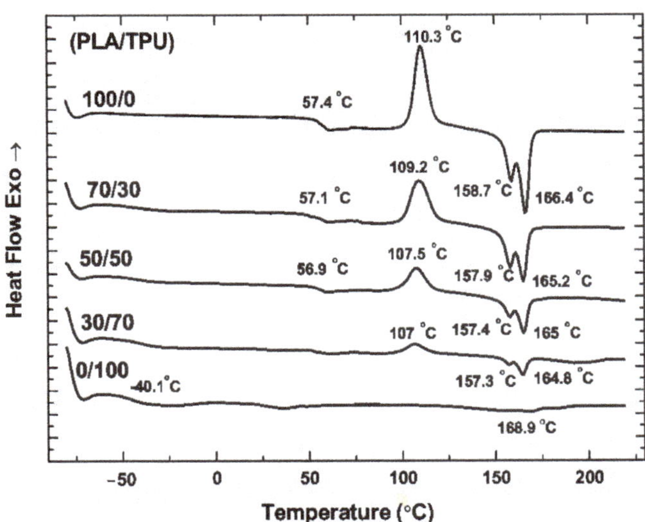


**Figure 2.** DSC curves of PLA, TPU, and PLA/TPU blends.

The dual melting behavior has been observed for many semi-crystalline polymers, including poly(butylene terephthalate), poly(ethylene terephthalate), and poly(ether ketone) [23–26]. The low-temperature endothermic peak represents partial melting of the "original" crystal, and the high-temperature endothermic peak represents the melting of "reorganized" crystals during the heating [27].

According to Figure 2, there is no exotherm peak of cold crystallization for the TPU in consistence with the amorphous structure of TPU. PLA is a semi-crystalline polymer with a slow crystallization rate. The rapid quenching of PLA from 220 °C before the DSC measurement does not allow the PLA to crystallize during cooling, i.e., the PLA polymer chains do not have enough time to migrate to equilibrium positions. During the heating, the polymer absorbs energy enabling the migration of polymer chains and leading to the presence of an exothermic peak of cold crystallization around 110.3 °C for the PLA. The exothermic peak of cold crystallization of the PLA/TPU blends decreases with the increase of the TPU fraction, and no exothermic peak of cold crystallization is present for the TPU, as expected.

Figure 3 and Figures S1–S5 in Supplementary Information present the creep curves and recovery curves of the PLA, TPU, and PLA/TPU blends under different tensile stresses at four temperatures of 10, 20, 30, and 40 °C. It is evident that the creep curves consist of two states—a transient state and a steady state (secondary creep). There is a recovery process after the complete removal of the stress/load for all the PLA, TPU, and PLA/TPU blends. It should be noted that the creep deformation of all the PLA, TPU, and PLA/TPU blends was confined to the secondary creep to avoid the presence of tertiary creep and the failure/breakage of the specimens.

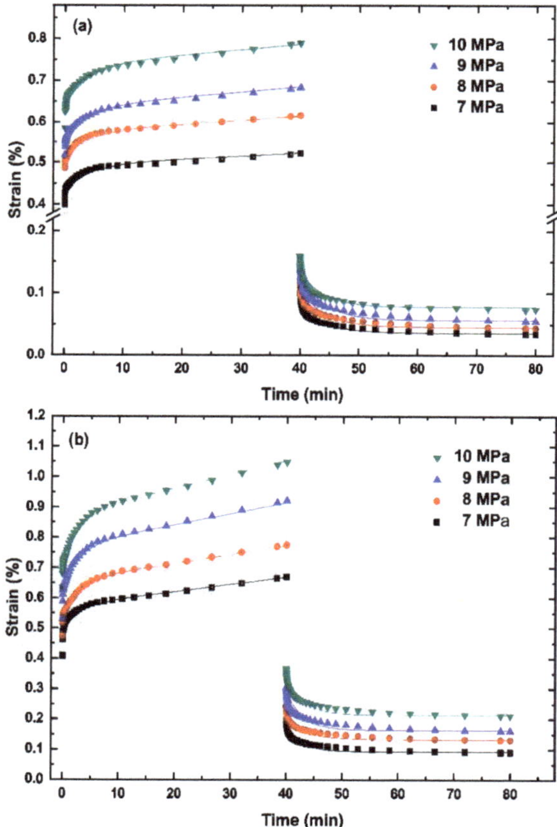


**Figure 3.** *Cont.*

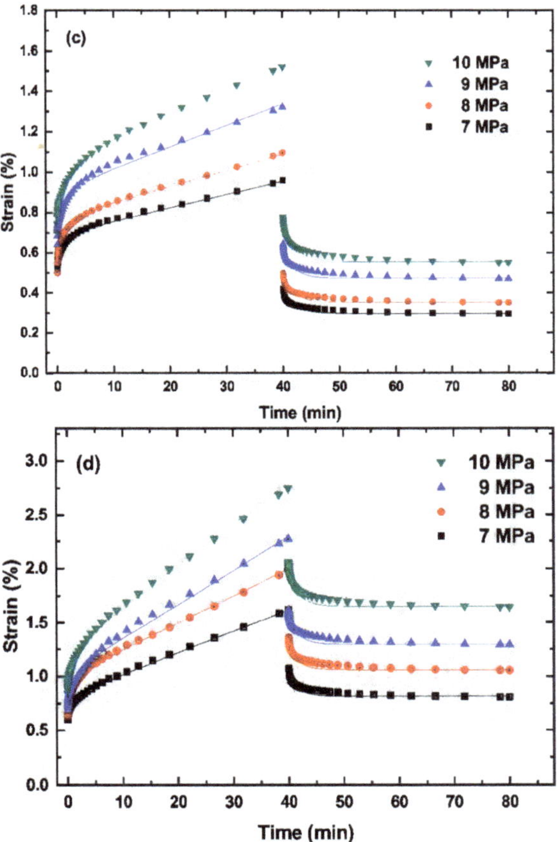


**Figure 3.** Creep curves of the PLA/TPU blend with 70:30 for the mass ratio of PLA to TPU under different tensile stresses at temperatures of (**a**) 10 °C, (**b**) 20 °C, (**c**) 30 °C, and (**d**) 40 °C.

## 4. Discussion

From Figure 3 and Figures S1–S5, we calculate the creep rate at the steady state. Figure 4 shows the variation of the creep rate with the applied stress for the steady state creep of the PLA, PLA/TPU blends, and TPU at different creep temperatures. The creep rate at the steady state increases with the increases in temperature and applied stress, as expected.

In general, the correlation between stress and creep rate at the steady-state creep can be expressed as a power-law law as [28]

$$\dot{\varepsilon} = A\sigma^n \tag{1}$$

where $\dot{\varepsilon}$ is the creep rate, A is a pre-exponential constant, $\sigma$ is the applied stress, and n is the stress exponent. Using Equation (1) to fit the experimental data in Figure 4, we obtain the stress exponent n. For comparison, the fitting curves are included in Figure 4. It is evident that Equation (1) describes well the correlation between the applied stress and the creep rate at the steady-state creep and there is no statistical difference in the stress exponent for the same material. Note that Equation (1) can be used to determine the activation energy for the creep at different temperatures under the same stress/load. For polymer, however, the creep deformation is better described by viscoelasticity as discussed below.

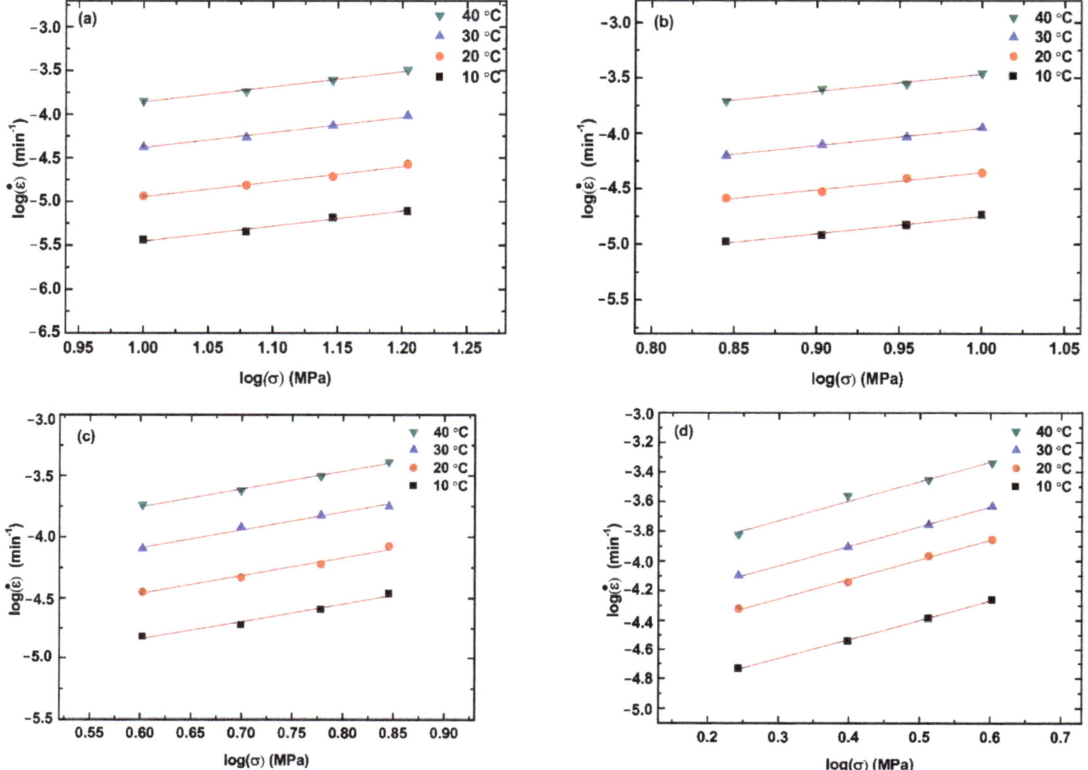


**Figure 4.** Stress dependence of the creep rate at the steady-state creep for different creep temperatures: (**a**) PLA, (**b**) PLA/TPU blend with 70:30 in the ratio of PLA to TPU, (**c**) PLA/TPU blend with 50:50 in the ratio of PLA to TPU and (**d**) PLA/TPU blend with 30:70 in the ratio of PLA to TPU.

Figure 5 displays the variation of the stress exponent with the mass fraction of TPU. The stress exponent decreases linearly from 1.73 for the PLA to 1.17 for the TPU. Such a trend suggests that increasing the fraction of TPU in the PLA/TPU blend reduces the resistance to the motion of the polymer chains. According to the result in Figure 5, the dependence of the stress exponent on the mass fraction of TPU in the PLA/TPU blend can be expressed as

$$n = 1.73 - 1.17m_{TPU}/(m_{PLA} + m_{TPU}) \quad (2)$$

where $m_{PLA}$ and $m_{TPU}$ are the masses of the PLA and TPU, respectively. Such a result is consistent with that PLA and TPU are thermodynamically immiscible.

Figure 6 shows the schematic of Kelvin representation of the non-linear Burgers model. Here, $E_1$ and $E_2$ are elastic constants of the corresponding spring elements, $\eta_1$ is the viscosity for nonlinear dashpot I, $\eta_2$ is the viscosity of the linear dashpot II, $\sigma$, $\sigma_A$, and $\sigma_B$ are the stresses acting on the corresponding elements, and $\varepsilon$, $\varepsilon_1$ and $\varepsilon_2$ are the strain of the corresponding elements. The springs I and II and dashpot II are linear elements.

To analyze the creep deformation of the PLA, TPU, and PLA/TPU blends with the power-law relation between the stress and creep rate at the steady-state creep, we introduce the Kelvin representation of the non-linear Burgers model, as shown in Figure 6, in which the stress dependence of the creep rate of the dashpot I follows a power-law relation similar to Equation (1) as

$$\sigma^n = \eta_1 \dot{\varepsilon}_3 \quad (3)$$

The strain/strain rate that other elements experience is proportional to the corresponding applied stress. Under the action of constant stress (creep deformation), we can follow the same approach as the Kelvin representation of the linear Burgers model to obtain the time dependence of the resultant strain, ε, of the non-linear Burgers model as

$$\varepsilon = \frac{\sigma}{E_1} + \frac{\sigma^n t}{\eta_1} + \frac{\sigma}{E_2}\left[1 - \exp\left(-\frac{E_2 t}{\eta_2}\right)\right] = \frac{\sigma}{E_1} + \frac{\sigma^n t}{\eta_1} + \frac{\sigma}{E_2}[1 - \exp(-\beta_c t)] \quad (4)$$

with t being the creep time, and $\beta_c = E_2/\eta_2$. Here, $E_1$ and $\eta_1$ are the elastic constant of spring 1 and the viscosity of non-linear dashpot 1, respectively, $E_2$ and $\eta_2$ are the elastic constant of spring 2 and the viscosity of linear dashpot 2, and $\beta_c^{-1}$ is the characteristic time of the non-linear Burgers model for creep deformation. The first term represents instantaneous elastic deformation, the second term represents steady-state creep, and the third term corresponds to transient creep deformation.

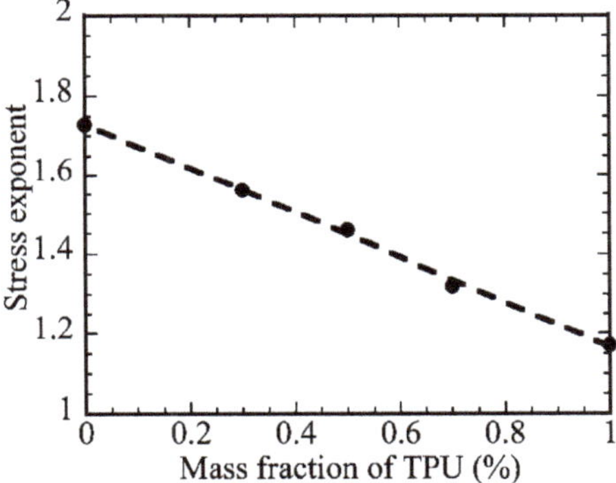


**Figure 5.** Variation of the stress exponent with the mass fraction of TPU.

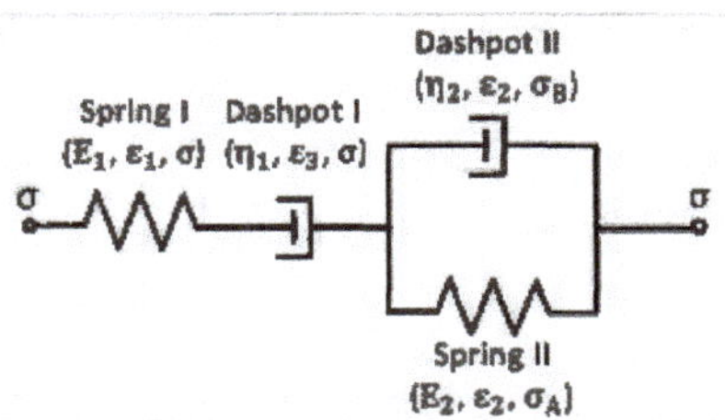


**Figure 6.** Kelvin representation of non-linear Burgers model.

Using the exponents presented in Figure 5 and Equation (4), we curve-fit the creep curves in Figure 3 and Figures S1–S5 and determine the parameters of $E_1$, $E_2$, $\eta_1$ and $\eta_2$. For comparison, the fitting curves are included in Figure 3 and Figures S1–S5. It is evident that Equation (4) describes well the creep deformation of the PLA, TPU, and PLA/TPU blends up to the steady-state creep.

Table 1 summarizes the temperature dependence of the elastic constants of the PLA, TPU, and PLA/TPU blends. Both $E_1$ and $E_2$ of the material decrease with the increase of the creep temperature, as expected, which is due to the increase in space allowing less constraint to the stretch of polymer chains. According to Table 1, increasing the mass fraction of TPU causes decreases in both $E_1$ and $E_2$, which is due to that the PLA has a higher modulus of 2.382 ± 0.114 GPa than 0.025 ± 0.003 GPa of the TPU.

**Table 1.** Temperature dependence of $E_1$ and $E_2$ of the PLA, TPU, and PLA/TPU blends.

| Material | Modulus | Temperature (°C) | | | |
|---|---|---|---|---|---|
| | | 10 | 20 | 30 | 40 |
| PLA | $E_1$ (GPa) | 2.65 ± 0.07 | 2.53 ± 0.06 | 2.41 ± 0.06 | 2.31 ± 0.07 |
| | $E_2$ (GPa) | 38.6 ± 2.0 | 23.4 ± 0.7 | 18.6 ± 0.9 | 11.2 ± 0.5 |
| PLA70/TPU30 | $E_1$ (GPa) | 1.59 ± 0.06 | 1.43 ± 0.05 | 1.27 ± 0.06 | 1.15 ± 0.08 |
| | $E_2$ (GPa) | 11.5 ± 0.7 | 7.09 ± 0.42 | 4.61 ± 0.35 | 2.99 ± 0.24 |
| PLA50/TPU50 | $E_1$ (GPa) | 1.05 ± 0.053 | 0.929 ± 0.055 | 0.772 ± 0.023 | 0.707 ± 0.026 |
| | $E_2$ (GPa) | 4.97 ± 0.2 | 3.36 ± 0.28 | 1.87 ± 0.13 | 1.28 ± 0.09 |
| PLA30/TPU70 | $E_1$ (GPa) | 0.425 ± 0.021 | 0.366 ± 0.026 | 0.345 ± 0.017 | 0.309 ± 0.02 |
| | $E_2$ (GPa) | 0.779 ± 0.04 | 0.516 ± 0.028 | 0.326 ± 0.021 | 0.232 ± 0.017 |
| TPU | $E_1$ (MPa) | 35.6 ± 0.9 | 30.8 ± 1.2 | 26.8 ± 0.7 | 23.9 ± 0.7 |
| | $E_2$ (MPa) | 103 ± 6 | 71.0 ± 4.6 | 51.4 ± 4.1 | 37.6 ± 2.4 |

According to the theory of the thermal activation process, the temperature dependence of $\eta_1$ and $\eta_2$ follows the Arrhenius relation as

$$\eta_1^{-1} = \eta_{10}^{-1} \exp\left(-\frac{Q_s}{RT}\right) \text{ and } \eta_2^{-1} = \eta_{20}^{-1} \exp\left(-\frac{Q_k}{RT}\right) \quad (5)$$

where $\eta_{10}$ and $\eta_{20}$ are two constants, $Q_s$ and $Q_k$ are the activation energies of the rate processes for the steady-state creep and transient creep, respectively, R is the gas constant, and T is the absolute temperature. Figure 7 shows the temperature dependence of $\eta_1$ and $\eta_2$. It is evident that both the $\eta_1^{-1}$ and $\eta_2^{-1}$ are exponentially decreasing functions of $T^{-1}$. Using Equation(5) to fit the data in Figure 7a,b, we obtain the activation energies for the creep deformation of the PLA, TPU, and PLA/TPU blends.

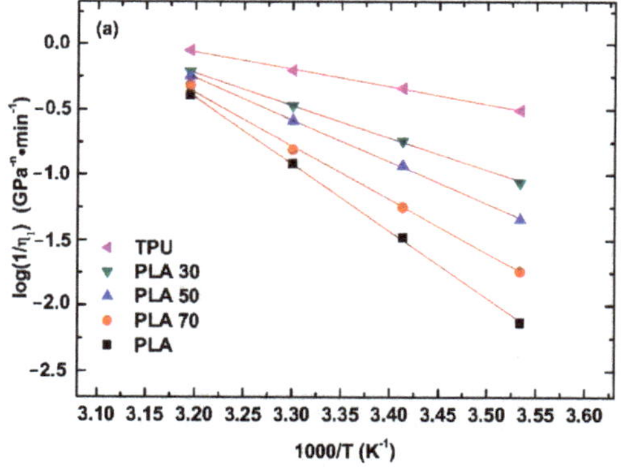


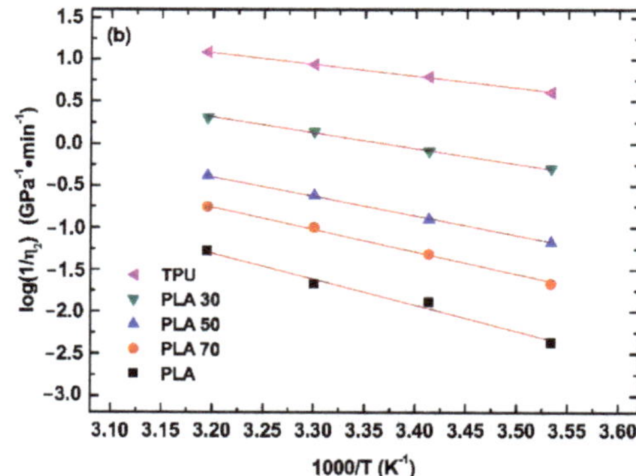


**Figure 7.** Temperature dependence of $\eta_1$ (**a**) and $\eta_2$ (**b**).

Figure 8 displays variations of $Q_s$ and $Q_k$ with the mass fraction of TPU. Both the $Q_s$ and $Q_k$ decrease linearly with the increase of the mass fraction of TPU, suggesting that the activation energies of the PLA/TPU blends can be expressed as

$$\left(Q_s^b, Q_k^b\right) = \left(Q_s^{PLA}, Q_k^{PLA}\right) - \left(Q_s^{TPU}, Q_k^{TPU}\right) \text{mTPU}/(\text{mPLA} + \text{mTPU}) \quad (6)$$

in which the superscript b represents the PLA/TPU blend, and the superscript and subscript of PLA and TPU represent the corresponding PLA and TPU. Such a result is consistent again with that PLA and TPU are thermodynamically immiscible. Note that $Q_s$ is larger than $Q_k$ for the same PLA/TPU blend. Such behavior is associated with the different states of the polymer chains. Before the onset of the steady-state creep, the polymer chains are in a relatively relaxed state with less resistance to migration. At the steady state creep, the polymer chains are under stretch with a large resistance to the migration. The polymer chains need to overcome a larger energy barrier to reach a new state at the steady state creep than at the transient state.

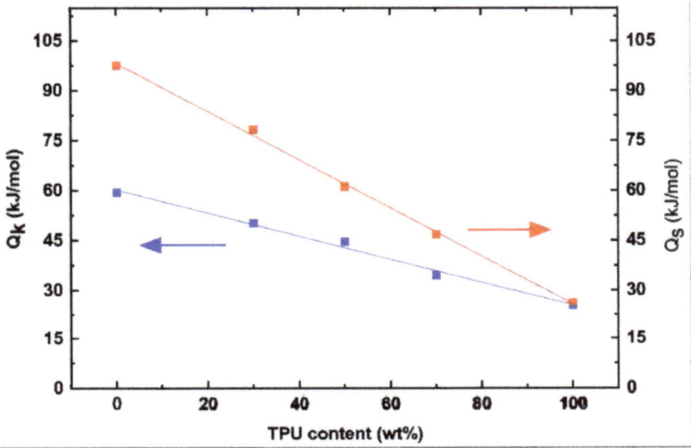


**Figure 8.** Variations of $Q_s$ and $Q_k$ with the mass fraction of TPU.

According to Figure 3 and Figures S1–S5, the PLA/TPU blends experienced an instantaneous elastic recovery, then a time-related recovery, and finally a permanent plastic deformation after the end of the creep deformation. Following the analysis of the creep deformation of the PLA/TPU blends, we use the Kelvin representation of the non-linear Burgers model to analyze the recovery behavior of the PLA/TPU blends. The time dependence of the recovery strain is expressed as

$$\varepsilon_r(t) = \left\{[\varepsilon(t_c)] - \frac{\sigma}{E_1} - \varepsilon_p\right\} \exp\left[-\frac{E_2(t - t_c)}{\eta_2}\right] + \varepsilon_p \text{ with } \varepsilon_p = B\sigma^m \exp\left(-\frac{Q_p}{RT}\right) \quad (7)$$

where $t_c$ is the creep time of the creep test, $\varepsilon(t_c)$ is the final strain of the creep deformation, and $\varepsilon_p$ is the plastic strain. For the plastic strain, B is a pre-exponential constant, m is the stress exponent, and $Q_p$ is the activation energy of the plastic deformation. Using the first equation in Equation (7) to curve-fit the recovery curves, we obtain the plastic strain. Figures S6 and S7 in Supplementary Information display the stress dependence of the plastic strain for the recovery of the PLA/TPU blends at different temperatures and the temperature dependence of the plastic strain for the recovery of the PLA/TPU blends under different stresses, respectively. Using the second equation in Equation (7) to curve-fit the results in Figures S6 and S7, we determine the stress exponents and activation energies, as listed in Table 2. Both the stress exponent (m) and the activation energy (Qp) are the same as the corresponding ones (n and $Q_s$) for the steady-state creep of the same PLA/TPU blends. Such results reveal the same rate mechanisms controlling the migration of polymer chains at the steady-state creep and the recovery after the creep deformation.

**Table 2.** Numerical values of m and $Q_p$ for the recovery of the PLA/TPU blends after the creep.

| Material | Temperature (10–40 °C) | |
|---|---|---|
| | m | $Q_p$ (kJ/mol) |
| PLA | 1.73 | 96.5 ± 3.1 |
| PLA70/TPU30 | 1.56 | 76.8 ± 3.3 |
| PLA50/TPU50 | 1.46 | 62.3 ± 2.5 |
| PLA30/TPU70 | 1.32 | 45.4 ± 2.7 |
| TPU | 1.17 | 27.4 ± 1.5 |

## 5. Conclusions

In summary, we have studied the thermal behavior, tensile creep deformation and recovery after complete unloading of the PLA, TPU, and PLA/TPU blends under different stresses in the temperature range of 10 to 40 °C. The thermal analysis has revealed that the PLA/TPU blends maintained the thermal characteristics of individual PLA and TPU, consistence with that PLA and TPU are thermodynamically immiscible. To avoid the presence of tertiary creep and the failure/breakage of the specimens, we have focused on the creep deformation of the PLA, TPU, and PLA/TPU blends on the transient and steady-state creep. The stress dependence of the creep rate of the PLA, TPU, and PLA/TPU blends for the steady-state creep under the conditions used in this work follows a power-law relation. The stress exponent of the power-law relation is a linearly decreasing function of the mass fraction of TPU in consistence with that PLA and TPU are thermodynamically immiscible.

On the base of the power-law relation between the tensile stress and creep rate at a steady state, we have proposed a Kelvin representation of the non-linear Burgers model for the analysis of the creep deformation and recovery of the PLA, TPU, and PLA/TPU blends. Such a Kelvin representation of the non-linear Burgers model can describe well the creep deformation and recovery of the PLA, TPU, and PLA/TPU blends. The results obtained from the curve-fitting of the creep and recovery curves reveal that the stress exponent and activation energies are linearly decreasing functions of the mass fraction of TPU, which are consistent with that PLA and TPU are thermodynamically immiscible. The activation energy of the transient creep is less than that of the steady state creep for the same PLA/TPU blend, revealing the increase in the resistance to the migration of polymer chains at the steady-state creep.

**Supplementary Materials:** The following supporting information can be downloaded at: https://www.mdpi.com/article/10.3390/polym14235276/s1, Figures S1–S5: Creep curves for PLA, PLA70/TPU30, PLA50/TPU50, PLA30/TPU70, and TPU, respectively. Figure S6: Stress dependence of the plastic strain at different creep temperatures for the recovery behavior. Figure S7: Temperature dependence of the plastic strain for the recovery behavior.

**Author Contributions:** Conceptualization, F.Y.; Data curation, Y.-S.J.; Investigation, Y.-S.J.; Methodology, Y.-S.J., H.O., F.Y. and S.L.; Project administration, H.O. and S.L.; Supervision, H.O. and S.L.; Validation, H.O. and F.Y.; Writing—original draft, Y.-S.J.; Writing—review & editing, F.Y. and S.L. All authors have read and agreed to the published version of the manuscript.

**Funding:** The authors thank the National Science and Technology Council, Taiwan for the financial support.

**Institutional Review Board Statement:** Not applicable.

**Data Availability Statement:** Not applicable.

**Conflicts of Interest:** The authors declare no conflict of interest.

## References


1. Drumright, R.E.; Gruber, P.R.; Henton, D.E. Polylactic acid technology. *Adv. Mater.* **2000**, *12*, 1841–1846. [CrossRef]
2. Matta, A.; Rao, R.U.; Suman, K.; Rambabu, V. Preparation and characterization of biodegradable PLA/PCL polymeric blends. *Procedia Mater. Sci.* **2014**, *6*, 1266–1270. [CrossRef]
3. Ho, C.-H.; Wang, C.-H.; Lin, C.-I.; Lee, Y.-D. Synthesis and characterization of TPO–PLA copolymer and its behavior as compatibilizer for PLA/TPO blends. *Polymer* **2008**, *49*, 3902–3910. [CrossRef]
4. Tokoro, R.; Vu, D.M.; Okubo, K.; Tanaka, T.; Fujii, T.; Fujiura, T. How to improve mechanical properties of polylactic acid with bamboo fibers. *J. Mater. Sci.* **2008**, *43*, 775–787. [CrossRef]
5. Bax, B.; Mussig, J. Impact and tensile properties of PLA/Cordenka and PLA/flax composites. *Compos. Sci. Technol.* **2008**, *68*, 1601–1607. [CrossRef]
6. Rozite, L.; Varna, J.; Joffe, R.; Pupurs, A. Nonlinear behavior of PLA and lignin-based flax composites subjected to tensile loading. *J. Thermoplast. Compos.* **2013**, *26*, 476–496. [CrossRef]
7. Couture, A.; Lebrun, G.; Laperriere, L. Mechanical properties of polylactic acid (PLA) composites reinforced with unidirectional flax and flax-paper layers. *Compos. Struct.* **2016**, *154*, 286–295. [CrossRef]
8. Papa, I.; Lopresto, V.; Simeoli, G.; Langella, A.; Russo, P. Ultrasonic damage investigation on woven jute/poly (lactic acid) composites subjected to low velocity impact. *Compos. Part B-Eng.* **2017**, *115*, 282–288. [CrossRef]
9. Gibeop, N.; Lee, D.W.; Prasad, C.V.; Toru, F.; Kim, B.S.; Song, J.I. Effect of plasma treatment on mechanical properties of jute fiber/poly (lactic acid) biodegradable composites. *Adv. Compos. Mater.* **2013**, *22*, 389–399. [CrossRef]
10. Yang, T.C.; Wu, T.L.; Hung, K.C.; Chen, Y.L.; Wu, J.H. Mechanical properties and extended creep behavior of bamboo fiber reinforced recycled poly(lactic acid) composites using the time-temperature superposition principle. *Constr. Build. Mater.* **2015**, *93*, 558–563. [CrossRef]
11. Georgiopoulos, P.; Kontou, E.; Christopoulos, A. Short-term creep behavior of a biodegradable polymer reinforced with wood-fibers. *Compos. Part B-Eng.* **2015**, *80*, 134–144. [CrossRef]
12. Morreale, M.; Mistretta, M.C.; Fiore, V. Creep Behavior of Poly(lactic acid) Based Biocomposites. *Materials* **2017**, *10*, 395. [CrossRef] [PubMed]
13. Ye, J.; Yao, T.Y.; Deng, Z.C.; Zhang, K.; Dai, S.; Liu, X.B. A modified creep model of polylactic acid (PLA-max) materials with different printing angles processed by fused filament fabrication. *J. Appl. Polym. Sci.* **2021**, *138*, 50270. [CrossRef]
14. Waseem, M.; Salah, B.; Habib, T.; Saleem, W.; Abas, M.; Khan, R.; Ghani, U.; Siddiqi, M.U.R. Multi-Response Optimization of Tensile Creep Behavior of PLA 3D Printed Parts Using Categorical Response Surface Methodology. *Polymers* **2020**, *12*, 2962. [CrossRef]
15. Guedes, R.M.; Singh, A.; Pinto, V. Viscoelastic modelling of creep and stress relaxation behaviour in PLA-PCL fibres. *Fiber Polym.* **2017**, *18*, 2443–2453. [CrossRef]
16. Niaza, K.V.; Senatov, F.S.; Stepashkin, A.; Anisimova, N.Y.; Kiselevsky, M.V. Long-term creep and impact strength of biocompatible 3d-printed pla-based scaffolds. In *Nano Hybrids and Composites*; Trans Tech Publication Ltd.: Zurich, Switzerland, 2017; Volume 13, pp. 15–20.
17. Jing, X.; Mi, H.Y.; Peng, X.F.; Turng, L.S. The morphology, properties, and shape memory behavior of polylactic acid/thermoplastic polyurethane blends. *Polym. Eng. Sci.* **2015**, *55*, 70–80. [CrossRef]
18. Li, Y.; Shimizu, H. Toughening of polylactide by melt blending with a biodegradable poly (ether) urethane elastomer. *Macromol. Biosci.* **2007**, *7*, 921–928. [CrossRef]
19. Lubis, M.A.R.; Handika, S.O.; Sari, R.K.; Iswanto, A.H.; Antov, P.; Kristak, L.; Lee, S.H.; Pizzi, A. Modification of Ramie Fiber via Impregnation with Low Viscosity Bio-Polyurethane Resins Derived from Lignin. *Polymers* **2022**, *14*, 2165. [CrossRef]
20. Aristri, M.A.; Lubis, M.A.R.; Iswanto, A.H.; Fatriasari, W.; Sari, R.K.; Antov, P.; Gajtanska, M.; Papadopoulos, A.N.; Pizzi, A.J.F. Bio-Based Polyurethane Resins Derived from Tannin: Source, Synthesis, Characterisation, and Application. *Forests* **2021**, *12*, 1516. [CrossRef]
21. Yasuniwa, M.; Tsubakihara, S.; Sugimoto, Y.; Nakafuku, C. Thermal analysis of the double-melting behavior of poly (L-lactic acid). *J. Polym. Sci. Part B Polym. Phys.* **2004**, *42*, 25–32. [CrossRef]
22. Shi, Q.; Mou, H.; Gao, L.; Yang, J.; Guo, W. Double-melting behavior of bamboo fiber/talc/poly (lactic acid) composites. *J. Polym. Environ.* **2010**, *18*, 567–575. [CrossRef]
23. Yasuniwa, M.; Tsubakihara, S.; Ohoshita, K.; Tokudome, S.I. X-ray studies on the double melting behavior of poly (butylene terephthalate). *J. Polym. Sci. Part B Polym. Phys.* **2001**, *39*, 2005–2015. [CrossRef]
24. Hobbs, S.; Pratt, C. Multiple melting in poly (butylene terephthalate). *Polymer* **1975**, *16*, 462–464. [CrossRef]
25. Holdsworth, P.; Turner-Jones, A. The melting behaviour of heat crystallized poly (ethylene terephthalate). *Polymer* **1971**, *12*, 195–208. [CrossRef]
26. Todoki, M.; Kawaguchi, T. Origin of double melting peaks in drawn nylon 6 yarns. *J. Polym. Sci. Polym. Phys. Ed.* **1977**, *15*, 1067–1075. [CrossRef]
27. Chen, H.L.; Porter, R.S. Melting behavior of poly (ether ether ketone) in its blends with poly (ether imide). *J. Polym. Sci. Part B Polym. Phys.* **1993**, *31*, 1845–1850. [CrossRef]
28. Mitchell, T.E.; Hirth, J.; Misra, A. Apparent activation energy and stress exponent in materials with a high Peierls stress. *Acta Mater.* **2002**, *50*, 1087–1093. [CrossRef]

MDPI

*Article*

# Blending PLA with Polyesters Based on 2,5-Furan Dicarboxylic Acid: Evaluation of Physicochemical and Nanomechanical Properties

**Zoi Terzopoulou [1,2,*], Alexandra Zamboulis [2], Lazaros Papadopoulos [2], Maria-Eirini Grigora [3], Konstantinos Tsongas [3], Dimitrios Tzetzis [3], Dimitrios N. Bikiaris [2] and George Z. Papageorgiou [1,4,*]**

1. Department of Chemistry, University of Ioannina, GR-45110 Ioannina, Greece
2. Laboratory of Polymer and Colors Chemistry and Technology, Department of Chemistry, University of Thessaloniki, GR-54124 Thessaloniki, Greece
3. Digital Manufacturing and Materials Characterization Laboratory, School of Science and Technology, International Hellenic University, 14km, N. Moudania, GR-57001 Thessaloniki, Greece
4. Institute of Materials Science and Computing, University Research Center of Ioannina (URCI), GR-45110 Ioannina, Greece

* Correspondence: terzozoi@chem.auth.gr (Z.T.); gzpap@uoi.gr (G.Z.P.)

**Abstract:** Poly(lactic acid) (PLA) is a readily available, compostable biobased polyester with high strength and toughness, and it is excellent for 3D printing applications. Polymer blending is an economic and easy way to improve its properties, such as its slow degradation and crystallization rates and its small elongation, and thus, make it more versatile. In this work, the effects of different 2,5-furan dicarboxylic acid (FDCA)-based polyesters on the physicochemical and mechanical properties of PLA were studied. Poly(butylene furan 2,5-dicarboxylate) (PBF) and its copolymers with poly(butylene adipate) (PBAd) were synthesized in various comonomer ratios and were blended with 70 wt% PLA using melt compounding. The thermal, morphological and mechanical properties of the blends are investigated. All blends were immiscible, and the presence of the dispersed phases improved the crystallization ability of PLA. Mechanical testing revealed the plasticization of PLA after blending, and a small but measurable mass loss after burying in soil for 7 months. Reactive blending was evaluated as a compatibilizer-free method to improve miscibility, and it was found that when the thermal stability of the blend components allowed it, some transesterification reactions occurred between the PLA matrix and the FDCA-based dispersed phase after 20 min at 250 °C.

**Keywords:** polymer blends; poly(lactic acid); aliphatic-aromatic copolyesters; 2,5-furan dicarboxylic acid



**Citation:** Terzopoulou, Z.; Zamboulis, A.; Papadopoulos, L.; Grigora, M.-E.; Tsongas, K.; Tzetzis, D.; Bikiaris, D.N.; Papageorgiou, G.Z. Blending PLA with Polyesters Based on 2,5-Furan Dicarboxylic Acid: Evaluation of Physicochemical and Nanomechanical Properties. *Polymers* **2022**, *14*, 4725. https://doi.org/10.3390/polym14214725

Academic Editors: José Miguel Ferri, Vicent Fombuena Borràs and Miguel Fernando Aldás Carrasco

Received: 19 October 2022
Accepted: 1 November 2022
Published: 4 November 2022



## 1. Introduction

Poly(lactic acid) (PLA) is a biobased and compostable aliphatic polyester that has been studied for use in many applications over the last decade. It is currently the most widely used biobased polymer in the world, and due to its excellent printability in 3D printing, its demand exceeds production. Many properties of PLA, such as strength, stiffness and gas permeability, were found to be comparable to those of traditional petrochemical-based polymers. However, PLA-based materials have a significant number of limitations for specific applications, such as slow biodegradation, high cost and low hardness [1]. The modification of PLA by blending it with other polymers to achieve suitable properties for different applications has received considerable attention in recent years. When compared with copolymerization, polymer blending is an easier and more cost-effective method of fabricating polymer-based materials for a wide range of applications.

Polymer blends containing PLA are offered in the market by several companies, including Bioflex® from FKuR Kunststoff GmbH (Willich, Germany) and Ecovio® from BASF (Ludwigshafen, Germany) [2]. Ecovio® is a compostable, partially biobased blend of

PLA with petroleum-based poly(butylene adipate-co-terephthalate) (PBAT) that is suitable for the fabrication of bags and mulch films. The addition of PBAT improves the elongation and toughness of PLA.

An integral part of the efforts to transition to a more sustainable society is the development and scaling up of the production of new biobased polymers. 2,5-Furan dicarboxylic acid (FDCA)-based polyesters and poly(ethylene 2,5-furan dicarboxylate) (PEF) in particular are expected to play a dominant role in the ever-growing biobased plastics market [3–7]. PEF is considered one of the most promising biobased polymers to replace the petroleum-based poly(ethylene terephthalate), either fully or partially [3]. Avantium has already started constructing a new plant with the capacity to produce 5 kilotonnes of FDCA per year, which is expected to reduce the price of biomass-derived FDCA significantly and, therefore, make the production of cost-competitive PEF possible as well. Besides PEF, a plethora of polyesters of FDCA and their copolymers were reported, with a wide range of properties that can be tuned by tuning their composition [6,8,9]. In the efforts to impart biodegradability to FDCA homopolyesters, adipic acid was used. PBF-co-poly(butylene adipate) (PBAd) copolymers are compostable and can behave either as thermoplastics or as elastomers, depending on the comonomer composition [10–14].

Because of the sustainable character, as well as the good barrier and mechanical properties of FDCA-based polyesters, the number of reports in the literature concerning their blends with PLA is increasing [15–23]. In our previous work, we prepared blends with PLA and PEF, poly(propylene 2,5-furan dicarboxylate) (PPF), or poly(butylene 2,5-furan dicarboxylate (PBF) using solution casting [20,23]. These blends were immiscible in all compositions tested. Regardless of the immiscibility, Long et al. [22] found that uncompatibilized PBF/PLA blends with small PBF content showed that the elongation of the blends was 17 times greater than PLA, while the measure of elasticity and the strength at the breaking point remained the same. In addition, impact resistance was improved compared with the two neat polymers. This peculiar behavior was attributed to the glass–amorphous transition and stretch-induced crystallization in the PBF phase. Uncompatibilized poly(alkylene 2,5-furan dicarboxylate) (PAF)/PLA, including PBF/PLA blends prepared with solvent casting, had improved elongation in comparison with neat PLA [15]. Blending poly(pentylene 2,5-furan dicarboxylate) (PPeF) with PLA improved its UV-shielding properties, its barrier properties and its ductility despite their immiscibility [19]. Additionally, blends of PLA with PPeF, POF and PDoF are suitable for producing textile fibers using wet-spinning [18,21].

The scope of this work was to prepare PLA/PBF and PLA/PBF-co-PBAd blends using melt blending and evaluate their physicochemical properties. The miscibility was assessed with SEM and DSC, while mechanical properties were tested with nanoindentation testing. Nanoindentation tests were assisted by a finite element analysis (FEA) process to curve fit the experimental load–depth curves and extract the materials' stress–strain behavior. Structural properties were characterized with FTIR and XRD and the effect of the blending on the soil degradation of PLA was studied using mass loss quantification and surface observation with microscopy. Finally, reactive blending was tested as a compatibilizer-free approach to improve blend miscibility.

## 2. Materials and Methods

### 2.1. Materials

The PLA used was Ingeo$^{TM}$ Biopolymer 3052D (NatureWorks, Plymouth, MN, USA), which was designed for injection molding [24], and it was kindly donated by Plastika Kritis S.A., Heraklion, Greece. It contains ~96% of L- and ~4% of D-lactide and has Mn = 81,700 g/mol. Its intrinsic viscosity is $[\eta]$ = 1.24 dL/g.

### 2.2. Synthesis of the Polyesters

For the synthesis of PBAd, the following protocol was used: First, for the esterification step, adipic acid and 1,4-butanediol in a molar ratio 1:1.1 were charged into a three-necked round bottom flask equipped with a condenser, a mechanical stirrer and a nitrogen inlet. The mixture was gradually heated until homogenization, and then the temperature was increased to 190 °C for 4 h. Then, for the polycondensation step, 400 ppm TBT was added to the reaction mixture, and a vacuum of 0.05 mbar was applied gradually over 30 min. The reaction temperature was increased to 220 °C for an additional 3 h.

For the synthesis of PBF, the following protocol was used: First, for the transesterification step, 2,5-dimethyl furan dicarboxylate (DMFD) and 1,4-butanediol in a molar ratio 1:2.2 were charged into the apparatus that was previously described. Furthermore, 400 ppm of TBT was added to act as a catalyst for transesterification. The mixture was gradually heated until homogenization, and then the temperature was set at 160 °C for 1 h, 170 °C for 1 h, 180 °C for another hour and, finally, 190 °C for an additional hour. For the polycondensation step, a vacuum of 0.05 mbar was applied over 30 min and the reaction temperature was set to 210 °C for 1 h, 220 °C for 1 h and 230 °C for 1 h.

For the synthesis of the copolymers, the product from the esterification step of PBAd synthesis, and the product from the transesterification step of PBF synthesis were used. 6-(4-Hydroxybutoxy)-6-oxohexanoic acid and bis(4-hydroxylbutyl) furan 2,5-dicarboxylate at molar ratios of 3:1, 1:1 and 1:3 were charged into the previously described apparatus. Then, 400 ppm of TBT was added to the flask, and the temperature was set at 160 °C for 1 h, 170 °C for 1 h, 180 °C for 1 h and 190 °C for an additional hour. For the polycondensation step, a vacuum of 0.05 mbar was applied over 30 min and the reaction temperature was set to 210 °C for 1 h, 220 °C for 1 h and 230 °C for 1 h. Through this protocol, the materials PBF-PBAd 75-25, PBF-PBAd 50-50 and PBF-PBAd 25-75 were synthesized.

### 2.3. Blend Preparation

Blends containing 70 wt% PLA and 30 wt% of each FDCA-based polyester were prepared via melt blending. Prior to this, the components were dried overnight under a vacuum. To prepare the blends, the components were introduced in a twin screw co-rotating extruder, operating at 190 °C and 35 rpm for 5 min.

### 2.4. Characterization

The intrinsic viscosity of the produced polyester was measured with an Ubbelohde viscometer (Schott Gerate GMBH, Hofheim, Germany) at 25 °C using a phenol/1,1,2,2-tetrachloroethane (60/40 $w/w$) solution. The sample was heated in the solvent mixture at 80 °C for 20 min until complete dissolution. After cooling, the solution was filtered through a disposable Teflon filter to remove possible solid residues. The calculation of the intrinsic viscosity value of the polymer was performed by applying the Solomon–Cuita Equation (1) of a single point measurement:

$$[\eta] = \frac{\left[2\left\{\frac{t}{t_0} - ln\left(\frac{t}{t_0}\right) - 1\right\}\right]^{1/2}}{c}, \tag{1}$$

where $c$ is the solution concentration, $t$ is the flow time of the solution and $t_0$ is the flow time of the solvent. The experiment was performed three times and the average value was estimated.

Molecular weight was measured with an Agilent Technologies 1260 Infinity II LC Gel Permeation Chromatography (GPC) System consisting of an Isocratic Pump, a PL gel MIXED Guard column and two PLgel 5 μm MIXED-C columns, and an Agilent RID detector. For the calibration, 10 polystyrene (PS) standards of molecular weights between 600 and 1,000,000 g/mol were employed. The prepared solutions had a concentration of 1 mg/mL and the injection volume was 25 μL with a flow of 1 mL/min at a temperature of 40 °C.

Nuclear magnetic resonance (NMR) spectra were recorded in deuterated chloro-form for the structural study of the synthesized polymers. An Agilent 500 spectrometer was utilized (Agilent Technologies, Santa Clara, CA, USA) at room temperature. Spectra were calibrated using the residual solvent peaks.

The morphology of cryofractured cross-sections of the samples was studied with a JEOL (Tokyo, Japan) JSM 7610F field emission scanning electron microscope (SEM) operating at 5 kV.

ATR spectra of the samples were recorded using an IRTracer-100 (Shimadzu, Kyoto, Japan) equipped with a QATR™ 10 Single-Reflection ATR Accessory with a Diamond Crystal. The spectra were collected in the range from 450 to 4000 $cm^{-1}$ at a resolution of 2 $cm^{-1}$ (a total of 16 co-added scans), while the baseline was corrected and converted into absorbance mode.

XRD diffractograms were recorded using a MiniFlex II XRD system (Rigaku, Co., Tokyo, Japan) with Cu K$\alpha$ radiation (0.154 nm) over the 2$\theta$ range from 5° to 50° with a scanning rate of 1°/min. Melt-quenched films prepared with compression molding that were first annealed at the peak of their $T_{cc}$ for 1 h were used.

Differential scanning calorimetry (DSC) analysis was performed using a Perkin Elmer Pyris Diamond DSC differential scanning calorimeter (Solingen, Germany) calibrated with pure indium and zinc standards. The system included a PerkinElmer Intracooler 2 (Solingen, Germany) cooling accessory. Samples of 5 ± 0.1 mg sealed in aluminum pans were used to test the thermal behavior of the polymers.

Thermogravimetric analysis (TGA) was carried out using a SETARAM (Caluire, France) SETSYS TG-DTA 16/18 instrument. The samples (6 ± 0.5 mg) were placed in alumina crucibles and heated from 25 °C to 600 °C in a 50 mL/min flow of $N_2$ at heating rates of 20 °C/min.

Mechanical properties measurements were performed at room temperature (25 °C). The nanoindentation measurements were carried out on a DUH-211S Shimadzu (Kyoto, Japan) device with a force resolution of 0.196 μN. A diamond triangular tip Berkovich indenter (angle of 65°, tip radius is 100 nm) was used and three points were selected and measured using an optical microscope integrated into the device. The modulus and hardness were determined based on the method of Oliver and Pharr [25] and previous work [26–30]. The maximum load was 10 mN and was achieved at a rate of 1.66 mN/s. Due to the material's viscoelastic nature, a dwell time of 50 s was implemented to allow for sufficient time at peak load for the creep effects to saturate. The additional depth induced during the dwell time at constant load was recorded to provide insight into the creep response of the material. In order to calculate the nanomechanical properties, the average value of ten measurements taken at different locations was used. A finite element analysis (FEA) process was developed in order to fit the nanoindentation test curves and extract the stress–strain behavior of the specimens. The interface between the indenter and the surface of the sample was simulated with contact elements and assumed to be frictionless. The nanoindentation experiments were computationally generated by considering the simulation of the loading stage of the indenter penetrating the surface. Other works [26,27,31] showed that kinematic hardening leads to a rapid convergence in the corresponding FEA calculations, and thus, this method was utilized in the developed curve-fitting procedure.

The water contact angle was measured with an Ossila (Sheffield, United Kingdom) contact angle goniometer L2004A1 at room temperature (25 °C). The contact angle was measured by gently placing a water droplet (5 μL) on the surface of the films of the samples prepared via compression molding. At least three measurements were performed and the mean value is reported herein.

For the preliminary evaluation of degradability in soil, the samples were buried in a pot containing a 1:1 by weight mixture of soil and digested sheep manure at an outdoors temperature in Thessaloniki, Greece, during March to September 2021. The soil was collected from a horticulture farm in the rural region of Thessaloniki (40°40′35.3″ N 23°15′37.9″ E). The climate of Thessaloniki is Mediterranean, with hot and dry summers and warm and temperate springs. The average daily temperature exported from a local weather station is presented in Figure S1. Moisture in the soil was replenished at regular intervals with tap water. After predetermined time intervals, the samples were removed from the soil, dried under a vacuum for 3 days and weighed.

The mass loss percentage was calculated according to the following equation:

$$Mass\ loss\ \% = \frac{W_0 - W_i}{W_0}, \tag{2}$$

where $w_0$ is the initial sample of the sample and $w_i$ is the final weight after a certain time interval.

Each sample was buried in triplicate and the mean value was calculated. The surface of the samples was examined using a Jenoptik (Jena, Germany) ProgRes GRYPHAX ARKTUR camera attached to a ZEISS (Oberkochen, Germany) SteREO Discovery V20 microscope, and Gryphax image capturing software and the scanning electron microscope described above were also used.

## 3. Results

### 3.1. Synthesis of the FDCA-Based Polyesters

The synthetic procedure applied is presented in Scheme 1 and the obtained molecular intrinsic viscosity, Mn and PDI are shown in Table 1. Butylene adipate (BAd) and butylene furanoate (BF) oligomers were first prepared, and then each one was added in the proper amounts to produce the copolymers in question through polycondensation. The use of dihydroxybutylene furanoate and dihydroxybutylene adipate assured that the monomer ratio was very close to the feed ratio, as only 1,4-butanediol molecules can be removed during the polycondensation step. Furthermore, materials of high molecular weight up to 23,000 g/mol were obtained, confirming that the protocol followed was suitable for the synthesis of copolyesters. Three materials with different compositions were prepared to examine the effect of each comonomer on the properties of the resulting material.

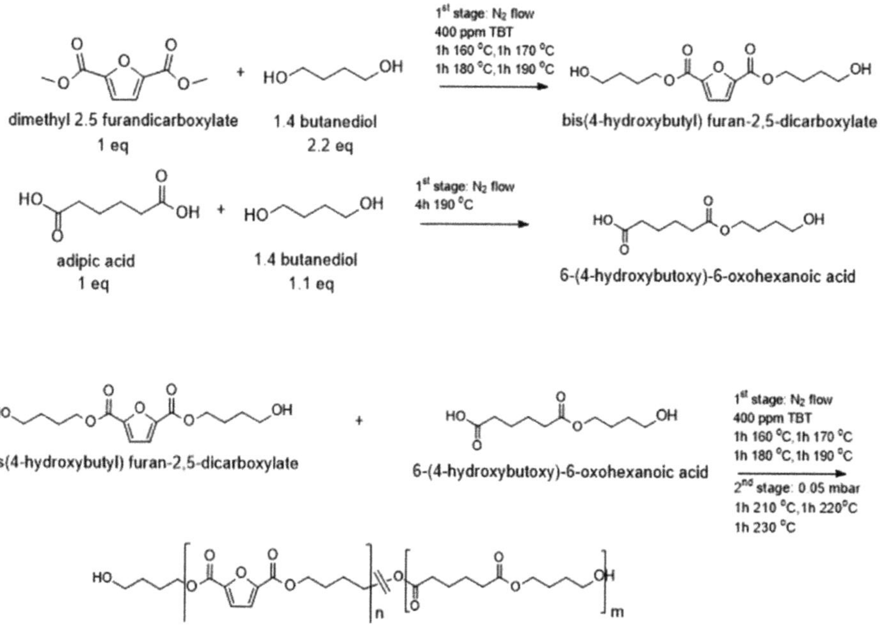

**Scheme 1.** Synthesis of PBF, PBAd and their copolymers.

**Table 1.** Intrinsic viscosity values of the synthesized oligomers and polymers.

| Sample | [η] (dL/g) | Mn (g/mol) | PDI |
|---|---|---|---|
| PBAd oligo | 0.26 | 3600 | 2.18 |
| PBAd | 0.47 | 15,400 | 2.28 |
| PBF oligo | 0.06 | N.D. | N.D. |
| PBF | 0.6 | N.D. | N.D. |
| PBF-PBAd 75 25 | 0.54 | 23,600 | 2.59 |
| PBF-PBAd 50 50 | 0.41 | 9900 | 2.25 |
| PBF-PBAd 25 75 | 0.63 | 15,790 | 2.25 |

N.D.: not determined.

The NMR spectra of the copolymers are presented in Figure S2. For the PBF-PBAd copolymers, resonance signals corresponding to the BF segments were observed at 7.28 ppm CH C, 4.44 ppm $CH_2$ D and 1.92 ppm $CH_2$ E. The BAd peaks were observed at 4.15 ppm $OCH_2$ 4, 2.41 ppm $CH_2C(O)$ 2, 1.73 ppm $CH_2$ 5 and 1.65 ppm $CH_2$ 3. Additionally, signals corresponding to butanol units linked to both FDCA and adipic acids were also observable: 4.41 ppm $OCH_2$ D′, 4.15 ppm $OCH_2$ 4′, 1.85 ppm E′ and 1.80 ppm 5′ (Figure S2c). The $^{13}C$ spectra (Figure S2b) confirmed the $^{1}H$ NMR spectra. Two ester peaks were observed at 174.5 (C=O 1) and 158.3 (C=O A) ppm, along with all the other expected signals: 146.5 and 118.8 ppm CH B and C, 65.1 ppm $OCH_2$ D, 64.3 ppm $OCH_2$ 4, 33.8 ppm $CH_2C(O)$ 2, 25.01 ppm $CH_2$ E, 24.98 ppm $CH_2$ 5 and 24.2 ppm $CH_2$ 3.

The composition of the copolymers was calculated using the integrations of the peaks at 7.28 ppm and 2.41 ppm (Table 2). The microstructure of the copolymers was deduced from peaks at 4.44 to 4.15 ppm corresponding to the butanol methylene groups adjacent to the ester linkages. As depicted in Figure S3, there were three possible structures according to the acid that is linked to each side of butanol: FBF, FBAd (equivalent to AdBF) and AdBAd. The corresponding resonance signals (Figure S2c) were used to calculate the average sequence length of the butylene furanoate ($L_{BF}$) and butylene adipate ($L_{BAd}$) segments in the copolymers and the degree of randomness ($R$) according to Equations (3)–(5). The degree of randomness was slightly over 1 for all copolymers (Table 2), suggesting not only a completely random structure but even alternating sequences.

$$L_{BF} = 1 + \frac{2 \times I_{FBF}}{I_{FBAd} + I_{AdBF}}, \quad (3)$$

$$L_{BAd} = 1 + \frac{2 \times I_{AdBAd}}{I_{FBAd} + I_{AdBF}}, \quad (4)$$

$$R = \frac{1}{L_{BF}} + \frac{1}{L_{BAd}}, \quad (5)$$

where $I_{BF}$ is the integration of the butylene furanoate peak, $I_{FBF}$ is the integration of the peak of the FBF segments, $I_{AdBAd}$ is the integration of the peak of the AdBAd segments and $I_{AdBF}$ is the integration of the peak of the AdBF segments in the $^{1}H$ NMR spectrum.

**Table 2.** Comonomer ratio in the feed calculated with NMR, as well as the block length and degree of randomness of the PBF-co-PBAd copolyesters.

| PBF-PBAd Feed Ratio (mol%) | Ratio Calculated with NMR (mol%) | $L_{BF}$ | $L_{BAd}$ | $R$ |
|---|---|---|---|---|
| PBF-PBAd 75 25 | 74–26 | 2.0 | 1.1 | 1.41 |
| PBF-PBAd 50 50 | 60–40 | 1.7 | 1.4 | 1.31 |
| PBF-PBAd 25 75 | 31–69 | 1.4 | 3.7 | 1.00 |

### 3.2. Characterization of the FDCA-Based Polyesters

Poly(butylene adipate-co-butylene 2,5-furandicarboxylate)s in several comonomer ratios were reported in the literature [6,10,13,14,32]. PBF-co-PBAd with low FDCA content (≤60 mol%) are degradable via hydrolysis and composting and are remarkably elastic. The DSC thermograms of PBF, PBAd and their copolymers are shown in Figure S4. PBAd is a fast-crystallizing polyester with a glass transition $T_g = -57.3$ °C, a double melting peak at 55 and 60 °C, and a melt crystallization $T_c = 31.4$ °C. PBF melts at 170 °C and has $T_g = 37.7$ °C and $T_{cc} = 103.4$ °C, in agreement with previous reports [33–36]. The copolymers PBF-PBAd 75 25 and PBF-PBAd 50 50 show weak cold crystallization and melting during heating after quenching (Figure S4d), and PBF-PBAd 25 75 does not crystallize during heating. The effect of the composition (calculated using NMR) on the thermal transitions of the copolymers is shown in Figure S5. As expected, when increasing the butylene adipate content, the $T_m$, $T_{cc}$ and $T_g$ are reduced because the incorporation of the flexible BAd segments increases the chain mobility [6]. The presence of a single $T_g$ for all copolymers complies with a random sequence structure, in agreement with the calculations from the NMR (Table 2).

The XRD patterns of PBF, PBAd and their copolymers are presented in Figure S6. PBF has a triclinic unit cell, and the diffraction peaks of the planes (010) and (100) appear at 17.5° and 24.5° [37]. PBAd has diffractions peaks at $2\theta = 17.9°$ (002), 19.7°, 20.3°, 21.8° (110), 22.5° (020) and 24.2° (021) [38]. PBAd can crystallize in different crystal forms depending on the crystallization conditions, namely, α- and β-form crystals [39]. Herein, PBAd was melt-quenched at room temperature and shows peaks from both α- and β-form crystals. While the α-form has a monoclinic unit cell, the β-form has an orthorhombic one [39]. The copolymer PBF-PBAd 25 75 has diffraction peaks of both PBF and PBAd, indicating both units can crystallize. Increasing the BF content to 50 and 75% turns the copolymers amorphous, showing weak diffraction peaks of the PBF moiety; therefore, the crystallization of both comonomers is suppressed due to incompatibility of the crystal lattices of PBF and PBAd. This observation supports the findings of the DSC data.

### 3.3. Microstructural Features and Spectroscopic Analysis of the Blends

The miscibility of polymer blends is controlled by many factors: viscosity ratio, composition, shear forces, elasticity, the characteristics of the interface, molecular weight and crystallinity are among them. Blends of PLA/PBAd with PBAd content ≥30 wt% are immiscible, which was attributed to the crystallization of PBAd in the PLA matrix [40]. Blends of PLA/PBF prepared using melt blending were reported in the literature, with PBF content ranging from 5 to 75 wt% [16,22,23]. All of them were immiscible, as the typical sea-island morphology was observed with SEM.

The miscibility of the blends was first evaluated using SEM observations of cryofractured surfaces. The microphotographs are shown in Figure 1 at two different magnifications: ×1000 and ×5000. All blends show the sea-island morphology, which is typical for immiscible blends. The blends PBF PLA, PBF-PBAd 75 25 PLA and PBF-PBAd 50 50 PLA had small spherical particles of the second component with diameters from approximately 0.3 to 6 μm. On the blend PBF-PBAd 25 75 PLA, which contained the largest amount of PBAd among all the blends, the droplets had an irregular platelet-like shape rather than a spherical shape. All samples had empty cavities on their surfaces and evidence of debonding of the dispersed phase, but numerous domains remained attached, especially smaller ones, which can be attributed to transesterification reactions during melt blending [41]. The domain sizes were measured and fitted with a lognormal distribution. The lognormal distribution curves and the histograms, along with the lognormal mean domain sizes, are presented in Figure 2. The mean domain size varied between the blends; however, the standard deviation is quite large, suggesting that there were no significant variations between the samples. While there was a trend that showed an increase in the PBAd content in the dispersed phase also increased the mean domain size, the blend PBF-PBAd 50 50 PLA did not follow this trend, as a large number of smaller spherical droplets (≤0.5 μm) existed on its cryofractured surface. This smaller domain size hinted at slightly improved compatibility and could be attributed to the smaller intrinsic viscosity of the copolymer PBF-PBAd 50 50. The viscosity ratio of the dispersed phase to the matrix influences the blend miscibility and combining a high-viscosity matrix, such as PLA, with a low-viscosity dispersed phase yields a homogeneous and fine dispersion [42].

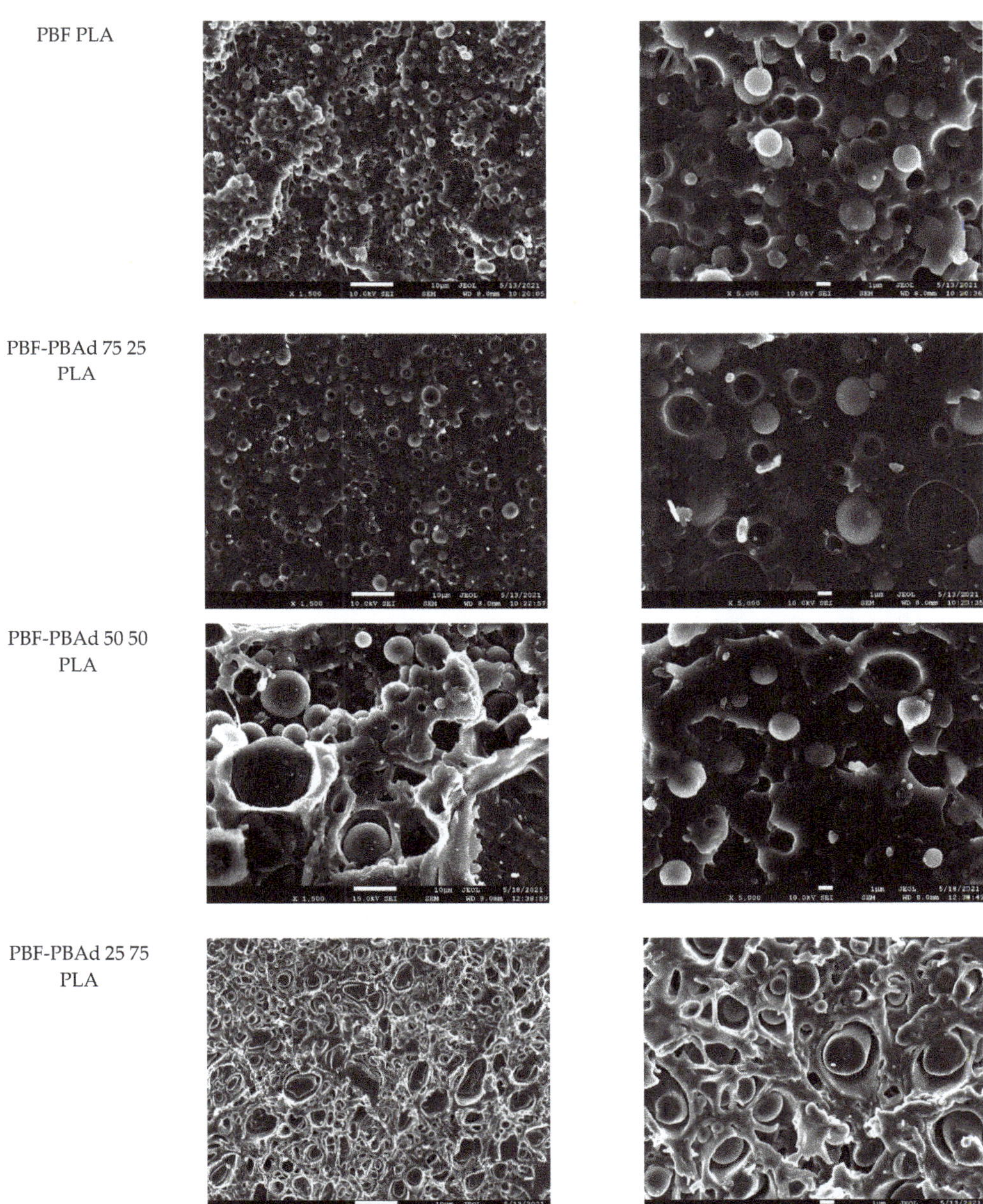


**Figure 1.** SEM micrographs of cryofractured cross-sections of the PLA-based blends. Left column magnification ×1000, right column magnification ×5000.

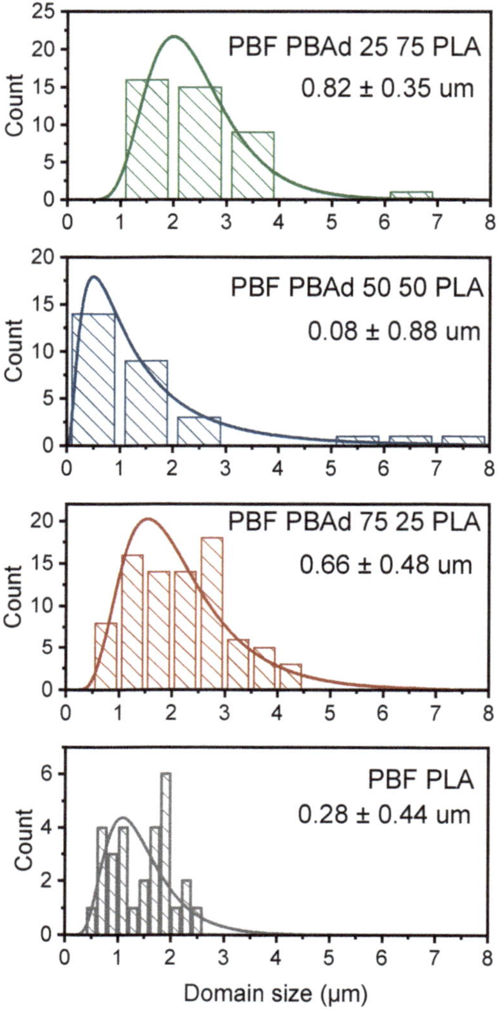


**Figure 2.** Histograms of the domain size and lognormal distribution curve.

The FTIR spectra of the homopolymers, the copolymers and their blends are shown in Figure 3. PLA, due to its high molecular weight, had weak absorption bands of terminal hydroxyls at 3500 cm$^{-1}$. The C-H stretching appeared at ~2900 cm$^{-1}$ and ~2945 cm$^{-1}$, the C=O stretching at 1752 cm$^{-1}$, the -$CH_3$ asymmetric stretching at 1453 cm$^{-1}$ and the C-O-C stretching at 1185–1080 cm$^{-1}$. The spectra of PBF and its blend with PLA are presented in Figure 3a. In the spectrum of PBF, the terminal -OH peak at ~3600 cm$^{-1}$ was visibly stronger than in PLA. The asymmetric and symmetric stretching of the furan ring can be seen at 3150 cm$^{-1}$ and 3120 cm$^{-1}$, respectively [10,15,43]. The bands in the range of 2900–2800 cm$^{-1}$ corresponded to the vibrations of the methyl groups of butanediol, and the band at 1732 cm$^{-1}$ to the C=O stretching. The bands of the C-H and C=H vibrations of the furan ring were visible at 1580 and 1530 cm$^{-1}$, respectively [10,36]. After blending PLA with the FDCA-based polyesters, a combination of the bands of the two components appeared on the spectra (Figure 3, middle spectra). It is noteworthy that the band of the terminal -OH groups was weaker, and the position of the C=O band of the FDCA-based component shifted to smaller wavenumbers, indicating that some limited transesterification reactions might have taken place during the melt blending. In contrast, no such reactions could be hypothesized when blends of PLA and FDCA-based polyesters were prepared via solvent casting [15]. The C=O absorption band did not shift in the spectrum of the blend PBF-PBAd 25 75 PLA (Figure 3d middle spectrum), which could have been due to the higher viscosity of the copolymer PBF-PBAd 25 75 in comparison with PBF-PBAd 50 50 and PBF-PBAd 75 25.

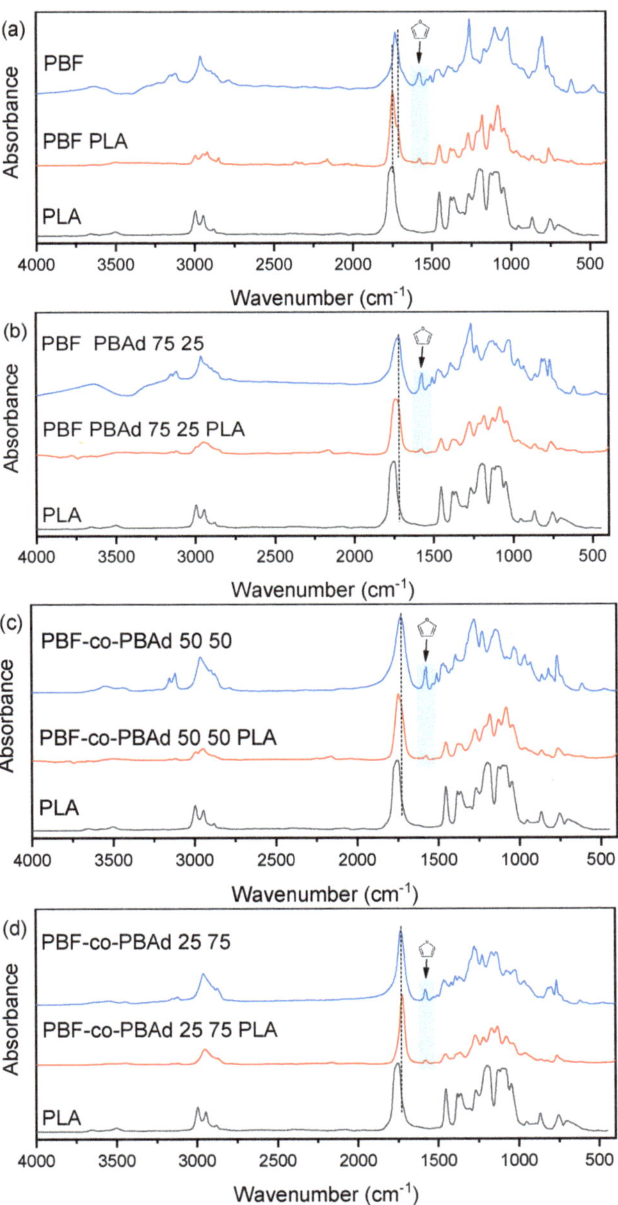


**Figure 3.** FTIR spectra of (**a**) PLA, PBF and PBF PLA blend (**b**) PLA, PBF-PBAd 75 25 and PBF-PBAd 75 25 PLA blend, (**c**) PLA, PBF-PBAd 50 50 and PBF-PBAd 50 50 PLA blend and (**d**) PLA, PBF-PBAd 25 75 and PBF-PBAd 25 75 PLA blend.

### 3.4. Thermal Properties and Crystallization of the Blends

The most common method used to assess whether two polymers are miscible is measuring the changes in their glass transition temperature ($T_g$) via DSC. That is feasible only when the two $T_g$ values differ significantly such that they can be detected. When a blend exhibits a single $T_g$, miscibility is assumed, while two $T_g$ values indicate partial miscibility or immiscibility [44].

The DSC traces of the prepared blends during heating after quenching are shown in Figure 4a. PLA has a $T_g$ = 62.6 °C and very weak cold crystallization at $T_{cc}$ = 126.5 °C. The formed crystals melted at $T_m$ = 153.3 °C. The two components of the blends had significantly different $T_g$ values, with $\Delta T_g$ ranging from ~25 to 100 °C (Figure 4b, gray columns). After mixing, the two $T_g$ values were still present, confirming the immiscibility of the blends, and the $\Delta T_g$ reduced after blending. All blends crystallized during heating, and the crystallization of PLA in the blends was a lot more pronounced than in neat PLA as the $\Delta H_{cc}$ increased, suggesting the dispersed phases facilitated the cold crystallization of PLA. The blending of a polymer with an incompatible component can improve its nucleation rate, and therefore its crystallization [45]. This improvement in crystallization could be the result of heterogeneous nucleation occurring along the interface of the two phase-separated domains [46]. It was previously concluded that blending PBF with PLA led to increased PLA crystallinity [23]. Both PBF and PLA crystallized in the PBF PLA blend since two distinct cold crystallization peaks were detected at 97.4 °C and 136.2 °C, as well as two melting peaks at 170.8 °C and 157.5 °C corresponding to PBF and PLA, respectively. The $T_{cc}$ and the $T_m$ of PLA shifted to higher temperatures, but the $T_{cc}$ of PBF shifted slightly toward lower temperatures in comparison with the neat polyesters. The blends with the PBF-PBAd copolymers did not show detectable cold crystallization of the dispersed phase, as the copolymers alone did not crystallize significantly.

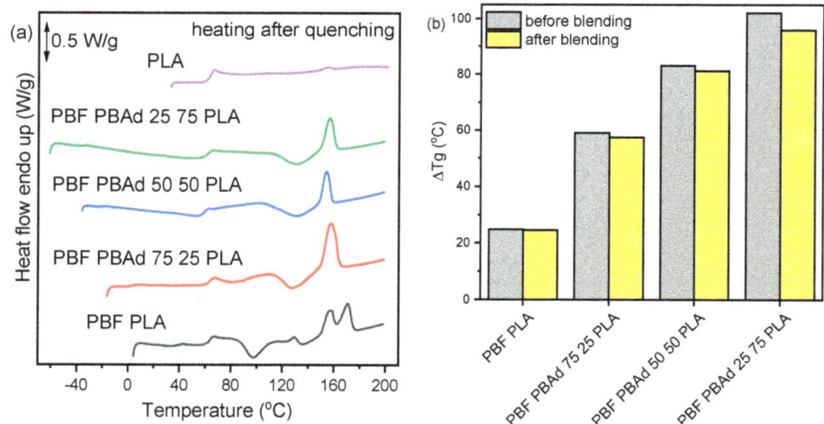


**Figure 4.** (**a**) DSC scans of the blends during heating after quenching at a rate of 20 °C/min and (**b**) $\Delta T_g$ values of the blend's components before and after the melt blending.

The XRD patterns of PLA, the FDCA-based polyesters and their blends are presented in Figure 5. The unit cell of PLA is orthorhombic, with the main diffraction peaks at 2θ = 15, 16.5 and 19°, which correspond to the (010), (200/110) and (203) crystal planes, respectively. The small peak at 22.3° is attributed to the (210) plane [47]. After annealing at the $T_{cc}$ for 1 h, all the prepared blends showed the diffraction peaks of PLA, and some smaller peaks of the dispersed phases, suggesting that both components crystallized during isothermal cold crystallization (annealing). The peaks of the dispersed phase were clearly visible for the blends PBF PLA and PBF-PBAd 25 75, as those polymers had strong diffraction peaks. In the blends PBF-PBAd 75 25 and PBF-PBAd 50 50, no safe conclusions could be drawn, as the weak diffraction peaks of these copolymers might have overlapped with the weak diffraction peak of PLA at about 25°. In our previous study, both PLA and PBF were able to crystallize in a wide range of PLA PBF compositions [16].

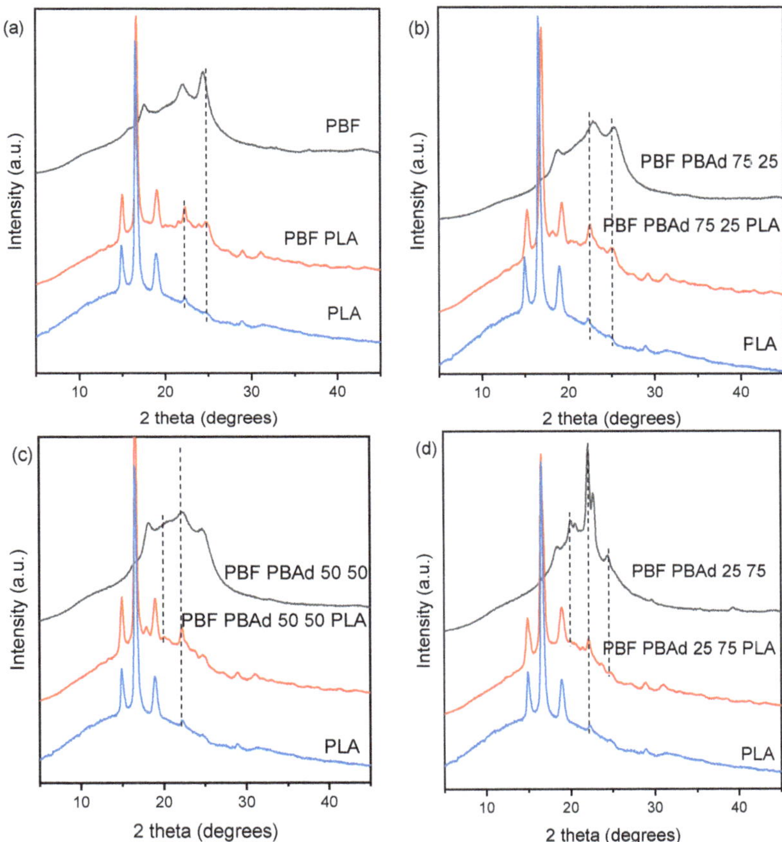


**Figure 5.** XRD patterns of isothermally cold-crystallized (**a**) PLA, PBF and PBF PLA blend (**b**) PLA, PBF-PBAd 75 25 and PBF-PBAd 75 25 PLA blend, (**c**) PLA, PBF-PBAd 50 50 and PBF-PBAd 50 50 PLA blend and (**d**) PLA, PBF-PBAd 25 75 and PBF-PBAd 25 75 PLA blend.

### 3.5. Nanomechanical Properties

The nanomechanical properties of the copolymers and their blends with PLA were measured through nanoindentation testing. In Figure S7, the representative indentation load–depth curves are illustrated, along with the FEA force–depth data that fit the experimental nanoindentation loading curve.

For the FE analysis, an initial value was introduced into the model for the first tangent modulus of the sample's stress–strain curve. This value was related to the elastic modulus that was derived from the nanoindentation tests. The measured indentation depth was applied in steps to the FE model (on the indenter), and then the force reaction was computed and compared with the measured value. The FEA force–depth data should fit the experimental nanoindentation curve; else, the value of the tangent modulus has to be computed again. For the cases where the solutions returned a computational force matching the measured force, the value of the tangent modulus was considered to be accepted, and the next couple of values of force and depth were applied to the model. The following calculation steps started with the previous indentation depth value, considering the already existing stress status and the previously obtained tangent modulus. This process was repeated until the last couple of load–depth values converged and the loop ended. At least 20 simulation steps were considered sufficient to achieve converged FEA solutions and proceed to a satisfactory curve fitting of the nanoindentation curves. The potential to calculate the stress–strain curves of polymers based on the force–depth nanoindentation testing under varied conditions allowed for the estimation of the materials' constitutive laws. The computationally generated stress–strain curves are presented in Figure 6. The resulting elastic modulus, ultimate stress and strain from the FEA-assisted nanoindentation testing (analytical–experimental method), as well as the elastic modulus and hardness from nanoindentation testing, are summarized in Table 3.

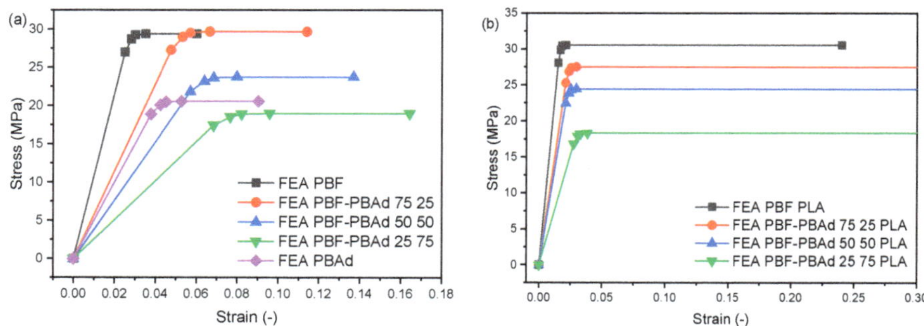


**Figure 6.** FEA generated stress–strain curves of (**a**) PBF, PBAd and PBF-co-PBAd, (**b**) PBF-PBAd PLA blends.

**Table 3.** Mechanical properties of the homopolymers, copolymers and blends obtained using nanoindentation and FEA-assisted nanoindentation.

| Property | Elastic Modulus | FEA Elastic Modulus | FEA Ultimate Stress | Nanoindentation Hardness |
|---|---|---|---|---|
| Units | MPa | MPa | MPa | MPa |
| PLA | 3572 ± 260 | 3600 | 60 | 142 ± 7.4 |
| PBF | 774 ± 43 | 1080 | N.D. | 39.2 ± 3.1 |
| PBF PLA | 2260 ± 107 | 1873 | 30.569 | 125.5 ± 11.4 |
| PBF-PBAd 75 25 | 217.7 ± 4.51 | 574 | 29.679 | 14.5 ± 0.7 |
| PBF-PBAd 75 25 PLA | 1262 ± 38 | 1204 | 27.512 | 72.9 ± 0.4 |
| PBF-PBAd 50 50 | 35 ± 1 | 383 | 23.743 | 2.6 ± 0.08 |
| PBF-PBAd 50 50 PLA | 1004 ± 16.4 | 1070 | 24.455 | 66.7 ± 2.8 |
| PBF-PBAd 25 75 | 31 ± 0.43 | 255 | 18.994 | 1.9 ± 0.08 |
| PBF-PBAd 25 75 PLA | 561 ± 19.3 | 617 | 18.342 | 37.7 ± 1.5 |
| PBAd | 506.03 ± 57 | 502 | 20.570 | 29.72 ± 5.5 |

N.D.: not determined.

The results showed a good correlation between the measured nanoindentation tests and the computational data for all specimens that contained PLA (Figure 6b). For the PBF-PBAd copolymer specimens (Figure 6a), the FEA seemed to overpredict the elastic modulus results. This was due to the inherent nature of calculation from the Oliver–Pharr approach [26], which measures the elastic modulus from the unloading section of the load–depth curve, while the FEA approach calculates the response from the loading curve up to the maximum force. Overall, the results showed a significant increase in strength for the PLA-containing blends. Considering these results, it can be concluded that the PBF-PBAd copolymers affected the specimen's overall stress–strain behavior by reducing the elastic modulus values; however, from the elasto-plastic type stress–strain curves, the ultimate strain increased due to the flexible macromolecular chain of PBAd.

Furthermore, the experimental nanoindentation technique assisted by FEA was shown to be a very successful method for determining the mechanical behavior of the materials. The values of elastic modulus obtained from the nanoindentation testing and FEA of the blends were almost similar and always in the same order of magnitude. A decreasing trend in the elastic modulus values of PBF-PBAd was observed with the increase of PBAd content in the copolymers, which was expected due to the flexible behavior of PBAd. A similar drop is observed in the values of hardness.

The elastic properties of the PLA matrix differed significantly from those of the dispersed phases since the PBF-PBAd copolymers had a much smaller elastic modulus when compared with PLA, which is a stiff polymer with small elongation. Thus, the blends had intermediate elastic modulus and stress at break values that decreased as the softness of the dispersed phase increased. Similar behavior was observed when PBAT was added in the PLA for PBAT content up to 40 wt% [48]. It was reported that when using only a small amount of PBF in the preparation of immiscible blends of PLA/PBF, the elongation and impact toughness of PLA significantly increased [22]. Herein, the amount of PBF used might have been too large to obtain a similar improvement in elongation. Overall, the materials PBF-PBAd 75 25 and PBF-PBAd 75 25 PLA presented the best nanomechanical properties during nanoindentation testing in comparison with all the PBF-PBAd copolymers and blends. However, it should be noted that since the blends are all immiscible, the cavities from the dispersed phase particles are expected to act as stress concentration points upon mechanical loading.

### *3.6. Wettability and Soil Burial Test*

The effect of blending on the wettability of PLA was investigated by measuring the water contact angle. The values obtained are shown in Figure 7. Water contact angle values are related both to the composition of the material, as well as the roughness of the substrate [49]. The contact angle values of all the materials were <90°, which is considered hydrophilic and was attributed to the numerous oxygen species of the repeating units of the polyesters. The contact angle of PBF and its copolymers with PBAd are represented by the green bars of Figure 7. PBF has a contact angle of 55.8 ± 3.5°, and it was increased after copolymerization with PBAd, as well as with increased PBAd content. This increase was due to the introduction of the eight methylene groups of the PBAd repeating unit into the macromolecular chain. The contact angle of PBF reported in the literature was larger than in this study [43,50], which was likely a result of the higher molecular weight and, as a result, fewer polar terminal -OH and -COOH groups. PLA was the least hydrophilic of all the polymers in this work, as it had the highest molecular weight. After introducing the FDCA-based polymers in the PLA matrix, its contact angle decreased and, therefore, its hydrophilicity increased, and all the blends had contact angle values in between those of their components.

PLA degrades under industrial composting conditions (58 °C, 90% biodegradation up to 6 months) and is likely to degrade in thermophilic anaerobic digestion at 52 °C. PBAT is also degradable under industrial composting conditions and certain grades are susceptible to home composting and soil degradation [51]. PBF does not degrade after 8 weeks of composting [52]. PBF-co-PBAd with 40–60% BF was found to be compostable under standard conditions regulated in ISO 14855-1:2005 and GB/T 19277.2-2013, as they reached 90% biodegradation in 110 days [32]. As expected, increasing the BF content decelerated the biodegradation.

The hydrolysis rate of aliphatic–aromatic copolyesters depends on the sequence length of the aromatic unit, the hydrophilicity and the inherent sensitivity of the ester bonds to hydrolytic attack. Ester bonds adjacent to aliphatic moieties are more prone to hydrolysis [53,54]. In addition, in soil degradation, the soil type and quality affect degradation rates [55]. In the scope of this work, a preliminary evaluation of the degradability of the polymers in real environmental conditions was performed. The cumulative mass loss of the polymers studied is shown in Figure 8. As seen in Figure 8a, both PBF and

PBF-PBAd 75 25 did not lose significant mass after 7 months of burying. Increasing the BAd content to 50 and 75 mol% enabled measurable mass loss, which reached ~25% in 7 months. In fact, PBF-PBAd 50 50 degraded similarly to PBAd, which contained only aliphatic units in its macromolecular chain. The accelerated degradation of this copolymer can be attributed to its lower molecular weight. PLA and its blends (Figure 8a) showed a different mass loss pattern since PLA and the matrix of the blends were not degradable in soil. Both PBF and the blend PBF PLA seemed to gain weight, which was attributed to the contamination of their surface with soil and a simultaneous lack of mass loss. PLA showed an insignificant mass loss with a large standard deviation; therefore, it was considered completely non-degradable under the testing conditions. After blending it with the FDCA-based copolyesters, a small mass loss was observed, reaching ~6% for the blend PBF-PBAd 25 75 PLA. This mass loss might be limited but it is an indication that having only a small fraction of the biodegradable repeating BAd units in the blend composition could accelerate the degradation of PLA. The properties that dictate degradation rates of the materials are molecular weight and hydrophilicity. Overall, for the samples that degraded, mass loss was increasing faster after 3 months of burying, which could be associated with the higher temperatures in that timeframe (Figure S1).

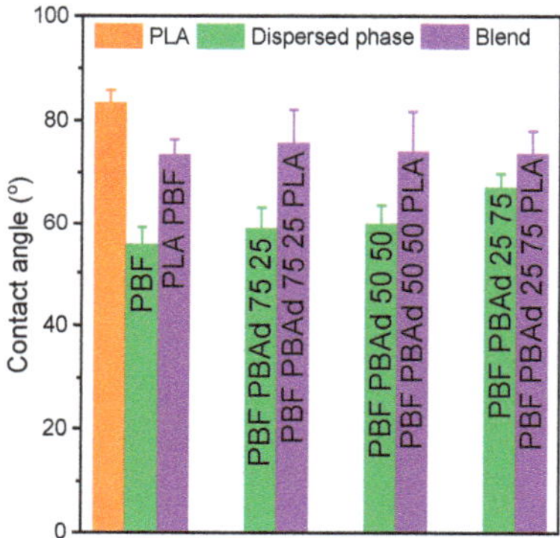


**Figure 7.** Water contact angle of the polymers before and after blending.

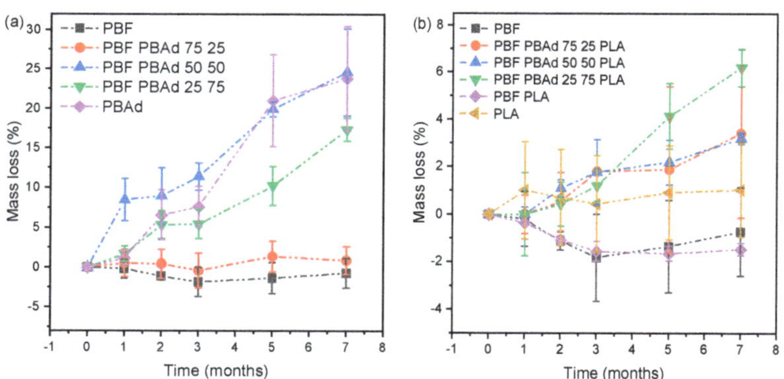


**Figure 8.** Cumulative mass loss of (**a**) PBF, PBAd and their copolymers, and (**b**) PBF, PLA and the blends during soil burial.

The appearance of the samples after being buried for 7 months is shown in Figure 9. PLA was not significantly affected macroscopically, except for the contamination of its surface by soil residues (Figure 9a). PBAd broke into several fragments and its color turned from white to brown (Figure 9b). Among the PBF-PBAd copolymers, macroscopic changes were noticed for PBF-PBAd 50 50 (Figure 9g) and PBF-PBAd 25 75 (Figure 9i), which had the lowest molecular weights and the highest BAd contents. These two films had holes visible to the naked eye and started breaking after 5 months of burying. The appearance of the blends was affected the most in terms of coloring with contamination from the soil, and some holes started appearing at the thinnest parts of the films of the blend PBF-PBAd 25 75 PLA, Figure 9j, which agreed with the mass loss results and revealed the important role of film thickness on degradation rates. The microscopic changes in the surface of the materials were examined with SEM. Besides PLA and PBF (Figure 9a,c), all polymers had very rough surfaces with defects and holes, suggesting surface erosion, as well as deposited particles. These could be traces of soil, mineral deposits or biofilms. In Figure 9j, a fiber-like structure that resembled microbial biofilms formed on the PBF-PBAd 25 75 PLA blend.

**Figure 9.** The surfaces of the materials after 7 months of soil degradation. From left to right: macroscopic photograph, stereoscope image with ×7.5 magnification and SEM image with ×2000 magnification. Reactive blending. (**a**) PLA; (**b**) PBAd; (**c**) PBF; (**d**) PBF PLA; (**e**) PBF-PBAd 75 25; (**f**) PBF-PBAd 75 25 PLA; (**g**) PBF-PBAd 50 50; (**h**) PBF-PBAd 50 50 PLA; (**i**) PBF-PBAd 25 75; (**j**) PBF-PBAd 25 75 PLA.

### 3.7. Reactive Blending

Reactive blending was performed to examine whether it can improve the compatibility between the components of the blends. In general, reactive blending concerns the mixing of the blend at elevated temperatures for prolonged times to enable transesterification reactions to occur. Increasing the blending time leads to the formation of block copolymers initially, and further increasing it yields random copolymers. Reactive blending was simulated inside the DSC pan to make a preliminary estimation of its effect on miscibility improvement. To do so, the blends were held isothermally at 220 °C or 250 °C for different times ranging from 10 to 50 min. After each isothermal step, a quenching and a heating scan with a heating rate of 20 °C/min were performed to measure the $T_g$. The start of thermal degradation during reactive blending was concluded from the DSC curves of the isothermal steps.

The effect of the reactive blending time on the $T_g$ of the blends is shown in Figure 10. In general, when $T_g$ values gradually approach each other, it indicates compatibilization [56]. Another indication for improved miscibility is the suppression of cold crystallization and a decrease in the $T_m$ since the ability of copolymers to crystallize is limited in comparison with blends.

In the PBF PLA blend, only the $T_g$ of PBF increased during reactive blending at 220 °C (Figure 10a); the cold crystallization (Figure S9a) shifted slightly toward higher temperatures, indicating reduced macromolecular mobility; and the positions of the $T_m$ peaks were not significantly affected. The total melting enthalpy $\Delta H_m$ reduced with time, which was reflected by the smaller area of the melting peaks, showing the reduced crystallinity after reactive blending. At 250 °C, the $T_g$ of both PLA and PBF shifted toward each other (Figure 10b), cold crystallization was suppressed significantly, the $T_m$ values shifted to lower temperatures and the two distinct melting peaks merged into one ((Figure S9c). The reduction in the $T_m$ could be a result of less perfected crystallites due to reduced chain mobility as transesterification reactions take place. TGA measurements of the PBF PLA blend did not show any significant mass loss up to 300 °C (Figure S8).

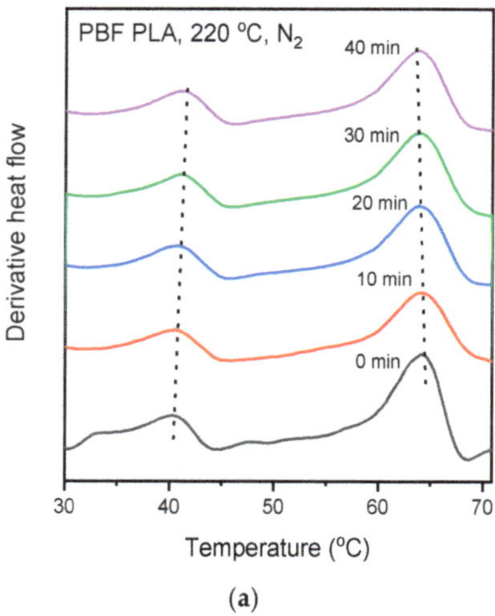


(a)

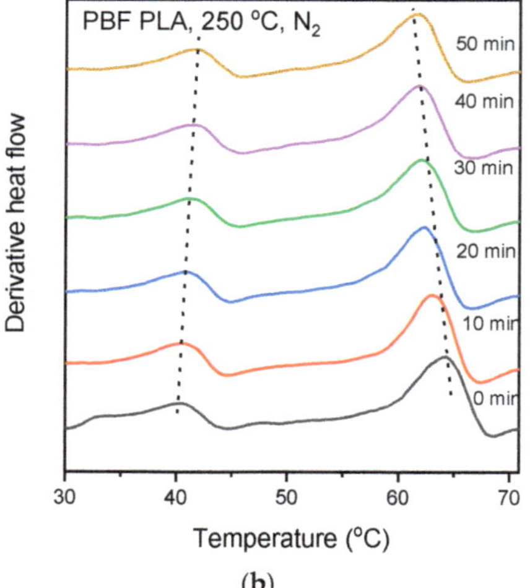


(b)

**Figure 10.** *Cont.*

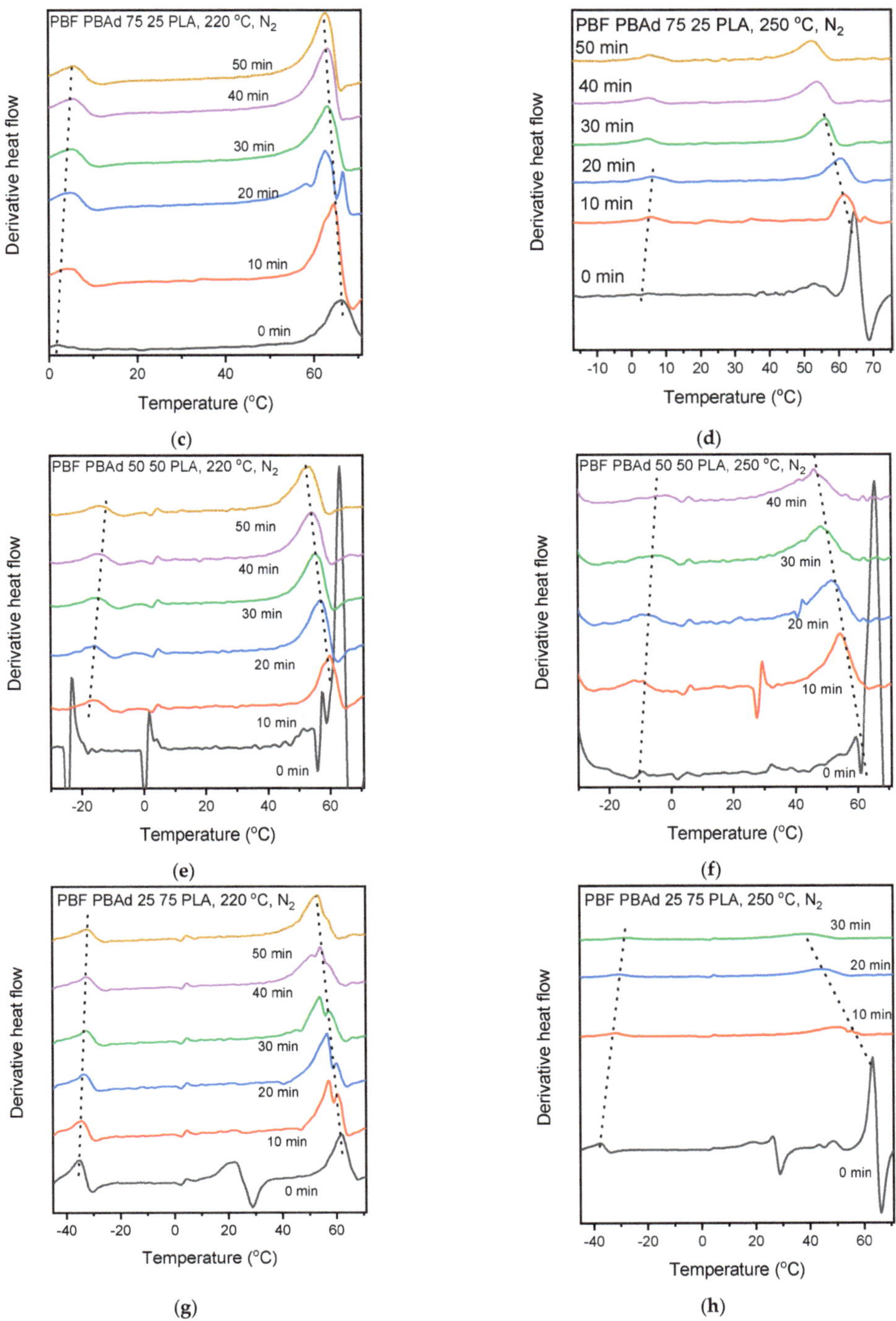


**Figure 10.** Study of reactive blending with DSC: derivative heat flow and change in the $T_g$ of the (**a**,**b**) PBF PLA, (**c**,**d**) PBF-PBAd 75 25 PLA, (**e**,**f**) PBF-PBAd 50 50 PLA and (**g**,**h**) PBF-PBAd 25 75 PLA blends.

The change in the $T_g$ of the blend PBF-PBAd 75 25 PLA during the reactive blending is shown in Figure 10c,d. At 220 °C, the two $T_g$ values approached each other up to 20 min, and after that, their changes remained insignificant. Signs of degradation started showing up on the DSC curves after 40 min of reactive blending as a small exothermic decline of the baseline (data not shown). The $T_m$ and the $T_{cc}$ were reduced with blending time, and the extent of cold crystallization remained similar. At 250 °C, reactive blending seemed to be efficient again until 20 min, with reduced $\Delta T_g$ and $T_m$ values, as well as suppressed cold crystallization, but thermal degradation occurred for prolonged times. Indeed, the TGA curves (Figure S8) showed that the blend PBF-PBAd 75 25 PLA lost 1.5% at 220 °C and ~5% of its initial mass at 250 °C.

The PBF-PBAd 50 50 PLA blend was thermally stable up to 275 °C (Figure S8); therefore, extensive degradation was not expected to occur during reactive blending. At 220 °C and with 40 min of reactive blending, the $T_g$ of PLA reduced by 9.2 °C, and the $T_g$ of the dispersed phase increased by 7 °C (Figure 10e). Based on the shift of the $T_g$ values, transesterification reactions might have occurred for the first 30 min. After that, the changes in both $T_g$ were subtle. The $T_{cc}$ and $T_m$ of PLA shifted to lower temperatures with time, but the overall crystallinity was sustained after reactive blending. As the $T_{cc}$ of PLA becomes smaller, the $T_m$ splits into two peaks because PLA crystallizes into both $\alpha$ and $\alpha'$ form crystals when the $T_{cc}$ decreases [57]. Despite the good thermal stability of the blend PBF-PBAd 50 50 PLA, thermal degradation was detected using DSC after the reactive blending at 250 °C for times $\geq$20 min. After 20 min, the shift in the $T_g$ values (Figure 10f) became less pronounced, marking the competition between potential transesterification and degradation reactions. The $T_{cc}$ increased and crystallization of PLA was suppressed, which are additional indications that transesterification reactions might have occurred during the reactive blending for 10 and 20 min at 250 °C.

The blend PBF-PBAd 25 75 PLA was thermally stable at 220 °C in TGA and started losing mass right after 250 °C (Figure S8). During reactive blending at 220 °C (Figure 10g), the $T_g$ of the blend shifted toward each other and, like the rest of the blends, the shift was more intense during the first 20 min. The $T_{cc}$ increased and the extent of the crystallization remained unaffected. At 250 °C (Figure 10h), degradation had occurred already during the first 10 min; therefore, reactive blending of the blend PBF-PBAd 25 75 PLA was not possible at this temperature.

The largest shift in $T_g$ values after 20 min of reactive blending occurred for the blend PBF-PBAd 50 50 PLA, both at 220 °C and 250 °C. Thus, reactive blending had the most prominent effect on the improvement of the compatibility of this specific blend. This observation was in line with the observed smaller domain size of PBF-PBAd 50 50 PLA, which was correlated to the smaller viscosity of the dispersed phase PBF-PBAd 50 50.

To further investigate the effect of reactive blending on the chemical structure of the blends, NMR spectra were recorded. Indicative NMR spectra of the PBF-PBAd 75-25 PLA blends obtained after 20 min at 220 °C and 250 °C are displayed in Figure 11, while the $^1$H NMR spectra of the other blends are presented in Figure S13. Resonance signals attributed to PLA were observed at 5.19 ppm (OCH) and 1.59 ppm ($CH_3$), in addition to the PBF-PBAd copolymer peaks (vide supra). New peaks indicative of the formation of bonds between PLA and PBF-PBAd copolymers could not be observed. Nevertheless, the $-CH_2OH$ end groups of PBF-PBAd copolymers decreased during the blending, strongly suggesting some transesterification reactions occurred with PLA. Furthermore, this decrease was proportional to the temperature, confirming a more effective blending at 250 °C. The blends composition was calculated using the integrals of the peaks at 5.19 ppm and 4.43 ppm for PBF, or 4.14–4.49 and 4.38–4.43 ppm for PBF-PBAd copolymers and was found in accordance with the feed ratio.

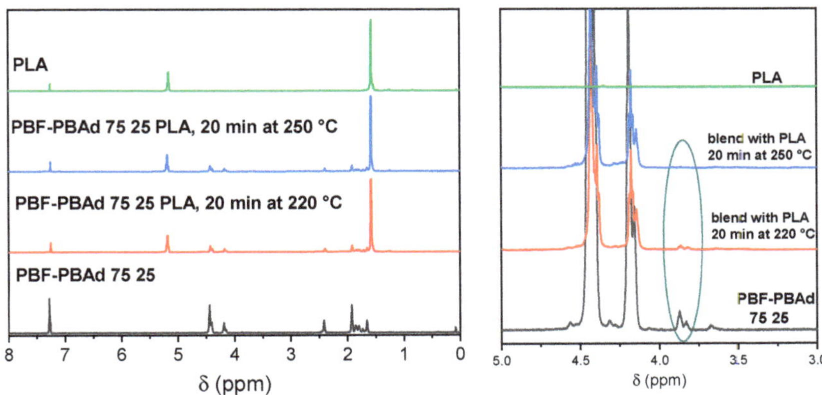


**Figure 11.** $^{1}$H NMR spectra of the PBF-PBAd 75 25 PLA blends.

## 4. Conclusions

PLA/PBF and PLA/PBF-co-PBAd blends with 70 wt% PLA were prepared using melt blending and their physicochemical properties were evaluated. The blends were immiscible (confirmed using SEM and DSC) and the domain size had an increasing trend with increasing PBAd content, except for the blend PBF-PBAd 50 50 PLA, which had the smallest domain size due to the smaller intrinsic viscosity of the copolymer PBF-PBAd 50 50. During the melt blending, some transesterification reactions might have occurred, as detected in the FTIR spectra. In all blends, the PLA matrix was able to crystallize during annealing. Nanomechanical characterization revealed that the elastic modulus, stress at break and hardness of the PLA reduced after the addition of the dispersed phases. FEA allowed for the generation of stress–strain curves that showed a good correlation between the measured nanoindentation tests and the computational data and confirmed the improvement in the case of the presence of PLA. Soil degradation was studied using mass loss quantification and surface observation with microscopy. The blends showed a small but detectable mass loss of the blends, which was accompanied by contaminated surfaces and microscopic holes, among other defects. Finally, reactive blending was found to be a possible compatibilizer-free approach to improve blend miscibility since it reduced the $\Delta T_g$ of the blend's components.

**Supplementary Materials:** The following supporting information can be downloaded from https://www.mdpi.com/article/10.3390/polym14214725/s1. Figure S1: Average daily temperature in Thessaloniki, Greece, during the soil degradation testing. Figure S2: NMR spectra of the synthesized copolymers. (a) $^{1}$H NMR, (b) $^{13}$C NMR, (c) $^{1}$H NMR zoom in the 1.5–4.5 ppm region and (d) numbered structures of PBF-PBAd copolymers. The peaks with the asterisk are assigned to the deuterated solvent ($CDCl_3$ and TFA-d1). Figure S3: Possible triads of PBF-PBAd copolymers. Figure S4: DSC scans of the homopolymers and copolymers during (a) 1st heating with a rate of 20 °C/min, (b) 2nd heating with a rate of 20 °C/min, (c) cooling with a rate of 10 °C/min and (d) heating after quenching with a rate of 20 °C/min. Figure S5: Thermal characteristics of the homopolymers PBF, PBAd and their copolymers. Figure S6: XRD patterns of PBF, PBAd and their copolymers. Figure S7: Load–depth nanoindentation curves of the PBF, PLA, PBF-PBAd and PBF-PBAd PLA specimens, along with the curve-fitted FEA response. Figure S8: (a) TGA and (b) DTG curves of the PLA-based blends. Figure S9: Study of reactive blending with DSC: heating curves after different times, with a zoomed-in view of the $T_g$ region of PBF PLA at 220 °C (a,b) and (c,d) 250 °C. Figure S10: Study of reactive blending with DSC: heating curves after different times, with a zoomed-in view of the Tg region of PBF-PBAd 75 25 PLA at 220 °C (a,b) and (c,d) 250 °C. Figure S11: Study of reactive blending with DSC: heating curves after different times, with a zoomed-in view of the Tg region of PBF-PBAd 50 50 PLA at 220 °C (a,b) and (c,d) 250 °C. Figure S12: Study of reactive blending with DSC: heating curves after different times, with a zoomed-in view of the $T_g$ region of PBF-PBAd

25 75 PLA at 220 °C (a,b) and (c,d) 250 °C. Figure S13: $^1$H NMR spectra of (a) PBF PLA, (b) PBF-PBAd 50-50 PLA, (c) PBF-PBAd 25-75 PLA blends, at 220 °C and 250 °C.

**Author Contributions:** Conceptualization, Z.T.; formal analysis, Z.T., K.T. and M.-E.G.; funding acquisition, Z.T., D.N.B. and G.Z.P.; investigation, Z.T., A.Z., M.-E.G. and L.P.; project administration, Z.T.; resources, D.N.B. and D.T.; validation, Z.T. and M.-E.G.; visualization, Z.T., A.Z. and M.-E.G.; writing—original draft, Z.T., A.Z., K.T. and M.-E.G.; writing—review and editing, Z.T., D.T., D.N.B. and G.Z.P. All authors have read and agreed to the published version of the manuscript.

**Funding:** This research is co-financed by Greece and the European Union (European Social Fund—ESF) through the Operational Programme "Human Resources Development, Education and Lifelong Learning" in the context of the project "Reinforcement of Postdoctoral Researchers—2nd Cycle" (MIS-5033021) implemented by the State Scholarships Foundation (IKY).

**Institutional Review Board Statement:** Not applicable.

**Informed Consent Statement:** Not applicable.

**Data Availability Statement:** Data supporting reported results can be found within the article and the Supplementary Materials.

**Acknowledgments:** This publication is based upon work from COST Action FUR4Sustain, CA18220, supported by COST (European Cooperation in Science and Technology). Z.T. would also like to thank Chrysanthi Papoulia and Eleni Pavlidou for the SEM observations, Dimitra Kourtidou and Konstantinos Chrissafis for the TGA measurements, and Nikolaos Bikiaris for performing the GPC measurements. A.Z. would like to acknowledge the Center of Interdisciplinary Research and Innovation of Aristotle University of Thessaloniki (CIRI-AUTH), Greece, for access to the Large Research Infrastructure and Instrumentation of the Nuclear Magnetic Resonance Laboratory at the Center for Research of the Structure of Matter in the Chemical Engineering Department.

**Conflicts of Interest:** The authors declare no conflict of interest.

## References

1. Hamad, K.; Kaseem, M.; Ayyoob, M.; Joo, J.; Deri, F. Polylactic acid blends: The future of green, light and tough. *Prog. Polym. Sci.* **2018**, *85*, 83–127. [CrossRef]
2. Papadopoulos, L.; Klonos, P.; Terzopoulou, Z.; Psochia, E.; Sanusi, O.M.; Hocine, N.A.; Benelfellah, A.; Giliopoulos, D.; Triantafyllidis, K.; Kyritsis, A.; et al. Comparative study of crystallization, semicrystalline morphology, and molecular mobility in nanocomposites based on polylactide and various inclusions at low filler loadings. *Polymer* **2021**, *217*, 123457. [CrossRef]
3. Sousa, A.F.; Patrício, R.; Terzopoulou, Z.; Bikiaris, D.N.; Stern, T.; Wenger, J.; Loos, K.; Lotti, N.; Siracusa, V.; Szymczyk, A.; et al. Recommendations for replacing PET on packaging, fiber, and film materials with biobased counterparts. *Green Chem.* **2021**, *23*, 8795–8820. [CrossRef]
4. Fei, X.; Wang, J.; Zhang, X.; Jia, Z.; Jiang, Y.; Liu, X. Recent Progress on Bio-Based Polyesters Derived from 2,5-Furandicarbonxylic Acid (FDCA). *Polymers* **2022**, *14*, 625. [CrossRef]
5. Papageorgiou, G.Z.; Papageorgiou, D.G.; Terzopoulou, Z.; Bikiaris, D.N. Production of bio-based 2,5-furan dicarboxylate polyesters: Recent progress and critical aspects in their synthesis and thermal properties. *Eur. Polym. J.* **2016**, *83*, 202–229. [CrossRef]
6. Terzopoulou, Z.; Papadopoulos, L.; Zamboulis, A.; Papageorgiou, D.G.; Papageorgiou, G.Z.; Bikiaris, D.N. Tuning the properties of furandicarboxylic acid-based polyesters with copolymerization: A review. *Polymers* **2020**, *12*, 1209. [CrossRef]
7. Loos, K.; Zhang, R.; Pereira, I.; Agostinho, B.; Hu, H.; Maniar, D.; Sbirrazzuoli, N.; Silvestre, A.J.D.; Guigo, N.; Sousa, A.F. A Perspective on PEF Synthesis, Properties, and End-Life. *Front. Chem.* **2020**, *8*, 585. [CrossRef]
8. Papadopoulos, L.; Magaziotis, A.; Nerantzaki, M.; Terzopoulou, Z.; Papageorgiou, G.Z.; Bikiaris, D.N. Synthesis and characterization of novel poly(ethylene furanoate-co-adipate) random copolyesters with enhanced biodegradability. *Polym. Degrad. Stab.* **2018**, *156*, 32–42. [CrossRef]
9. Pandey, S.; Dumont, M.J.; Orsat, V.; Rodrigue, D. Biobased 2,5-furandicarboxylic acid (FDCA) and its emerging copolyesters' properties for packaging applications. *Eur. Polym. J.* **2021**, *160*, 110778. [CrossRef]
10. Wu, B.; Xu, Y.; Bu, Z.; Wu, L.; Li, B.G.; Dubois, P. Biobased poly(butylene 2,5-furandicarboxylate) and poly(butylene adipate-co-butylene 2,5-furandicarboxylate)s: From synthesis using highly purified 2,5-furandicarboxylic acid to thermo-mechanical properties. *Polymer* **2014**, *55*, 3648–3655. [CrossRef]
11. Zhou, W.; Wang, X.; Yang, B.; Xu, Y.; Zhang, W.; Zhang, Y.; Ji, J. Synthesis, physical properties and enzymatic degradation of bio-based poly(butylene adipate-co-butylene furandicarboxylate) copolyesters. *Polym. Degrad. Stab.* **2013**, *98*, 2177–2183. [CrossRef]

12. Little, A.; Pellis, A.; Comerford, J.W.; Naranjo-Valles, E.; Hafezi, N.; Mascal, M.; Farmer, T.J. Effects of Methyl Branching on the Properties and Performance of Furandioate-Adipate Copolyesters of Bio-Based Secondary Diols. *ACS Sustain. Chem. Eng.* **2020**, *8*, 14471–14483. [CrossRef]
13. Peng, S.; Wu, B.S.; Wu, L.; Li, B.G.; Dubois, P. Hydrolytic degradation of biobased poly(butylene succinate-co-furandicarboxylate) and poly(butylene adipate-co-furandicarboxylate) copolyesters under mild conditions. *J. Appl. Polym. Sci.* **2017**, *134*, 44674. [CrossRef]
14. Kim, H.; Kim, T.; Choi, S.; Jeon, H.; Oh, D.X.; Park, J.; Eom, Y.; Hwang, S.Y.; Koo, J.M. Remarkable elasticity and enzymatic degradation of bio-based poly(butylene adipate-: Co-furanoate): Replacing terephthalate. *Green Chem.* **2020**, *22*, 7778–7787. [CrossRef]
15. Fredi, G.; Rigotti, D.; Bikiaris, D.N.; Dorigato, A. Tuning thermo-mechanical properties of poly(lactic acid) films through blending with bioderived poly(alkylene furanoate)s with different alkyl chain length for sustainable packaging. *Polymer* **2021**, *218*, 123527. [CrossRef]
16. Poulopoulou, N.; Kantoutsis, G.; Bikiaris, D.N.; Achilias, D.S.; Kapnisti, M.; Papageorgiou, G.Z. Biobased engineering thermoplastics: Poly(butylene 2,5-furandicarboxylate) blends. *Polymers* **2019**, *11*, 937. [CrossRef]
17. Fredi, G.; Karimi Jafari, M.; Dorigato, A.; Bikiaris, D.N.; Pegoretti, A. Improving the Thermomechanical Properties of Poly(lactic acid) via Reduced Graphene Oxide and Bioderived Poly(decamethylene 2,5-furandicarboxylate). *Materials* **2022**, *15*, 1316. [CrossRef]
18. Perin, D.; Rigotti, D.; Fredi, G.; Papageorgiou, G.Z.; Bikiaris, D.N.; Dorigato, A. Innovative Bio-based Poly(Lactic Acid)/Poly(Alkylene Furanoate)s Fiber Blends for Sustainable Textile Applications. *J. Polym. Environ.* **2021**, *29*, 3948–3963. [CrossRef]
19. Rigotti, D.; Soccio, M.; Dorigato, A.; Gazzano, M.; Siracusa, V.; Fredi, G.; Lotti, N. Novel Biobased Polylactic Acid/Poly(pentamethylene 2,5-furanoate) Blends for Sustainable Food Packaging. *ACS Sustain. Chem. Eng.* **2021**, *9*, 13742–13750. [CrossRef]
20. Poulopoulou, N.; Smyrnioti, D.; Nikolaidis, G.N.; Tsitsimaka, I.; Christodoulou, E.; Bikiaris, D.N.; Charitopoulou, M.A.; Achilias, D.S.; Kapnisti, M.; Papageorgiou, G.Z. Sustainable plastics from biomass: Blends of polyesters based on 2,5-furandicarboxylic acid. *Polymers* **2020**, *12*, 225. [CrossRef]
21. Perin, D.; Fredi, G.; Rigotti, D.; Soccio, M.; Lotti, N.; Dorigato, A. Sustainable textile fibers of bioderived polylactide/poly(pentamethylene 2,5-furanoate) blends. *J. Appl. Polym. Sci.* **2022**, *139*, 51740. [CrossRef]
22. Long, Y.; Zhang, R.; Huang, J.; Wang, J.; Jiang, Y.; Hu, G.H.; Yang, J.; Zhu, J. Tensile Property Balanced and Gas Barrier Improved Poly(lactic acid) by Blending with Biobased Poly(butylene 2,5-furan dicarboxylate). *ACS Sustain. Chem. Eng.* **2017**, *5*, 9244–9253. [CrossRef]
23. Long, Y.; Zhang, R.; Huang, J.; Wang, J.; Zhang, J.; Rayand, N.; Hu, G.H.; Yang, J.; Zhu, J. Retroreflection in binary bio-based PLA/PBF blends. *Polymer* **2017**, *125*, 138–143. [CrossRef]
24. NatureWorks Ingeo$^{TM}$ Biopolymer 3052D Technical Data Sheet. NatureWorks 2018, 001, 1–4. Available online: https://www.natureworksllc.com/~{}/media/Files/NatureWorks/Technical-Documents/Technical-Data-Sheets/TechnicalDataSheet_3052D_injection-molding_pdf.pdf (accessed on 25 October 2022).
25. Oliver, W.C.; Pharr, G.M. An improved technique for determining hardness and elastic modulus using load and displacement sensing indentation experiments. *J. Mater. Res.* **1992**, *7*, 1564–1583. [CrossRef]
26. Tzetzis, D.; Tsongas, K.; Mansour, G. Determination of the mechanical properties of epoxy silica nanocomposites through FEA-supported evaluation of ball indentation test results. *Mater. Res.* **2017**, *20*, 1571–1578. [CrossRef]
27. Tsongas, K.; Tzetzis, D.; Karantzalis, A.; Banias, G.; Exarchos, D.; Ahmadkhaniha, D.; Zanella, C.; Matikas, T.; Bochtis, D. Microstructural, surface topology and nanomechanical characterization of electrodeposited Ni-P/SiC nanocomposite coatings. *Appl. Sci.* **2019**, *9*, 2901. [CrossRef]
28. Mansour, G.; Tzetzis, D. Nanomechanical Characterization of Hybrid Multiwall Carbon Nanotube and Fumed Silica Epoxy Nanocomposites. *Polym.-Plast. Technol. Eng.* **2013**, *52*, 1054–1062. [CrossRef]
29. Mansour, G.; Tzetzis, D.; Bouzakis, K.D. A nanomechanical approach on the measurement of the elastic properties of epoxy reinforced carbon nanotube nanocomposites. *Tribol. Ind.* **2013**, *35*, 190–199.
30. Grigora, M.E.; Terzopoulou, Z.; Tsongas, K.; Bikiaris, D.N.; Tzetzis, D. Physicochemical Characterization and Finite Element Analysis-Assisted Mechanical Behavior of Polylactic Acid-Montmorillonite 3D Printed Nanocomposites. *Nanomaterials* **2022**, *12*, 2641. [CrossRef]
31. Mansour, G.; Zoumaki, M.; Tsongas, K.; Tzetzis, D. Microstructural and finite element analysis—Assisted nanomechanical characterization of maize starch nanocomposite films. *Mater. Res.* **2021**, *24*, e20200409. [CrossRef]
32. Peng, S.; Wu, L.; Li, B.G.; Dubois, P. Hydrolytic and compost degradation of biobased PBSF and PBAF copolyesters with 40–60 mol% BF unit. *Polym. Degrad. Stab.* **2017**, *146*, 223–228. [CrossRef]
33. Papageorgiou, G.Z.; Tsanaktsis, V.; Papageorgiou, D.G.; Exarhopoulos, S.; Papageorgiou, M.; Bikiaris, D.N. Evaluation of polyesters from renewable resources as alternatives to the current fossil-based polymers. Phase transitions of poly(butylene 2,5-furan-dicarboxylate). *Polymer* **2014**, *55*, 3846–3858. [CrossRef]
34. Ma, J.; Yu, X.; Xu, J.; Pang, Y. Synthesis and crystallinity of poly(butylene 2,5-furandicarboxylate). *Polymer* **2012**, *53*, 4145–4151. [CrossRef]

35. Papadopoulos, L.; Xanthopoulou, E.; Nikolaidis, G.N.; Zamboulis, A.; Achilias, D.S.; Papageorgiou, G.Z.; Bikiaris, D.N. Towards high molecular weight furan-based polyesters: Solid state polymerization study of bio-based poly(propylene furanoate) and poly(butylene furanoate). *Materials* **2020**, *13*, 4880. [CrossRef]
36. Papadopoulos, L.; Terzopoulou, Z.; Vlachopoulos, A.; Klonos, P.A.; Kyritsis, A.; Tzetzis, D.; Papageorgiou, G.Z.; Bikiaris, D. Synthesis and characterization of novel polymer/clay nanocomposites based on poly (butylene 2,5-furan dicarboxylate). *Appl. Clay Sci.* **2020**, *190*, 105588. [CrossRef]
37. Zhu, J.; Cai, J.; Xie, W.; Chen, P.H.; Gazzano, M.; Scandola, M.; Gross, R.A. Poly(butylene 2,5-furan dicarboxylate), a biobased alternative to PBT: Synthesis, physical properties, and crystal structure. *Macromolecules* **2013**, *46*, 796–804. [CrossRef]
38. Nikolic, M.S.; Djonlagic, J. Synthesis and characterization of biodegradable poly(butylene succinate-co-butylene adipate)s. *Polym. Degrad. Stab.* **2001**, *74*, 263–270. [CrossRef]
39. Gan, Z.; Abe, H.; Doi, Y. Temperature-induced polymorphic crystals of poly(butylene adipate). *Macromol. Chem. Phys.* **2002**, *203*, 2369–2374. [CrossRef]
40. Siafaka, P.I.; Barmbalexis, P.; Bikiaris, D.N. Novel electrospun nanofibrous matrices prepared from poly(lactic acid)/poly(butylene adipate) blends for controlled release formulations of an anti-rheumatoid agent. *Eur. J. Pharm. Sci.* **2016**, *88*, 12–25. [CrossRef]
41. Farias da Silva, J.M.; Soares, B.G. Epoxidized cardanol-based prepolymer as promising biobased compatibilizing agent for PLA/PBAT blends. *Polym. Test.* **2021**, *93*, 106889. [CrossRef]
42. Wang, L.; Feng, L.F.; Gu, X.P.; Zhang, C.L. Influences of the Viscosity Ratio and Processing Conditions on the Formation of Highly Oriented Ribbons in Polymer Blends by Tape Extrusion. *Ind. Eng. Chem. Res.* **2015**, *54*, 11080–11086. [CrossRef]
43. Jiang, M.; Liu, Q.; Zhang, Q.; Ye, C.; Zhou, G. A series of furan-aromatic polyesters synthesized via direct esterification method based on renewable resources. *J. Polym. Sci. Part A Polym. Chem.* **2012**, *50*, 1026–1036. [CrossRef]
44. Poulopoulou, N.; Kasmi, N.; Bikiaris, D.N.; Papageorgiou, D.G.; Floudas, G.; Papageorgiou, G.Z. Sustainable Polymers from Renewable Resources: Polymer Blends of Furan-Based Polyesters. *Macromol. Mater. Eng.* **2018**, *303*, 1800153. [CrossRef]
45. Xu, J.; Reiter, G.; Alamo, R.G. Concepts of nucleation in polymer crystallization. *Crystals* **2021**, *11*, 304. [CrossRef]
46. Mitra, M.K.; Muthukumar, M. Theory of spinodal decomposition assisted crystallization in binary mixtures. *J. Chem. Phys.* **2010**, *132*, 184908. [CrossRef]
47. Zhang, J.; Tashiro, K.; Tsuji, H.; Domb, A.J. Disorder-to-order phase transition and multiple melting behavior of poly(L-lactide) investigated by simultaneous measurements of WAXD and DSC. *Macromolecules* **2008**, *41*, 1352–1357. [CrossRef]
48. Su, S.; Duhme, M.; Kopitzky, R. Uncompatibilized pbat/pla blends: Manufacturability, miscibility and properties. *Materials* **2020**, *13*, 4897. [CrossRef]
49. Liu, J.; Mei, Y.; Xia, R. A new wetting mechanism based upon triple contact line pinning. *Langmuir* **2011**, *27*, 196–200. [CrossRef]
50. Hu, H.; Zhang, R.; Sousa, A.; Long, Y.; Ying, W.B.; Wang, J.; Zhu, J. Bio-based poly(butylene 2,5-furandicarboxylate)-b-poly(ethylene glycol) copolymers with adjustable degradation rate and mechanical properties: Synthesis and characterization. *Eur. Polym. J.* **2018**, *106*, 42–52. [CrossRef]
51. Nova Institute Biodegradable Polymers in Various Environments. Available online: https://renewable-carbon.eu/publications/product/biodegradable-polymers-in-various-environments-$-$-graphic-pdf/ (accessed on 8 June 2022).
52. Dong, Y.; Wang, J.; Yang, Y.; Wang, Q.; Zhang, X.; Hu, H.; Zhu, J. Bio-based poly(butylene diglycolate-co-furandicarboxylate) copolyesters with balanced mechanical, barrier and biodegradable properties: A prospective substitute for PBAT. *Polym. Degrad. Stab.* **2022**, *202*, 110010. [CrossRef]
53. Kasmi, N.; Terzopoulou, Z.; Chebbi, Y.; Dieden, R.; Habibi, Y.; Bikiaris, D.N. Tuning thermal properties and biodegradability of poly(isosorbide azelate) by compositional control through copolymerization with 2,5-furandicarboxylic acid. *Polym. Degrad. Stab.* **2022**, *195*, 109804. [CrossRef]
54. Marten, E.; Müller, R.J.; Deckwer, W.D. Studies on the enzymatic hydrolysis of polyesters. II. Aliphatic-aromatic copolyesters. *Polym. Degrad. Stab.* **2005**, *88*, 371–381. [CrossRef]
55. Qi, R.; Jones, D.L.; Liu, Q.; Liu, Q.; Li, Z.; Yan, C. Field test on the biodegradation of poly(butylene adipate-co-terephthalate) based mulch films in soil. *Polym. Test.* **2021**, *93*, 107009. [CrossRef]
56. Han, Y.; Shi, J.; Mao, L.; Wang, Z.; Zhang, L. Improvement of Compatibility and Mechanical Performances of PLA/PBAT Composites with Epoxidized Soybean Oil as Compatibilizer. *Ind. Eng. Chem. Res.* **2020**, *59*, 21779–21790. [CrossRef]
57. Saeidlou, S.; Huneault, M.A.; Li, H.; Park, C.B. Poly(lactic acid) crystallization. *Prog. Polym. Sci.* **2012**, *37*, 1657–1677. [CrossRef]

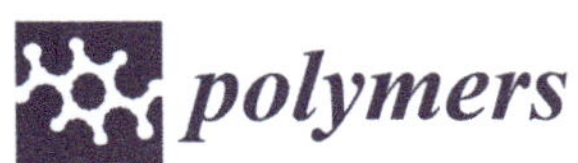

Article

# Nano-Hydroxyapatite from White Seabass Scales as a Bio-Filler in Polylactic Acid Biocomposite: Preparation and Characterization

Preeyaporn Injorhor [1,2], Tatiya Trongsatitkul [1,2], Jatuporn Wittayakun [2,3], Chaiwat Ruksakulpiwat [1,2,*] and Yupaporn Ruksakulpiwat [1,2,*]

1 School of Polymer Engineering, Institute of Engineering, Suranaree University of Technology, Nakhon Ratchasima 30000, Thailand
2 Research Center for Biocomposite Materials for Medical Industry and Agricultural and Food Industry, Nakhon Ratchasima 30000, Thailand
3 School of Chemistry, Institute of Science, Suranaree University of Technology, Nakhon Ratchasima 30000, Thailand
* Correspondence: charuk@sut.ac.th (C.R.); yupa@sut.ac.th (Y.R.); Tel.: +66-44-22-4430 (C.R.); +66-44-22-3033 (Y.R.)

**Abstract:** Nano-hydroxyapatite (nHAp) as a bio-filler used in PLA composites was prepared from fish by acid deproteinization (1DP) and a combination of acid-alkali deproteinization (2DP) followed by alkali heat treatment. Moreover, the PLA/nHAp composite films were developed using solution casting method. The mechanical and thermal properties of the PLA composite films with nHAp from different steps deproteinization and contents were compared. The physical properties analysis confirmed that the nHAp can be prepared from fish scales using both steps deproteinization. 1DP-nHAp showed higher surface area and lower crystallinity than 2DP-nHAp. This gave advantage of 1DP-nHAp for use as filler. PLA composite with 1DP-nHAp gave tensile strength of 66.41 ± 3.63 MPa and Young's modulus of 2.65 ± 0.05 GPa which were higher than 2DP-nHAp at the same content. The addition of 5 phr 1DP-nHAp into PLA significantly improved the tensile strength and Young's modulus. PLA composite solution with 1DP-nHAp at 5 phr showed electrospinnability by giving continuous fibers without beads.

**Keywords:** fish scales; hydroxyapatite; deproteinization; polylactic acid (PLA); physical properties; mechanical properties; thermal properties; electrospinnability

**Citation:** Injorhor, P.; Trongsatitkul, T.; Wittayakun, J.; Ruksakulpiwat, C.; Ruksakulpiwat, Y. Nano-Hydroxyapatite from White Seabass Scales as a Bio-Filler in Polylactic Acid Biocomposite: Preparation and Characterization. *Polymers* **2022**, *14*, 4158. https://doi.org/10.3390/polym14194158

Academic Editors: José Miguel Ferri, Vicent Fombuena Borràs and Miguel Fernando Aldás Carrasco

Received: 30 August 2022
Accepted: 30 September 2022
Published: 4 October 2022



## 1. Introduction

Fish waste of more than 7.2 million tons is annually produced and discarded around the world, leading to environmental problems. Fish scale is one of the wastes from the aquaculture sector and fish markets which has not been used much commercially. Fish scales comprise functional materials such as collagen and hydroxyapatite (HAp) and could be the sources of sustainable biomaterials in various applications, especially biomedical applications [1–6]. HAp from fish scales is an attractive biomaterial with excellent bioactivity, osteointegration, and osteoconductivity [7–9]. It has been used as biosorbent for dyes and metal ions [9]. Therefore, the utilization of fish scales by converting them into high-value materials reduces the waste that causes environmental problems and develops low-cost medical materials. Fish scales comprise HAp and type I collagen [9]. Notably, teleost fish such as sea bass have elasmoid scale, which is similar to the bone composed of extracellular matrix, mainly type I collagen fibers and needle-like hydroxyapatite. Moreover, important anions such as $Cl^-$ and $F^-$ and cations such as $Mg^{2+}$, $Al^{3+}$, $Sr^{2+}$, $Zn^{2+}$, $K^+$, and $Na^+$ are presence as trace elements [10–14]. HAp is an interesting type of calcium phosphate with the theoretical chemical formula $Ca_{10}(PO_4)_6(OH)_2$. It is used in biomedical applications such as bone replacement material, dental implants, and bone tissue engineering [15,16].

HAp could be prepared or extracted from fish scales by various methods. There are two main methods to eliminate collagen constituents: calcination and alkali heat treatment [3,9,17–20]. Kongsri et al. [19] prepared nanocrystalline HAp from fish scale waste using the alkali heat treatment method. The method consists of deproteinization by acid and alkali, followed by alkali heat treatment. The products are nanocrystals HAp with high porosity. Although nano-sized HAp (nHAp) can be prepared using alkali heat treatment it requires two steps deproteinization or acid-alkali deproteinization (2DP). Some reports say HAp could be prepared from fish scales by chemical deproteinization [3,8,9] but it takes two times to remove residue protein composition. Hence, this study aims to reduce the steps of deproteinization to obtain nHAp. One-step of deproteination or acid deproteinization (1DP) is used before alkali heat treatment. White seabass is an economically significant species. They are widely cultivated in Thailand. In 2021, white seabass production presented 97.43% of all marine fish farming production. The total commodity value was 5.05 billion baht [21]. Its huge production makes the scales abundant and could be considered a sustainable source of HAp.

Polylactic acid or PLA is a thermoplastic polyester that has advantageous characteristics such as renewability, biocompatibility and inherent biodegradability, ease of preparation, non-toxic nature, and the ability to form fibers. It has been utilized in medical applications extensively [22,23]. However, the disadvantages of PLA such as low thermal stability, hydrophobic nature, and high brittleness, still limit some applications [22,24]. The incorporation of HAp could overcome the hydrophobicity of PLA, improve its poor properties, and stimulate the properties of osteoconduction and osseointegration of the implanted scaffold [7].

This study aims to reduce the step of nHAp preparation and to compare the physical properties of nHAp from 1DP and 2DP. The nHAp samples from 1DP and 2DP with 5 phr are used to form a composite with PLA. The mechanical and thermal properties of both composites are compared. The sample with better performance is further investigated at 2.5, 5, and 10 phr to determine the optimum amount for PLA composite. The PLA/nHAp composite films are fabricated by solvent casting. Effects of deproteinization steps and 1DP-nHAp content on mechanical and thermal properties of the PLA/nHAp composite films are investigated. Moreover, the electrospinnability of PLA/1DP-$nHAp_5$ is studied. The obtained nHAp is expected to be useful as a bio-filler in PLA for medical applications.

## 2. Materials and Methods

### 2.1. Preparation of nHAp Powder from White Sea Bass Scales

White seabass scales of approximately 10 kg were collected from the local market in Rayong, Thailand, washed with deionized water, and oven-dried at 80 °C. Polylactic acid (PLA, Ingeo™ Biopolymer 4043D-General Purpose Grade) was used as the matrix of composites supplied by NatureWorks Llc. (Minnetonka, MN, USA). Hydrochloric acid (HCl) 37% RPE, sodium hydroxide (NaOH) 99% RPE-ACS, and dichloromethane (DCM) RPE were purchased from Carlo Erba (Milano, Italy).

The preparation method of nHAp was adapted from Kongsri et al. [19]. The steps of preparation are shown in Figure 1. The dried fish scales were treated with 0.1 M HCl for 1 h at room temperature to eliminate collagen, non-collagen proteins, and limiting layers of fish scales. Then, they were washed with deionized water several times before oven-drying at 80 °C. The remaining was treated with 5% ($w/v$) NaOH under stirring at 250 rpm for 3 h at 60 °C. The obtained fish scale powder was washed until the pH = 6.5–7 and oven-dried at 80 °C. The alkali heat treatment method was performed on the fish scale powder from the previous step by treating with 50% ($w/v$) NaOH under stirring at 250 rpm for 3 h at 80 °C to produce a slurry. The nHAp slurry was washed with deionized water until the pH = 6.5–7 and oven-dried at 80 °C. The final product was nHAp powder. This process is two-steps deproteinization (2DP-nHAp). For one-step of deproteinization (1DP-nHAp), the process was carried out without 5% (*w*/*v*) NaOH treatment.

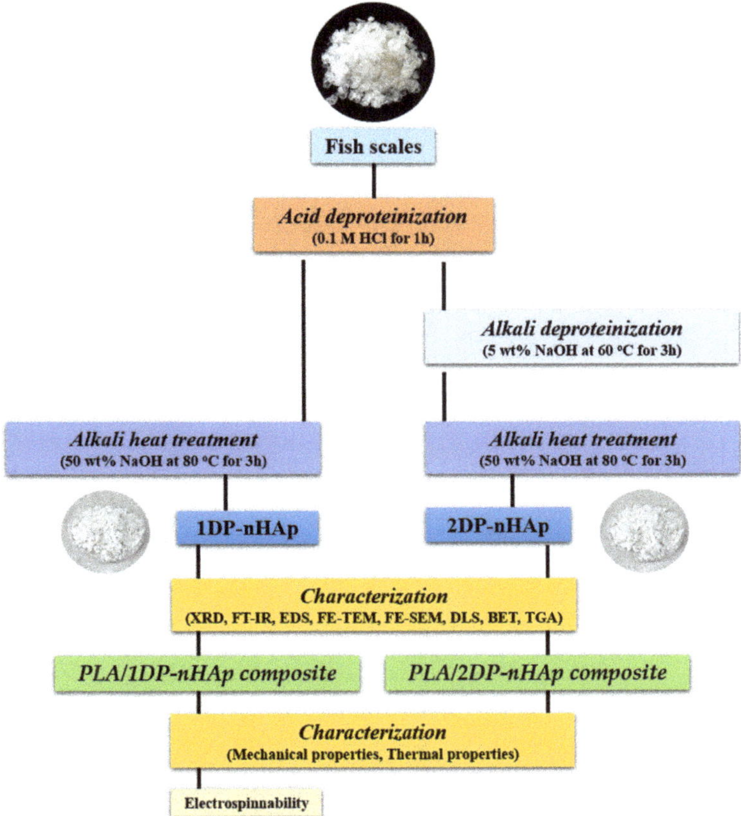


**Figure 1.** Schematic diagram showing the preparation of nHAp from white seabass scales by alkali heat treatment method.

### 2.2. Preparation of PLA/nHAp Composite Films

To compare nHAp from one-step and two-steps preparation, the powder of 1DP-nHAp and 2DP-nHAp at the same content (5 phr) was dispersed using a magnetic stirrer in DCM for 24 h. To study the effect of nHAp contents, 1DP-nHAp at 2.5 and 10 phr was dispersed with the same procedure. Then, the dispersed nHAp was poured into PLA solution which was dissolved in DCM at a concentration of 10 wt%. The mixture was mixed for 72 h using a magnetic stirrer until a homogenous solution was obtained. The composite films were prepared by casting PLA/nHAp solution into a Petri glass and drying at room temperature for 24 h before evaporating the residue solvent at 40 °C for 72 h in an oven. After that, the PLA/nHAp composite films were stored in a desiccator for further characterization. The effect of nHAp contents on mechanical properties and thermal properties was studied by adding 1DP-nHAp at 2.5, 5, and 10 phr to fabricate PLA/nHAp composite films.

### 2.3. Characterization of nHAp Powder

The crystalline phase and degree of crystallinity ($\chi_c$) of nHAp powder were determined by X-ray diffraction (XRD, D2 PHASER, Bruker, Billerica, MA, USA) with Cu K$\alpha$ radiation source operated at 30 kV and current 10 mA. Bragg's angle of diffraction (2$\theta$)

was measured from 10° to 60° at a scan rate of 2°/min and a step size of 0.02°. The $\chi_c$ was calculated by the Equation (1) [25]:

$$\chi_c = (A_C \ / \ A_T) \times 100\% \tag{1}$$

where $A_C$ is the area of crystalline peaks and $A_T$ is the total area of amorphous and crystalline peaks, respectively.

The crystallite size ($D_{hkl}$) of the samples was calculated from the Scherrer equation (Equation (2)) [26]:

$$D_{hkl} = \frac{k\lambda}{\beta \cos \theta} \tag{2}$$

where λ is the wavelength of the X-ray radiation, k is the Scherrer constant generally taken to be 0.9, θ is the diffraction angle, and β refers to the full width at half maximum (FWHM).

The functional groups of nHAp powder were analyzed by Fourier transform infrared (FT-IR) spectroscopy on Tensor 27 (Bruker, Billerica, MA, USA) with 64 scans and a resolution of 4 $cm^{-1}$. Each sample was oven-dried at 110 °C for 24 h, mixed with dried potassium bromide (KBr), mashed in an agate mortar and pressed into a disk. Percentage by weight of carbonate content (wt% $CO_3$) of HAp structure was calculated by Equation (3) [27]:

$$\text{wt\% } CO_3 = 28.62 \times r_{c/p} + 0.0843 \tag{3}$$

where $r_{c/p}$ is the ratio of the integrated the area between area of $v_3(CO_3)$ and area of $v_1v_3(PO_4)$ from the absorbance spectra calculated from OriginPro software (OriginPro 2022, OriginLab, Northampton, MA, USA).

The elemental composition of nHAp was analyzed by energy dispersive X-ray spectroscopy (EDS) in a scanning transmission electron microscope (STEM, Talos F200X, Thermo Fisher Scientific, Waltham, MA, USA).

Micrographs of nano-sized HAp from fish scales were acquired using field emission transmission electron microscopy (FE-TEM, Talos F200X, Thermo Fisher Scientific, Waltham, MA, USA). The microstructure of nHAp powder was observed using field emission scanning electron microscopy (FE-SEM, Carl Zeiss Auriga, Oberkochen, Germany) at 3 kV. The samples were sputter-coated with gold for 3 min at 10 mA.

Particle size distribution was analyzed with dynamic light scattering (DLS) using Zetasizer-ZS (Malvern Panalytical, Malvern, UK). The samples were dispersed in ethanol and analyzed at 25 °C. The average particle size in this study was obtained from Z-average.

Nitrogen adsorption–desorption analysis was performed on BelSorp-Mini II (Micro-tracBEL, Osaka, Japan). The nHAp was degassed at 150 °C for 24 h before the analysis. The specific surface area was calculated by Brunauer–Emmett–Teller (BET) method.

Thermal properties, including decomposition temperatures, weight loss, and remaining residue were investigated by thermogravimetric analysis (TGA, TGA/DSC1, Mettler Toledo, Schwerzenbach, Switzerland). Approximately 10 mg of sample was placed on the alumina crucible and heated at a rate of 10 °C/min from 30 to 1000 °C for the analysis under an air atmosphere.

### 2.4. Characterization of PLA/nHAp Composite Films

Tensile properties of PLA/nHAp composite films were measured according to ASTM D882-10 using a universal testing machine (INSTRON/5565, Norwood, MA, USA) with a load cell of 5 kN and a crosshead speed of 250 mm/min at room temperature. The specimens with 1 cm width and 10 cm length were analyzed. Tensile strength, elongation at break, and Young's modulus of PLA/nHAp composite films were obtained from the average results of five test specimens.

Thermal properties of PLA/nHAp composite films such as enthalpy of melting ($\Delta H_m$), enthalpy of cold crystallization ($\Delta H_{cc}$), glass transition temperature ($T_g$), cold crystallization temperature ($T_{cc}$), melting temperature ($T_m$) was evaluated by using differential scanning calorimeter (DSC, Pyris Diamond DSC, Perkin Elmer, Waltham, MA, USA). The samples were characterized under nitrogen flow rate at 20 mL/min from 25 to 200 °C and a heating rate of 10 °C/min. The degree of crystallinity was calculated according to Equation (4) [28]:

$$\chi_c = \left[ \frac{(\Delta H_m - \Delta H_{cc})}{\left(\Delta H_m^0 \times w\right)} \right] \times 100\% \quad (4)$$

where $\Delta H_m^0$ is the heat of melting of purely crystalline PLA (93 $J \cdot g^{-1}$), w is the weight fraction of PLA in the sample.

*2.5. Preparation of PLA/nHAp Fibers by Electrospinning Technique and Their Electrospinnability*

The aggregation of nHAp may affect the viscosity of PLA/nHAp composite solution which is one of the factors that affect the electrospinning process. To determine electrospinnability, PLA/1DP-nHAp$_5$ composite solution was fabricated by electrospinning. Nanofibers were spun at 150 mm distance to a drum collector, which was covered with aluminum foil. The collector rotation speed was set at 300 rpm. The high voltage between the needle tip and the drum collector was set to 25 and 30 kV. The PLA/1DP-nHAp$_5$ solution was fed at a constant flow rate of 1.0 mL/h. Electrospun fibers were performed by SEM (JSM-6010LV, JOEL, Akishima, Tokyo, Japan) with EDS (EDAX Genesis 2000, AMETEX, Berwyn, PA, USA) to observe their morphology and check the distribution nHAp particles in fibers. The fiber diameter was measured from SEM images using image analysis software (Image J 1.53k, Wayne Rasband and contributors, National Institutes of Health, Bethesda, MD, USA).

## 3. Results and Discussion

*3.1. Characterization of nHAp Powder*

The XRD patterns of 1DP-nHAp and 2DP-nHAp are shown in Figure 2. The characteristic peaks of nHAp powder are compared with Crystallography Open Database (COD 9003552) for hexagonal HAp structure. The results show that diffraction peaks of both samples correspond to the HAp planes [19,29–31]. They only show the crystalline phase of HAp without other phases. All diffraction peaks agree with the standard XRD pattern of hydroxyapatite in Crystallography Open Database (COD 9003552) and the standard of JCPDS card no. 09-0432. Both samples are hexagonal structures with $\alpha = \beta = 90°$, and $\gamma = 120°$. Corresponding to Sathiskumar et al. [6] nHAp from *Cirrhinus mrigala* fish scales using the same method (2DP) have similar characteristic peaks; this corroborates the method of purity nano-sized HAp preparation. However, their nHAp crystallinity is close to the crystallinity of 1DP-nHAp. It indicated that the deproteinization reduction method could provide nHAp. The crystallinity of 2DP-nHAp is slightly higher than 1DP-nHAp, indicating that 2DP increased the crystallinity of nHAp powder. As a result, 1DP-nHAp with lower crystallinity may be more suitable for enhancing biodegradation behavior and higher metabolic [29]. The degree of crystallinity and crystallite size of nHAp powder are included in Table 1.

FT-IR spectra of 1DP-nHAp and 2DP-nHAp in transmission mode are shown in Figure 3. Both samples show a broad peak of associating hydroxyl stretching of adsorbed water around 3500 $cm^{-1}$. The bending mode of the water molecule appears at 1639 $cm^{-1}$. The bands at 1458, 1417, and 874 $cm^{-1}$ are assigned to carbonate groups in nHAp powder. It indicated the nHAp powder is B-type carbonated hydroxyapatite with carbonate ions substituting phosphate ions in the hydroxyapatite structure. The strong broadband between 1083–1042 $cm^{-1}$ and the bands at 962, 603, and 565 $cm^{-1}$ correspond to phosphate groups [8,32,33]. The percent weight of carbonate in 1DP-nHAp and 2DP-nHAp, calculated from the band in absorption mode as shown in Figure 4, are 6.57 and 15.64 wt%, respectively. It suggested the presence of carbonate ions substituting phosphate ions more than 1DP-nHAp. In addition, the spectrum of 2DP-nHAp showed the stretching of free hydroxyl groups at 3642 $cm^{-1}$. It is in agreement with the report by Gergely et al. [34] that the free hydroxyl groups may connect to calcium oxide on the surface. In addition, The FT-IR results of 1DP-nHAp and 2DP-nHAp are similar to Gopalu et al. [35]. They presented FT-IR and Raman results of pure HAp. It can be assumed that 1DP-nHAp and 2DP-nHAp have the same characteristics as pure HAp.

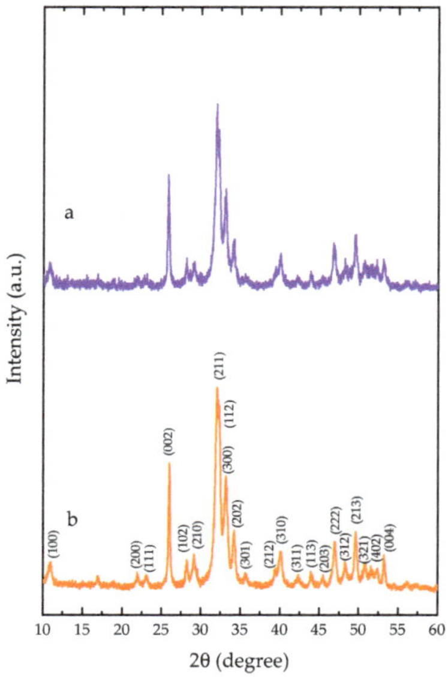


**Figure 2.** XRD patterns of nHAp powder (a) 1DP-nHAp and (b) 2DP-nHAp.

**Table 1.** The analysis information of n-HAp powder.

| Properties | Materials | |
|---|---|---|
| | **1DP-nHAp** | **2DP-nHAp** |
| Crystallinity (%$\chi_c$) | 71.41 | 80.99 |
| Crystallite size, $D_{hkl}$(nm) | 19.41 | 13.87 |
| BET surface area (cm$^3$/g) | 50 | 41 |
| Total pore volume (cm$^3$/g) | 0.26 | 0.17 |
| Mean pore diameter (nm) | 21.25 | 16.32 |

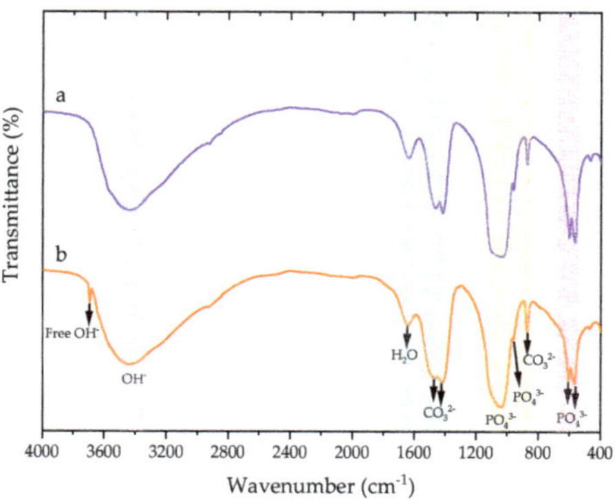


**Figure 3.** FT-IR transmittance spectra of nHAp powder (a) 1DP-nHAp and (b) 2DP-nHAp.

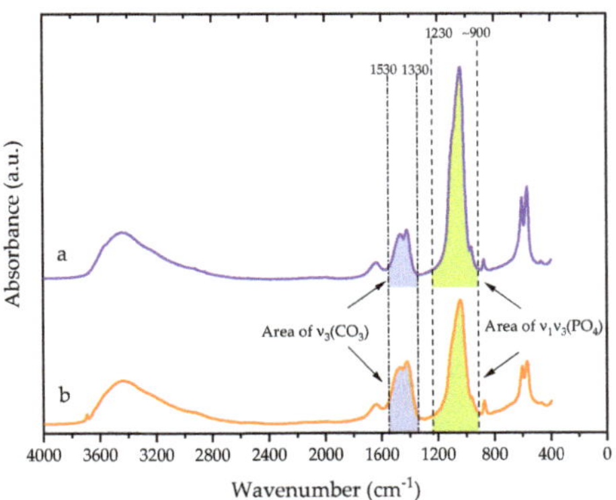


**Figure 4.** FT-IR absorbance spectra of nHAp powder (a) 1DP-nHAp and (b) 2DP-nHAp.

EDS spectra of the nHAp with element composition are shown in Figure 5. Both 1DP-nHAp and 2DP-nHAp have constituents carbon (C), oxygen (O), magnesium (Mg), calcium (Ca), and phosphorous (P). Typically, the presence of Mg constituent is a significant factor in bone and teeth growth [8]. Hydroxyapatite is a type of calcium phosphate ceramic which is classified by calcium/phosphorus atomic ratio (Ca/P); for example, hydroxyapatite $Ca_{10}(PO_4)_6(OH)_2$, HAp, Ca/P = 1.667) and β-tricalcium phosphate ($\beta - Ca_3(PO_4)_2$, β-TCP, Ca/P = 1.5) [36]. The Ca/P has illustrated the molar ratio from nHAp. The Ca/P of 1DP-nHAp and 2DP-nHAp are 1.63 and 2.01, respectively. The Ca/P of 1DP-nHAp is close to the theoretical value (1.67) [8]. The Ca/P ratio of 2DP-nHAp is higher than the theoretical value due to the substitution of phosphate ions with carbonate ions. These results are the same as the report by Deb and Deoghare [29] and consistent with the FT-IR results.

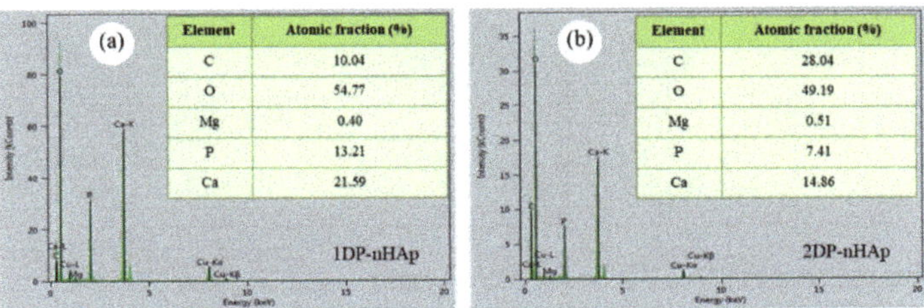


| Element | Atomic fraction (%) |
|---|---|
| C | 10.04 |
| O | 54.77 |
| Mg | 0.40 |
| P | 13.21 |
| Ca | 21.59 |

| Element | Atomic fraction (%) |
|---|---|
| C | 28.04 |
| O | 49.19 |
| Mg | 0.51 |
| P | 7.41 |
| Ca | 14.86 |

**Figure 5.** Results from elemental analysis by EDS of nHAp powder (**a**) 1DP-nHAp and (**b**) 2DP-nHAp.

The nano-sized particles of nHAp from 1DP-nHAp and 2DP-nHAp were confirmed using FE-TEM, as shown in Figure 6a,b. The particle sizes of both samples in the range of nano-scale. The microstructure is observable in the FE-SEM images. FE-SEM in Figure 6. 1DP-nHAp has rod-like shapes with different widths and lengths (Figure 6c), while 2DP-nHAp has an irregular shape (Figure 6d). Typically, the external elasmodine layer of fish scales is composed of needle-like HAp crystals hybridized with randomly arranged collagen fibers. However, the HAp products from fish scales have various shapes, such as irregular, rod-like, spherical, and needle-like. According to Qin et al. [9], the rod-like shape or needle-like shape of HAp is frequently seen from HAp extracted from natural sources. However, the extraction method or the source has no effect on the shape of HAp particles. The same extraction method can provide a different HAp shape. It can be concluded that 1DP-HAp has the natural shape that is frequently found. Moreover, 2DP-nHAp shows an aggregation of particles that leads to non-homogenous distribution in the matrix and deteriorates mechanical properties [37].

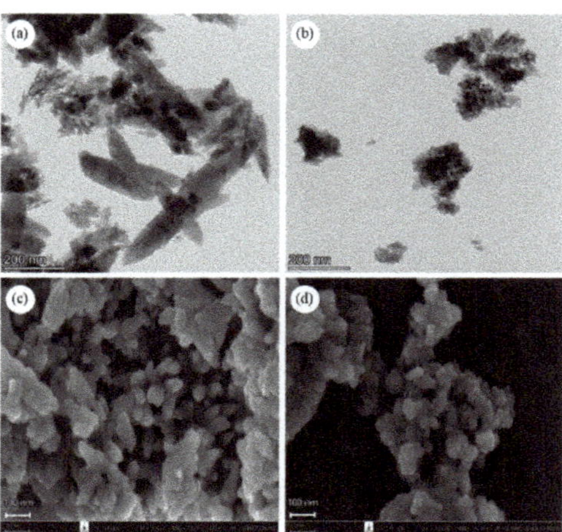

**Figure 6.** FE-TEM images of (**a**) 1DP-nHAp and (**b**) 2DP-nHAp and FE-SEM images of (**c**) 1DP-nHAp and (**d**) 2DP-nHAp.

Particle size distribution of 1DP-nHAp and 2DP-nHAp measured by DLS technique is shown in Figure 7. The particle size was reported by Z-average particle size, which is the intensity weighted harmonic mean size. The particle size distribution of both samples is a mono-modal distribution with the Z-average particle size of 223.6 and 172.9 nm, respectively. 1DP-nHAp shows a narrower distribution than 2DP-nHAp. According to Raita et al. [38], the size from the DLS technique is larger than that observed from the electron microscopy because DLS measures a hydrodynamic size, rather than a physical one.

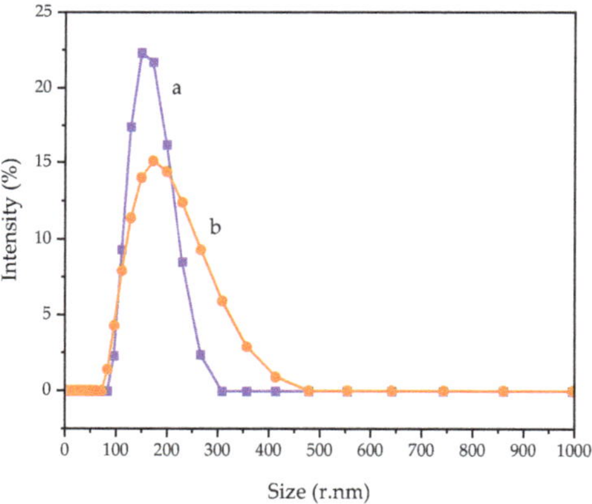


**Figure 7.** DLS measurements of particle size distribution of nHAp powder (a) 1DP-nHAp and (b) 2DP-nHAp.

The adsorption–desorption isotherms of 1DP-nHAp and 2DP-nHAp are shown in Figure 8. Both samples demonstrate reversible Type II isotherms which are the physisorption on nonporous adsorbents according to the IUPAC classification. Moreover, their adsorption and desorption lines do not overlap, forming a type H3 hysteresis loop which is found on materials with non-rigid aggregates of plate-like particles [39,40]. Their BET surface area, total pore volume, and mean pore diameter are included in Table 1. 1DP-nHAp has a larger surface area than 2DP-nHAp. For tissue engineering, 1DP-nHAp, the sample with a larger surface area, could interact better with osteoblast cells to promote cell growth and proliferation [1,9].

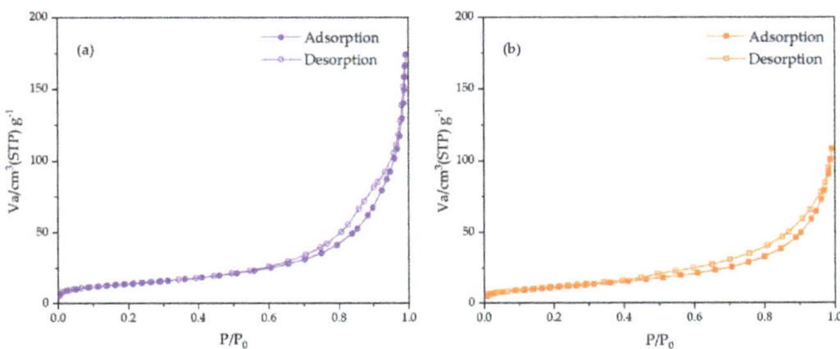


**Figure 8.** Nitrogen adsorption–desorption isotherms of nHAp powder (**a**) 1DP-nHAp and (**b**) 2DP-nHAp.

TGA analysis was used to confirm the composition of HAp in 1DP-nHAp and 2DP-nHAp. Figure 9a,b exhibits the TGA results showing the weight loss of 1DP-nHAp and 2DP-nHAp in the temperature range from 30–1000 °C, respectively. The figure also includes their DTG curves, the derivative of the weight loss, which indicate three stages of weight loss from both samples. The first stage at 30–200 °C is the weight loss from water evaporation. The second stage around 375–500 °C corresponds to the weight loss from the combustion of hydrocarbons which are the organic residue. The final stage, at around 600–800 °C corresponds to the loss of carbonate groups in the nHAp structure [19,41]. According to the FT-IR results, 2DP-nHAp has carbonate groups more than 1DP-nHAp. So, the carbonate weight loss of 2DP-nHAp is greater. The HAp residue of 1DP-nHAp and 2DP-nHAp were 86.10 and 76.12 wt%, respectively. In addition, maximum degradation temperature ($Td_{max}$) of 1DP-nHAp and 2DP-nHAp were 716.67 °C and 704.50 °C, respectively. It indicated that 1DP-nHAp has better thermal stability than 2DP-nHAp.

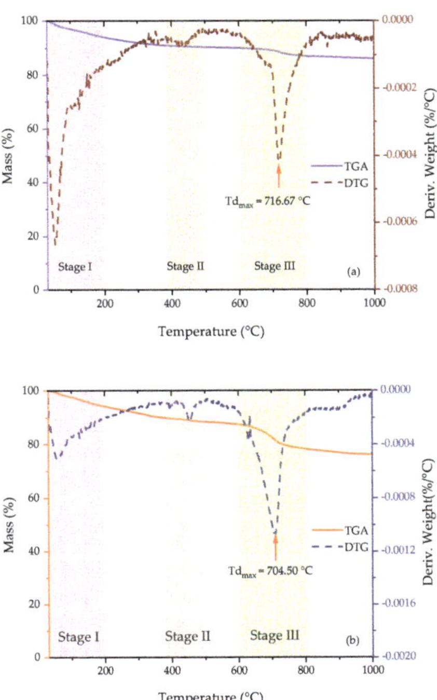


**Figure 9.** TGA and DTG curves of thermal decomposition of (**a**) 1DP-nHAp and (**b**) 2DP-nHAp.

### 3.2. Characterization of PLA/nHAp Composite Films

The PLA/nHAp composite films were fabricated by the solution casting method. 5phr of 1DP-nHAp and 2DP-nHAp was added for comparison. Figure 10 shows the tensile stress–strain curve of the PLA/nHAp composite films with nHAp from different preparation steps and with different 1DP-nHAp contents. The results of the tensile properties are shown in Table 2. It shows Young's modulus, tensile strength, and elongation at break of PLA and PLA/nHAp composite films. The Young's modulus of PLA was 1.73 ± 0.18 GPa and the value for the PLA/1DP-nHAp$_5$ and PLA/2DP-nHAp$_5$ increased up to 2.65 ± 0.05 MPa and 2.38 ± 0.11 MPa, respectively. However, elongation at break of PLA/2DP-nHAp$_5$ is lower than PLA/1DP-nHAp$_5$. According to Kamarudin et al. [42] and Boey et al. [43], the modulus and strength of the composite depend on mechanical interlocking or chemical interaction between the filler and the matrix. Thus, adhesion or bonding between the filler and the matrix is an important factor. Mechanical interlocking is a form of physical force that holds filler and matrix together, whereas chemical interaction is the formation of chemical bonding via functional groups between filler and matrix. In this work, the surface area of nHAp that is available for the mechanical interlocking and the chemical bonding between carbonyl (−COO) of PLA and $Ca^{2+}$ ions on the surface of nHAp were considered to affect the mechanical properties of the PLA composite [44,45]. The composite with 1DP-nHAp$_5$ addition showed the highest tensile strength (66.41 ± 3.63 MPa). This result corresponds to the surface properties of 1DP-nHAp which has larger surface area and could cause more mechanical interlocking with matrix than 2DP-nHAp. So, the interlocking between filler and matrix was expected that greater. According to EDS results, 1DP-nHAp has atomic fraction of $Ca^{2+}$ more than 2DP-nHAp, indicating that 1DP-nHAp has more interaction sites. This assumption supplemented that the strength of PLA/1DP-nHAp$_5$ could occur from the chemical bonding of −COO and $Ca^{2+}$. However, both samples have mechanical properties which correspond to the tensile strength of human skeletal bones (ranging from 40–200 MPa) and the critical mechanical modulus of bone replacement material in non-load bearing sites (ranging from 10–1500 MPa) [46].

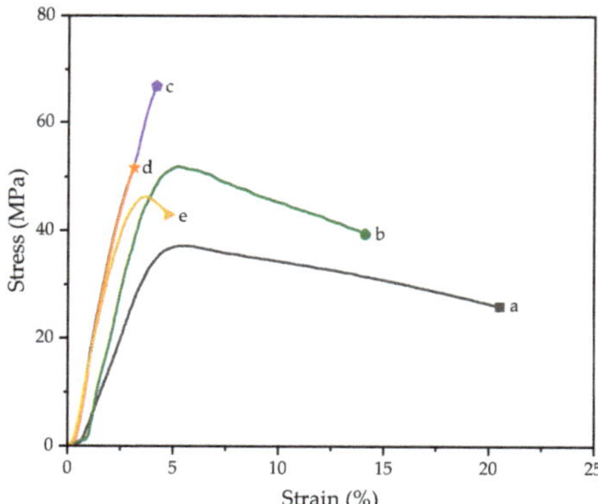


**Figure 10.** Tensile stress–strain curve of PLA/nHAp composite films with nHAp from different preparation steps and 1DP-nHAp at various contents (a) PLA, (b) PLA/1DP-nHAp$_{2.5}$, (c) PLA/1DP-nHAp$_5$, (d) PLA/2DP-nHAp$_5$, and (e) PLA/1DP-nHAp$_{10}$.

The mechanical properties of the PLA/nHAp composite films with various 1DP-nHAp content are also given in Table 2. The results indicated that PLA/nHAp composite films show more tensile strength than neat PLA films. The tensile strength of PLA/nHAp composite films increases with increasing 1DP-nHAp content up to 5 phr as well as the result of Young's modulus. It indicated good dispersion of nHAp into PLA matrix and strong interfacial actions between the PLA and nHAp [46]. Therefore, the optimum content of nHAp is 5phr. However, elongation at break is the lowest. Tensile strength decreased with adding nHAp content at 10 phr. The high content of nHAp would agglomerate, causing poor dispersion of nHAp in the PLA matrix leading to the deterioration of tensile strength as previously reported by Li et al. [47]. According to Boey et al. [43], the mechanical properties of composites are found to improve linearly with increasing filler content up to a certain optimum value. Moreover, the addition of filler above that limit adversely affects the mechanical strength due to the formation of agglomerates. Elongation at break of PLA/nHAp composite films decreased by the addition of nHAp up to 5 phr. The results are suggested that the optimum addition of nHAp into the PLA matrix can be improved the rigidity of the composite film. Nevertheless, all composite samples have enough strength for developing medical materials that can be degraded. This strength is an outgrowth from inorganic filler that the materials should maintain sufficient strength while providing specific cell-surface receptors during the tissue remodeling process [48].

**Table 2.** Mechanical properties of PLA/nHAp composite films with nHAp from different steps of preparation and 1DP-nHAp at various contents.

| Designation | nHAp Content (phr) | Young's Modulus (GPa) | Tensile Strength (MPa) | Elongation at Break (%) |
|---|---|---|---|---|
| PLA | – | 1.73 ± 0.18 | 38.21 ± 0.95 | 23.39 ± 1.97 |
| $PLA/1DP-nHAp_{2.5}$ | 2.5 | 1.94 ± 0.27 | 54.45 ± 1.42 | 14.74 ± 2.92 |
| $PLA/1DP-nHAp_{5}$ | 5 | 2.65 ± 0.05 | 66.41 ± 3.63 | 4.32 ± 0.34 |
| $PLA/2DP-nHAp_{5}$ | 5 | 2.38 ± 0.11 | 52.21 ± 4.67 | 3.44 ± 0.66 |
| $PLA/1DP-nHAp_{10}$ | 10 | 2.02 ± 0.18 | 45.80 ± 1.78 | 4.72 ± 0.59 |

The thermal properties of PLA and PLA/nHAp composite films was studied by differential scanning calorimetry (DSC). Their DSC thermograms are shown in Figure 11. The effect of the nHAp from different preparation steps on the thermal properties of PLA/nHAp composite films is shown in Table 3. The glass transition temperature ($T_g$) of all samples was slightly different, all in the range of 60–61 °C. It indicates that the interaction between matrix and filler is low, and it can slightly change the mobility of polymer chains related to the glass transition. An exothermic peak corresponds to the crystallinity of the PLA. The addition of nHAp particles in the PLA matrix affects the temperature of cold crystallization ($T_{cc}$) values tend to decrease due to the nHAp particles acting as nucleation centers for PLA crystals [49]. So, the PLA crystallinity was enhanced by loading nHAp due to the exothermic peak, as observed in PLA/nHAp composites being sharper than neat PLA and the inorganic filler could promote the polymer crystallization on their surface [50]. As compared between 1DP-nHAp and 2DP-nHAp at the same filler contents, the $T_g$, $T_{cc}$, and $T_m$ of both composites seem slightly different. Still, their degree of crystallinity is dramatically different because the 2DP-nHAp has smaller particles. The smaller size of 2DP-HAp can generate more cross-linking points in the PLA matrix than 1DP-nHAp, which restricts the movement of the PLA chain [51].

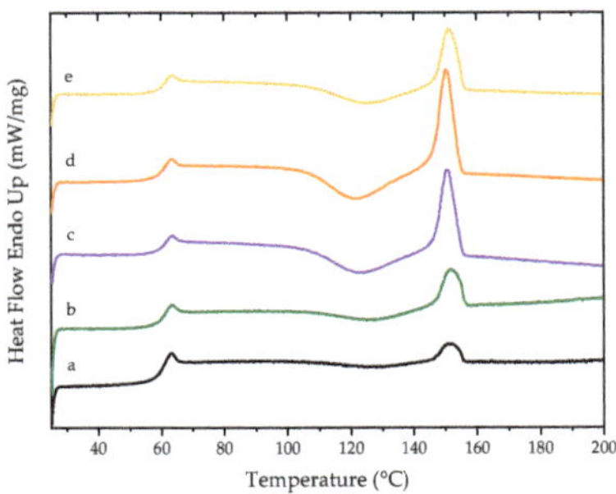


**Figure 11.** The second heating DSC thermograms of PLA and PLA/nHAp composite films with nHAp from different preparation steps and 1DP-nHAp at various contents (a) PLA, (b) PLA/1DP-nHAp$_{2.5}$, (c) PLA/1DP-nHAp$_{5}$, (d) PLA/2DP-nHAp$_{5}$, and (e) PLA/1DP-nHAp$_{10}$.

**Table 3.** Thermal characteristics of PLA and PLA/nHAp composite films with nHAp from different preparation steps and 1DP-nHAp at various contents.

| Designation | $T_g$ (°C) | $T_{cc}$ (°C) | $\Delta H_{cc}$ ($J \cdot g^{-1}$) | $T_m$ (°C) | $\Delta H_m$ ($J \cdot g^{-1}$) | $X_c$ (%) |
|---|---|---|---|---|---|---|
| PLA | 60.10 | 127.89 | 3.57 | 151.60 | 3.16 | 7.23 |
| $PLA/1DP - nHAp_{2.5}$ | 60.81 | 128.39 | 6.71 | 151.59 | 6.51 | 14.50 |
| $PLA/1DP - nHAp_5$ | 61.00 | 122.40 | 12.98 | 150.61 | 15.69 | 32.38 |
| $PLA/2DP - nHAp_5$ | 60.77 | 121.90 | 19.72 | 150.44 | 16.74 | 41.17 |
| $PLA/1DP - nHAp_{10}$ | 61.14 | 126.07 | 11.22 | 151.28 | 10.21 | 25.35 |

The melting temperature peak of PLA is 151.60 °C while the peak of PLA/nHAp composite films is decreased to a lower temperature. However, it is the same as the neatPLA. Meanwhile, the effect of the 1DP-nHAp content on the thermal properties of PLA/nHAp composite films is shown in Table 3. The $T_g$ of PLA was 60.10 °C, while $T_g$ of the PLA/nHAp composites slightly increased with increasing 1DP-nHAp content. This change indicated that the interactions of the 1DP-nHAp slightly interfere with the mobility of polymer chains related to the glass transition [52]. The $T_{cc}$ value of the PLA and PLA/nHAp composites are 127.89, 128.39, 122.40, and 126.07 °C when 0, 2.5, 5, and 10 phr of 1DP-nHAp were added, respectively. It indicated that the 1DP-nHAp at 5 and 10 phr accelerated the cold crystallization of PLA due to the ability of 1DP-nHAp to induce heterogeneous nucleation into the PLA matrix. However, adding 1DP-nHAp enhanced the PLA's crystallinity compared with neat PLA. The crystallinity corresponds to $T_m$; typically, the polymers melt at a higher temperature when they form fewer perfect crystals [53,54].

SEM images of PLA/1DP-nHAp$_5$ composite fibers with applied high voltage at 25 and 30 kV are shown in Figure 12. The continuous PLA/1DP-nHAp$_5$ composite fibers with rough surfaces without the formation of beads were successfully fabricated. The average diameter of the fibers obtained from the high voltage of 25 and 30 kV is 3.83 ± 1.09 and 2.62 ± 0.35 μm, respectively. The fiber diameter from 30 kV high voltage shows a slightly smaller fiber diameter than that from 25 kV. The increase in the applied voltage leads to the stretching of the polymer chains in correlation with the charge repulsion within the polymer jet [55].

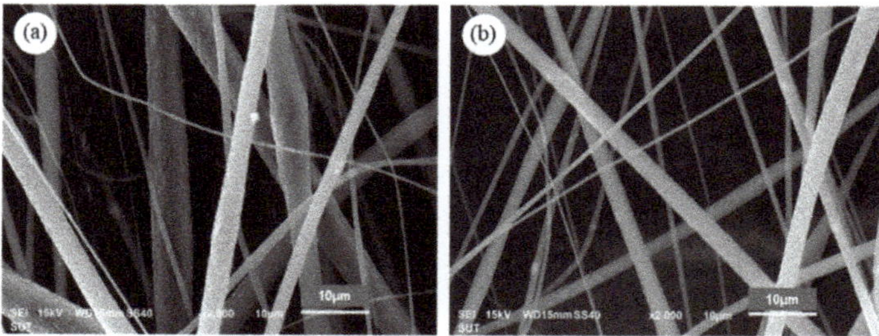


**Figure 12.** SEM images of PLA/1DP-nHAp$_5$ composite fibers operated different applied high voltage (**a**) 25 kV and (**b**) 30 kV.

EDS analyzed area and mappings of PLA/1DP-nHAp$_5$ at 25 and 30 kV high voltage are shown in Figure 13. The EDS analyzed area of each sample was shown by pink frame. The existence of calcium was indicated to nHAp that disperse in fibers. The EDS mappings show a good dispersion of nHAp in both samples. This indicated that nHAp could be incorporated into PLA solution without phase separation and aggregation.

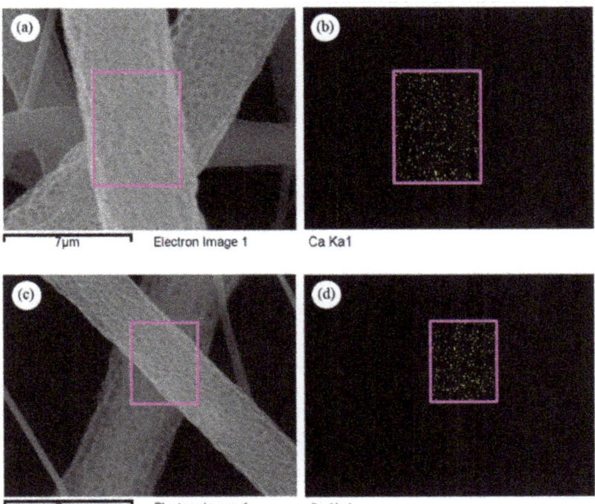


**Figure 13.** EDS analyzed area and EDS mappings of PLA/1DP-nHAp$_5$ composite fibers operated different applied high voltage (**a**,**b**) 25 kV and (**c**,**d**) 30 kV.

## 4. Conclusions

Nano-hydroxyapatite (nHAp) was prepared from white seabass scales by two methods: (1) acid deprotonization (1DP) and (2) combination of acid-alkali deproteinization (2DP) followed by alkali heat treatment. Physicochemical properties of 1DP-nHAp and 2DP-nHAp are compared by several techniques. Both samples are B-type carbonated hydroxyapatite. Their particles are non-porous with irregular shapes, and nano-sized diameters. 1DP-nHAp has lower crystallinity, narrower particle size distribution, lower Ca/P ratio and larger surface area. 1DP-nHAp and 2DP-nHAp are used as bio-fillers for polylactic acid (PLA). The composite films PLA/1DP-nHAp and PLA/2DP-nHAp are compared in terms of mechanical strength and thermal behavior. PLA/1DP-nHAp$_5$ is the better composite. Both composites have higher Young's modulus and tensile strength than the neat PLA but shorter elongation at break. At the same nHAp content (5 phr), the composite film PLA/1DP-nHAp$_5$ shows higher Young's modulus and tensile strength with higher elongation at break than the composite film PLA/2DP-nHAp$_5$. The strength of PLA/1DP-nHAp decreases by increasing or decreasing the nHAp content. The thermal behaviors of all PLA/nHAp composite films are slightly different from the neat PLA. The interaction between matrix and filler is low, and it slightly changes the mobility of polymer chains. The nHAp can induce heterogeneous nucleation into the PLA matrix via accelerated cold crystallization. Moreover, $PLA/1DP - nHAp_5$ demonstrates good eletrospinnability, producing continuous fibers without beads. The nHAp dispersed well in PLA without phase separation and aggregation.

**Author Contributions:** Conceptualization, C.R. and Y.R.; methodology, P.I., Y.R. and C.R.; validation, T.T., J.W., C.R., and Y.R.; formal analysis, P.I.; investigation, P.I.; resources, Y.R. and C.R.; data curation, P.I.; writing—original draft preparation, P.I.; writing—review and editing, P.I., T.T., J.W., C.R. and Y.R.; visualization, C.R. and Y.R.; supervision, Y.R. and C.R.; project administration, Y.R. and C.R.; funding acquisition, Y.R. and C.R. All authors have read and agreed to the published version of the manuscript.

**Funding:** This research was funded by Thailand Science Research and Innovation (TSRI), Grant No. 42853.

**Institutional Review Board Statement:** Not applicable.

**Data Availability Statement:** Not applicable.

**Acknowledgments:** The authors are grateful to Suranaree University of Technology (SUT), Thailand Science Research and Innovation (TSRI), to the National Science, Research and Innovation Fund (NSRF), and to the Research Center for Biocomposite Materials for the Medical Industry and Agricultural and Food Industry for the financial support.

**Conflicts of Interest:** The authors declare no conflict of interest.

## References

1. Kodali, D.; Hembrick-Holloman, V.; Gunturu, D.R.; Samuel, T.; Jeelani, S.; Rangari, V.K. Influence of Fish Scale-Based Hydroxyapatite on Forcespun Polycaprolactone Fiber Scaffolds. *ACS Omega* **2022**, *7*, 8323–8335. [CrossRef] [PubMed]
2. Ideia, P.; Degli Esposti, L.; Miguel, C.C.; Adamiano, A.; Iafisco, M.; Castilho, P.C. Extraction and characterization of hydroxyapatite-based materials from grey triggerfish skin and black scabbardfish bones. *Int. J. Appl. Ceram.* **2021**, *18*, 235–243. [CrossRef]
3. Mohd Pu'ad, N.A.S.; Abdul Haq, R.H.; Mohd Noh, H.; Abdullah, H.Z.; Idris, M.I.; Lee, T.C. Nano-size hydroxyapatite extracted from tilapia scale using alkaline heat treatment method. *Mater. Today Proc.* **2020**, *29*, 218–222. [CrossRef]
4. Athinarayanan, J.; Periasamy, V.S.; Alshatwi, A.A. Simultaneous fabrication of carbon nanodots and hydroxyapatite nanoparticles from fish scale for biomedical applications. *Mater. Sci. Eng. C* **2020**, *117*, 111313. [CrossRef] [PubMed]
5. Predoi, D.; Iconaru, S.L.; Predoi, M.V. Dextran-Coated Zinc-Doped Hydroxyapatite for Biomedical Applications. *Polymers* **2019**, *11*, 886. [CrossRef]
6. Kaimonov, M.; Safronova, T.; Shatalova, T.; Filippov, Y.; Tikhomirova, I.; Sergeev, N. Composite Ceramics in the in the $Na_2O$–$CaO$–$SiO_2$–$P_2O_5$ System Obtained from Pastes including Hydroxyapatite and an Aqueous Solution of Sodium Silicate. *Ceramics* **2022**, *5*, 550–561. [CrossRef]
7. Bernardo, M.P.; da Silva, B.C.R.; Hamouda, A.E.I.; de Toledo, M.A.S.; Schalla, C.; Rütten, S.; Sechi, A. PLA/Hydroxyapatite scaffolds exhibit in vitro immunological inertness and promote robust osteogenic differentiation of human mesenchymal stem cells without osteogenic stimuli. *Sci. Rep.* **2022**, *12*, 2333. [CrossRef]
8. Sathiskumar, S.; Vanaraj, S.; Sabarinathan, D.; Bharath, S.; Sivarasan, G.; Arulmani, S.; Ponnusamy, V.K. Green synthesis of biocompatible nanostructured hydroxyapatite from Cirrhinus mrigala fish scale—A biowaste to biomaterial. *Ceram. Int.* **2019**, *45*, 7804–7810. [CrossRef]
9. Qin, D.; Bi, S.; You, X.; Wang, M.; Cong, X.; Yuan, C.; Chen, X.G. Development and application of fish scale wastes as versatile natural biomaterials. *J. Chem. Eng.* **2022**, *428*, 131102. [CrossRef]
10. Ibañez, A.; Cowx, I.; O'Higgins, P. Variation in elasmoid fish scale patterns is informative with regard to taxon and swimming mode. *Zool. J. Linn. Soc.* **2009**, *155*, 834–844. [CrossRef]
11. Kara, A.; Tamburaci, S.; Tihminlioglu, F.; Havitcioglu, H. Bioactive fish scale incorporated chitosan biocomposite scaffolds for bone tissue engineering. *Int. J. Biol. Macromol.* **2019**, *130*, 266–279. [CrossRef] [PubMed]
12. Irwansyah, F.S.; Noviyanti, A.R.; Eddy, D.R.; Risdiana, R. Green Template-Mediated Synthesis of Biowaste Nano-Hydroxyapatite: A Systematic Literature Review. *Molecules* **2022**, *27*, 5586. [CrossRef] [PubMed]
13. Negrila, C.C.; Predoi, M.V.; Iconaru, S.L.; Predoi, D. Development of Zinc-Doped Hydroxyapatite by Sol-Gel Method for Medical Applications. *Molecules* **2018**, *23*, 2986. [CrossRef] [PubMed]
14. Huang, B.; Li, M.; Mo, H.; Chen, C.; Chen, K. Effects of Substitution Ratios of Zinc-Substituted Hydroxyapatite on Adsorption and Desorption Behaviors of Bone Morphogenetic Protein-2. *Int. J. Mol. Sci.* **2022**, *23*, 10144. [CrossRef] [PubMed]
15. Santos, C.; Luklinska, Z.B.; Clarke, R.L.; Davy, K.W.M. Hydroxyapatite as a filler for dental composite materials: Mechanical properties and in vitro bioactivity of composites. *J Mater Sci Mater Med* **2001**, *12*, 565–573. [CrossRef]
16. Gibson, I.R. 1.3.4A—Natural and Synthetic Hydroxyapatites. In *Biomaterials Science*, 4th ed.; Wagner, W.R., Sakiyama-Elbert, S.E., Zhang, G., Yaszemski, M.J., Eds.; Academic Publisher: Cambridge, MA, USA, 2020; pp. 307–317.
17. Mohd Pu'ad, N.A.S.; Koshy, P.; Abdullah, H.Z.; Idris, M.I.; Lee, T.C. Syntheses of hydroxyapatite from natural sources. *Heliyon* **2019**, *5*, e01588. [CrossRef]
18. Chittara, Y. Effect of Calcination Temperature on Quality of Hydroxyapatite that fabricated from Fish Scale Biowaste. In Proceedings of the Twenty Third International Conference on Processing and Fabrication of Advanced Materials XXIII, Indian Institute of Technology Roorkee, Roorkee, India, 5–7 December 2014.
19. Kongsri, S.; Janpradit, K.; Buapa, K.; Techawongstien, S.; Chanthai, S. Nanocrystalline hydroxyapatite from fish scale waste: Preparation, characterization and application for selenium adsorption in aqueous solution. *Chem. Eng. J.* **2013**, *215–216*, 522–532. [CrossRef]
20. Majhool, A.; Zainol, I.; Jaafar, C.; Mudhafar, M.; Alsailawi, H.A.; Asaad, A.; Mezaal, F. Preparation of Fish Scales Hydroxyapatite (FsHAp) for Potential Use as Fillers in Polymer. *J. Chem. Chem. Eng.* **2019**, *13*, 62–75.
21. Department of Fisheries. Available online: https://www4.fisheries.go.th/local/file_document/20220916111633_1_file.pdf (accessed on 21 September 2022).
22. Ranjbar Mohammadi Bonab, M.; Shakoori, P.; Arab, Z. Design and characterization of keratin/PVA-PLA nanofibers containing hybrids of nanofibrillated chitosan/ZnO nanoparticles. *Int. J. Biol. Macromol.* **2021**, *187*, 554–565. [CrossRef]
23. Wardhono, E.Y.; Kanani, N.; Alfirano; Rahmayetty. Development of polylactic acid (PLA) bio-composite films reinforced with bacterial cellulose nanocrystals (BCNC) without any surface modification. *J. Dispers. Sci. Technol.* **2020**, *41*, 1488–1495. [CrossRef]

24. Farah, S.; Anderson, D.G.; Langer, R. Physical and mechanical properties of PLA, and their functions in widespread applications—A comprehensive review. *Adv. Drug Deliv. Rev.* **2016**, *107*, 367–392. [CrossRef] [PubMed]
25. Sa, Y.; Guo, Y.; Feng, X.; Wang, M.; Li, P.; Gao, Y.; Jiang, T. Are different crystallinity-index-calculating methods of hydroxyapatite efficient and consistent? *New J. Chem.* **2017**, *41*, 5723–5731. [CrossRef]
26. Rattan, S.; Fawcett, D.; Poinern, G.E.J. Williamson-Hall based X-ray peak profile evaluation and nano-structural characterization of rod-shaped hydroxyapatite powder for potential dental restorative procedures. *AIMS Mater. Sci.* **2021**, *8*, 359–372. [CrossRef]
27. Grunenwald, A.; Keyser, C.; Sautereau, A.M.; Crubézy, E.; Ludes, B.; Drouet, C. Revisiting carbonate quantification in apatite (bio)minerals: A validated FTIR methodology. *J. Archaeol. Sci.* **2014**, *49*, 134–141. [CrossRef]
28. Mysiukiewicz, O.; Barczewski, M. Crystallization of polylactide-based green composites filled with oil-rich waste fillers. *J. Polym. Res.* **2020**, *27*, 374. [CrossRef]
29. Deb, P.; Deoghare, A.B. Effect of Acid, Alkali and Alkali–Acid Treatment on Physicochemical and Bioactive Properties of Hydroxyapatite Derived from Catla catla Fish Scales. *Arab. J. Sci. Eng.* **2019**, *44*, 7479–7490. [CrossRef]
30. Rouhani, P.; Taghavinia, N.; Rouhani, S. Rapid growth of hydroxyapatite nanoparticles using ultrasonic irradiation. *Ultrason. Sonochem.* **2010**, *17*, 853–856. [CrossRef]
31. Karim, B.; LaKrat, M.; Elansari, L.L.; Mejdoubi, E. Synthesis of B-type carbonated hydroxyapatite by a new dissolution-precipitation method. *Mater. Today Proc.* **2020**, *31*, S83–S88.
32. Prasad, A.; Mohan Bhasney, S.; Sankar, M.R.; Katiyar, V. Fish Scale Derived Hydroxyapatite Reinforced Poly (Lactic acid) Polymeric Bio-films: Possibilities for Sealing/locking the Internal Fixation Devices. *Mater. Today Proc.* **2017**, *4*, 1340–1349. [CrossRef]
33. Deb, P.; Deoghare, A.B. Effect of pretreatment processes on physicochemical properties of hydroxyapatite synthesized from Puntius conchonius fish scales. Bull. *Mater. Sci.* **2019**, *42*, 3.
34. Gergely, G.; Wéber, F.; Lukács, I.; Tóth, A.L.; Horváth, Z.E.; Mihály, J.; Balázsi, C. Preparation and characterization of hydroxyapatite from eggshell. *Ceram. Int.* **2010**, *36*, 803–806. [CrossRef]
35. Gopalu, K.; Cho, E.-B.; Thirumurugan, K.; Govindan, S.; Kumar, G.; Kolesnikov, E.; Selvakumar, R. Mesoporous Mn-doped hydroxyapatite nanorods obtained via pyridinium chloride enabled microwave-assisted synthesis by utilizing Donax variabilis seashells for implant applications. *Mater. Sci. Eng. C* **2021**, *126*, 112170. [CrossRef]
36. Tariq, U.; Haider, Z.; Chaudhary, K.; Hussain, R.; Ali, J. Calcium to phosphate ratio measurements in calcium phosphates usinglibs. *J. Phys. Conf. Ser.* **2018**, *1027*, 012015. [CrossRef]
37. Samal, S. Effect of shape and size of filler particle on the aggregation and sedimentation behavior of the polymer composite. *Powder Technol.* **2020**, *366*, 43–51. [CrossRef]
38. Raita, M.S.; Iconaru, S.L.; Groza, A.; Cimpeanu, C.; Predoi, G.; Ghegoiu, L.; Predoi, D. Multifunctional Hydroxyapatite Coated with Arthemisia absinthium Composites. *Molecules* **2020**, *25*, 413. [CrossRef] [PubMed]
39. Thommes, M.; Kaneko, K.; Neimark, A.V.; Olivier, J.P.; Rodriguez-Reinoso, F.; Rouquerol, J.; Sing, K.S.W. Physisorption of gases, with special reference to the evaluation of surface area and pore size distribution (IUPAC Technical Report). *Pure Appl. Chem.* **2015**, *87*, 1051–1069. [CrossRef]
40. Chen, K.; Zhang, T.; Chen, X.; He, Y.; Liang, X. Model construction of micro-pores in shale: A case study of Silurian Longmaxi Formation shale in Dianqianbei area, SW China. *Pet. Explor. Dev.* **2018**, *45*, 412–421. [CrossRef]
41. Muhammad, N.; Gao, Y.; Iqbal, F.; Ahmad, P.; Ge, R.; Nishan, U.; Ullah, Z. Extraction of biocompatible hydroxyapatite from fish scales using novel approach of ionic liquid pretreatment. *Sep. Purif. Technol.* **2016**, *161*, 129–135. [CrossRef]
42. Kamarudin, S.; Luqman Chuah, A.; Aung, M.M.; Ratnam, C.; Jusoh, E. A study of mechanical and morphological properties of PLA based biocomposites prepared with EJO vegetable oil based plasticiser and kenaf fibres. *Mater. Res. Express* **2018**, *5*, 085314. [CrossRef]
43. Boey, J.Y.; Lee, C.K.; Tay, G.S. Factors Affecting Mechanical Properties of Reinforced Bioplastics: A Review. *Polymers* **2020**, *14*, 3737. [CrossRef]
44. Jodeh, S.; Azzaoui, K.; Mejdoubi, E.; Lamhamdi, A.; Hammouti, B.; Akartasse, N.; Abidi, N. Novel Tricomponenets composites Films From Polylactic Acid/ Hydroxyapatite/ Poly- Caprolactone Suitable For Biomedical Applications. *J. Mater. Environ. Sci.* **2016**, *7*, 761–769.
45. Goreke, M.D.; Alakent, B.; Soyer-Uzun, S. Comparative Study on Factors Governing Binding Mechanisms in Polylactic Acid–Hydroxyapatite and Polyethylene–Hydroxyapatite Systems via Molecular Dynamics Simulations. *Langmuir* **2020**, *36*, 1125–1137. [CrossRef] [PubMed]
46. Rakmae, S.; Lorprayoon, C.; Ekgasit, S.; Suppakarn, N. Influence of Heat-Treated Bovine Bone-Derived Hydroxyapatite on Physical Properties and in vitro Degradation Behavior of Poly (Lactic Acid) Composites. *Polym. Plast. Technol. Eng.* **2013**, *52*, 1043–1053. [CrossRef]
47. Li, J.; Lu, X.; Zheng, Y. Effect of surface modified hydroxyapatite on the tensile property improvement of HA/PLA composite. *Appl. Surf. Sci.* **2008**, *255*, 494–497. [CrossRef]
48. Xiao, L.; Wang, B.; Yang, G.; Gauthier, M. Poly (lactic acid)-based biomaterials: Synthesis, modification and applications. *Biomed.Sci. Eng. Technol.* **2012**, *11*, 247–282.
49. Pandele, A.; Constantinescu, A.; Radu, C.; Miculescu, F.; Voicu, Ş.I.; Ciocan, L. Synthesis and Characterization of PLA-Microstructured Hydroxyapatite Composite Films. *Materials* **2020**, *13*, 274. [CrossRef] [PubMed]

50. Kamarudin, S.H.; Abdullah, L.C.; Aung, M.M.; Ratnam, C.T. Thermal and Structural Analysis of Epoxidized Jatropha Oil and Alkaline Treated Kenaf Fiber Reinforced Poly(Lactic Acid) Biocomposites. *Polymers* **2020**, *12*, 2604. [CrossRef] [PubMed]
51. Zhang, S.; Liang, Y.; Qian, X.; Hui, D.; Sheng, K. Pyrolysis kinetics and mechanical properties of poly(lactic acid)/bamboo particle biocomposites: Effect of particle size distribution. *Nanotechnol. Rev.* **2020**, *9*, 524–533. [CrossRef]
52. Chinh, N.T.; Manh, V.Q.; Trung, V.Q.; Trang, T.D.M.; Hoang, T. Extraction of hydroxyapatite and collagen from the Vietnamese tilapia scales. *Vietnam J. Chem.* **2019**, *57*, 225–228. [CrossRef]
53. Liu, S.; Zheng, Y.; Liu, R.; Tian, C. Preparation and characterization of a novel polylactic acid/hydroxyapatite composite scaffold with biomimetic micro-nanofibrous porous structure. *J. Mater. Sci. Mater. Med.* **2020**, *31*, 74. [CrossRef]
54. Akindoyo, J.O.; Beg, M.D.H.; Ghazali, S.; Heim, H.P.; Feldmann, M. Impact modified PLA-hydroxyapatite composites—Thermo-mechanical properties. *Compos. A Appl. Sci. Manuf.* **2018**, *107*, 326–333. [CrossRef]
55. Haider, A.; Haider, S.; Kang, I.-K. A comprehensive review summarizing the effect of electrospinning parameters and potential applications of nanofibers in biomedical and biotechnology. *Arab. J. Chem.* **2018**, *11*, 1165–1188. [CrossRef]

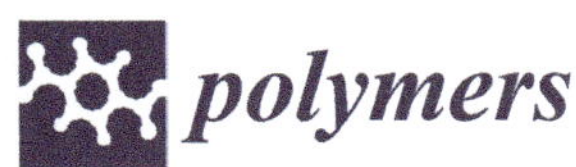

*Article*

# A Feature of the Crystalline and Amorphous Structure of Ultra Thin Fibers Based on Poly(3-hydroxybutyrate) (PHB) Containing Minor Concentrations of Hemin and a Complex of Tetraphenylporphyrin with Iron

**Svetlana G. Karpova [1,*], Ivetta A. Varyan [1,2], Anatoly A. Olkhov [1,2], Polina M. Tyubaeva [1,2] and Anatoly A. Popov [1,2]**

1 Department of Biological and Chemical Physics of Polymers, Emanuel Institute of Biochemical Physics, Russian Academy of Sciences, 4 Kosygina Street, 119334 Moscow, Russia
2 Academic Department of Innovational Materials and Technologies Chemistry, Plekhanov Russian University of Economics, 36 Stremyanny Lane, 117997 Moscow, Russia
* Correspondence: karpova@sky.chph.ras.ru

**Abstract:** Comprehensive studies combining X-ray diffraction analysis, thermophysical, dynamic measurements by probe method and scanning electron microscopy have been carried out. The peculiarity of the crystalline and amorphous structure of ultra-thin fibers based on poly(3-hydroxybutyrate) (PHB) containing minor concentrations (0–5%) of a gene and a tetraphenylporphyrin (TFP) complex with iron (in the form of FeCl) are considered. When these complexes are added to the PHB fibers, the morphology of the fibers change: a sharp change in the crystallinity and molecular mobility in the amorphous regions of PHB is observed. When adding a gel to the fibers of PHB, a significant decrease in the degree of crystallinity, melting enthalpy, and correlation time can be observed. The reverse pattern is observed in a system with the addition of FeCl-TFP—there is a significant increase in the degree of crystallinity, melting enthalpy and correlation time. Exposure of PHB fibers with gemin in an aqueous medium at 70 °C leads to a decrease in the enthalpy of melting in modified fibers—to an increase in this parameter. The molecular mobility of chains in amorphous regions of PHB/gemin fibers increases at the same time, a nonlinear dependence of changes in molecular dynamics is observed in PHB/FeCl-TFP fibers. Ozonolysis has a complex effect on the amorphous structure of the studied systems. The obtained fibrous materials have bactericidal properties and should be used in the creation of new therapeutic systems of antibacterial and antitumor action.

**Keywords:** hemin; TEMPO stable radical; correlation times; ultrathin fibers; poly(3-hydroxybutyrate); ozo-nation; amorphous phase

**Citation:** Karpova, S.G.; Varyan, I.A.; Olkhov, A.A.; Tyubaeva, P.M.; Popov, A.A. A Feature of the Crystalline and Amorphous Structure of Ultra Thin Fibers Based on Poly(3-hydroxybutyrate) (PHB) Containing Minor Concentrations of Hemin and a Complex of Tetraphenylporphyrin with Iron. *Polymers* **2022**, *14*, 4055. https://doi.org/10.3390/polym14194055

Academic Editors: José Miguel Ferri, Dimitrios Bikiaris, Vicent Fombuena Borràs and Miguel Fernando Aldás Carrasco

Received: 2 September 2022
Accepted: 23 September 2022
Published: 27 September 2022
Corrected: 25 April 2023



## 1. Introduction

Since the diffusion of the drug substance is largely dependent on the structure of the polymer matrix, structural changes in the polymer can significantly affect the kinetics of drug release, and, consequently, the effectiveness of the dosage form. Structural changes in fibrillar mats during storage and operation can be caused by water absorption, heating, the action of oxygen, ozone and UV radiation on the polymer, as well as the action of microorganisms. These factors may act simultaneously or sequentially depending on the operating conditions of the dosage form and climatic features. One of the most common natural polyesters for medical use is poly(3-hydroxybutyrate) (PHB). An effective method of directed influence on the structure of a polymeric material is its doping with low molecular weight substances of various nature. In our earlier work, we considered the structure formation of fibrous materials based on PHB containing dipyridamole [1–9], chitosan [10], nanoparticles of titanium dioxide and silicon [11], complexes:

iron (III)-chloroporphyrin [12], zinc porphyrin [13], manganese-chloroporphyrin [14], tin-chloroporphyrin [15]. A strong influence of low molecular-weight substances on the structure of the crystalline and amorphous phases of PHB fibers was shown. These substances, due to the presence of chemically active polar functional groups in them, enter into intermolecular interactions with the biopolymer. Depending on the nature of the metal, metalloporphyrins show a different tendency to aggregation, and this determines their ability to act as nucleating agents during polymer crystallization. In addition, porphyrins and their metal complexes have several binding centers in their structure, which contribute to the emergence of coordination interactions with the molecular environment [16,17]. First of all, the coordinating activity of the central-complexing metal ion should be singled out. Thus, metal cations with chloride extraligands contained in the structure of porphyrin metal complexes can exchange ligands for polar fragments of the environment, for example, polymeric, or such as oxygen-containing groups such as carboxyls or hydroxyls.

One of the most promising PHB dopants for the creation of polymeric materials for medical purposes are complexes of tetraphenylporphyrin (TPP) with tin (IV) (Sn(TPP)$Cl_2$) and iron (III) (Fe(TPP)Cl). Sn(TPP)$Cl_2$ is currently used to create photocatalysts that promote the destruction of organic toxicants and photosensitizers for medical diagnostics and therapy. A significant advantage of porphyrin complexes with tin(IV) is the presence of two extraligands directed on opposite sides of the plane of the porphyrin macrocycle. This structural feature causes the practical absence of the ability of these metal complexes to aggregate in comparison with the complexes of porphyrins with metals in the +2 and +3 oxidation states [18–20]. Fe(TPP)Cl molecules have a strong interaction with PHB macromolecules, however, the presence of only one extraligand chloride leads to intermolecular repulsion, so the molecules are capable of aggregation. Due to the unique geometric and electronic structure, the molecules of porphyrin metal complexes have a significant effect on the crystallization and segmental mobility of polymer macromolecules during the formation of composite matrices based on them. As a result of such an interaction, a sharp deceleration or acceleration of crystallization processes, a change in the structure of amorphous substances, and relaxation of polymer macromolecules can occur. Various complexes of porphyrins of natural and synthetic origin are used to modify the polymer to achieve specific properties of the entire composite system. Metal complexes with tetraphenylporphyrins have unique photocatalytic and antimicrobial properties [21,22].

Our aim was a comprehensive study of the properties of the new biocompatible composites based on a system of polymer and hemin for biomedical applications. One of the most promising areas for these materials is a wound-healing bandage: a biopoylmer-hemin-protein that provides regeneration. Also we found that there are different materials based on hemin: "hemin can be used in various biomedical materials, as a basis for binding proteins to a polymer, for container molecules (such as cavitands and capsules) for delivering systems, for constructing new biocatalysts tailored to specific functions, for creation of the innovative anticoagulants and others" [4–6].

In view of the fact that hemin is widely used in medicine as an independent drug complex in the treatment of porphyria, requiring a carrier polymer, and in various fields to create combined drugs based on proteins and peptides for drug delivery, combinations of PHB-hemin should definitely be recommended for use in biomedicine in view of the stability and high physical and mechanical properties of this composite.

The purpose of this work is to obtain new materials based on natural porphyrin—hemin and to compare the obtained regularities with the results obtained earlier for the PHB/Fe(TPP)Cl system. Due to its properties, hemin can be used in various biomedical materials, as a basis for binding proteins to a polymer [23], for container molecules (such as cavitands and capsules), for system delivery [24], to create new biocatalysts adapted for specific functions [25], which opens up prospects for the development of new biocompatible composites based on the system of nanopolymer and hemin for biomedical applications.

## 2. Materials and Methods

Polymer nanofibrous materials were obtained by electrospinning (ES) on a single-capillary laboratory setup with a capillary diameter of 0.1 mm. Polymer-molding solutions were prepared from 16F series semi-crystalline biodegradable polymer poly(3-hydroxybutyrate) (PHB) (BIOMER®, Germany) with a molecular weight of 206 kDa, a density of 1.248 g/cm$^3$, and a crystallinity level of 59% (Figure 1a). Finely dispersed PHB powder was dissolved in chloroform at a temperature of 60 °C to obtain molding solutions. Hemin is an iron coordination complex (oxidation state: III) (Figure 1b), obtained by organic synthesis [26]. Hemin was dissolved in N,N-dimethylformamide at 25 °C and homogenized with PHB solution. The content of PHB in the solution was 7% wt.; the content of hemin was 1, 3, and 5% wt. The conditions of the ES process were voltage 17–20 kV, distance between electrodes 190–200 mm, and gas pressure on the solution 10–15 kg(f)/cm$^2$.

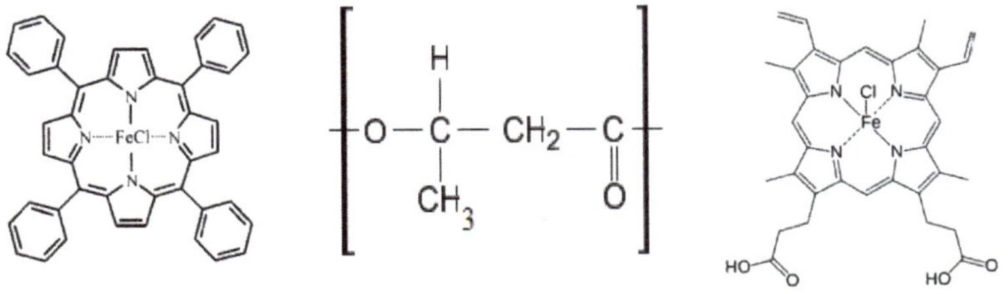

(a) (b) (c)

**Figure 1.** Structural formulas of (**a**) Fe(TPP)Cl, (**b**) PHB, and (**c**) hemin.

X-ray diffraction analysis (XRD) of the samples was performed using transmission imaging. High-resolution two-dimensional scattering patterns were obtained using an S3-Micropix small- and wide-angle X-ray scattering system manufactured by Hecus (Cu(Ka) radiation, l = 1.542 Å). A Pilatus 100 K detector and also a PSD 50M linear detector were used in Ar/Me current at 8 bar argon pressure, high voltage on the Xenocs Genix source tube and current were 50 kV and 1 mA, respectively. The X-ray beam was formed using Fox 3D X-ray optics; the diameters of the forming slots in the collimator were 0.1 and 0.2 mm, respectively. The measurement range of diffraction angles was from 0.003 to 1.9 Å$^{-1}$. To eliminate X-ray scattering in the air, the block of X-ray mirrors and the chamber were placed in a vacuum system (pressure $\approx 2.3 \times 10^{-2}$ Torr). The accumulation time was varied in the range of 600–5000 s. This pattern of intermolecular interactions determined the structural and dynamic characteristics of ultrathin fibers. Hemin molecules have polar groups, which causes their attraction among themselves and, as a result, the formation of large particles.

Small-angle X-ray diffraction was used to study PHB/hemin fibers of various compositions. The average effective crystallite size $L_{hkl}$ in the crystallographic direction hkl was determined from the integral half-width of the line of the corresponding X-ray reflection using the Scherrer equation:

$$\Delta_{hkl}(2\theta) = \lambda / L_{hkl} \cos\theta \quad (1)$$

The value of the long period was calculated by the equation:

$$d = n\lambda / 2\theta_m \quad (2)$$

where $d$—the long period, $\lambda$—1.542 Å wavelength of CuK$\alpha$—radiation, $\theta_m$—diffraction angle, $n$—reflection order.

Electron paramagnetic resonance (EPR) X-band spectra were recorded on an EPR-V automated spectrometer (Federal Research Center for Chemical Physics, Russian Academy of Sciences, Moscow). The value of the microwave power to avoid saturation effects did not exceed 1 mW. The modulation amplitude was always much smaller than the resonance line width and did not exceed 0.5 G. The stable nitroxide radical TEMPO was used as a spin probe. The radical was introduced into the fibers from the gas phase at a temperature of 50 °C for an hour. The concentration of the radical in the polymer was determined by double integration of the EPR spectra; the reference was an evacuated solution of TEMPO in $CCl_4$ with a radical concentration of ~$1 \times 10^{-3}$ mol/L.

The samples were studied by differential scanning calorimetry (DSC) on a DSC 204 F1 instrument (Netzsch) in argon at a heating rate of 10 K/min. The average statistical error in the measurement of thermal effects was ±3%. The enthalpy of fusion was calculated using the NETZSCH Proteus program. Thermal analysis was carried out according to the standard procedure [27]. Peak separation was carried out using the software "NETZSCH Peak Separation 2006.01".

The fibers were obtained by electrospinning using a single-capillary laboratory setup with the following parameters: capillary diameter, 0.1 mm; electric current voltage, 12 kV; distance between electrodes, 18 cm; solution electrical conductivity, 10 μS/cm.

The effect on samples of distilled water was studied at 70 ± 1 °C. Before introducing the radical, the samples exposed to water were dried to constant weight.

The ozonation of the samples was carried out in the gas phase. The ozone concentration was $3 \times 10^{-5}$ mol/L.

It should be noted that dopant molecules are localized only in amorphous regions of the polymer, and that the proportion of such regions decreases with the increasing concentration of the additive. For this reason, the local dopant concentration in the amorphous regions of the polymer was significantly higher than the set values of 1, 3 and 5%.

## 3. Results

### 3.1. Effect of Hemin Concentration on the Geometric Parameters of the Fiber

The PHB/hemin materials were obtained by double molding. The essence of the electrospinning process ensured the curing of the fibers of the polymer matrix after complete evaporation of the solvent [28]. This method made it possible to uniformly arrange additives that occupied the free space between crystallites in the amorphous regions of the polymer [29]. The introduction of hemin led to the formation of uniform and thin fibers, which might be due to the presence of metal, which increased the electrical conductivity of the spinning solution, and which, of course, improved the quality of the resulting fibers [30].

The introduction of hemin into the PHB structure accelerated the enzymatic hydrolysis of ultrathin PHB/hemin fibers compared to PHB/FeCl-TFP fibers due to a looser amorphous phase and a lower degree of crystallinity. When porphyrin metal complexes were added to the PHB molding solution, the morphology of the fibrous material changed dramatically. When 1–5% of FeCl-TFP complexes were added, elliptical elements in the fiber structure disappeared completely. When 1% of the FeCl-TFP complex was added, fibers were formed mainly with average diameters of 1.5–2.0; 3.0–4.0 and 5.0–6.0 microns. The presence of thin fibers of less than 3 microns was a consequence of the splitting effect of the primary jet of the molding solution in the field of electrostatic forces. With an increase in the concentration of FeCl-TFP from 3 to 5%, fibers with a diameter of 3 microns predominated.

Scanning electron microscopy was used to study in detail the morphology of fibrous materials. The images of the most characteristic parts of the samples are shown in Figure 1. The structure of the initial fibers of PHB and with a low content of hemin (up to 3%) is heterogeneous. The fibers are characterized by a high degree of tortuosity, adhesions and the presence of large formations in the form of thickenings of an elliptical shape. The average size of these structures is from 20 to 30 microns in the longitudinal direction and 15–25 microns in the transverse direction. An increase in the concentration of hemin leads to a decrease in the diameter and an increase in the uniformity of the fibers in diameter. The microrelief, with characteristic pores (Figure 2a) changes with the introduction of even 1% additive. Defects in the form of glues and thickenings are extremely rare, the fibers become thinner, the number of defects is reduced, and with the introduction of 5%, defects in the form of thickenings are almost completely absent. Roughness on the surface of the fibers almost completely disappears at 5% addition, as do elliptical thickenings.

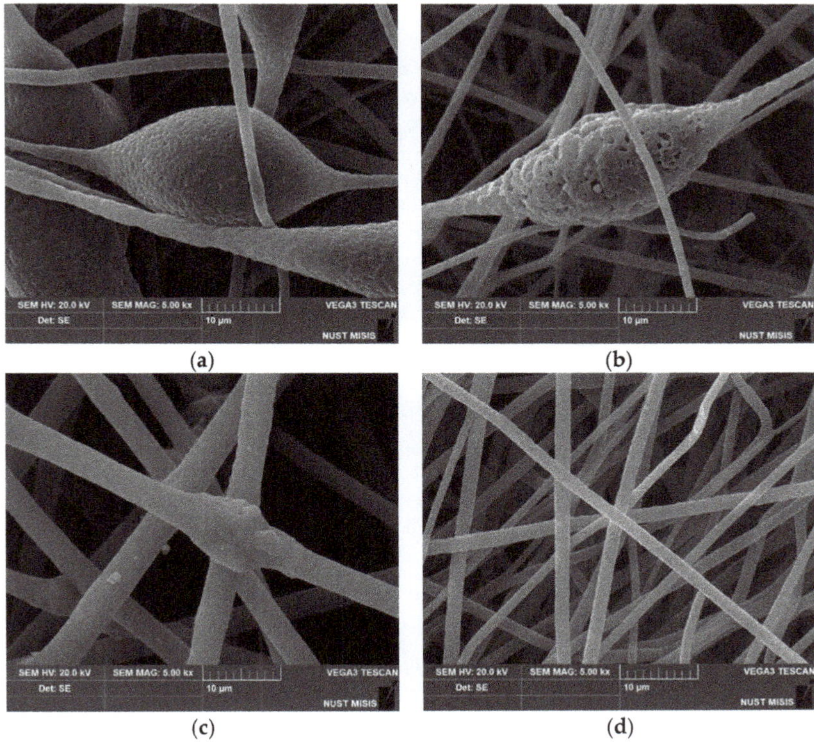


**Figure 2.** Micrographs of PHB with different hemin content: (**a**) 0%, (**b**) 1%, (**c**) 3% and (**d**) 5%.

The deepening of pores on such a thickening (Figure 2b) can be explained by more intensive evaporation of the solvent due to the technology of electrospinning using two solvents, which is used, among other things, to obtain highly porous fibers [30], however, for a 5% hemin content, this factor did not add to the contribution in the form of a more optimal balance of electrical conductivity-viscosity of the casting solution. In all likelihood, the presence of large hemin molecules in the intermolecular space of PHB led to an increase in the free volume of amorphous regions, which increased the rate of solvent desorption. To determine the uniformity of distribution of hemin in the fibers, we used the EDX method.

To ensure that the hemin was indeed incorporated into the fiber structure and not on the surface, an EDX analysis was performed. The results of the analysis of the composition of the material by the energy-dispersive X-ray method showed that hemin in the studied concentration range was distributed fairly evenly in the material, and not only concentrated in inclusions on the surface. Iron and chlorine atoms were chosen as the atom identifying hemin, since they were included in the central part of the tetrapyrrole ring.

### 3.2. X-ray Diffraction Study of PHB/Hemin Fibers of Different Composition

Along with a change in the geometry of ultrathin fibers with the introduction of an additive, one should also expect changes in the structural and dynamic parameters of the compositions.

It is known that the intermolecular interaction between porphyrin-metal complex particles, as well as between these particles and PHB molecules, depends significantly on the nature of the metal complex. Thus, it was shown in [31–33] that porphyrin molecules were attracted and, as a result, rather large particles of porphyrin were formed in PHB fibers. The intermolecular interaction of these particles with PHB macromolecules was extremely small. Although the molecules of the porphyrin complex with FeCl had a strong interaction with PHB macromolecules, the presence of the loride extraligand in the metal complex led to intermolecular repulsion, i.e., the molecules aggregated into smaller particles. Finally, $Sn(TPP)Cl_2$ molecules were distributed in the system at the molecular level, due to the presence of two chlorine anions, repulsive forces between the complexes acted. The interaction of these complexes was weaker than that of Fe(TPP)Cl complexes with PHB macromolecules.

Figure 3a shows the dependence of the degree of crystallinity $\chi$ on the composition of the PHB/hemin system. For comparison, Figure 3b shows such a dependence for the PHB/Fe(TPP)Cl system. As seen, the dependencies point in different directions. Fe(TPP)Cl particles, being crystallization nuclei, caused an increase in the degree of crystallinity, and hemin, aggregating into large particles, prevented the crystallization of the PHB/hemin mixture and, as a result, the fraction of crystallites decreased with increasing hemin concentration in the polymer. In a system with 5% hemin, crystalline hemin particles 55–100 nm in size were formed. If the large period d increased with the addition of Fe(TPP)Cl particles to PHB (from 58 to 63 nm), then this parameter practically did not change in the PHB/hemin system.

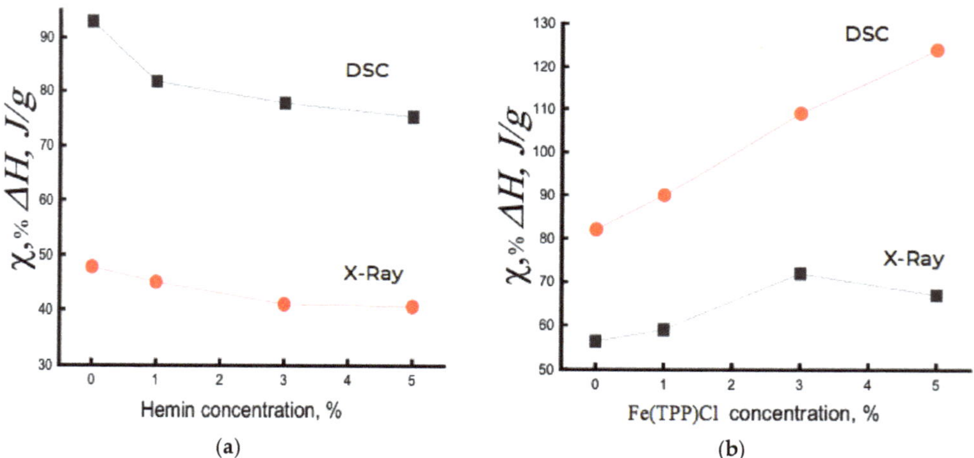


**Figure 3.** Dependence of the degree of crystallinity $\chi$ and enthalpy of melting $\Delta H$ on the composition of the mixed composition of (**a**) PHB/hemin and (**b**) PHB/Fe(TPP)Cl.

It should be noted that with an increase in the concentration of Fe(TPP)Cl from 3% to 5%, a slight decrease in the degree of crystallinity was observed, apparently due to the aggregation of complexes into large particles, which did not lead to a further increase in the proportion of crystallites during fiber spinning.

Thus, X-ray diffraction study indicated a decrease in the degree of crystallinity in the PHB/hemin system, and in PHB/Fe(TPP)Cl fibers, an increase in the longitudinal dimensions (along the fibril) of crystal structures, a large period, and also the degree of crystallinity with increasing concentration of the additive was observed. Along with the presented changes, the transverse sizes of crystallites significantly decreased upon the addition of porphyrin: for example, $L_{020}$ changed from 20.2 Å in PHB to $\approx$ 16 Å in PHB/Fe(TPP)Cl, and $L_{110}$ changed from 20.1 Å in PHB to $\approx$14.5 Å in PHB with an additive. Therefore, it can be concluded that the addition of porphyrin to PHB fibers caused a decrease in the transverse dimensions of fibrils, and the longitudinal size of crystallites increased significantly. The observed changes, in our opinion, took place due to the straightening of the folded sections of the chains in the crystallites and their partial crystallization during the electrospinning of the fibers. The presence of small Fe(TPP)Cl particles contributed to the processes of intensive plasticization of the polymer, since it was in the PHB/Fe(TPP)Cl samples that the strongest increase in the degree of crystallinity occurred with increasing additive concentration. In PHB/hemin samples, however, the degree of crystallinity decreased significantly with an increase in the additive, which may be due to the formation of large particles, and with an increase in their concentration, the process of crystallite deformation took place to an increasing extent.

The presented data indicate that the addition of Fe(TPP)Cl particles to PHB fibers caused a significant increase in the degree of crystallinity with a higher longitudinal crystallite size ($L_{002}$). When hemin was added, due to its aggregation into large particles, crystallization processes were inhibited, which led to a significant decrease in the degree of fiber crystallinity.

### 3.3. *Thermophysical Characteristics of the Crystalline Phase of PHB/Hemin and PHB/Fe(TPP)Cl Compositions*

Let us consider the effect of hemin on the crystalline phase of PHB fibers using the DSC method (Figure 3, Table 1). It should be noted that the concentration of the porphyrin complex in the amorphous regions of the polymer varied significantly beyond the specified values of 1, 3, and 5%. The latter were calculated for the entire mass of the sample, whereas the particles of the porphyrin complex were only in the amorphous interlayer, the proportion of which decreased significantly with an increase in the concentration of Fe(TPP)Cl and increased with the addition of hemin to the system. For example, in PHB fibers containing 5% Fe(TPP)Cl complex, its concentration in amorphous regions would be ~15%.

**Table 1.** Enthalpy ($\Delta H$) and melting point ($T_m$) of ultrathin fibers of PHB/hemin mixtures studied by DSC.

| Characteristic | PHB | PHB + 1% | PHB + 3% | PHB + 5% |
|---|---|---|---|---|
| **$\Delta H$, 1 scan** | 93.05 | 81.8 | 77.8 | 75.3 |
| **$\Delta H$, 2 scan** | 90.8 | 78.7 | 75.3 | 72.7 |
| **$T_m$ °C, 1 scan** | 174.5 | 172 | 172.8 | 174 |
| **$T_m$ °C, 2 scan** | 170.7 | 168.5 | 170.2 | 170.3 |
| | **Annealing in an aqueous medium at 70 °C for 0.5 h** | | | |
| **$\Delta H$, 1 scan** | 88 | 84.5 | 81.6 | 67.3 |
| **$\Delta H$, 2 scan** | 81.3 | 74.8 | 76 | 64 |
| **$T_m$ °C, 1 scan** | 175 | 172 | 172.8 | 172.5 |
| **$T_m$ °C, 2 scan** | 171 | 160 | 172 | 172 |
| | **Annealing in an aqueous medium at 70 °C for 1.5 h** | | | |
| **$\Delta H$, 1 scan** | 89.2 | 84 | 76.7 | 74.3 |
| **$\Delta H$, 2 scan** | 77.8 | 87.7 | 78.3 | 77 |
| **$T_m$ °C, 1 scan** | 173.4 | 173.3 | 172.7 | 170.7 |
| **$T_m$ °C, 2 scan** | 168 | 170.3 | 172.7 | 168 |
| | **Annealing in an aqueous medium at 70 °C for 5 h** | | | |
| **$\Delta H$, 1 scan** | 90.2 | 80.7 | 80.3 | 81 |
| **$\Delta H$, 2 scan** | 92 | 75 | 80.9 | 78.9 |
| **$T_m$ °C, 1 scan** | 174.5 | 172.5 | 174 | 174 |
| **$T_m$ °C, 2 scan** | 168.5 | 166.4 | 171.7 | 170.5 |

Depending on the chemical structure of the additive, the indicators of ΔH and Tm varied greatly, due to the different interaction of PHB and the additive [34].

It is important to note that the enthalpy of melting ΔH obtained by DSC provided information both on the fraction of the crystalline phase and on linear structures (structures of straightened chains with a two-dimensional order). An increase in hemin concentration, aggregating into large particles, increasingly loosened the PHB structure, slowing down the process of crystallization and the formation of linear structures. As a result, ΔH decreased with increasing hemin concentration in PHB (Figure 3a). Moreover, there was an imbalance between the DSC and XRD data of ~40%, which was explained by the presence of linear systems in the polymer.

For PHB/Fe(TPP)Cl fibers, the opposite was observed. Having Fe(TPP)Cl particles of small size and being crystallization nuclei, due to a fairly strong interaction with PHB, they predetermined an increase in the degree of crystallinity and the proportion of linear structures in the composition (Figure 3b).

The thermograms of the PHB/hemin samples had an asymmetric shape with a low-temperature shoulder, which was due to the formation of linear structures. The general view of the thermograms of PHB/hemin samples subjected to repeated temperature scanning (heating) differed fundamentally from the thermograms of fibers of the same composition obtained during the first heating: single melting maxima transformed into bimodal maxima. We also observed a similar melting pattern for films and fibers with chitosan and dipyridomole [8–10]. The change in thermophysical characteristics during repeated temperature scanning was most likely associated with a deterioration in the organization of its crystalline phase due to a sufficiently high cooling rate when the fiber structure did not have time to return to its original state (the fibrous material passed into a film material).

### *3.4. Exposure of Ultrathin PHB Fibers in an Aqueous Medium in the Presence of Hemin and Porphyrin Complex Fe(TPP)Cl*

Sorption and diffusion transport of water in PHB films were considered in detail earlier in a number of works, in which the diffusion coefficients and water absorption isotherms were estimated, the segmental mobility of PHB molecules was studied, and micrographs of the surface of samples obtained by SEM were presented, as well as their structural characteristics [15,35–38]. It was interesting to study the effect of exposure in an aqueous medium (distilled water) at a temperature of 70 °C on the structural and dynamic parameters of the PHB/hemin mixture composition and compare it with the regularities for the PHB/Fe(TPP)Cl system.

The structure and dynamics of medical materials to a large extent affect the diffusion processes of drugs introduced into polymers and, consequently, the kinetic patterns of drug release into the body's environment. Since the polymer in the body is in the aquatic environment, it is important to identify changes in its structural and dynamic parameters during swelling. The choice of temperature for processing fibrous materials in water at 70 °C is associated with the need to increase the hydrophilicity of PHB, since after removing most of the water by drying, a smaller part of it remains in the polymer in a bound form (hydrogen bonds). Even a slight increase in the hydrophilicity of polymer matrices affects the diffusion processes of drug transport, which, as a rule, are polar. At room temperature, the diffusion of water into the polymer takes a long time, up to several days. The plasticizing action of water, which is most intense in the amorphous regions of the fibers, in combination with elevated temperature creates favourable conditions for the completion of additional crystallization of the polymer, leads to a more ordered state of the intergranular regions and to a redistribution of the ratio between the "loose" and dense phases. Earlier studies of the effect of an aqueous medium on the structure of polymers were carried out [12,14,36]. At 70 °C, hydrogen bonds are destroyed and the process of additional orientation of straightened chains in the fiber is realized. It was necessary to establish the degree of these structural changes for the system studied in the work.

Crystallites and amorphous regions generally do not correspond to the free energy minimum. The system tends to a minimum of free energy, and this process is facilitated when the intermolecular interaction decreases (for example: the structure loosens at elevated temperature, when additives are introduced into the polymer, etc.). Loose, amorphous regions relax to the state of a supercooled melt, and crystalline regions relax to a minimum of free energy through a change in size and, first of all, longitudinal dimensions ($\Delta G = \Delta H - T\Delta S$). It should be noted that amorphous regions in polymers are heterogeneous, there are regions with a high degree of straightening and regions with folded macromolecules. Under conditions where steric hindrance is removed (e.g., at elevated temperature), highly straightened chains assume more straightened conformations.

Water molecules act on the polymer structure in two ways: on the one hand, they have a plasticizing effect, which allows the crystallites to increase their size, on the other hand, after the removal of water, hydrated complexes remain in the fiber, loosening the polymer structure. Water molecules, hydrated complexes penetrate into the end surfaces of crystallites and linear structures, partially destroying them and, as a result, the degree of crystallinity decreases, as was shown in [15]. The accumulation of hydrated complexes in the amorphous component leads to its decompaction. It was necessary to establish the extent of these structural changes. At different exposure times in samples containing the addition of hemin and Fe(TPP)Cl at different concentrations, the intensity of the listed processes is different.

Let us first consider the effect of water treatment at 70 °C on the crystal structure of the PHB/hemin fiber (Table 1). Figure 4 presents the data on the change in the enthalpy of melting of this system on the concentration of hemin at different exposure times in an aqueous medium. An increase in ΔH in mixed compositions compared to the original untreated in an aqueous medium was observed at all studied times of water-temperature exposure to the polymer (with the exception of fibers with 5% hemin). In PHB fibers, the enthalpy of melting decreased. The research results indicated the contradictory nature of the impact of the aquatic environment on the fibers of PHB and PHB with complexes.

As previously noted, water molecules affected the polymer structure in two ways: on the one hand, they had a plasticizing effect, which allowed crystallites to increase in size. On the other hand, after the removal of water, hydrated complexes remained in the fiber, which are formed between the polar groups of PHB and hemin, as well as between hemin molecules, PHB groups. As a result, the fiber structure became looser and the enthalpy of the system decreased.

The increase in ΔH in PHB fibers with hemin after exposure can be explained by the predominance of plasticization processes and, as a consequence, an increase in the proportion of straightened chains in the fiber and, upon longer exposure, the formation of hydrated complexes that loosened the structure of crystalline regions (end surfaces of crystallites, linear structures). Hemin, due to large particle sizes, greatly loosens the structure of PHB, which leads to a significant concentration of water molecules in the fiber, the plasticizing factor also increases and, as a result, ΔH increases. In samples with 5% hemin, loosening was maximal and represented the highest increase in the melting enthalpy in them(Figure 5) It can be assumed that hemin, due to large particle sizes, does not deform the end surfaces of crystallites and linear systems; therefore, at a 5% additive in the polymer, the maximum opportunity for a certain proportion of macromolecules to straighten is realized; stress relaxation occurs in the nonequilibrium supramolecular structure of the fiber, which leads to additional densification of amorphous intercrystalline regions of their packing into crystallites.

Let us now consider the dynamic characteristics of the amorphous phase of ultrathin PHB/hemin fibers after water-temperature treatment. The plasticizing effect of water, which is most intense in the amorphous regions of the polymer, in combination with elevated temperature, created favorable conditions for the completion of its additional crystallization and for a more ordered state of the intercrystalline regions. After the removal of water, hydrated complexes remained in the fiber, which loosened the amorphous structure to a greater extent than non-hydrated particles; therefore, a sharp decrease in the correlation time was observed.

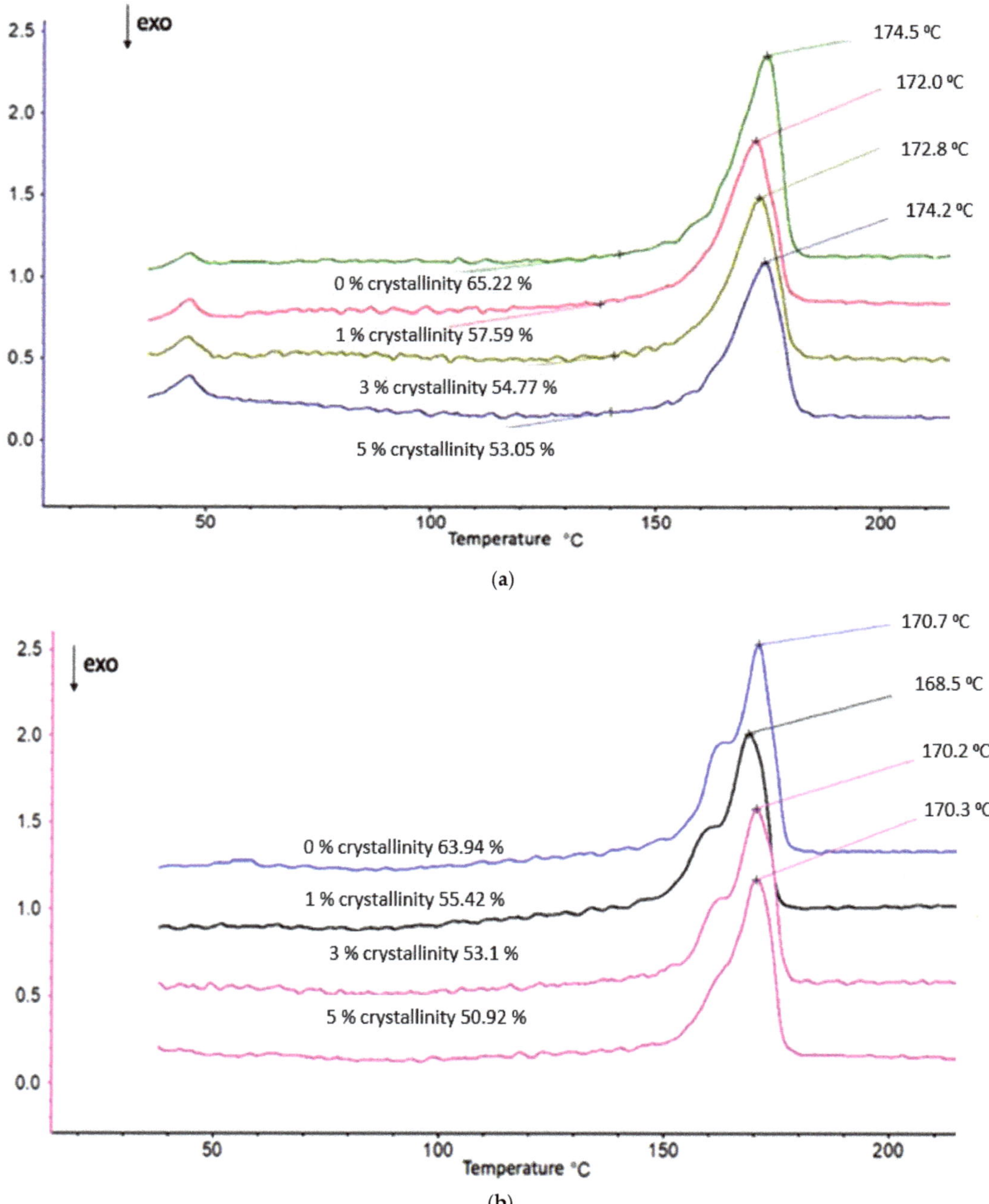


**Figure 4.** Thermograms of PHB/hemin melting (**a**) 1st and (**b**) 2nd scanning.

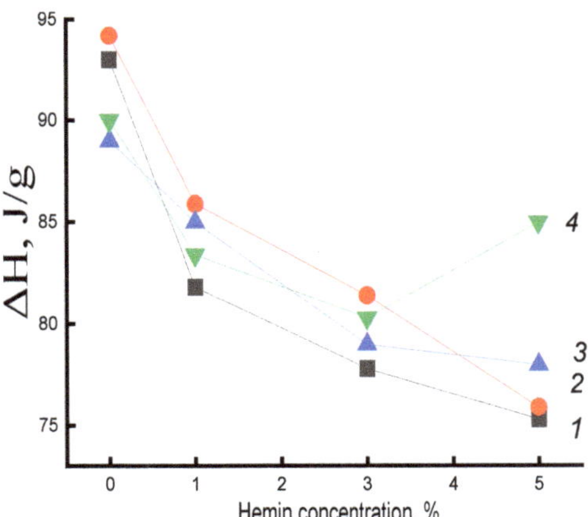


**Figure 5.** Dependence of the enthalpy of melting ΔH on the composition of the system for PHB fibers with hemin. Exposure time in the aquatic environment at 70 °C: 1—0, 2—30, 3—90, 4—300 min.

The specificity of this work was associated with a qualitative analysis of changes in the mobility of radicals during one or another treatment of the polymer matrix, and not with the interpretation of the exact values of the parameters of rotational mobility. For this reason, we needed to introduce a parameter that would qualitatively characterize the rotational mobility of radicals, be determined without a numerical analysis of all points of the spectrum, and have sufficient sensitivity to small changes in the shape of the spectrum, such as the changes shown in Figure 6. It can be seen that the greatest difference in the spectra was observed in the region of the high-field and low-field components; therefore, as the parameter of the spectrum shape, we chose the characteristic rotational correlation time calculated by the well-known formula [38,39]. This formula was proposed to calculate the rotational correlation times of nitroxyl radicals of the piperidine series in the case of their isotropic rotation in the region of rotational correlation times $5 \times 10^{-11} \leq \tau \leq 1 \times 10^{-9}$ s. In this region, the EPR spectrum consisted of three well-resolved components, the width of which was described in terms of the Redfield theory. In our case, this parameter qualitatively characterized the line shape of the EPR spectrum and could be considered as some characteristic correlation time.

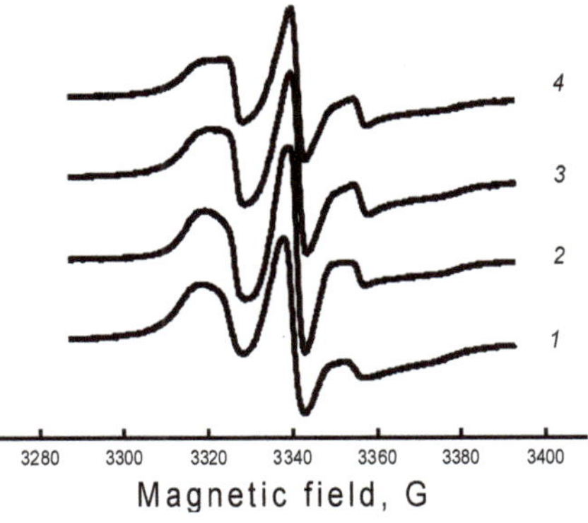


**Figure 6.** EPR spectra of the TEMPO nitroxide radical for PHB samples with hemin concentration: 1—0%, 2—1%, 3—3%, 4—5%.

Figure 7a shows the change in the time of rotational correlation of radicals with an increase in the concentration of hemin complexes. The greatest intensity of the decrease in τ in the system was in the concentration up to 1% hemin. A further increase in the concentration of hemin in the fiber also caused an increase in molecular mobility, but not so significant. A similar pattern of change in χ and ΔH was observed with an increase in hemin concentration, which indicated the decompression of amorphous and crystalline regions upon the addition of hemin to the PHB structure.

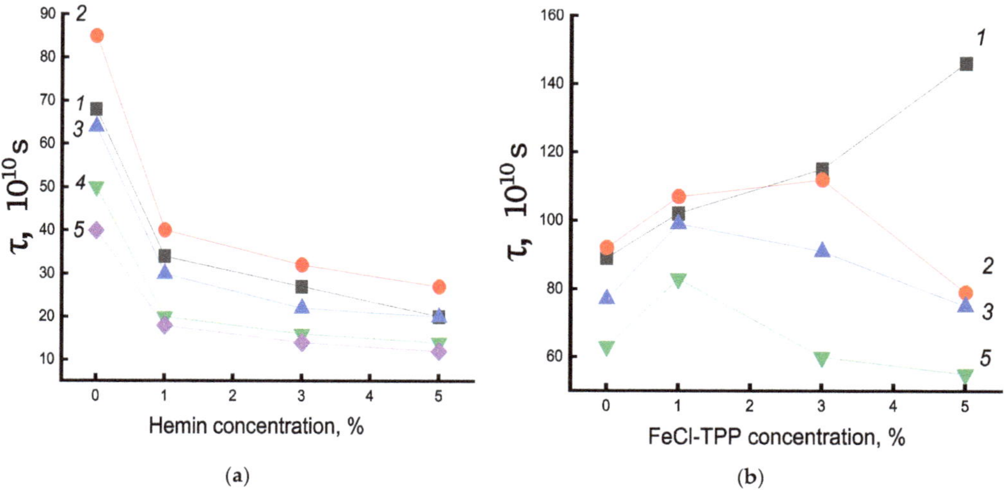


**Figure 7.** Dependence of τ on the composition of (**a**) PHB/hemin and (**b**) PHB/Fe(TPP)Cl. Exposure time in the aquatic environment at 70 °C: 1—0, 2—30, 3—50, 4—90, 5—300 min.

In contrast to the observed changes in the structure of the amorphous component of the fiber with hemin, in the PHB/Fe(TPP)Cl system, in which the additive particles are small, opposite patterns of change in τ are observed (Figure 7b). Fe(TPP)Cl particles, being crystallization nuclei, promoted an increase in the degree of crystallinity and, as a result, a significant increase in the proportion of straightened macromolecules, which led to a decrease in molecular dynamics [12,35].

Let us now consider the effect of exposure to an aqueous medium at 70 °C on the molecular mobility of the PHB/hemin system. After 30 min exposure in an aqueous medium, τ increased for all systems, which indicated a slowdown in molecular mobility. However, with an increase in the time of water-temperature exposure, the mobility of chains in the amorphous phase increased significantly for all compositions of the system, although the enthalpy of melting remained higher than for untreated fibers. It can be assumed that as the density of amorphous regions (with increasing crystallinity) also increased, the concentration of the radical in them also decreased and, as a result, the molecular mobility increased. Characteristically, the most dramatic change in molecular dynamics and melting enthalpy was observed after the addition of 1% hemin at all times of water-temperature exposure. Apparently, the concentration of hemin increased not so sharply due to the formation of ever larger particles with an increase in the concentration of the additive.

The explanation of the obtained data was the fact that hemin, due to the large particle size, strongly loosened the PHB structure, which led to a significant concentration of water molecules in the fiber and to two opposite processes: not only the plasticizing factor increases, leading to the growth of straightened chains in the fiber, but to an increase in the concentration of hydrated complexes and, as a result, the structure of the amorphous phase was loosened. In our case, at all processing times, as shown by DSC and τ data (up to 30 min), the first factor prevailed. The decrease in the correlation time for all compositions of the system at treatment times of more than 30 min was explained by the ever lower concentration of the radical in the dense regions of the polymer, which were formed under water-temperature exposure, as well as by the loosening of the amorphous phase after the formation of hydrated complexes, the concentration of which increased with the time of fiber treatment. To investigate how the density of the fiber changed, data were obtained on the equilibrium concentration of the radical adsorbed in samples of the studied compositions of the same mass using the software from Bruker (winer). Figure 8 shows the dependences of the concentration of the radical (C) on the composition of the fibers for a series of exposure times. As expected, the concentration of the radical increased with increasing additive. After water-temperature treatment in PHB samples, the concentration increased with increasing exposure time, which was consistent with the increase in the fraction of the amorphous component according to DSC data. In fibers with hemin, the concentration of the radical decreased (deviation 3%). The addition of 5% hemin was accompanied by the highest increase in ΔH with increasing exposure time; the strongest changes in the radical concentration are observed precisely in these fibers. Therefore, it can be argued that despite the formation of hydrated complexes that loosened the polymer structure, the processes of plasticization of the structure leading to an increase in the enthalpy of melting, proceeding with the compaction of the material, prevailed.

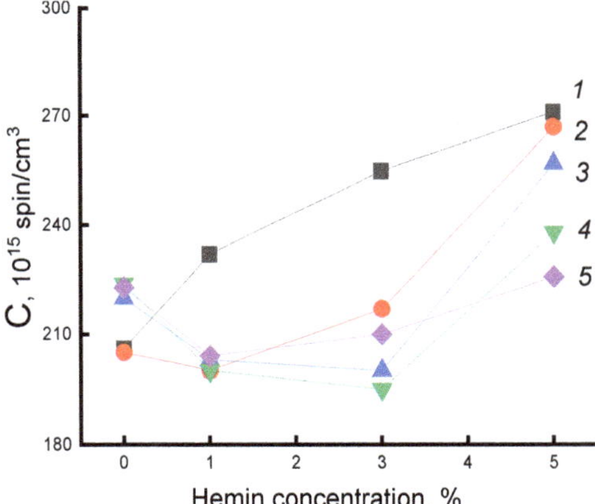


**Figure 8.** Dependence of the C radical concentration in PHB/hemin on the concentration of the additive. Exposure time in the aquatic environment at 70 °C: 1—0, 2—30, 3—50, 4—90, 5—300 min.

Let us now consider the change in molecular dynamics under water-temperature treatment of PHB/Fe(TPP)Cl fibers. The dependence of molecular dynamics on the additive concentration after exposure to an aqueous medium was extreme. At an additive concentration of 1%, there was a sharp increase in $\tau$. It was in these polymers that a high proportion of straightened chains was evidenced by a sharp increase in $\chi$ and $\Delta H$ with an increase in the concentration of the porphyrin complex. Therefore, at a low concentration of this additive, the plasticizing effect prevailes. As a result, mobility slowed down. At higher concentrations of Fe(TPP)Cl, the processes loosening the structure due to the formation of hydrated complexes began to predominate to an increasing extent, and the molecular mobility increased. Similar processes took place with an increase in the time of water-temperature treatment. The concentration of the radical in the fibers after exposure was significantly lower than in the original samples. For example, in the initial sample of PHB-porphyrin 5% concentration of the radical $9.3 \times 10^{15}$, and after annealing—$3.1 \times 10^{15}$.

Thus, the hydrophilization of the studied systems depended both on the composition of the composition and on the type of additive in PHB, as well as on the time of water-temperature treatment. The sorption capacity, due to the high affinity of water for Fe(TPP)Cl and hemin molecules, increased sharply compared to PHB, in this case, the diffusion coefficients of water and low molecular weight compounds increased exponentially.

Studies of the structural and dynamic characteristics of fibrous materials PHB/hemin and PHB/FeClTPP after exposure to an aqueous medium at 70 °C showed that:

1. The enthalpy of melting of PHB/hemin fibers at all times of water-temperature treatment (up to 5 h) and for all concentrations of hemin increased, which was due to the predominance of plasticization processes and structure compaction over loosening processes due to the formation of hydrated complexes in the polymer. The molecular mobility of the probe, and hence the mobility of macromolecules, slowed down after holding the fiber in an aqueous medium for 30 min due to an increase in the enthalpy of melting (according to DSC data). At higher times of such an impact on the system, the molecular dynamics began to increase, which was caused by the predominance of structure loosening processes during the formation of hydrated complexes. The radical was sorbed at a lower and lower concentration in amorphous regions with increasing exposure time in an aqueous medium due to an increase in density.
2. Opposite regularities are observed when the water-temperature effect iwas applied to the PHB/Fe(TPP)Cl system. The presence of the complex determined the complex dependence of $\tau$ on its concentration in the fiber after exposure to water. The segregation of Fe(TPP)Cl molecules into small particles determined the structural and dynamic changes not only with an increase in the concentration of the additive, but also during water-temperature treatment of the polymer.

It is well known that water, penetrating into the polymer matrix, can affect both the physicochemical and mechanical properties of polymers and the flows of the third low molecular weight component (electrolyte, drug substance) [38].

Additional information on the dynamic behaviour of the PHB/hemin system of various compositions was obtained by studying the temperature dependence of the radical rotation rate and determining the corresponding activation energies $E_\tau$. A characteristic feature of $E_\tau$ as a function of the percentage of hemin in PHB was a sharp difference between these values in mixed compositions. This parameter decreased with increasing hemin concentration: 0%—50, 1%—28, 3%—22, 5%—18 kJ/mol. Such a sharp decrease in the activation energy of probe rotation upon the introduction of hemin was associated with a change in the state of the intercrystalline polymer phase; the density of the amorphous phase decreased significantly with an increase in the hemin concentration.

### *3.5. Effect of Ozonation on the Dynamics of Radical Rotation in PHB/Hemin and PHB/FeClTPP Composition Fibers*

When biomedical materials are used, along with mechanical and thermal effects, their structure and segmental mobility are affected by ozone. Two sources of its appearance in the atmosphere should be indicated here. Firstly, ozone is formed during the operation of powerful electrical devices that ensure the vital activity of patients both during surgical operations and during therapy or monitoring in a hospital. Secondly, ozone in some special cases continues to be used in the sterilization of medical devices. Despite the daily contact of this aggressive compound with polymers, its effect on their morphological and dynamic characteristics remains a little-studied area of polymer materials science.

In the process of ozonation, two opposite processes take place that affect the molecular dynamics in the polymer: they are the breakage of macromolecules, which contributes to an increase in chain mobility and the formation of oxygen-containing groups on the side chains of macromolecules, which contributes to an increase in the rigidity of macromolecules and, as a result, a slowdown in molecular dynamics. The loops of macromolecules on the end surfaces of the crystallites are oxidized at the highest rate, since the highest stresses and, consequently, the highest oxidation rate are concentrated in these areas [12,35]. The rupture of these loops is accompanied by the straightening of regions of macromolecules, which slows down the molecular mobility in these regions. The formation of oxygen-containing groups in the side chains of macromolecules leads to an increase in intermolecular interaction and, as a result, the molecular mobility of the chains also slows down, whereas the rigidity of the chains increases and the processes of chain reorientation take place, which also leads to a slowdown in their mobility. In parallel, the processes of chain destruction proceed, which, as a result, can cause both the reorientation of chains and their folding into coils, depending on the initial degree of straightening of macromolecules. The tendency of a macromolecule to unfold and reorient itself or to coil into a ball is determined by the conformational criterion $k^* = h/L = \sqrt{2}/k$ (where h is the average distance between the ends of the chains, L is the contour length of the chain, k is the number of segments). Chains in which $k > k^*$ tend to reorientate, macromolecules in which $k < k^*$ tend to take the coil conformation.

Figure 8 shows the dependences of the rotational correlation time of the spin probe on the time of ozonation of the PHB/hemin and PHB/Fe(TPP)Cl systems. It can be seen that, firstly, the dependences were extreme, and, secondly, the change in the mobility of radicals with increasing ozonation time largely depended on the concentration of the additive and on the time of ozone-oxygen exposure of the polymer.

The extreme nature of the dependencies was due to the following factors. In PHB/hemin fibers (up to 30 min) and in PHB/Fe(TPP)Cl (up to 100 min), processes leading to a slowdown in molecular dynamics predominated. Moreover, if such changes in the PHB/hemin fiber were not so significant, then in the PHB/Fe(TPP)Cl system this effect was much greater, which indicated the predominance of processes leading to an increase in the rigidity of the polymer matrix. Let us consider the reason for the sharp difference in the patterns of change in molecular dynamics from the time of ozonation. The addition of Fe(TPP)Cl, as noted earlier, caused a significant increase in the degree of crystallinity and linear systems (these particles are the nuclei of crystallization), as a result, the proportion of straightened chains in the fiber greatly increased and, with an increase in the proportion of the composite, the $\chi$ and $\Delta H$. The accumulation of oxygen-containing groups on such chains was accompanied by a significant increase in their rigidity and an increase in intermolecular interaction, as a result of which the coefficient k* increased and some of the macromolecules straightened, as a result, the molecular dynamics slowed down. Rupture of sufficiently straightened chains, which at the same time reorient themselves and their mobility also decreased. Another aspect leading to an increase in $\tau$ was the breaking and straightening of the chains on the end surfaces of the crystallites (due to the highest stresses on the chain folds). All these components caused a sharp increase in $\tau$ at the initial stage of ozonation of PHB/Fe(TPP)Cl fibers. With a longer exposure to ozone, destruction processes began to predominate with an increase in molecular mobility. With an increase in the concentration of the additive, the proportion of straightened chains, the proportion of crystallites increased, which caused an increasingly stronger growth effect $\tau$ with a change in the concentration of the additive (Figure 9).

In contrast to the Fe(TPP)Cl additive, hemin loosened the polymer structure to an increasing extent with increasing concentration. Against the background of a loose structure, the growth of intermolecular interaction during the accumulation of oxygen-containing groups did not have such a significant effect, and the fraction of end surfaces of crystallites also decreased with an increase in the fraction of hemin. As a result, a weak growth of $\tau$ is observed at the initial stage of ozonation. At deeper oxidation states, the molecular dynamics began to increase, and only after ozonation for more than 60 min did the chain mobility begin to decrease, which was apparently caused by the formation of a high concentration of oxygen-containing groups.

In general, the effect of ozone on the structural and dynamic characteristics of PHB doped with porphyrin metal complexes largely depended on the nature of the additive [15,36,38,40,41].

Thus, when comparing the results of ozonation of two polymers PHB/hemin and PHB/Fe(TPP)Cl, it follows that the structure that was formed upon addition of the complex determined to a large extent the ability of the system to change the dynamic parameters from the time of ozonation.

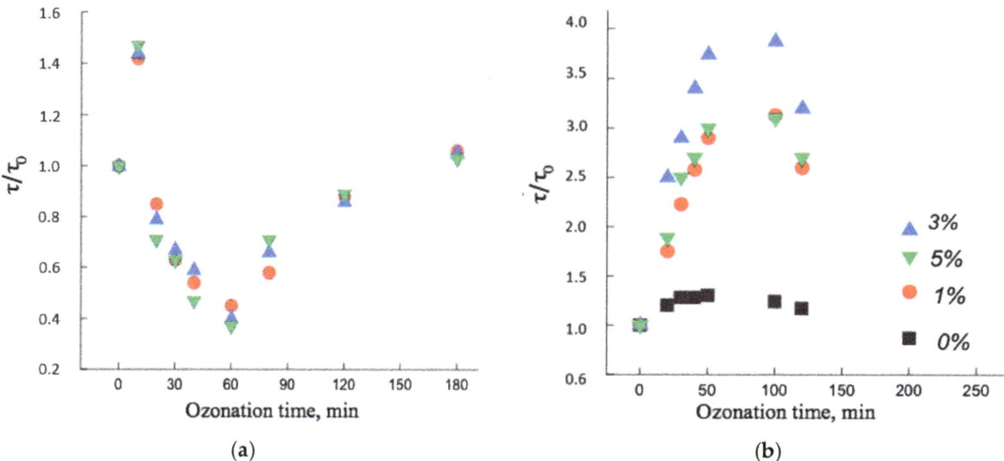


**Figure 9.** Change in molecular dynamics $\tau/\tau_0$ as a function of ozonation time in PHB fibers with metal complexes: (**a**) hemin, (**b**) Fe(TPP)Cl: 1—0%, 2—1%, 3—3%, 4—5%.

## 4. Conclusions

In this work, based on structural-dynamic studies, combining EPR, DSC and X-ray diffraction analysis, the results of the effect of low concentrations (1, 3, 5%) of hemin and the Fe(TPP)Cl complex on the degree of crystallinity, melting enthalpy and molecular dynamics of chains in amorphous regions of ultrafine fibers are presented. It is shown that the particle size of the additive and its chemical structure play an important role in the structuring of the fiber material. The intermolecular interaction between porphyrin complexes and the interaction of complexes with PHB macromolecules determine the degree of structural changes in the fiber. The addition of hemin to PHB fibers causes a strong loosening effect on both crystalline and amorphous regions of the polymer. The mutual influence of crystalline and amorphous regions in biodegradable highly crystalline polymers and their compositions remains a rather complex and little-studied problem of modern polymer materials science. These first studies have made it possible to interpret at the molecular level the effect of a number of aggressive factors (such as water-temperature and ozone effects) on the structural and dynamic characteristics of PHB–porphyrin complex fibers. Exposure to the aquatic environment and exposure to ozone in the gas phase greatly changes the structure of the fibers, and the observed changes depend on the nature and amount of the additive.

**Author Contributions:** Conceptualization, S.G.K. and I.A.V.; investigation, S.G.K., P.M.T. and I.A.V.; resources, I.A.V. and P.M.T.; data curation, A.A.O.; writing—original draft preparation, S.G.K. and A.A.P.; writing—review and editing, A.A.P. and I.A.V.; visualization, S.G.K.; supervision, S.G.K.; project administration, I.A.V.; funding acquisition, I.A.V. All authors have read and agreed to the published version of the manuscript.

**Funding:** This research received no external funding.

**Institutional Review Board Statement:** Not applicable.

**Acknowledgments:** The study was carried out using scientific equipment of the Center of Shared Usage «New Materials and Technologies» of Emanuel Institute of Biochemical Physics and the Common Use Centre of Plekhanov Russian University of Economics. The authors express their gratitude for the research conducted by P.M. Tyubaeva.

**Conflicts of Interest:** The authors declare no conflict of interest.

## References


1. Almaguer-Flores, A.; Silva-Bermúdez, P.; Rodi, S.E. Nanostructured biomaterials with antimicrobial activity for tissue engineering. In *Nanostructured Biomaterials for Regenerative Medicine*; Woodhead Publishing: Sawston, UK, 2020; pp. 81–137. [CrossRef]
2. Chifiriuc, M.C.; Ficai, A.; Grumezescu, A.M.; Ditu, L.M.; Popa, M.; Iordache, C.; Lazar, V. Chapter 1—Soft tissue engineering and microbial infections: Chal-lenges and perspectives. In *Nanobiomaterials in Soft Tissue Engineering*; William Andrew: Norwich, NY, USA, 2016; pp. 1–29. [CrossRef]
3. Lund, A.; van der Velden, N.M.; Persson, N.-K.; Hamedi, M.M.; Müller, C. Electrically conducting fibres for e-textiles: An open playground for conjugated polymers and carbon nanomaterials. *Mater. Sci. Eng. R Rep.* **2018**, *126*, 1–29. [CrossRef]
4. Tabassum, R.; Kant, R. Recent trends in surface plasmon resonance based fiber–optic gas sensors utilizing metal oxides and carbon nanomaterials as functional entities. *Sens. Actuators B Chem.* **2020**, *310*, 127813. [CrossRef]
5. Laskowska, M.; Kityk, L.; Pastukh, O.; Dulski, M.; Zubko, M.; Jedryka, J.; Laskowski, E. Nanocomposite for photonics—Nickel pyrophosphate nanocrystals synthesised in silica nanoreactors. *Microporous Mesoporous Mater.* **2020**, *306*, 110435. [CrossRef]
6. Chen, Z.; Mai, B.; Tan, H.; Chen, X. Nucleic acid based nanocomposites and their applications in biomedicine. *Compos. Commun.* **2018**, *10*, 194–204. [CrossRef]
7. Ghosal, K.; Agatemor, C.; Špitálsky, Z.; Thomas, Z.S.; Kny, E. Electrospinning Tissue Engineering and Wound Dressing Scaffolds from Polymer–Titanium Dioxide Nanocomposites. *Chem. Eng. J.* **2019**, *358*, 1262–1278. [CrossRef]
8. Karpova, S.G.; Ol'khov, A.A.; Shilkina, N.G.; Kucherenko, E.L.; Iordanskii, A.L. Influence of Drug on the Structure and Segmental Mobility of Poly(3-Hydroxybutyrate) Ultrafine Fibers. *Polym. Sci. Ser. A* **2017**, *59*, 58–66. [CrossRef]
9. Iordanskii, A.L.; Ol'khov, A.A.; Karpova, S.G.; Kucherenko, E.L.; Kosenko, R.Y.u.; Rogovina, S.Z.; Chalykh, A.E.; Berlin, A.A. Influence of the Structure and Morphology of Ultrathin Poly(3-hydroxybutyrate) Fibers on the Diffusion Kinetics and Transport of Drugs. *Polym. Sci. Ser. A* **2017**, *59*, 352–362. [CrossRef]
10. Karpova, S.G.; Ol'khov, A.A.; Iordanskii, A.L.; Gumargalieva, K.Z.; Berlin, A.A. Nonwoven Blend Composites Based on Poly(3-hydroxybutyrate)–Chitosan Ultrathin Fibers Prepared via Electrospinning. *Polym. Sci. Ser. A* **2016**, *58*, 76–86. [CrossRef]
11. Siracusa, V.; Ingrao, C.; Karpova, S.G.; Olkhov, A.A.; Iordanskii, A.L. Gas transport and characterization of poly(3-hydroxybutyrate) films. *Eur. Polym. J.* **2017**, *91*, 149–161. [CrossRef]
12. Karpova, S.G.; Olkhov, A.A.; Chvalun, S.N.; Bakirov, A.V.; Shilkina, N.G.; Popov, A.A. Poly(3-hydroxybutyrate) Matrices Modified with Iron(III) Comple-xes with Tetraphenylporphyrin. Analysis of the Structural Dynamic Parameters. *Russ. J. Phys. Chem. B* **2018**, *12*, 142–153. [CrossRef]
13. Karpova, S.G.; Ol'khov, A.A.; Krivandin, A.V.; Shatalova, O.V.; Lobanov, A.V.; Popov, A.A.; Iordanskii, A.L. Effect of Zinc–Porphyrin Complex on the Structure and Properties of Poly(3-hydroxybutyrate) Ultrathin Fibers. *Polym. Sci. Ser. A* **2019**, *61*, 70–84. [CrossRef]
14. Karpova, S.G.; Ol'khov, A.A.; Lobanov, A.V.; Popov, A.A.; Iordanskii, A.L. Biological Compositions of Ultrathin Poly-3-Hydroxybutyrate Fibers with MnCl–Tetraphenylporphyrin Complexes. Dynamics, Structure and Properties. *Nanotechnol. Russ.* **2019**, *14*, 132–143. [CrossRef]
15. Karpova, S.G.; Ol'khov, A.A.; Zhul'kina, A.L.; Popov, A.A.; Iordanskii, A.L. Nonwoven Materials Based on Electrospun Ultrathin Fibers of Poly(3-hydro-xybutyrate) and Complex Tin Chloride–Porphyrin. *Polym. Sci. Ser. A* **2021**, *63*, 369–381. [CrossRef]
16. Altaee, N.; El-Hiti, G.A.; Fahdil, A.; Sudesh, K.; Yousif, E. Biodegradation of different formulations of polyhydroxybutyrate films in soil. *Springer Plus* **2016**, *5*, 762. [CrossRef] [PubMed]
17. Gradova, M.A.; Zhdanova, K.A.; Bragina, N.A.; Lobanov, A.V.; Mel'nikov, M.Y. Aggregation state of amphiphilic cationic tetraphenylporphyrin derivatives in aqueous microheterogeneous systems. *Russ. Chem. Bull.* **2015**, *64*, 806–811. [CrossRef]
18. Amati, A.; Cavigli, P.; Demitri, N.; Natali, M.; Indelli, T.M.; Iengo, E. Sn(IV) Multiporphyrin Arrays as Tunable Photoactive Systems. *Am. Chem. Soc. Inorg. Chem.* **2019**, *58*, 4399–4411. [CrossRef]
19. Kim, H.; Kim, W.; Mackeyev, Y.; Lee, G.-S.; Kim, H.-J.; Tachikawa, T. Selective Oxidative Degradation of Organic Pollutants by Singlet Oxygen-Mediated Photosensitization: Tin Porphyrin versus C60 Aminofullerene Systems. *Environ. Sci. Technol.* **2012**, *46*, 9606–9613. [CrossRef] [PubMed]
20. Huang, H.; Chauhan, S.; Geng, J.; Qin, Y.; Watson, D.F. Implantable Tin Porphyrin-PEG Hydrogels with pH-Responsive Fluorescence. *Biomacromolecules* **2017**, *18*, 562–567. [CrossRef] [PubMed]
21. Nitzan, Y.; Ladan, H.; Gozansky, S.; Malik, Z. Characterization of hemin antibacterial action on Staphylococcus aureus. *FEMS Microbiol. Lett.* **1987**, *48*, 401–406. [CrossRef]
22. Walker, B.W.; Lara, R.P.; Mogadamd, E.; Yub, C.H.; Kimball, W.; Annabi, N. Rational design of microfabricated electroconductive hydrogels for biomedical applications. *Prog. Polym. Sci.* **2019**, *92*, 135–157. [CrossRef] [PubMed]
23. Lu, Y.; Berry, S.M.; Pfister, T.D. Engineering Novel Metalloproteins: Design of Metal-Binding Sites into Native Protein Scaffolds. *Chem. Rev.* **2001**, *101*, 3047–3080. [CrossRef] [PubMed]
24. Zhang, Y.; Xu, C.; Li, B. Self-assembly of hemin on carbon nanotube as highly active peroxidase mimetic and its application for biosensing. *RSC Adv.* **2013**, *3*, 6044–6055. [CrossRef]
25. Qu, R.; Shen, L.; Chai, Z.; Jing, C.; Zhang, Y.; An, Y.; Shi, L. Hemin-Block Copolymer Micelle as an Artificial Peroxidase and Its Applications in Chromo-genic Detection and Biocatalysis. *ACS Appl. Mater. Interfaces* **2014**, *6*, 19207–19216. [CrossRef] [PubMed]
26. Adler, A.D.; Longo, F.R.; Kampas, F.; Kim, J. On the preparation of metalloporphyrins. *J. Inorg. Nucl. Chem.* **1970**, *32*, 2443–2445. [CrossRef]

27. Stephen, Z.D. *Handbook of Thermal Analysis and Calorimetry, Applications to Polymers and Plastics*; Elsevier: Amsterdam, The Netherlands; Boston, MA, USA; London, UK, 2002; Volume 3.
28. Katsogiannis, K.A.; Vladisavljević, G.T.; Georgiadou, S. Porous electrospun polycaprolactone (PCL) fibres by phase separation. Katsogiannis, K.A.G.; Vladi-savljević, G.T.; Georgiadou, S. *Eur. Polym. J.* **2015**, *69*, 284–295. [CrossRef]
29. Renekera, D.H.; Yarinbc, A.L.; Zussmanb, E.E.; Xuad, H. Electrospinning of Nanofibers from Polymer Solutions and Melts. *Adv. Appl. Mech.* **2007**, *41*, 43–195. [CrossRef]
30. Baumgarten, P. Electrostatic spinning of acrylic microfibers. *J. Colloid Interface Sci.* **1971**, *36*, 71–91. [CrossRef]
31. Lobanov, A.V.; Kholuiskaya, S.N.; Komissarov, G.G. The $H_2O_2$ as donor of electrons in catalytic reduction of inorganic carbon. *Russ. J. Phys. Chem. B* **2004**, *23*, 44–48.
32. Lobanov, A.V.; Kholuiskaya, S.N.; Komissarov, G.G. Photocatalytic synthesis of formaldehyde from $C_{O2}$ and $H_2O_2$. *Dokl. Phys. Chem.* **2004**, *399*, 266–268. [CrossRef]
33. Lobanov, A.V.; Gromova, G.A.; Gorbunova, Y.G.; Tsivadze, A.Y. Supramolecular associates of double-decker lanthanide phthalocyanines with macromolecular structures and nanoparticles as the basis of biosensor devices. *Prot. Met. Phys. Chem. Surf.* **2014**, *50*, 570–577. [CrossRef]
34. Bossu, J.; Angellier-Coussy, H.; Totee, C.; Matos, M.; Reis, M.; Guillard, M. Effect of the molecular structure of Poly(3-hydroxybutyrate-co-3-hydroxyvalerate)(P (3HB-3HV)) produced from mixed bacterial cultures on its crystallization and mechanical properties. *Biomacromolecules* **2020**, *21*, 4709–4723. [CrossRef] [PubMed]
35. Karpova, S.G.; Ol'khov, A.A.; Chvalun, S.N.; Tyubaeva, P.M.; Popov, A.A.; Iordanskii, A.L. Comparative Structural Dynamic Analysis of Ultrathin Fibers of Poly-(3-Hydroxybutyrate) Modified by Tetraphenyl–Porphyrin Complexes with the Metals Fe, Mn, and Zn. *Nanotechnol. Russ.* **2019**, *14*, 367–377. [CrossRef]
36. Karpova, S.G.; Ol'khov, A.A.; Tyubaeva, P.M.; Shilkina, N.G.; Popov, A.A.; Iordanskii, A.L. Composite Ultrathin Fibers of Poly-3-Hydroxybutyrate and a Zinc Porphyrin: Structure and Properties. *Russ. J. Phys. Chem. B* **2019**, *13*, 313–327. [CrossRef]
37. Karpova, S.G.; Olkhova, A.A.; Popov, A.A.; Iordanskii, A.L.; Shilkina, N.G. Characteristics of the Parameters of Superfine Fibers of Poly(3-hydroxybutyrate) Modified with Tetraphenylporphyrin. *Inorg. Mater. Appl. Res.* **2021**, *12*, 44–54. [CrossRef]
38. Olkhov, A.A.; Tyubaeva, P.M.; Vetcher, A.A.; Karpova, S.G.; Kurnosov, A.S.; Rogovina, S.Z.; Iordanskii, A.L.; Berlin, A.A. Aggressive Impacts Affecting the Biodegradable Ultrathin Fibers Based on Poly(3-Hydroxybutyrate), Polylactide and Their Blends: Water Sorption, Hydrolysis and Ozonation. *Polymers* **2021**, *13*, 941. [CrossRef] [PubMed]
39. Buchachenko, A.L.; Wasserman, A.M. *Stable Radicals*; Chemistry Publishing: Moscow, Russia, 1973; p. 408.
40. Karpova, S.G.; Iordanskii, A.L.; Klenina, N.S.; Popov, A.A.; Lomakin, S.M.; Shilkina, N.G.; Rebrov, A.V. Changes in the structural parameters and molecular dynamics of polyhydroxybutyrate-chitosan mixed compositions under external influences. *Russ. J. Phys. Chem. B* **2013**, *7*, 225–231. [CrossRef]
41. Karpova, S.G.; Iordanskii, A.L.; Popov, A.A.; Shilkina, N.G.; Lomakin, S.M.; Shcherbin, M.A.; Chvalun, S.N.; Berlin, A.A. Effect of external influences on the structural and dynamic parameters of polyhydroxybutyrate-hydroxyvalerate-based biocomposites. *Russ. J. Phys. Chem. B* **2012**, *6*, 72–80. [CrossRef]

MDPI

Review

# A Review on Fully Bio-Based Materials Development from Polylactide and Cellulose Nanowhiskers

**Purba Purnama [1,2,*], Muhammad Samsuri [3] and Ihsan Iswaldi [1]**

1 School of Applied STEM, Universitas Prasetiya Mulya, Tangerang 15339, Banten, Indonesia
2 Vanadia Utama Science and Technology, PT Vanadia Utama, Jakarta 14470, Indonesia
3 Chemical Engineering Department, Universitas Bhayangkara Jakarta Raya, Bekasi 17121, West Java, Indonesia
* Correspondence: purbapur@gmail.com

**Abstract:** This review covers the development of eco-friendly, bio-based materials based on polylactide (PLA) and cellulose nanowhiskers (CNWs). As a biodegradable polymer, PLA is one of the promising materials to replace petroleum-based polymers. In the field of nanocomposites, CNWs offer many advantages; they are made from renewable resources and exhibit beneficial mechanical and thermal properties in combination with polymer matrix. A wide range of surface modifications has been done to improve the miscibility of CNW with the PLA homopolymer, which generally gives rise to hydrophobic properties. PLA–CNW nanocomposite materials are fully degradable and sustainable and also offer improved mechanical and thermal properties. Limitations pertaining to the miscibility of CNWs with PLA were solved through surface modification and chemical grafting on the CNW surfaces. Further development has been done by combining PLA-based material via stereocomplexation approaches in the presence of CNW particles, known as bio-stereo-nanocomposite PLA–CNW. The combination of stereocomplex crystalline structures in the presence of well-distributed CNW particles produces synergetic effects that enhance the mechanical and thermal properties, including stereocomplex memory (melt stability). The bio-based materials from PLA and CNWs may serve as eco-friendly materials owing to their sustainability (obtained from renewable resources), biodegradability, and tunability properties.

**Keywords:** polylactide; cellulose nanowhiskers; biopolymer; nanocomposite; stereocomplex; interfacial compatibility; nucleating agent

**Citation:** Purnama, P.; Samsuri, M.; Iswaldi, I. A Review on Fully Bio-Based Materials Development from Polylactide and Cellulose Nanowhiskers. *Polymers* **2022**, *14*, 4009. https://doi.org/10.3390/polym14194009

Academic Editors: José Miguel Ferri, Vicent Fombuena Borràs and Miguel Fernando Aldás Carrasco

Received: 29 August 2022
Accepted: 19 September 2022
Published: 25 September 2022



## 1. Introduction

Petroleum-based and bio-based polymers are widely used for various purposes. Petroleum-based polymers, including polyethylene (PE), polypropylene (PP), polystyrene (PS), high-impact polystyrene (HIPS), polyvinyl chloride (PVC)polyethylene terephthalate (PET), polycarbonate (PC); are widely used in the packaging, electronics, and automotive industries, among others. Petroleum-based polymers are cheaper than bio-based polymers but are non-degradable. The high consumption of petroleum-based polymers in various fields causes tremendous environmental issues including air and soil pollution. Incineration of non-degradable polymers always produces large amounts of carbon dioxide, thereby contributing to global warming; moreover, it sometimes produces toxic gases that also contribute to air pollution worldwide. Thus, recycling petroleum-based polymers may help protect the environment. Petroleum sources are finite, non-renewable, non-sustainable, and on the brink of exhaustion.

In contrast, bio-based polymers are defined as polymeric materials derived from renewable resources. Biodegradable polymers are polymers that can fully degrade by microorganisms through physical and chemical deterioration processes [1]. Bio-based polymers are not the same as biodegradable polymers, which means that not all bio-based polymers are biodegradable and vice versa. Therefore, in this review bio-based polymers refer to bio-based polymers with biodegradable properties.

Because bio-based polymers are biodegradable and made from renewable resources, they have the potential to reduce environmental problems and preserve sustainability. However, bio-based polymers have certain limitations in their physical, thermal, and mechanical properties as compared with petroleum-based polymers [2]. Consequently, the properties of bio-based polymers should be modified before their use as a replacement for conventional polymers in various applications.

Environmental legislation drives the development of various biodegradable polymers as an alternative to petroleum-based polymers, including polylactide (PLA), polyglycolic acid (PGA), polycaprolactone (PCL), polyhydroxyalkanoates (PHAs), starches, poly(butylene succinate) (PBS), and polyhydroxy butyrates (PHBs). Of these, PLA is a bio-based polymer that offers outstanding properties. It is a well-known "eco-friendly polymer" produced from biodegradable agricultural resources [3,4]. It has remarkable properties, including biodegradability, biocompatibility, UV stability, and processability; moreover, it has great potential in various advanced applications such as the development of high-performance material for use in the automotive industry and biomedical applications. However, PLA has some limitations; it requires certain modifications to overcome its thermal and mechanical property-related shortcomings. PLA can be combined with other materials to achieve acceptable properties. Many researchers have reported the development of PLA-based materials through the addition of other polymers [5–7], stereocomplexation [8–14], and through the addition of inorganic materials as reinforcing fillers [15–18]. The morphology, thermal behavior, interfacial, and rheological properties were studied from PBS/PLA and PCL/PLA blends, which strongly depend on polymer miscibility, were studied [5,6]. Interfacial adhesion and the morphology of the polymer blends were responsible for the improvements in their properties [7]. Stereocomplexation led to the improved thermal and mechanical properties of PLA-based materials via the formation of stereocomplex crystalline structures [8,9]. Functional hybrid materials offer a synergetic assembly of biopolymers and nano-sized inorganic particles [15]. With a combination of organic polymers with nano-sized inorganic materials, polymer nanocomposites showed significant improvements in their thermal, mechanical, and barrier properties [17]. Nano-sized cellulose particles with good interfacial interactions with polymer matrix are responsible for the enhancement of the properties of polymer nanocomposites [18]. Nature provides an unlimited source of plant- or animal-derived biomaterials, including biopolymers and bio-based nanoparticles. To avoid environmental damage, the use of degradable fillers derived from renewable resources such as cellulose-based fillers is preferred. The use of cellulose nanowhiskers (CNWs) attracted great interest owing to their advantages, including their processability, recyclability, high aspect ratio, high strength and stiffness, degradability, and renewable characteristics [18–22].

In the future, the development of biomaterials should focus on sustainability, degradability, and the tunability of properties. Of all the bio-based materials, PLA and CNW comply with the future materials development requirements; moreover, these materials have specific characteristics. PLA can be developed as an advanced material through nanocomposites [22–28] and stereocomplexation [29,30]. Furthermore, the combination of PLA-based materials and CNWs offers the synergetic properties enhancement of these materials [18,22,29,30], preserving their renewable and degradable characteristics. The molecular structure adaptability of PLA and CNW through molecular and structure modifications helps overcome the main problems concerning their miscibility. Herein, we reviewed the development of fully bio-based materials derived from PLA and CNW blends and assessed the over process and property enhancements. We first present a discussion of some general aspects of PLA and CNWs. The synergetic effects of the PLA and CNW blend on the improvement of the properties of fully bio-based PLA and CNW materials are also discussed.

## 2. Synthesis and Properties of PLA and CNWs

### 2.1. Polylactide (PLA)

PLA is an important aliphatic polyester comprising repeating units of lactide. It is considered a "green polymer" because it is produced from renewable resources. Lactide is a cyclic dimer prepared from lactic acid dehydration through a controlled depolymerization process. Lactic acid is a chiral molecule and exists as L and D isomers. It can be obtained through the fermentation of corn, sugar cane, sugar beet, etc [31,32]. Lactide has three different stereoisomers: L-lactide, D-lactide and meso-lactide (D,L-lactide); thus, the ring-opening polymerization of lactide produces poly(L-lactide) (PLLA), poly(D-lactide) (PDLA) and poly(D,L-lactide) (PDLLA), respectively.

PLA is formed either through the direct polycondensation of lactic acid monomers or by cyclic intermediate dimers (lactide) followed by the ring-opening polymerization process, as shown in Figure 1 [32]. In the polycondensation process, water is removed through condensation and the use of a solvent under high vacuum and temperature conditions. This process produces low-to-intermediate molecular weight PLA owing to the difficulties associated with the removal of water and its impurities. However, this process has some disadvantages: it requires a large reactor, an evaporation step, and solvent recovery, and has a high chance of racemization. Mitsui Toatsu used a solvent-based process to produce high-molecular-weight PLA through direct condensation with azeotropic distillation to continuously remove water from the condensation process [33].

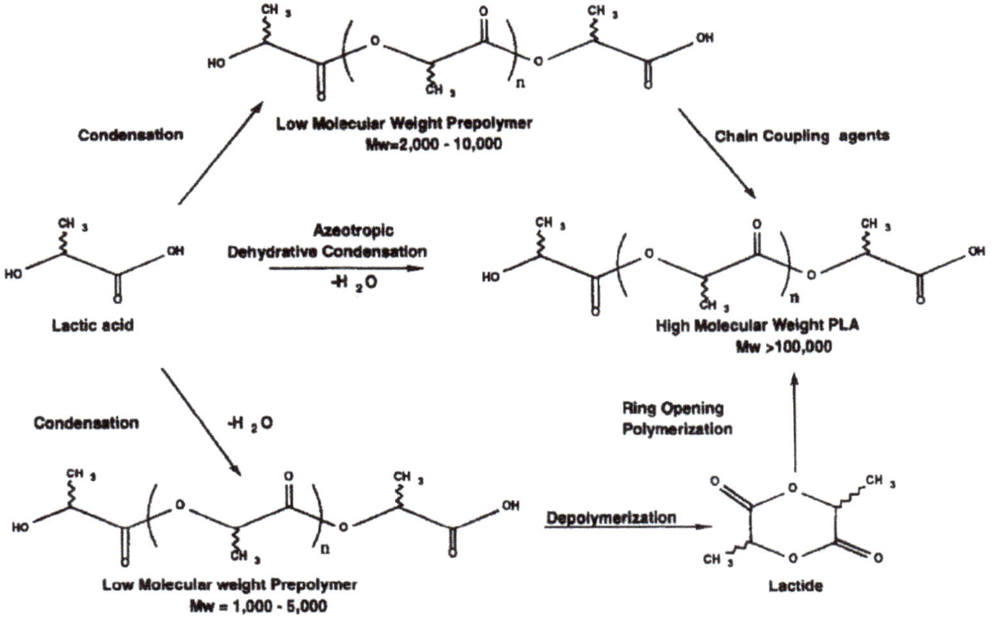

**Figure 1.** Polymerization route of polylactide. (Copyright and permission, [32]).

Ring-opening polymerization is superior to better than polycondensation in that it produces high-molecular-weight PLA. This process removes water using mild conditions and without the need for an organic solvent, thereby producing a lactide intermediate that can be subjected to vacuum distillation for purification. NatureWorks LLC has developed and patented a low-cost continuous process for the production of PLA with a controlled molecular weight [34]. The molecular weight range can be predicted by controlling the lactide dimers purity and catalyst ratio. In the ring-opening polymerization process, many catalysts can be used for polymerization, including stannous octoate [35–39],

Al-alkoxides [40,41], yttrium and lanthanide alkoxides [42], and iron alkoxides [43]. Stannous octoate is a common catalyst used to produce high-molecular-weight PLA with stereoregularity through a coordination–insertion mechanism [36]. Yttrium and lanthanide alkoxide support good polymerization activity [42]. Aluminum alkoxide can be used to produce high-molecular-weight PLA with a homogenous chain length or narrow molecular-weight distribution [41]. Generally, ring-opening polymerization can be performed under heat [34,35], the presence of solvent [33,37,40,41], or supercritical fluid conditions [38].

PLA is a potential replacement for petroleum-based polymers (PE, PET, PS, PC) owing to its biodegradability and biocompatibility properties [32,44] and is being produced on a large scale since 2001 [3]. PLA is a thermoplastic polymer and has properties similar to those of PS and PET as shown in Table 1 [45–47]. At room temperature, PLA is a stiff and brittle biopolymer with a glass transition temperature of approximately 55 °C and a melting temperature of 180 °C, depending on its crystallinity and molecular structures. Isotactic PLLA and PDLA are mostly high crystalline polymers, whereas PDLLA is an amorphous material [48]. Although PLA-based materials with high crystallinity have good physical, mechanical, and thermal properties, including biocompatibility, they have a slow degradation process. PLA is now being used as an alternative to petroleum-based polymers such as PET, PVC, HIPS in the packaging industry. PLA can also be combined with other materials for use in specific applications.

**Table 1.** Properties of PLA, PS, and PET.

| Properties | PLLA | PS | PET |
|---|---|---|---|
| Density (kg $m^{-3}$) | 1.26 | 1.05 | 1.40 |
| Tensile strength (MPa) | 59 | 45 | 57 |
| Elastic modulus (GPa) | 3.8 | 3.2 | 2.8–4.1 |
| Elongation at break (%) | 4–7 | 3 | 300 |
| Notched izod (J $m^{-1}$) | 26 | 21 | 59 |
| Heat deflection (°C) | 55 | 75 | 67 |

### 2.2. Cellulose Nano-Whiskers (CNW)

Cellulose is a bio-based material derived from renewable resources, such as plants, bacteria, trees, and tunicate, and is synthesized from glucose through condensation polymerization. During this process, a long chain of anhydro-glucose units joins together through the formation of β-1,4-glycosidic bonds [49]. Cellulose generally exists in the form of cellulose fiber, microcrystalline cellulose, and nanosized cellulose [49,50]. Of all these, as nanosized cellulose crystals, CNWs are attracting great interest, as evidenced by the increasing number of publications over the past decades [18–30,51–62].

Cellulose sources are not completely crystalline but are a combination of crystalline and amorphous. The amorphous nature renders the cellulose structure sensitive to acid hydrolysis followed by breakdown into single crystallites. CNWs can be produced from various cellulosic materials through acid hydrolysis [18,51–53]. The dimensions of the CNWs depend on the initial source and hydrolysis conditions [54]. For some cellulosic materials, different hydrolysis times result in different dimensions of the CNWs [55,56]. The dimensions of the CNWs obtained from the acid hydrolysis of cellulose range from 200 to 500 nm in length and from 3 to 30 nm in diameter [57]. Rod-like CNWs can be generated by controlling the concentrations of the cellulosic materials and sulfuric acid, temperature, and ultrasonic treatment time [55]. The success of CNW synthesis can be confirmed using transmission electron microscopy (TEM) [18,29,53], atomic force microscopy (AFM) [30], and functional group analysis using Fourier transform infrared (FTIR) spectroscopy [18,58]. Typical structures of some CNWs obtained from the acid hydrolysis of cellulose are shown in Figure 2 [29].

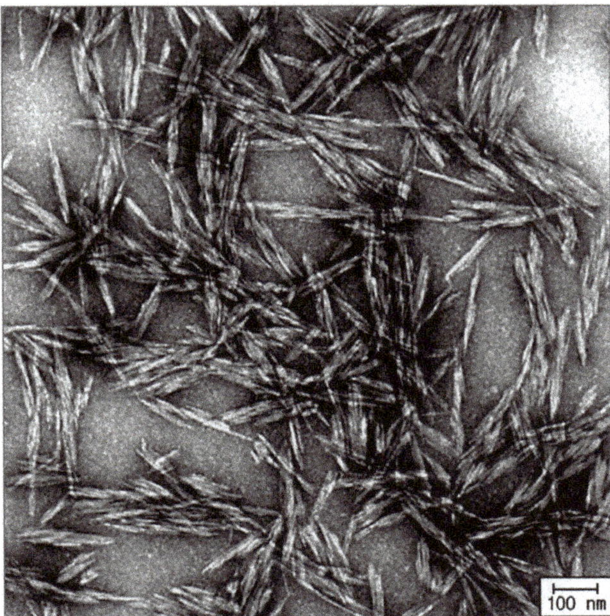


**Figure 2.** Transmission Electron Microscopy (TEM) image of acetylated CNW isolated from hydrolysis microcrystalline cellulose. (Copyright and permission, [29]).

CNWs are generally semicrystalline, and their properties depend on their source. Using X-ray diffraction, Sakurada et al. determined that the crystal modulus of CNWs from cellulose-I is 138 GPa; a similar value was noted for CNWs from plant cellulose fibers [60]. Sturcova et al. isolated CNWs from tunicin with a tensile modulus of approximately 143 GPa [61], whereas Rusli and Eichhorn isolated CNWs from cotton with a tensile modulus of approximately 57–105 GPa [62]. Determining the mechanical properties of very small particles is really difficult. By theoretical calculation, modulus values ranging from 100–160 GPa can be obtained through molecular dynamics/mechanics methods [63]. These modulus values are indicative that the particles may be used as fillers or additives.

### 2.3. PLA and CNW Compatibility

The main things to consider for future materials include supply continuity (use of renewable resources), properties (physical and mechanical), and processability. PLA and CNWs are potential bioresources with future prospects. As biodegradable and biocompatible material, PLA has some shortcomings in its mechanical and thermal properties. CNWs are interesting materials owing to their excellent mechanical and thermal properties; moreover, they are the most abundant materials on the planet and are renewable, inexpensive, and easy to process. Combining PLA and CNWs helps enhance the individual materials' properties without losing their biodegradability [64]. Material compatibility is the key to ensuring the formation of homogenous blends with other materials and is strongly dependent on surface functionality. As derivatives of polysaccharides, CNWs are not miscible with hydrophobic polymers [65]. When hydrophilic functional groups are present on CNWs surfaces, the interfacial compatibility of PLA with CNWs is limited. Consequently, hydrophilic CNWs exhibit weak interfacial bonding and dispersion in the PLA matrix, which affects their thermal and mechanical properties [66].

The dispersion of the CNWs in the solvent or polymer matrix is also affected by the type of acid used in the hydrolysis process [67]. When sulfuric acid hydrolysis is used, CNWs have better dispersion in aqueous solvents, owing to electrostatic repulsion from the negatively charged (anionic) sulfonate ester groups [55]. CNWs are highly reactive and functionalizable, owing to their high surface-to-volume ratios, and contain many surface hydroxyl groups (–OH) [67]. Moreover, surface modifications are required to improve the CNW-polymer interface in polymer nanocomposite systems. Substitution of the surface hydroxyl groups on the CNW surfaces with organic molecules and polymer grafting can help improve the hydrophobic ability of the CNWs or their interfacial compatibility with the PLA matrix [66].

Many studies have utilized certain special chemicals to improve CNW dispersion in organic solvents, including poly(ethylene glycol) (PEG) [63], silanes [68], n-octadecyl isocyanate [69], hydrochloric acid combine with acetic acid [18,29,30,70,71], sulfuric acid [58], and polyvinyl alcohol [72]. Some surface modifications have been performed to improve the miscibility of CNW with the PLA matrix, as illustrated in Figure 3. Moreover, the –OH functional group can be substituted with other hydrophobic molecules through acetylation [18,29,30], esterification with organic acid (benzoic acid, valeric acid) [73,74], and lactic acid [75]. Substitution of these –OH functional groups with hydrophobic chains diminishes the hydrophilicity of the CNWs surface and increases its hydrophobicity based on the replacement portion. These strategies improve the dispersion of CNWs in the PLA matrix. PLA molecules can also be grafted on the CNW surface by controlling the number of –OH functional groups as a co-initiator during polymerization [28,58]. Unmodified CNWs containing many –OH functional groups result in low-molecular-weight grafted-PLA chains. Blocking the surface hydroxyl groups of CNWs support the ring-opening polymerization of lactide to obtain high-molecular-weight PLA [18,29,30,58]. The surface modifications of CNWs improve their dispersibility in the PLA matrix and preserves their high mechanical properties, crystallinity, and degradability [76].

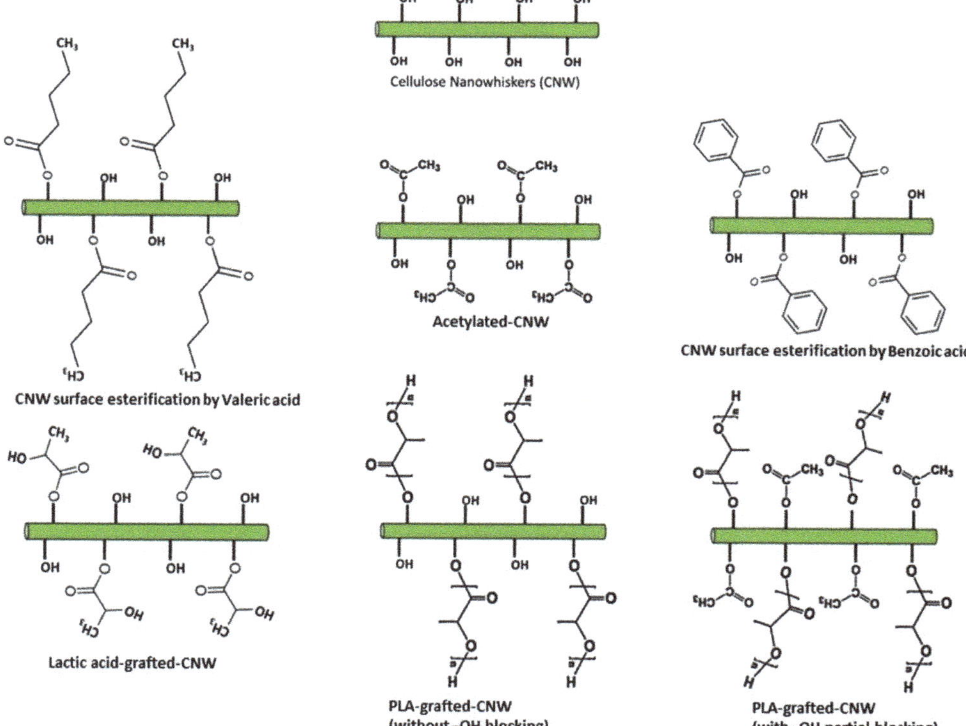


**Figure 3.** Illustration of surface modifications of CNW particles.

## 3. PLA–CNW Nanocomposites

Nanocomposites reportedly improve the mechanical and physical properties of the polymer matrix [15–17]. Among all the available nanoparticles, CNWs are natural organic-based nanoparticles and are considered promising nanomaterials owing to their cheap price and abundance in nature. PLA and CNW blends may replace petroleum-based polymers and provide advantages such as property tunability, degradability, and resource sustainability.

Some studies have demonstrated the direct utilization of cellulose-based materials to improve the properties of PLA [70,77–79]. The biocomposite PLA and natural fibers almost achieved the mechanical properties of glass fiber-reinforced polymers [77]. Oksman et al. reported on a promising composite of PLA and flax fibers with a significant improvement in stiffness at 30% of the flax fiber content [78]. The excess –OH groups on the cellulose fiber surface increase its reactive compatibility with the PLA matrix but restrict the molecular weight of PLA [79]. The portion of the substitution of microfibrillated cellulose –OH groups with acetyl functional groups affects the mechanical properties of PLA-microfibrillated cellulose composites [70]. Hydrophilic surfaces limit cellulose dispersion in the PLA matrix. Furthermore, a high cellulose particle content affects the optical properties of the blended materials. It is thus important to control the surface hydrophilicity and minimize the particle loading of cellulose-based materials.

CNWs are nano-sized and whisker-like crystalline particles isolated from cellulose through the process of acid hydrolysis. Nano-sized CNWs with excellent dispersion in polymer matrix results in impressive improvement in the mechanical properties at low loading levels through a percolation mechanism supported by hydrogen bonds [80].

Many researchers have studied the combination of PLA matrix with CNWs (neat and modified surfaces) to improve their physical, mechanical, barrier-related, and thermal properties, as presented in Table 2. The addition of CNWs with abundant surface hydroxyl groups from the acid hydrolysis of cellulose enhanced the mechanical, thermal, and barrier-related properties of hydrophobic PLA matrix with a high molecular weight ($M_w$ > 100,000 gr/mol) through solution casting [22,26,27]. This enhancement of properties was attributed to the char formation and increased crystallinity [22].

**Table 2.** Polymer matrix, surface modification, processing, and properties of PLA–CNW nanocomposites.

| Polymer Matrix | CNW Modifications/Synthesis | Nanocomposites Processing | Properties Improvement | Ref. (Year) |
|---|---|---|---|---|
| PLA ($M_w$ = 100,000 g/mol) | Acid hydrolysis from cotton fabric | Solution casting in chloroform/N,N'dimethylformamide | Thermal stability, tensile strength, and Young's modulus | [22] (2021) |
| PLA ($M_n$ = 130,000 g/mol) | Acid hydrolyzed from cellulose | Solution casting in chloroform | Water permeability and oxygen permeability | [26] (2010) |
| PLA | Acid hydrolyzed from flax cellulose | Solution process in N,N'dimethylformamide to form nanofibrous mat | Crystallinity and water absorption | [27] (2018) |
| PDLLA | Acid hydrolyzed from eucalyptus kraft wood pulp | Solution casting in dimethylformamide | Hydrolytic degradation, thermal stability | [81] (2011) |
| PLA | Acetylation CNW | Solution polymerization<br>Bulk Polymerization<br>Solution Blending | Thermal stability and Crystallinity | [18] (2012) |
| PLA ($M_w$ = 52,000 g/mol) | Acetylation using acetic anhydride | Solution casting in chloroform | Tensile strength, thermal stability, dimensional stability, and dynamic mechanical properties | [25] (2014) |
| PLA ($M_w$ = 180,000 g/mol) | Acetylation CNW | Solution casting in dichloromethane | Stress transfer between CNW and PLA matrix | [82] (2017) |
| PLA ($M_n$ = 1.0 × $10^5$ g/mol) | Surface esterification by formic acid | Solution casting in chloroform | Barrier properties | [83] (2017) |
| PLA ($M_w$ = 87,000 g/mol) | Grafted toluene diisocyanate | Solution casting in chloroform | Tensile strength | [84,85] (2016) |

**Table 2.** *Cont.*

| Polymer Matrix | CNW Modifications/Synthesis | Nanocomposites Processing | Properties Improvement | Ref. (Year) |
|---|---|---|---|---|
| PLA ($M_w$ = 100,000 g/mol) | Surface modification by triazine derivative | Hot compression process 170oC 40 MPa | Breaking strength, elongation, compatibility, and thermal properties | [23,86] (2017, 2018) |
| PLA ($M_n$ = 98,000 g/mol; $M_w$ = 199,590 g/mol) | Surface esterification by benzoic acid | Masterbatch followed by extrusion process | The Young's modulus and ultimate tensile stress | [73] (2018) |
| PLA ($M_n$ = 98,000 g/mol; $M_w$ = 199,590 g/mol) | Surface esterification by valeric acid | Masterbatch followed by extrusion process | Thermal decomposition, mechanical properties, and crystallinity growth | [74] (2019) |
| PLA ($M_w$ = 2.1 × $10^5$ g/mol) | Graft modification by 3-aminopropyltriethoxysilane | Solution casting in dichloromethane | Air permeability, light resistance, thermal stability, and mechanical properties | [87] (2020) |
| PLA ($M_n$ = 150,000 g/mol) | Addition radical initiator with dicumyl peroxide | Reactive extrusion by Twin-screw extruder | Mechanical properties, crystallinity. processability, melt-strength, rheological behavior | [88,89] (2016, 2017) |
| PLLA ($M_w$ = 100,000 g/mol) | Grafted lactic acid | Solution casting in chloroform | Tensile strength and Young's modulus | [75] (2018) |
| PLA ($M_n$ = 95,000 g/mol) PLA ($M_n$ = 130,000 g/mol) | Grafting PLLA by ring-opening polymerization in toluene | Melt-blending in mini extruder | Compatibility, thermal behavior, and mechanical properties | [58] (2011) |
| PLA | Grafting PLLA by ring-opening polymerization in toluene | Twin-screw extruder | Thermal, mechanical, optical, and morphological properties | [84] (2016) |
| PLA | Grafting PLLA by ring-opening polymerization in dimethyl sulfoxide | Solution casting and co-extrusion | Barrier and dynamic mechanical properties | [90] (2016) |
| PLA ($M_n$ = 110,000 g/mol) | Grafting PLLA by ring-opening polymerization in toluene | Melt spinning using twin-screw micro-compounder | Thermal stability, degree of crystallinity, and mechanical properties | [91] (2016) |

Note: $M_n$: number average molecular weight; $M_w$: weight average molecular weight.

Sanchez-Garcia and Lagaron used CNWs as a reinforcing agent in the polylactide matrix to reduce the gas and vapor permeability: optimum barrier enhancement was obtained with CNWs at low loadings, as shown in Figure 4a,b [26]. The addition of 3% well-distributed CNW particles in the PLA matrix reduced the water and oxygen permeability by approximately 82% and 90%, respectively. The presence of crystalline CNW materials affects the barrier properties of the PLA–CNW materials due to filler-induced nucleation and results in increased crystallinity [26], whereas the surface hydroxyl groups improve the water absorption of the PLA nanocomposites [27]. A degradation study was performed; compare with the original PLA homopolymers, low-loading unmodified CNWs showed better stability at a temperature of 310 °C (Figure 4c) [27]. The presence of CNWs without surface modifications delays the hydrolytic degradation of the PDLLA matrix (Figure 4d) because these CNWs function as a physical barrier and prevent water diffusion to polymer molecules, thereby re-routing the kinetics of hydrolytic degradation [83].

The surface hydroxyl group on CNWs limits their compatibility and dispersion in the PLA matrix. Controlling the number of the –OH groups on the CNWs helps increase the miscibility of the CNWs with PLA matrix and control their hydrophilicity. Therefore, CNW surface modifications are required to achieve acceptable dispersion and compatibility levels of CNWs in the PLA matrix. The polarity of CNWs can be decreased by the blocking effect of the acetyl group on the CNW surface, leading to enhanced CNW dispersion in PLA and accordingly improving the physical and mechanical properties of the PLA–CNW nanocomposites [18,25]. The interaction between the acetyl groups on the CNW surface and PLA was evidenced by the shift in the carbonyl stretching spectra at 1748 $cm^{-1}$ [84]. Other organic CNW surface modifiers mainly focus on reducing the surface hydroxyl groups through esterification to successfully increase the CNW dispersion in the PLA matrix [73,74,83–86]. The shielding effect of triazine derivatives on the CNW surface supports the improvement of the thermal degradation properties of the PLA–CNW nanocomposites [23,86]. Well-dispersed silanized CNWs in the PLA matrix are a result of the strong hydrogen bonding of –COOH groups in PLA and the –OH groups in the silanized CNWs [87]. The functionalization of the CNW surface was also enhanced by the addition of a radical initiator dicumyl peroxide through grafting between the PLA methine (–CH) groups with methylene groups of the CNWs [88].

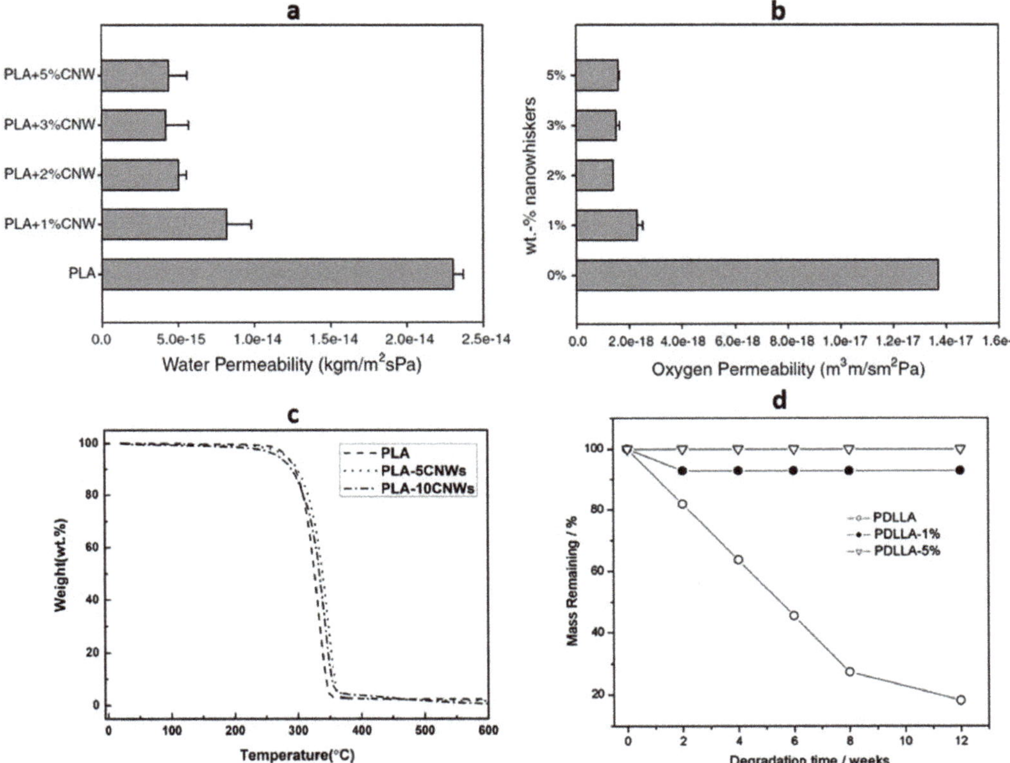


**Figure 4.** (**a**). Water permeability of PLA and their nanocomposites with various CNW content.; (**b**). Oxygen permeability of PLA and their nanocomposites with various CNW content. (Copyright and permission, [26]); (**c**). Thermogravimetric analysis (TGA) thermogram of the PLA and PLA–CNW; (**d**). Residual mass of the neat PDLLA and its nanocomposites as a function of degradation time. (Copyright and permission, [83]).

Compared with the functional groups, the grafted-PLA chains on the CNWs reportedly increases the miscibility. It can be obtained via in situ ring-opening polymerization [18,28,58,90,91]. Because –OH groups on the CNW surface can be controlled by the partial blocking of organic molecules, these –OH groups can be used as co-initiators to graft a PLA chain onto the CNW surface. The in-situ polymerization of lactide onto unmodified CNWs results in low-molecular-weight PLA chains owing to the excess –OH groups present on the CNW particles [58]. Partially-blocked –OH groups on the CNW surface led to effective ring-opening polymerization of lactide, resulting in the formation of high-molecular-weight PLA chains [18]. High-molecular-weight PLA–graft–CNW tackles the hydrophobic/hydrophilic incompatibilities of the PLA–CNW nanocomposites and consequently, leads to tremendous improvements in the material properties: thermal stability, crystallinity, mechanical, barrier-related, optical, and morphological properties [28,58,90,91]. The PLA–graft–CNW improves the miscibility of the CNWs with PLA matrix, resulting in the homogenous dispersion of the CNW particles. Dispersed CNW particles accelerate the formation of nucleating sites and enhance the crystallization ability of PLA matrix [58]. PLA–CNW nanocomposites can be easily obtained in the presence of organic solvents [18,22,25–27,75,81–85,87,90] or be produced using thermal mixing process [23,28,58,73,74,86,88,89], which also suitable for high-molecular-weight PLA matrix [25,74,90].

## 4. Bio-Stereo-Nanocomposites PLA–CNW

Nanocomposite is not the only method that can be used to improve the thermal and mechanical properties of PLA. Stereocomplex PLA (s-PLA) is another method to improve the properties of PLA. S-PLA is an enantiomeric polymer blend of PLLA and PDLA [8]. The crystal structure of s-PLA is different from that of its homopolymer as shown in wide-angle x-ray spectroscopy (WAXS) profiles. The main peaks of the PLA homopolymer ($X_D$ = 1) appeared at 2$\theta$ values of 15°, 17°, and 19°, whereas the main peaks of s-PLA ($X_D$ = 0.5) were observed at 2$\theta$ values of 12°, 21°, and 24° [8]. Okihara et al. proposed the crystal structure of s-PLA (Figure 5) [9]. The thermal and mechanical properties of s-PLA improved because s-PLA is strongly bounded by the crystal network and exhibits decreased motion as compared with PLA homopolymers.

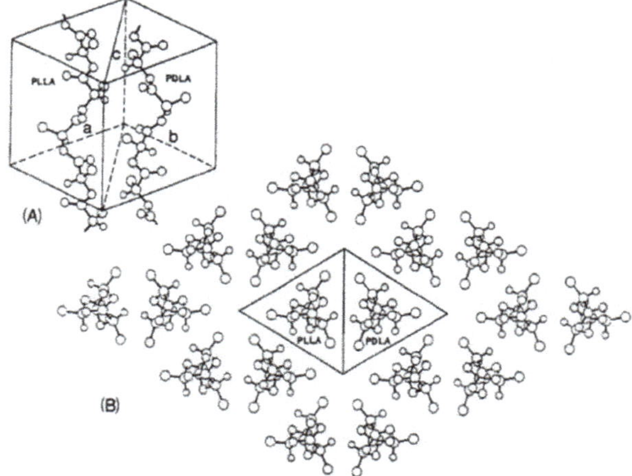


**Figure 5.** Crystal structure of s-PLA. (**A**) Structural model of the stereocomplex of PLLA and PDLA with the space group p 1. In the present case, the pointing direction of methyl groups is upward. (**B**) Molecular arrangement projected on the plane normal to the chain axis. (Copyright and permission, [9]).

S-PLA can be produced using the solution [92–94], melt blending [95–97], supercritical fluid [13,14,98], and microwave irradiation method [99], which is mostly similar to the synthesis route of PLA–CNW nanocomposites. Therefore, both stereocomplexation and nanocomposite approaches can be applied using the same method for the combination of PLA and CNW materials. The molecular weight of PLA (critical value: approximately 100,000 g/mol) is an additional hindrance to the formation of stereocomplex crystallites [100].

Bio-stereo-nanocomposites are defined as advanced materials developed from biodegradable materials by combining stereocomplexation and nanocomposite approaches. The bio-stereo-nanocomposite PLA–CNW is the combination of fully bio-based PLA and CNW through stereocomplexation and nanocomposite approaches. As an advanced biomaterial, bio-stereo-nanocomposite PLA–CNW offers many benefits for wider application in the future: biodegradability, sustainability, and tune-ability of the thermal-mechanical properties.

Bio-stereo-nanocomposite PLA–CNW can be synthesized by combining enantiomeric PLA mixtures with pristine or surface-modified CNWs or by mixing graft structures such as CNW-grafted PDLA and PLLA.

Bio-stereo-nanocomposite PLA–CNW was first developed through solution casting of PLLA and CNW-grafted PDLA at a 50:50 weight ratio and was found to contain excellent stereocomplex crystallites [101]. The addition of CNW-grafted PDLA at a low amount (up to 10% weight ratio) also supports the formation of stereocomplex crystallites concurrently with PLA homocrystallites [102,103]. The formation of stereocomplex crystallites was confirmed using differential scanning calorimeter (DSC) analysis, the results of which are shown in Figure 6a,b [102]. A single melting temperature ($T_m$) of 220 °C for the equimolar ratio indicated the presence of stereocomplex crystallites, whereas a lower CNW-grafted PDLA indicated the presence of homocrystallites [102]. Cao et.al. produced a bio-stereo-nanocomposite PLA–CNW with partial stereocomplex crystallites from a mixture of PLLA and PDLA containing 5% acetylated CNW particles [104]. However, the generated bio-stereo-nanocomposite did not adequately support the formation of s-PLA crystallites from high-molecular-weight PLLA and PDLA blends because the CNW particles could not prevail in the formation of homocrystallites during the crystallization process for high molecular weight (Mw > 100,000). The bio-stereo-nanocomposites PLA–CNW was also successfully developed by mixing CNW-grafted PLLA and PDLA at a 50:50 weight ratio using solution casting [30] and supercritical carbon dioxide–dichloromethane [29]. The formation of a perfect stereocomplex in the bio-stereo-nanocomposite PLA–CNW was confirmed using the DSC and X-ray diffraction (XRD) analyses, as shown in Figure 6c,d [29]. The DSC thermograms of the bio-stereo-nanocomposite PLA–CNW with various CNW loadings (s-PLAC-CNW1, s-PLA–CNW2, and s-PLA–CNW3) exhibited a single $T_m$ of approximately ~230 °C without the homopolymer $T_m$ (approximately ~180 °C). Perfect stereocomplexation was also confirmed through characteristic diffraction peaks of s-PLA at 11.86°, 20.56°, and 23.86° of $2\theta$ value (Figure 6d).

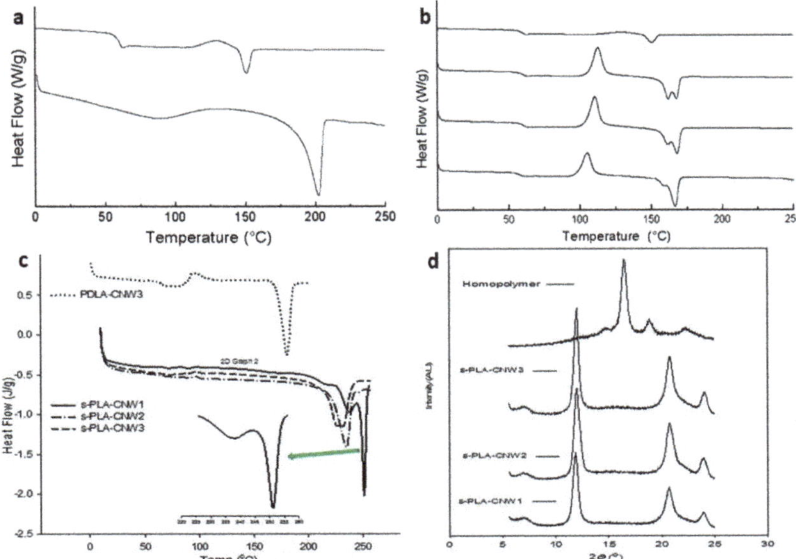


**Figure 6.** (**a**). DSC thermogram of neat PLLA (**top**) and PLLA/CNW-grafted PDLA at 50:50 ratio; (**b**). DSC thermogram of neat PLLA (**top**) and PLLA/CNW-grafted PDLA at different content 2.5, 5, and 10% (**top** to **bottom**).; (**c**). DSC thermogram of the PDLA-CNW3 (homopolymer) and generated bio-stereo-nanocomposite PLA–CNW materials; (**d**). XRD diffractogram of homopolymer PLA–CNW and generated bio-stereo-nanocomposite PLA–CNW materials. (Copyright and permission, [29]).

Bio-stereo-nanocomposites PLA–CNW from the CNW-grafted PLLA and PDLA showed unique processing characteristics [29]. The formation of stereocomplex crystallites in the solution is strongly related to the solubility of the polymer matrix in the solvent. The graft structure of PLA–CNW nanocomposites magnifies the number of functional groups that reduce the polymer viscosity in organic solvents. Contrary to the formation of linear s-PLA, unusual stereocomplex crystallite formation was found for bio-stereo-nanocomposites PLA–CNW. The well-distributed CNW particles and the reducing solution viscosity resulted in a higher possibility of grafted-PLLA and grafted-PDLA chains interacting and immediately forming stereocomplex crystallites. Bio-stereo-nanocomposites PLA–CNW can be obtained using the solution method in a relatively short time (approximately 5 min) from various molecular weights (low, medium, and high) and concentrations of CNW-grafted PLLA and PDLA [29].

Neat s-PLA materials have improved mechanical and thermal properties compare with PLA homopolymers. The addition of unmodified or surface-modified CNW particles in the stereocomplex matrix can enhance its properties to a greater extent. As mentioned in a previous report, s-PLA crystallites increase the tensile strength and Young's modulus up to 25% more than neat homopolymers do [13]. In a study focusing on polymer nanocomposites, well-distributed nanoparticles also improved the mechanical properties of polymeric materials [105]. Bio-stereo-nanocomposite PLA–CNW showed enhanced mechanical and thermal properties despite the low content of the CNW particles. The presence of CNW particles (3% content) as a core structure in bio-stereo-nanocomposites PLA–CNW (s-PLA–CNW3) enhances its mechanical properties more than that in neat s-PLA materials (up to 2.70 GPa and 62.96 MPa for Young's modulus and tensile strength, respectively) [30]. Moreover, the addition of 10% CNW-grafted PDLA in the PLLA matrix also improves the storage modulus E′ at approximately 4.15 GPa (at 25 °C) and 0.66 GPa (at 100 °C) as compared to neat PLLA (approximately 2.93 GPa and 0.41 GPa at 25 °C and 100 °C, respectively) [102]. These property enhancements are most probably caused by the reinforcing effect of CNW crystallites through hydrogen bonding and the stabilizing percolation network by stereocomplex crystallization [102]. Wu et al. reported that the addition of 10% CNW-grafted-PDLA in the PLLA matrix shows significant s-PLA crystallites which act as nucleation sites for PLLA homopolymers; accordingly, notable improvement was observed in the storage modulus at high temperatures as shown in Figure 7a [103].

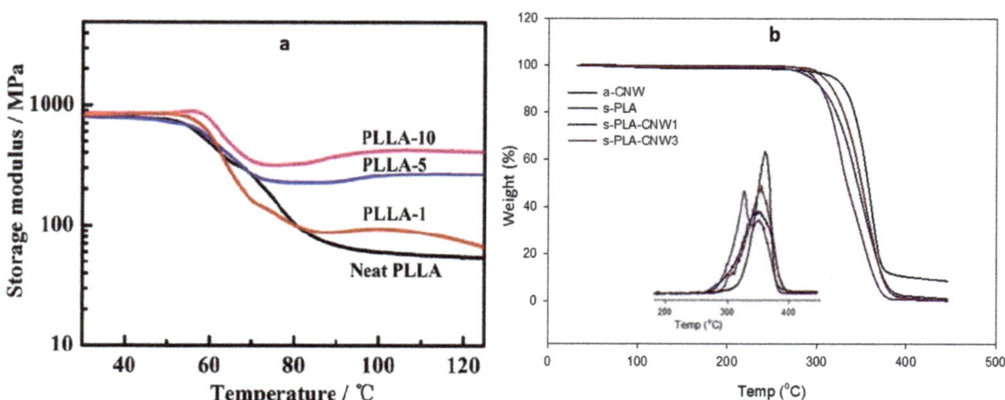


**Figure 7.** (**a**). Storage modulus 9E′—temperature curve of neat PLLA and CNW-grafted-PLLA at different content (PLLA-1; PLLA-5; and PLLA-10) [103]; (**b**). TGA traces of a-CNW, s-PLA, and bio-stereo-nanocomposite with different CNW contents (s-PLA–CNW1, and s-PLA–CNW3) under nitrogen atmosphere. The insert shows the first derivative of the weight loss. (Copyright and permission, [30]).

Bio-stereo-nanocomposite PLA–CNW also shows significant enhancements in its thermal properties. A comparison of acetylated CNW, neat s-PLA, and bio-stereo-nanocomposite PLA–CNW (s-PLA–CNW) showed that the presence of CNW particles improves the thermal decomposition of the materials (Figure 7b) [30]. S-PLA formation increases the thermal degradation temperature owing to the strong interactions between the L- and D-lactide chains, which significantly reduces molecular mobility and disturbs thermal degradation [106,107]. The bio-stereo-nanocomposites PLA–CNW materials s-PLA–CNW1 and s-PLA–CNW3 showed enhanced thermal degradation properties as compared with the original s-PLA, indicating that the CNW particles can act as a superior insulator and mass transport barrier during the thermal degradation process. The enhancement of these valuable properties represents a synergetic effect of the strong interaction among the molecules in s-PLA crystallites and well-distributed CNW particles in bio-stereo-nanocomposite PLA–CNW.

One of the important properties of stereocomplex materials related to processing materials is stereocomplex memory or melt stability. Stereocomplex memory is the ability of stereocomplex materials to reform stereocomplexes after they are melted. Because its unzipped fragment cannot re-zip to form s-PLA after melting, stereocomplex memory is the main limitation of linear s-PLA [14,106]. Some structural modifications were developed to solve this limitation [14,106,108]. Biela et al. reported that a star-shaped s-PLA with a minimum of 13 arms could maintain stereocomplex memory through parallel or anti-parallel orientations of hardlock-type interactions [106]. Star-shaped PLA with a smaller arm number (8 arms) and a three-dimensional core structure also shows excellent stereocomplex memory because a precise position and space distribution minimizes the movement of the lactide chains [108].

The bio-stereo-nanocomposites from CNW-grafted PLLA and PDLA show excellent stereocomplex memory, which is supported by the graft structure of supramolecules similar to the star-shaped structure and the presence of the CNW particles' three-dimensional core structures. Figure 8a shows the mechanism of stereocomplex memory through anti-parallel interaction, as supported by the highly distributed CNW particles [29]. The stereocomplex memory was confirmed through the presence of a single peak during the cooling process and by performing a second scanning of DSC analysis (Figure 8b) [29]. The grafted structure of the PLA–CNW materials minimized the freedom of the PLLA and PDLA chains during the melting process. Consequently, the unzipped PLLA and PDLA chains could easily re-zip after melting. This is also supported by the well-distributed CNW particles as nucleating agents.

The combination of stereocomplex crystallization and nanocomposite approaches brings simultaneous improvements in the mechanical and thermal properties of PLA, including stereocomplex memory and an unusual stereocomplex crystallization process. The simultaneous property enhancement is attributed to the synergistic effect derived from the nucleating agent support of well-dispersed CNW particles and chain freedom of PLLA and PDLA. PLA chains can be modified using various polymer chains and structures; thus, the bio-stereo-nanocomposites PLA–CNW has numerous general and specific applications and found an important role in food packaging, automotive, engineering, and biomedical applications owing to their unique properties [109–113]. Controlling CNW miscibility in PLA matrix helps enhance the mechanical, optical, and thermal properties, which are the main requirements for food packaging applications [109–111]. As a bio-based material, PLA–CNW with its biodegradability and cytocompatibility fulfills the basic requirements for use in biomedical applications [112,113] and is a potential candidate for the development of eco-friendly materials in the future owing to its sustainability (from renewable resources), biodegradability, and property tune-ability. Highly functional nanomaterials can be generated from PLA and CNW, thereby expanding the advanced and highly specific applications, such as biomedical and tissue engineering applications.

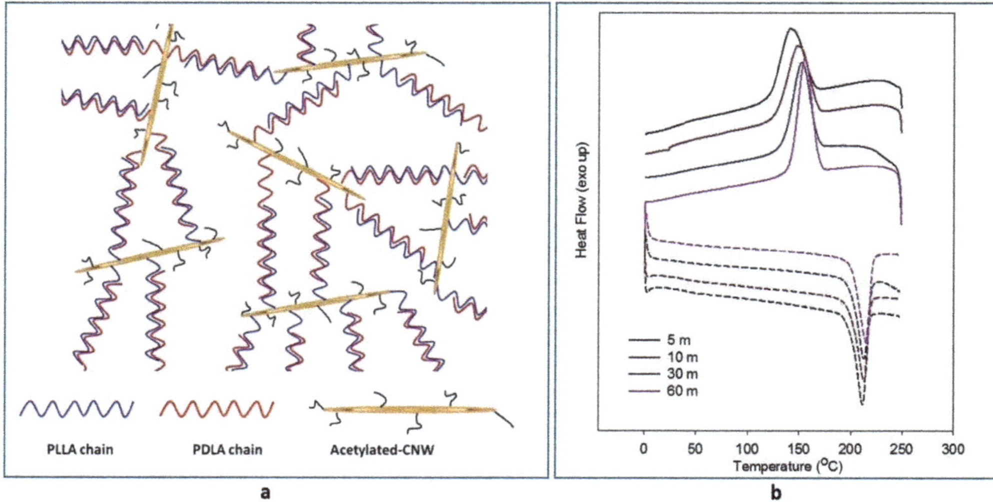


**Figure 8.** (**a**) Schematic structure of s-PLA–CNW via antiparallel re-assembles supported by graft-structure and well-distributed acetylated-CN; (**b**) DSC thermograms of bio-stereo-nanocomposites PLA–CNW at various processing times; cooling (solid-line) and second heating (dash-line). (Copyright and permission, [29]).

## 5. Conclusions and Prospective

PLA and CNW materials are a promising material for future applications as they are derived from sustainable resources and degradability. PLA and CNW can be developed as advanced materials through nanocomposite and stereocomplex approaches. The combination of PLA and CNW attracted great interest because they are fully bio-based materials. The development of fully-bio-based materials utilizing PLA and CNW materials through nanocomposite and stereocomplexation approaches. PLA–CNW nanocomposites can be obtained through solution casting and melt blending processes. The critical point of development of the PLA–CNW nanocomposites is mainly associated with the inherent hydrophilicity of CNW particles. Unmodified CNWs improve the PLA properties with limited CNW dispersibility in the PLA matrix. Chemical modifications on the CNW surface to control the –OH groups improved the dispersion and compatibility of CNWs in the PLA matrix. Acetyl groups, esterification, and grafted-PLA chains on the CNW surface increase the interfacial interaction with the polymer matrix through hydrogen bonding. Well-distributed CNW particles in PLA–CNW nanocomposites can act as a nucleating agent to improve the PLA crystallinity, thereby affecting the thermal, mechanical, and barrier-related properties.

Bio-stereo-nanocomposite PLA–CNW is an advanced material developed using stereocomplexation and nanocomposite approaches. The presence of CNW particles in high-molecular-weight linear PLLA and PDLA homopolymer blends constrained by the molecular weight resulted in the formation of partial stereocomplex crystallites. Bio-stereo-nanocomposite PLA–CNW from CNW-grafted PLLA and PDLA shows unusual stereocomplex formation in a very short time and excellent stereocomplex memory after melting. These graft structures presumably provided a three-dimensional core structure and minimize the chain mobility of PDLA and PLLA. The stereocomplex crystallites and nucleating agents from CNW particles contribute to the simultaneous improvement of the properties of bio-stereo-nanocomposites PLA–CNW. The simultaneous enhancement of the properties opens the window to be applied in a wide range of industrial applications. Bio-based materials from PLA and CNW can be used in the development of future materials owing to their properties of biodegradability, sustainability, and property tunability. Moreover,

PLA and CNW have the ability to be modified with specific functional structures. With its tacticity and functionality, PLA has a great chance for molecular modifications and stereocomplexation. The surface hydroxyl groups on the CNW surface can be controlled to adjust the hydrophobicity and surface functionality. These properties of PLA and CNW bring bio-based PLA–CNW promising for the development of future materials for specific and advanced applications.

**Author Contributions:** Conceptualization, P.P.; resources, P.P., M.S. and I.I.; data curation, M.S. and I.I.; writing—original draft preparation, P.P., M.S. and I.I.; writing—review and editing, P.P., M.S. and I.I.; visualization, P.P., M.S. and I.I. All authors have read and agreed to the published version of the manuscript.

**Funding:** This research was funded by Kementerian Pendidikan, Kebudayaan, Riset dan Teknologi Republik Indonesia, World Class Research Grant.

**Institutional Review Board Statement:** Not applicable.

**Informed Consent Statement:** Not applicable.

**Data Availability Statement:** Data is contained within the article.

**Conflicts of Interest:** The authors declare no conflict of interest.

## References

1. Babu, R.P.; O'Connor, K.; Seeram, R. Current progress on bio-based polymers and their future trends. *Prog. Biomater.* **2013**, *2*, 8. [CrossRef] [PubMed]
2. Williams, C.K.; Hillmyer, M.A. Polymer from renewable resources: A perspective for a special issue of polymer reviews. *Polym. Rev.* **2008**, *45*, 1–10. [CrossRef]
3. Garlotta, D. A literature review of poly(lactic acid). *Polym. Environ.* **2001**, *9*, 63–84. [CrossRef]
4. Ikada, Y.; Tsuji, H. Biodegradable polyesters for medical and ecological applications. *Macromol. Rapid Commun.* **2000**, *21*, 117–132. [CrossRef]
5. Wu, D.; Yuan, L.; Laredo, E.; Zhang, M.; Zhou, W. Interfacial properties, viscoelasticity, and thermal behaviours of poly(butylene succinate)/polylactide blend. *Ind. Eng. Chem. Res.* **2012**, *51*, 2290–2298. [CrossRef]
6. Noroozi, N.; Schafer, L.L.; Hatzikiriakos, S.G. Thermorheological properties of poly(ε-caprolactone)/polylactide blends. *Polym. Eng. Sci.* **2012**, *52*, 2348–2359. [CrossRef]
7. Ojijo, V.; Ray, S.S. Super toughened biodegradable polylactide blends with non-linear copolymer interfacial architechture obtained via facile in-situ reactive compatibilization. *Polymer* **2015**, *80*, 1–17. [CrossRef]
8. Ikada, Y.; Jamshidi, K.; Tsuji, H.; Hyon, S.-H. Stereocomplex formation between enantiomeric poly(lactides). *Macromolecules* **1987**, *20*, 904–906. [CrossRef]
9. Okihara, T.; Tsuji, M.; Kawaguchi, A.; Katayama, K.; Tsuji, H.; Hyon, S.-H.; Ikada, Y. Crystal structure of stereocomplex of poly(L-lactide) and poly(D-lactide). *J. Macromol. Sci. B Phys.* **1991**, *B30*, 119–140. [CrossRef]
10. Brochu, S.; Prud'homme, R.E.; Barakat, I.; Jerome, R. Stereocomplexation and Morphology of Polylactides. *Macromolecules* **1995**, *28*, 5230–5239. [CrossRef]
11. Zhang, J.; Sato, H.; Tsuji, H.; Noda, I.; Ozaki, Y. Infrared spectroscopic study of CH3—O = C interaction during poly(L-lactide)/poly(D-lactide) stereocomplex formation. *Macromolecules* **2005**, *38*, 1822–1828. [CrossRef]
12. Fukushima, K.; Kimura, Y. Stereocomplexed polylactides (Neo-PLA) as high-performance bio-based polymers: Their formation, properties, and applications. *Polym. Int.* **2006**, *55*, 626–642. [CrossRef]
13. Purnama, P.; Kim, S.H. Stereocomplex formation of high-molecular-weight polylactide using supercritical fluid. *Macromolecules* **2010**, *43*, 1137–1142. [CrossRef]
14. Purnama, P.; Jung, Y.; Kim, S.H. Stereocomplexation of poly(L-lactide) and Random Copolymer Poly(D-lactide-co-ε-caprolactone) to enhance melt stability. *Macromolecules* **2012**, *45*, 4012–4014. [CrossRef]
15. Darder, M.; Aranda, P.; Ruiz-Hitzky, E. Bionanocomposites: A new concept of ecological, bioinspired, and functional hybrid materials. *Adv. Mater.* **2007**, *19*, 1309–1319. [CrossRef]
16. Darder, M.; Aranda, P.; Ferrer, M.L.; Gutierrez, M.C.; del Monte, F.; Ruiz-Hitzky, E. Progress in bionanocomposite and bioinspired foams. *Adv. Mater.* **2011**, *23*, 5262–5267. [CrossRef]
17. Bitinis, N.; Hernandez, M.; Verdejo, R.; Kenny, J.M.; Lopez-Manchado, M.A. Recent advances in clay/polymer nanocomposites. *Adv. Mater.* **2011**, *23*, 5229–5236. [CrossRef]
18. Braun, B.; Dorgan, J.R.; Hollingsworth, L.O. Supra-molecular ecobionanocomposites based on polylactide and cellulosic nanowhiskers: Synthesis and properties. *Biomacromolecules* **2012**, *13*, 2013–2019. [CrossRef]
19. Park, W.I.; Kang, M.; Kim, H.S.; Jin, H.J. Electrospinning of poly (ethylene oxide) with bacterial cellulose whiskers. *Macromol. Symp.* **2007**, *249*, 289–294. [CrossRef]

20. Peresin, M.S.; Habibi, Y.; Zoppe, J.O.; Pawlak, J.; Rojas, O.J. Nanofiber composites of polyvinyl alcohol and cellulose nanocrystals: Manufacture and characterization. *Biomacromolecules* **2010**, *11*, 674–681. [CrossRef]
21. Zoppe, J.O.; Peresin, M.S.; Habibi, Y.; Venditi, R.; Rojas, O.J. Reinforcing poly (ε-caprolactone) nanofibers with cellulose nanocrystals. *Appl. Mater. Interfaces* **2009**, *1*, 1996–2004. [CrossRef] [PubMed]
22. Shi, Q.F.; Zhou, C.J.; Yue, Y.Y.; Guo, W.; Wu, Y.; Wu, Q. Mechanical properties and in vitro degradation of electrospun bio-nanocomposite mats from PLA and cellulose nanocrystals. *Carbohyd. Polym.* **2012**, *90*, 301–308. [CrossRef] [PubMed]
23. Yin, Y.; Zhao, L.; Jiang, X.; Wang, H.; Gao, W. Poly(lactic acid)-based biocomposites reinforced with modified cellulose nanocrystals. *Cellulose* **2017**, *24*, 4773–4784. [CrossRef]
24. Mokhena, T.C.; Sefadi, J.S.; Sadiku, E.R.; John, M.J.; Mochane, M.J.; Mtibe, A. Thermmoplastic processing of PLA/Cellulose nanomaterials composites. *Polymers* **2018**, *10*, 1363. [CrossRef]
25. Lee, J.H.; Park, S.H.; Kim, S.H. Surface modification of cellulose nanowishkers and their reinforcing effect in polylactide. *Macromol. Res.* **2014**, *22*, 424–430. [CrossRef]
26. Sanchez-Garcia, M.D.; Lagaron, J.M. On the use of plant cellulose nanowishkers to enhance the barrier properties of polylactic acid. *Cellulose* **2010**, *17*, 987–1004. [CrossRef]
27. Liu, W.; Dong, Y.; Liu, D.; Bai, X.; Lu, X. Polylactid acid (PLA)/cellulose nanowishkers (CNWs) composite nanofibers: Microstructural and properties analysis. *J. Compos. Sci.* **2018**, *2*, 4. [CrossRef]
28. Lizundia, E.; Fortunati, E.; Dominici, F.; Vilas, J.L.; Leon, L.M.; Armentano, I.; Torre, L.; Kenny, J.M. PLLA-grafted cellulose nanocrystals: Role of the CNC content and grafting on the PLA bionanocomposite film properties. *Carbohydr. Polym.* **2016**, *142*, 105–113. [CrossRef]
29. Purnama, P.; Kim, S.H. Synergism of cellulosic nanowhiskers and graft structure in stereocomplex-based materials: Formation in solution and a stereocomplex memory study. *Cellulose* **2014**, *21*, 2539–2548. [CrossRef]
30. Purnama, P.; Kim, S.H. Bio-based composite of stereocomplex polylactide and cellulose nanowhiskers. *Polym. Degrad. Stab.* **2014**, *109*, 430–435. [CrossRef]
31. Drumright, R.E.; Gruber, P.R.; Henton, D.E. Polylactic acid technology. *Adv. Mater.* **2000**, *12*, 1841. [CrossRef]
32. Lunt, J. Large-scale production, properties and commercial applications of polylactic acid polymers. *Polym. Degrad. Stab.* **1998**, *59*, 145–152. [CrossRef]
33. Enomoto, K.; Ajioka, M.; Yamaguchi, A. Polyhydroxycarboxylic Acid and Preparation Process Thereof. U.S. Patent 5,310,865, 10 May 1994.
34. Gruber, P.R.; Hall, E.S.; Kolstad, J.J.; Iwen, M.L.; Benson, R.D.; Borchardt, R.L. Continuous Process for Manufacture of Lactide polymers with Controlled Optical Purity. U.S. Patent 5,142,023, 25 August 1992.
35. Gruber, P.; O'Brien, M. Biopolymer. In *Polylactides: NatureWorks® PLA*; Wiley-VCH: Weinheim, Germany, 2001; Volume 6, Chapter 8.
36. Kricheldorf, H.R. Syntheses and application of polylactides. *Chemosphere* **2001**, *43*, 49. [CrossRef]
37. Kowalski, A.; Duda, A.; Penczek, S. Kinetics and mechanicm of cyclic esters polymerization initiated with tin(II) octoate. 3. Polymerization L,L-lactide. *Macromolecules* **2000**, *33*, 7359–7370. [CrossRef]
38. Urbanczyk, L.; Noundjo, F.; Alexandre, M.; Jérôme, C.; Detrembleur, C.; Calberg, C. Synthesis of polylactide/clay nanocomposites by in situ intercalative polymerization in supercritical carbon dioxide. *Eur. Polym. J.* **2009**, *45*, 643–648. [CrossRef]
39. Purnama, P.; Jung, Y.; Hong, C.H.; Han, D.S.; Kim, S.H. Synthesis of poly(D-lactide) with different molecular weight via melt-polymerization. *Macromol. Res.* **2012**, *20*, 515–519. [CrossRef]
40. Dubois, P.; Jacobs, C.; Jérôme, R.; Teyssié, P. Macromolecular engineering of polylactones and polylactides. 4. Mechanism and kinetics of lactide homopolymerization by aluminum isopropoxide. *Macromolecules* **1991**, *24*, 2266–2270. [CrossRef]
41. Mecerreyes, D.; Jerome, R. From living to controlled aluminium alkoxide mediated ring-opening polymerization of (di)lactones, a powerful tool for the macromolecular engineering of aliphatic polyesters. *Macromol. Chem. Phys.* **1999**, *200*, 2581. [CrossRef]
42. Stevels, W.M.; Dijkstra, P.J.; Feijen, J. New initiators for the ring-opening polymerization of cyclic esters. *Trends Polym. Sci.* **1997**, *5*, 300.
43. O'Keefe, B.J.; Monnier, S.M.; Hillmyer, M.A.; Tolman, W.B. Rapid and controlled polymerization of lactide by structurally characterized ferric alkoxides. *J. Am. Chem. Soc.* **2001**, *123*, 339. [CrossRef]
44. Vink, E.T.H.; Rabago, K.R.; Glassner, D.A.; Gruber, P.R. Applications of life cycle assessment to NatureWorks™ polylactide (PLA) production. *Polym. Degrad. Stab.* **2003**, *80*, 403–419. [CrossRef]
45. Anderson, K.S.; Schreck, K.M.; Hillmyer, M.A. Toughening polylactide. *Polym. Rev.* **2008**, *48*, 85–108. [CrossRef]
46. Kaawashima, N.; Ogawa, S.; Obuchi, S.; Matsuo, M.; Yagi, T. Polylactic acid LACEA. In *Biopolymers*; Doi, Y., Steinbuchel, A., Eds.; Wiley Online Library: Hoboken, NJ, USA, 2002; pp. 251–2744.
47. Auras, R.; Harte, B.; Slke, S. An overview of polylactide as packaging materials. *Macromol. Biosci.* **2004**, *4*, 835–864. [CrossRef]
48. Bouapao, L.; Tsuji, H.; Tashiro, K.; Zhang, J.; Hanesaka, M. Crystallization, spherulite growth, and structure of blends of crystalline and amorphous poly(lactide)s. *Polymer* **2009**, *50*, 4007–4017. [CrossRef]
49. Klemm, D.; Heinze, T.; Heinze, U.; Edgar, K.J.; Philip, B.; Zugenmaier, P. *Comprehensive Cellulose Chemistry*, 2nd ed.; Wiley-VCH: Weinheim, Germany, 2009.
50. Dufresne, A. Nanocellulose: A new ageless biomaterials. *Mater. Today* **2013**, *16*, 220–227. [CrossRef]

51. Gousse, C.; Chanzy, H.; Excoffier, G.; Soubeyrand, L.; Fleury, E. Stable suspension of partially silylated cellulose whiskers dispersed in organic solvents. *Polymer* **2002**, *43*, 2645–2651. [CrossRef]
52. Grunert, M.; Winter, W.T. Nanocomposites of cellulose acetate butyrate reinforced with cellulose Nanocrystals. *J. Polym. Environ.* **2002**, *10*, 27–30. [CrossRef]
53. Habibi, Y.; Goffin, A.L.; Schiltz, N.; Duquesne, E.; Dubois, P.; Dufresne, A. Bionanocomposites based on poly(ε-caprolactone)-grafted cellulose nanocrystals by ring-opening polymerization. *J. Mater. Chem.* **2008**, *18*, 5002–5010. [CrossRef]
54. Azizi Samir, M.A.S.; Alloin, F.; Dufresne, A. Review of recent research into cellulosic whiskers, their properties and their application in nanocomposite field. *Biomacromolecules* **2005**, *6*, 612–626. [CrossRef]
55. Bondeson, D.; Mathew, A.; Oksman, K. Optimization of the isolation of nanocrystals from microcrystalline cellulose by acid hydrolysis. *Cellulose* **2006**, *13*, 171–180. [CrossRef]
56. Beck-Candanedo, S.; Roman, M.; Gray, D.G. Effect of reaction conditions on the properties and behavior of wood cellulose nanocrystal suspensions. *Biomacromolecules* **2005**, *6*, 1048–1054. [CrossRef] [PubMed]
57. Elazzouzi-Hafraoui, S.; Nishiyama, Y.; Putaux, J.L.; Heux, L.; Dubreuil, F.; Rochas, C. The shape and size distribution nanoparticles prepared by acid hydrolysis of native cellulose. *Biomacromolecules* **2008**, *9*, 57–65. [CrossRef] [PubMed]
58. Goffin, A.-L.; Raquez, J.-M.; Duquesne, E.; Siqueira, G.; Habibi, Y.; Dufresne, A.; Dubois, P. From interfacial ring-opening polymerization to melt processing of cellulose nanowhiskers-filled polylactide-based nanocomposites. *Biomacromolecules* **2011**, *12*, 2456–2465. [CrossRef] [PubMed]
59. Sakurada, I.; Nukushina, Y.; Ito, T. Experimental determination of the elastic modulus of crystalline regions in oriented polymers. *J. Polym. Sci.* **1962**, *57*, 651–660. [CrossRef]
60. Nishino, T.; Takano, K.; Nakamae, K. Acetylation of plant cellulose fiber in supercritical carbon dioxide. *J. Polym. Sci. Part B Polym. Phys.* **1995**, *33*, 1647–1651. [CrossRef]
61. Sturcova, A.; Davies, G.R.; Eichhorn, S.J. Elastic modulus and stress-transfer properties of tunicate cellulose whiskers. *Biomacromolecules* **2005**, *6*, 1055–1061. [CrossRef]
62. Rusli, R.; Eichhorn, S.J. Determination of the stiffness of cellulose nanowhiskers and the fiber-matrix interface in a nanocomposite using raman spectroscopy. *Appl. Phys. Lett.* **2008**, *93*, 331111–331113. [CrossRef]
63. Araki, J.; Wada, M.; Kuga, S. Steric stabilization of a cellulose microcrystal suspension by poly(ethylene glycol) grafting. *Langmuir* **2001**, *17*, 21–27. [CrossRef]
64. Zhu, L.; Qiu, J.; Liu, W.; Sakai, E. Mechanical and thermal properties of rice Straw/PLA modified by nano attapulgite/PLA interfacial layer. *Compos. Commun.* **2019**, *13*, 18–21. [CrossRef]
65. Dufresne, A. Interfacial phenomena in nanocomposites based on polysaccharide nanocrystals. *Compos. Interfaces* **2003**, *10*, 369–387. [CrossRef]
66. Ren, Z.; Guo, R.; Bi, H.; Jia, X.; Xu, M.; Cai, L. Interfacial Adhesion of Polylactic Acid on Cellulose Surface: A Molecular Dynamics Study. *ACS Appl. Mater. Interfaces* **2020**, *12*, 3236–3244. [CrossRef] [PubMed]
67. Eichhorn, S.J. Cellulose nanowhiskers: Promising materials for advanced applications. *Soft Matter* **2011**, *7*, 303–315. [CrossRef]
68. Tomé, L.C.; Pinto, R.J.B.; Trovatti, E.; Freire, C.S.R.; Silvestre, A.J.D.; Neto, C.P.; Gandini, A. Transparent bionanocomposites with improved properties prepared from acetylated bacterial cellulose and poly(lactic acid) through a simple approach. *Green Chem.* **2011**, *13*, 419–427. [CrossRef]
69. Siqueira, G.; Bras, J.; Dufresne, A. New process of chemical grafting of cellulose nanoparticles with a long chain isocyanate. *Langmuir* **2010**, *26*, 402–411. [CrossRef]
70. Tingaut, P.; Zimmermann, T.; Lopez-Suevos, F. Syntheis and characterization of bionanocomposites with tunable properties from poly(lactic acid) and acetylated microfibrillated cellulose. *Biomacromolecules* **2010**, *11*, 454–464. [CrossRef]
71. Braun, B.; Dorgan, J.R. Single-step method for the isolation and surface functionalization of cellulosic nanowhiskers. *Biomacromolecules* **2009**, *2*, 334–341. [CrossRef]
72. Bondeson, D.; Oksman, K. Polylactic acid/cellulose whisker nanocomposites modified by polyvinyl alcohol. *Compos. Part A* **2007**, *38*, 2486–2492. [CrossRef]
73. Shojaeiarani, J.; Bajwa, D.S.; Stark, N.M. Green esterification: A new approach to improve thermal and mechanical properties of poly(lactic acid) composites reinforced by cellulose nanocrystals. *J. Appl. Polym. Sci.* **2018**, *135*, 46468. [CrossRef]
74. Shojaeiarani, J.; Bajwa, D.S.; Hartman, K. Esterified cellulose nanocrystals as reinforcement in poly(lactic acid) nanocomposites. *Cellulose* **2019**, *26*, 2349–2362. [CrossRef]
75. Wu, H.; Nagarajan, S.; Shu, J.; Zhang, T.; Zhou, L.; Duan, Y.; Zhang, J. Green and facile surface modification of cellulose nanocrystal as the route to produce poly(lactic acid) nanocomposites with improved properties. *Carbohydr. Polym.* **2018**, *197*, 204–214. [CrossRef]
76. Zhou, L.; Ke, K.; Yang, M.-B.; Yang, W. Recent progress on chemical modification of cellulose for high mechanical-performance Poly(lactic acid)/Cellulose composite: A short review. *Compos. Commun.* **2020**, *23*, 100548. [CrossRef]
77. Riedel, U.; Nickel, J. Natural fibre-reinforced biopolymers as construction materials-new discoveries. *Die Angew. Macromol. Chem.* **1999**, *272*, 34–40. [CrossRef]
78. Oksman, K.; Skrifvars, M.; Selin, J.F. Natural fiber as reinforcement in polylactic acid (PLA) composites. *Compos. Sci. Technol.* **2003**, *9*, 1317–1324. [CrossRef]

79. Braun, B.; Dorgan, J.R.; Knauss, D.M. Reactively compatibilized cellulosic polylactide microcomposites. *J. Polym. Environ.* **2006**, *1*, 49–58. [CrossRef]
80. Favier, V.; Cavaille, J.Y.; Canova, G.R.; Shrivastava, S.C. Mechanical percolation in cellulose whisker nanocomposites. *Polym. Eng. Sci.* **1997**, *10*, 1732–1739. [CrossRef]
81. De Paula, E.L.; Mano, V.; Pereira, F.V. Influence of cellulose nanowhiskers on the hydrolytic degradation behavior of poly(D,L-lactide). *Polym. Degrad. Stab.* **2011**, *96*, 1631–1638. [CrossRef]
82. Mukherjee, T.; Tobin, M.J.; Puskar, L.; Sani, M.-A.; Kao, N.; Gupta, R.K.; Pannirselvam, M.; Quazi, N.; Bhattacharya, S. Chemically imaging the interaction of acetylated nanocrystalline cellulose (NCC) with a polylactic acid (PLA) polymer matrix. *Cellulose* **2017**, *24*, 1717–1729. [CrossRef]
83. Yu, H.Y.; Zhang, H.; Song, M.L.; Zhou, Y.; Yao, J.; Ni, Q.Q. From Cellulose Nanospheres, Nanorods to Nanofibers: Various Aspect Ratio Induced Nucleation/Reinforcing Effects on Polylactic Acid for Robust-Barrier Food Packaging. *ACS Appl. Mater. Interfaces* **2017**, *9*, 43920–43938. [CrossRef]
84. Gwon, J.-G.; Cho, H.-J.; Chun, S.-J.; Lee, S.; Wu, Q.; Lee, S.-Y. Physiochemical, optical and mechanical properties of poly(lactic acid) nanocomposites filled with toluene diisocyanate grafted cellulose nanocrystals. *RSC Adv.* **2016**, *6*, 9438–9445. [CrossRef]
85. Gwon, J.-G.; Cho, H.-J.; Chun, S.-J.; Lee, S.; Wu, Q.; Li, M.; Lee, S.-Y. Mechanical and thermal properties of toluene diisocyanate-modified cellulose nanocrystal nanocomposites using semi-crystalline poly(lactic acid) as base matrix. *RSC Adv.* **2016**, *6*, 73879–73886. [CrossRef]
86. Yin, Y.; Zhao, L.; Jiang, X.; Wang, H.; Gao, W. Cellulose nanocrystals modified with a triazine derivative and their reinforcement of poly(lactic acid)-based bionanocomposites. *Cellulose* **2018**, *25*, 2965–2976. [CrossRef]
87. Jin, K.; Tang, Y.; Zhu, X.; Zhou, Y. Polylactic acid based biocomposite films reinforced with silanized nanocrystalline cellulose. *Int. J. Biol. Macromol.* **2020**, *162*, 1109–1117. [CrossRef] [PubMed]
88. Dhar, P.; Tarafder, D.; Kumar, A.; Katiyar, V. Thermally recyclable polylactic acid/cellulose nanocrystal films through reactive extrusion process. *Polymer* **2016**, *87*, 268–282. [CrossRef]
89. Dhar, P.; Gaur, S.S.; Soundararajan, N.; Gupta, A.; Bhasney, S.M.; Milli, M.; Kumar, A.; Katiyar, V. Reactive Extrusion of Polylactic Acid/Cellulose Nanocrystal Films for Food Packaging Applications: Influence of Filler Type on Thermomechanical, Rheological, and Barrier Properties. *Ind. Eng. Chem. Res.* **2017**, *56*, 4718–4735. [CrossRef]
90. Miao, C.; Hamad, W.Y. In-situ polymerized cellulose nanocrystals (CNC)-poly(l-lactide) (PLLA) nanomaterials and applications in nanocomposite processing. *Carbohydr. Polym.* **2016**, *153*, 549–558. [CrossRef]
91. Mujica-Garcia, A.; Hooshmand, S.; Skrifvars, M.; Kenny, J.M.; Oksman, K.; Peponi, L. Poly(lactic acid) melt-spun fibers reinforced with functionalized cellulose nanocrystals. *RSC Adv.* **2016**, *6*, 9221–9231. [CrossRef]
92. Tsuji, H.; Ikada, Y. Stereocomplex formation between enantiomeric poly(lactic acid)s. 6. Binary blends from copolymers. *Macromolecules* **1992**, *25*, 5719–5723. [CrossRef]
93. Yamane, H.; Sasai, K. Effect of the addition of poly(D-lactic acid) on the termal property of poly(L-actic acid). *Polymer* **2003**, *44*, 2569–2575. [CrossRef]
94. Fukushima, K.; Chang, Y.H.; Kimura, Y. Enhanced stereocomplex formation of poly(L-lactic acid) and poly(D-lactic acid) in the presence of stereoblock poly(lactic acid). *Macromol. Biosci.* **2007**, *7*, 829–835. [CrossRef]
95. Tsuji, H.; Ikada, Y. Stereocomplex formation between enantiomeric poly(lactic acid)s. 9. Stereocomplexation from the melt. *Macromolecules* **1993**, *26*, 6918–6926. [CrossRef]
96. Fukushima, K.; Kimura, Y. An efficient solid-state polycondensation method for synthesizing stereocomplexed poly(lactic acid)s with high molecular weight. *J. Polym. Sci. A Polym. Chem.* **2008**, *46*, 3714–3722. [CrossRef]
97. Anderson, K.S.; Hillmyer, M.A. Melt preparation and nucleation efficiency of polylactide stereocomplex crystallites. *Polymer* **2006**, *47*, 2030–2035. [CrossRef]
98. Purnama, P.; Kim, S.H. Rapid stereocomplex formation of polylactide using supercritical fluid technology. *Polym. Int.* **2012**, *61*, 939–942. [CrossRef]
99. Purnama, P.; Kim, S.H. Stereocomplex formation of polylactide using microwave irradiation. *Polym. Int.* **2014**, *63*, 741–745. [CrossRef]
100. Tsuji, H.; Ikada, Y. Stereocomplex formation between enantiomeric poly(lactic acid)s. XI. Mechanical properties and morphology of solution-cast films. *Polymer* **1999**, *40*, 6699–6708. [CrossRef]
101. Dorgan, J.R.; Braun, B. Polymer Composites Incorporating Stereocomplexation. US Patent 2011/0319509 A1, 29 December 2011.
102. Habibi, Y.; Aouadi, S.; Raquez, J.-M.; Dubois, P. Effect of interfacial stereocomplexation in cellulose nanocrystal-filled polylactide nanocomposites. *Cellulose* **2013**, *20*, 2877–2885. [CrossRef]
103. Wu, H.; Nagarajan, S.; Zhou, L.; Duan, Y.; Zhang, J. Synthesis and characterization of cellulose nanocrystal-graft-poly(D-lactide) and its nanocomposite with poly(L-lactide). *Polymer* **2016**, *103*, 365–375. [CrossRef]
104. Cao, X.; Wang, Y.; Chen, H.; Hu, J.; Cui, L. Preparation of different morphologies cellulose nanocrystals from waste cotton fibers and its effect on PLLA/PDLA composites films. *Compos. Part B* **2021**, *217*, 108934. [CrossRef]
105. Raquez, J.-M.; Murena, Y.; Goffin, A.-L.; Habibi, Y.; Ruelle, B.; DeBuyl, F.; Dubois, P. Surface-modification of cellulose nanowhiskers and their use as nanoreinforcers into polylactide: A sustainability-integrated approach. *Compos. Sci. Technol.* **2012**, *72*, 544–549. [CrossRef]

106. Biela, T.; Duda, A.; Penczek, S. Enhanced melt stability of star-shaped stereocomplexes as compared with linear stereocomplexes. *Macromolecules* **2006**, *39*, 3710–3713. [CrossRef]
107. Tsuji, H. Poly(lactide) stereocomplex: Formation, structure, properties, degradation, and applications. *Macromol. Biosci.* **2005**, *5*, 569–597. [CrossRef] [PubMed]
108. Purnama, P.; Jung, Y.; Kim, S.H. Melt stability of 8-arms star-shaped stereocomplex polylactide with three-dimensional core structures. *Polym. Degrad. Stab.* **2013**, *96*, 1097–1101. [CrossRef]
109. Oksman, K.; Mathew, A.P.; Sain, M. Novel bionanocomposites: Processing, properties and potential applications. *Plast. Rubber Compos.* **2009**, *38*, 396–405. [CrossRef]
110. Spunella, S.; Samuel, C.; Raquez, J.-M.; McCallum, S.A.; Gross, R.; Dubois, P. Green and efficient sytnthesis of dispersible cellulose nanocrystals in biobased polyesters for engineering applications. *ACS Sustain. Chem. Eng.* **2016**, *4*, 2517–2527. [CrossRef]
111. Sucinda, E.F.; Abdul Majid, M.S.; Ridzuan, M.J.M.; Cheng, E.M.; Alshahrani, H.A. Development and characterization of packaging film from napier cellulose nanowhisker reinforced polylactic acid (PLA) bionanocomposites. *Int. J. Biol. Macromol.* **2021**, *18*, 43–53. [CrossRef] [PubMed]
112. Pal, N.; Banerjee, S.; Roy, P.; Pal, K. Melt-blending of unmodified and modified cellulose nanocrystals with reduced graphene oxide into PLA matrix for biomedical application. *Polym. Adv. Technol.* **2019**, *30*, 3049–3060. [CrossRef]
113. Pal, N.; Banerjee, S.; Roy, P.; Pal, K. Reduced graphene oxide and PEG-grafted TEMPO-oxidized cellulose nanocrystal reinforced poly-lactic acid nanocomposite film for biomedical application. *Mater. Sci. Eng. C* **2019**, *104*, 109956. [CrossRef]

*Article*

# Bio-Based Poly(lactic acid)/Poly(butylene sebacate) Blends with Improved Toughness

**Adriana Nicoleta Frone *, Marius Stelian Popa, Cătălina Diana Uşurelu, Denis Mihaela Panaitescu, Augusta Raluca Gabor, Cristian Andi Nicolae, Monica Florentina Raduly, Anamaria Zaharia and Elvira Alexandrescu**

National Institute for Research and Development in Chemistry and Petrochemistry—ICECHIM, 202 Splaiul Independentei, 060021 Bucharest, Romania
* Correspondence: adriana.frone@icechim.ro

**Abstract:** A series of poly(butylene sebacate) (PBSe) aliphatic polyesters were successfully synthesized by the melt polycondensation of sebacic acid (Se) and 1,4-butanediol (BDO), two monomers manufactured on an industrial scale from biomass. The number average molecular weight ($M_n$) in the range from 6116 to 10,779 g/mol and the glass transition temperature ($T_g$) of the PBSe polyesters were tuned by adjusting the feed ratio between the two monomers. Polylactic acid (PLA)/PBSe blends with PBSe concentrations between 2.5 to 20 wt% were obtained by melt compounding. For the first time, PBSe's effect on the flexibility and toughness of PLA was studied. As shown by the torque and melt flow index (MFI) values, the addition of PBSe endowed PLA with both enhanced melt processability and flexibility. The tensile tests and thermogravimetric analysis showed that PLA/PBSe blends containing 20 wt% PBSe obtained using a BDO molar excess of 50% reached an increase in elongation at break from 2.9 to 108%, with a negligible decrease in Young's modulus from 2186 MPa to 1843 MPa, and a slight decrease in thermal performances. These results demonstrated the plasticizing efficiency of the synthesized bio-derived polyesters in overcoming PLA's brittleness. Moreover, the tunable properties of the resulting PBSe can be of great industrial interest in the context of circular bioeconomy.

**Keywords:** poly(butylene sebacate); biopolymer blends; thermal analysis; dynamic mechanical analysis

**Citation:** Frone, A.N.; Popa, M.S.; Uşurelu, C.D.; Panaitescu, D.M.; Gabor, A.R.; Nicolae, C.A.; Raduly, M.F.; Zaharia, A.; Alexandrescu, E. Bio-Based Poly(lactic acid)/Poly(butylene sebacate) Blends with Improved Toughness. *Polymers* **2022**, *14*, 3998. https://doi.org/10.3390/polym14193998

Academic Editors: José Miguel Ferri, Vicent Fombuena Borràs and Miguel Fernando Aldás Carrasco

Received: 5 September 2022
Accepted: 20 September 2022
Published: 24 September 2022



## 1. Introduction

The persistent consumption of the already scarce global oil reserves and the increasing public awareness of the environmental problems caused by the improper management of plastic waste have encouraged the development of bio-based polymers as alternatives for the conventional non-biodegradable petroleum-based plastics. A very attractive substitute for the oil-based plastics is polylactic acid (PLA), an aliphatic polyester [1] whose synthesis is currently achieved by two main routes: the ring-opening polymerization (ROP) of the *L*-lactide dimer in the presence of a metal-based or organic catalyst, and the direct polycondensation of lactic acid [2]. Of the two methods, ROP is preferred, since it allows for the obtaining of PLA with high molecular weights in a shorter time [3], as well as good control over the molecular weight and polydispersity of the resulting polymer [4]. Regardless of the synthesis route, a great benefit is the possibility of obtaining the starting monomer—lactic acid—by the bacterial fermentation of sugars from renewable sources, such as corn, potato, wheat, rice, sugarcane, sugar beet pulp [5], or agricultural waste [6]. The most appealing characteristics of PLA reside in its origin from renewable sources, as well as its biodegradability, biocompatibility, and non-toxicity [7]. Additionally, its high tensile strength and elastic modulus, which are comparable to those of conventional oil-based polymers [8], and the possibility of processing PLA using all techniques characteristic of petroleum-based polymers [9], make PLA one of the most in-demand bioplastics on the market [10]. Today, PLA is used in a wide range of applications, from disposable packaging and food service ware [11]; to implants, drug delivery systems, scaffolds, sutures, wound dressings, and coatings in biomedicine [12]; toys and home appliances [13]; textiles [11]; and mulch films in agriculture [14]. Consequently, the global production of PLA is expected to grow from the almost 200,000 tons/year reported in 2019, to 370,000 tons/year by 2023 [15]. Nevertheless, the use of PLA on a larger scale is restrained due to its unsatisfactory properties, such as its inherent brittleness, low elongation at break, poor impact strength, low thermal stability, low heat distortion temperature, narrow processing window, and slow crystallization rate [16]. To increase the flexibility, ductility, toughness, and processability of PLA, several methods have been applied, plasticization, copolymerization, and physical blending being some of the most popular choices [17]. In order not to compromise the bio-based origin of PLA, in the last two decades, researchers have focused on finding effective agents for improving the flexibility and toughness of PLA that are entirely, or at least partially, derived from renewable sources [18].

An interesting solution to overcome the brittleness of PLA is the use of poly(*L*,*D*-lactic acid) (PLDLA), a copolymer obtained by the polymerization of a racemic mixture of *L*-lactide and *D*-lactide enantiomers. Unlike poly(*L*-lactic acid) (denoted here by PLA), which is a semicrystalline polymer, PLDLA is an amorphous, atactic polymer with higher structural flexibility [19] and thus, higher processability as compared to PLA.

The addition of plasticizers to PLA is known to induce an increase in the mobility of the PLA chains, along with a decrease in glass transition temperature ($T_g$) and melt viscosity, which improves the ductility and processability of PLA [20]. Lactide and oligomers of lactic acid, citrate and adipate esters, poly(ethylene glycol), poly(propylene glycol), glycerol, and glycerol derivatives are just a few bio-based plasticizing agents that have been proposed for adjusting PLA's flexibility. However, cardanol and derivatives of cardanol [21,22], levulinic acid esters [23], maleic acid and its esters [21], and vegetable oils and their epoxidized/maleinized/acrylated/hydroxylated derivatives [24] are also plasticizers originating from renewable resources that have been studied in the last decade for enhancing PLA properties. Although many of them have proven to be extremely effective in improving the flexibility and toughness of PLA, they still exhibit some limitations [23,25,26]. For example, by employing acetylated malic acid butyl ester as a bio-based plasticizer for PLA, Park et al. [25] obtained an increase in the elongation at break of PLA from around 16 to 648%, with a simultaneous decrease in the $T_g$ from 60 °C to 38 °C. However, the increase in flexibility was accompanied by a significant decrease in the tensile strength and elastic modulus of PLA. Moreover, the plasticizer began to vaporize from the material at about 200 °C, which is not convenient, considering that neat PLA's processing is usually conducted at 160–220 °C [27]. By the addition of 20 wt% of renewable sources-derived glycerol dilevulinate (ED) in a PLA formulation, Xuan et al. [23] reported a remarkable increase in the elongation at break of PLA to 546% (from 5% for pure PLA) and a significant decrease in the $T_g$ to 15 °C (from 59 °C for neat PLA). In another work, Zych et al. [26] reported an improvement in the elongation at break of PLA from around 2 to 368% and 783%, respectively, along with a decrease in the $T_g$ value from 71.3 °C to 51.2 °C and 47.0 °C, respectively, for PLA formulations plasticized with 10 and 40 wt% epoxidized soybean oil methyl ester (ESOME). Moreover, about 192-fold increase in the PLA toughness was achieved at a content of ESOME of 10 wt%, while above this concentration, phase separation processes were observed. Nevertheless, both ED and ESOME caused drastic decreases in the tensile strength and elastic modulus of PLA and showed an increased tendency of migration in different food simulants [23,26].

The copolymerization of *L*-lactide with different monomers emerged as another attractive technique to obtain PLA-based materials with increased flexibility and toughness [28]. In this sense, comonomers such as glycolide, ethylene glycol, ε-caprolactone, δ-valerolactone [29], β-methyl-δ-valerolactone [30], 1,5-dioxepan-2-one [31], 1,3-trimethylene carbonate [29], and ethylene carbonate [32] have been studied over time. The limitations of the copolymerization method are related to the long reaction times, high costs, difficulty in establishing and controlling the reaction conditions (reaction temperature and time, comonomer feed ratio, type of catalyst, amount of catalyst, etc.) in order to obtain copolymers with the desired properties, and the more laborious transfer of the process to an industrial scale as compared to the plasticization method [28]. Zhang et al. [33] reported the synthesis of poly(lactic acid-co-ε-caprolactone) copolymers with different flexibilities by the ROP of *L*-lactide and ε-caprolactone at 150 °C, using stannous octoate as a catalyst and varying the feed molar ratio between the two monomers and polymerization time. An elongation at break as high as 2661.3 ± 575.9% and improved toughness were obtained at a molar feed ratio of 1:1 between *L*-lactide and ε-caprolactone and a reaction time of 30 h. In another study, Xi et al. [34] obtained a flexible poly(*L*-lactide-co-trimethylene carbonate) copolymer with an elongation at break of 512% and a $T_g$ of 34.1 °C by the ROP of *L*-lactide and trimethylene carbonate at 130 °C for 72 h, using stannous octoate as catalyst and a molar ratio of 60/40 between the two monomers.

A less costly, more time-efficient, and simple technique to obtain PLA-based materials with increased flexibility and toughness is blending PLA [35] with "softer" polymers that possess lower $T_g$, higher flexibility, and increased ductility as compared to PLA [36]. In this regard, the use of polymers entirely or partially derived [36] from renewable sources, such as starch [37], poly(butylene succinate), poly(butylene succinate-co-adipate) [38], poly(glycerol sebacate) [37], poly(butylene succinate-co-terephthalate), poly(butylene adipate-co-terephthalate) [38], etc. has been explored. An essential requirement in this technique is the existence of strong interfacial interactions, i.e., a good compatibility between the constituent polymers of the blend [39]. Poor compatibility between polymers leads to the aggregation of the particles of the more flexible polymer (the dispersed phase) in the PLA matrix, and an insufficient improvement in the elongation at break and toughness of PLA [40]. Unfortunately, most bio-based flexible polymers show poor compatibility with PLA. However, promising results were obtained by Hu et al. [41], who prepared blends using PLA and a series of self-synthesized bio-based poly(lactate/butanediol/sebacate/itaconate) (PLBSI) elastomers. An elongation at break as high as 324% and an impact strength of 35.7 kJ/m$^2$ (approximately 50 and 15-fold higher than those of neat PLA) were reported at a content of PLBSI in the blend of 15 wt%. These results indicated a high compatibility between the two polymers, which was attributed to the structural similarities (lactate units, ester groups) between PLA and PLBSI. In another study, Kang synthesized a bio-sourced elastomer (BE) [42] based on sebacic acid, itaconic acid, succinic acid, 1,3-propanediol, and 1,4-butanediol as toughener for PLA. With a BE content of 11.5 vol% in the PLA/BE blend, a maximum elongation at break of 179% and a notched impact strength of 10.3 kJ/m$^2$ were obtained, which are 25-fold and 4-fold higher, respectively, than those of neat PLA.

Another promising candidate for improving the flexibility and toughness of PLA may be poly(butylene sebacate) (PBSe), a bio-based and biodegradable aliphatic polyester that can be synthesized by the polycondensation of sebacic acid (Se) with 1,4-butanediol (BDO) [43]. The bio-based character of PBSe comes from the fact that both of its precursor monomers can be obtained from renewable sources, Se being produced by the alkaline pyrolysis of castor oil [44] and BDO being obtainable via the microbial fermentation of sugars from renewable sources [45]. Similar to PLA, PBSe is a biodegradable polymer in composting conditions [46]. With a $T_g$ of −29.8 °C [43], PBSe possesses a good flexibility, which could be attributed to the relatively long methylene (-$CH_2$-) chain segments in its structure. The presence of ester bonds in both PBSe and PLA structures may ensure some compatibility between the two polymers, while the flexible chains of PBSe are expected to cause an increase in the flexibility of PLA.

In this work, a series of bio-based polyesters aimed to improve the ductility and toughness of PLA were synthesized by the polycondensation of sebacic acid and 1,4-butanediol, two monomers that are obtained at a commercial scale from renewable raw materials. The polycondensation reaction was carried out by varying the Se: BDO molar feed ratio, and the resulting polyesters were characterized in terms of molecular weight, polydispersity, structure, and thermal properties. Subsequently, the synthesized polyesters were employed in the preparation of PLA/polyester blends via melt compounding using different concentrations of PBSe and carefully established processing parameters. Since PBSe is bio-sourced and biodegradable, the renewability and biodegradability of PLA were not compromised. Moreover, the effects of the synthesized polyesters on the thermal properties, mechanical properties, morphology, and rheological behavior of PLA were investigated and discussed. To the best of our knowledge, this is the first time when PBSe was used as a plasticizer for tuning the ductility of PLA. The resulting PLA/PBSe blends are promising materials for the fabrication of sustainable children's toys or other goods which require increased flexibility, in addition to other properties.

## 2. Materials and Methods

### 2.1. Materials

Polylactic acid (PLA) Ingeo 4043D, with an *L*-lactide content of 98%, an average molecular weight of 111 kDa, and a density of 1.24 g/cm$^3$, was supplied by Nature Works (Blair, NE, USA). Sebacic acid (Se) (purity 99%) and titanium (IV) butoxide (TBT) (purity 97%) were purchased from Aldrich (Milwaukee, WI, USA), while 1,4-butanediol (BDO) (purity 99%) was procured from Merck Co. (Darmstadt, Germany). All chemicals were used as received without further purification.

### 2.2. Synthesis of PBSe

PBSe was synthesized via a two-step melt polycondensation reaction, as illustrated in Scheme 1. For this, the calculated amounts of Se and BDO were charged in a 250 mL three-neck round-bottom reaction flask equipped with a thermometer, a nitrogen inlet, and a condenser for collecting the water that results as a condensation by-product. The molar feed ratio between Se and BDO was varied so that different excesses of BDO and Se, respectively, were ensured. The amounts of Se and BDO employed in each experiment and the code name for the poly(butylene sebacate) obtained in each case are listed in Table 1. Each of the samples was denoted by "PBSey," where y is an indicator of the ratio between the two reactants in the feed mixture, as shown in Table 1. In the first stage of the reaction, called "esterification," the reaction mixture was heated at 190 °C for 1 h, under rigorous stirring and nitrogen atmosphere, in the absence of any catalyst. During this step, sebacic acid reacted with 1,4-butanediol, with the formation of short chain oligoesters and water as a by-product. In the second stage, called "polycondensation," TBT (0.75 wt% relative to the Se amount) was added as a transesterification catalyst, and the reaction mixture was heated at 200 °C for 4 h, under stirring and nitrogen atmosphere. During this stage, the oligomers resulted in the "esterification stage" condensed to form polyesters with higher molecular weights. Toward the end of the reaction, a significant increase in the viscosity of the reaction mixture and hindered stirring were observed, indicating that an important increase in the molecular weight and chain length of the reaction products had taken place. At the end of the reaction, the resulting molten mass of polyesters was poured onto a glass surface to solidify. Subsequently, the obtained polyesters were used in the obtaining of PLA-based blends. Similar work concerning the synthesis of poly(butylene sebacates) by polycondensation was reported by Kim et al. [43].

All the synthesized PBSe, regardless of the molar composition of the feed mixture, were solid and tough at room temperature, having a wax-like appearance. The PBSe sample prepared using the highest excess of BDO (PBSe2) was slightly sticky as compared with the other polyesters from the series. The polyesters obtained by employing a BDO excess (PBSe1, PBSe2) exhibited a pale white color, while the ones prepared using an Se excess (PBSe3, PBSe4) were a pale cream color.

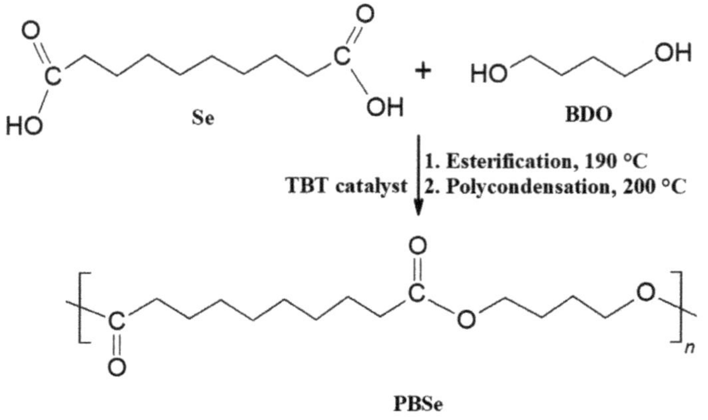

Scheme 1. Reaction scheme for the synthesis of PBSe.

**Table 1.** Code name, feed molar ratio, molecular weight, polydispersity index (PDI), and retention time ($R_t$) for the synthesized PBSe samples.

| Sample Code | Se:BDO Molar Ratio | $M_n$ (g/mol) | PDI | $R_t$ (min) |
|---|---|---|---|---|
| PBSe1 | 1:1.1 | 10,779± 490 | 1.28 | 1.8 |
| PBSe2 | 1:1.5 | 7620 ± 22 | 1.15 | 1.4 |
| PBSe3 | 1.1:1 | 7812 ± 37 | 1.18 | 1.6 |
| PBSe4 | 1.5:1 | 6116 ± 47 | 1.13 | 1.2 |

### 2.3. Preparation of PLA/PBSe Blends

PLA/PBSe blends were prepared by melt mixing PLA and PBSe in a 30 $cm^3$ chamber of a Brabender mixer (Brabender GmbH & Co. KG, Duisburg, Germany), for 13 min, using carefully established temperature/rotor speed/time profiles in order to ensure a good incorporation of the PBSe in the PLA matrix. Based on repeated experimental trials, the temperature/rotor/speed/time profile was set to include three stages: (i) temperature: 165 °C; rotor speed: 100 rpm; mixing time: 2 min; (ii) 155 °C; 50 rpm; 5 min; (iii) 155 °C; 100 rpm; 5 min. These conditions ensured both a rapid melting of PLA (first stage) and a good incorporation of PBSe. The concentration of PBSe in the blends was varied in the range between 2.5 and 20 wt%. The code names and compositions of the PLA/PBSey (y being 1, 2, 3, or 4) blends are summarized in Table 2. Films of the PLA/PBSe blends having a uniform thickness of 1.00 mm (±0.03 mm) were obtained by compression-molding using a laboratory P200E platen press (Dr. Collin, Maitenbeth, Germany) at 170 °C under the following conditions: pre-heating with 0.5 MPa for 150 s, pressing at 10 MPa for 60 s, and cooling in a cooling cassette at 0.5 MPa for 60 s. A reference sample based on pristine PLA was processed under the same conditions.

**Table 2.** Code names and compositions of the prepared PLA/PBSe blends.

| Sample | PLA (wt%) | PBSey (wt%) |
|---|---|---|
| PLA | 100 | 0 |
| PLA/2.5PBSey | 97.5 | 2.5 |
| PLA/5PBSey | 95 | 5 |
| PLA/7.5PBSey | 92.5 | 7.5 |
| PLA/10PBSey | 90 | 10 |
| PLA/20PBSey | 80 | 20 |

### *2.4. Characterization Methods*

#### 2.4.1. Fourier-Transform Infrared Spectroscopy (FTIR) Analysis

The FTIR absorbance spectra of the obtained polyesters were recorded using a Jasco FTIR 6300 instrument (JASCO Int. Co., Ltd., Tokyo, Japan). The PBSe polyesters and Se were firstly ground with KBr powder and pressed to form disc samples. Then, the FTIR data were collected between 400 and 4000 $cm^{-1}$, with a 4 $cm^{-1}$ resolution and 32 scans.

#### 2.4.2. Size Exclusion Chromatography (SEC-GPC)

The molecular weight distributions of the PBSe polyesters were determined using an Agilent Technologies 1200 series gel permeation chromatograph equipped with a PLgel Mixed-C column (300 × 7.5 mm) and an Agilent 1200 differential refractometer, in dimethylformamide (DMF), at 30 °C and a flow rate of 1 mL/min. The calibration was made using polystyrene standards. The PBSe samples were dissolved in HPLC grade DMF (0.1 g/5 mL) and filtered before analysis.

#### 2.4.3. Thermal Properties Analysis

Thermogravimetric analysis (TGA) was used to characterize the thermal stability of both the synthesized PBSe and the PLA/PBSe blends. TGA measurements were carried out according to ISO 11358. TA Q5000 equipment (TA Instruments Inc., New Castle, DE, USA) using nitrogen as a purge gas at a flow rate of 40 mL/min was utilized for these measurements. All the samples, polyesters, and blends were placed in platinum pans and heated from 25 to 700 °C at a heating rate of 10 °C/min.

Modulated differential scanning calorimetry (MDSC) analysis of the PBSe synthesized polyesters was carried out using a DSC Q2000 from TA Instruments (New Castle, DE, USA) under helium flow (25 mL/min), with the base rate of 10 °C/min, amplitude of 0.8 °C/min, and a period of 30 s, as follows: cooling from 30 °C to −90 °C; heating from −90 °C to 105 °C and equilibrating for 3 min for erasing the thermal history, cooling down to −90 °C, isothermal for 3 min, and reheating to 105 °C. Instrument calibration was performed according to ASTM E 967 using indium reference.

#### 2.4.4. Scanning Electron Microscopy (SEM)

The compression molded sheets of pristine PLA and PLA/PBSe blends were fractured in liquid nitrogen and then sputter-coated (Q150R Plus, Quorum, SXE, Lewes, UK) with a 5 nm layer of gold. The surface of the fractured blends was analyzed by SEM using an environmental scanning electron microscope (ESEM-FEI Quanta 200, Eindhoven, The Netherlands) working in low vacuum, large field detector mode, at 15 kV accelerating voltage.

#### 2.4.5. Mechanical Property Analysis

Tensile tests on the PLA/PBSe blends were conducted according to ISO 527-3:2018 (applicable to films) at room temperature using an Instron 3382 universal testing machine (Norwood, MA, USA) with a load cell of 10 kN. The tensile tests were performed on rectangular samples of 60 mm × 10 mm × 1 mm (length × width × thickness), which were cut from the previously compression-molded sheets. For each sample, more than 5 specimens were measured with a crosshead speed of 2 mm/min. The average values and the standard deviations for Young's modulus (YM), maximum tensile strength ($\sigma_{max}$), and elongation at break ($\varepsilon_B$) were calculated using the Bluehill 2 Software (Instron, Norwood, MA, USA).

A DMA Q800 instrument (TA Instruments, New Castle, DE, USA) was used to carry out the dynamic mechanical analysis (DMA) of pristine PLA and PLA/PBSe blends, following the guidance of ASTM D 5206 (standard test method for plastics: dynamic mechanical properties: in tension). Bar specimens of 12 mm × 6 mm × 0.5 mm (length × width × thickness), cut from the pristine PLA and PLA/PBSe blends compressed plates, were measured in multi-frequency-strain mode (tension clamp) from room temperature to

145 °C, with a heating rate of 3 °C/min and a frequency of 1 Hz. The storage modulus and the tan δ or damping factor were plotted against temperature.

2.4.6. Melt Rheology Evaluation of PLA/PBSe Blends

The melt rheological properties were measured by melt flow index (MFI) using a plastometer capillary rheometer (LMI 4003 Melt Indexer, Dynisco Polymer Testing, Franklin, MA, USA). The measurements were carried out at temperature of 190 °C, using a load of 2.16 kg (ASTM 1238), in five replicates.

## 3. Results and Discussion

### 3.1. SEC-GPC Analysis

The number average molecular weight ($M_n$), polydispersity index (PDI), and retention time ($R_t$) of the synthesized PBSe polyesters, determined by SEC-GPC, are given in Table 1. As shown in Table 1, the polyesters obtained at a Se:BDO molar ratio of 1:1.1 showed the highest $M_n$ and $R_t$ values. Increasing the content of Se or BDO in the feed mixture led to lower $M_n$ and $R_t$ values. Indeed, the PBSe polyester samples prepared using a higher amount of Se or BDO in the polymerization substrate displayed smaller $M_n$ values, which correlates with the retention time values [47]. Moreover, all the synthesized PBSe polyesters, regardless of the Se:BDO ratio, showed PDI values close to 1.

The lower $M_n$ of the polymers obtained at different excesses of Se in the feed mixture can be attributed to the formation in the "esterification" stage of a larger number of oligoesters possessing strictly COOH groups at both of their ends. These oligomers did not have at their disposal sufficient oligoesters with at least one OH end group for condensation and growth in the second stage of the reaction, and thus, their chain extension was limited, and a PBSe with lower $M_n$ was formed [48]. The same explanation is valid for the lower $M_n$ obtained when a molar excess of 50% diol was added to the polycondensation reaction, i.e., for the PBSe2 sample. However, it should be noted that a higher $M_n$ was achieved when a molar excess of 10% diol was employed, i.e., for the PBSe1 sample. This can be explained by considering the volatility of BDO, which could have evaporated, to a small extent, at the high temperatures at which the polycondensation was conducted [49]. Thus, the excess of BDO was likely lost by evaporation, and an equimolar ratio between BDO and Se may have actually remained in the polymerization substrate. An equimolar ratio between Se and BDO provides the opportunity for the formation, in the "esterification stage," of oligoesters that possess COOH groups at both ends, OH groups at both ends, or an OH group at one end and a COOH group at the other, which increases the possibilities of chain growth in the second stage of the polycondensation reaction, leading to PBSes with superior $M_n$.

According to the DSC measurements, the synthesized PBSe polyesters had a $T_g$ between −15 and −27 °C and a melting transition in the temperature range from 49 to 52 °C, depending on the molar ratio between Se and BDO (Table 3). All the polyester samples also displayed a crystallization event, which indicates them as semi-crystalline materials.

**Table 3.** Thermal properties of the synthesized PBSe.

| Sample | $T_{d,10\%}$, °C | $T_{d,max1}$, °C | $T_{d,max2}$, °C | $WL_{165^\circ C}$ % | $R_{700^\circ}$, % | $T_g$ *, °C | $T_m$ **, °C | $T_c$ ***, °C |
|---|---|---|---|---|---|---|---|---|
| PBSe1 | 284. | 411 | 459 | 3. 65 | 1.70 | −27 | 52 | 38 |
| PBSe2 | 291 | 418 | 464 | 2.46 | 1.05 | −23 | 51 | 37 |
| PBSe3 | 362 | 415 | 463 | 0.61 | 1.22 | −16 | 49 | 38 |
| PBSe4 | 279 | 418 | 463 | 0.31 | 1.02 | −14 | 47 | 40 |

* $T_g$ from DSC 1st cooling cycle. ** $T_m$ from DSC 2nd heating cycle. *** $T_c$ from DSC 2nd cooling cycle.

### 3.2. FTIR Analysis

FTIR analysis was used to characterize the molecular structure of the PBSe polyesters. Figure 1 shows the FTIR spectra of the synthesized PBSe, and the Se used as a reference. It can be observed that all PBSe spectra showed similar absorption peaks. The absorptions at

2930 $cm^{-1}$ and 2852 $cm^{-1}$ were assigned to the symmetrical and the antisymmetric stretching vibration of methylene groups in the polyester structures ($-CH_2$), respectively [50,51]. The intensity of these two bands was greater for the PBSe samples with a higher Se content in their composition. The spectra of all PBSes also displayed a band at around 3450 $cm^{-1}$, which can be attributed to the O-H bonds from the hydroxyl end groups of the PBSes chains which, as expected, are more intense for the PBSes obtained at an excess of BDO in the polymerization substrate [50].

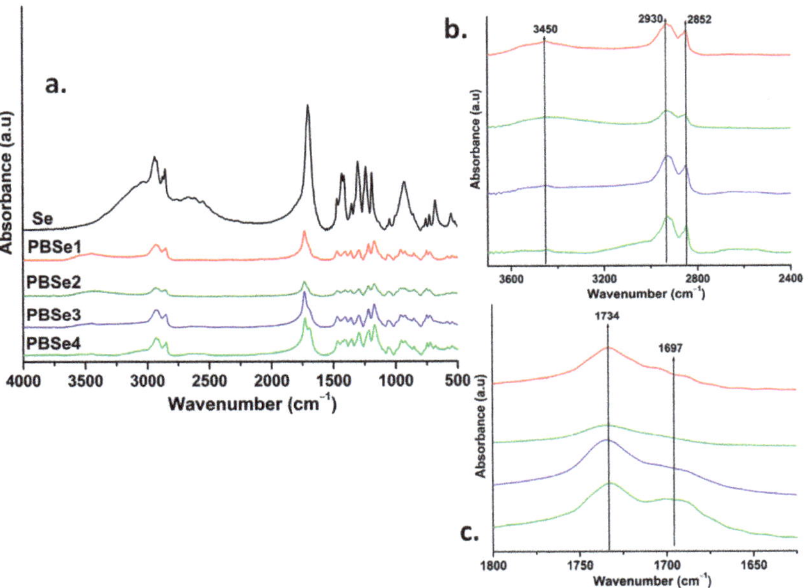


**Figure 1.** FTIR spectra of sebacic acid (Se) and PBSe polyesters (**a**) and FTIR spectra of PBSes in the range from 3600 to 2500 $cm^{-1}$ (**b**) and 1800 to 1675 $cm^{-1}$ (**c**).

An intense absorption band was seen at 1734 $cm^{-1}$ in the FTIR spectra of all synthesized PBSe polyesters, which was associated with the stretching vibration of the carbonyl group (C=O), thus indicating the formation of ester bonds. Furthermore, the absorption at 1173 $cm^{-1}$ confirms the presence of the C–O–C=O groups of ester bonds in the structure of all synthesized PBSe [50–52]. The characteristic absorption bands from the structure of Se situated at 1697, 1300, and 930 $cm^{-1}$, which are specific to carboxylic acid groups, were not detected in the spectra of PBSe. All these results reveal the successful formation of polyester molecular structures.

### 3.3. TGA Analysis of PBSe

The TGA and DTG curves of PBSe are shown in Figure 2. The related thermal parameters, i.e., the temperatures at 10% weight loss ($T_{d,10\%}$) and at the maximum degradation rate ($T_{d,max1}$, $T_{d,max2}$), the weight loss at the PLA/PBSe processing temperature of 165 °C, and the residue at 700 °C ($R_{700}$) are listed in Table 3.

Similar two-step degradation processes were observed for all PBSe polyesters, with some small differences (Figure 2). Thus, two additional small degradation shoulders situated at around 255 °C and 349 °C were observed before the main decomposition peak in the DTG curve of the PBSe prepared with the largest excess of Se, and a broad degradation peak with a maximum at 264 °C was noted in the DTG curve of the PBSe having the highest BDO content (Figure 2). These differences could indicate the degradation at lower temperatures of the polyester with an excess of sebacic terminal moieties (PBSe4) in the first case and of the PBSe with an excess of BDO as in the case of PBSe2. Indeed,

Bikiaris et al. stated that the low degradation temperatures of some aliphatic polyesters obtained from succinic acid and diols with different lengths (2, 4, 6, 8, and 10 methylene groups) could be due to the existence in these products of some oligomers with lower degradation temperatures than those of the larger macromolecules, which were difficult to remove during polycondensation, since they were prepared from diols with high boiling points, as in our case (the boiling point of BDO being at 230 °C) [53]. The previously mentioned PBSe polyester samples had the lowest $M_n$ and $R_t$ values, meaning that they contain higher amounts of products with lower molecular weights, which are the first to decompose during heating [47].

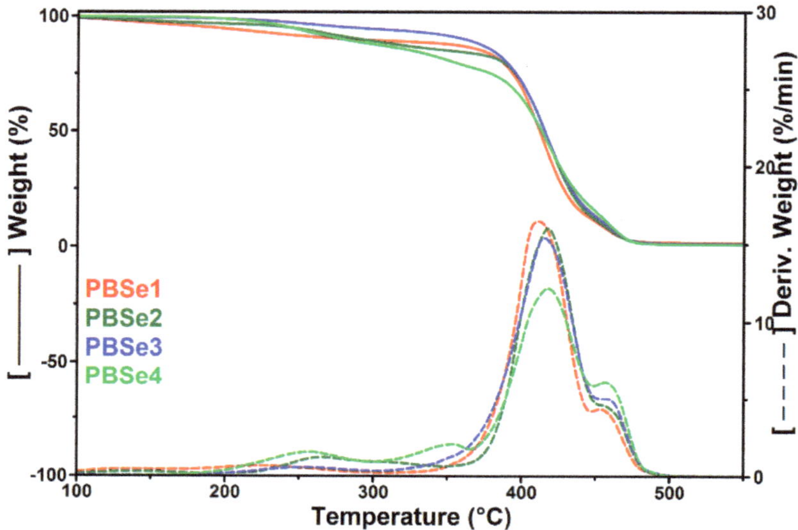


**Figure 2.** TGA and DTG curves of PBSe with different Se:BDO molar ratios.

The TGA results showed that the variation in the molar ratios between the components of PBSe did not significantly influence the thermal stability of the resulting polyesters: $T_{d,max1}$ and $T_{d,max2}$ values of the resulting PBSe are similar (Table 3). Similar $T_{d,10\%}$ values were obtained for all PBSe samples, except for the PBSe3 sample, which presented better thermal stability at lower temperatures, likely because of the lower content of volatile products. Moreover, no notable differences were found between the char yields of the polyesters, a slightly higher residue being observed for PBSe1, the polyester with the highest molecular weight. In summary, considering their thermal decomposition, which took place at high temperatures ranging from 411 to 418 °C, and the small mass loss values at 165 °C (Table 3), it can be stated that all synthesized PBSe polyesters are suitable for engineering applications, in terms of processing temperature requirements.

### 3.4. TGA Analysis of PLA/PBSe Blends

TGA and DTG curves of pristine PLA and its blends with different content of PBSe polyesters (from 2.5 wt% to 20 wt%) are shown in Figure 3. The degradation temperatures at which 10 wt% mass loss occurred ($T_{d,10\%}$) and at the maximum degradation rate ($T_{d,max}$), the weight loss at 200 °C ($WL_{200}$), and the residue determined at 700 °C ($R_{700}$) for the same samples are listed in Table S1.

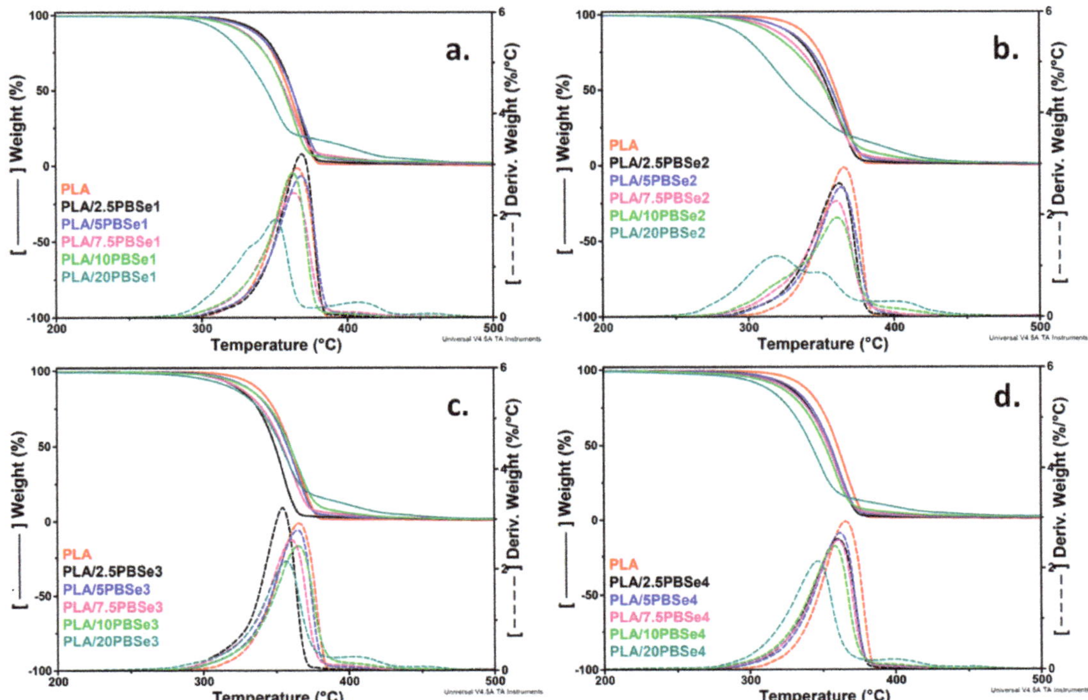


**Figure 3.** TGA/DTG curves for pristine PLA and PLA blends containing different amounts of PBSe1 (**a**), PBSe2 (**b**), PBSe3 (**c**), and PBSe4 (**d**).

PLA degradation occurred in a single step with a $T_{d,max}$ of 366 °C and a percentage of carbon residue equal to 0%, thus showing 100% degradation at 700 °C. Overall, the TGA and DTG curves of the PLA/PBSe blends showed one predominant degradation process similar to the degradation profile of pristine PLA, regardless of the PBSe composition. In general, the incorporation of PBSe polyesters promoted a small reduction in the thermal stability of the PLA blends, visible by a slight shift of the $T_{d,10\%}$ and $T_{d,max}$ to lower values. It is obvious that the thermal stability of the PLA blends decreases proportionally with the increase in the PBSe content. As compared to PLA, the $T_{d,10\%}$ of PLA/10PBSe1 was lower by 10 °C, that of PLA/10PBSe2 by 24 °C, that of PLA/10PBSe3 by 8 °C and that of PLA/10PBSe4 by 17 °C while the $T_{d,max}$ decreased only by a few degrees at this concentration of PBSes in the blends. Interestingly, the addition of low amounts (2.5 and 5 wt%) of PBSe obtained at a BDO excess of 10% led to a slight enhancement in the thermal stability of PLA (Table S1). This effect can be related to the higher $M_n$ and longer molecular chains of this polyester (PBSe1), which render it harder to degrade, and to its higher melting and crystallization temperatures (Table 3).

Further, it was observed that the thermal degradation of PLA/PBSe2, containing the PBSe polyester prepared using a BDO molar excess of 50%, occurred over a broader range of temperatures. Thus, with the increase in PBSe2 content, the degradation peak of PLA decreased in intensity and was shifted to lower values, while additional small degradation shoulders appeared in the DTG curves of the PLA/PBSe2 blends. This effect was also observed in the PLA/PBSe1 blend with 20 wt% PBSe1, but at a lower intensity. In general, the PLA blends with 20 wt% PBSes had a lower thermal stability compared to PLA. This suggests that at high PBSe contents, a possible involvement of PBSe in the degradation mechanism of PLA occurred [54]. Moreover, a new small degradation shoulder was identified in the DTG curves of all PLA/PBSe blends, regardless of the PBSe type, at temperatures above 400 °C. This may be determined by the formation of cyclic or cross-linked products in polyesters at elevated temperatures [55].

However, the main degradation temperature for PLA/PBSe blends was diminished by only 5% as compared with that of pristine PLA. This means that these new materials can be processed in the molten state, without the cleavage of the polymeric chains or other decomposition processes.

### 3.5. Morphology of Fracture Surface of PLA/PBSe Blends

The investigation of the phase morphology of the PLA/PBSe blends can provide important information regarding the microstructure/compatibility/mechanical performances relationships in these materials. The size of the dispersed PBSe particles in blends is governed by many factors, such as compatibility, viscosity match, and shear rate [42]. The micrographs of the cryo- fractured surfaces of PLA/PBSe blends at different magnifications are shown in Figures S1, S2 and 4.

A smooth surface was observed in the PLA micrograph showing its well-known brittle fracture (Figure S1). From the SEM micrographs taken at different magnifications, the following aspects were noted for the PLA/PBSe blends: (i) the PBSes form particles with spherical and irregular geometry in the blends; (ii) the PBSe particles are homogeneously dispersed in the PLA matrix, especially at low concentrations; (iii) brittle-to-ductile transition occurs with an increase in the concentration of PBSes.

The synthesized PBSe seems to have a low affinity for the PLA matrix, and they tend to bond with themselves to reduce the surface tension, which leads to the formation of spherical PBSe particles with different diameters inside the PLA matrix. This behavior was especially notable in the case of PBSe1—having a higher molecular weight—which exhibited a stronger tendency to self-assemble. Due to the limited adhesion at the PLA-PBSe interface, upon cryo-fracture, the spherical PBSe particles were pulled out causing the appearance of cavities, or even voids [42] (Figures S2 and 4). Still, these features were more obvious for the PLA blends containing the PBSe1 polyester, while in all other PLA blends, this effect was visible only at the highest PBSe concentration. Indeed, the PBSe1 polyester had the highest $M_n$, $T_m$, and $T_c$ values, which influenced its dispersion in the PLA matrix.

Interestingly, at low PBSe concentrations (2.5 wt.%), the PLA blends showed a homogeneous phase microstructure, indicating a rather good compatibility between PBSes and PLA (Figure S1, 1000×). Moreover, the number average particle diameter of the dispersed PBSe particles in all the blends showed values between 0.4 and 0.7 μm, except for the PLA/PBSe1 blend, which displayed PBSe1 particles with an average diameter of 0.92 ± 0.13 μm. According to Hu et al., good compatibility between components in a binary material leads to a uniform dispersion of the dispersed phase, with small particle size distribution, as in the case of most of the PLA/PBSe blends with a small content of PBSes [52].

At a higher PBSe content (10 and 20 wt%), the fractured surface of the PLA blends showed a more ductile fracture, having more PLA matrix fibrils and a highly deformed matrix (Figures S2 and 4, magnifications 2500× and 5000×). When the polyester reached the maximum concentration in the PLA blends (20 wt%), a change in the size of the PBSe particles was observed. The number average particle diameter of the dispersed PBSe particles in the PLA/PBSe blends displayed a broader range of 0.7–4 μm. A few particles having an average diameter of 14 μm were detected only in the case of blends containing the PBSe1 and PBSe4 polyesters, but only above 10 wt%. Even at higher concentration, PBSe polyesters were well dispersed in the PLA matrix (Figure 4). Noteworthy, the PBSe2 formed particles with irregular geometry, which become visible as agglomerates at 20 wt% PBSe2 in the PLA matrix (Figure 4, magnification 5000×).

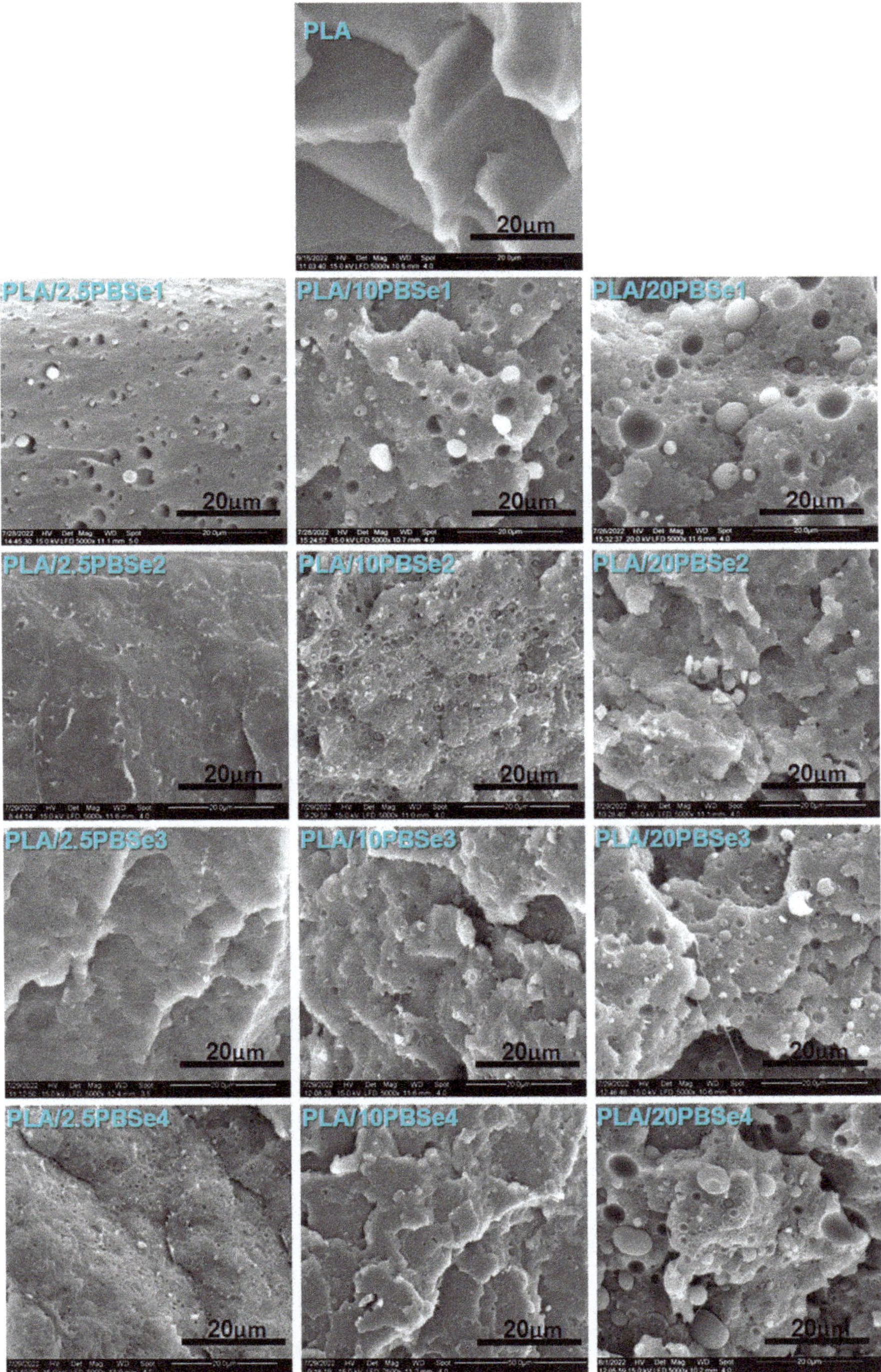


**Figure 4.** Micrographs of neat PLA and PLA blends containing different amounts of PBSe1, PBSe2, PBSe3, and PBSe4 polyesters at 5000× magnification.

On the other hand, the SEM cryo-fractured surfaces of the PLA blends containing PBSe types 2, 3, and 4 exhibited fewer and smaller cavities and detached particles; the particles seemed more attached to the matrix, suggesting a higher interfacial adhesion and compatibility with the PLA matrix when compared with the PLA/PBSe1 blends. This may be a consequence of the lower $M_n$ values of these polyesters, which may assure an enhanced solubility and dispersibility in the PLA matrix. Moreover, PLA blends with the above mentioned polyesters displayed lower torque values and higher melt flow index values (lower viscosity in the melt) as compared with the PLA blends containing the PBSe1 type polyester (Figure 5a,b).

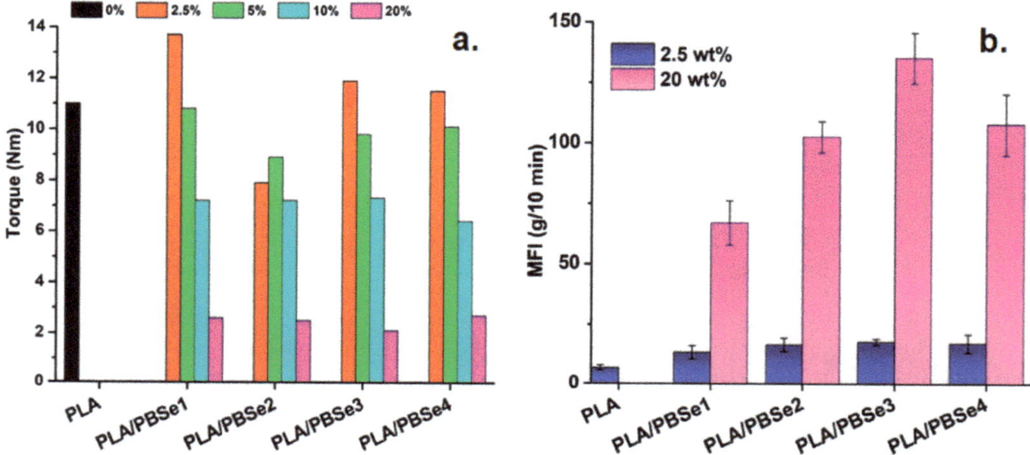


**Figure 5.** Torque (**a**) and melt flow index (**b**) values for pristine PLA and PLA/PBSe blends.

Therefore, a higher compatibility may be presumed between the PLA and PBSe types 2, 3, and 4 due to the lower melt viscosity of these blends as compared with the PLA/PBSe1 blends. The effect of the PBSe polyesters was enhanced at higher content in the blends: a decrease in the torque values was observed with increasing PBSe content (Figure 5a), while higher MFI values were attained at higher PBSe concentrations (Figure 5b). These results also indicated a better melt processing behavior of the PLA/PBSe blends as compared to PLA. Thus, lower shear forces were involved and lower mechanical energy was consumed in the melt processing of PLA/PBSe blends than for neat PLA [56].

### *3.6. Mechanical Behavior of PLA/PBSe Blends*

#### 3.6.1. Dynamic Mechanical Properties

Temperature dependence of the storage modulus (E′) and tan $\delta$ for pristine PLA and PLA/PBSe blends as determined from the DMA analysis are presented in Figures 6 and 7. The E′ curves for pristine PLA and the PLA/PBSe blends showed a typical behavior of thermoplastic polymers, with a drastic decrease in the E′ values when approaching the glass transition temperature ($T_g$) of PLA, and an increase starting from around 100 °C due to cold crystallization. Below $T_g$, the $E'_{40}$ of all PLA blends dropped gradually to lower values with an increase in the PBSes concentration, regardless of the PBSe type (Table 4, Figure 6a–d). Interestingly, cold crystallization started at lower temperatures with an increase in the PBSe content of the blend. Therefore, the addition of PBSe improved the cold crystallization ability of the PLA matrix, thus lowering its cold crystallization temperature in the resulting blends.

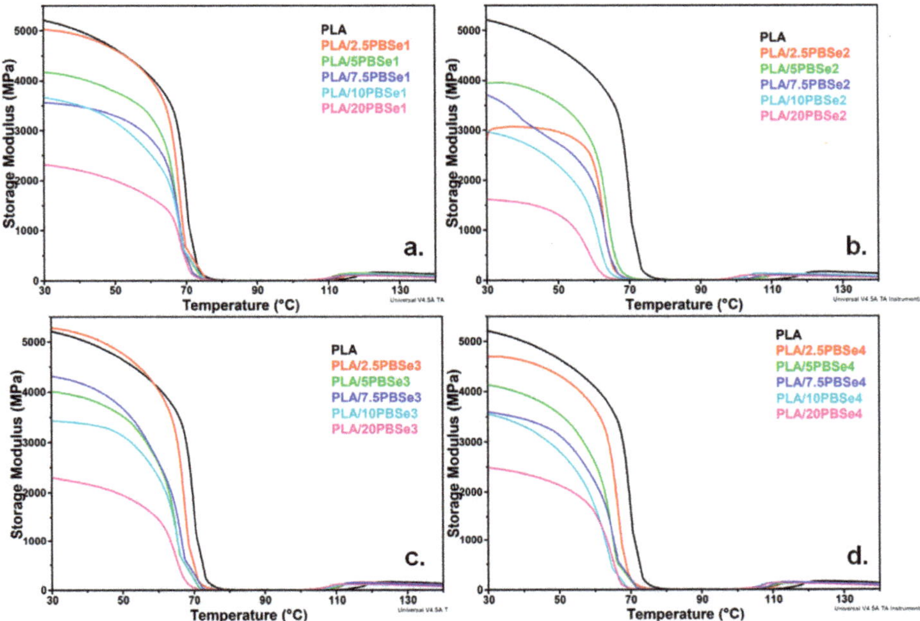


**Figure 6.** E′ curves of neat PLA and PLA blends containing different amounts of PBSe1 (**a**), PBSe2 (**b**), PBSe3 (**c**), and PBSe4 (**d**) polyesters.

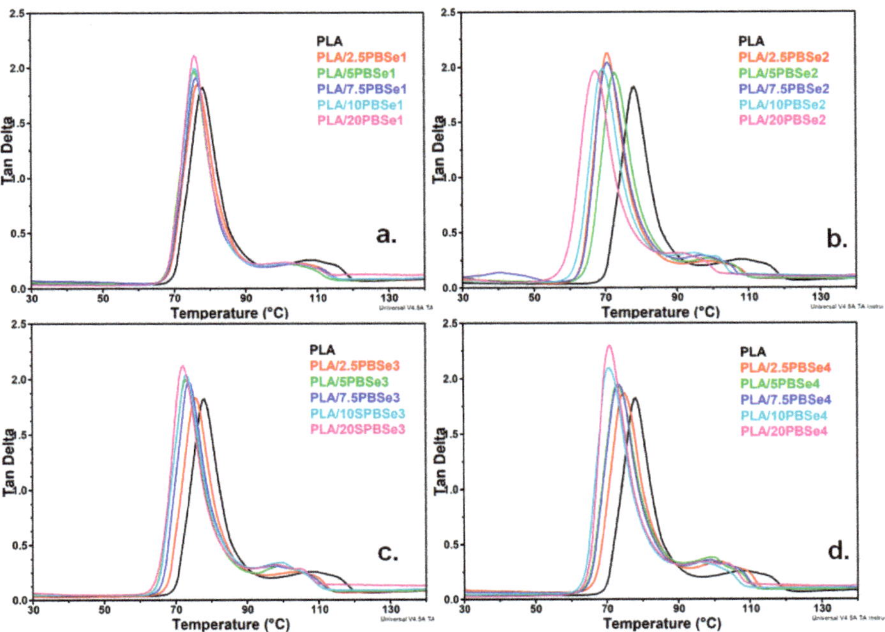


**Figure 7.** Tan δ curves of neat PLA and PLA blends containing different amounts of PBSe1 (**a**), PBSe2 (**b**), PBSe3 (**c**), and PBSe4 (**d**) polyesters.

**Table 4.** Storage modulus (E′) at temperatures of 40 and 80 °C ($E'_{40}$, $E'_{80}$), damping factor (tan $\delta$) values, and height.

| Sample | $E'_{40}$, MPa | $E'_{80}$, MPa | $T_g$, °C | tan $\delta$ ($T_g$) |
|---|---|---|---|---|
| PLA | 5001 | 13.89 | 78.16 | 1.824 |
| PLA/2.5PBSe1 | 4917 | 9.76 | 76.78 | 1.856 |
| PLA/5PBSe1 | 4056 | 8.62 | 75.83 | 1.968 |
| PLA/7.5PBSe1 | 3490 | 8.21 | 76.26 | 1.910 |
| PLA/10PBSe1 | 3514 | 8.08 | 75.99 | 1.997 |
| PLA/20PBSe1 | 2208 | 5.82 | 75.80 | 2.111 |
| PLA/2.5PBSe2 | 3065 | 4.76 | 70.65 | 2.128 |
| PLA/5PBSe2 | 3891 | 5.65 | 72.74 | 1.954 |
| PLA/7.5PBSe2 | 3219 | 4.67 | 70.76 | 2.048 |
| PLA/10PBSe2 | 2765 | 4.24 | 69.38 | 1.988 |
| PLA/20PBSe2 | 1545 | 2.48 | 67.32 | 1.967 |
| PLA/2.5PBSe3 | 5128 | 9.08 | 75.71 | 1.829 |
| PLA/5PBSe3 | 3881 | 5.72 | 72.97 | 1.998 |
| PLA/7.5PBSe3 | 4135 | 6.29 | 73.84 | 1.987 |
| PLA/10PBSe3 | 3369 | 5.07 | 73.05 | 2.043 |
| PLA/20PBSe3 | 2175 | 4.36 | 72.18 | 2.125 |
| PLA/2.5PBSe4 | 4617 | 7.40 | 74.91 | 1.864 |
| PLA/5PBSe4 | 3955 | 4.73 | 73.18 | 1.935 |
| PLA/7.5PBSe4 | 3460 | 5.44 | 73.39 | 1.946 |
| PLA/10PBSe4 | 3320 | 3.46 | 70.78 | 2.100 |
| PLA/20PBSe4 | 2368 | 3.57 | 70.88 | 2.300 |

As shown in Figure 7, PLA exhibited a sharp tan $\delta$ peak at 78 °C, which corresponds to its glass transition temperature (Table 4). A depression in the $T_g$ of PLA was noticed with increasing amount of PBSe polyesters in the blends, indicating an enhancement of PLA chain segment mobility (Table 4). In particular, $T_g$ decreased from 78 °C for PLA to 76 °C for PLA/PBSe1, 67 °C for PLA/PBSe2, 72 °C for PLA/PBSe3, and 71 °C for PLA/PBSe4. The most important plasticizing effect was observed by the addition of PBSe2, and the most insignificant effect was noticed for the PLA/PBSe1 blends (Figure 7). Moreover, an increase in the height of the tan δ peak as a function of PBSe concentration was observed in all PLA blends, thus confirming the increase in segmental mobility (Table 4, Figure 7a–d).

### 3.6.2. Tensile Properties

The effect of the synthesized PBSe on the mechanical properties of pristine PLA was investigated by using tensile tests. The tensile strength ($\sigma_{max}$), Young's modulus (YM), and elongation at break ($\varepsilon_B$) of PLA blends containing the synthesized PBSe polyesters in different amounts are shown in Figure 8.

Pristine PLA is a stiff and brittle polymer and has a poor ductility of about 2.9%, due to extensive intermolecular forces [52,54]. The mechanical properties of PLA were strongly affected by the addition of PBSes, which showed a satisfactory affinity for the PLA matrix and manifested plasticizing effect by increasing its elongation at break (Figure 8c). This correlates with the decreasing $T_g$ values observed when increasing the PBSe content in the blends, as indicated before by the DMA results (Table 4). As observed from the $\varepsilon_B$ variation, the increase in PBSe concentration promoted a proportional increase in the chain mobility of the PLA macromolecules. The PLA blends containing PBSe2 showed the highest elongation at break (107.5% for 20 wt% PBSe2), a 37-fold improvement over pristine PLA. These results were in agreement with the DMA results, which indicated that the highest decrease in the $T_g$ values occurs for the PLA/20PBSe2 blends (Table 4). However, a good improvement in the $\varepsilon_B$ of the PLA blends was also recorded when PBSe1 was used as a toughener, despite the modest decrease in the $T_g$ of PLA inflicted over the entire concentration range (2.5 to 20 wt%) (Figure 8, Table 4). It is worth mentioning that this polyester had the lowest $T_g$ (−27 °C), as well as the lowest $T_c$ (Table 3), which likely contributed to an improved flexibility of the blends. The important enhancement in the PLA's ductility is sustained by the brittle-to-ductile transition observed in the SEM images with the increase in the PBSes content in the blends, as well as by the MFI and torque results.

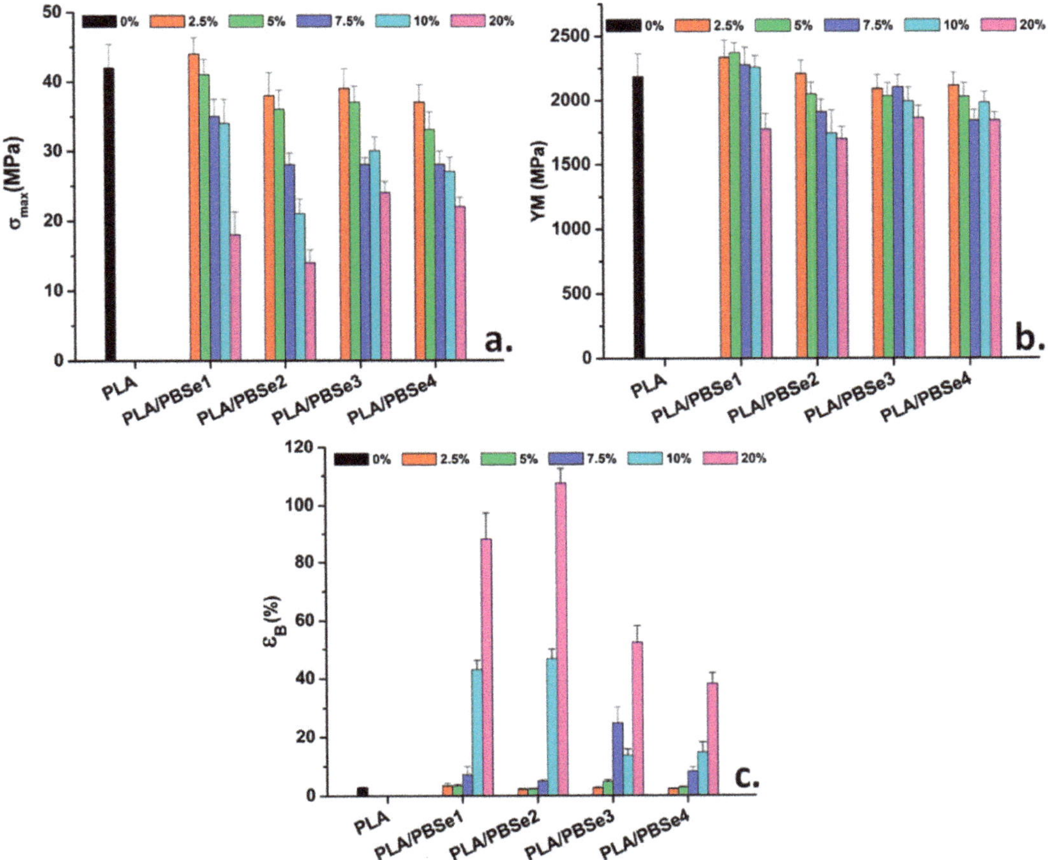


**Figure 8.** Mechanical properties determined from the tensile tests: $\sigma_{max}$ (**a**), YM (**b**), and $\varepsilon_B$ (**c**).

PBSes were able to increase the free volume between the polymer chains of PLA, resulting in higher $\varepsilon_B$, while the $\sigma_{max}$ and YM values decreased (Figure 8). Of all the PLA/PBSe blends, the PLA blends containing PBSe1 showed the best tensile strength and modulus, at a PBSe1 concentration of up to 10 wt% (Figure 8a,b). This can be explained by the limited miscibility between PBSe1 and PLA, which caused a lower plasticizing effect at these PBSe concentrations. As observed for other immiscible binary blends, the immiscible plasticizer may promote the crystallization of the PLA matrix through the formation of nucleation sites at their interface [52,54].

Contrarily, the PBSe2 polyester led to slightly lower $\sigma_{max}$ and YM values for the PLA blends due to the presence of some irregular particle agglomerates which were detected in the SEM images, especially at higher concentrations. Moreover, the more pronounced decrease in the tensile strength of the PLA blends with PBSe2, PBSe3, or PBSe4 could be due to the miscibility of these polyesters with the PLA matrix, which weakened the interactions between the PLA macromolecules, and finally led to a decrease in the PLA's tensile strength. These results correlated well with the decrease in the $T_g$ values of PLA in these blends, as determined by DMA.

It is important to note that although a considerable improvement in the ductility and mobility of PLA chains was attained by the addition of PBSe polyesters, the YM did not show a significant decrease. A maximum decrease in YM of only 22% was obtained at the highest content of PBSe of 20 wt% in the PLA blends (Figure 8). This is much lower than other reported results, especially when considering the remarkable improvement in the elongation at break that was attained.

## 4. Conclusions

Bio-derived poly(butylene sebacate) (PBSe) polyesters were synthesized via the direct melt polycondensation of sebacic acid and 1,4-butanediol and these were subsequently characterized. It was found that both the $M_n$ and $T_g$ of the resulting PBSes can be tailored by varying the monomer ratio. FTIR results confirmed the presence of ester groups in the structure of the PBSes, while TGA showed a high thermal decomposition temperature (from 411 to 418 °C) for the prepared PBSe polyesters, accompanied by small mass loss values at 165 °C. For the first time, PBSe polyesters were used as plasticizers for tuning PLA's ductility in concentrations from 2.5 to 20 wt%. The incorporation of PBSes into the PLA matrix led to a slight decrease in the thermal stability of PLA, which was proportional to the increase in the PBSe content in the blends. SEM results indicated a limited affinity between PBSe and the PLA matrix, which led to the formation of spherical particles of PBSe with different diameters in the PLA/PBSe blends. The PBSe particles were homogeneously dispersed in the PLA matrix, especially at low concentrations, and a brittle-to-ductile transition of the PLA matrix occurred with an increase in the PBSe concentration in the blends. An improvement in the ductility and mobility of the PLA chains was achieved by the addition of PBSe polyesters, without a drastic decrease in the Young's modulus values. These findings demonstrated that the PBSes alone, or in blends with other biopolymers, can be promising candidates for engineering applications that require renewability in addition to good thermal and mechanical characteristics.

**Supplementary Materials:** The following supporting information can be downloaded at: https://www.mdpi.com/article/10.3390/polym14193998/s1, Table S1: TGA/DTG data for pristine PLA and PLA/PBSe blends; Figure S1: Micrographs of neat PLA and PLA blends containing different amounts of PBSe1, PBSe2, PBSe3 and PBSe4 polyesters at 1000× magnification.; Figure S2: Micrographs of neat PLA and PLA blends containing different amounts of PBSe1, PBSe2, PBSe3 and PBSe4 polyesters at 2500× magnification.

**Author Contributions:** Conceptualization and writing—original draft preparation, A.N.F.; methodology, C.D.U. and M.S.P.; investigation, A.R.G., C.A.N., M.F.R., A.Z. and E.A.; writing—review and editing, A.N.F. and D.M.P.; funding acquisition, A.N.F. All authors have read and agreed to the published version of the manuscript.

**Funding:** This research was funded by a grant from the Ministry of Research, Innovation, and Digitization, CNCS/CCCDI–UEFISCDI, project number 67TE/2020, within PNCDI III.

**Institutional Review Board Statement:** Not applicable.

**Informed Consent Statement:** Not applicable.

**Data Availability Statement:** Not applicable.

**Conflicts of Interest:** The authors declare no conflict of interest.

## References

1. Naser, A.Z.; Deiab, I.; Darras, B.M. Poly(lactic acid) (PLA) and polyhydroxyalkanoates (PHAs), green alternatives to petroleum-based plastics: A review. *RSC Adv.* **2021**, *11*, 17151–17196. [CrossRef] [PubMed]
2. Masutani, K.; Kimura, Y. PLA Synthesis. From the Monomer to the Polymer. In *Poly(Lactic Acid) Science and Technology: Processing, Properties, Additives and Applications*; RSC Publishing: Cambridge, UK, 2014; pp. 1–36.
3. Shrivastava, A. Polymerization. In *Introduction to Plastics Engineering*; William Andrew Publishing: Norwich, NY, USA, 2018; pp. 17–49.

4. Walton, M.J.; Lancaster, S.J.; Redshaw, C. Highly Selective and Immortal Magnesium Calixarene Complexes for the Ring-Opening Polymerization of *rac*-Lactide. *ChemCatChem* **2014**, *6*, 1892–1898. [CrossRef]
5. Nwamba, M.C.; Sun, F.; Mukasekuru, M.; Song, G.; Harindintwali, J.D.; Boyi, S.A.; Sun, H. Trends and hassles in the microbial production of lactic acid from lignocellulosic biomass. *Environ. Technol. Innov.* **2021**, *21*, 101337. [CrossRef]
6. Panaitescu, D.M.; Frone, A.N.; Chiulan, I.; Gabor, R.A.; Spataru, C.I.; Căşărică, A. Biocomposites from Polylactic Acid and Bacterial Cellulose Nanofibers Obtained by Mechanical Treatment. *Bioresources* **2017**, *12*, 662–672. [CrossRef]
7. Nofar, M.; Sacligil, D.; Carreau, P.J.; Kamal, M.R.; Heuzey, M.-C. Poly (lactic acid) blends: Processing, properties and applications. *Int. J. Biol. Macromol.* **2018**, *125*, 307–360. [CrossRef] [PubMed]
8. Brebu, M. Environmental Degradation of Plastic Composites with Natural Fillers—A Review. *Polymers* **2020**, *12*, 166. [CrossRef] [PubMed]
9. Frone, A.N.; Batalu, D.; Chiulan, I.; Oprea, M.; Gabor, A.R.; Nicolae, C.-A.; Raditoiu, V.; Trusca, R.; Panaitescu, D.M. Morpho-Structural, Thermal and Mechanical Properties of PLA/PHB/Cellulose Biodegradable Nanocomposites Obtained by Compression Molding, Extrusion, and 3D Printing. *Nanomaterials* **2020**, *10*, 51. [CrossRef]
10. Valvez, S.; Santos, P.; Parente, J.M.; Silva, M.P.; Reis, P.N.B. 3D printed continuous carbon fiber reinforced PLA composites: A short review. *Procedia Struct. Integr.* **2020**, *25*, 394–399. [CrossRef]
11. Castro-Aguirre, E.; Iñiguez-Franco, F.; Samsudin, H.; Fang, X.; Auras, R. Poly(lactic acid)—Mass production, processing, industrial applications, and end of life. *Adv. Drug Deliv. Rev.* **2016**, *107*, 333–366. [CrossRef]
12. Liu, S.; Qin, S.; He, M.; Zhou, D.; Qin, Q.; Wang, H. Current applications of poly(lactic acid) composites in tissue engineering and drug delivery. *Compos. Part B Eng.* **2020**, *199*, 108238. [CrossRef]
13. Nedelcu, D.; Mazurchevici, S.-N.; Popa, R.-I.; Lohan, N.-M.; Maldonado-Cortés, D.; Carausu, C. Tribological and Dynamical Mechanical Behavior of Prototyped PLA-Based Polymers. *Materials* **2020**, *13*, 3615. [CrossRef]
14. França, D.C.; Almeida, T.G.; Abels, G.; Canedo, E.L.; Carvalho, L.H.; Wellen, R.M.R.; Haag, K.; Koschek, K. Tailoring PBAT/PLA/Babassu films for suitability of agriculture mulch application. *J. Nat. Fibers* **2019**, *16*, 933–943. [CrossRef]
15. China Aims to Go as Big in Bioplastics as It Did in Solar Panels. Available online: https://asia.nikkei.com/Spotlight/Environment/China-aims-to-go-as-big-in-bioplastics-as-it-did-in-solar-panels (accessed on 23 August 2022).
16. Coiai, S.; Di Lorenzo, M.L.; Cinelli, P.; Righetti, M.C.; Passaglia, E. Binary Green Blends of Poly(lactic acid) with Poly(butylene adipate-co-butylene terephthalate) and Poly(butylene succinate-co-butylene adipate) and Their Nanocomposites. *Polymers* **2021**, *13*, 2489. [CrossRef] [PubMed]
17. Si, W.-J.; Yang, L.; Weng, Y.-X.; Zhu, J.; Zeng, J.-B. Poly(lactic acid)/biobased polyurethane blends with balanced mechanical strength andtoughness. *Polym.Test.* **2018**, *69*, 9–15. [CrossRef]
18. Yang, Y.; Zhang, L.; Xiong, Z.; Tang, Z.; Zhang, R.; Zhu, J. Research progress in the heat resistance, toughening and filling modification of PLA. *Sci. China Chem.* **2016**, *59*, 1355–1368. [CrossRef]
19. Kim, K.; Lee, J.W.; Chang, T.; Kim, H.I. Characterization of Polylactides with Different Stereoregularity Using Electrospray Ionization Ion Mobility Mass Spectrometry. *J. Am. Soc. Mass Spectrom.* **2014**, *25*, 1771–1779. [CrossRef]
20. Kang, H.; Li, Y.; Gong, M.; Guo, Y.; Guo, Z.; Fang, Q.; Li, X. An environmentally sustainable plasticizer toughened polylactide. *RSC Adv.* **2018**, *8*, 11643–11651. [CrossRef]
21. Greco, A.; Maffezzoli, A. Cardanol derivatives as innovative bio-plasticizers for poly-(lactic acid). *Polym. Degrad. Stab.* **2016**, *132*, 213–219. [CrossRef]
22. Mele, G.; Bloise, E.; Cosentino, F.; Lomonaco, D.; Avelino, F.; Marcianò, T.; Massaro, C.; Mazzetto, S.E.; Tammaro, L.; Scalone, A.G.; et al. Influence of Cardanol Oil on the Properties of Poly(lactic acid) Films Produced by Melt Extrusion. *ACS Omega* **2019**, *4*, 718–726. [CrossRef]
23. Xuan, W.; Hakkarainen, M.; Odelius, K. Levulinic acid as a versatile building-block for plasticizer design. *ACS Sustain. Chem. Eng.* **2019**, *7*, 12552–12562. [CrossRef]
24. Carbonell-Verdu, A.; Boronat, T.; Quiles-Carrillo, L.; Fenollar, O.; Dominici, F.; Torre, L. Valorization of Cotton Industry Byproducts in Green Composites with Polylactide. *J. Polym. Environ.* **2020**, *28*, 2039–2053. [CrossRef]
25. Park, M.; Choi, I.; Lee, S.; Hong, S.; Kim, A.; Shin, J.; Kang, H.-C.; Kim, Y.-W. Renewable Malic Acid-based Plasticizers for Both PVC and PLA Polymers. *J. Ind. Eng. Chem.* **2020**, *88*, 148–158. [CrossRef]
26. Zych, A.; Perotto, G.; Trojanowska, D.; Tedeschi, G.; Bertolacci, L.; Francini, N.; Athanassiou, A. Super Tough Polylactic Acid Plasticized with Epoxidized Soybean Oil Methyl Ester for Flexible Food Packaging. *ACS Appl. Polym. Mater.* **2021**, *3*, 5087–5095. [CrossRef]
27. Volpe, V.; De Filitto, M.; Klofacova, V.; De Santis, F.; Pantani, R. Effect of mold opening on the properties of PLA samples obtained by foam injection molding. *Polym. Eng. Sci.* **2017**, *58*, 475–484. [CrossRef]
28. García-Campo, M.J.; Boronat, T.; Quiles-Carrillo, L.; Balart, R.; Montanes, N. Manufacturing and Characterization of Toughened Poly(lacticacid) (PLA) Formulations by Ternary Blends with Biopolyesters. *Polymers* **2018**, *10*, 3. [CrossRef]
29. Puthumana, M.; Krishnan, P.S.G.; Nayak, S.K. Chemical modifications of PLA through copolymerization. *Int. J. Polym. Anal. Charact.* **2020**, *25*, 634–648. [CrossRef]
30. Siehr, A.; Flory, C.; Callaway, T.; Schumacher, R.J.; Siegel, R.A.; Shen, W. Implantable and Degradable Thermoplastic Elastomer. *ACS Biomater. Sci. Eng.* **2021**, *7*, 5598–5610. [CrossRef]

31. Dånmark, S.; Finne-Wistrand, A.; Schander, K.; Hakkarainen, M.; Arvidson, K.; Mustafa, K.; Albertsson, A.-C. In vitro and in vivo degradation profile of aliphatic polyesters subjected to electron be amsterilization. *Acta Biomater.* **2011**, *7*, 2035–2046. [CrossRef]
32. Plichta, A.; Florjańczyk, Z.; Kundys, A.; Frydrych, A.; Dębowski, M.; Langwald, N. On the copolymerization of monomers from renewable resources: L-lactide and ethylene carbonate in the presence of metal alkoxides. *Pure Appl. Chem.* **2014**, *86*, 733–745. [CrossRef]
33. Zhang, M.; Chang, Z.; Wang, X.; Li, Q. Synthesis of Poly(l-lactide-co-ε-caprolactone) Copolymer: Structure, Toughness, and Elasticity. *Polymers* **2021**, *13*, 1270. [CrossRef]
34. Xi, L.; Wang, Y.; Su, F.; Zhu, Q.; Li, S. Biocompatibility and degradation studies of poly(L-lactide-co-trimethylene carbonate) copolymers as cardiac occluders. *Materialia* **2019**, *7*, 100414. [CrossRef]
35. Mahmud, M.S.; Buys, Y.F.; Anuar, H.; Sopyan, I. Miscibility, Morphology and Mechanical Properties of Compatibilized Polylactic Acid/Thermoplastic Polyurethane Blends. *Mater. Today* **2019**, *17*, 778–786. [CrossRef]
36. Zhao, X.; Hu, H.; Wang, X.; Yu, X.; Zhou, W.; Peng, S. Super tough poly(lactic acid) blends: A comprehensive review. *RSC Adv.* **2020**, *10*, 13316–13368. [CrossRef] [PubMed]
37. Saini, P.; Arora, M.; Kumar, M.N.V.R. Poly(lactic acid) blends in biomedical applications. *Adv. Drug Deliv. Rev.* **2016**, *107*, 47–59. [CrossRef] [PubMed]
38. Wang, X.; Peng, S.; Chen, H.; Yu, X.; Zhao, X. Mechanical properties, rheological behaviors, and phase morphologies of high-toughness PLA/PBAT blends by in-situ reactive compatibilization. *Compos. Part B Eng.* **2019**, *173*, 107028. [CrossRef]
39. Aliotta, L.; Vannozzi, A.; Canesi, I.; Cinelli, P.; Coltelli, M.-B.; Lazzeri, A. Poly(lactic acid) (PLA)/Poly(butylene succinate-co-adipate) (PBSA) Compatibilized Binary Biobased Blends: Melt Fluidity, Morphological, Thermo-Mechanical and Micromechanical Analysis. *Polymers* **2021**, *13*, 218. [CrossRef]
40. Salehiyan, R.; Ray, S.S. Influence of Nanoclay Localization on Structure-Property Relationships of Polylactide-Based Biodegradable Blend Nanocomposites. *Macromol. Mater. Eng.* **2018**, *30*, 1800134. [CrossRef]
41. Hu, X.; Li, Y.; Li, M.; Kang, H.; Zhang, L. Renewable and Supertoughened Polylactide-Based Composites: Morphology, Interfacial Compatibilization, and Toughening. *Mechanism. Ind. Eng. Chem. Res.* **2016**, *55*, 9195–9204. [CrossRef]
42. Kang, H.; Qiao, B.; Wang, R.; Wang, Z.; Zhang, L.; Ma, J.; Coates, P. Employing a novel bioelastomer to tough enpolylactide. *Polymer* **2013**, *54*, 2450–2458. [CrossRef]
43. Kim, S.J.; Kwak, H.W.; Kwon, S.; Jang, H.; Park, S.-I. Synthesis, Characterization and Properties of Biodegradable Poly(ButyleneSebacate-*Co*-terephthalate). *Polymers* **2020**, *12*, 2389. [CrossRef]
44. Jeon, W.-Y.; Jang, M.-J.; Park, G.-Y.; Lee, H.-J.; Seo, S.-H.; Lee, H.-S.; Han, C.; Kwon, H.; Lee, H.-C.; Lee, J.-H.; et al. Microbial production of sebacic acid from a renewable source: Production, purification, and polymerization. *Green Chem.* **2019**, *21*, 6491–6501. [CrossRef]
45. De Bari, I.; Giuliano, A.; Petrone, M.T.; Stoppiello, G.; Fatta, V.; Giardi, C.; Razza, F.; Novelli, A. From Cardoon Lignocellulosic Biomass to Bio-1,4 Butanediol: An Integrated Biorefinery Model. *Processes* **2020**, *8*, 1585. [CrossRef]
46. Siotto, M.; Zoia, L.; Tosin, M.; Degli Innocenti, F.; Orlandi, M.; Mezzanotte, V. Monitoring biodegradation of poly(butylene sebacate) by Gel Permeation Chromatography, 1H-NMR and 31P-NMR techniques. *J. Environ. Manag.* **2013**, *116*, 27–35. [CrossRef] [PubMed]
47. Raghunanan, L.; Narine, S.S. Influence of Structure on Chemical and Thermal Stability of Aliphatic Diesters. *J. Phys. Chem. B* **2013**, *117*, 14754–14762. [CrossRef] [PubMed]
48. Wu, Y.; Xie, Q.; Gao, C.; Wang, T.; Wang, C. Synthesis and characterization of a novel aliphatic polyester based on itaconic acid. *Polym. Eng. Sci.* **2013**, *54*, 2515–2521. [CrossRef]
49. Sokolsky-Papkov, M.; Langer, R.; Domb, A.J. Synthesis of aliphatic polyesters by polycondensation using inorganic acid as catalyst. *Polym. Adv. Technol.* **2011**, *22*, 502–511. [CrossRef]
50. Inácio, E.M.; Lima, M.C.P.; Souza, D.H.S.; Sirelli, L.; Dias, M.L. Crystallization, thermal and mechanical behavior of oligosebacate plasticized poly(lactic acid) films. *Polímeros* **2018**, *28*, 381–388. [CrossRef]
51. Hu, X.; Kang, H.; Li, Y.; Li, M.; Wang, R.; Xu, R.; Qiao, H.; Zhang, L. Direct Copolycondensation of Biobased Elastomers Based on Lactic Acid with Tunable and Versatile Properties. *Polym. Chem.* **2015**, *6*, 8112–8123. [CrossRef]
52. Hu, X.; Li, Y.; Gao, Y.; Wang, R.; Wang, Z.; Kang, H.; Zhang, L. Renewable and super-toughened poly (butylene succinate) with bio-based elastomers: Preparation, compatibility and performances. *Eur. Polym. J.* **2019**, *116*, 438–444. [CrossRef]
53. Bikiaris, R.D.; Ainali, N.M.; Christodoulou, E.; Nikolaidis, N.; Lambropoulou, D.A.; Papageorgiou, G.Z. Thermal Stability and Decomposition Mechanism of Poly(alkylene succinate)s. *Macromol* **2022**, *2*, 58–77. [CrossRef]
54. Athanasoulia, I.G.; Tarantili, P.A. Preparation and characterization of polyethylene glycol/poly(L-lactic acid) blends. *Pure Appl. Chem.* **2017**, *89*, 141–152. [CrossRef]
55. Kervran, M.; Vagner, C.; Cochez, M.; Ponçot, M.; Saeb, M.R.; Vahabi, H. Thermal degradation of polylactic acid (PLA)/polyhydroxybutyrate (PHB) blends: A systematic review. *Polym. Degrad. Stab.* **2022**, *201*, 10999. [CrossRef]
56. Panaitescu, D.M.; Nicolae, C.A.; Frone, A.N.; Chiulan, I.; Stanescu, P.O.; Draghici, C.; Iorga, M.; Mihailescu, M. Plasticized poly(3-hydroxybutyrate) with improved melt processing and balanced properties. *J. Appl. Polym. Sci.* **2017**, *134*, 44810. [CrossRef]

 polymers

MDPI

Article

# Development of a New Eco-Friendly Copolymer Based on Chitosan for Enhanced Removal of Pb and Cd from Water

Iolanda-Veronica Ganea [1,2], Alexandrina Nan [2,*], Carmen Roba [1], Iulia Neamțiu [1,3], Eugen Gurzău [1,3,4], Rodica Turcu [2], Xenia Filip [2] and Călin Baciu [1,*]

[1] Faculty of Environmental Science and Engineering, Babes-Bolyai University, 30 Fantanele, 400294 Cluj-Napoca, Romania
[2] Development of Isotopic and Molecular Technologies, National Institute for Research, 67-103 Donath, 400293 Cluj-Napoca, Romania
[3] Environmental Health Center, 58 Busuiocului, 400240 Cluj-Napoca, Romania
[4] Cluj School of Public Health, College of Political, Administrative and Communication Sciences, Babeș-Bolyai University, 7 Pandurilor, 400095 Cluj-Napoca, Romania
* Correspondence: alexandrina.nan@itim-cj.ro (A.N.); calin.baciu@ubbcluj.ro (C.B.)

**Abstract:** Worldwide, concerns about heavy metal contamination from manmade and natural sources have increased in recent decades. Metals released into the environment threaten human health, mostly due to their integration into the food chain and persistence. Nature offers a large range of materials with different functionalities, providing also a source of inspiration for scientists working in the field of material synthesis. In the current study, a new type of copolymer is introduced, which was synthesized for the first time by combining chitosan and poly(benzofurane-*co*-arylacetic acid), for use in the adsorption of toxic heavy metals. Such naturally derived materials can be easily and inexpensively synthesized and separated by simple filtration, thus becoming an attractive alternative solution for wastewater treatment. The new copolymer was investigated by solid-state nuclear magnetic resonance, thermogravimetric analysis, scanning electron microscopy, Fourier transform infrared spectroscopy, and X-ray photon electron microscopy. Flame atomic absorption spectrometry was utilized to measure heavy metal concentrations in the investigated samples. Equilibrium isotherms, kinetic 3D models, and artificial neural networks were applied to the experimental data to characterize the adsorption process. Additional adsorption experiments were performed using metal-contaminated water samples collected in two seasons (summer and winter) from two former mining areas in Romania (Roșia Montană and Novăț-Borșa). The results demonstrated high (51–97%) adsorption efficiency for Pb and excellent (95–100%) for Cd, after testing on stock solutions and contaminated water samples. The recyclability study of the copolymer indicated that the removal efficiency decreased to 89% for Pb and 58% for Cd after seven adsorption–desorption cycles.

**Keywords:** eco-friendly copolymer; poly(benzofurane-*co*-arylacetic acid); chitosan; heavy metals removal; wastewater; Roșia Montană; adsorption mechanism

**Citation:** Ganea, I.-V.; Nan, A.; Roba, C.; Neamțiu, I.; Gurzău, E.; Turcu, R.; Filip, X.; Baciu, C. Development of a New Eco-Friendly Copolymer Based on Chitosan for Enhanced Removal of Pb and Cd from Water. *Polymers* **2022**, *14*, 3735. https://doi.org/10.3390/polym14183735

Academic Editors: José Miguel Ferri, Vicent Fombuena Borràs and Miguel Fernando Aldás Carrasco

Received: 31 July 2022
Accepted: 2 September 2022
Published: 7 September 2022



## 1. Introduction

It is well-known that anthropogenic impact causes water pollution, habitat loss or degradation, and spread of invasive species, thus affecting marine ecosystems, wildlife, and human health, and contributing to climate change and quantitative as well as qualitative decrease of freshwater resources [1–3]. These effects are harmful not only to individual species and populations but also to entire communities [4–6]. Nowadays, heavy metal pollution has become a serious problem due to metals' difficult natural degradation processes and persistence in the environment, from where they are gradually released into water bodies which serve as sinks for contaminant discharge [7]. Moreover, human exposure to toxic concentrations of cadmium and lead can cause acute symptoms (e.g., irritation, abdominal pain, diarrhea, headache, nausea) and long-term effects (e.g., "itai-itai" disease, renal tubular dysfunction, encephalopathy, hypertensive disorders, cancer, coma, and death) [8,9].

Nowadays, scientists find inspiration from the enormous variety of materials that nature offers and their diverse uses. Adsorbents derived from biopolymers, including polysaccharides, have the benefits of being biocompatible and biodegradable [10]. For instance, chitosan is a polysaccharide synthesized through the deacetylation of chitin (the second most abundant polymer in nature after cellulose) and is used in a variety of fields including catalysis, biomedicine, veterinary medicine, pharmaceuticals, drug delivery, decontamination, membrane and film synthesis, food science, and enzyme immobilization [11–14]. Chitin can be found in significant quantities in seafood processing waste produced in many eastern and southeastern Asian countries [15]. Compared with conventional adsorbents, chitosan is biocompatible, antibacterial, biodegradable, easily separable through filtration, involves low costs, and has amino and hydroxyl functional groups that provide effective binding sites for contaminants, especially heavy metals (through chelation, i.e., ion exchange) [16]. The nitrogen atom in amino groups is the donor of electrons, while metal ions act as acceptors [17]. Chemical or physical modifications can improve chitosan's poor solubility and small surface area. Physical modification methods can enable processing into membranes, beads, nanofibers, gels, nanoparticles, honeycomb, etc. [18,19]. The common chemical modifications that can be applied to chitosan include N-alkylation, acylation, carboxylation, esterification using inorganic oxygen acids or anhydrides, grafting on polymers such as poly(ethylene imine), polyaniline, poly(vinyl amine), poly(alkyl methacrylate), poly(vinyl alcohol), triethylenetetramine, or polyacrylamide, crosslinking with glyoxal, tripolyphosphate, ethylene glycol diglycidyl ether, formaldehyde, epichlorohydrin, glutaraldehyde, dimethyloldihydroxy ethylene urea, or isocyanates, etc. [20–33]. Several studies have reported that pure chitosan has affinity for metals in the following order: Hg > Cu > Fe > Ni > Ag > Cd > Mn > Pb > Co > Cr, while others have stated that after cross-linking, this turns to Cu > Pb > Zn [34–38].

The present study introduces a new hybrid material, **CHIT-PAAA**, which was synthesized through the modification of chitosan (**CHIT**) with poly(benzofurane-*co*-arylacetic acid) (**PBAAA**), using simple green chemical methods. The obtained copolymer was structurally and morphologically investigated by solid-state nuclear magnetic resonance (ss-NMR), scanning electron microscopy (SEM), X-ray photon electron spectroscopy (XPS), and Fourier transform infrared spectroscopy (FTIR). Thermogravimetric analysis (TGA) was used to evaluate the material's thermal stability. Furthermore, the material was put to use to remove heavy metals from contaminated water samples. Flame atomic absorption spectrometry (FAAS) measurements were performed to determine the metal concentrations. A preliminary adsorption study was conducted to check the suitability of this material for environmental applications [39]. Pb and Cd were chosen for applying various isotherms and kinetic models, and to observe the adsorption behavior of the material based on the initial contaminant concentration and contact time. Furthermore, 3D adsorption rate models and artificial neural networks (ANNs) were also generated to characterize the adsorption process.

## 2. Materials and Methods

### 2.1. Chemical Reagents

**PBAAA** was synthesized by closely following a previously reported procedure [40], being highly soluble in most commonly used solvents. Chitosan (medium molecular weight), cadmium chloride hydrate 98% ($CdCl_2 \times H_2O$), and lead chloride 98% ($PbCl_2$) were purchased from Sigma-Aldrich (St. Louis, MO, USA). All reagents used were of analytical grade, commercially available, and no further purification was involved.

### 2.2. *Synthesis of the Adsorbent Material*

Preparation of the new copolymer through the modification of chitosan with **PBAAA** is shown in Scheme 1. **PBAAA** (2 g) and chitosan (1 g) were placed in a 250 mL flask and dissolved in a mixed solution of deionized water (150 mL) and methanol (20 mL). The prepared solution was ultrasonicated for 1 h and then refluxed for 2 days. After the reaction finished, the solvents were evaporated using a rotary evaporator. Furthermore, a mixture of methanol and deionized water (2:1) was added to the remaining solid, and the solution was ultrasonicated for 30 min. The solid material was filtered off and subsequently washed with methanol. Finally, the product **CHIT-PAAA** was dried at room temperature and then analyzed. The homopolymer was removed with the water–methanol mixture to evaluate the efficiency of the graft copolymerization. Although there are no unified definitions for calculating the parameters of the graft copolymerization, herein we report the use of the grafting yield (G) and the copolymerization yield (Y) (Equations (1) and (2)) [41]:

$$G(\%) = \frac{W_{CHIT-PAAA} - W_{CHIT}}{W_{CHIT}} \cdot 100 \quad (1)$$

$$Y\,(\%) = \frac{W_{CHIT-PAAA}}{W_{CHIT} + W_{PBAAA}} \cdot 100 \quad (2)$$

where $W_{CHIT\text{-}PAAA}$ is the mass of the copolymer after grafting, $W_{CHIT}$ is the initial weight of chitosan added in the copolymerization reaction, and $W_{PBAAA}$ is the initial weight of **PBAAA** added in the copolymerization reaction.

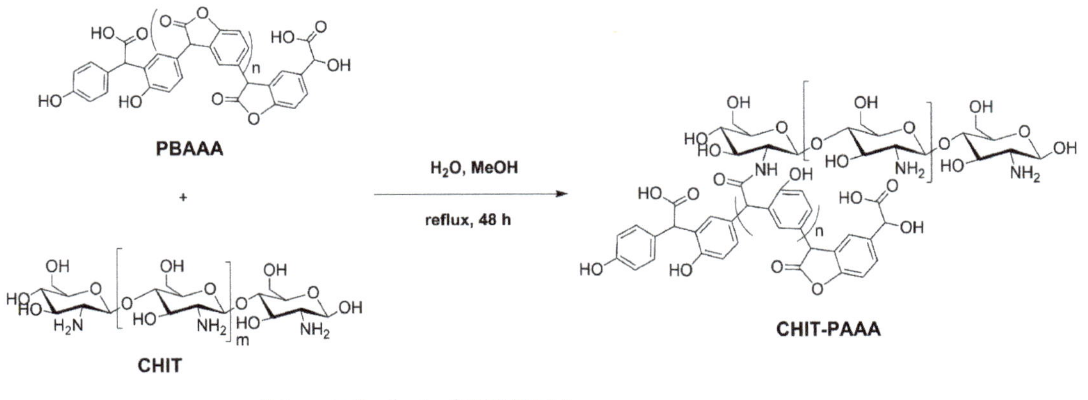

**Scheme 1.** Synthesis of **CHIT-PAAA**.

### 2.3. *Characterization Methods*

#### 2.3.1. Solid-State Nuclear Magnetic Resonance

A Bruker Advance III 500 MHz wide-bore NMR spectrometer operating at room temperature was used, with a 4 mm double resonance (1H/X) MAS probe. The material was packed in 4 mm zirconia rotors, and the solid-state $^{13}C$ and $^{15}N$ NMR spectra were recorded at 125.73 and 50.66 MHz Larmor frequencies. Standard RAMP $^{13}C/^{15}N$ CP-MAS spectra were acquired at 14/7 kHz spinning frequencies, 2/4 ms contact times, and proton decoupling under TPPM. For $^{13}C$ spectra, the acquisition parameters were optimized to the following values of relaxation delay and number of transients: 2 s/30,000 transients for **PBAAA** and **CHIT**, and 2 s/50,000 transients for sample **CHIT-PAAA**. For $^{15}N$ spectra the relaxation delay and number of transients were: 2 s/31,000 transients for **CHIT** and 2 s/60,000 transients for **CHIT-PAAA**. The recorded spectra were calibrated relative to the $CH_3$ line in tetramethylsilane (TMS) and the $^{15}NO_2$ line in nitromethane, through an indirect procedure that used L-Glycine as an external reference (C=O of glycine at 176.5 ppm for $^{13}C$ and −347.6 ppm for $^{15}N$), and line broadening was applied at 20 Hz (for $^{13}C$ spectra) and 150 Hz (for $^{14}N$ spectra).

#### 2.3.2. Fourier Transform Infrared Spectroscopy

FTIR investigation was conducted on a Jasco FTIR-6100 spectrophotometer (JASCO Deutschland GmbH, Pfungstadt, Germany), recording the material's spectra in the 400–4000 $cm^{-1}$ spectral range. Pressed pellets prepared from polymer powder embedded in KBr were used for this purpose.

#### 2.3.3. Scanning Electron Microscopy

SEM analysis was performed on a Hitachi SU8230 High-Resolution Scanning Electron Microscope (Hitachi Ltd., Tokyo, Japan) equipped with a cold field-emission gun. The samples were placed on aluminum stubs and covered with a 10 nm gold coating for morphological analysis.

#### 2.3.4. Thermo-Gravimetric Analysis

TGA was conducted in air, using TA Instruments SDT Q 600 equipment (TA Instruments Inc., New Castle, DE, USA), in the temperature range 30 °C–800 °C, with a heating rate of 10 °C $min^{-1}$ in air.

#### 2.3.5. X-ray Photon Electron Spectroscopy

An XPS spectrometer SPECS (SPECS Surface Nano Analysis GmbH, Berlin, Germany) equipped with a dual-anode X-ray source Al/Mg, a PHOIBOS 150 2D CCD hemispherical energy analyzer, and a multi-channeltron detector with vacuum maintained at $1 \times 10^{-9}$ torr was used to record XPS spectra. XPS investigations were conducted using the Al K$\alpha$ X-ray source (1486.6 eV) operating at 200 W. The XPS survey spectra were captured at 30 eV pass energy, 0.5 eV/step. The high-resolution spectra for individual elements were recorded by accumulating 30 scans at 30 eV pass energy and 0.1 eV/step. The powder samples were pressed on an indium foil to allow the XPS measurements. The sample surface was cleaned by argon ion bombardment (300 V) and the spectra were recorded before and after the cleaning. Data analysis and curve fitting were performed using CasaXPS software (Casa Software Ltd., Teignmouth, UK) with Gaussian-Lorentzian product functions and a non-linear Shirley background subtraction.

#### 2.3.6. Brunauer-Emmett-Teller Surface Area Analysis

The total surface area (St), pore volume (Vp), and pore radius (Rm) of $N_2$ adsorption–desorption isotherms (recorded at −196 °C) were determined using the Brunauer–Emmett–Teller (BET) technique for measuring St, and the Dollimore–Heal model. A Sorptomatic 1990 device (Thermo Electron Corporation, Waltham, MA, USA) was used to record the isotherms. Prior to analysis, samples were degassed at 70 °C for 5 h at a pressure of 1 Pa to eliminate any physisorbed contaminants from the surface.

#### 2.3.7. Flame Atomic Absorption Spectrometry

FAAS was used in batch experiments to determine the heavy metal concentrations. The samples were atomized using an atomic absorption spectrophotometer AAS Spectra AA110 (Varian, Australia) in a flame of air and acetylene. The analysis method closely followed the protocol described in detail in the standard SR ISO 8288/2001. In brief, the samples were digested in nitric acid, and five-point calibration curves were drawn for each metal, with the range of concentrations between 0.05 mg $L^{-1}$ and 0.4 mg $L^{-1}$ for Cd, and between 0.25 mg $L^{-1}$ and 2.50 mg $L^{-1}$ for Pb. Dilutions were made for samples that had concentrations exceeding the intervals previously mentioned. The reference material used for standard preparation (for the calibration curves) was 1000 mg $L^{-1}$ Spectro Econ Chem Lab Stock Solution (Chem Lab, Zedelgem, Belgium), while 1000 mg $L^{-1}$ Merck Stock Solution (Merck KGaA, Darmstadt, Germany) was used for quality control. The method's detection limits were 0.03 mg $L^{-1}$ for Cd and 0.25 mg $L^{-1}$ for Pb.

### 2.4. Batch Adsorption Experiments

Stock solutions of metal contaminants (Pb, Cd) were prepared at six different concentrations (10, 20, 40, 60, 80, and 100 mg $L^{-1}$) using Pb and Cd salts ($PbCl_2$, $CdCl_2 \times H_2O$) and Milli-Q ultrapure water (Millipore, Bedford, MA, USA) with pH adjusted to 5.0. The effects of two parameters (initial metal concentration and contact time) were investigated to study the adsorptive behavior of **CHIT-PAAA**. Pb and Cd adsorption assays were performed on the synthesized material **CHIT-PAAA** under magnetic agitation (at 600 rpm rotational speed) and normal atmospheric conditions (room temperature). Afterwards, samples were filtered off (on Rotilabo folded filters, type 113 P, membrane Ø 150 mm, Macherey-Nagel GmbH, Dueren, Germany), and heavy metals in the supernatant were analyzed by FAAS. An AAS Spectra AA110 atomic absorption spectrophotometer was used to determine the metal concentrations in the solutions.

The removal efficiencies (adsorption percentages) and sorption capacities were calculated based on the following equations:

$$R\ (\%) = \frac{C_i - C_f}{C_i} \cdot 100 \tag{3}$$

$$q\left(mg\ g^{-1}\right) = \frac{(C_i - C_f) \cdot V}{w} \tag{4}$$

where R is the removal efficiency (%); $C_i$ is the initial concentration (before adsorption) (mg $L^{-1}$); $C_f$ is the final concentration (after adsorption) (mg $L^{-1}$); q is the sorption capacity (mg $g^{-1}$); V is the volume of solution (L); w is the amount of sorbent (material) used (g).

### 2.5. Equilibrium Adsorption Isotherms

Pb and Cd adsorption equilibrium studies were carried out using 0.04 L metal stock solutions of six different concentrations (10, 20, 40, 60, 80, and 100 mg $L^{-1}$) and 0.02 g adsorbent material for 24 h contact time. Linear and nonlinear forms of Langmuir [42], Freundlich [43], Dubinin–Radushkevich [44], Temkin [45], Khan [46], Redlich–Peterson [47], Sips [48], Toth [49] and Koble–Corrigan [50] isotherm models were applied to fit the **CHIT-PAAA** experimental adsorption data. Table S1 summarizes the equations of the isotherms used in the current study. For the linear forms, the values of each isotherm constant were obtained from the slope and intercept of various plots: $C_e/q_e$ versus $C_e$ (Langmuir 1st type), $\ln(q_e)$ versus $\ln(C_e)$ (Freundlich), $\ln(q_e)$ versus $\varepsilon^2$ (Dubinin–Radushkevich), $q_e$ versus $\ln(C_e)$ (Temkin), $\ln(C_e/q_e)$ versus $\ln(C_e)$ (Redlich–Peterson), $\ln(q_e)/(q_{max}-q_e))$ versus $\ln(C_e)$ (Sips).

The separation factor was also determined, because it highlights the essential characteristics of Langmuir isotherm (the shape of the isotherm and the nature of the adsorption process):

$$R_L = \frac{1}{1 + K_L\, C_i} \tag{5}$$

where $R_L$ is the separation factor, $C_i$ is the initial concentration of the metal ion solution (mg $L^{-1}$), and $K_L$ is the Langmuir constant (L $mg^{-1}$). The nature of the adsorption process can be categorized as unfavourable ($R_L > 1$), linear ($R_L = 1$), favourable ($0 < R_L < 1$), or irreversible ($R_L = 0$) [51,52].

### 2.6. Kinetic Studies

The prediction of batch adsorption kinetics was essential to describe the adsorption rates and sorbate interactions. The **CHIT-PAAA** kinetic experiments were conducted using 0.04 g adsorbent material and 0.08 L metal solutions of 10, 20, 40, 60, 80 and 100 mg $L^{-1}$ concentrations. Samples were collected from each solution after 1 min, 5 min, 10 min, 20 min, 30 min, 45 min, 1 h, 3 h, 6 h, 9 h, 12 h, and 24 h of contact time, and heavy metals were determined via FAAS. Four types of kinetic models were applied to characterize the adsorption behaviour of **CHIT-PAAA**, namely the pseudo-first order [53], pseudo-second order [54], Weber–Morris intra-particle diffusion [55], and Elovich model [56]. Table S2 summarizes the equations of the models used to determine the adsorption kinetics of Pb and Cd onto **CHIT-PAAA**. Another useful kinetic parameter is the adsorption half-time ($\tau_{\frac{1}{2}}$) which represents the amount of time needed to attain half the adsorption progress or half the equilibrium value [57,58]. This parameter can be calculated as follows:

$$\tau_{1/2}[\text{min}] = \frac{1}{k_2 \cdot q_e} \quad (6)$$

*2.7. Statistics*

The results of all equilibrium and kinetic models used in this study were evaluated through the least-square method and correlation coefficient ($R^2$) analysis. The statistical evaluation was performed using Origin v.2018 (OriginLab Corporation, Northampton, MA, USA). The root mean square error function was also determined in order to establish the best-fitting models [59]:

$$\text{RMSE} = \sqrt{\frac{1}{n} \cdot \sum_{i=1}^{n} (q_{calc(i)} - q_{exp\,(i)})^2} \quad (7)$$

where RMSE represents the root mean square error; $q_{calc}$ is the calculated amount of pollutant adsorbed per unit mass of material (mg $g^{-1}$); $q_{exp}$ is the measured amount of pollutant adsorbed per unit mass of material (mg $g^{-1}$).

*2.8. Recyclability Studies*

Seven adsorption–desorption cycles were conducted to check the reusability of **CHIT-PAAA**. For this purpose, 10 mg $L^{-1}$ Pb and Cd aqueous solutions were shaken at 600 rpm for 1 h with specific amounts of copolymer. Afterwards, the material was separated from the contaminated solutions through filtration and subsequently washed with 30 mL 0.1 M $HNO_3$ solution and distilled water. The filtrated solutions were analysed using FAAS.

*2.9. 3D Adsorption Rate Models*

A 3D adsorption rate model is a representation providing a clear overview of the adsorption process and the factors that influence the sorption capacities of a studied material [60]. Herein, contact time and initial metal concentration were considered the main parameters affecting **CHIT-PAAA** adsorption rates. The 3D adsorption rate models were generated with a resolution of 1 mg $L^{-1}$, over a range of initial metal concentrations from 1 to 100 mg $L^{-1}$ for each pollutant.

*2.10. Artificial Neural Networks Models*

The high complexity of the adsorption process makes it difficult to model only through statistical methods. Therefore, computational intelligence models such as adaptive fuzzy inference systems (ANFIS), least square support vector regression (LSSVR), random forest (RF), or artificial neural networks (ANNs), which rely on artificial intelligence (AI) prediction, represent some of the best methods for modeling complex datasets [61–68]. ANNs were first introduced by McCulloch and Pitts [69], inspired by the structure and functions of biological neural networks, and have become a powerful tool for predicting system behaviors and for analyzing processes [69–72]. In general, ANNs consist of artificial neurons with specific weights, placed in various layers, interconnected through a system of artificial synapses that train interrelationships between inputs and outputs [73].

A multilayer perceptron (MLP), one of the most common types of ANNs, includes an input layer, an output layer, and one or more hidden (intermediate) layers. The numbers of layers and neurons, the networks' structure, the transfer function, and the training component form the architecture of an ANN [74]. ANNs undergo a training algorithm to enable them to predict the correlation between inputs and outputs and to reproduce known and unknown data. The MLP network is a feed-forward ANN, because data are processed from the input to the output layers [60]. The most common ANN structures used for adsorption experiments are multilayer feed-forward neural networks (MLFFN) [75]. Collected data is usually divided into 70–80% training data (for generating the output values) and 20–30% testing data (for examining the parameters of the trained ANN). The performance of the ANN model can be checked and improved by adjusting the mean squared error function (MSE) and the correlation coefficient defined by the following equations [76]:

$$\mathrm{MSE} = \frac{1}{n}\sum\nolimits_{i=1}^{n}\left(\left|\hat{Y}_i - Y\right|\right)^2 \tag{8}$$

$$R^2 = 1 - \frac{\sum_{i=1}^{n}\left(\hat{Y}_i - Y\right)}{\sum_{i=1}^{n}\left(\hat{Y}_i - Y_{av}\right)} \tag{9}$$

where n is the number of data, $\hat{Y}$ represents the predicted data, Y is the actual output data, and $Y_{av}$ is the average of the experimental values.

For the current study, a three-layer ANN (two inputs and one output) was developed by using the Neural Network Toolbox of MATLAB 7.6 (R2008a) mathematical software (MathWorks, Natick, MA, USA). The three layers consisted of two neurons in the input layer represented by the initial metal concentration (10, 20, 40, 60, 80, or 100 mg $L^{-1}$) and contact time (0–1440 min), and one neuron in the output layer (the amount of metal adsorbed). The ANN was trained with 840 data points and validated with 180. Algorithms involved 1000 iterations with tangent sigmoid transfer functions (*tansig*) and linear transfer functions (*purelin*) for training the MLFFN.

### 2.11. Adsorption Assays on Metal-Polluted Water Samples

Four water samples (Table S3) were collected in two seasons (summer and winter 2020) from two former mining areas in Romania (Novăț-Borșa and Roșia Montană, Figure 1). The map with the sampling points was developed using ArcGIS 10.6.1 (ESRI, Redlands, CA, USA). Novăț-Borșa is located in northern Romania, in the Maramureș Mountains (Maramureș County), close to the border with Ukraine. It has been an important source of lead and zinc ores (associated with copper, antimony, bismuth, cadmium, gold, and silver). Roșia Montană is situated in western Romania, in the Apuseni Mountains (Alba County) and has been exploited for its gold and silver ores. These two sampling areas were selected because they are considered among the most polluted sites in Romania, with soil, groundwater, and surface waters often reported to contain significant loads of heavy metals.

The **CHIT-PAAA** adsorption experiments using the collected metal-polluted water samples were carried out in similar conditions as the stock solution assays. After following the adsorption protocol, the metal content in the water samples was determined with FAAS.

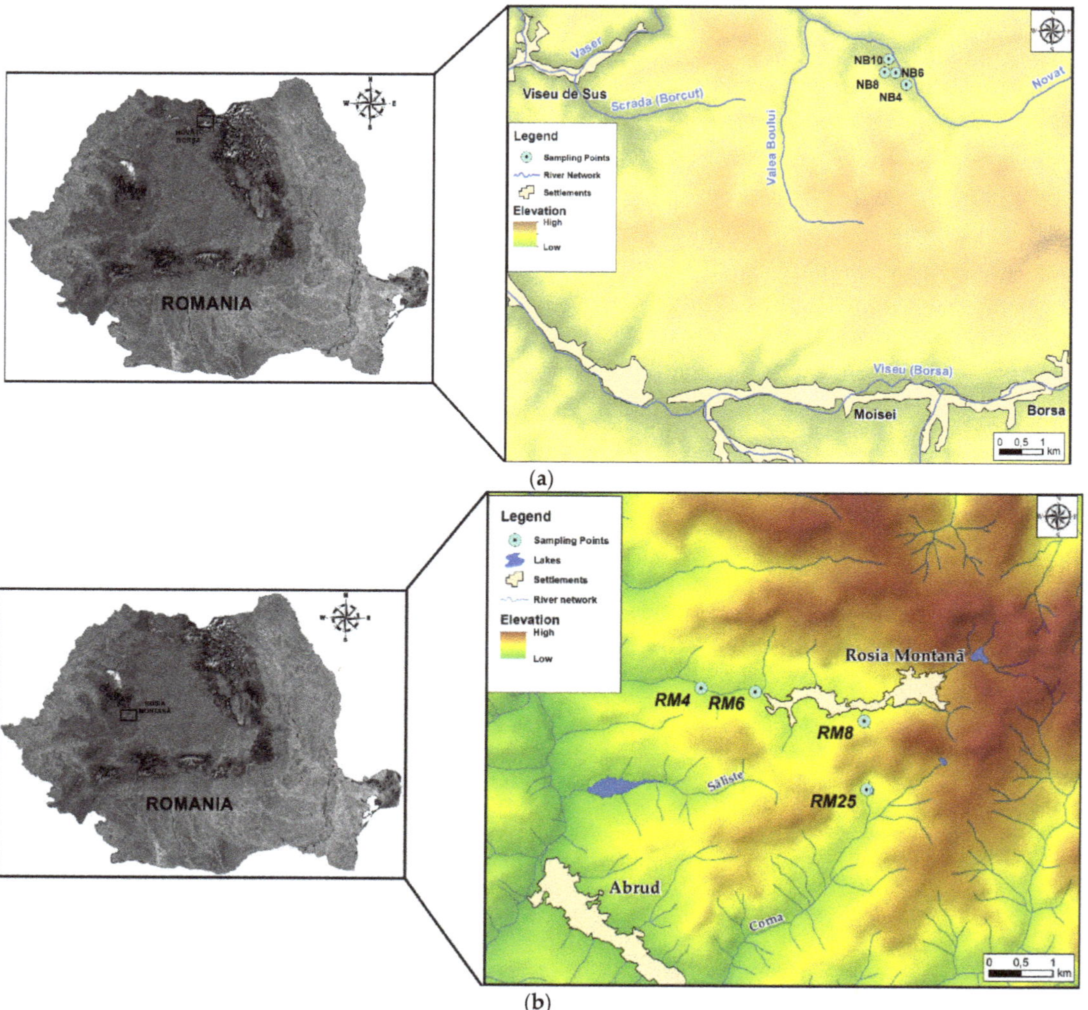


**Figure 1.** Map of the collected water samples: (**a**) Novăț-Borșa mining area, (**b**) Roșia Montană mining area.

## 3. Results and Discussion

### 3.1. Synthesis and Characterization of the Adsorbent Material

Graft polymerization represents one approach to the fabrication of chemically bonded natural–synthetic copolymer compositions. Grafting has been utilized as an important technique for modifying the chemical and physical properties of polymers. Graft copolymers are increasingly gaining importance due to their tremendous industrial potential. The current study explored the possibility of obtaining a new eco-friendly material, insoluble in water and easily separable via filtration, by opening the lactone rings in the **PBAAA** polymer chain [40] with the free amino groups of chitosan in an uncatalyzed reaction. The grafting yield determined for **CHIT-PAAA** copolymer was 140% and the copolymerization reaction yield was 83%.

Considering the fact that **CHIT-PAAA** is solid and insoluble in water or organic solvents, NMR-spectra were recorded in solid state, i.e., as $^{13}C$ *ss*-NMR and $^{15}N$ *ss*-NMR spectra. As shown in Figure 2, there is only one signal in the $^{15}N$ *ss*-NMR spectrum of chitosan, whereas two signals for nitrogen atoms can be observed in the spectrum of the final copolymer. The peak at −347.4 ppm of the copolymer is attributed to the $-NH_2$ group of the chitosan chain, and the weaker broad peak at −259.5 ppm is assigned to the new amide bond (-NH-C=O), which appears after the covalent linkage of chitosan to the **PBAAA** by opening the lactone ring.

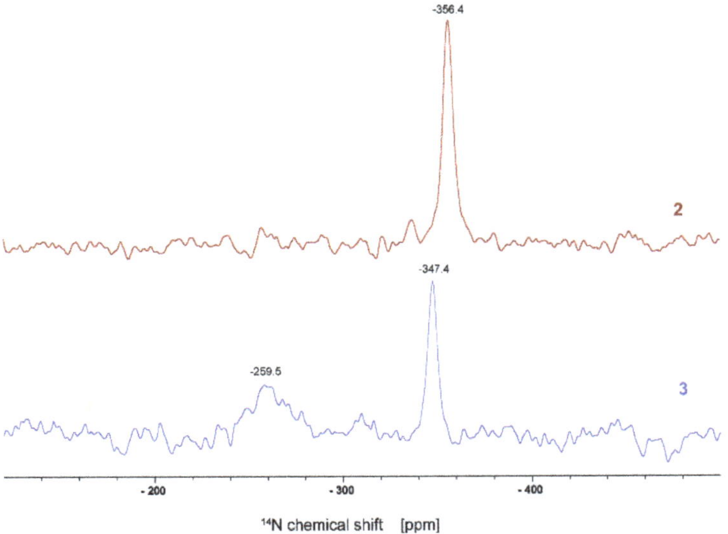


**Figure 2.** $^{15}N$ *ss*-NMR spectra of the: **CHIT** (red line) and **CHIT-PAAA** (blue line).

Significant changes were also seen in the $^{13}C$ *ss*-NMR spectrum of the final copolymer compared with the starting materials (Figure 3). As a result of the attachment of the chitosan to the **PBAAA** chain, the peaks at around 50 ppm in the $^{13}C$ *ss*-NMR spectra of **PBAAA,** typical for the –CH in the lactone units, do not appear in the $^{13}C$ *ss*-NMR spectra of copolymer **CHIT-PAAA**. The peak in the 15–25 ppm region indicates -CH groups belonging to the chitosan chain. It was also present in the spectrum of the copolymer but was hidden under the sideband. Moreover, a significant additional change in the $^{13}C$ *ss*-NMR spectrum of the copolymer was also observed in the aromatic part, where the signals for the peaks belonging to the benzene ring of **PBAAA** appeared due to the covalent linkage of the chitosan on the polymer chain.

Figure 4 shows the FTIR spectra of **PBAAA, CHIT** and **CHIT-PAAA**. The latter contains both sets of bands of the starting materials **CHIT** and **PBAAA,** showing the linkage of both moieties. In addition, the absorption band around 1620 $cm^{-1}$ belonging to the amide bond of **CHIT-PAAA** was more intensive than in the case of **CHIT**. This demonstrates the covalent attachment of the polymer chain to chitosan. Furthermore, a decrease in band intensity at 1800 $cm^{-1}$ attributable to the lactone C=O bond was observed in comparison with the copolymer with r, indicating the opening of lactone rings. Also, this is due to the linkage of the amino group of chitosan to the lactone ring of r. The FTIR bands located between 1381–1457 $cm^{-1}$ are attributed to the –C-H bond of the -CHOH-group, while those between 991–1078 $cm^{-1}$ for the copolymer are typical for the –C-O-bond in the –COH group.

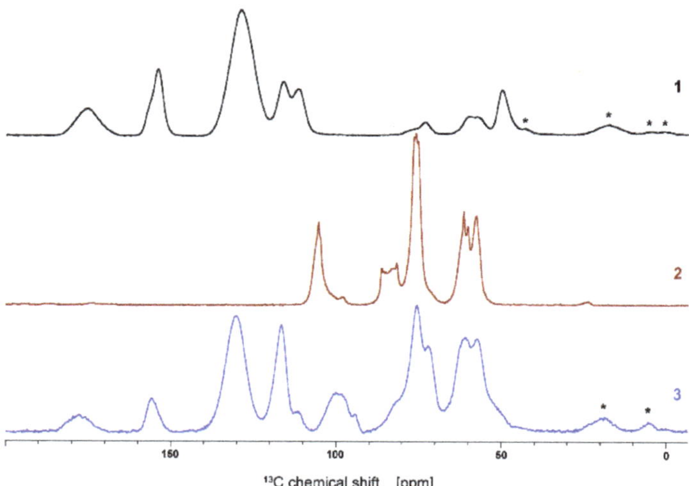


**Figure 3.** $^{13}C$ ss-NMR spectra of **PBAAA** (black), **CHIT** (dark red) and **CHIT-PAAA** (blue). The asterisk * indicates a sideband.

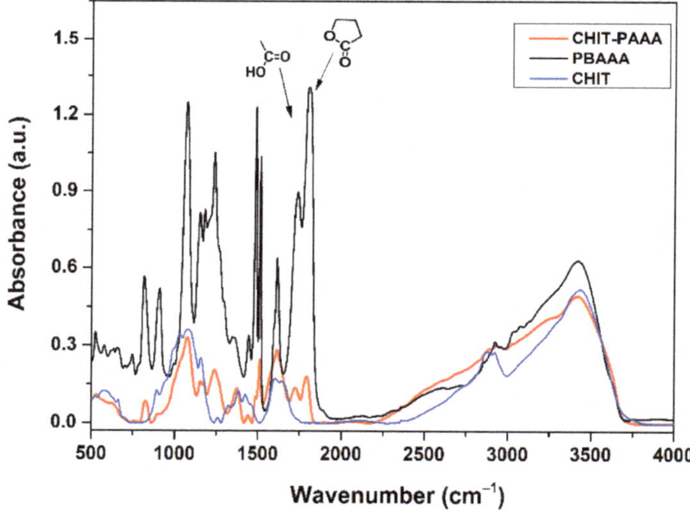


**Figure 4.** FTIR spectra of **PBAAA** (black), **chitosan** (blue), and copolymer **CHIT-PAAA** (red).

Figure 5 presents the TGA curves of **CHIT**, **PBAAA** and **CHIT-PAAA,** ranging from room temperature to 800 °C. **CHIT** underwent three degradation phases; first, between 39 °C–151°C (9.9% weight loss) corresponding to the elimination of water adsorbed in the polysaccharide structure; second, between 230 °C–360 °C (39% weight loss); and third, above 360 °C, attributed to total degradation. An increment in the decomposition trend of organic matter was observed in the case of **CHIT-PAAA** compared to **PBAAA**, due to the higher number of hydroxyl groups present after the covalent linkage of the polymer to the chitosan chain. Thus, an initial weight loss of 6.9% at 284 °C was recorded for **PBAAA**, associated with a decarboxylation process, followed by continuous degradation until 570 °C. Regarding the new material **CHIT-PAAA**, a 61% weight loss was observed in two steps; first, between 45 °C–105 °C (13% weight loss) and then between 200 °C–550 °C

(48% weight loss), corresponding to the decomposition of the polysaccharide structure of the chitosan. The whole copolymer structure collapsed at 570 °C.

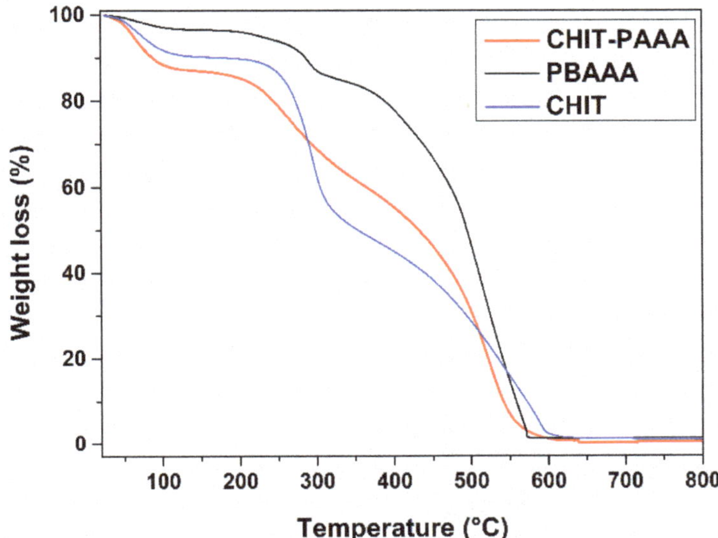


**Figure 5.** TGA curves of **PBAAA** (black), **CHIT** (blue) and **CHIT-PAAA** (red).

**PBAAA**, **CHIT**, and **CHIT-PAAA** were also investigated through SEM (Figure 6). Major morphological differences can be seen: **PBAAA** resembles an arboreal self-assembling structure, while **CHIT** has a flat, uniform folded surface, and **CHIT-PAAA** forms aggregates with a cauliflower-like aspect. The rough, heterogeneous surface of **CHIT-PAAA** with many pores makes it suitable as an adsorbent for different applications.

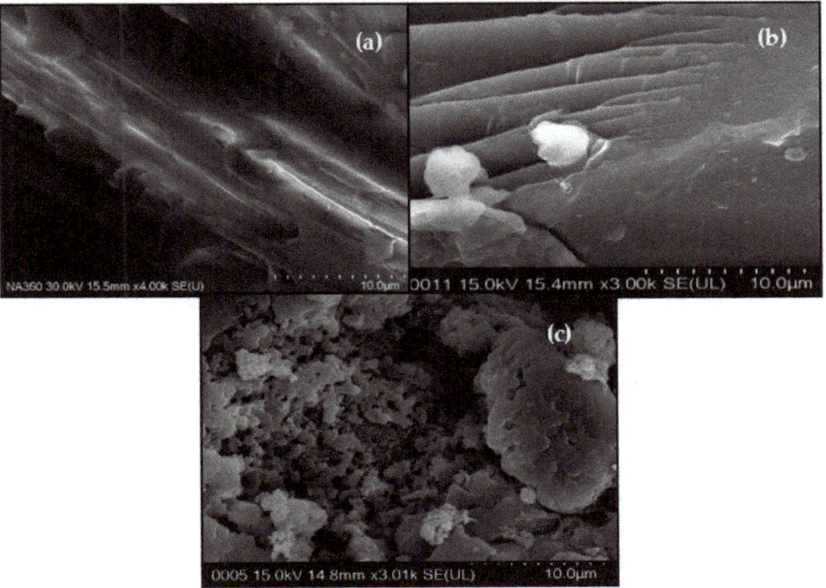


**Figure 6.** SEM images of (**a**) **PBAAA**, (**b**) **CHIT** and (**c**) **CHIT-PAAA**.

### 3.2. Metal Removal Assays on Stock Solutions

The experiments on stock solutions started with investigation of the effects induced by contact time and Pb and Cd initial concentrations on the adsorption efficiency and sorption capacity of **CHIT-PAAA**. As indicated in Figure 7a, Pb recorded very high removal efficiencies (90.63–96.07%) for a range of metal concentrations between 10–60 mg $L^{-1}$. However, a decreasing trend (76.71–84.60%) can be noticed at high concentrations (80–100 mg $L^{-1}$). Cd also recorded better adsorption efficiencies (62.20–68.60%) at low concentrations, compared with those obtained with elevated concentrations. As expected, sorption capacity increased proportionally to metal concentration (153.42 mg $g^{-1}$ and 102.26 mg $g^{-1}$ were obtained at 100 mg $L^{-1}$ for Pb and Cd, respectively; Figure 7b).

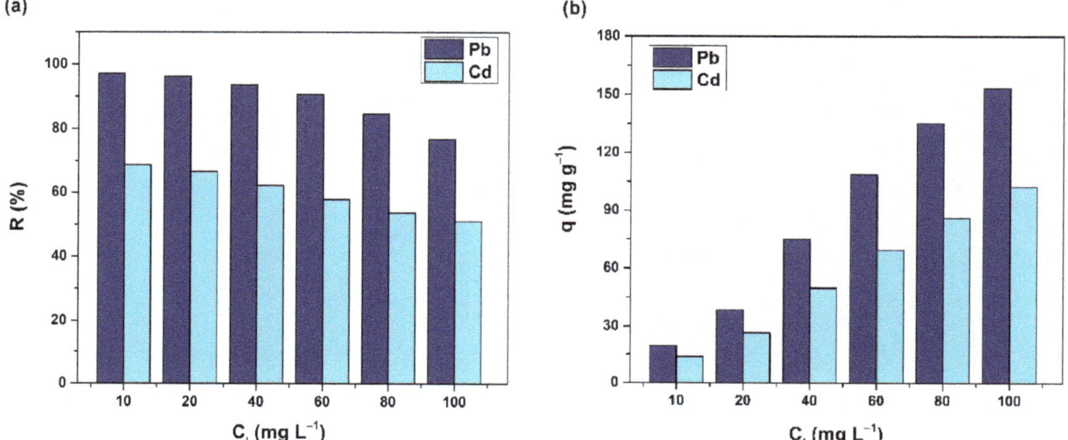


**Figure 7.** (**a**) Removal efficiency and (**b**) sorption capacity of **CHIT-PAAA** for Pb and Cd stock solutions.

The Pb and Cd sorption capacities of the newly synthesized copolymer determined in the current study are comparable to the results reported in the literature for other chitosan-based materials, as listed in Table 1. **CHIT-PAAA** recorded excellent adsorption capacities for both investigated metals compared with most adsorbents tested by other researchers [77–87]. Moreover, the copolymer registered sorption capacities up to 18 times higher than CHIT, the parent component, as seen in Table 1.

The changes occurring in the **CHIT-PAAA** structure after Pb and Cd sorption could easily be noticed in the infrared spectra of the copolymer (Figure S1). Equal amounts of material were used to make the pellets used for FTIR determination, and all obtained spectra were normalized after recording. Important modifications emerged in the 350–600 $cm^{-1}$ wavenumber range, namely the intensification of Pb-O, Cd-O bands at 462 $cm^{-1}$ and 505 $cm^{-1}$. Furthermore, other rises were seen in the bands belonging to C-O, C=C bonds from the benzene ring, and in the C-N, -N-H-, C=O bonds, located at 1072 $cm^{-1}$, 1156 $cm^{-1}$, 1245 $cm^{-1}$, 1377–1435 $cm^{-1}$, and 1513–1617 $cm^{-1}$.

Figure 8 and Figure S2 show the effects of contact time on the evolution of Pb and Cd adsorption onto **CHIT-PAAA**. For both investigated metals, adsorption occurred quickly in the first hour of interaction with the new material. Maximum adsorption efficiency was registered after 45 min of contact time for Pb and 60 min for Cd, after which equilibrium was reached.

Based on the results obtained from the adsorption experiments, 3D adsorption models were developed for each metal. As can be seen in Figure 9, sorption capacity was directly influenced by contact time and initial metal concentrations, highlighting proportional dependencies between these variables.

**Table 1.** Comparison of Pb and Cd adsorption capacities reported in the literature for different adsorbent materials.

| Heavy Metal | $q_{max}$ (mg g$^{-1}$) | Adsorbent Material | Reference |
|---|---|---|---|
| Cd | 1.06 | Chitosan | [88] |
| Cd | 9.9 | | [89] |
| Cd | 94 | | [90] |
| Pb | 7.64 | | [77] |
| Pb | 34.98 | | [79] |
| Pb | 55.5 | | [91] |
| Pb | 34.13 | Epichlorohydrin crosslinked chitosan | [78] |
| Pb | 63.33 | Chitosan–magnetite | [80] |
| Pb | 112.98 | Magnetic chitosan nanocomposites | [81] |
| Pb | 142.67 | Geopolymer–alginate– chitosan | [82] |
| Pb | 189.60 | Magnetic chitosan functionalized with EDTA | [92] |
| Pb | 334.90 | Crosslinked carboxylated chitosan–carboxylated nanocellulose hydrogel beads | [93] |
| Pb | 441.20 | Polydopamine-modified chitosan | [94] |
| Cd | 344.00 | Chitosan–activated-carbon–iron bio-nanocomposite | [95] |
| Pb<br>Cd | 11.98<br>9.73 | Chitosan bead-supported $MnFe_2O_4$ nanoparticles | [83] |
| Pb<br>Cd | 13.23<br>12.87 | Polyaniline-grafted chitosan | [84] |
| Pb<br>Cd | 86.09<br>14.14 | Chitosan-coated cotton fibers | [85] |
| Pb<br>Cd | 96.62<br>80.57 | Chitosan-*g*-methylenebisacrylamide/poly(acrylic acid) | [86] |
| Pb<br>Cd | 125.40<br>69.40 | Activated carbon–chitosan complex | [87] |
| Pb<br>Cd | 395.00<br>269.00 | Crosslinked glucan–chitosan | [96] |
| Pb<br>Cd | 447.00<br>177.00 | Chitosan–sulfhydryl-functionalized graphene oxide composites | [97] |
| **Pb**<br>**Cd** | **170.07**<br>**180.51** | **CHIT-PAAA** | **Current Study** |

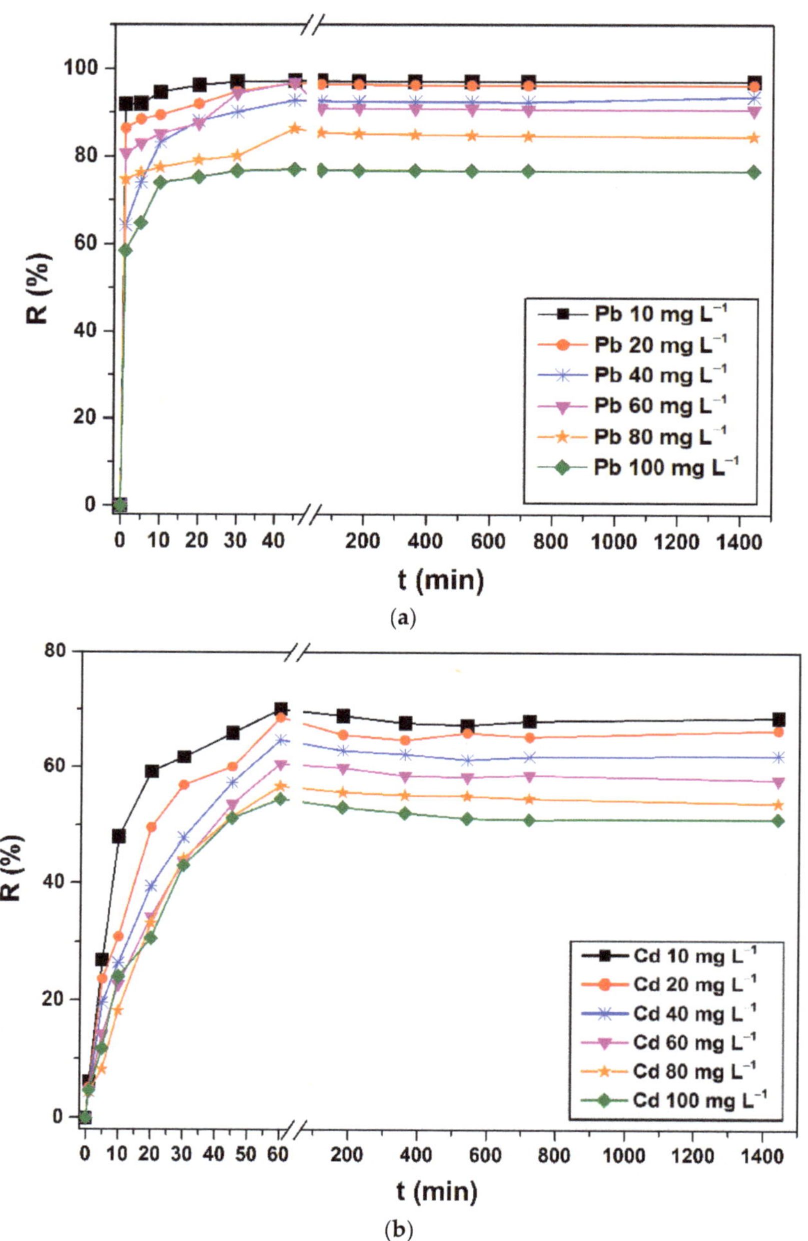


**Figure 8.** The effect of contact time on (**a**) Pb and (**b**) Cd removal efficiencies of **CHIT-PAAA**.

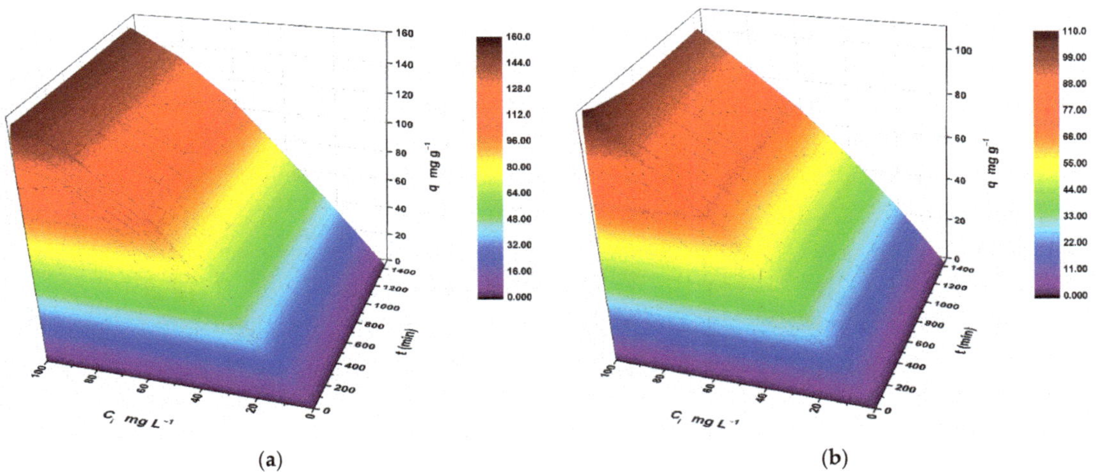


(a) (b)

**Figure 9.** The 3D adsorption models of combined effects of variables on (**a**) Pb and (**b**) Cd adsorption capacity of **CHIT-PAAA**.

The results obtained from the equilibrium study (Figure 10) show that the linear form of the Langmuir isotherm best fitted the ($R^2$ = 0.998) Pb experimental sorption data. In contrast, linear Sips (which is a combination of the Langmuir and Freundlich models) was more suitable ($R^2$ = 0.999) for Cd. The constants and correlation coefficients resulting from the linear plots of the applied isotherms are summarized in Table S4. The maximum sorption capacities of **CHIT-PAAA** were 170.068 mg $g^{-1}$ for Pb and 180.505 mg $g^{-1}$ for Cd. Pb and Cd adsorption processes could both be categorized as favorable based on the calculated values of the separation factors ($R_L$ ranged from 0.00006 to 0.00058) (Figure S3). The Sips model indicated that at low $C_e$ values ($SS_S$ between 0–1), the isotherm reduced to Freundlich characteristics, while at high $C_e$ values ($SS_S$ approaches 1), it highlighted the Langmuir monolayer sorption features [98].

The nonlinear fitting of the experimental data is presented in Figure 11. Best fit was recorded with Sips and Koble–Corrigan isotherms ($R^2$ = 0.999; RMSE = 1.339) for Pb adsorption data and with the Khan isotherm ($R^2$ = 0.999; RMSE = 0.636) for Cd. Pb and Cd isotherm shapes were generally attributed to the class L subgroup 2 type, with and without strict plateaus, respectively, based on Giles et al. [99] and Essington [100] categorizations.

Kinetics is one of the most important characteristics of the adsorption process and describes the uptake rate depending on the contact time. Linear regression was performed on the four kinetic models applied to the Pb and Cd adsorption data (Figures 12 and 13). The correlation coefficients ($R^2$) and the differences between the calculated ($q_e$) and experimental amounts of metals adsorbed ($q_{e1}$, $q_{e2}$) were taken into consideration to determine the kinetic model that best described the sorption process onto **CHIT-PAAA** (Table S5). Comparing the correlation coefficients obtained, the data followed the sequence: Pseudo-second order kinetics ($R^2$ varied between 0.999–1) > Elovich kinetics ($R^2$ varied between 0.774–0.898) > Weber–Morris intra-particle diffusion ($R^2$ varied between 0.450–0.670) > pseudo-first order kinetics ($R^2$ varied between 0.387–0.715). Pseudo-second order kinetics provided the most appropriate model for the characterization of both Pb and Cd sorption mechanisms, indicating that chemisorption was the process influencing the rates of adsorption. This fact was also supported by the BET measurements (St < 1 $m^2$ $g^{-1}$), highlighting that for **CHIT-PAAA** the mechanism of adsorption relies mainly on the chelation of metal ions by the functional groups present in the structure of the copolymer. Nevertheless, the results obtained for $\tau_{\frac{1}{2}}$ showed that **CHIT-PAAA** required between 1s–2 min and 4–7 min to reach half the adsorption capacities for Pb and Cd, respectively.

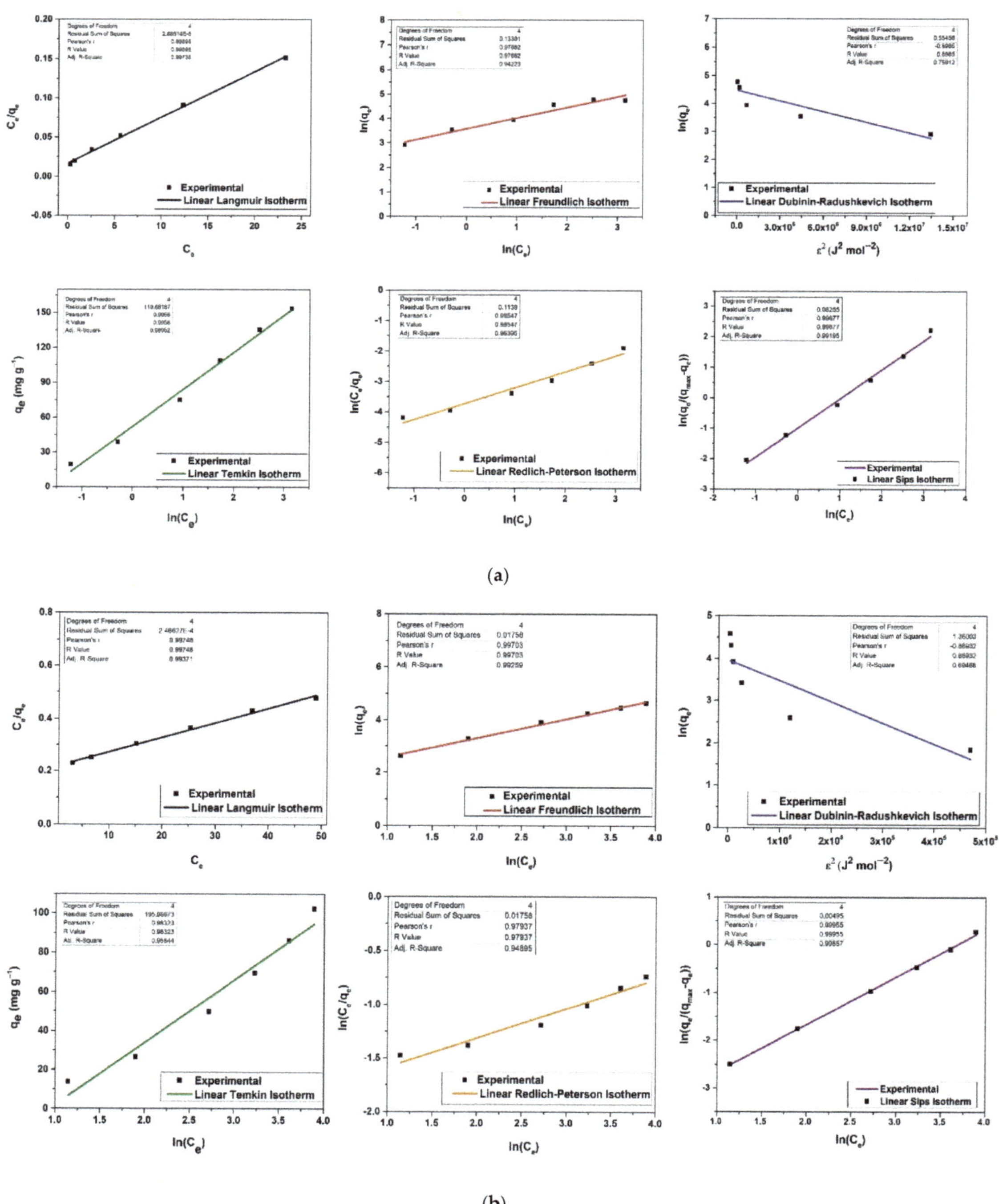


**Figure 10.** Linear forms of the equilibrium isotherms obtained for (**a**) Pb and (**b**) Cd adsorption onto **CHIT-PAAA**.

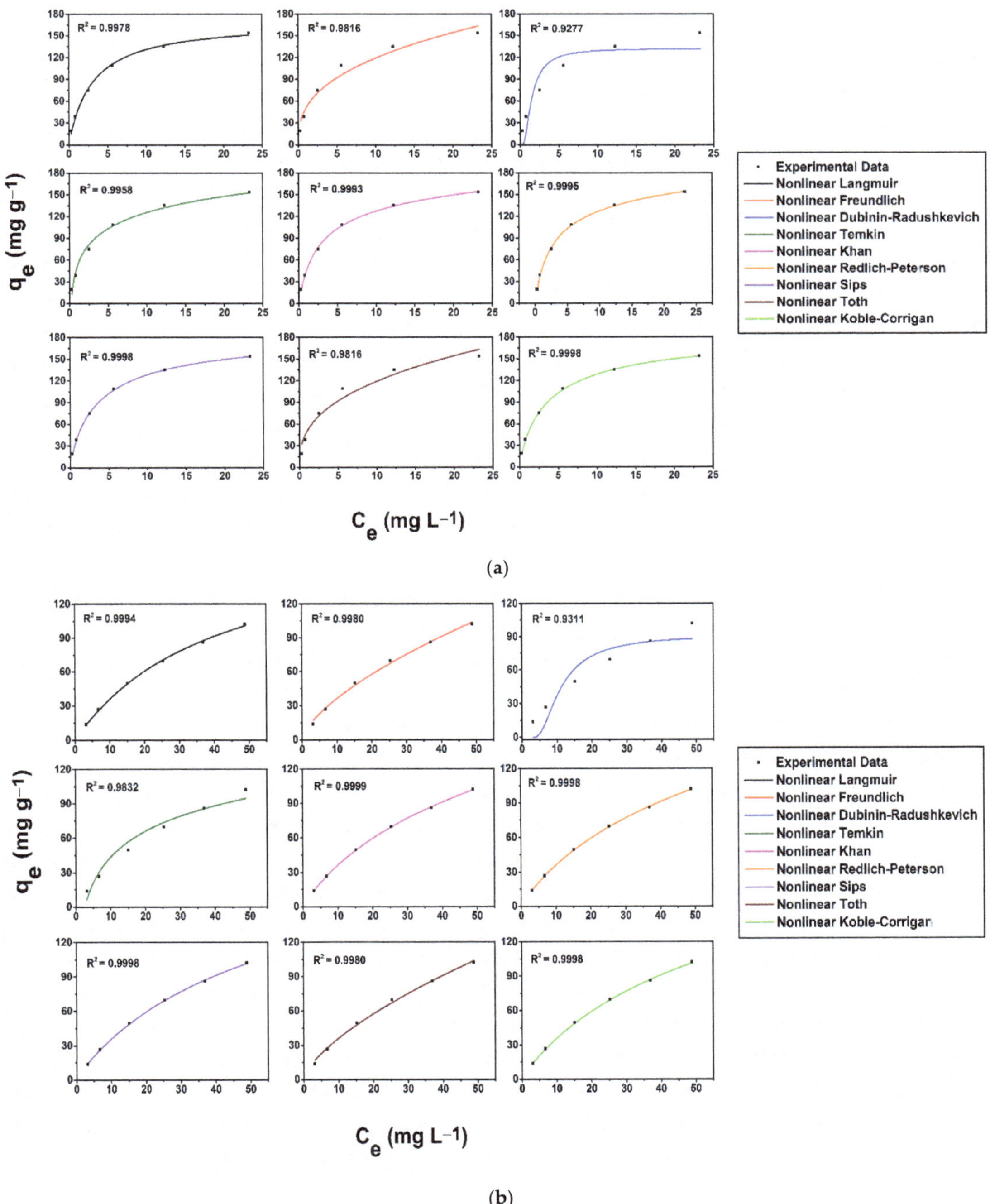


**Figure 11.** Nonlinear forms of the equilibrium isotherms obtained for (**a**) Pb and (**b**) Cd adsorption onto **CHIT-PAAA**.

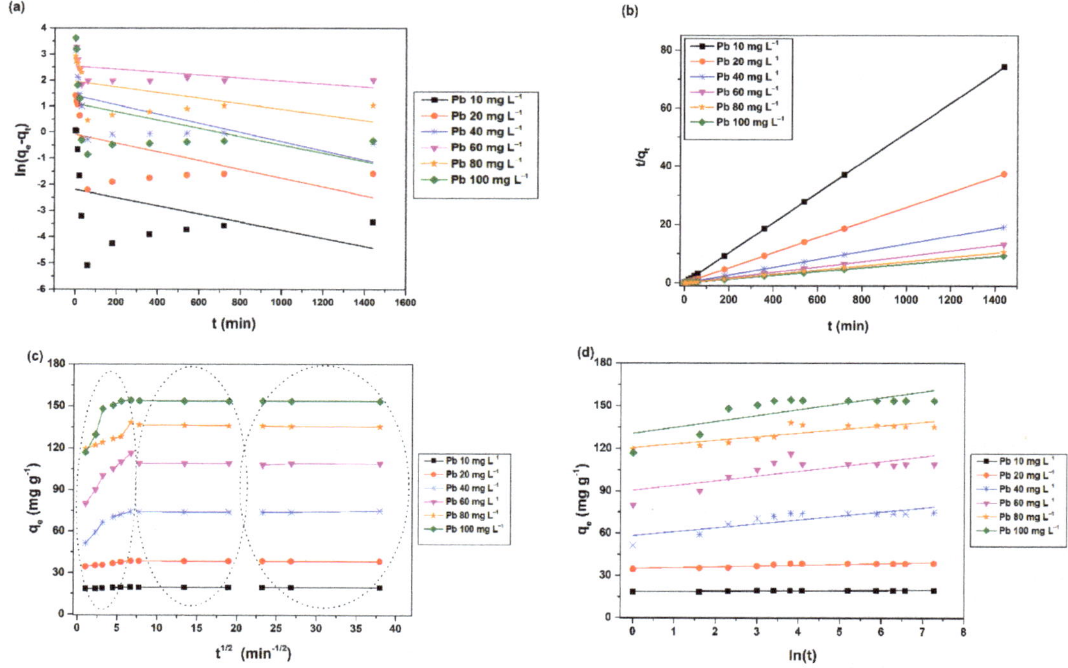


**Figure 12.** (**a**) Pseudo-first order, (**b**) pseudo-second order, (**c**) Weber–Morris intra-particle diffusion, and (**d**) Elovich kinetic models used to describe Pb adsorption on **CHIT-PAAA**.

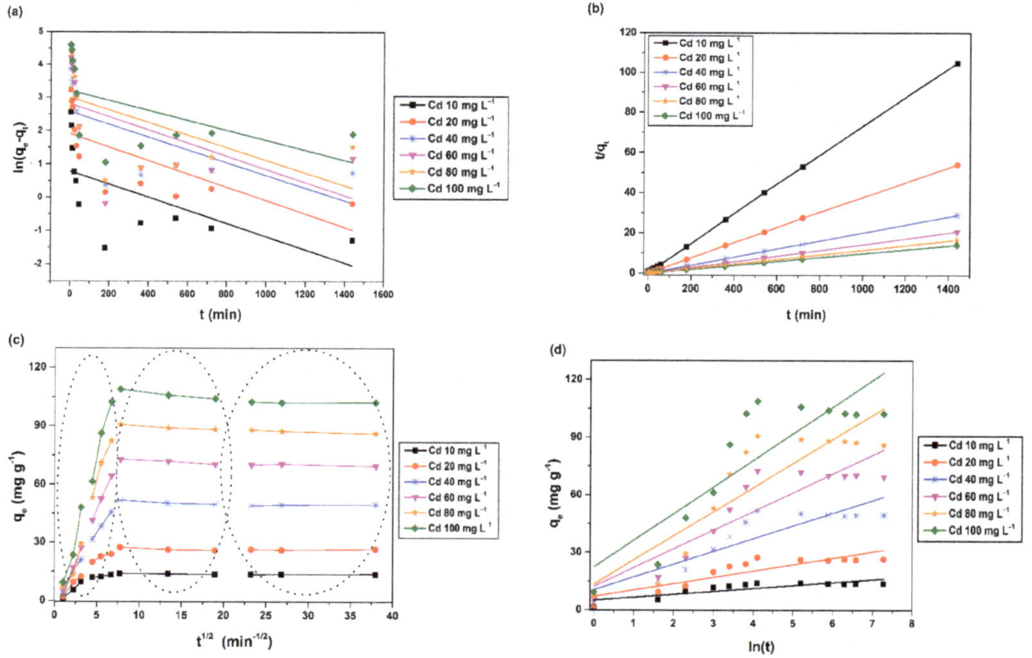


**Figure 13.** (**a**) Pseudo-first order, (**b**) pseudo-second order, (**c**) Weber–Morris intra-particle diffusion, and (**d**) Elovich kinetic models used to describe Cd adsorption on **CHIT-PAAA**.

The conducted recyclability study showed that **CHIT-PAAA** removal efficiency decreased from 97.18% to 89% for Pb and from 70% to 58% for Cd after seven cycles of adsorption–desorption (Figure 14).

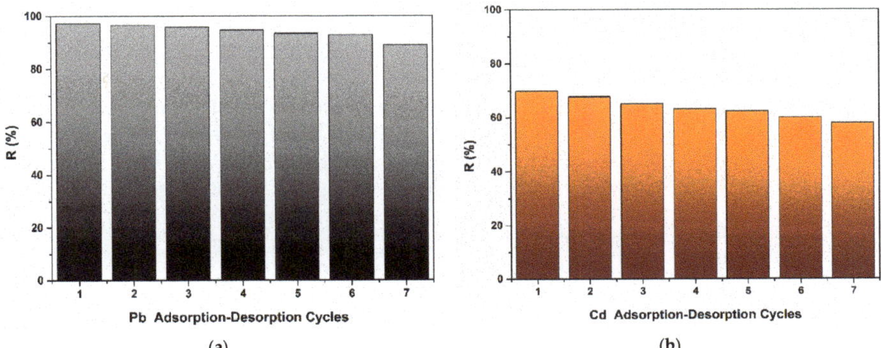


**Figure 14. CHIT-PAAA** recyclability study performed for (**a**) Pb and (**b**) Cd adsorption.

The ANN architectures developed for models of Pb and Cd adsorption onto **CHIT-PAAA** are presented in Figure 15. Eight algorithms were trained for each ANN: Levenberg–Marquardt backpropagation, resilient backpropagation, Fletcher–Reeves conjugate gradient backpropagation, Polak–Ribiére conjugate gradient backpropagation, Powell–Beale conjugate gradient backpropagation, scaled conjugate gradient backpropagation, BFGS Quasi-Newton gradient backpropagation and one-step secant backpropagation (Table S6). It was determined that the Levenberg–Marquardt design was the most suitable algorithm to model both metals' adsorption processes (Figure S4). The algorithm selection was achieved by checking the highest $R^2$ values and the lowest MSEs for Pb $R^2$ = 0.999 and MSE = 8.88 $10^{-2}$, and for Cd $R^2$ = 0.999 and MSE = 7.89 $10^{-2}$. The optimum number of hidden neurons was five in the case of Pb and six for Cd sorption. The predicted ANN results were very close to the experimental data, highlighting a good fit and a low MSE (Table S7).

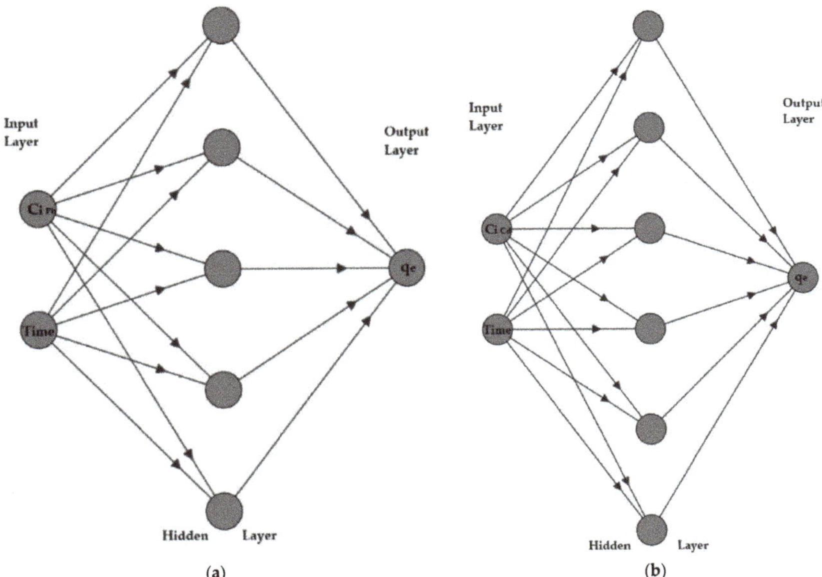


**Figure 15.** ANN architectures for (**a**) Pb and (**b**) Cd adsorption modeling onto **CHIT-PAAA**.

### 3.3. Assays on Metal-Polluted Water Samples

Adsorption assays were performed on the collected contaminated water samples after establishing the initial metal composition. As can be seen in Figure 16, Roșia Montană water samples registered significant amounts of Fe (11.48–460.81 mg $L^{-1}$), Mn (26.34–185.22 mg $L^{-1}$), Cu (0.06–1.55 mg $L^{-1}$), and Ni (0.24–0.89 mg $L^{-1}$), while Novăț-Borșa samples were rich in Fe (35.21–70.2 mg $L^{-1}$), Zn (1.77–50 mg $L^{-1}$), Cu (0.11–0.94 mg $L^{-1}$) and Pb (0.01–0.40 mg $L^{-1}$). In general, high concentrations were measured for Fe, Mn, and Zn in all investigated samples. Important metal inputs could be noticed during the winter season due to the increase in water flow from rain and snow. Each mining area had a specific metal composition profile as a result of its local geology and geochemistry, and Roșia Montană water samples recorded higher metal concentrations compared with Novăț-Borșa samples.

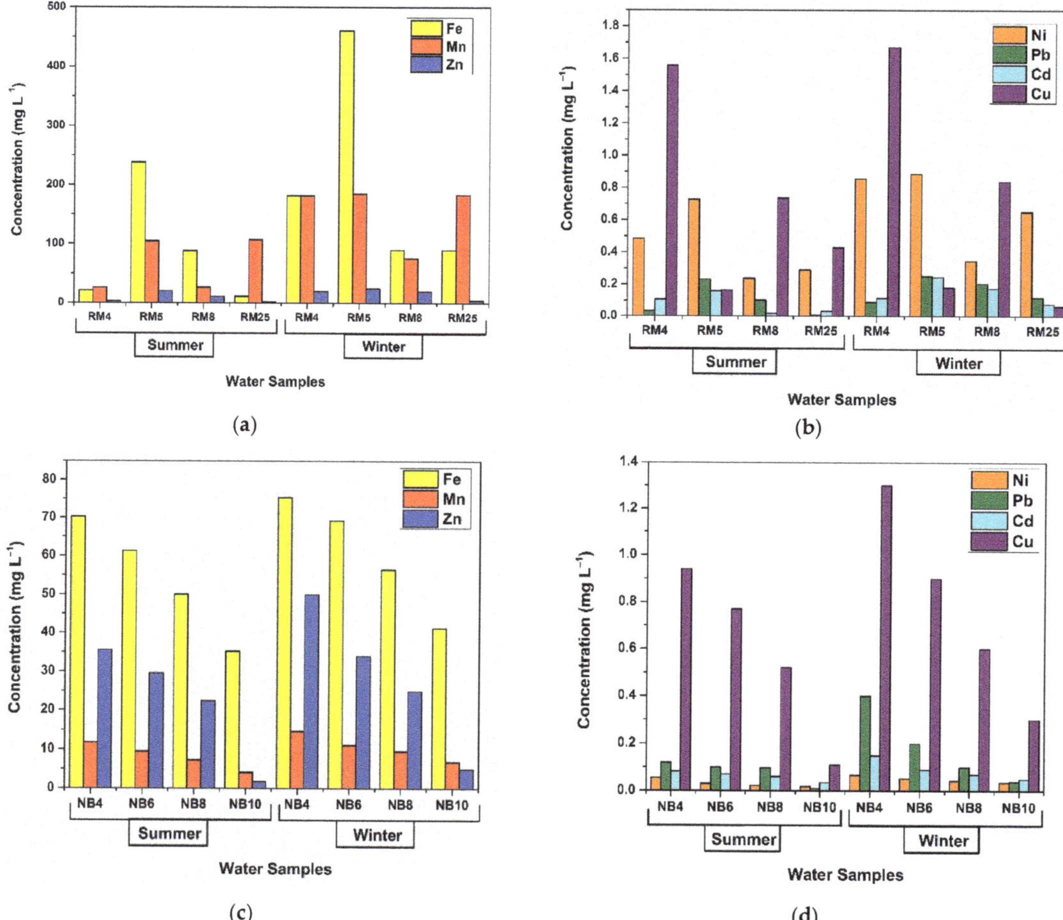


**Figure 16.** Heavy metals concentrations measured in water samples from (**a**,**b**) Roșia Montană and (**c**,**d**) Novăț-Borșa mining areas, Romania.

On the other hand, excellent removal efficiencies (100%) were obtained for Ni, Pb, Cd, and Cu in water samples collected from both locations (Figure 17). This fact indicates a better adsorption performance of **CHIT-PAAA** at low metal concentrations. Nonetheless, very good adsorption efficiencies were also determined for Fe (up to 95%) in Roșia Montană water samples and Zn (up to 85%) in Novăț-Borșa samples.

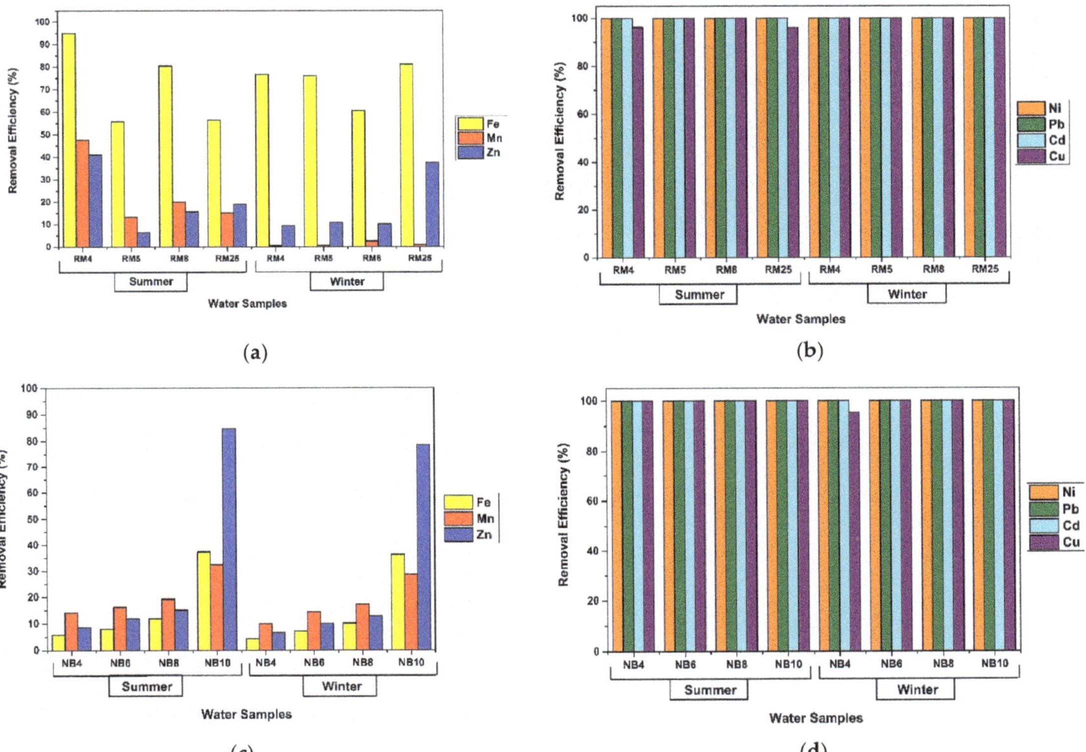


**Figure 17.** **CHIT-PAAA** removal efficiencies for heavy metals in water samples from (**a**,**b**) Roșia Montană and (**c**,**d**) Novăț-Borșa mining areas, Romania.

### 3.4. XPS Results

XPS analysis evidences the formation of the copolymer **CHIT-PAAA** and the adsorption of Pb into this copolymer. Figure 18 shows the high-resolution XPS spectra for C1s, O1s, and N1s for the copolymer **CHIT-PAAA**. The best fit for the C1s spectrum was obtained with four components; the component located at 284.8 eV corresponded to C-C, C-H; that at 285.78 eV corresponded to C-N, C-O; the component at 287.6 eV corresponded to the amide group N-C=O which demonstrates the copolymer formation; the higher binding energy component located at 289.2 eV corresponded to the O-C=O group. The N1s spectrum exhibited three components assigned to the nitrogen atoms, from $NH_2$, N-C=O groups, and protonated nitrogen $NH_3^+$.

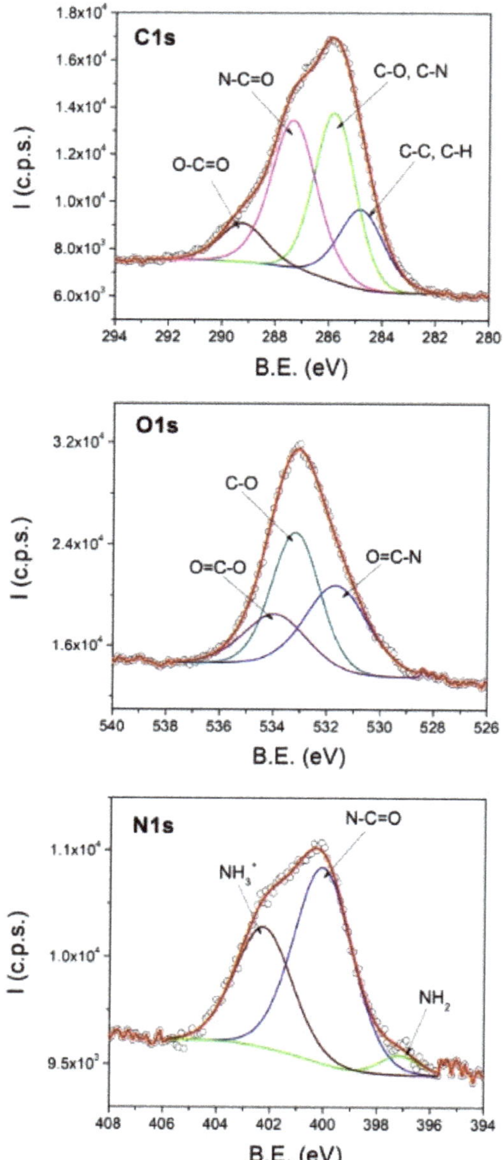


**Figure 18.** High resolution XPS spectra of C1s, O1s and N1s core levels from **CHIT-PAAA**.

The adsorption of Pb on **CHIT-PAAA** is evidenced by the high-resolution XPS spectra shown in Figure 19. The Pb spectrum from Figure 19 exhibits the doublet Pb4f5/2 and Pb 4f7/2 located at 143.7 eV and 138.8 eV, corresponding to Pb. A comparison of the XPS spectra from Figures 18 and 19 shows changes in the relative intensities of the component peaks, especially for O1s and N1s. This fact suggests that the mechanism of Pb adsorption on **CHIT-PAAA** involved the interaction of O and N atoms with metallic ions.

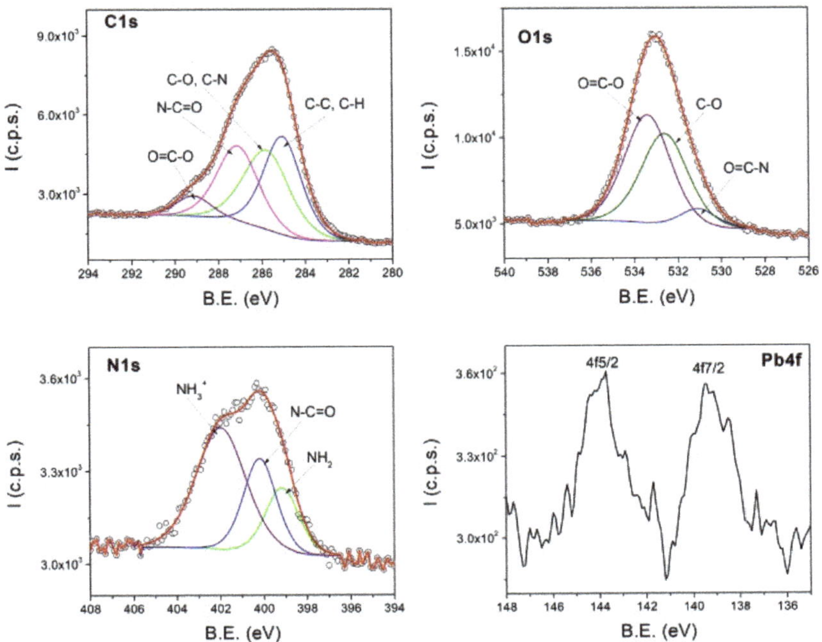


**Figure 19.** High-resolution XPS spectra of C1s, O1s, N1s and Pb4f core levels from **CHIT-PAAA** after adsorption of $Pb^{2+}$.

## 4. Conclusions

In conclusion, a new hybrid material was synthesized by green methods through the modification of **CHIT** with **PBAAA**. Successful results were achieved regarding the material's efficiency and selectivity in retaining Pb (96.07–100%), Cd (76.71–100%), Fe (95%), Zn (85%), Ni (100%), Cu (100%) from batch solutions and contaminated mining water samples. Maximum adsorption was reached quickly, after only 45 min contact time for Pb and 60 min for Cd. The applied 3D, equilibrium, and kinetic models suggested that the sorption capacity of **CHIT-PAAA** was directly dependent on the contact time and initial metal concentrations, and chemisorption was the rate-limiting process. Similar results were generated with the neural network architectures developed, highlighting a high level of trust in the ANN models for both Pb and Cd adsorption. The reciclability study of the copolymer indicated that the removal efficiency decreased to 89% for Pb and 58% for Cd after seven adsorption–desorption cycles. The results of the present investigation suggest that the newly synthesized material is cost-effective, eco-friendly, and has excellent performance in removing metal ions, being suitable for applications in the field of water and wastewater treatment technologies. Therefore, this new copolymer can be used to remediate the issue of contaminated waters, reducing heavy metal pollution, and promoting sustainable development.

**Supplementary Materials:** The following are available online at https://www.mdpi.com/article/10.3390/polym14183735/s1, Table S1: Isotherm models used on **CHIT-PAAA** adsorption data, Table S2: Kinetic models used on **CHIT-PAAA** adsorption data, Table S3: Geographic coordinates of the collected metal-polluted water samples, Figure S1: FTIR of **CHIT-PAAA** before and after adsorption of Pb and Cd from 40 mg $L^{-1}$ (**a**) and 100 mg $L^{-1}$ (**b**) stock solutions, Figure S2: The effect of contact time on (**a**) Pb and (**b**) Cd sorption capacities of **CHIT-PAAA,** Table S4. Results of the Pb and Cd adsorption equilibrium study performed on **CHIT-PAAA.** (Ci = 10–100 mg $L^{-1}$, 0.02 g material, 298 K, 600 rpm, 24 h), Figure S3. Separation factors determined for Pb (**a**) and Cd (**b**) adsorption onto **CHIT-PAAA,** Table S5: Results of the kinetic models applied for Pb and Cd sorption onto **CHIT-PAAA** (Ci = 10–100 mg L−1, 0.04 g material, 298 K, 600 rpm), Table S6: Performance of algorithms applied for ANN modeling of Pb and Cd adsorption onto **CHIT-PAAA**, Figure S4: Error histograms and regression plots of Pb (a,b) and Cd (c,d) ANN adsorption models generated with Levenberg–Marquardt algorithm, Table S7: Comparison between calculated and ANN predicted values of the metal amount adsorbed onto r.

**Author Contributions:** Conceptualization, methodology and validation I.-V.G., A.N. and C.R.; investigation I.-V.G. and A.N.; resources R.T., E.G. and C.B.; software, I.-V.G., C.R. and X.F.; data curation, A.N., X.F., I.N. and R.T.; formal analysis, I.-V.G. and C.R.; writing—original draft preparation I.-V.G., A.N., C.B. and R.T.; writing—review and editing I.-V.G., A.N., I.N. and C.R.; supervision and project administration and funding acquisition, R.T., E.G. and C.B. All authors have read and agreed to the published version of the manuscript.

**Funding:** This work was supported by the Ministry of Research, Innovation and Digitalisation through Core Project PN-19-35-02-03 and Programme 1—Development of the National Research and Development System, Subprogramme 1.2—Institutional Performance—Funding Projects for Excellence in RDI, Contract No. 37PFE/30.12.2021.

**Institutional Review Board Statement:** Not applicable.

**Informed Consent Statement:** Not applicable.

**Data Availability Statement:** No new data were created or analyzed in this study. Data sharing is not applicable to this article.

**Acknowledgments:** The authors would like to acknowledge Sebastian Alin Porav, Monica Dan and Diana Lazăr, from INCDTIM Cluj-Napoca for SEM, TGA and BET investigations and Cristian Malos for the support provided with GIS analysis.

**Conflicts of Interest:** Iolanda-Veronica Ganea is an employee of MDPI; however, she did not work for the journal Polymers at the time of submission and publication.

## References

1. Schwarzenbach, R.P.; Escher, B.I.; Fenner, K.; Hofstetter, T.B.; Johnson, C.A.; Von Gunten, U.; Wehrli, B. The Challenge of Micropollutants in Aquatic Systems. *Science* **2006**, *313*, 1072–1077. [CrossRef] [PubMed]
2. Delpla, I.; Jung, A.V.; Baures, E.; Clement, M.; Thomas, O. Impacts of Climate Change on Surface Water Quality in Relation to Drinking Water Production. *Environ. Int.* **2009**, *35*, 1225–1233. [CrossRef]
3. Kallenborn, R. Persistent Organic Pollutants (POPs) as Environmental Risk Factors in Remote High-Altitude Ecosystems. *Ecotoxicol. Environ. Saf.* **2006**, *63*, 100–107. [CrossRef] [PubMed]
4. Dudgeon, D.; Arthington, A.H.; Gessner, M.O.; Kawabata, Z.-I.; Knowler, D.J.; Lévêque, C.; Naiman, R.J.; Ne Prieur-Richard, A.-H.; Soto, D.; Stiassny, M.L.J.; et al. Freshwater Biodiversity: Importance, Threats, Status and Conservation Challenges. *Cambridge Philos. Soc.* **2006**, *81*, 163–182. [CrossRef] [PubMed]
5. Schindler, D.E.; Hilborn, R. Prediction, Precaution, and Policy under Global Change. Emphasize Robustness, Monitoring, and Flexibility. *Science* **2015**, *347*, 953–954. [CrossRef]
6. Malmqvist, B.; Rundle, S. Threats to the Running Water Ecosystems of the World. *Environ. Conserv.* **2002**, *29*, 134–153. [CrossRef]
7. Narayan, S.S.; Dipak, P. Heavy Metal Tolerance and Accumulation by Bacterial Strains Isolated from Waste Water. *J. Chem. Biol. Phys. Sci.* **2014**, *4*, 812–817.
8. Agency for Toxic Substances and Disease Registry (ATSDR). *Toxicological Profile for Lead*; U.S. Department of Health and Human Services: Washington, DC, USA, 2007.
9. Agency for Toxic Substances and Disease Registry (ATSDR). *Toxicological Profile for Cadmium*; U.S. Department of Health and Human Services: Washington, DC, USA, 2012.

10. Wan Ngah, W.S.; Teong, L.C.; Hanafiah, M.A.K.M. Adsorption of Dyes and Heavy Metal Ions by Chitosan Composites: A Review. *Carbohydr. Polym.* **2011**, *83*, 1446–1456. [CrossRef]
11. Reddy, D.H.K.; Lee, S.M. Application of Magnetic Chitosan Composites for the Removal of Toxic Metal and Dyes from Aqueous Solutions. *Adv. Colloid Interface Sci.* **2013**, *201–202*, 68–93. [CrossRef] [PubMed]
12. Varma, A.J.; Deshpande, S.V.; Kennedy, J.F. Metal Complexation by Chitosan and Its Derivatives: A Review. *Carbohydr. Polym.* **2004**, *55*, 77–93. [CrossRef]
13. Guibal, E. Interactions of Metal Ions with Chitosan-Based Sorbents: A Review. *Sep. Purif. Technol.* **2004**, *38*, 43–74. [CrossRef]
14. Fang, C.; Xiong, Z.; Qin, H.; Huang, G.; Liu, J.; Ye, M.; Feng, S.; Zou, H. One-Pot Synthesis of Magnetic Colloidal Nanocrystal Clusters Coated with Chitosan for Selective Enrichment of Glycopeptides. *Anal. Chim. Acta* **2014**, *841*, 99–105. [CrossRef] [PubMed]
15. Li, N.; Bai, R. Highly Enhanced Adsorption of Lead Ions on Chitosan Granules Functionalized with Poly(Acrylic Acid). *Ind. Eng. Chem. Res.* **2006**, *45*, 7897–7904. [CrossRef]
16. Charpentier, T.V.J.; Neville, A.; Lanigan, J.L.; Barker, R.; Smith, M.J.; Richardson, T. Preparation of Magnetic Carboxymethylchitosan Nanoparticles for Adsorption of Heavy Metal Ions. *ACS Omega* **2016**, *1*, 77–83. [CrossRef] [PubMed]
17. Mahaweero, T. Extraction of Heavy Metals from Aqueous Solutions Using Chitosan/Montmorillonite Hybrid Hydrogels. Master's Thesis, Case Western Reserve University, Cleveland, OH, USA, 2013.
18. Zhang, L.; Zeng, Y.; Cheng, Z. Removal of Heavy Metal Ions Using Chitosan and Modified Chitosan: A Review. *J. Mol. Liq.* **2016**, *214*, 175–191. [CrossRef]
19. Gokila, S.; Gomathi, T.; Sudha, P.N.; Anil, S. Removal of the Heavy Metal Ion Chromiuim(VI) Using Chitosan and Alginate Nanocomposites. *Int. J. Biol. Macromol.* **2017**, *104*, 1459–1468. [CrossRef] [PubMed]
20. Medeiros Borsagli, F.G.L.; Mansur, A.A.P.; Chagas, P.; Oliveira, L.C.A.; Mansur, H.S. O-Carboxymethyl Functionalization of Chitosan: Complexation and Adsorption of Cd (II) and Cr (VI) as Heavy Metal Pollutant Ions. *React. Funct. Polym.* **2015**, *97*, 37–47. [CrossRef]
21. Sun, X.; Li, Q.; Yang, L.; Liu, H. Chemically Modified Magnetic Chitosan Microspheres for Cr(VI) Removal from Acidic Aqueous Solution. *Particuology* **2016**, *26*, 79–86. [CrossRef]
22. Kulkarni, P.S.; Deshmukh, P.G.; Jakhade, A.P.; Kulkarni, S.D.; Chikate, R.C. 1,5 Diphenyl Carbazide Immobilized Cross-Linked Chitosan Films: An Integrated Approach towards Enhanced Removal of Cr(VI). *J. Mol. Liq.* **2017**, *247*, 254–261. [CrossRef]
23. Gopal Reddi, M.R.; Gomathi, T.; Saranya, M.; Sudha, P.N. Adsorption and Kinetic Studies on the Removal of Chromium and Copper onto Chitosan-g-Maliec Anhydride-g-Ethylene Dimethacrylate. *Int. J. Biol. Macromol.* **2017**, *104*, 1578–1585. [CrossRef] [PubMed]
24. Anitha, T.; Senthil Kumar, P.; Sathish Kumar, K.; Ramkumar, B.; Ramalingam, S. Adsorptive Removal of Pb(II) Ions from Polluted Water by Newly Synthesized Chitosan–Polyacrylonitrile Blend: Equilibrium, Kinetic, Mechanism and Thermodynamic Approach. *Process Saf. Environ. Prot.* **2015**, *98*, 187–197. [CrossRef]
25. Sobahi, T.R.A.; Abdelaal, M.Y.; Makki, M.S.I. Chemical Modification of Chitosan for Metal Ion Removal. *Arab. J. Chem.* **2014**, *7*, 741–746. [CrossRef]
26. Dubey, R.; Bajpai, J.; Bajpai, A.K. Chitosan-Alginate Nanoparticles (CANPs) as Potential Nanosorbent for Removal of Hg (II) Ions. *Environ. Nanotechnol. Monit. Manag.* **2016**, *6*, 32–44. [CrossRef]
27. Gupta, V.K.; Gupta, D.; Agarwal, S.; Kothiyal, N.C.; Asif, M.; Sood, S.; Pathania, D. Fabrication of Chitosan-g-Poly(Acrylamide)/Cu Nanocomposite for the Removal of Pb(II) from Aqueous Solutions. *J. Mol. Liq.* **2016**, *224*, 1319–1325. [CrossRef]
28. Miretzky, P.; Cirelli, A.F. Hg(II) Removal from Water by Chitosan and Chitosan Derivatives: A Review. *J. Hazard. Mater.* **2009**, *167*, 10–23. [CrossRef] [PubMed]
29. Mousa, N.E.; Simonescu, C.M.; Pătescu, R.E.; Onose, C.; Tardei, C.; Culiţă, D.C.; Oprea, O.; Patroi, D.; Lavric, V. Pb2 + Removal from Aqueous Synthetic Solutions by Calcium Alginate and Chitosan Coated Calcium Alginate. *React. Funct. Polym.* **2016**, *109*, 137–150. [CrossRef]
30. Ji, G.; Bao, W.; Gao, G.; An, B.; Zou, H.; Gan, S. Removal of Cu (II) from Aqueous Solution Using a Novel Crosslinked Alumina-Chitosan Hybrid Adsorbent. *Chin. J. Chem. Eng.* **2012**, *20*, 641–648. [CrossRef]
31. Lalhmunsiama; Lalchhingpuii;; Nautiyal, B.P.; Tiwari, D.; Choi, S.I.; Kong, S.H.; Lee, S.M. Silane Grafted Chitosan for the Efficient Remediation of Aquatic Environment Contaminated with Arsenic(V). *J. Colloid Interface Sci.* **2016**, *467*, 203–212. [CrossRef]
32. Pal, P.; Pal, A. Surfactant-Modified Chitosan Beads for Cadmium Ion Adsorption. *Int. J. Biol. Macromol.* **2017**, *104*, 1548–1555. [CrossRef] [PubMed]
33. Kyzas, G.Z.; Kostoglou, M. Swelling–Adsorption Interactions during Mercury and Nickel Ions Removal by Chitosan Derivatives. *Sep. Purif. Technol.* **2015**, *149*, 92–102. [CrossRef]
34. Hu, C.; Zhu, P.; Cai, M.; Hu, H.; Fu, Q. Comparative Adsorption of Pb(II), Cu(II) and Cd(II) on Chitosan Saturated Montmorillonite: Kinetic, Thermodynamic and Equilibrium Studies. *Appl. Clay Sci.* **2017**, *143*, 320–326. [CrossRef]
35. Shankar, P.; Gomathi, T.; Vijayalakshmi, K.; Sudha, P.N. Comparative Studies on the Removal of Heavy Metals Ions onto Cross Linked Chitosan-g-Acrylonitrile Copolymer. *Int. J. Biol. Macromol.* **2014**, *67*, 180–188. [CrossRef] [PubMed]
36. Sutirman, Z.A.; Sanagi, M.M.; Abd Karim, K.J.; Wan Ibrahim, W.A. Preparation of Methacrylamide-Functionalized Crosslinked Chitosan by Free Radical Polymerization for the Removal of Lead Ions. *Carbohydr. Polym.* **2016**, *151*, 1091–1099. [CrossRef]
37. Liu, Q.; Yang, B.; Zhang, L.; Huang, R. Adsorptive Removal of Cr(VI) from Aqueous Solutions by Cross-Linked Chitosan/Bentonite Composite. *Korean J. Chem. Eng.* **2015**, *32*, 1314–1322. [CrossRef]

38. Seyedmohammadi, J.; Motavassel, M.; Maddahi, M.H.; Nikmanesh, S. Application of Nanochitosan and Chitosan Particles for Adsorption of Zn(II) Ions Pollutant from Aqueous Solution to Protect Environment. *Model. Earth Syst. Environ.* **2016**, *2*, 165. [CrossRef]
39. Ganea, I.-V.; Nan, A.; Neamt, I.; Baciu, C.; Serrano, A.R. Neoteric Material Based on Renewable Resources for Metal-Contaminated Waters. *Environ. Sci. Proc.* **2021**, *9*, 3. [CrossRef]
40. Nan, A.; Bunge, A.; Cîrcu, M.; Petran, A.; Hădade, N.D.; Filip, X. Poly(Benzofuran-Co-Arylacetic Acid)—A New Type of Highly Functionalized Polymers. *Polym. Chem.* **2017**, *8*, 3504–3514. [CrossRef]
41. Ortega, A.; Sánchez, A.; Burillo, G. Binary Graft of Poly(N-Vinylcaprolactam) and Poly(Acrylic Acid) onto Chitosan Hydrogels Using Ionizing Radiation for the Retention and Controlled Release of Therapeutic Compounds. *Polymers* **2021**, *13*, 2641. [CrossRef]
42. Langmuir, I. The Constitution and Fundamental Properties of Solids and Liquids. Part II.-Liquids. *J. Franklin Inst.* **1917**, *184*, 721. [CrossRef]
43. Freundlich, H. Über Die Absorption in Lösungen. *Z. Phys. Chem. Stöch. Verwand.* **1907**, *57*, 385–470. [CrossRef]
44. Dubinin, M.M.; Radushkevich, L.V. The Equation of the Characteristic Curve of Activated Charcoal. *Proc. Acad. Sci. USSR Phys. Chem. Sect.* **1947**, *55*, 331–337.
45. Temkin, M.J.; Pyzhev, V. Kinetics of Ammonia Synthesis on Promoted Iron Catalysts. *Acta Physicochim. URSS* **1940**, *12*, 217–222.
46. Khan, A.R.; Ataullah, R.; Al-Haddad, A. Equilibrium Adsorption Studies of Some Aromatic Pollutants from Dilute Aqueous Solutions on Activated Carbon at Different Temperatures. *J. Colloid Interface Sci.* **1994**, *194*, 154–165. [CrossRef] [PubMed]
47. Redlich, O.; Peterson, D.L. A Useful Adsorption Isotherm. *J. Phys. Chem.* **1959**, *63*, 1024. [CrossRef]
48. Sips, R. Combined Form of Langmuir and Freundlich Equations. *J. Phys. Chem.* **1948**, *16*, 490–495. [CrossRef]
49. Toth, J. State Equation of the Solid Gas Interface Layer. *Acta Chim.* **1971**, *69*, 311–317.
50. Koble, R.A.; Corrigan, T.E. Adsorption Isotherms for Pure Hydrocarbons. *Ind. Eng. Chem.* **1952**, *44*, 383–387. [CrossRef]
51. Hamzaoui, M.; Bestani, B.; Benderdouche, N. The Use of Linear and Nonlinear Methods for Adsorption Isotherm Optimization of Basic Green 4-Dye onto Sawdust-Based Activated Carbon. *J. Mater. Environ. Sci.* **2018**, *9*, 1110–1118.
52. Kocadagistan, B.; Kocadagistan, E. The Effects of Sunflower Seed Shell Modifying Process on Textile Dye Adsorption: Kinetic, Thermodynamic and Equilibrium Study. *Desalination Water Treat.* **2014**, *57*, 3168–3178. [CrossRef]
53. Lagergren, S.; Sven, K. Zur Theorie Der Sogennanten Adsorptiongeloster Stoffe. *K. Sevenska Vetenskapsakademiens. Handl.* **1898**, *24*, 1–39.
54. Ho, Y.S.; McKay, G. The Kinetics of Sorption of Divalent Metal Ions onto Sphagnum Moss Peat. *Water Res.* **2000**, *34*, 735–742. [CrossRef]
55. Weber, W.J.; Morris, J.C. Kinetic of Adsorption on Carbon from Solution. *Am. Soc. Civ. Eng.* **1963**, *89*, 31–59. [CrossRef]
56. Zeldowitsch, J. Über Den Mechanismus Der Katalytischen Oxidation Von CO a MnO2. *URSS Acta Physiochim.* **1934**, *1*, 364–449.
57. Marczewski, A.W. Application of Mixed Order Rate Equations to Adsorption of Methylene Blue on Mesoporous Carbons. *Appl. Surf. Sci.* **2010**, *256*, 5145–5152. [CrossRef]
58. Derylo-Marczewska, A.; Marczewski, A.W.; Winter, S.; Sternik, D. Studies of Adsorption Equilibria and Kinetics in the Systems: Aqueous Solution of Dyes–Mesoporous Carbons. *Appl. Surf. Sci.* **2010**, *256*, 5164–5170. [CrossRef]
59. Tvrdík, J.; Křivý, I.; Mišík, L. Adaptive Population-Based Search: Application to Estimation of Nonlinear Regression Parameters. *Comput. Stat. Data Anal.* **2007**, *52*, 713–724. [CrossRef]
60. Roman, T.; Asavei, R.L.; Karkalos, N.E.; Roman, C.; Virlan, C.; Cimpoesu, N.; Istrate, B.; Zaharia, M.; Markopoulos, A.P.; Kordatos, K.; et al. Synthesis and Adsorption Properties of Nanocrystalline Ferrites for Kinetic Modeling Development. *Int. J. Appl. Ceram. Technol.* **2019**, *16*, 693–705. [CrossRef]
61. Ghaedi, M.; Ansari, A.; Nejad, P.A.; Ghaedi, A.; Vafaei, A.; Habibi, M.H. Artificial Neural Network and Bees Algorithm for Removal of Eosin B Using Cobalt Oxide Nanoparticle-Activated Carbon: Isotherm and Kinetics Study. *Environ. Prog. Sustain. Energy* **2015**, *34*, 155–168. [CrossRef]
62. Ghaedi, M.; Hosaininia, R.; Ghaedi, A.M.; Vafaei, A.; Taghizadeh, F. Adaptive Neuro-Fuzzy Inference System Model for Adsorption of 1,3,4-Thiadiazole-2,5-Dithiol onto Gold Nanoparticales-Activated Carbon. *Spectrochim. Acta Part A Mol. Biomol. Spectrosc.* **2014**, *131*, 606–614. [CrossRef] [PubMed]
63. Ghaedi, M.; Ghaedi, A.M.; Negintaji, E.; Ansari, A.; Vafaei, A.; Rajabi, M. Random Forest Model for Removal of Bromophenol Blue Using Activated Carbon Obtained from Astragalus Bisulcatus Tree. *J. Ind. Eng. Chem.* **2014**, *20*, 1793–1803. [CrossRef]
64. Elemen, S.; Akçakoca Kumbasar, E.P.; Yapar, S. Modeling the Adsorption of Textile Dye on Organoclay Using an Artificial Neural Network. *Dye. Pigment.* **2012**, *95*, 102–111. [CrossRef]
65. Despagne, F.; Massart, D.L. Neural Networks in Multivariate Calibration. *Analyst* **1998**, *123*, 157R–178R. [CrossRef]
66. Khan, T.; Mustafa, M.R.U.; Isa, M.H.; Manan, T.S.B.A.; Ho, Y.C.; Lim, J.W.; Yusof, N.Z. Artificial Neural Network (ANN) for Modelling Adsorption of Lead (Pb (II)) from Aqueous Solution. *Water. Air. Soil Pollut.* **2017**, *228*, 426. [CrossRef]
67. Narayana, P.L.; Maurya, A.K.; Wang, X.S.; Harsha, M.R.; Srikanth, O.; Alnuaim, A.A.; Hatamleh, W.A.; Hatamleh, A.A.; Cho, K.K.; Paturi, U.M.R.; et al. Artificial Neural Networks Modeling for Lead Removal from Aqueous Solutions Using Iron Oxide Nanocomposites from Bio-Waste Mass. *Environ. Res.* **2021**, *199*, 111370. [CrossRef]
68. Olanrewaju, R.F.; Mariam, R.; Ahmed, A.A. Modeling of ANN to Determine Optimum Adsorption Capacity for Removal of Pollutants in Wastewater. In Proceedings of the 2017 IEEE 4th International Conference on Smart Instrumentation, Measurement and Application (ICSIMA), Putrajaya, Malaysia, 28–30 November 2017; pp. 1–5. [CrossRef]
69. McCulloch, W.S.; Pitts, W. A Logical Calculus of the Ideas Immanent in Nervous Activity. *Bull. Math. Biophys.* **1943**, *5*, 115–133. [CrossRef]

70. Mohanraj, M.; Jayaraj, S.; Muraleedharan, C. Applications of Artificial Neural Networks for Thermal Analysis of Heat Exchangers—A Review. *Int. J. Therm. Sci.* **2015**, *90*, 150–172. [CrossRef]
71. Sha, W.; Edwards, K.L. The Use of Artificial Neural Networks in Materials Science Based Research. *Mater. Des.* **2007**, *28*, 1747–1752. [CrossRef]
72. Mjalli, F.S.; Al-Asheh, S.; Alfadala, H.E. Use of Artificial Neural Network Black-Box Modeling for the Prediction of Wastewater Treatment Plants Performance. *J. Environ. Manag.* **2007**, *83*, 329–338. [CrossRef] [PubMed]
73. Cavas, L.; Karabay, Z.; Alyuruk, H.; Doğan, H.; Demir, G.K. Thomas and Artificial Neural Network Models for the Fixed-Bed Adsorption of Methylene Blue by a Beach Waste *Posidonia oceanica* (L.) Dead Leaves. *Chem. Eng. J.* **2011**, *171*, 557–562. [CrossRef]
74. Chowdhury, S.; Saha, P. Das Artificial Neural Network (ANN) Modeling of Adsorption of Methylene Blue by NaOH-Modified Rice Husk in a Fixed-Bed Column System. *Environ. Sci. Pollut. Res. Int.* **2013**, *20*, 1050–1058. [CrossRef] [PubMed]
75. Ghaedi, A.M.; Vafaei, A. Applications of Artificial Neural Networks for Adsorption Removal of Dyes from Aqueous Solution: A Review. *Adv. Colloid Interface Sci.* **2017**, *245*, 20–39. [CrossRef] [PubMed]
76. Yildiz, S. Artificial Neural Network (ANN) Approach for Modeling Zn(II) Adsorption in Batch Process. *Korean J. Chem. Eng.* **2017**, *34*, 2423–2434. [CrossRef]
77. Luu, T.T.; Dinh, V.P.; Nguyen, Q.H.; Tran, N.Q.; Nguyen, D.K.; Ho, T.H.; Nguyen, V.D.; Tran, D.X.; Kiet, H.A.T. Pb(II) Adsorption Mechanism and Capability from Aqueous Solution Using Red Mud Modified by Chitosan. *Chemosphere* **2021**, *287*, 132279. [CrossRef] [PubMed]
78. Chen, A.H.; Liu, S.C.; Chen, C.Y.; Chen, C.Y. Comparative Adsorption of Cu(II), Zn(II), and Pb(II) Ions in Aqueous Solution on the Crosslinked Chitosan with Epichlorohydrin. *J. Hazard. Mater.* **2008**, *154*, 184–191. [CrossRef] [PubMed]
79. Ngah, W.S.W.; Fatinathan, S. Pb(II) Biosorption Using Chitosan and Chitosan Derivatives Beads: Equilibrium, Ion Exchange and Mechanism Studies. *J. Environ. Sci.* **2010**, *22*, 338–346. [CrossRef]
80. Tran, H.V.; Tran, L.D.; Nguyen, T.N. Preparation of Chitosan/Magnetite Composite Beads and Their Application for Removal of Pb(II) and Ni(II) from Aqueous Solution. *Mater. Sci. Eng. C* **2010**, *30*, 304–310. [CrossRef]
81. Rasoulzadeh, H.; Dehghani, M.H.; Mohammadi, A.S.; Karri, R.R.; Nabizadeh, R.; Nazmara, S.; Kim, K.H.; Sahu, J.N. Parametric Modelling of Pb(II) Adsorption onto Chitosan-Coated Fe3O4 Particles through RSM and DE Hybrid Evolutionary Optimization Framework. *J. Mol. Liq.* **2020**, *297*, 111893. [CrossRef]
82. ALSamman, M.T.; Sánchez, J. Recent Advances on Hydrogels Based on Chitosan and Alginate for the Adsorption of Dyes and Metal Ions from Water. *Arab. J. Chem.* **2021**, *14*, 103455. [CrossRef]
83. Li, H.; Ji, H.; Cui, X.; Che, X.; Zhang, Q.; Zhong, J.; Jin, R.; Wang, L.; Luo, Y. Kinetics, Thermodynamics, and Equilibrium of As(III), Cd(II), Cu(II) and Pb(II) Adsorption Using Porous Chitosan Bead-Supported $MnFe_2O_4$ Nanoparticles. *Int. J. Min. Sci. Technol.* **2021**, *31*, 1107–1115. [CrossRef]
84. Karthik, R.; Meenakshi, S. Removal of Pb(II) and Cd(II) Ions from Aqueous Solution Using Polyaniline Grafted Chitosan. *Chem. Eng. J.* **2015**, *263*, 168–177. [CrossRef]
85. Zhang, G.; Qu, R.; Sun, C.; Ji, C.; Chen, H.; Wang, C.; Niu, Y. Adsorption for Metal Ions of Chitosan Coated Cotton Fiber. *J. Appl. Polym. Sci.* **2008**, *110*, 2321–2327. [CrossRef]
86. Paulino, A.T.; Belfiore, L.A.; Kubota, L.T.; Muniz, E.C.; Almeida, V.C.; Tambourgi, E.B. Effect of Magnetite on the Adsorption Behavior of Pb(II), Cd(II), and Cu(II) in Chitosan-Based Hydrogels. *Desalination* **2011**, *275*, 187–196. [CrossRef]
87. Ge, H.; Fan, X. Adsorption of Pb2+ and Cd2+ onto a Novel Activated Carbon-Chitosan Complex. *Chem. Eng. Technol.* **2011**, *34*, 1745–1752. [CrossRef]
88. Babakhani, A.; Sartaj, M. Competitive Adsorption of Nickel(II) and Cadmium(II) Ions by Chitosan Cross-Linked with Sodium Tripolyphosphate. *Chem. Eng. Commun.* **2021**, *209*, 1348–1366. [CrossRef]
89. Bassi, R.; Prasher, S.O.; Simpson, B.K. Removal of Selected Metal Ions from Aqueous Solutions Using Chitosan Flakes. *Sep. Sci. Technol.* **2000**, *35*, 547–560. [CrossRef]
90. Zielińska, K.; Chostenko, A.; Truszkowski, S. Adsorption of Cadmium Ions on Chitosan Membranes: Kinetics and Equilibrium Studies. *Prog. Chem. Appl. Chitin Deriv.* **2010**, *15*, 73–78.
91. Sobhanardakani, S.; Zandipak, R.; Parvizimosaed, H.; Khoei, A.J.; Moslemi, M.; Tahergorabi, M.; Hosseini, S.M.; Delfieh, P. Efficiency of Chitosan for the Removal of Pb (II), Fe (II) and Cu (II) Ions from Aqueous Solutions. *Iran. J. Toxicol.* **2014**, *8*, 1145–1151.
92. Chen, B.; Zhao, H.; Chen, S.; Long, F.; Huang, B.; Yang, B.; Pan, X. A Magnetically Recyclable Chitosan Composite Adsorbent Functionalized with EDTA for Simultaneous Capture of Anionic Dye and Heavy Metals in Complex Wastewater. *Chem. Eng. J.* **2019**, *356*, 69–80. [CrossRef]
93. Xu, X.; Ouyang, X.-K.; Yang, L.-Y. Adsorption of Pb(II) from Aqueous Solutions Using Crosslinked Carboxylated Chitosan/Carboxylated Nanocellulose Hydrogel Beads. *J. Mol. Liq.* **2021**, *322*, 114523. [CrossRef]
94. Guo, D.-M.; An, Q.-D.; Xiao, Z.-Y.; Zhai, S.-R.; Yang, D.-J. Efficient Removal of Pb(II), Cr(VI) and Organic Dyes by Polydopamine Modified Chitosan Aerogels. *Carbohydr. Polym.* **2018**, *202*, 306–314. [CrossRef] [PubMed]
95. Sharififard, H.; Shahraki, Z.H.; Rezvanpanah, E.; Rad, S.H. A Novel Natural Chitosan/Activated Carbon/Iron Bio-Nanocomposite: Sonochemical Synthesis, Characterization, and Application for Cadmium Removal in Batch and Continuous Adsorption Process. *Bioresour. Technol.* **2018**, *270*, 562–569. [CrossRef] [PubMed]
96. Jiang, C.; Wang, X.; Wang, G.; Hao, C.; Li, X.; Li, T. Adsorption Performance of a Polysaccharide Composite Hydrogel Based on Crosslinked Glucan/Chitosan for Heavy Metal Ions. *Compos. Part B Eng.* **2019**, *169*, 45–54. [CrossRef]

97. Li, X.; Zhou, H.; Wu, W.; Wei, S.; Xu, Y.; Kuang, Y. Studies of Heavy Metal Ion Adsorption on Chitosan/Sulfydryl-Functionalized Graphene Oxide Composites. *J. Colloid Interface Sci.* **2015**, *448*, 389–397. [CrossRef] [PubMed]
98. Kumara, N.T.R.N.; Hamdan, N.; Petra, M.I.; Tennakoon, K.U.; Ekanayake, P. Equilibrium Isotherm Studies of Adsorption of Pigments Extracted from Kuduk-Kuduk (*Melastoma malabathricum* L.) Pulp onto $TiO_2$ Nanoparticles. *J. Chem.* **2014**, *2014*, 468975. [CrossRef]
99. Giles, C.H.; Smith, D.; Huitson, A. A General Treatment and Classification of the Solute Adsorption Isotherm. *J. Colloid Interface Sci.* **1974**, *47*, 755–765. [CrossRef]
100. Essington, M.E. *Soil and Water Chemistry: An Integrative Approach*; CRC Press: Boca Raton, FL, USA, 2004; ISBN 9772081415.

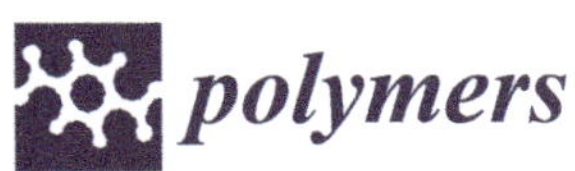

*Article*

# Chitin Nanocrystals Provide Antioxidant Activity to Polylactic Acid Films

**Murat Yanat *, Ivanna Colijn and Karin Schroën**

Laboratory of Food Process Engineering, Wageningen University and Research, Bornse Weilanden 9, 6708 WG Wageningen, The Netherlands; ivanna.colijn@wur.nl (I.C.); karin.schroen@wur.nl (K.S.)
* Correspondence: murat.yanat@wur.nl; Tel.: +31-647308241

**Abstract:** About 1/3rd of produced food goes to waste, and amongst others, advanced packaging concepts need to be developed to prevent this from happening. Here, we target the antioxidative functionality of food packaging to thus address food oxidation without the need for the addition of antioxidants to the food product, which is not desirable from a consumer point of view. Chitin nanocrystals (ChNC) have been shown to be promising bio-fillers for improving the mechanical strength of biodegradable plastics, but their potential as active components in plastic films is rather unexplored. In the current study, we investigate the antioxidant activity of chitin nanocrystals as such and as part of polylactic acid (PLA) films. This investigation was conducted using DPPH (1,1-diphenyl-2-picrylhydrazyl) radical scavenging activity. Chitin nanocrystals produced via acid hydrolysis showed five times higher activity compared to crude chitin powder. When using these crystals as part of a polylactic acid film (either inside or on top), in both scenarios, antioxidant activity was found, but the effect was considerably greater when the particles were at the surface of the film. This is an important proof of the principle that it is possible to create biodegradable plastics with additional functionality through the addition of ChNC.

**Keywords:** chitin nanocrystals; antioxidant activity; polylactic acid; nanocomposite; active packaging; DPPH

**Citation:** Yanat, M.; Colijn, I.; Schroën, K. Chitin Nanocrystals Provide Antioxidant Activity to Polylactic Acid Films. *Polymers* **2022**, *14*, 2965. https://doi.org/10.3390/polym14142965

Academic Editors: José Miguel Ferri, Vicent Fombuena Borràs and Miguel Fernando Aldás Carrasco

Received: 23 June 2022
Accepted: 20 July 2022
Published: 21 July 2022



## 1. Introduction

Oxidation is one of the greatest issues leading to food quality deterioration and, ultimately, food waste. This chain reaction is induced and propagated by radicals that react with atmospheric oxygen, thus directly affecting oxygen-sensitive food compounds, including proteins, vitamins, pigments, and especially unsaturated lipids [1–4]. Once initiated, the oxidation reaction will proceed, leading to food quality deterioration via nutrient and sensory losses such as off-odors, off-flavors, color or texture changes, and even the formation of toxic components. There are indications that products from oxidation reactions can react with biomolecules in the human body and thus impair human health due to cytotoxic, mutagenic, and carcinogenic effects [5,6]. As a consequence of all mentioned effects, a substantial reduction in food shelf life, and related to that, food loss, takes place [4]. From this, it is clear that it is crucial to reduce and ideally prevent food oxidation.

Conventionally, food oxidation is influenced by the addition of antioxidant compounds, for example, tocopherols, plant phenolics, or synthetic antioxidants such as butylated hydroxytoluene (BHT) and butylated hydroxyanisole (BHA). Although these methods can be very effective, there has been increasing (consumer) concern towards the addition of antioxidant compounds which are often presented as E-numbers on the food package. This is known to influence consumer acceptance and promotes so-called 'clean-label strategies' in food production. This makes it immediately clear that other strategies to prevent food oxidation are relevant and worthy of exploration.

An alternative approach is the incorporation of bioactive compounds in the food package instead of in the food products themselves. Bioactive packaging has become quite a trend: in essence, a bioactive compound is used in food packages to maintain the quality and extend the shelf life of food [7–9]. The antioxidant action of bioactive films relies on the release of bioactive compounds such as organic acids, essential oils, fruit and plant extracts, etc., from the package to the food matrix. These antioxidants can be package-incorporated [10,11] or encapsulated by another substance [12–14] to achieve controlled release. Edible bioactive films are also available [15–17]. However, their applications are mostly limited to fruit and vegetable coating or short-term packaging.

Considering the widespread use of bioactive films in food packaging, one of the main challenges is the heat sensitivity of the added antioxidant compounds. Commonly, high temperatures between 170–250 °C are used during plastic production, which results in the degradation of antioxidant compounds. For example, the degradation of essential oils can be as high as 90% during plastic extrusion [18]. Alternatively, it has been suggested to add small pouches containing antioxidant components after package material preparation, which in itself is a valid approach, but also comes with challenges (e.g., recycling of the pouches and recontamination of the food package through the pouch).

Heat-stable nanoparticles with antioxidative properties have been considered as part of the packaging material. In particular, metal nanoparticles such as silver, zinc oxide, copper, and titanium dioxide have been comprehensively studied in the literature [19–21]. They show great potential in terms of additional functionality, including antimicrobial and antioxidant activity, and UV-blocking properties, while simultaneously enhancing the material's mechanical strength [20,22]. On the other hand, there are growing concerns about the migration of these nanoparticles to the food during storage, and also into the natural environment at the end of the lifetime of the package if not disposed of appropriately, or after incineration when not considered as part of appropriate waste management [23]. The impact on living organisms and natural ecosystems is potentially huge considering the non-biodegradable and long-lasting nature of these nanoparticles. This immediately makes it clear that biodegradable particles would be preferred, that is, if they can also supply the required functionality.

Others have shown the importance of enhancing circularity in food production and, more generally, in processing systems [24,25]. Creating high-value products by utilization of waste stream by-products is one of the key points. In the current paper, we develop nanocrystals from a biomaterial that is currently considered waste or, at most, a low-value by-product. In this way, we create not only added functionality but also added value. We do this using chitin, a linear copolymer consisting of 2-acetamido-2-deoxy-D-glucose linked together with $\beta(1\rightarrow4)$ glycosidic linkage. It is the second most abundant biopolymer in nature and is present in exoskeletons of crustaceans such as shrimps and crabs or cell walls of fungi [26]. Chitin powder and its derivative chitosan are well-known to scavenge free radicals in aqueous medium, thus illustrating their antioxidant activity [27]. The antioxidant action of chitin and chitosan is attributed to their hydroxyl and amino groups, which donate hydrogen to unstable free radicals [28,29] and thus terminate the radical chain reaction.

Chitin nanocrystals (ChNC) can be produced via the simple acid hydrolysis treatment of chitin powder and are known to improve the mechanical strength of biodegradable plastic (e.g., polylactic acid) and limit oxygen and water vapor transfer [30–33]. These beneficial attributes of nanocrystals are related to their large surface-to-volume ratio. In addition, in relation to the prevention of oxidative reactions, a high surface area is expected to be beneficial. In earlier work, it was already shown that the addition of ChNC to polylactic acid improved the mechanical strength of these bioplastics. These particles can also be used to create additional functionality in polymer films, which is rather unexplored, and here we take a major step toward using the particles to provide polymer films with antioxidant properties.

In the present study, we focused on the antioxidant property of chitin nanocrystals as well as composite materials containing chitin nanocrystals and polylactic acid (the most prominent biobased plastic). This study differs from current bioactive packages in terms of antioxidative mechanisms that do not rely on an active ingredient transferring from the packaging material to the food matrix. We prepared chitin nanocrystals via acid hydrolysis and studied the nanocrystals as such or within nanocrystals/polylactic acid films using DPPH (1,1-diphenyl-2-picrylhydrazyl) radical scavenging. We studied the effect of particle size, ζ-potential, and degree of acetylation on antioxidant activity in the current manuscript. We demonstrated that the antioxidant activity of the nanocrystals and nanocomposite films is considerably higher compared to the initial chitin powder, which opens new routes to active packaging with additional functionality.

## 2. Materials and Methods

### 2.1. Materials

Crude chitin powder (≥98% purity) was purchased from Glentham Life Sciences (Corsham, UK). Polylactic acid was obtained from Natureworks LLC (Ingeo Biopolymer 4043D, Plymouth, MN, USA). Calcofluor white stain (calcofluor white M2R 1 g/L and Evans blue 0.5 g/L) and all other chemicals or solvents were of analytical grade and supplied by Sigma-Aldrich (St. Louis, MO, USA). All dilutions were prepared with ultrapure Milli-Q water (Q-POD with Millipak Express 40 0.22 µm filter, MilliporeSigma, Burlington, MA, USA).

### 2.2. Sample Preparation

#### 2.2.1. Chitin Nanocrystals Preparation

ChNC were prepared by acid hydrolysis of crude chitin powder in 3 M hydrochloric acid (HCl) at 80 °C for 90 min under magnetic stirring. The hot mixture was cooled on ice to stop the reaction. The HCl was removed by centrifuging the samples at 4500× $g$ for 5 min (Sorvall Lynx 4000, Thermo Fisher Scientific, Waltham, MA, USA), after which the supernatant was discarded, and the pellet was redispersed in an equal amount of MilliQ. This step was repeated three times. Subsequently, a 30 mL sample was added to a 50 mL reaction tube, and a sonification step was applied (5000 J energy input with an SFX150, Branson Ultrasonics, Brookfield, CT, USA) to increase chitin nanocrystal yield. Samples were diluted 10 times and centrifuged at 1200× $g$ for 20 min. The supernatant containing the chitin nanocrystals was collected, and the pellet containing the amorphous chitin was discarded. A final centrifugation step was applied at 4500× $g$ for 5 min to concentrate the chitin nanocrystal solution. The chitin nanocrystals were freeze dried before further usage (Christ Epsilon 2-6D, Martin Christ Gefriertrocknungsanlagen GmbH, Osterode am Harz, Germany). For this, samples were frozen at −40 °C for 12 h. The freeze-drying process was performed under a 1030 mbar vacuum. The temperature was increased from −20 °C to +20 °C in 8 stages during the freeze-drying process, which lasted for a total of 48 h.

#### 2.2.2. Nanocomposite Preparation

The solvent casting method, which is commonly used for the production of nanocomposites on a lab scale, was employed to produce PLA nanocomposites containing ChNC [34,35]. An amount of 5% ($w/v$) PLA chloroform solution was prepared and gently stirred for >12 h to ensure full solubilization. Likewise, the required amounts of nanocrystals were mixed in chloroform and vigorously stirred for 12 h. To create mixed matrix samples, PLA and ChNC solutions were mixed in desired amounts and left to stir for 3 h. The PLA/ChNC mixtures were poured into aluminum trays, and chloroform was allowed to evaporate in a fume hood. For ChNC positioned on top, first PLA−chloroform solution was poured into aluminum trays and left in the fume hood for 2 days to form a film. Afterwards, ChNC−chloroform suspension was added, which resulted in a ChNC layer on top of the PLA. All films were then placed in a vacuum oven at 40 °C for a week in order to remove chloroform fully.

#### 2.2.3. Deacetylation of ChNC

Alkaline treatment was used to increase the degree of deacetylation of ChNC [36]; ChNC was added to 50% NaOH (*w*/*v*) at 90 °C, and this mixture was continuously stirred for 1 h. After deacetylation, the samples were washed with Milli-Q water and centrifuged three times (4500× *g*, 5 min) to remove NaOH. The pH of the samples was adjusted to 5 using 0.1 M HCl. The deacetylated ChNC (D-ChNC) were freeze-dried using the same method described before for ChNC before further use.

### 2.3. *Sample Characterization*

#### 2.3.1. Particle Size and Zeta (ζ)-Potential

Laser diffraction (Mastersizer 3000, Malvern Instruments Ltd., Malvern, UK) and dynamic light scattering (Zetasizer Ultra, Malvern Instruments Ltd., Malvern, UK) were used to determine the size of chitin powder and chitin nanocrystals, respectively. The absorption index was set at 0.01, and refractive indices of 1.560, 1.360, and 1.330 were used for chitin, ethanol, and water, respectively.

The ζ-potential of all samples was measured through laser Doppler electrophoresis (Zetasizer Ultra). Size and ζ-potential analyses were conducted at 25 °C using capillary cells (DTS1070, Malvern Instruments Ltd., Malvern, UK). Prior to measurement, the pH of the sample was adjusted to 5 using 0.1 M HCl and 0.1 M NaOH.

#### 2.3.2. Viscosity Measurement

A stress-controlled rotational rheometer (MCR 301, Anton Paar, Austria) was used to investigate the behavior of ChNC in ethanol at various concentrations. A 50 mm cone plate (CP50-4, Anton Paar, Graz, Austria) was used in a shear sweep from 0.1 to 100 1/s.

#### 2.3.3. Degree of Deacetylation

A slightly modified version of the first derivative method described by Hein et al. (2008) [37] was used to determine the degree of deacetylation of samples. The reference substances, N-acetyl glucosamine (NAG) and D-glucosamine hydrochloride (GlcN.HCl), were dissolved in 85% (*w*/*w*) phosphoric acid to prepare 0.05 M reference solutions. Thereafter, the produced solutions were diluted with Milli-Q water to 1 mM, and a calibration curve was established (Appendix A, Figure A3). The powder and ChNC samples were dissolved in 85% (*w*/*w*) phosphoric acid under vigorous stirring at 55 °C for 60 min. After this, the sample solutions were diluted 40 times, and UV-VIS absorbance was measured at 210 nm (Beckman DU720, Beckman Coulter Inc., USA), with Milli-Q water as a blank solution. Experiments were carried out in triplicate, and values were averaged. Equation (1) was used to calculate the degree of deacetylation (DD, %).

$$DD = 100 \times (1 - \frac{\mu\text{mole NAG}}{\mu\text{mole NAG} - \mu\text{mole GlcN.HCl}}). \quad (1)$$

For μmole NAG, the calibration curve was used, and Equation (2) followed μmole GlcN.HCl, with *W* article weight (mg) per mL, and 0.203 and 0.161 conversion factors.

$$\mu\text{mole GlcN.HCl} = \frac{W - (\mu\text{mole NAG} \times 0.203)}{0.161}. \quad (2)$$

#### 2.3.4. Nanocomposite Morphology

The distribution of ChNC in the polymer film was visualized using confocal laser scanning microscopy (CLSM) (SP5X-SMD, Leica Microsystems, Wetzlar, Germany) using calcofluor white for fluorescence labelling. ChNC and calcofluor white were mixed in dark conditions at room temperature for 30 min, after which 10% potassium hydroxide was added and left to stir for another 30 min. The mixture was diluted 10 times with Milli-Q water and centrifuged at 4000× *g* for 10 min, after which the supernatant was discarded, and the pellet was redispersed in an equal amount of Milli-Q water. This step was repeated three times. The particles were freeze-dried, after which they were ready for use, as explained before.

For microscopy, nanocomposite samples were placed on an objective glass and attached by heating briefly. A 63x (water) objective (Leica Microsystems, Wetzlar, Germany) was used, in combination with a 480 nm filter, to observe the excited calcofluor white labelled-ChNC within the plastic matrix. Three hundred z-stack microscopic images were created per ~140 nm thick sample, and these images were processed with the software ImageJ (Fiji ImageJ 1.52, National Institutes of Health, Bethesda, MD, USA).

### 2.4. Antioxidant Activity

The antioxidant activity of ChNC and nanocomposites was determined following the DPPH (2,2-diphenyl-1-picrylhydrazyl) radical scavenging assay described by Brand-Williams et al. (1995) [38]. In short, a 0.5 mM DPPH/ethanol solution was prepared by stirring under dark conditions at room temperature until DPPH was completely dissolved. Next, particle/ethanol solutions were prepared by adding 375 mg particles to 25 mL ethanol. Approximately 0.5 mL of DPPH solution and 1.5 mL of particle solution were added to 2 mL tubes and stirred at 1400 rpm at 30 °C (Thermomixer C, Eppendorf AG, Hamburg, Germany) for 30 min. To prevent disturbance, samples were centrifuged for 5 min at 15,000 g (Eppendorf Centrifuge 5424, Eppendorf AG, Hamburg, Germany), after which the antioxidant activity of the supernatant was determined by measuring absorbance at 517 nm using a UV-VIS spectrophotometer (Beckman DU720, Beckman Coulter Inc., Brea, CA, USA), with ethanol as a blank and 0.125 mM DPPH in ethanol solution was a control. Each sample was measured in triplicate. Equation (3) was used to calculate radical scavenging activity in terms of DPPH inhibition, $I$ (%).

$$I\ (\%) = \left( \frac{Abs_{control} - Abs_{sample}}{Abs_{control}} \right) \times 100. \tag{3}$$

In which, $Abs_{control}$ and $Abs_{sample}$ indicate the absorbance values of DPPH + ethanol mixtures without and with the sample at 517 nm, respectively. To determine the antioxidant activity of nanocomposites, the films were immersed in DPPH/ethanol mixture for 8 h using the neat PLA film as control.

## 3. Results

### 3.1. Particle Characterization

Table 1 provide the size, ζ-potential, degree of deacetylation (DD%), and specific surface area of chitin nanocrystals (ChNC), deacetylated chitin nanocrystals (D-ChNC), and the starting chitin powder. Crude chitin powder possesses an average size of ~120 µm and a ζ-potential of +16.8 mV. Acid hydrolysis resulted in chitin nanocrystals (ChNC) with an average size of ~400 nm, which is a stark reduction compared to the starting material, as was also expected. Additionally, the harsh process conditions resulted in an increase in DD% of ~10% compared to the base material, and remarkably the ζ-potential almost doubled to +32 mV. Upon further alkaline treatment, the DD and ζ-potential increased strongly to 69% and sightly to +35.9 mV, respectively. These results are in line with the literature [39,40].

**Table 1.** Particle characteristics of ChNC, D-ChNC, and chitin powder.

| Sample | Size | ζ-Potential | Degree of Deacetylation | Specific Surface Area |
|---|---|---|---|---|
| | (µm) | (mV) | (%) | ($m^2$/kg) |
| ChNC | 0.4 ± 0.0 | +32.0 ± 1.0 | 16.3 ± 1.6 | 285 ± 2 |
| D-ChNC | 0.8 ± 0.0 | +35.9 ± 0.4 | 69.2 ± 3.8 | 184 ± 1 |
| Chitin powder | 122 ± 5 | +16.8 ± 2.7 | 6.8 ± 1.1 | 35.2 ± 1 |

* Size measurements of ChNC and D-ChNC were carried out with Zetasizer Ultra, Mastersizer was used for chitin powder. ζ-potential values were measured at pH 5. For specific surface area measurement, samples were added to ethanol at 15 mg/mL.

It is good to point out that the specific surface area of the ChNC may have been larger [41,42]. It is known that freeze drying results in strong aggregates that are difficult to redisperse and break up [43].

### 3.2. The Effect of Particle Size and Concentration on Radical Scavenging Activity

The antioxidant activity was determined through DPPH radical scavenging assay; Figure 1 shows the DPPH inhibition (%) of crude chitin powder and ChNC at different concentrations after 30 min of incubation time in ethanol. Compared to crude chitin powder, the radical scavenging activity of ChNC was typically four to five times higher, and the activity did not increase completely proportionally with concentration. This may be caused by the difference in ratio between available surface area and substrate concentration, potentially going from a surface to a substrate-limited situation. Alternatively, crude chitin powder contains pores that can reach 250 nm [44,45]. This may lead to the underestimation of the specific surface area when measured by laser diffraction. Furthermore, the viscosity of the liquid may have influenced the reaction by slowing down the mass transfer. Especially the latter effect seems to be prominent in the ChNC systems that we investigated.

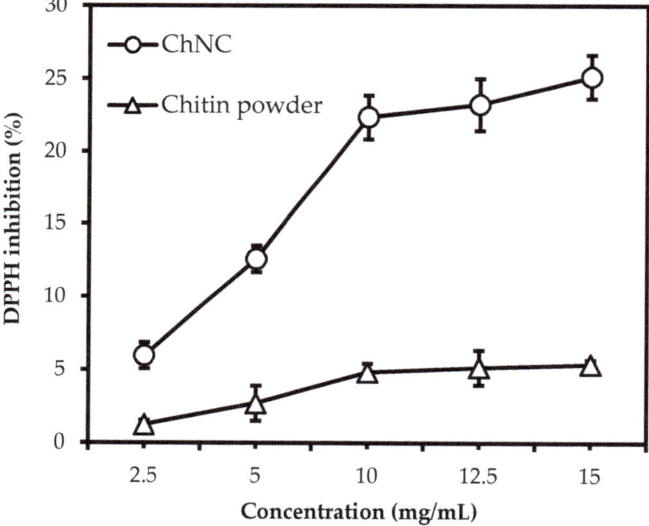


**Figure 1.** DPPH inhibition by ChNC (o) and chitin powder (Δ) for different concentrations.

Figure 2 shows the rheological behavior of ChNC/ethanol mixtures for shear rates between 0.1 1/s and 100 1/s. The viscosity of ChNC/ethanol suspensions drastically increases with ChNC concentration and shows complex behavior. This was most probably due to interparticle interactions, including hydrogen bonds and van der Waals interactions which result in a gel-like structure at high concentrations [46,47]. The viscosity at 0.1 1/s shear rate was 5, 181, and 1383 mPa.s for 2.5, 5.0, and 15.0 mg/mL particle concentrations, respectively. Moreover, the response at high particle concentration at shear rate region 1, −6 1/s, may be related to the disruption of the gel structure into sub-micrometer-sized elongated tactoids [48]. When tested, 10 mg/mL and 12.5 mg/mL ChNC/ethanol suspensions showed similar behavior (Appendix A, Figure A1). This gel-like structure is expected to result in limitations in the interaction between nanocrystals and the DPPH radical, leading to the levelling-off of antioxidant activity at concentrations above 10 mg/mL.

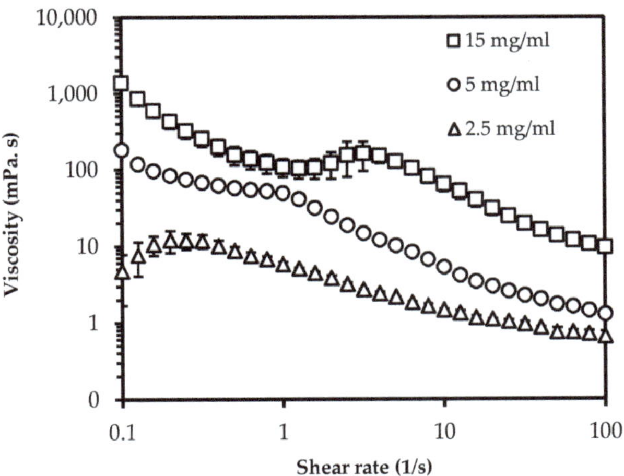


**Figure 2.** Viscosity of ChNC dispersions as function of shear rate.

*3.3. The Effect of pH and Degree of Deacetylation on Radical Scavenging Activity*

The charge of the ChNC is expected to influence radical scavenging activity; therefore, the ζ-potential (Figure 3a) and DPPH inhibition were measured as a function of pH (Figure 3b). The average ζ-potential of ChNC was highest at pH 4 (+34 ± 1 mV). The high ζ-potential values at lower pH are the result of amino groups being highly protonated at acidic pH. At higher pH values, the average ζ-potential ultimately decreased to −3 ± 1 mV at pH 8.

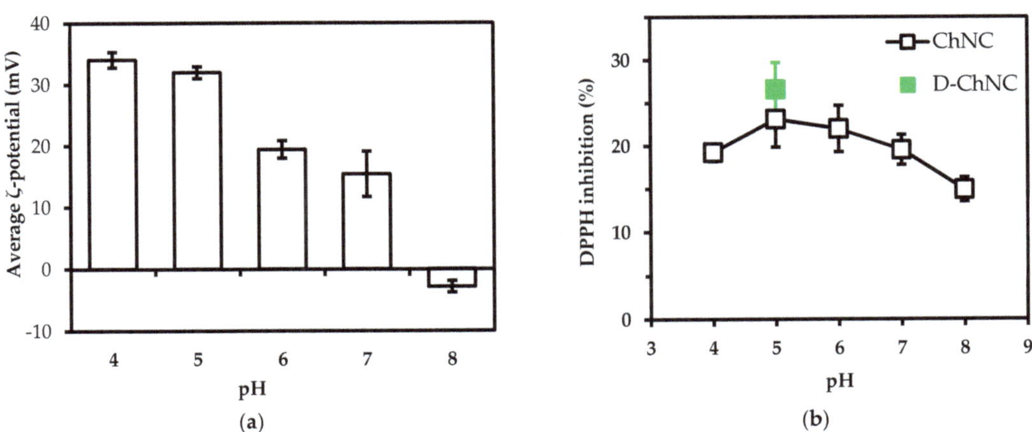


**Figure 3.** (**a**) The influence of pH on average ζ-potential of ChNC; (**b**) The DPPH inhibition (%) values at pH 4–8. 15 mg/mL concentration of ChNC was used during experiments.

Radical scavenging activity was much less pH dependent and remained over 15% for pH values 4–8 (pH-values as expected for foods (mostly < pH 7)). The radical scavenging activity for D-ChNC at pH 5 was similar to that for regular ChCN and for the ζ-potential (36 mV). In Figure 3b, we report on deacetylated chitin nanocrystals with ~70%DD, and it is good to share that at different DD (30–70%), we found no significant difference in radical scavenging activity (Appendix A, Table A1). It is good to point out that at pH 8, ChNC contains a distribution of charges; the distribution can be found in the Appendix A, Figure A2.

### 3.4. The Antioxidant Activity in Polylactic Acid Films

ChNC were positioned inside and on top of polylactic acid films (Figure 4b), and radical scavenging activity was measured [38] at 1, 2.5, and 5% (*w*/*w*) (Figure 4a). The CLSM micrographs show that ChNC could be both distributed in PLA (bottom image) and positioned on top of a PLA film (top image). From Figure 4b, it is clear that chitin nanocrystal aggregation occurred during nanocomposite preparation, but it is good to point out that 90% of the particles were below 6 μm. When placed on top of the PLA film, the particles were found within a thickness of 35 μm.

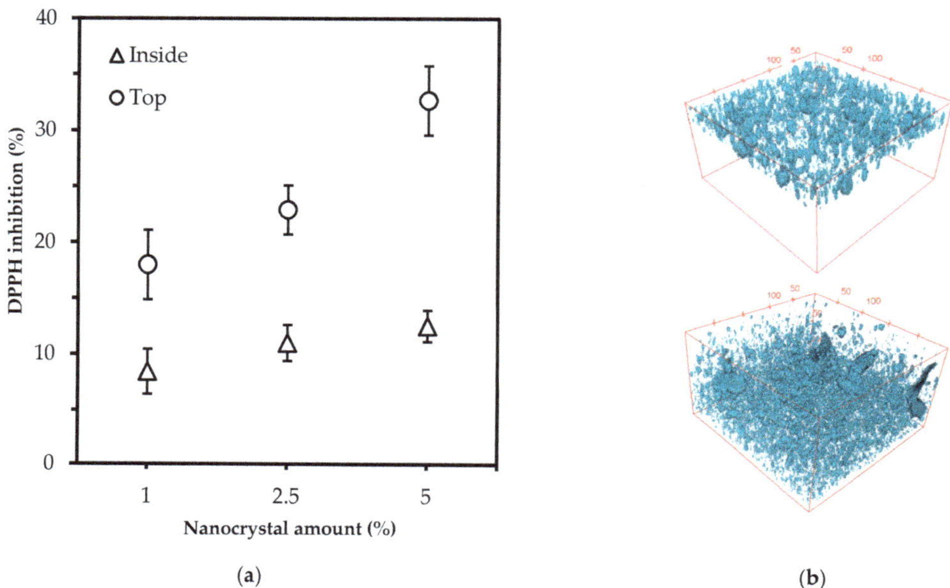


(a) (b)

**Figure 4.** (**a**) DPPH inhibition (%) of ChNC/PLA nanocomposites after 8 h incubation time; (Δ) particles dispersed inside the plastic, (o) particles at the surface of the plastic; (**b**) CLSM micrographs of calcofluor white labelled ChNC in PLA (dimensions; 250 × 250 × 140 μm), and on top of PLA (1% on top, 5% inside the matrix, respectively).

The position of the ChNC clearly affected antioxidant activity. Overall, 1% ChNC placed on the surface showed higher activity (18%) than 5% ChNC dispersed inside the plastic (12.5%).Additionally, the radical scavenging activity increased much more strongly with particles deposited on top of the PLA film instead of inside. Still, both options can contribute to increasing the shelf life of food products through the proven radical scavenging effects.

## 4. Discussion

Protonation of amine groups ($NH_3^+$) has been related to antioxidant activity in chitin-derived products [49–51], and to date, most research is focused on chitosan, but our results indicate that chitin nanocrystals can be used as well. For example, we found DPPH scavenging activity of 22.5% at a concentration of 10 mg/mL ChNC after 30 min of incubation, and Yen et al. (2008) [52] reported 28.4% for chitosan at the same concentration and conditions, which is a comparable result. The ζ-potential of the particles is positive and more positive than the starting material, which will contribute to colloidal stability [53,54], as well as radical scavenging activity [55,56].

The antioxidant activity of ChNC is rather low compared to conventional antioxidants such as ascorbic acid, essential oils, tocopherols, or phenolics that commonly show a DPPH inhibition up to 75–90% at low concentrations (0.1–1 mg/mL) [57,58]. However, if these antioxidants were to be used as part of packing concepts, they would need to be resistant to the processing conditions used during the production of thermoplastic food packages. As these substances are sensitive to heat and high pressure [18,59,60], these conventional antioxidants will be degraded, mostly to a very great extent, during plastic production at ~170–250 °C [61] or need to be overdosed. In contrast, chitin nanocrystals are heat stable up to 250–300 °C [62,63] and, therefore, are better candidates for the production of active packages. When comparing chitosan with chitin nanoparticles, chitin has better miscibility than hydrophobic polylactic acid (chitosan is more hydrophilic) [64,65], which has been linked to improved mechanical and barrier properties [66–69].

For our envisioned application, active packaging, it is important that nanoparticles retain their antioxidant activity. Our results clearly show that ChNC possesses antioxidant activity, even when they are dispersed in the polylactic acid film, although they are more effective when positioned on top of the polymer film, which is both interesting leads for the design of active packages. Additionally, the activity occurred for the entire pH range investigated, meaning that ChNC can be used to prevent oxidation in pretty much any food ranging from animal-origin foods such as meat (raw), cheese (cottage, cheddar), and fermented dairy (yoghurt) to seafood (oysters, sardines, and tuna) [70]. These foods are commonly stored in plastic packages for retail sale.

We expect that various approaches can be used to increase the antioxidant activity of ChNC further. In particular, the reduction of aggregation, which is expected to be possible through industrial-scale extrusion, can improve the dispersion of the particles in the plastic matrix [71,72], thus increasing the surface area available for antioxidative action. An alternative route is utilized to increase the antioxidant activity of ChNC by surface modification with a phenolic compound [73,74]; for instance, using Steglich esterification with caffeic acid [75]. This is expected to improve the dispersion and antioxidant activity simultaneously due to the hydrophobic nature of the added groups and their functionality.

## 5. Conclusions

In this study, we produced chitin nanocrystals with an average size of 390 nm. These particles showed significant radical scavenging activity (DPPH assay) over a pH range from 4 to 8. When added to polylactide films, the chitin nanocrystals kept their antioxidant properties and transferred them to the polylactic acid film as a whole. The position of the nanocrystals in the film is an important design parameter, with particles on top of the package exhibiting significantly higher radical scavenging activity than those embedded in the film.

The findings in this paper will contribute to the development of active packaging concepts for biodegradable plastics that can be applied to different foods that require reduction of oxidation during storage and thereby contribute to extending shelf-life.

**Author Contributions:** M.Y. Conceptualization, Formal analysis, Investigation, Writing—original draft, Visualization. I.C. Conceptualization, Investigation, Writing—review and editing. K.S. Conceptualization, Writing—review and editing, Funding acquisition. All authors have read and agreed to the published version of the manuscript.

**Funding:** This work was carried out under project number A17020 as part of the research program of the Materials Innovation Institute (M2i), which is supported by the Dutch government. Funding for this research was obtained from Yparex, Oerlemans Packaging B.V., Koninklijke Peijnenburg B.V., Heineken, and Jacobs Douwe Egberts. The authors are thankful for the financial support for the PhD training of Murat Yanat by the Republic of Turkey Ministry of National Education.

**Institutional Review Board Statement:** Not applicable.

**Informed Consent Statement:** Not applicable.

**Data Availability Statement:** Data available on request.

**Acknowledgments:** We thank Maurice Strubel, Jos Sewalt, and Wouter de Groot for their technical support during laboratory experiments. We thank Jan Willem Borst for CLSM advice. We also thank thesis students Iris Luiten, Pien van Streun, and Kirsten Versluis, who took part in the preliminary experiments.

**Conflicts of Interest:** The authors declare no conflict of interest.

## Appendix A

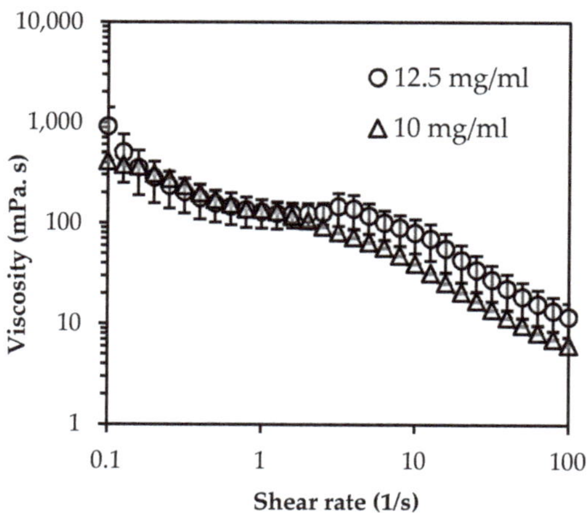


**Figure A1.** The viscosity of ChNC dispersions as function of shear rate.

**Table A1.** DPPH inhibition (%) values of deacetylated chitin samples after 30 min of incubation with DPPH radical solution.

| Sample | Degree of Deacetylation (%) | DPPH Inhibition (%) |
|---|---|---|
| D-ChNC | 69.2 ± 3.8 | 26.63 ± 3.10 |
| D2-ChNC | 53.1 ± 5.2 | 22.7 ± 1.91 |
| D3-ChNC | 33.4 ± 2.8 | 20.8 ± 2.89 |

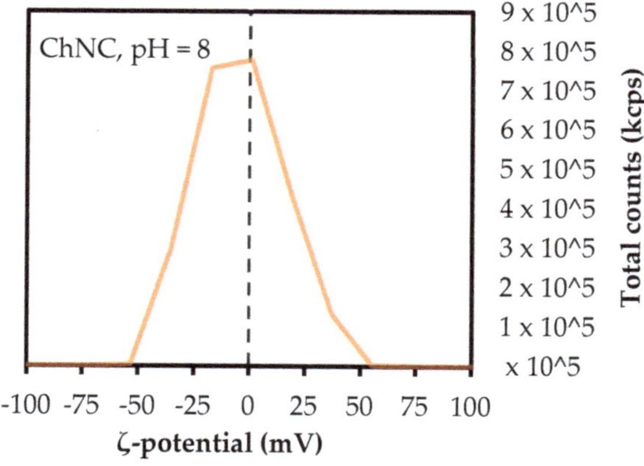


**Figure A2.** The ζ-potential (mV) distribution of ChNC at pH 8.

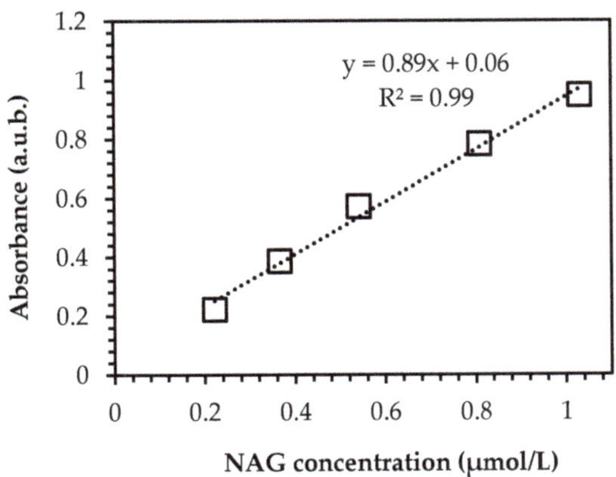


**Figure A3.** The calibration curve of NAG that was used to calculate the degree of acetylation.

## References


1. Jacobsen, C.; Meyer, A.S.; Adler-Nissen, J. Oxidation Mechanisms in Real Food Emulsions: Oil-Water Partition Coefficients of Selected Volatile off-Flavor Compounds in Mayonnaise. *Eur. Food Res. Technol.* **1999**, *208*, 317–327. [CrossRef]
2. Farhoosh, R. A Reconsidered Approach Providing Kinetic Parameters and Rate Constants to Analyze the Oxidative Stability of Bulk Lipid Systems. *Food Chem.* **2020**, *327*, 127088. [CrossRef] [PubMed]
3. Berton-Carabin, C.C.; Schröder, A.; Rovalino-Cordova, A.; Schroën, K.; Sagis, L. Protein and Lipid Oxidation Affect the Viscoelasticity of Whey Protein Layers at the Oil–Water Interface. *Eur. J. Lipid Sci. Technol.* **2016**, *118*, 1630–1643. [CrossRef]
4. Tian, F.; Decker, E.A.; Goddard, J.M. Controlling Lipid Oxidation of Food by Active Packaging Technologies. *Food Funct.* **2013**, *4*, 669–680. [CrossRef] [PubMed]
5. Ahmed Laskar, A.; Younus, H. Aldehyde Toxicity and Metabolism: The Role of Aldehyde Dehydrogenases in Detoxification, Drug Resistance and Carcinogenesis. *Drug Metab. Rev.* **2019**, *51*, 42–64. [CrossRef] [PubMed]
6. Estévez, M.; Li, Z.; Soladoye, O.P.; Van-Hecke, T. Health Risks of Food Oxidation. *Adv. Food Nutr. Res.* **2017**, *82*, 45–81. [CrossRef]
7. Zhang, L.; Yu, D.; Regenstein, J.M.; Xia, W.; Dong, J. A Comprehensive Review on Natural Bioactive Films with Controlled Release Characteristics and Their Applications in Foods and Pharmaceuticals. *Trends Food Sci. Technol.* **2021**, *112*, 690–707. [CrossRef]
8. de Oliveira Filho, J.G.; Braga, A.R.C.; de Oliveira, B.R.; Gomes, F.P.; Moreira, V.L.; Pereira, V.A.C.; Egea, M.B. The Potential of Anthocyanins in Smart, Active, and Bioactive Eco-Friendly Polymer-Based Films: A Review. *Food Res. Int.* **2021**, *142*, 110202. [CrossRef]
9. Yang, Y.; Jiao, Q.; Wang, L.; Zhang, Y.; Jiang, B.; Li, D.; Feng, Z.; Liu, C. Preparation and Evaluation of a Novel High Internal Phase Pickering Emulsion Based on Whey Protein Isolate Nanofibrils Derived by Hydrothermal Method. *Food Hydrocoll.* **2022**, *123*, 107180. [CrossRef]
10. Chan-Matú, D.I.; Toledo-López, V.M.; Vargas, M.d.L.V.y.; Rincón-Arriaga, S.; Rodríguez-Félix, A.; Madera-Santana, T.J. Preparation and Characterization of Chitosan-Based Bioactive Films Incorporating Moringa Oleifera Leaves Extract. *J. Food Meas. Charact.* **2021**, *15*, 4813–4824. [CrossRef]
11. Ribeiro, A.C.B.; Cunha, A.P.; da Silva, L.M.R.; Mattos, A.L.A.; de Brito, E.S.; de Souza Filho, M.d.S.M.; de Azeredo, H.M.C.; Ricardo, N.M.P.S. From Mango By-Product to Food Packaging: Pectin-Phenolic Antioxidant Films from Mango Peels. *Int. J. Biol. Macromol.* **2021**, *193*, 1138–1150. [CrossRef] [PubMed]
12. Roy, S.; Rhim, J.W. Fabrication of Chitosan-Based Functional Nanocomposite Films: Effect of Quercetin-Loaded Chitosan Nanoparticles. *Food Hydrocoll.* **2021**, *121*, 107065. [CrossRef]
13. Siddiqui, S.A.; Bahmid, N.A.; Taha, A.; Khalifa, I.; Khan, S.; Rostamabadi, H.; Jafari, S.M. Recent Advances in Food Applications of Phenolic-Loaded Micro/Nanodelivery Systems. *Crit. Rev. Food Sci. Nutr.* **2022**, 1–21. [CrossRef] [PubMed]
14. Lu, W.; Chen, M.; Cheng, M.; Yan, X.; Zhang, R.; Kong, R.; Wang, J.; Wang, X. Development of Antioxidant and Antimicrobial Bioactive Films Based on Oregano Essential Oil/Mesoporous Nano-Silica/Sodium Alginate. *Food Packag. Shelf Life* **2021**, *29*, 100691. [CrossRef]
15. Sarker, A.; Grift, T.E. Bioactive Properties and Potential Applications of Aloe Vera Gel Edible Coating on Fresh and Minimally Processed Fruits and Vegetables: A Review. *J. Food Meas. Charact.* **2021**, *15*, 2119–2134. [CrossRef]
16. Chaudhary, S.; Kumar, S.; Kumar, V.; Sharma, R. Chitosan Nanoemulsions as Advanced Edible Coatings for Fruits and Vegetables: Composition, Fabrication and Developments in Last Decade. *Int. J. Biol. Macromol.* **2020**, *152*, 154–170. [CrossRef]
17. Shivangi, S.; Dorairaj, D.; Negi, P.S.; Shetty, N.P. Development and Characterisation of a Pectin-Based Edible Film That Contains Mulberry Leaf Extract and Its Bio-Active Components. *Food Hydrocoll.* **2021**, *121*, 107046. [CrossRef]

18. Nerín, C. Antioxidant Active Food Packaging and Antioxidant Edible Films. In *Oxidation in Foods and Beverages and Antioxidant Applications*; Woodhead Publishing: Sawston, UK, 2010; pp. 496–515. [CrossRef]
19. Munteanu, B.S.; Aytac, Z.; Pricope, G.M.; Uyar, T.; Vasile, C. Polylactic Acid (PLA)/Silver-NP/VitaminE Bionanocomposite Electrospun Nanofibers with Antibacterial and Antioxidant Activity. *J. Nanoparticle Res.* **2014**, *16*, 2643. [CrossRef]
20. Llorens, A.; Lloret, E.; Picouet, P.A.; Trbojevich, R.; Fernandez, A. Metallic-Based Micro and Nanocomposites in Food Contact Materials and Active Food Packaging. *Trends Food Sci. Technol.* **2012**, *24*, 19–29. [CrossRef]
21. Garcia, C.V.; Shin, G.H.; Kim, J.T. Metal Oxide-Based Nanocomposites in Food Packaging: Applications, Migration, and Regulations. *Trends Food Sci. Technol.* **2018**, *82*, 21–31. [CrossRef]
22. Ghozali, M.; Fahmiati, S.; Triwulandari, E.; Restu, W.K.; Farhan, D.; Wulansari, M.; Fatriasari, W. PLA/Metal Oxide Biocomposites for Antimicrobial Packaging Application. *Polym. Technol. Mater.* **2020**, *59*, 1332–1342. [CrossRef]
23. Hoseinnejad, M.; Jafari, S.M.; Katouzian, I. Inorganic and Metal Nanoparticles and Their Antimicrobial Activity in Food Packaging Applications. *Crit. Rev. Microbiol.* **2018**, *44*, 161–181. [CrossRef] [PubMed]
24. Koppelmäki, K.; Helenius, J.; Schulte, R.P.O. Nested Circularity in Food Systems: A Nordic Case Study on Connecting Biomass, Nutrient and Energy Flows from Field Scale to Continent. *Resour. Conserv. Recycl.* **2021**, *164*, 105218. [CrossRef]
25. Tseng, M.L.; Chiu, A.S.F.; Chien, C.F.; Tan, R.R. Pathways and Barriers to Circularity in Food Systems. *Resour. Conserv. Recycl.* **2019**, *143*, 236–237. [CrossRef]
26. Shamshina, J.L.; Berton, P.; Rogers, R.D. Advances in Functional Chitin Materials: A Review. *ACS Sustain. Chem. Eng.* **2019**, *7*, 6444–6457. [CrossRef]
27. Ngo, D.H.; Kim, S.K. *Antioxidant Effects of Chitin, Chitosan, and Their Derivatives*, 1st ed.; Elsevier: Amsterdam, The Netherlands, 2014; Volume 73, ISBN 9780128002681.
28. Hromis, N.; Lazic, V.; Popovic, S.; Suput, D.; Bulut, S. Antioxidative Activity of Chitosan and Chitosan Based Biopolymer Film. *Food Feed Res.* **2017**, *44*, 91–100. [CrossRef]
29. Younes, I.; Rinaudo, M. Chitin and Chitosan Preparation from Marine Sources. Structure, Properties and Applications. *Mar. Drugs* **2015**, *13*, 1133–1174. [CrossRef]
30. Herrera, N.; Salaberria, A.M.; Mathew, A.P.; Oksman, K. Plasticized Polylactic Acid Nanocomposite Films with Cellulose and Chitin Nanocrystals Prepared Using Extrusion and Compression Molding with Two Cooling Rates: Effects on Mechanical, Thermal and Optical Properties. *Compos. Part A Appl. Sci. Manuf.* **2016**, *83*, 89–97. [CrossRef]
31. Herrera, N.; Roch, H.; Salaberria, A.M.; Pino-Orellana, M.A.; Labidi, J.; Fernandes, S.C.M.; Radic, D.; Leiva, A.; Oksman, K. Functionalized Blown Films of Plasticized Polylactic Acid/Chitin Nanocomposite: Preparation and Characterization. *Mater. Des.* **2016**, *92*, 846–852. [CrossRef]
32. Broers, L.; van Dongen, S.; de Goederen, V.; Ton, M.; Spaen, J.; Boeriu, C.; Schroën, K. Addition of Chitin Nanoparticles Improves Polylactic Acid Film Properties. *Nanotechnol. Adv. Mater. Sci.* **2018**, *1*, 1–8.
33. Coltelli, M.B.; Cinelli, P.; Gigante, V.; Aliotta, L.; Morganti, P.; Panariello, L.; Lazzeri, A. Chitin Nanofibrils in Poly(Lactic Acid) (PLA) Nanocomposites: Dispersion and Thermo-Mechanical Properties. *Int. J. Mol. Sci.* **2019**, *20*, 504. [CrossRef] [PubMed]
34. Liu, D.Y.; Yuan, X.W.; Bhattacharyya, D.; Easteal, A.J. Characterisation of Solution Cast Cellulose Nanofibre—Reinforced Poly(Lactic Acid). *Express Polym. Lett.* **2010**, *4*, 26–31. [CrossRef]
35. Petersson, L.; Kvien, I.; Oksman, K. Structure and Thermal Properties of Poly(Lactic Acid)/Cellulose Whiskers Nanocomposite Materials. *Compos. Sci. Technol.* **2007**, *67*, 2535–2544. [CrossRef]
36. Yanat, M.; Schroën, K. Preparation Methods and Applications of Chitosan Nanoparticles; with an Outlook toward Reinforcement of Biodegradable Packaging. *React. Funct. Polym.* **2021**, *161*, 104849. [CrossRef]
37. Hein, S.; Ng, C.H.; Stevens, W.F.; Wang, K. Selection of a Practical Assay for the Determination of the Entire Range of Acetyl Content in Chitin and Chitosan: UV Spectrophotometry with Phosphoric Acid as Solvent. *J. Biomed. Mater. Res.—Part B Appl. Biomater.* **2008**, *86*, 558–568. [CrossRef]
38. Brand-Williams, W.; Cuvelier, M.E.; Berset, C. Use of a Free Radical Method to Evaluate Antioxidant Activity. *LWT—Food Sci. Technol.* **1995**, *28*, 25–30. [CrossRef]
39. Nam, Y.S.; Park, W.H.; Ihm, D.; Hudson, S.M. Effect of the Degree of Deacetylation on the Thermal Decomposition of Chitin and Chitosan Nanofibers. *Carbohydr. Polym.* **2010**, *80*, 291–295. [CrossRef]
40. Liu, L.; Wang, R.; Yu, J.; Jiang, J.; Zheng, K.; Hu, L.; Wang, Z.; Fan, Y. Robust Self-Standing Chitin Nanofiber/Nanowhisker Hydrogels with Designed Surface Charges and Ultralow Mass Content via Gas Phase Coagulation. *Biomacromolecules* **2016**, *17*, 3773–3781. [CrossRef]
41. Muzzarelli, R.A. Chitin Nanostructures in Living Organisms. In *Chitin*; Gupta, N.S., Ed.; Topics in Geobiology; Springer: Dordrecht, The Netherlands, 2011; Volume 34, ISBN 978-90-481-9683-8.
42. Cai, J.; Kuga, S. Aerogels of Cellulose and Chitin Crystals. In *Handbook of Green Materials*; World Scientific: Singapore, 2014; pp. 139–161. [CrossRef]
43. Colijn, I.; Fokkink, R.; Schroën, K. Quantification of Energy Input Required for Chitin Nanocrystal Aggregate Size Reduction through Ultrasound. *Sci. Rep.* **2021**, *11*, 17217. [CrossRef]
44. Tian, Z.; Wang, S.; Hu, X.; Zhang, Z.; Liang, L. Crystalline Reduction, Surface Area Enlargement and Pore Generation of Chitin by Instant Catapult Steam Explosion. *Carbohydr. Polym.* **2018**, *200*, 255–261. [CrossRef]

45. Phongying, S.; Aiba, S.-i.; Chirachanchai, S. Direct Chitosan Nanoscaffold Formation via Chitin Whiskers. *Polymer* **2007**, *48*, 393–400. [CrossRef]
46. Tzoumaki, M.V.; Moschakis, T.; Biliaderis, C.G. Metastability of Nematic Gels Made of Aqueous Chitin Nanocrystal Dispersions. *Biomacromolecules* **2010**, *11*, 175–181. [CrossRef] [PubMed]
47. Tzoumaki, M.V.; Moschakis, T.; Kiosseoglou, V.; Biliaderis, C.G. Oil-in-Water Emulsions Stabilized by Chitin Nanocrystal Particles. *Food Hydrocoll.* **2011**, *25*, 1521–1529. [CrossRef]
48. Pignon, F.; Challamel, M.; De Geyer, A.; Elchamaa, M.; Semeraro, E.F.; Hengl, N.; Jean, B.; Putaux, J.L.; Gicquel, E.; Bras, J.; et al. Breakdown and Buildup Mechanisms of Cellulose Nanocrystal Suspensions under Shear and upon Relaxation Probed by SAXS and SALS. *Carbohydr. Polym.* **2021**, *260*, 117751. [CrossRef]
49. Sun, T.; Zhou, D.; Xie, J.; Mao, F. Preparation of Chitosan Oligomers and Their Antioxidant Activity. *Eur. Food Res. Technol.* **2007**, *225*, 451–456. [CrossRef]
50. Saddam, N.S.; Hadi, A.G.; Mohammed, A.A. Enhanced Antioxidant Activity of Chitosan Extracted from Fish Shells Materials. *Mater. Today Proc.* **2021**, *49*, 2793–2796. [CrossRef]
51. Tu, Y.; Xu, Y.; Ren, F.; Zhang, H. Characteristics and Antioxidant Activity of Maillard Reaction Products from α-Lactalbumin and 2′-Fucosyllactose. *Food Chem.* **2020**, *316*, 126341. [CrossRef]
52. Yen, M.T.; Yang, J.H.; Mau, J.L. Antioxidant Properties of Chitosan from Crab Shells. *Carbohydr. Polym.* **2008**, *74*, 840–844. [CrossRef]
53. Pochapski, D.J.; Carvalho Dos Santos, C.; Leite, G.W.; Pulcinelli, S.H.; Santilli, C.V. Zeta Potential and Colloidal Stability Predictions for Inorganic Nanoparticle Dispersions: Effects of Experimental Conditions and Electrokinetic Models on the Interpretation of Results. *Langmuir* **2021**, *37*, 13379–13389. [CrossRef]
54. Cano-Sarmiento, C.; Téllez-Medina, D.I.; Viveros-Contreras, R.; Cornejo-Mazón, M.; Figueroa-Hernández, C.Y.; García-Armenta, E.; Alamilla-Beltrán, L.; García, H.S.; Gutiérrez-López, G.F. Zeta Potential of Food Matrices. *Food Eng. Rev.* **2018**, *10*, 113–138. [CrossRef]
55. Liu, S.; Chen, Y.; Liu, C.; Gan, L.; Ma, X.; Huang, J. Polydopamine-Coated Cellulose Nanocrystals as an Active Ingredient in Poly(Vinyl Alcohol) Films towards Intensifying Packaging Application Potential. *Cellulose* **2019**, *26*, 9599–9612. [CrossRef]
56. Zhang, M.; Li, Y.; Wang, W.; Yang, Y.; Shi, X.; Sun, M.; Hao, Y.; Li, Y. Comparison of Physicochemical and Rheology Properties of Shiitake Stipes-Derived Chitin Nanocrystals and Nanofibers. *Carbohydr. Polym.* **2020**, *244*, 116468. [CrossRef] [PubMed]
57. Julianus Sohilait, H.; Kainama, H. Free Radical Scavenging Activity of Essential Oil of *Eugenia caryophylata* from Amboina Island and Derivatives of Eugenol. *Open Chem.* **2019**, *17*, 422–428. [CrossRef]
58. Di Mambro, V.M.; Azzolini, A.E.C.S.; Valim, Y.M.L.; Fonseca, M.J.V. Comparison of Antioxidant Activities of Tocopherols Alone and in Pharmaceutical Formulations. *Int. J. Pharm.* **2003**, *262*, 93–99. [CrossRef]
59. Oey, I.; Verlinde, P.; Hendrickx, M.; Van Loey, A. Temperature and Pressure Stability of L-Ascorbic Acid and/or [6s] 5-Methyltetrahydrofolic Acid: A Kinetic Study. *Eur. Food Res. Technol.* **2006**, *223*, 71–77. [CrossRef]
60. Ali, A.; Chong, C.H.; Mah, S.H.; Abdullah, L.C.; Choong, T.S.Y.; Chua, B.L. Impact of Storage Conditions on the Stability of Predominant Phenolic Constituents and Antioxidant Activity of Dried Piper Betle Extracts. *Molecules* **2018**, *23*, 484. [CrossRef]
61. Wagner, J.R.; Mount, E.M.; Giles, H.F. Processing Conditions. In *Extrusion*; William Andrew: Norwich, NY, USA, 2014; pp. 181–192. [CrossRef]
62. Ehrlich, H.; Rigby, J.K.; Botting, J.P.; Tsurkan, M.V.; Werner, C.; Schwille, P.; Petrášek, Z.; Pisera, A.; Simon, P.; Sivkov, V.N.; et al. Discovery of 505-Million-Year Old Chitin in the Basal Demosponge *Vauxia gracilenta*. *Sci. Rep.* **2013**, *3*, 17–20. [CrossRef]
63. Wysokowski, M.; Petrenko, I.; Stelling, A.L.; Stawski, D.; Jesionowski, T.; Ehrlich, H. Poriferan Chitin as a Versatile Template for Extreme Biomimetics. *Polymers* **2015**, *7*, 235–265. [CrossRef]
64. Akopova, T.A.; Demina, T.S.; Khavpachev, M.A.; Popyrina, T.N.; Grachev, A.V.; Ivanov, P.L.; Zelenetskii, A.N. Hydrophobic Modification of Chitosan via Reactive Solvent-Free Extrusion. *Polymers* **2021**, *13*, 2807. [CrossRef]
65. Ioelovich, M. Crystallinity and Hydrophility of Chitin and Chitosan. *Res. Rev. J. Chem.* **2014**, *3*, 7–14.
66. Lu, T.; Liu, S.; Jiang, M.; Xu, X.; Wang, Y.; Wang, Z.; Gou, J.; Hui, D.; Zhou, Z. Effects of Modifications of Bamboo Cellulose Fibers on the Improved Mechanical Properties of Cellulose Reinforced Poly(Lactic Acid) Composites. *Compos. Part B Eng.* **2014**, *62*, 191–197. [CrossRef]
67. Martínez-Sanz, M.; Abdelwahab, M.A.; Lopez-Rubio, A.; Lagaron, J.M.; Chiellini, E.; Williams, T.G.; Wood, D.F.; Orts, W.J.; Imam, S.H. Incorporation of Poly(Glycidylmethacrylate) Grafted Bacterial Cellulose Nanowhiskers in Poly(Lactic Acid) Nanocomposites: Improved Barrier and Mechanical Properties. *Eur. Polym. J.* **2013**, *49*, 2062–2072. [CrossRef]
68. Zhang, Q.; Wei, S.; Huang, J.; Feng, J.; Chang, P.R. Effect of Surface Acetylated-Chitin Nanocrystals on Structure and Mechanical Properties of Poly(Lactic Acid). *J. Appl. Polym. Sci.* **2014**, *131*, 2–9. [CrossRef]
69. Lin, N.; Huang, J.; Chang, P.R.; Feng, J.; Yu, J. Surface Acetylation of Cellulose Nanocrystal and Its Reinforcing Function in Poly(Lactic Acid). *Carbohydr. Polym.* **2011**, *83*, 1834–1842. [CrossRef]
70. Shafiur Rahman, M.; Rahman, M.R.T. pH in Food Preservation. In *Handbook of Food Preservation*; CRC Press: Boca Raton, FL, USA, 2007; pp. 287–298. [CrossRef]
71. Yang, W.; Fortunati, E.; Dominici, F.; Kenny, J.M.; Puglia, D. Effect of Processing Conditions and Lignin Content on Thermal, Mechanical and Degradative Behavior of Lignin Nanoparticles/Polylactic (Acid) Bionanocomposites Prepared by Melt Extrusion and Solvent Casting. *Eur. Polym. J.* **2015**, *71*, 126–139. [CrossRef]

72. Sarul, D.S.; Arslan, D.; Vatansever, E.; Kahraman, Y.; Durmus, A.; Salehiyan, R.; Nofar, M. Effect of Mixing Strategy on the Structure-Properties of the Pla/Pbat Blends Incorporated with Cnc. *J. Renew. Mater.* **2022**, *10*, 149–164. [CrossRef]
73. Tan, W.; Zhang, J.; Mi, Y.; Li, Q.; Guo, Z. Synthesis and Characterization of α-Lipoic Acid Grafted Chitosan Derivatives with Antioxidant Activity. *React. Funct. Polym.* **2022**, *172*, 105205. [CrossRef]
74. Woranuch, S.; Yoksan, R. Preparation, Characterization and Antioxidant Property of Water-Soluble Ferulic Acid Grafted Chitosan. *Carbohydr. Polym.* **2013**, *96*, 495–502. [CrossRef]
75. Colijn, I.; Yanat, M.; Terhaerdt, G.; Molenveld, K.; Boeriu, C.G.; Schroën, K. Chitin Nanocrystal Hydrophobicity Adjustment by Fatty Acid Esterification for Improved Polylactic Acid Nanocomposites. *Polymers* **2022**, *14*, 2619. [CrossRef]

MDPI
St. Alban-Anlage 66
4052 Basel
Switzerland
www.mdpi.com

*Polymers* Editorial Office
E-mail: polymers@mdpi.com
www.mdpi.com/journal/polymers

www.ingramcontent.com/pod-product-compliance
Lightning Source LLC
LaVergne TN
LVHW071609170726
843515LV00008B/2355
*9783036590639*